ENGINEERING, CONSTRUCTION, AND OPERATIONS IN SPACE II : VOLUME 2

Proceedings of Space 90

Albuquerque, New Mexico
April 22-26, 1990

Edited by Stewart W. Johnson and John P. Wetzel

Sponsored by:

Aerospace Division of the American Society of Civil Engineers
Aerospace & Electronics Society of IEEE
American Astronautical Society
American Institute of Aeronautics and Astronautics
American Society of Macro Engineering
American Society of Mechanical Engineers
Lunar and Planetary Institute
NASA Office of Exploration—Code Z
National Space Society
Society of American Military Engineers
Space Studies Institute
The Planetary Society
U.S. Air Force Engineering & Services
U.S. Army Corps of Engineers
U.S. International Space Year Association
U.S. Naval Facilities Engineering Command

Published by the
American Society of Civil Engineers
345 East 47th Street
New York, New York 10017-2398

ABSTRACT

This proceedings, *Engineering, Construction and Operations in Space II*, consists of the 165 papers presented at Space 90, the Second International Conference, in Albuquerque, New Mexico, April 22-26, 1990. The papers are divided into four sections. The first section presents the keynote addresses covering such topics as the legalities of extraterrestrial mining, NASA initiatives in technology, and advanced engineering design. Section two Extraterrestrial Mission, includes discussions on all aspects of lunar engineering and construction from the use of lunar metals for rocket fuels to design criteria for lunar structures. Orbital Mission, the third section of the proceedings, explores such subjects as space station operations, space debris, lunar base site constraints, optimum structural design, and gravity effect on beam behavior. The final section, Support Infrastructure, consists of papers on such topics as commercial spaceport considerations, limits of radiation exposure, cultural diversity and education.

Library of Congress Cataloging-in-Publication Data

Engineering, construction, and operations in space II: proceedings of Space 90/sponsored by Aerospace Division of the American Society of Civil Engineers . . . [et al.]. Albuquerque, New Mexico, April 22-26, 1990; edited by Stewart W. Johnson and John P. Wetzel.

p. cm.

ISBN 0-87262-752-7

1. Space stations—Design and construction—Congresses. 2. Extraterrestrial bases—Design and construction—Congresses. 3. Lunar bases—Design and construction—Congresses. I. Johnson, Stewart W. II. Wetzel, John P. III. American Society of Civil Engineers. Aerospace Division.

TL797.E553 1990

620'.419—dc20 90-373

CIP

Library of Congress Catalog Card No: 90-373
ISBN 0-87262-752-7
Manufactured in the United States of America

Cover illustration: On its way to the Moon, the Lunar Landing Vehicle is transferred from the shuttle to integration facilities aboard Space Station Freedom. Courtesy of Griffin/Hudson.

CONTENTS

VOLUME I

1.0 KEYNOTE ADDRESSES

2.0 EXTRATERRESTRIAL MISSION

2.1 SURFACE PROPERTIES

2.2 PROCESSING LUNAR SOILS

2.3 AUTOMATED PROCESSING OF EXTRATERRESTRIAL MATERIALS

2.4 SITE CHARACTERIZATION AND PROSPECTING

2.5 EXCAVATION AND MINING

2.6 LUNAR OXYGEN

2.7 LUNAR BASE DEVELOPMENT

2.8 LUNAR SURFACE STRUCTURES

2.9 LUNAR SUBSURFACE STRUCTURES

2.10 PLANETARY SURFACE TRANSPORTATION

2.11 SUPPLYING A LUNAR BASE

2.12 ASTRONOMY: SCIENCE, ENGINEERING, AND CONSTRUCTION

2.13 MARS MISSIONS

VOLUME II

3.0 ORBITAL MISSION

3.1 SPACE STATION

3.2 SPACE DEBRIS

3.3 ORBITAL SUPPORT FOR EXTRATERRESTRIAL MISSIONS

3.4 ROBOTICS AND AUTOMATED PROCESSING

3.5 MATERIALS APPLICATIONS

3.6 STRUCTURES AND STRUCTURAL RESPONSE

3.7 MISSION TO PLANET EARTH

4.0 SUPPORT INFRASTRUCTURE

4.1 EARTH ANALOGS AND TERRESTRIAL TESTING

4.2 LIFE SUPPORT SYSTEMS

4.3 LAUNCH FACILITIES AND SUPPORT

4.4 SPACE POWER

4.5 HUMAN FACTORS

4.6 ARTIFICIAL GRAVITY

4.7 EDUCATION

4.8 RESOURCE MANAGEMENT, ECONOMICS, AND LOGISTICS

4.9 STUDENT PAPERS

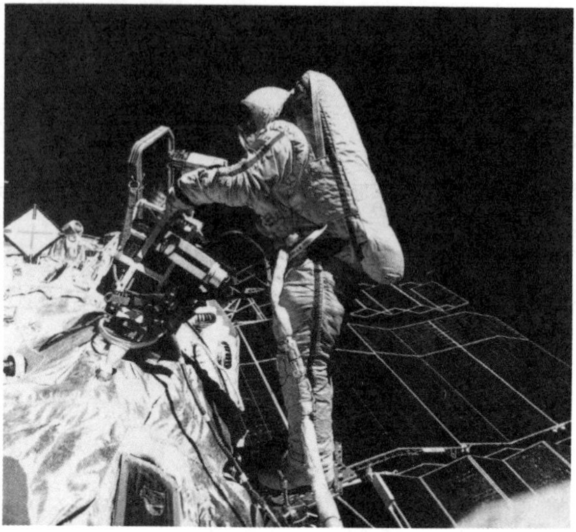

Successful use of space welding by Soviet Cosmonaut on Soyuz T12 Mission. (from Anderson, Pearson, and Drewry, this volume)

SECTION 3

ORBITAL MISSION

The Space Station *Freedom* concept continues to evolve and be shaped and driven by not only the innovations and imaginations of planners and engineers and scientists, but also by national and international political processes. The orbital mission section which follows presents the role of orbital operations in the study of Earth, in reaching out to reestablish a presence on the Moon, and in going to Mars. Also considered are the debris environments and aspects of orbital operations involving telerobotics deployment, maintenance, payload accommodation, and manufacturing in orbit. The papers in this section are subdivided into 7 sections:

SPACE STATION: The plan to construct the first permanently manned international space station, *Freedom,* is discussed, as well as the solar flux and possible future initiatives.

SPACE DEBRIS: The hazards of space debris on orbital structures, including the space station are discussed as well as mitigation approaches.

ORBITAL SUPPORT OF EXTRATERRESTRIAL MISSIONS: Space stations and other orbital nodes are examined which could support maintenance of satellites and extraterrestrial basing.

ROBOTICS AND AUTOMATED PROCESSES: Various papers examine space manufacturing, space construction, robotic research, and design issues relating to the use of robotics.

MATERIALS APPLICATIONS: The use of various materials in space applications and recent advancements are presented.

STRUCTURES AND STRUCTURAL RESPONSE: Papers examine the dynamics and reliability of various materials and structures and response to the environmental considerations of space.

MISSION TO EARTH: The utilization of Earth orbiting satellites to better define planetary processes, investigate and mitigate global change, and meet various challenges are discussed.

Subsection 3.1: SPACE STATION

REQUIREMENTS AND CONCEPT FOR AN ON-ORBIT CONSTRUCTION FACILITY FOR PHASE 1 SPACE STATION FREEDOM

Robert W. Buchan*
Richard M. Gates**
Laura M. Waters***

Introduction

To meet the National Aeronautics and Space Administration's (NASA) long-range plan for manned exploration of the solar system[1] and to support other high priority science missions,[2] new technologies in the area of in-space construction will have to be developed. Currently, spacecraft delivered to orbit must be designed to fit within the launch vehicle's payload envelope; appendages outside the envelope such as solar arrays, antennas, and sensor booms must be automatically deployed and checked out in orbit. In addition to increasing the design complexity, this method of payload delivery severely restricts the size and nature of space structures. The large space structures necessary to support future missions will be difficult or perhaps impossible to build using this approach. Space Station Freedom (SSF) has the potential to allow larger spacecraft to be constructed in space with the aid of extravehicular activity (EVA) astronauts or robots. The delay, however, of the dual keel configuration to Phase 2 (figure 1) of the SSF Program seriously restricts the availability of suitable locations for construction of large external payloads during Phase 1 (figure 2) operation, and in particular restricts the capability to develop needed on-orbit construction technologies.

A construction facility attached to the SSF is required to accommodate the assembly of large payloads and to develop and demonstrate the techniques that will enable on-orbit construction of future spacecraft. The capabilities and accommodations provided by this facility must be flexible enough to address the wide variety of needs of future spacecraft without restricting the performance of the Phase 1 SSF. This paper describes future on-orbit construction needs, describes a candidate construction facility for Phase 1 SSF, and evaluates its influence on SSF control requirements.

*Robert W. Buchan, NASA Langley Research Center, Hampton, Virginia.
**Richard M. Gates, Boeing Aerospace and Electronics, Seattle, Washington.
***Laura M. Waters, Analytical Mechanics Associates, NASA Langley Research Center, Hampton, Virginia.

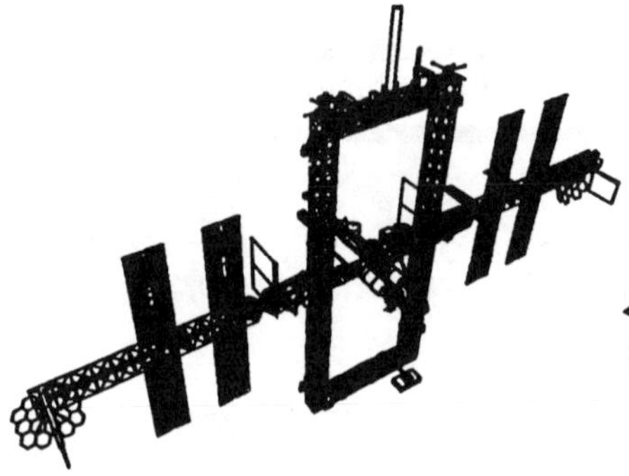

Figure 1. Dual Keel Phase 2 Space Station Freedom

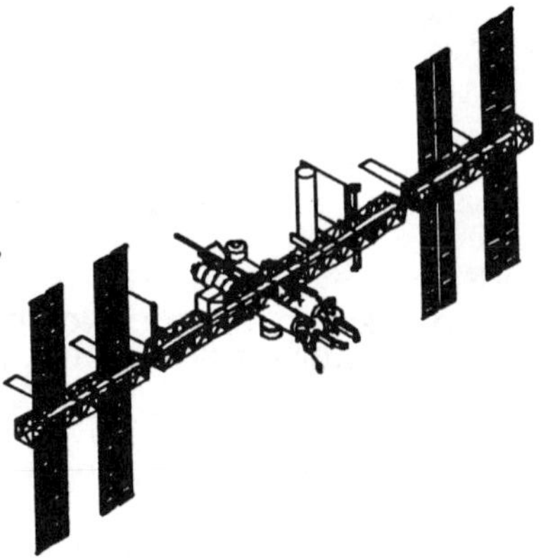

Figure 2. Phase 1 Space Station Freedom

In-Space Construction Plans and Facility Requirements

The construction of SSF will provide valuable experience in some of the techniques required for in-space construction; i.e., assembling truss structures, attaching utility trays, positioning and attaching modules, deploying solar arrays, assembling thermal radiators, etc. This will complement the experience gained from the Solar Array Flight Experiment (SAFE), the Assembly Concept for Construction of Erectable Space Structures (ACCESS), and the Experimental Assembly of Structures in EVA (EASE) truss assembly experiments which were conducted on previous Space Shuttle flights.

Looking beyond the construction of SSF, there are a number of projects, precursor experiments, and other technology experiments[3-11] envisioned that require in-space construction (figure 3). In addition, several configurations of interferometers and long baseline telescopes have been proposed that have overall dimensions of 26 to 100 meters. The size and precision required by these spacecraft are such that on-orbit assembly and checkout are necessary.

Program/Experiment

Large Deployable Reflector (LDR)
Pathfinder
Solar Dynamic Power System
Solar X-Ray Pinhole/Occulter Facility
Astromag
X-Ray Large Array
Geostationary Platform

Figure 3. Future Missions/Experiments Requiring In-Space Construction

Looking even farther into the future, interplanetary missions to explore the solar system will require very large spacecraft that will need to be assembled in space.[1] NASA's Pathfinder Program[2] is a focused technology program which is identifying and advancing the technologies and capabilities necessary to enable these missions; it will develop the basic technology required to construct large, massive structures and complex vehicles. One objective is to define and develop methodologies for constructing generic spacecraft components such as aerobrakes, backbone trusses, and pressurized modules that can be applied to many different missions. Another objective is to develop the processes (welding, bonding, and mechanically attaching) required to join components. A third objective is to develop the ability to manipulate and position large, massive vehicle components precisely so that they can be permanently joined. Concepts such as transporters, space cranes, and assembly fixtures will be fabricated and demonstrated to achieve this objective.

Although a facility large enough to accommodate the assembly of interplanetary exploration vehicles may evolve either as a major modification of SSF (figure 4) or as a separate orbital facility (figure 5), many construction methods and technologies could be demonstrated with precursor experiments attached to a modest space station construction facility. In addition, the projects identified in figure 3 could be assembled at such a facility.

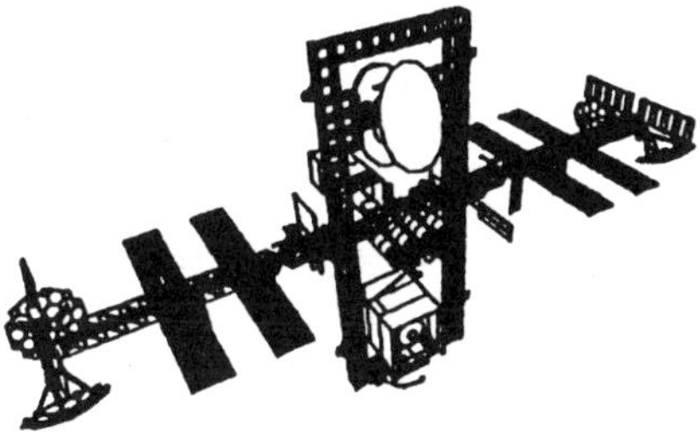

Figure 4. SSF Lunar Mars Configuration

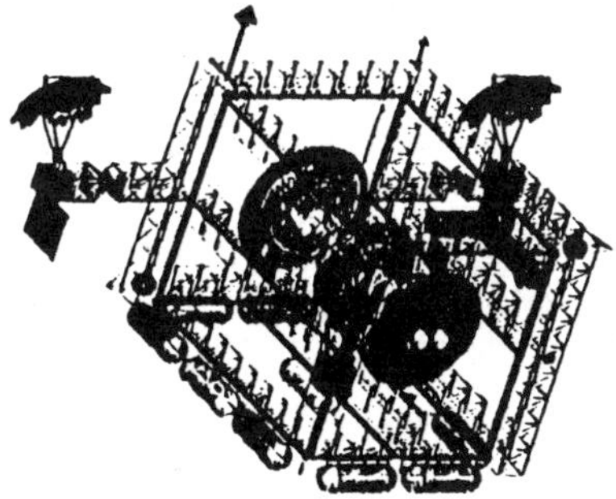

Figure 5. Man Tended Transportation Depot

One objective of an SSF construction facility is to provide an attachment location that permits construction without significant interference with space station operations. The Space Transportation System (STS) will deliver spacecraft parts and components ranging in size from small components packaged in containers to complete modules or assemblies. The construction facility, will, therefore, need a component storage location that will not be used for other activities. The preferred location for such a storage area is within close proximity to the orbiter bay so that the orbiter remote manipulator system (RMS) or SSF mobile service center (MSC) can place the container or module directly in the storage area without the need to translate along the space station truss. The facility should also provide a thermally controlled enclosed volume for containment of small items and thermal control of sensitive equipment.

The choice of location for a construction facility on SSF is governed by many factors. Neither the facility nor the construction projects can be allowed to interfere with space station modules, radiators, solar arrays, logistics modules, MSC operation, or the orbiter docking envelope. This includes not only physical interference, but also interference with the field of view of systems such as solar arrays, antennas, and experiments that require space or Earth viewing.

Another requirement is to locate the construction facility so that it does not significantly impact the microgravity (μg) environment within the laboratory modules. In fact, it is desirable to locate the facility such that the microgravity contour includes a larger portion of the laboratory modules.

The construction facility must also be designed such that when large payloads are attached, the control system is not overloaded and the torque equilibrium attitude (TEA) is within SSF program requirements. The desirable orientation of the facility is one that minimizes drag and other undesirable environmental effects such as solar radiation.

This facility must provide suitable attachment methods for construction projects. Depending upon the payload configuration, single or multiple attachment point locations may be required. For an axisymmetric structure, the facility should allow the payload to be attached to a turntable to allow the structure to rotate past a fixed assembly workstation.

Simulations on the ground and experiments in space have demonstrated that astronauts perform construction tasks more efficiently when using foot restraints. A fixed or mobile workstation with foot restraints should, therefore, be included in the facility. Some generic construction aids such as lighting, tools, video camera, and procedural displays should also be integrated into the workstation.

In-space construction often requires electrical power for workstations, robotic devices, rotating attachment mechanisms, general illumination, and payload test and checkout. Some projects also require data transmission and thermal control. The construction facility should, therefore, provide electrical power, data transmission capability, and thermal control.

Facility Concept

The construction facility proposed for Phase 1 SSF is shown in figure 6. It consists of two eight-bay portions of the Phase 2 SSF lower keel truss structure. Other equipment includes a storage module, a construction turntable, equipment attachment platforms, a surrogate payload bay structure, and a portable workstation. The facility provides electrical power, data transmission, and active thermal control consistent with that available to the Phase 1 SSF.

Facility Structure

The facility structure consists of two eight-bay beams located below the SSF keel truss. The forward face of each beam is equipped with MSC attachment posts to permit MSC translation. This modest structural concept is capable of accommodating most projects that identify a need for on-orbit construction as well as precursor technology demonstrations for those missions that require a more elaborate construction facility (e.g., interplanetary missions). The truss members and nodes are common to the SSF truss; in addition, SSF is already "scarred" to accommodate these truss beams. The facility structure is also coincident and consistent with SSF structural growth plans (figure 1).

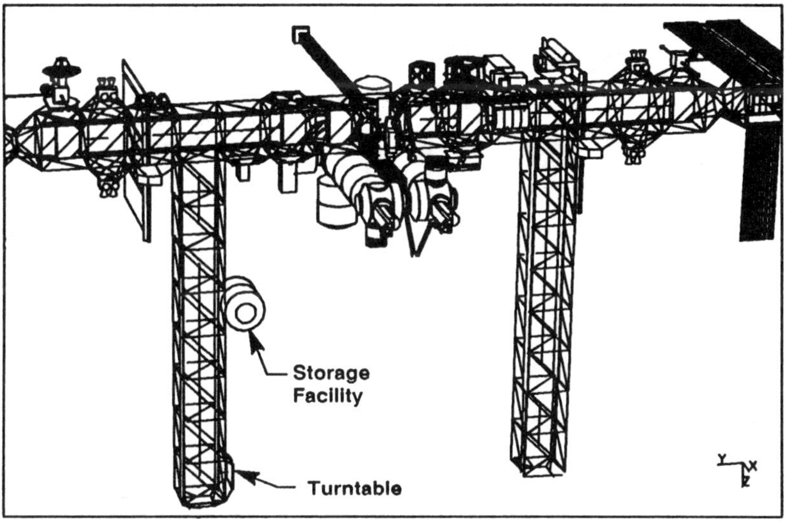

Figure 6. SSF Construction Facility Concept

The truss beams location was chosen so that the SSF center of mass would move toward the center of the module cluster. By lowering the center of mass, less of the laboratory module microgravity (10^{-6} g) envelope will be disturbed. This location also satisfies the requirement to locate construction projects for easy viewing from the modules.

Storage Module

The storage module (SM) is a modified SSF logistics module (figure 7) located on the inboard side of the starboard truss. It is attached at its longeron and keel trunnions by truss members. The keel attachment is a special X-truss that replaces the single diagonal member in the face of the standard 5-meter truss bay. The module contains a variety of storage racks and attachments for tools and other standard equipment. Access is gained through the personnel hatch on the end of the module or through a large cargo door built into the side. The SM is unpressurized, but is thermally controlled to provide a uniform temperature environment. It is oriented with its longitudinal axis aligned with the flight path to minimize drag. The personnel hatch is in the wake direction, and the cargo hatch is positioned to allow access by the MSC from either the main truss beam or from the construction facility. A container of spare SSF truss members and nodes is shown attached to the exterior of the SM. It can be easily removed by the MSC and transported to the site where truss members are needed.

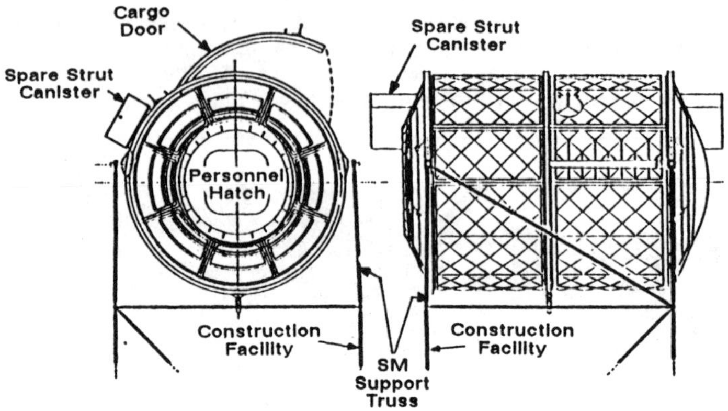

Figure 7. Storage Module (SM)

Construction Turntable

The ability to rotate a construction project, such as the Large Deployable Reflector (LDR), to facilitate construction is accomplished using a turntable. A modified alpha-joint, which is common to SSF equipment and accommodates electrical power, data transmission, and thermal control, is envisioned for this purpose. Figure 8 shows two possible turntable configurations, both of which are attached to the corner nodes of the truss in the same manner that the alpha-joint is attached to the main SSF boom. The low-profile turntable, however, brings the mounting interface closer to the truss by being attached internally to a cube of the facility truss.

Equipment Attachment Platforms

Temporary attachment of small equipment items can be accomplished using SSF structural interface adapter (SIA) hardware or modifications. For example, the SIA plate structure could be replaced with a gridwork plate equipped with quick-release, hold-down clamps or straps that can be inserted into any of the holes in the grid.

Surrogate Payload Bay Structure

Spacecraft elements, modules, or containers of parts that are attached to the orbiter longerons during delivery to SSF can be attached to the construction facility using a structure that duplicates the orbiter attachment method. This surrogate payload bay (figure 9) is equipped with structure that replicates the orbiter bridge fitting track and keel track and will include passive and active trunnion and keel fittings.

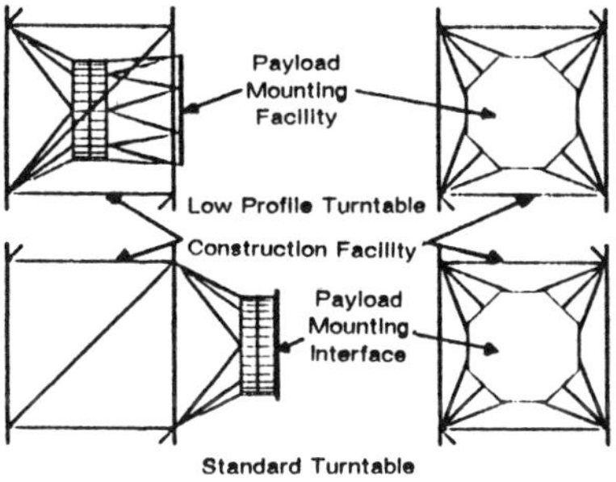

Figure 8. Turntable Configurations

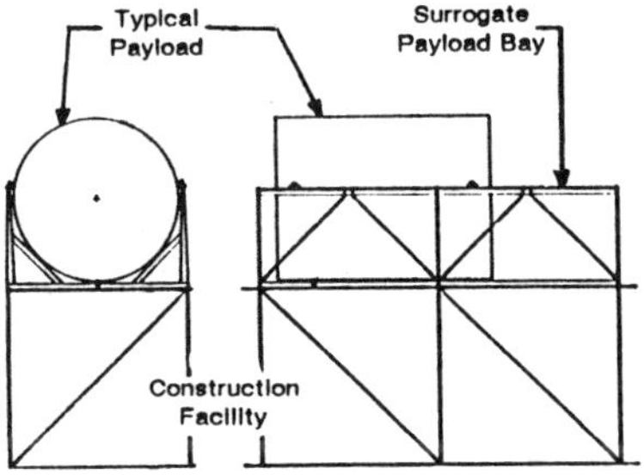

Figure 9. Surrogate Payload Bay Structure

Portable Workstation

The portable workstation is envisioned as a platform that can accommodate one or two EVA astronauts in foot restraints. It can be attached to any SIA to provide a fixed workstation, or it can be used while attached to the MSC robotic arm. It includes lighting, a video camera, accommodations for an assortment of tools, attachment locations for parts containers, controls for positioning the platform while attached to the MSC arm, and generic test and measurement equipment. It also includes a voice-activated "heads up" procedural display device. When not in use, it is attached to the SM or to a convenient SIA, and its equipment (lighting, video camera, etc.) is stored within the SM.

Performance Analysis

A performance analysis of the effects of accommodating the proposed construction facility of the Phase 1 SSF was performed. Payloads are located on the station such that they have a minimum impact on control requirements, microgravity (μg), and torque equilibrium attitude (TEA). Figure 10 depicts the payload locations of two of the larger payloads, LDR and Astromag. The Astromag was located opposite the LDR to help balance the effects of adding a large mass to one side of the transverse boom. This minimizes the center of mass migration along the y-axis, reducing aerodynamic-induced yaw torques about the z-axis.

Figure 11 compares the mass properties of the Phase 1 SSF to those of the construction facility configuration shown in figure 10. The mass increases by 19,000 kg, and the center of mass shifts 3.42 meters in the z direction. Although neither configuration has mass inertia properties which exhibit 3-

axis gravity gradient stability (i.e., Iyy > Ixx > Izz), the construction facility is pitch stabilized (Ixx > Izz).

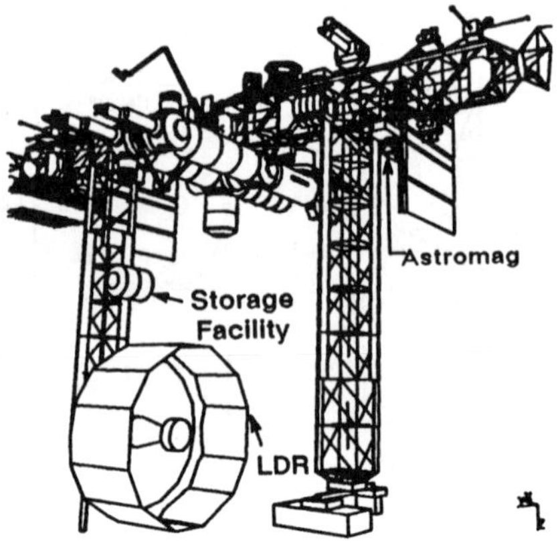

Figure 10. Construction Facility Configuration with LDR and Astromag Payloads

Configuration	Mass (Kg)	Center of Mass (m)			Moment of Inertia ($x10^6$ kg-m^2)		
		X	Y	Z	Ixx	Iyy	Izz
Phase 1 SSF	256000	-1.36	0.68	3.54	117.4	22.7	132.5
Construction Facility	275000	-1.57	0.00	6.96	156.1	56.3	140.0

Figure 11. Mass Properties

Figure 12, which compares control characteristics, indicates the construction facility has a smaller pitch TEA than Phase 1 SSF. The construction facility configuration also exhibits a -4.7° roll TEA due to the asymmetric mass distribution in the y-z plane. Overall, the construction facility improves the pitch and yaw TEA and does not violate the $\pm 5^{\circ}$ local vertical local horizontal (LVLH) space station requirement in the roll axis.

For an actively controlled station using control momentum gyroscopes (CMG), the maximum steady state attitude and the maximum steady state rate deviation increase $\pm 0.2^{\circ}$ per orbit, and $\pm 0.002^{\circ}$ per second, respectively, for the construction facility configuration. This remains within the space station limits as shown in figure 12. The steady state CMG angular momentum control required to hold the construction facility configuration at its TEA increases to

6000 Nms for the construction as seen in figure 12. This does not exceed the 28350 Nms CMG capability of the Phase 1 SSF.

Configuration	TEA (deg) Yaw (z)	Pitch (y)	Roll (x)	Attitude Delta (± deg)	Rate Delta (± deg/sec)	CMG Angular Momentum (Nms)	Max. μg at Lab. Center	Max. Drag (μg)
SSF Limits	±5	±5	±5	2.5	0.02	28350		
Phase 1	1.0	-7.9	-0.2	0.5	0.003	4000	1.4	0.8
Construction Facility	0.0	-6.6	-4.7	0.7	0.005	6000	1.3	0.9

Figure 12. Control Characteristics

The maximum μg at the laboratory center is also given in figure 12. The drag component increases for the construction facility configuration while the total μg magnitude decreases. The increase in drag is due to the increase in area associated with the facility and payloads whereas the decrease in μg magnitude is due to the change in the center of mass location; i.e., the center of the laboratory is closer to the center of mass, thus, the sensed acceleration is smaller.

Figures 13 and 14 show steady state μg profiles for the Phase 1 and construction facility configurations. In both configurations, the 1 μg contour contains only a small portion of the laboratory module while the 2 and 10 μg contours encompass the entire module pattern. As a result of the z-axis shift in the center of mass, the μg contours are lower on the construction facility configuration (figure 14).

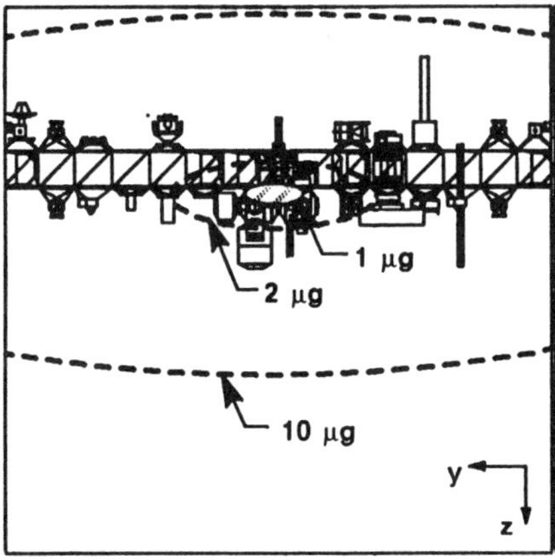

Figure 13. Phase 1 SSF Steady State μg Profile

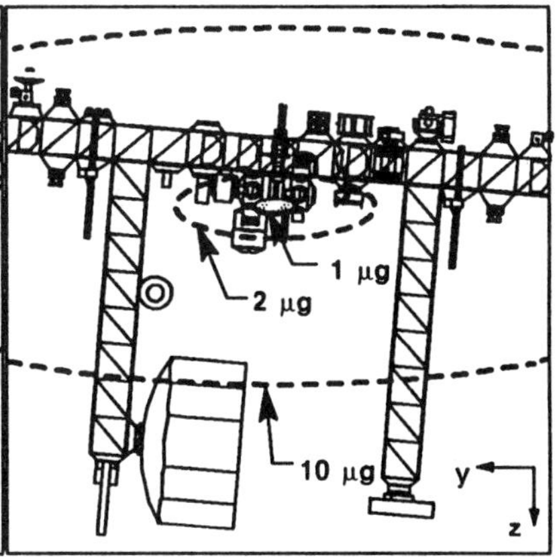

Figure 14. Construction Facility Steady State μg Profile

Conclusion

Future NASA plans for exploring the solar system and programs such as the Large Deployable Reflector and Pathfinder require in-space construction for their success. The technologies required by these and other programs can be demonstrated by experiments conducted while attached to SSF if a construction facility is incorporated in the SSF design. The construction facility concept and associated equipment defined in this paper could provide Phase 1 SSF with the capability to accommodate many on-orbit assembly projects and other experiments without compromising SSF operations, control requirements, or future growth plans.

References

1. NASA Office of Exploration: Beyond Earth's Boundaries, Human Exploration of the Solar System in the 21st Century. 1988 Annual Report to the Administrator.
2. Pathfinder, Research and Technology to Enable Future Space Missions. Program Overview, 1988.
3. Mahoney, M. J.; and Ibbott, A. C.: A Large Deployable Reflector Assembly Scenario. JPL D-5942, November 1988.
4. Russell, R. A.; and Gates, R. M.: LDR Structural Experiment Definition. NASA TM-100618, June 1988.
5. Migra, R. P.: Conceptual Definition of a Technology Development Mission for Advanced Solar Dynamic Power Systems. NASA CR-179482, July 1986.
6. NASA Marshall Space Flight Center: Pinhole Occulter Facility. Phase A Study, Program Development. January 1984.
7. General Electric: Astromag Facility Study. Status Report. April 1989.
8. NASA Marshall Space Flight Center: XLA The X-Ray Large Array. A Pre-Phase A Feasibility Study, Program Development. October 1987.
9. Garrett, L. B.: Langley Research Center Geostationary Platform System Study. Presented at the NASA Langley Research Center Global Change Technology Initiative Workshop, Williamsburg, Virginia, April 4-6, 1989.
10. Ford Aerospace Corporation: Geostationary Platform Bus Study for Earth Observation Sciences. WDL TR-11066, September 1987.
11. Boeing Aerospace: Engineering Analysis for Assembly and Checkout of Space Transportation Vehicles in Orbit. February 10, 1989.

An Astronaut Positioning System for Space Station Assembly and Maintenance Support

R. W. Adkisson[1]
D. R. Barron [2]
R. P. Velgos[3]

Abstract

Large space structures require special systems to support assembly, operations, and maintenance. These systems must be designed to support many tasks throughout the life of the space facility. This requires design solutions that provide maximum operational flexibility while limiting cost and maintaining reliability. This paper focuses on a method of positioning astronauts during extravehicular (EV) assembly, servicing, and maintenance of Space Station Freedom.

McDonnell Douglas has developed an astronaut positioning system (APS) concept for use on Space Station. The APS positions two EV astronauts within a worksite, as required, for assembly and maintenance tasks. During initial assembly, the APS is an integral part of the assembly work platform and provides an efficient method for EV assembly of truss and installation of attached hardware.

This paper discusses three areas related to the design and development of the APS: design requirements, system description, and operations.

Introduction

NASA has proposed to establish a permanently manned Space Station in space during the 1990s to provide a platform for mounting experiments, servicing satellites, and staging advanced missions to the moon and beyond.

---

[1]Design Engineer, Mechanical Systems,
McDonnell Douglas Space Systems Company (MDSSC),
5301 Bolsa Ave, Huntington Beach, CA 92647
[2]Manager-Mechanical Systems, MDSSC
[3]Design Engineer, EVA Systems, MDSSC

Services offered to Space Station users include assembly, maintenance, and resource supply. Extravehicular (EV) activity and telerobotics have been chosen by the aerospace community as the way to perform assembly and maintenance. The APS is a positioning tool for EV astronauts to access locations requiring these services.

The APS is an integral part of the Space Station mobile transporter (MT) shown in Figure 1. The MT is the mobility system for the Space Station Mobile Servicing Center (MSC) [Adkisson, 1988]. The MSC provides key assembly and maintenance support functions for EV astronauts and the flight telerobotic servicer (FTS) through the use of its Space Station remote manipulator system.

The following discussion focuses on the development and design of the APS for Space Station. It also addresses uses of the APS and suggests additional applications that may enhance Space Station operations.

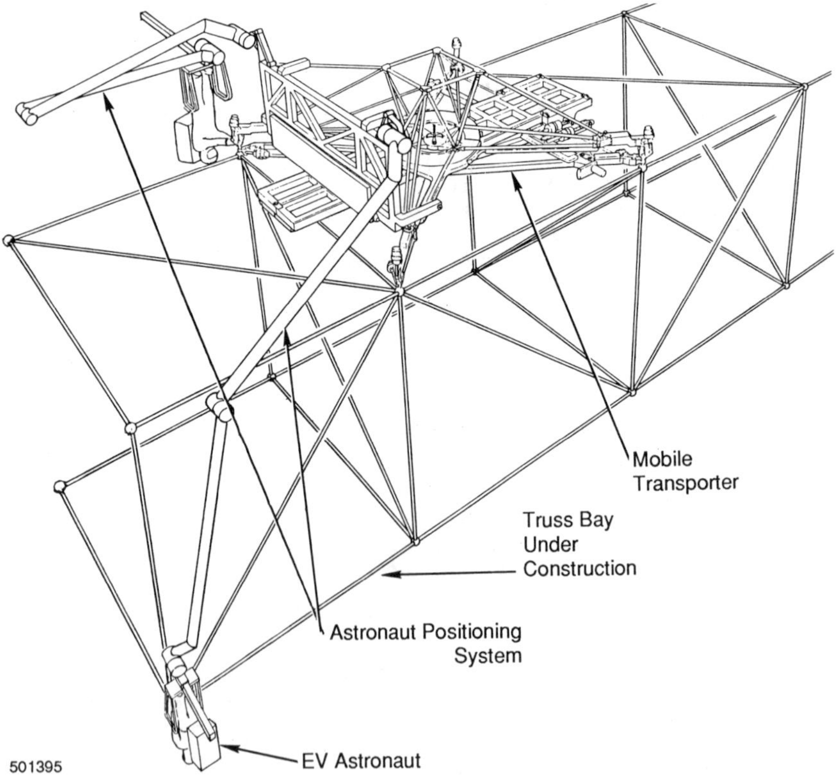

Figure 1. Mobile Transporter and Astronaut Positioning System

Design Requirements

APS design requirements can be divided into functional and interface requirements. The functional requirements are derived from the tasks required to assemble and maintain Space Station. Interface requirements are driven by the interfaces with the MT and the portable, powered foot restraint. In addition to these requirements, program requirements (e.g., safety, commonality, reliability, and maintainability) have also been considered in the APS design.

Functional Requirements. The primary functional requirement for the APS is to position EV astronauts at assembly and maintenance locations. A discussion of assembly and maintenance tasks is included in the Operations section of this paper. Repositioning must be controllable from both the foot restraint platform on the end of the APS and from the workstation within Space Station modules.

Two extreme access positions for building truss (Figure 1) drive the requirements for the APS configuration. Both requirements are derived from the MT and truss designs. The APS structural and mechanical performance is defined by these access requirements. By using the translation and rotation capabilities of the MT, the APS can access a majority of the Space Station hardware.

The APS design must be optimized for EV astronaut requirements e.g., a firm, steady support for fixed location activities, and a smooth comfortable ride between work locations. Excessive motion or accelerations will add to astronaut fatigue and must be avoided. Table 1 provides a summary of the APS performance requirements driven by EV astronaut operations. These requirements are defined by NASA documents (Space Station EVA User Interface Design Guidelines and Space Station Projects Description and Requirements). Velocity, steady-state acceleration, and bending stiffness have been derived by McDonnell Douglas and will be validated by a NASA flight experiment in June 1990.

Interface Requirements. The APS is attached to the MT. This provides a fixed support for APS operation. Attachment is accommodated on the MT by a strongback structure. Loads from the APS (Table 1) are reacted through the MT and into the truss. Power and data attachments, as well as supervisory control, are provided by the MT/MSC system. The APS must stow on the MT between scheduled EV operations as shown in Figure 2. This stowage position minimizes interference with other MSC operations and provides clearance with truss-mounted equipment during translation.

Table 1. APS Performance Requirement

Velocity (maximum)	1.0 ft/sec
Steady-state accelerations (maximum)	0.25 ft/sec^2
Emergency stop distance	1.0 ft
Reacted load* (maximum)	45 lb
Astronaut inertia mass	800 lbm
Bending stiffness** (minimum)	10 lb/in.
Operational lifetime	30 yrs

*Applied at the astronaut's chest; no load reaction tether, APS extended
**Deflection at the tip of the APS due to reacted load

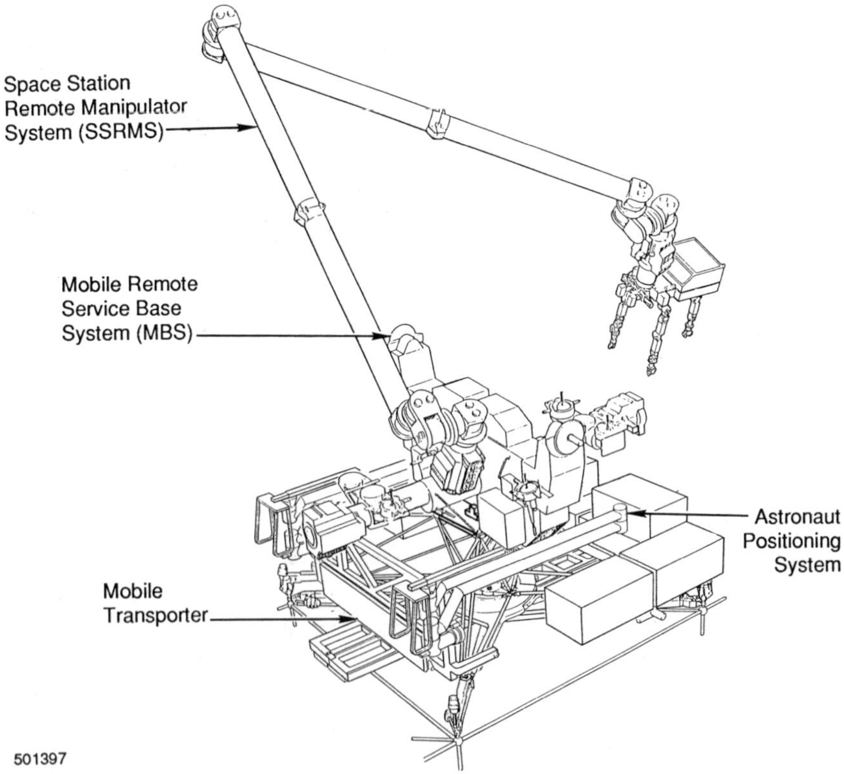

Figure 2. APS Stowed on Mobile Servicing Center

The APS provides a structural attachment for the universal foot restraint described in the next section. The interface also includes power, data, and a dedicated direct circuit for the emergency stop function.

System Description

The APS includes two astronaut positioning devices (Figure 3) that are approximately 25-ft long and contain three primary positioning joints. These three joints (two rotary and one linear) are arranged to provide three dimensional orthogonal positioning capability for an EV foot restraint attached at the end of the foot restraint boom. The foot restraint is removable and contains its own pitch, yaw, and roll joints. These foot restraint joints allow the EV astronaut to change his angular orientation after initial positioning at the worksite. The three primary positioning joints can be controlled locally by the EV astronaut or from a remote workstation. The remote operation is from the space shuttle aft flight deck during early space shuttle-based assembly operations, and from a Space Station cupola workstation when it becomes available later in the assembly process. For all operations, the foot restraint reorientation is locally controlled by the astronaut.

The joint arrangement selected for the APS, when compared to other options studied, provides advantages in operational flexibility, offers improved accessibil-

501396

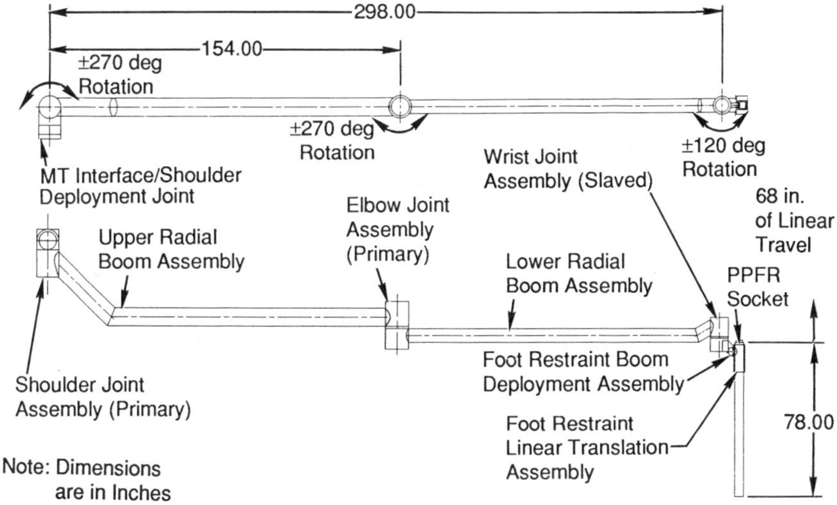

Figure 3. Astronaut Positioning Device

ity to the interior of the truss bays, and allows the use of simple control algorithms. In addition to the two primary positioning joints, the foot restraint boom is mounted to a rotary joint that is slaved to rotate counter to the two primary rotary joints. This slaved joint maintains the astronaut's body axis angular orientation during repositioning. The positioning devices are also equipped with two deployment joints that permit remotely controlled deployment and stowage.

Control Architecture. The control architecture for the APS is illustrated in Figure 4. For remote operation, control signals originate from the crew at the interior workstation in the form of a directional rate command defined for the EV foot restraint. This rate command is transmitted to the data processor on the MT where a control algorithm calculates rates for each of the drive motors. The individual drive motor rates are, in turn, distributed by data bus from the data processor to the motor control electronics at each joint. In addition to driving the APS, the crew will alternatively have the capability to select programmed positions whereby the APS will automatically relocate to the desired position. During all relocation operations, the EV astronaut has the capability to stop the APS by activating an emergency stop switch in the EV control panel. The emergency stop switch is a safety feature that can be used if obstacles are inadvertently encountered or if emergency shutdown is required.

Operations

The assembly of the 5-m cube truss requires a mechanism to translate an EV crew member between the truss node points. The APS solution not only satisfies this requirement, but also provides a convenient method for positioning the EV crew member during other assembly, servicing, and maintenance operations. This additional capability of the APS reduces the EVA support burden on the shuttle and Space Station remote manipulator systems. These systems are then able to place hardware simultaneously with APS positioning of the EV crew, thereby reducing the task duration.

Assembly and Maintenance. The APS is launched, attached to the MT as part of Mission Building Flight 1. The MT, along with a deployable structure, is called the assembly work platform and provides the interface between the space shuttle and partially assembled Space Station (See Figure 5). At the completion of Mission Building Flight 3, the MSC assembly is completed and subsequent assembly operations are Station-based, rather than shuttle-based. A summary of the assembly tasks supported by the APS is listed in Table 2.

A benefit of a manned Space Station is the availability of crew members to perform required repairs. External maintenance activities that include an EV crew member are excellent candidates for including the APS. These tasks will include scheduled hardware servicing, standard repairs, inspection, and failed component replacement. A summary of potential, APS-supported maintenance activities is included in Table 3.

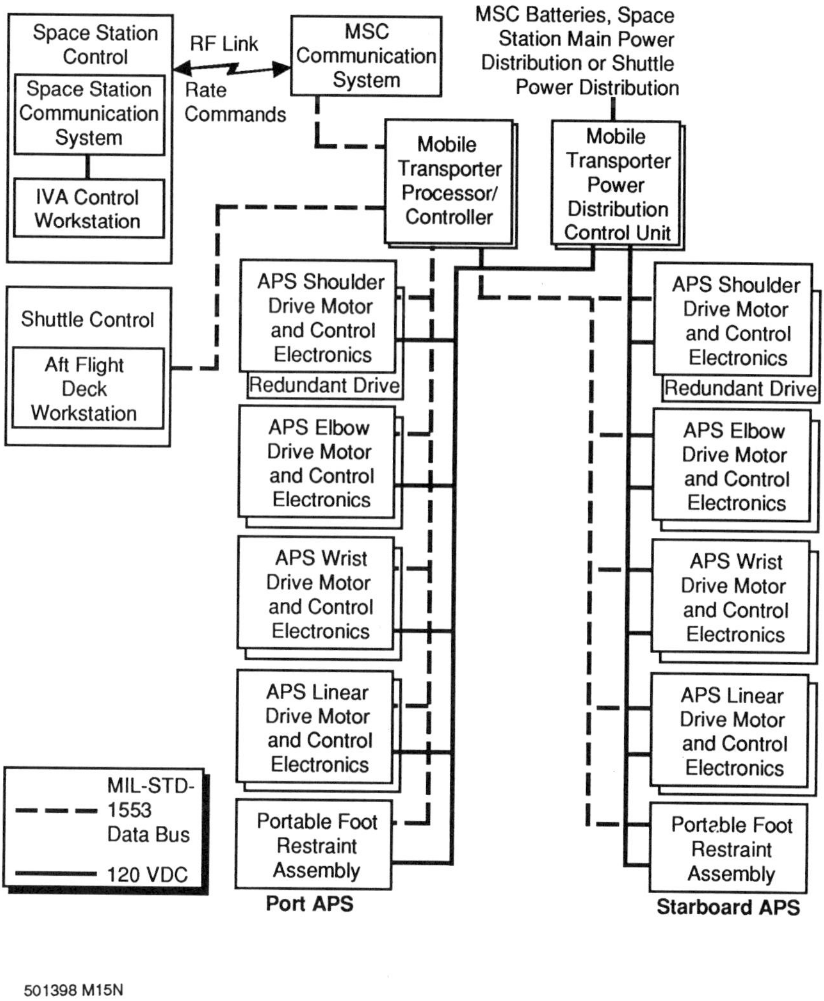

Figure 4. APS Control Architecture Block Diagram

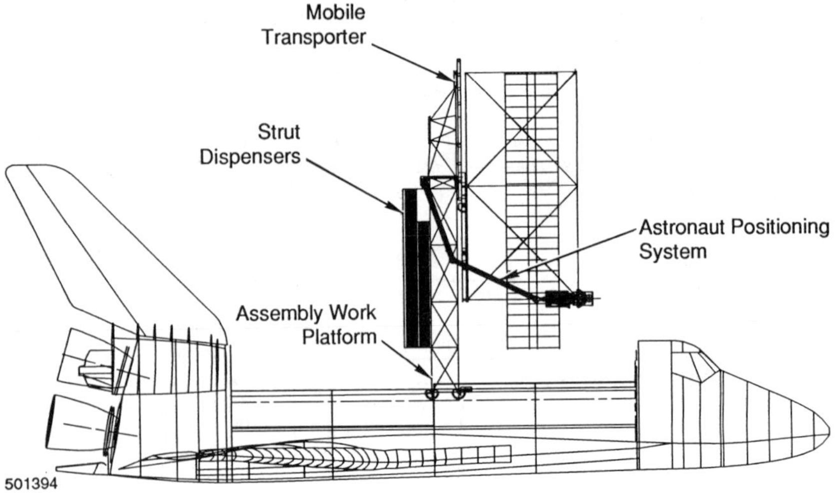

Figure 5. Shuttle-Based Assembly of the Space Station

Table 2. APS-Supported Assembly Tasks

- Installing of longeron, diagonal, and batten struts for shuttle-based truss assembly operations
- Installing of longeron, diagonal, and batten struts for Space Station based-truss assembly operations
- Installing of resource pallets and other equipment within the truss bays
- Installing of resource pallets and other equipment outside the truss bays
- Securing utility trays to the truss structure
- Connecting utility lines to utility trays and resource pallets
- Installing Solar Alpha Rotary Joints
- Attaching other truss-mounted equipment requiring installation as part of the Space Station assembly sequence
- Retrieving longeron, diagonal, and batten struts from containers

Table 3. APS-Supported Maintenance Activities

- Utility tray/utility line repair
- Pallet-mounted hardware maintenance
- Radiator panel replacement
- Pressurized module inspection
- Orbital replaceable unit change out
- Externally mounted experiment servicing

Potential Support of Telerobotic Operations. The capability of the APS to accommodate a flight telerobotic servicer (FTS) is being studied. The FTS would be positioned and restrained at the worksite by the APS where it would perform selected tasks in place of an astronaut. To accommodate the FTS, the APS requires provisions for the FTS structural attachment, as well as additional power, data, and video services. The APS also has to react to potentially higher inertial loads resulting from an FTS mass greater than an astronaut.

The FTS is under development and, as currently defined, will be able to telerobotically perform such tasks as replacement of robot friendly components, attachment and detachment of utility connectors, attachment of experiment packages, assembly of truss structure, and general servicing, inspection, and maintenance operations. The FTS telerobot has two, 7 degree-of-freedom manipulator arms: one, 6 degree-of-freedom attachment/stabilization arm, and associated cameras, control electronics, batteries, and end effectors.

The FTS is being designed to operate either while attached to stationary structure in the vicinity of the task to be performed or, while attached to the end of a manipulator arm. The Space Station and space shuttle manipulator systems have been identified as the primary manipulators to support the FTS. However, using these manipulators to support the FTS has the disadvantage of occupying them when it may be more productive to use them for other manipulating tasks. For example, by using the APS for positioning and restraining the FTS, the Space Station or space shuttle manipulator arms would then be free to retrieve and deliver cargo from the Space Shuttle payload bay or to hold equipment for hand off to the FTS.

Assembly of the Space Station thermal control system radiators is a specific example of potential EVA reduction and productivity improvement benefits through coordinated use of the APS-based FTS and the space shuttle manipulator arm. For this operation, the FTS would be positioned by the APS along the condenser assembly. It would then perform the repetitive task of inserting the radiator panels into the condenser as they are removed from their container (retrieved from the cargo bay and positioned by the space shuttle manipulator arm).

Combined Operations With Remote Manipulator Systems. The constraints on available time for EV assembly operations require simultaneous use of at least one equipment manipulator and two crew positioning devices. The equipment manipulation tasks can be performed by either the shuttle remote manipulator system or the Space Station remote manipulator system. The APS provides the crew positioning capability and is compatible with both remote manipulator systems. During assembly operations, the manipulator retrieves equipment and delivers it to the worksites where astronauts perform assembly tasks while being positioned and restrained by the APS.

Summary

The astronaut positioning system concept described in this paper was developed to satisfy Station configuration-driven requirements. Attached to the MT, the APS is a powerful tool for the EV crew during assembly and maintenance. As an operations tool, the APS provides EVA time-savings, making it a cost-effective solution to astronaut positioning. The addition of FTS support and positioning would enhance the Space Station's telerobotic operations capability. Future studies will identify operational activities associated with a maturing Space Station and the supporting role of APS in these activities.

Appendix

Adkisson, Robert W., The Role of a Mobile Transporter in Large Space Structures Assembly and Maintenance, Space 88, Albuquerque, New Mexico, 29-31 August, 1988.

Telerobotic Capabilities for Space Station Operations

David L. Akin[†]

Abstract

Over the past decade, the M.I.T. Space Systems Laboratory has accumulated an extensive data base on space operations typical of those planned for the Space Station, based on integrated systems tested in neutral buoyancy. This paper presents a summary of those tests involving either pure telerobotics, or telerobots in conjunction with extravehicular activity. The primary areas of application are structural assembly, maneuvering and docking tasks, and satellite servicing.

Configuration overviews are presented on three telerobotic vehicles of the Space Systems Laboratory: the Beam Assembly Teleoperator, used for structural assembly and satellite servicing; the Multimode Proximity Operations Device, used for maneuvering and docking tasks; and the Apparatus for Space Telerobotic Operations, a second-generation device used primarily to date for cooperative EVA/telerobotic operations testing. All of these vehicles are self-contained, with full six-degree-of-freedom closed-loop flight control capabilities. The Beam Assembly Teleoperator has three manipulator arms, including a dexterous general-purpose manipulator designed to replicate the range, speed, and force of motion of the human arm. Results are presented for four series of BAT tests: assembly of the EASE structure, with correlation to the EVA assembly data from Shuttle flight 61-B in 1985; assembly of a Space Station-type truss structure; selected servicing tasks performed on the Hubble Space Telescope; and rescue and return to the orbiter of an incapacitated astronaut. The station truss structure assembly and the HST servicing tasks were done in both pure telerobotic mode, and with cooperative EVA/telerobotic operations. The Multimode Proximity Operations Device is primarily a mobility vehicle, replicating the functions of the Orbital Maneuvering Vehicle and incorporating a cockpit for direct comparison of manned and remote control of free-flying vehicles in mobility tasks, such as docking with satellite mockups. MPOD results include the flight control tests, as well as the use of the vehicle as an Astronaut Support Vehicle for both structural assembly and HST servicing. The Apparatus for Space TeleRobotic Operations is the newest SSL telerobot, and test results presented on this vehicle focus on the use of the system as an assistant to EVA crewmen in structural assembly tasks, maintaining a supply of truss elements as the crew moves along the structure during assembly.

Introduction

By 1983, the M.I.T. Space Systems Laboratory (SSL) had compiled a significant experimental data base on the human assembly of space structures. A theoretical study for the NASA Marshall Space Flight Center had also been completed on the Space Applications of Automation, Robotics, and Machine Intelligence Systems (ARAMIS), which concluded that one of the seminal technologies to be developed was that of space telerobotics [Akin 1983]. As the logical extension of these two experiences, the SSL undertook the development of telerobotic devices capable of a variety of space operational tasks, in order to ex-

[†]Space Systems Laboratory, M. I. T., Cambridge, MA 02139

tend its experimental data base into the realm of automation and robotics.

A number of key decisions were made in the process of designing and developing these telerobot systems. These decisions profoundly affected the experimental approach of this research, and in turn have led to the capabilities and limitations of the present systems. These design decisions include:

• *The use of integrated systems* – to obtain the most complete and comprehensive data possible, the decision was made to develop complete telerobotic systems, including the capability for unlimited 6DOF motion control, manipulation (as required), and all necessary auxiliary systems, such as vision and other sensing. This decision grew out of prior experience in the development of research versions of the manned maneuvering unit, and reflected a serious interest in understanding the human control implications of full telerobotic systems in a high-fidelity simulation of space operations.

• *Neutral buoyancy simulation* – for meaningful simulations of manipulative space tasks, it was necessary to model the weightless dynamics of the space environment. The choice of neutral buoyancy simulation offered the potential for more meaningful correlation of the telerobotic operations results with that of pressure-suited subjects in the same environment, allowing direct comparison of humans and machines performing identical tasks. A good deal of high-fidelity hardware exists from satellite development efforts and crew training in neutral buoyancy; this hardware forms a ready-made set of future applications for the study of telerobotic satellite servicing.

• *Design for initial teleoperation, with provisions for system upgrades* – It was recognized immediately that one of the prime reasons for the existence of these research telerobot systems was to gain knowledge about the operational use of telerobots, rather than just how to design and build such a system. Since the operational implications were not apparent *a priori*, the decision was made to make the hardware as capable as possible initially, but with the potential to upgrade the software for increasing levels of system autonomy as experience indicated such a capability would be useful.

Three such systems have been developed over the last six years in the Space Systems Laboratory. An overview of each of these vehicles, along with a synopsis of important research results, is the focus of this paper.

Beam Assembly Teleoperator

Overview

The Beam Assembly Teleoperator (BAT) was designed to assemble the same structure used by the Space Systems Laboratory for the Experimental Assembly of Structures in EVA (EASE) program. EASE involved two pressure-suited subjects repeatedly assembling a six-element tetrahedral truss, and included both neutral buoyancy simulation and a shuttle flight experiment flown on STS 61-B in late 1985 [Akin 1986]. By choosing as a design case the assembly of this same structure, direct comparisons could be made between EVA and telerobotic assembly, as well as correlation to the flight experience. This structure was designed to be challenging for EVA assembly; no major modifications in the structure were allowed for simplifying the task for robotic assembly. Thus, BAT was designed from the outset to be as capable as EVA for this one specific assembly task, and generically capable of a variety of other EVA tasks as well.

The basic design of BAT (Figure 1) is based on a self-contained mobility base, with vision and manipulation systems attached. The mobility base contains the control electronics, on-board power supplies, and other support systems, as well as eight electrically-powered ducted propellers for underwater mobility. Careful attention was paid to simulation fidelity in the neutral buoyancy environment, and flotation panels and trim weights are attached to the base unit to adjust the centers of buoyancy and gravity to be coincident, so that the vehicle has no preferred orientation. In the current configuration, BAT is equipped with two pairs of stereo monochrome video cameras, one dexterous general-purpose manipulator, a non-articulated grappling arm for grasping the structure under assembly, and a specialized manipulator for performing the coarse alignment of the long struts for the truss assembly.

Figure 1

Configuration Studies

Initial research efforts with BAT were concerned with determining the most favorable system configuration for structural assembly, particularly with respect to manipulators and vision systems. Designed around a single dexterous manipulator, the original design had a grappling arm with single degree-of-freedom (DOF) articulation for handling the struts of the EASE structure. This proved in testing to be difficult to use. Moreover, it was found that the use of the dexterous arm to provide all of the necessary torque for rotating the struts into alignment with the structural connector was fatiguing to the operator and extremely wearing on the motors and drive trains of the manipulator. For this reason, the grappling arm was changed to a non-articulated structure with an end effector capable of a rigid grasp, and a third appendage was added to perform the coarse alignment of the struts prior to handing the strut off to the dexterous manipulator for fine alignment and connection. This combination of "pick-and-place" automation and flexible dexterous manipulation under teleoperation has proven to be a powerful approach to accomplishing a pre-planned task, such as repetitive structural assembly.

The initial vision system for BAT consisted of a single monochrome video camera in a pan and tilt unit, and a second camera with close-up lens slaved to the position of the dexterous end effector. This configuration proved inadequate for both manipulation and vehicle mobility control. Significant in all control regimes was the lack of depth cues to the operator. In addition, the use of a camera on the pan and tile unit proved to be confusing to the operator, as the test subjects universally tended to attempt to "fly" the vehicle along the camera axes, regardless of the amount of information provided as to camera orientation. It seems that, without much greater effort, the operators were unable to make the sequential conversions between visual, camera, and vehicle axes. Tests showed that the most advantageous camera configuration was a pair of stereo cameras rigidly mounted on the vehicle base, located as close as possible to the centerline of the control axes and boresighted along the grappling arm, for use in docking the telerobot to the work site. A second stereo camera pair, fitted with close-up lenses, was used in the pan and tilt unit to provide operator feedback during manipulative tasks. The fixed-axis stereo camera pair was shown in trials to produce comparable grapple times to several other video configurations, but was the only configuration to produce a 100% grapple reliability. Both of these systems are presently low-resolution monochrome; future plans include the substitution of color and high-resolution camera systems to assess the effect on task performance.

EASE Structure Assembly

Once the reference configuration had evolved, BAT was used to study teleoperated assembly of the EASE truss structure (Figure 2). Each of the basic tasks (attaching a strut to the structure, attaching a node to a strut, connecting a node and strut to close a face of the truss structure) were performed separately, and the results used to derive an assembly process for the entire structure. The structure was then built in an end-to-end test, requiring one hour and thirty-three minutes (1:33:00) in a master-slave teleoperated mode. This may be compared to the best time for a single pressure-suited person in neutral buoyancy of 12:30, which shows a clear-cut advantage for the EVA assembly technique. However, it is perhaps more properly compared to the 1:11:00 time for the same EVA subject on his first assembly of the EASE structure. Since the times of 1:33:00 and 1:11:00 are at equal points on the learning curve, teleoperated structural assembly may be seen as competitive with EVA assembly. [Viggh 1988]

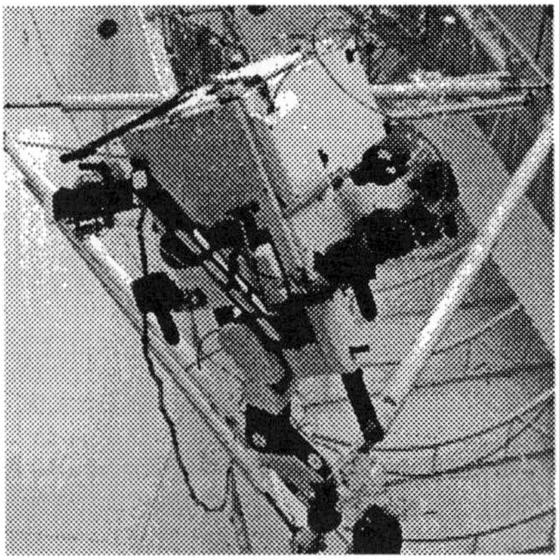

Figure 2

Subsequent telerobotic assemblies focused on the effect of supervisory control modes on assembly productivity. The dexterous activity of structural assembly (aligning and connecting the structural elements) was broken down into five tasks. Due to the nature of the BAT sensory system, one of these (aligning and inserting the strut into the connector) had to remain a master-slave task; the other four (grasping the beam, grasping the locking sleeve, locking the connector, and stowing the manipulator for subsequent free-flight) were automated, so that the BAT operator merely commanded the sequence, and monitored to insure proper operation. This increase in robotic capability was then assessed for its effect on the productivity of the structural assembly task. Of particular interest was the fact that supervisory control did not markedly affect the performance times, either of the individual connections or of the total structural assembly. It did make a marked difference in the level of operator training required (down from 20 hours in the master-slave case to 2 hours prior to the supervisory assembly test), and in the perceived levels of operator fatigue. From these results, it appears that increasing the level of automation tends to aid the operator, but not to increase the overall productivity. It should also be pointed out that, if the supervisory control routines fail, the operator is not generally skilled or practiced at performing the same task manually, in which case the effect is that of having a naive operator performing the operational task. This has some serious implications for system reliability and task assurance. [Anderson 1988]

Space Station Structure Assembly

BAT was also used to study telerobotic assembly of a space station truss mockup. As shown in Figure 3, the Space Systems Laboratory mockup of the space station truss used subscale struts (2.8 m as compared to the 5 m struts for the space station), based on theoretical models used to design neutral buoyancy structures with the maximum possible dynamic fidelity to structural assembly on-orbit. No complete end-to-end robotic

Figure 3

truss assembly was completed in this case; the times for the component tasks produced a projected assembly time for one cell of the truss of 1:39:17. This is much greater than the nominal time for 2 EVA subjects to assemble a cell of the same truss (7:00). This was also the first test of joint EVA/telerobotic activity in the same worksite. It proved to be no problem to have the EVA subjects perform the more demanding dexterous tasks, such as removing the struts from the launch restraints and handing them to BAT. Due to the wide disparity in assembly times, however, it was not found to be productive to have BAT and the EVA subjects jointly assemble a single cell, and the approach adopted was for BAT to assemble cells remote from the EVA assembly site, returning as necessary for more structural components.

Hubble Space Telescope Servicing

In January, 1989, the Space Systems Laboratory had the opportunity to perform telerobotic servicing tests on the crew training mockup of the Hubble Space Telescope (HST). Three types of test were performed: completely telerobotic (no EVA assistance); EVA replacement (BAT substituting for the second EVA crewman, and performing the tasks nominally assigned in the EVA timeline); and telerobotic assisted (BAT acting as an assistant to the prime EVA subject, and performing tasks to which it is best suited). At the request of the HST program office, these servicing tests focused primarily on the task of changing out the nickel-cadmium batteries, and replacing them with nickel-hydrogen batteries with different packaging and mounting configurations. Figure 4 shows BAT removing a nickel-hydrogen battery from the launch support in a completely telerobotic operation. The interface for this unit, as for all of the units tested, was completely unchanged from that developed for EVA servicing. The battery is held down with a large number of 7/16" hex-head bolts, which were torqued to levels beyond that achievable by the BAT power tool. In addition, the configuration of the unit required the use of an 18" extension on the 7/16" socket tool. The compliance of this total assembly reduced the positioning accuracy of the BAT dexterous manipulator to the point where engaging the bolt head was problematical. Due to this type of problem in each of the HST EVA interfaces, the results from both the completely telerobotic and EVA replacement approaches were disappointing.

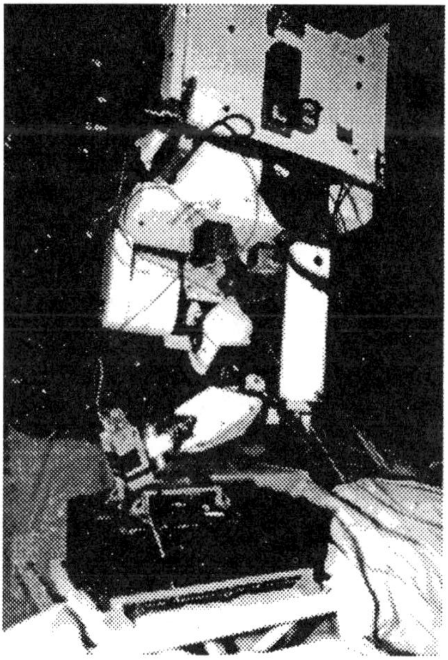

Figure 4

The best results came in the final series of tests, wherein the servicing procedures were freely adapted to allow the allocation of tasks to BAT for which it was best suited, and in which the EVA subject performed all of the dexterous or high force tasks beyond BAT's capabilities. One of the best configurations found is shown in Figure 5. It was found early on that the high complexity of the servicing tasks favored a stable platform for the manipulators, and therefore BAT was attached to the shuttle Remote Manipulator System (RMS), which provided all its subsequent mobility and positioning. One problem is that this prevented the use of the manipulator foot restraints (MFR), which provide a fixed work platform for the EVA subject. The approach taken was to have BAT mounted in the RMS, and

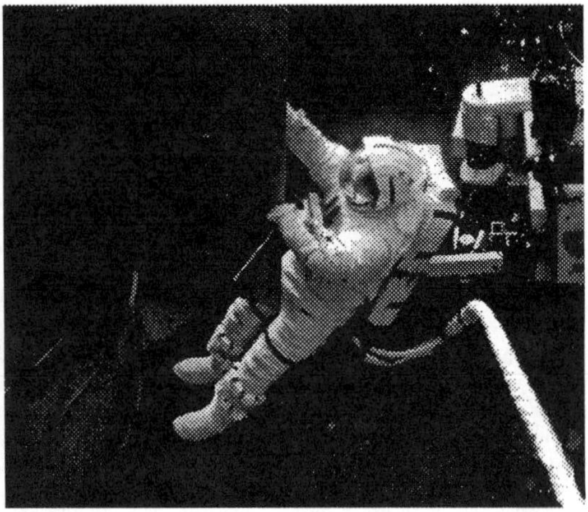

Figure 5

to hold the backpack of the EVA subject with the fixed grappling arm. In this way, the RMS still provided positioning for the EVA crewman, and BAT's dexterous arm was available to the right of the test subject to hold tools, receive packages (such as the removed battery packs) and stow them for transport while the EVA subject continued working, and perform other useful support tasks. The subjects reported that the backpack grasp was more comfortable than the long-term use of foot restraints, and the addition of degrees of freedom to the grappling arm would have allowed more intricate adjustment of EVA position than would be available through the RMS alone. These tests demonstrated that a dedicated telerobotic EVA support system is a powerful concept for enhancing EVA productivity.

EVA Rescue and Safety

Tests have demonstrated that telerobotic systems such as BAT can be a significant enhancement to EVA safety. In tests from 1987 (Figure 6), the free-flight capability of BAT was used to grapple an EVA subject pretending to be incapacitated, and to return the subject to the vicinity of the shuttle airlock. Tests showed this task to be even easier if the EVA subject is cooperative. Since the space station will not be able to navigate, some means of rescuing EVA crewmen who may become unattached from the station (for whatever reason) is critical; telerobotic freefliers would be on-site faster than if another EVA crewman had to ingress a manned maneuvering unit for the same task.

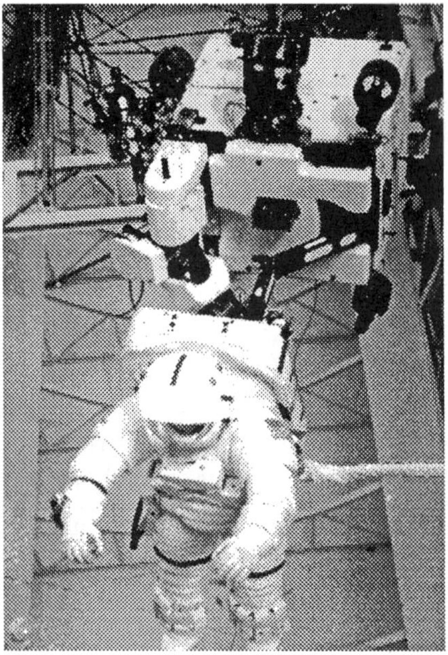

Figure 6

No discussion of cooperative EVA and telerobotic activities would be complete without consideration of safety issues. Many of the Space Systems Laboratory tests had the EVA subject in immediate contact with the telerobot, well within the operating envelope of the BAT dexterous manipulator. This would not be acceptable practice with a more restrictive safety rule, such as one requiring a total exclusion zone within the robotic reach envelope. However, this type of rule would also eliminate some of the most productive cooperative work strategies found in this research. Further research is necessary on methods of allowing safe access within the manipulator working envelope, such as depowered actuators, surface contact sensors, or other means of insuring that no inadvertent manipulator motion could imperil the EVA crewman. However, failure to make these accommodations will prevent the full integration of EVA and telerobots in the same worksite, and result in two competing, and lower productivity, approaches to performing the same task.

Multimode Proximity Operations Device

Overview

Figure 7

Although the Beam Assembly Teleoperator has a fully capable free-flight mobility unit, it was primarily designed to investigate teleoperated manipulation tasks. The Multimode Proximity Operations Device (MPOD), on the other hand, was specifically designed to investigate the teleoperated and automated control of a free-flying vehicle, similar in application to the planned Orbital Maneuvering Vehicle. As shown in Figure 7, MPOD is a large vehicle of approximately spherical configuration. The symmetry of the vehicle was designed to maintain equal drag properties in each direction, to maximize the fidelity of the flight dynamics in the neutral buoyancy environment. Since the initial research objective for MPOD was to investigate the role of different types and fidelities of sensory feedback to the human operator, it was designed around a cockpit for internal flight control. Thus, the same task (such as docking to a satellite mockup) can be performed onboard and teleoperated for direct comparison of task performance in the different control modes. Beside the redundant ducted fan thruster systems, the only external payload on MPOD in its reference configuration is an instrumented docking probe, used for rigidly docking with a matching drogue which may either be fixed, or mounted to a 4000 pound satellite mockup for unrestrained docking tests. [Vyhnalek 1985]

Operator Location Studies

The initial tests of MPOD dealt with operator location and sensory feedback. When the operator is on-board, he or she receives full vestibular feedback, as well as direct vision out the forward window. This window can be covered over, making the operator fly based on visual references from on-board and/or remote video cameras, displayed on a 4" monochrome monitor on the MPOD instrument panel. This monitor may also be used for supplementary views during tests with unobstructed direct vision. MPOD may also be controlled from an underwater control station, where direct external vision of MPOD and the docking target may be selectively controlled, and the same 4" monitor used to supply the same choice of video sources. By removing the operator from MPOD, the vestibular feedback can be eliminated. By the choice of whether or not direct vision is allowed, and the choice of video source on the monitor, the type and quality of visual feedback can likewise be controlled. The initial research application used these degrees of sensory feedback control, and looked at the effect on a standardized task of closing on and docking to a fixed target from an initial range of 5 m. Although testing time was too limited to run a statistically significant number of test subjects through the protocol, results to date indicate that remote teleoperated control is marginally better than on-board control. This is due largely to the difficulty in acquiring visual cues of closing velocity and distance, which are better formed from a direct exterior view than either a direct or video view on-board. At the maneuvering rates of this type of vehicle, vestibular sensory cues are not of noticeable value. [Tarrant 1987]

Astronaut Support Vehicle Tests

The Space Systems Laboratory has postulated the use of an astronaut support vehicle for extended-distance EVA operations. Analogous to the diver support vehicles of the undersea community, such a system would support the crew for extended duration missions away from the space station, and provide operations support for EVA tasks upon arrival at the work site. Figure 8 shows a proof-of-concept test utilizing MPOD as such a vehicle, with the docking probe removed and a standard manipulator foot restraint system attached to the front of the vehicle. The remote operator provides translation and positioning support for the EVA crewman. This concept has been tested for both structural assembly and HST servicing tasks. These tests have shown that vehicles of the size of MPOD can accurately maneuver in constrained spaces, such as around HST, and that the large mass of such a vehicle provides a great deal of inertia for stabilizing body motions, even without a rigid connection to the worksite. Further studies will examine the use of hand controllers for the EVA subject to directly control MPOD motion, as well as incorporation of the EVA assistance manipulators from the BAT HST servicing tests.

Figure 8

Apparatus for Space TeleRobotic Operations

Overview

The Apparatus for Space TeleRobotic Operations (ASTRO) is a second-generation telerobotic device, incorporating lessons learned from four years of operational experience with BAT and MPOD. ASTRO is a device the size of BAT, with a near-symmetrical shape derived from MPOD. It is designed to be modular, with interchangeable front end kits to support different configurations of manipulators. The current status of this vehicle is that the mobility unit has been completed, and the closed-loop control system has been installed and debugged.

EVA Logistical Support

The mobility unit of ASTRO was used with an open-loop control system as logistical support for an EVA subject assembling space structures. The ASTRO controller would bring structural struts and nodes to the EVA crewman, assembling a space station truss from the manipulator foot restraints on the end of the RMS (Figure 9). This test demonstrated that a free-flying vehicle could be accurately controlled, even open-loop, in the vicinity of EVA crew and the space station truss, and that the use of such a vehicle could increase assembly productivity by eliminating the requirement for the EVA subject to translate back to the launch support fixture to obtain parts. Further possibilities for this type of vehicle include remote inspection of hardware or monitoring of activities. Future plans for ASTRO include the development of two dexterous manipulators for general-purpose telerobotic manipulation studies.

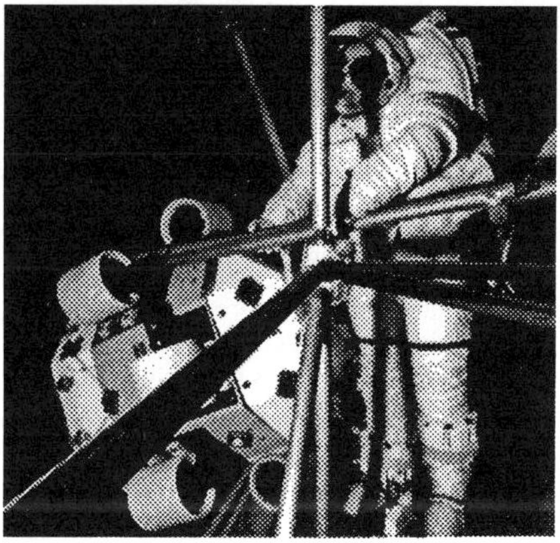

Figure 9

Future Efforts

This work will be continued and expanded in the future. The Space Systems Laboratory is in the final stages of developing a modular, high-torque positioning manipulator for use in further telerobot mobility studies. Using an innovative pneumatic/hydraulic hybrid drive system, this manipulator is composed of a two-link, 3 DOF coarse positioning arm, and a 6 DOF Stewart platform for fine positioning. The interchangeable links are 3, 6, 12, and 24 feet long, providing a system with total maximum reach lengths of between eight and fifty feet. The actuators are designed to provide ten pounds tip force at maximum (50') extension, thus providing an order of magnitude greater capability than any existing neutral buoyancy manipulator system. This manipulator will be used in conjunction with BAT in studying large positioning arms as an alternative to free-flight telerobotic mobility in the structural assembly and satellite servicing worksites.

The Space Systems Laboratory will move in its entirety from the Massachusetts Institute of Technology to the University of Maryland during the Summer of 1990. This move will allow greater expansion of laboratory activities. Most exciting is the development, already underway, of a neutral buoyancy research facility at the University of Maryland. With a water tank size of 50' diameter by 25' deep, this will be comparable to the neutral buoyancy facilities in NASA and industry, and will be the only facility designed from the outset and dedicated to fundamental research in telerobotic space operations.

This type of research can be easily and usefully extended into the development of space flight experiments. Using design techniques and hardware directly taken from the neutral buoyancy telerobotics research, the Space Systems Laboratory developed a spacecraft to test an innovative concept in reentry thermal protection. This spacecraft was developed in five months at a total cost of $80,000. It was onboard the American Rocket Company SET-1 launch vehicle, and is currently being refurbished following the failure of that booster. Extending this experience, design is currently proceeding on a low-cost telerobotic system (similar to BAT) for use in space telerobotics experiments.

Conclusions

The neutral buoyancy environment can be a difficult one for the simulation designer; however, it offers unique advantages for the integrated simulation of space telerobotic operations. A large and growing data base exists on EVA simulations in neutral buoyancy, with correlations to space flight experience. The development of full-scale telerobots for neutral buoyancy allows the use of existing hardware and mockups, direct comparison with EVA operational experience, and tests of EVA/telerobotic cooperative activities in a realistic simulation of the space environment. Results to date have demonstrated that telerobots, even if fairly simplistic from a technology viewpoint, can assemble space structures and service satellites in times that are generally comparable to those of EVA. More importantly, these tests have shown that a cooperative EVA/telerobot worksite in space can markedly enhance human productivity, while at the same time providing important safety features for the EVA crewmen. Further studies will be aimed at refining the data base on telerobots, and telerobots working in conjunction with humans, as well as developing and testing new concepts for telerobotic applications to the types of operations necessary in the space station era.

References

[Akin 1983] D.L. Akin, M.L. Minsky, E.D. Thiel, and C.R. Kurtzman, "Space Applications of Automation, Robotics and Machine Intelligence Systems (ARAMIS) - Phase II," Vol. 1,2, and 3 NASA Contractor Reports no. 3734, 3735, and 3736, for Contract NAS8-34381, October 1983

[Akin 1986] David L. Akin, "Experimental Assembly of Structures in EVA: Overview of Selected Results" Space Construction, Conference at NASA Langley Research Center, August 6-7, 1986. NASA Conference Publication 2490.

[Anderson 1988] David Elliott Anderson, "Supervisory Control Algorithms for Telerobotic Space Structure Assembly" S.M. Thesis, Department of Aeronautics and Astronautics, Massachusetts Institute of Technology, May 1988

[Tarrant 1987] Janice M. Tarrant, "Attitude Control and Human Factors Issues in the Maneuvering of an Underwater Space Simulation Vehicle" S.M. Thesis, Department of Aeronautics and Astronautics, Massachusetts Institute of Technology, Sept. 1987

[Viggh 1988] Herb Viggh, "Teleoperated Space Structure Assembly with a Task Planning Advisor" S.M. Thesis, Department of Aeronautics and Astronautics, Massachusetts Institute of Technology, Feb. 1988

[Vyhnalek 1985] Gary Gerard Vyhnalek, "A Digital Control System for an Underwater Space Simulation Vehicle using Rate Gyro Feedback," S.M. Thesis, Department of Aeronautics and Astronautics, Massachusetts Institute of Technology, June 1985

FLUID LINE DEPLOYMENT AND REPAIR FOR SPACE STATION

Raymond H. Anderson[1], Shirley J. Pearson[2], and J. Douglas Drewry[3]

Abstract

Deployment of the utility system for providing services to the NASA Space Station Freedom provides a unique challenge in design and material technology. The system must be lightweight, reliable, EVA and robotically compatible, and maintainable over the projected life of the station. The system must provide continuous operational environments at various pressures to users within the station.

The station will be built largely by IVA-controlled telerobots and EVA crew members. Deployment of the fluid system lines and the electrical power cables must be accomplished using simple, safe techniques which minimize EVA crew time. The system, once in orbit, must be maintainable and easily repaired over its projected life. Fluid line connections and electrical couplings, assembled in the hostile environment of space, must be reliable, easy to inspect, safe to prepare, and require a minimum of preparation and tooling.

On-orbit deployment of this system by a flexible metal tray meets these requirements. Deployment of the lines and power cable from a spool located within the Space Shuttle Orbiter bay and containing long sections of terrestrially connected metallic or nonmetallic lines provides a unique system for deploying the components while minimizing EVA time of the crew. The flexible aluminum trays with hinged metal covers provide ready access for crew or robotic inspection and maintenance. Possible use of flexible metallic or nonmetallic lines within the tray provides for either vertical or horizontal line adjustment during line maintenance or repair. Tube fittings and quick-disconnect couplings are being developed to provide a high degree of robotic and EVA compatibility for system maintainability. Continued development of automated tools for line maintenance and fitting repair will reduce the time required for either EVA crew or robots to perform this system maintenance.

[1] *Propulsion and Fluid Systems, Space Station Division, McDonnell Douglas Space Systems Company, 5301 Bolsa Ave., Huntington Beach, CA 92647*

[2] *Thermal Systems, Space Station Division, McDonnell Douglas Space Systems Company, 5301 Bolsa Ave., Huntington Beach, CA 92647*

[3] *Operations and Integration, Space Station Division, McDonnell Douglas Space Systems Company, 16055 Space Center Blvd., Houston, TX 77062*

On-orbit welding provides a viable alternate to mechanical methods for tube joint repair and preparation. Extensive on-orbit testing by the Soviet Union provides a data base for further miniaturization and developed reliability in space welding equipment. On-orbit joint preparation and inspection need further development to make the process viable for use on space structures such as the station.

Introduction

Future NASA space explorations to exploit Earth orbit, build a permanent outpost on the moon, or send personnel to explore Mars will involve large space structures that will act as permanent stations for satellite servicing, space experimentation, and rendezvous points for deep-space explorations. Such structures cannot be placed in orbit by a single launch vehicle. In order to prepare for these most ambitious goals, the United States is currently preparing to establish a permanent presence in low-Earth orbit with the development of the NASA Space Station Freedom. The station will be the first NASA structure to be fully assembled in space and will be placed in low-Earth orbit to serve as base for assembly and maintenance of other large space structures. Such space structures include earth-observing, free-flyer platforms and proposed large scientific instruments such as the lightweight 20-meter submillimeter wavelength telescope being proposed by NASA. Figure 1 shows a sketch of the currently proposed NASA Space Station Freedom.

Distribution of fluids and electrical power to the various systems on the station is vital to the successful operation of the structure. Over 2 miles of tubing and electrical power cabling will be required to distribute these services. Deployment procedures for

DAC81819 101129

Figure 1. NASA Space Station Freedom

these systems must be designed to minimize crew-assisted EVA time. Once deployed, the system must ensure that maintenance and repair procedures can be easily performed. Damage to line components from impact by space debris and meteoroids will require maintenance or repair to the system over their lifetime. Current station design defines tube outer diameters ranging from 1/4 to 1-1/4 inch. McDonnell Douglas has been active in developing leak-free, reliable components and EVA and telerobotic-compatible mate/demate techniques.[1-4]

System Deployment

On-orbit joining of each fluid line and electrical connector by EVA crew would be time consuming, crew fatiguing, hazardous, and would reduce system reliability and performance. Use of telerobotics to perform similar on-orbit joining operations would also be time consuming and reduce system reliability. To reduce EVA and robotic on-orbit assembly time, McDonnell Douglas is studying a system by which large sections of metallic, nonmetallic lines and electrical power cable segments are terrestrially preassembled, pressure checked for joint leak tightness, power cables checked for continuity, and clamped into flexible metal trays. These tray segments are coiled on a large, 14-ft-diameter spool (Figure 2) for storage in the Space Shuttle Orbiter. The power cables are isolated in the trays from the fluid lines.

DAC77640 101130

Figure 2. Spool for On-Orbit Fluid and Electrical Line Deployment

The flexible trays will be automatically uncoiled from the spool within the Shuttle Orbiter and attached to preassembled sections of the station truss (Figure 3) by a combination of EVA crew and telerobotics. This will result in less time for utility deployment and less fatigue for the EVA crew.

Attachment of individual line connectors at termination points of trays will be performed by a combined effort of EVA crew and telerobots. Couplings and fittings (both fluid and electrical) are being developed that have a high degree of both EVA and telerobotic compatibility.[5-7] Trays will contain breakout sections where system umbilical connections will be completed by either EVA or telerobots using EVA-compatible quick-disconnect couplings. Such couplings require a minimum assembly force and are sized to be easily handled by either EVA crew or telerobotic end effectors.

The spool and flexible metal tray concept for on-orbit line deployment is designed to save 10 minutes of EVA time per tray when compared with on-orbit joining of flat tray sections. It also provides for a safer system deployment. The spool concept of line deployment has been successfully demonstrated in a series of weightless simulation tests.[8] Flexible tray sections were successfully unwound from the spool, attached to preassembled sections of truss, and an umbilical connection was made by quick-disconnect coupling with an orbital replacement unit (ORU).

System Maintainability

Maintainability of the utility and electrical power lines and the supporting flexible tray segments is a major requirement in the system's design. Maintainability

101132

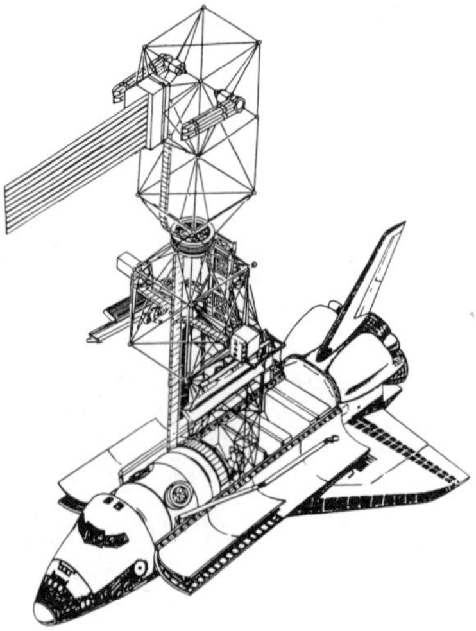

Figure 3. Preassembled Tray Segments Being Deployed On Orbit From the Space Shuttle

requirements such as (1) use of telerobots for line removal and replacement, (2) direct tray access for line and component inspection, and (3) tray design which minimizes special tools and equipment for repair and maintenance are important to ensure that the system can be adequately maintained by both EVA crew and telerobotic systems.

Extensive testing by McDonnell Douglas has demonstrated both the EVA and telerobotic compatibility of both mechanical tube fittings and quick-disconnect couplings.[1] Assembly, disassembly, and repair of these components both in a 1g and simulated weightless environment has been demonstrated (Figures 4, 5, and 6).

It has also been shown that it requires a longer time for a telerobot to perform operations such as nut torquing and tube cutting than EVA crew. The more precisely controlled Kraft telerobotic arm or NASA-developed flight telerobotic servicer (FTS) could provide greater flexibility and reduce mate/demate times based on single robotic systems (Figure 7). The Kraft arm has been utilized in 1-g tests at NASA-JSC to assemble and repair tubular components similar to those proposed for the station. Separable mechanical tube fittings can also be readily assembled and disassembled with automated or EVA-compatible manual tools as backup to telerobots. Such an automated, modular tool for assembly and disassembly of separable mechanical fittings is shown in Figure 8. The end effector can be removed from the tool body and manually EVA operated by a wrench. This tool has demonstrated its effectiveness in assembly and disassembly of separable tube fittings up to 1-1/4-inch in diameter by both EVA crew and the Kraft telerobotic arm.

SSC109859-3

101134

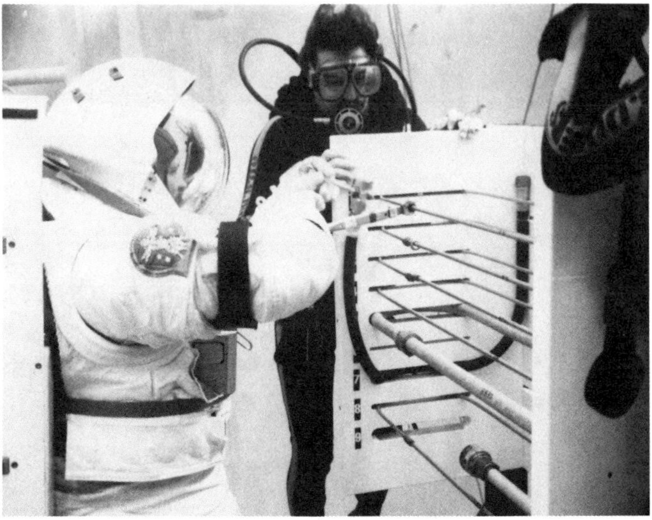

Figure 4. Assembly/Disassembly of Mechanical Tube Fitting in a Simulated Weightless Environment

SSC109831 101135

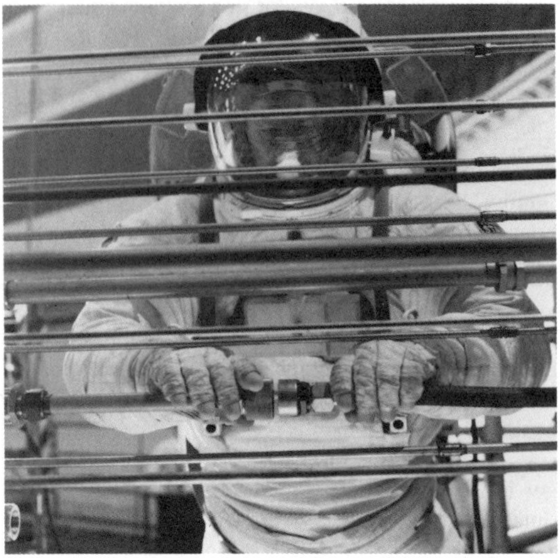

Figure 5. Mate/Demate of Quick-Disconnect Coupling in a Simulated Weightless Environment

SSC112261 101136

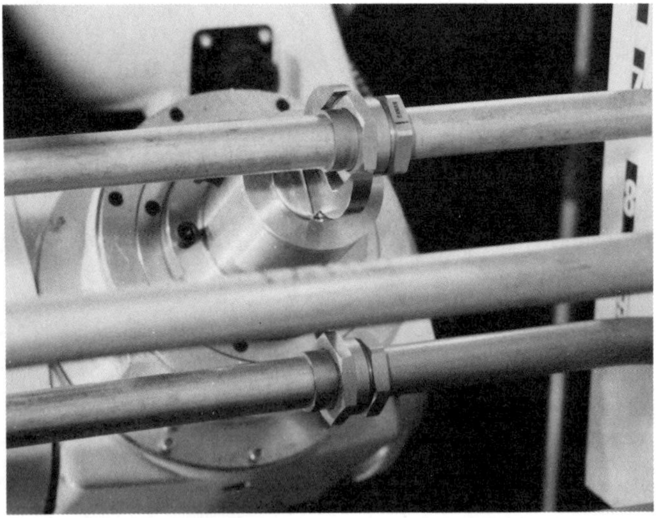

Figure 6. Robotic Assembly of a Separable Mechanical Tube Fitting in 1-g Environment

101137

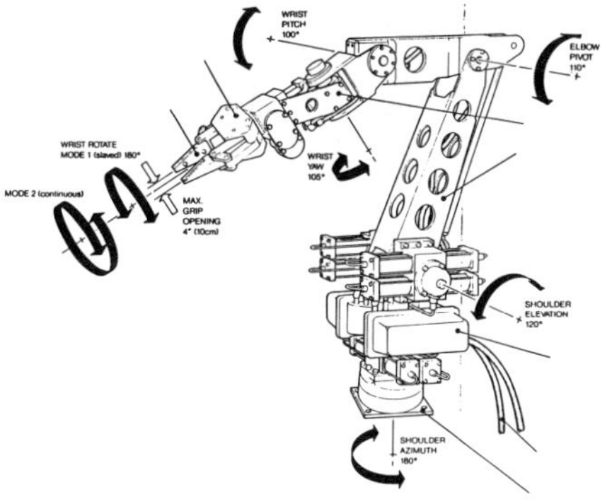

Figure 7. Kraft Telerobotic System

101138

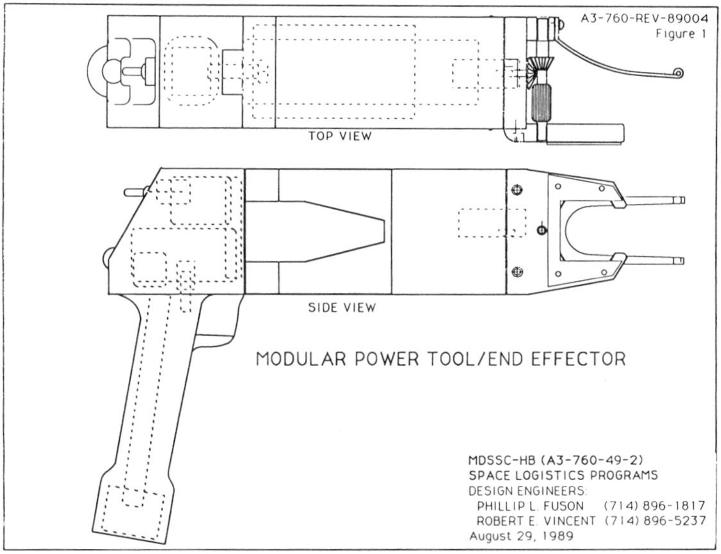

Figure 8. Automated Tool for On-Orbit Assembly and Disassembly of Mechanical Tube Fittings

The flexible metal tray is designed for easy access. Hinged tray covers can be opened either by telerobots or EVA crew. The covers open for visible access to the internal tray components. Distance between lines within trays remains small because of the large number of lines required to service all of the systems on the station. This will make glove and end effector access to the potentially defective line more difficult. Maintenance of the metal trays may require a different approach. Should a tray become heavily damaged due to micrometeoroid or space debris impact, it may require replacement of the entire tray section.

System Repair

On-orbit repair of lines or tray segments must be performed utilizing a process which is reliable, simple, safe, and provides for ease of final inspection. Mechanical tube fittings and quick-disconnect couplings to be used in these systems will be welded terrestrially to their mating tube sections and inspected before assembly into the flexible tray. This will result in the highest joint reliability. However, should a leak develop on-orbit in one of these joints, repair with high reliability is required. McDonnell Douglas and NASA are currently studying welding as a viable method to perform such repairs in space.[9,10] Welding processes being studied by NASA include (1) plasma arc, (2) electron beam, (3) laser, and (4) gas tungsten arc. The Soviet Union has demonstrated the ability to utilize a miniaturized welder in space on their Soyuz T12 mission (Figure 9). The Soviet electron beam weld head weighs 50 lb, is quite small (12 in. long x 6 in. wide x 3 in. thick) and requires low power for operation (0.5 kW). Increasing unit power demands could also adversely affect astronaut safety.

Repair of mechanical joints by both EVA crew and telerobots has been demonstrated by McDonnell Douglas as shown in Figure 10.[1] Tubular samples were cut and the entire section replaced and joined with the use of mechanical tube fittings. Both the standard NASA shuttle tube cutter and an automated EVA and telerobotic-

SSC113131

101139

Figure 9. Successful Use of Space Welding by Soviet Cosmonaut

SSC109811 101140

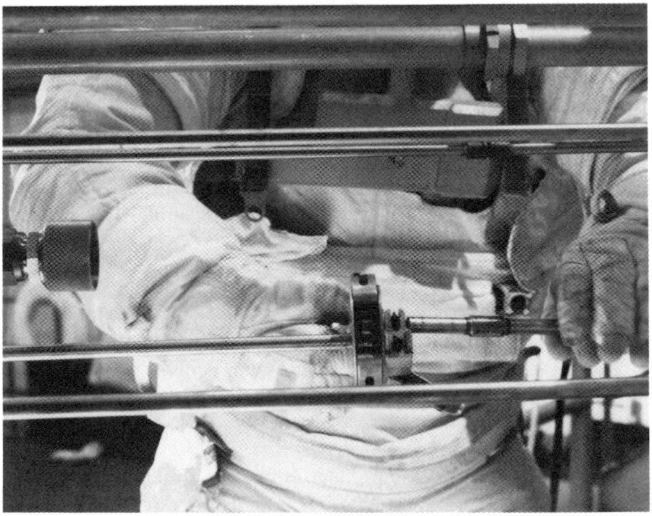

Figure 10. Manual Ratcheting Cutter Used in Tube Repair

compatible cutter were used to perform the operations. Pressure testing of the repaired tube samples indicated no leaks greater than $1x10^{-9}$ scc/sec when pressurized to 6000 psi with helium gas. Tube samples were repaired with the use of either permanently swaged or heatshrunk permanent one-piece fittings.

Conclusions

On-orbit deployment of the fluid and electrical power distribution systems for the NASA Space Station Freedom must be performed using reliable, safe procedures which minimize EVA crew time. Use of an automated spool and flexible hinged trays to deploy these lines is one option being considered to provide a unique concept which meets all these requirements. Further investigation is required to determine the type of fluid lines which can be successfully deployed with use of this spool-flexible tray concept with system pressures from 130 to 1000 psi and system environments from water to high-pressure oxygen. Stiffness of the trays containing the many unpressurized lines and its effect on line deployment must be investigated through demonstration testing of trays containing equivalent line sizes and quantities. The concept will minimize EVA crew time and provide for the optimum use of telerobotics and EVA crew assisted time to deploy the fluid and electrical systems for the station.

Maintainability of this utility system is achievable for its 30-year projected life. Components comprising the systems have been developed which have been demonstrated to be compatible with EVA crew and telerobotics in both a 1g and simulated weightless environment. The flexible hinged tray will provide ready access for maintenance by either EVA crew or telerobotic manipulators. Size, quantity, and spacing of lines must be controlled to ensure maintainability of the system throughout its lifetime.

Repair of the utility system over its projected 30 year life is a major design requirement. Joint leakage and damage from space debris are items which must be addressed. Repair of individual joints within the system has been demonstrated with the use of both EVA crew in a weightless environment and robots in a 1g environment. Joints can be repaired with the use of either automated tube cutting or the standard NASA ratcheting tube cutter. Removal of entire line sections may be the most appropriate method of repair for some damage areas. Tube fittings and couplings will be developed which are both EVA and telerobotic compatible. Use of flexible lines provides for line adjustment by either EVA crew or telerobotics to facilitate maintenance and repair.

On-orbit welding would provide for greater reliability of joint repair on the Space Station Freedom. The technique requires extensive development for on-orbit joint preparation and inspection. Success of the Soviet Union in past on-orbit weld preparation can be used as a basis for continued US development of the process.

References

1. R. H. Anderson and S. J. Pearson, On-Orbit Tube Joining and Repair for Space Station, McDonnell Douglas Report MDC H 5065, February 1989.

2. R. H. Anderson and J. J. Boman, EVA Construction and Assembly of Tubular Systems in Space Station, McDonnell Douglas Report MDC H 2829, June 1987.

3. S. J. Pearson, Advanced Development Project 20, Thermal Utility Connectors, Task 4.0, Space Station Definition and Preliminary Design, Work Package No. 2, McDonnell Douglas Report 20, January 1987.

4. D. E. Anderson, Telerobotic Application to EVA Activities, McDonnell Douglas Report MDC H 4121, October 1988.

5. Stanley Aviation Corporation, Aurora, Colorado, Mechanical Tube Fittings, Technical Proposal, July 1989.

6. Symetrics Inc, Newbury Park, CA, Quick Disconnect Couplings, Technical Proposal, May 1989.

7. Connector, Electrical, Zero-g, Specification For, MSFC Specification 40M 39580, April 1972.

8. S. J. Stock, Space Station Connectors, WETF Test, Utility System Integration, McDonnell Douglas Trip Report A93-J85-STS-M-8802475, November 1988.

9. Industry-University In-Space Technology Experiment Program (Outreach), NASA, Washington, DC, November 1987.

10. McDonnell Douglas Space Systems Company – Space Station Division, IRAD Program PD 12-100, MDC Q0931-12 and MDC Q0952-12.

SPACE STATION FREEDOM ATTACHED PAYLOADS AND COMMERCIALIZATION(1)

N. Furio, Ph.D.[2], B. Pyfer[3], R. B. Smith, Jr.[4], S. R. Hunt[5], Ph.D., and D. Suddeth[6]

SUMMARY

The Space Station Freedom (SSF) accommodates a class of payloads called external attached payloads. These payloads are attached to the main truss (Figure 1) of Space Station Freedom through the use of Attached Payload Accommodation Equipment (APAE). The APAE provides a reusable structural support for the payloads as well as access to station resources such as electrical power, command, and data handling. While in orbit, payloads are installed and maintained by a teleoperated robot or by an astronaut in a space suit performing extra vehicular activity (EVA).

---

1. This work was performed by GE AstroSpace Division, King of Prussia, PA, under contract NAS5-32000 to Goddard Space Flight Center, Greenbelt, MD
2,3. Sr. User Integration Engineer, GE AstroSpace, King of Prussia, PA
4. Mechanical System Engineer, GE AstroSpace, King of Prussia, PA
5. Manager, User Integration, GE AstroSpace, King of Prussia, PA
6. Manager, User's Systems Integration, NASA GSFC, Greenbelt, MD

Operational control of the payloads is performed by the SSF crew or from the ground by use of radio command channels. Provisions are made for payload transportation to orbit by the Space Shuttle or possibly by means of an expendable launch vehicle. Payload removal and return to Earth at the end of their mission are also provided.

The objectives of the User Integration Office are to: 1) provide information, coordination and support between the Space Station Freedom and assigned users; 2) assess user requirements to aid in the development of the APAE design; and 3) provide planning and analyses for the effective utilization of the APAE by its users.

The APAE User Integration office ensures that the interaction and coordination of the organizations required for successful utilization of external attached payloads on the SSF are as efficient and convenient as possible. The most important relationships will be between the assigned users and those who are responsible for flight implementation and operation.

Commercialization of APAE will be accomplished by private companies wishing to exploit the unique space environment by establishing research, development and production capabilities in Earth orbit. These ventures will have as their goal the development of profitable goods and services for consumer markets back on Earth. Among the areas to be commercially developed are: materials processing, Earth and oceanic monitoring, telecommunications, and industrial services.

This paper provides discussions of: 1) APAE design, capabilities and operations; 2) User integration activity related to APAE; and 3) commercialization for the APAE.

SPACE STATION FREEDOM OVERVIEW

The Space Station Freedom will be designed and built by the National Aeronautics and Space Administration (NASA) and its International Partners: the European Space Agency (ESA) composed of twelve European nations; the Canadian Space Agency (CSA); and the National Space Development Agency (NASDA) of Japan. The SSF Program encompasses the design, development, test, evaluation, verification, launch, assembly, operation and utilization of a habitable research facility based in low earth orbit (LEO) and the operation of supporting ground facilities. The program goals are to establish a permanent human presence in Space

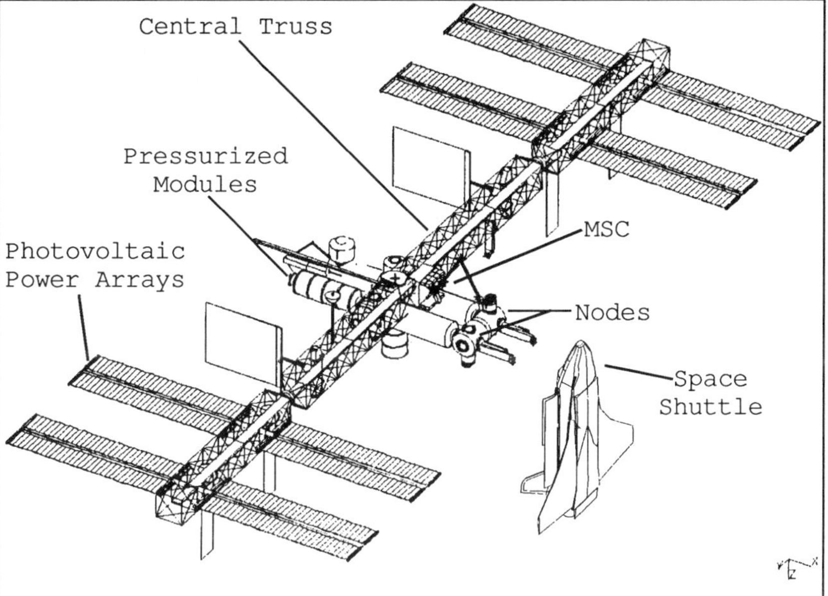

Figure 1. Space Station Freedom

and to perform a variety of scientific, technological and commercial activities.

The baseline SSF features a central transverse truss (see Figure 1) made up of uniform sequential bays. Located at each end of the truss are solar pointing photovoltaic arrays which provide power to the SSF subsystems and payloads. The overall length of the truss, including the photovoltaic arrays, is 155 meters. Located centrally are the pressurized modules and nodes providing habitation and research laboratory space. The central truss provides structural stiffness and dimensional stability to the entire SSF, which pitches about its longitudinal axis at a rate required to maintain a fixed attitude relative to the earth. To accommodate payload orientation requirements,the truss has provisions for the attachment of payloads along its length on the zenith (top or celestial viewing),nadir (bottom or earth viewing),and anti-velocity vector (back) faces. Limited provisions for payload servicing are available. A Mobile Servicing Center (MSC) and a Flight Telerobotic Servicer (FTS) will be used to assist in the remote assembly of payloads and payload equipment, and for a number of related servicing tasks.

The overall truss assembly includes: the central truss structure equipment to facilitate crew movement; an external lighting system; resource distribution trays and utility ports; avionics pallets; and drive mechanisms to orient the photovoltaic arrays.

The SSF is designed to operate at an altitude ranging from 150 to 270 nmi at a 28.5° inclination to the equator. The nominal operating altitude will be between 220 and 250 nmi.

The primary missions of external attached payloads are to carry out research in Astronomy, Earth Sciences, Plasma Physics,and Robotics, and to establish the practicality of commercializing space by investigating on-orbit material development, processing ,and fabrication techniques. These payloads include capabilities from short term (6 months) and completely passive (zero power) to semi-permanent test facilities having an extended multi-year operational on-orbit lifetime utilizing kilowatts of power.Payload size accommodation requirements range from about a 0.5 meter cube with a mass of 15 kg, up to a 4 meter diameter by 32 meter long cylinder with a 6000 kg mass.

APAE

Introduction

The diversity of payload requirements are accommodated by designing a modular APAE consisting of standard and optional items. Configurations can be created to meet the requirements of a wide variety of payloads constrained only by the total availability of resources. All interfaces between an attached payload and the SSF will be through the APAE. The APAE consists of the following hardware sub-elements (see Figure 2); the Payload Attachment System (PAS), the Payload Pointing System (PPS) and the Attitude Determination System (ADS). The PAS is the basic building block of the APAE, whereas the PPS and the ADS are both considerations for possible future expansion of APAE capability.

Standard APAE Equipment

The PAS provides the basic APAE structural and resource interface with the SSF (see Figure 2). The standard assemblies of the PAS are: the Station Interface Adapter (SIA); the Electric and Data interface System; and the Payload Interface Adapter (PIA). Optional assemblies of

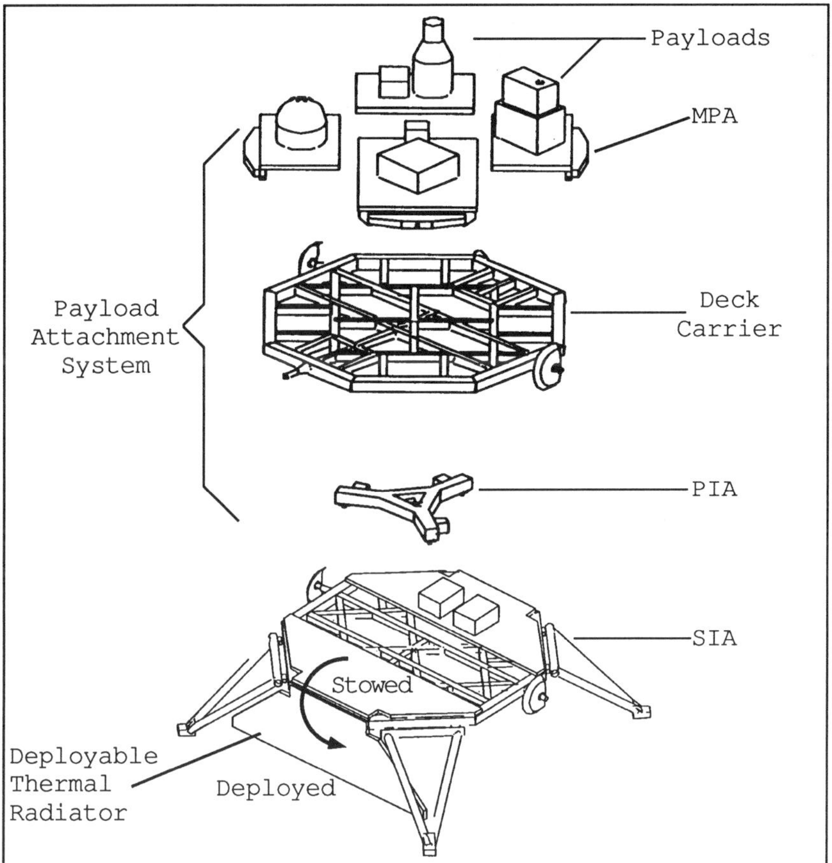

Figure 2. Exploded View of APAE Multiple Payload Configuration

the PAS are the Deck Carrier and the Multiple Payload Adapter (MPA). The SIA and PIA are required by all APAE Users. They provide the standard interfaces with the SSF for physical attachment, payload interchange and for access to and distribution of the power and data resources.Once installed, the SIA becomes a semi-permanent part of the SSF and remains in position when payloads are replaced.

The PIA provides the capability for remote mating with an SIA and incorporates the interfaces for power, command, control and data resources. The PIA is integrated with a payload on the ground and transported to the SSF as a single package. The PIA is also used to attach the Deck Carrier (Figure 2) to the SIA. The Deck Carrier is an optional means of accommodating payloads of limited size. The deck carrier spans the Space Shuttle cargo bay, mating with the shuttles keel and trunnion mounts, to provide the payload and associated APAE with structural support during launch and then attaches to the SIA on-orbit, simplifying the transition from transport to operation.

Optional APAE Equipment

Figure 3 illustrates five possible APAE configurations all of which utilize the SIA/PIA to couple payloads to nodes on the face of the truss bay.

The Deck Carrier is a carrier for a multiple payload set or a single small payload. The Deck Carrier with MPAs can accommodate up to four small payloads at a particular utility port location, allowing on-orbit removal and replacement of each individual MPA payload combination. Also the MPA, in conjunction with a Deck Carrier, provides Users with small payloads the structural support needed for launch.

Each MPA is integrated with its associated payload on the ground. The MPAs provide the capability to channel resources to each payload in the multiple payload set independently and to allow removal, or replacement, of a payload from the SSF. This may be done remotely with the MSC or FTS or as a backup by EVA. Attachment of multiple small payloads together with their associated MPAs is shown in an exploded view in Figure 2.

The pallet shown in Figure 3 is representative of a payload unique, User-supplied carrier to support the payload during launch and while on the SSF. In this case, the PIA is attached to the carrier during ground integration. Figure 3 also shows a large payload with PIA attached directly to the payload body structure.

The APAE has been designed to allow for future enhancements to accommodate payloads which require accurate pointing knowledge, articulation, environmental data, and active thermal control.

The Payload Pointing System (PPS) will provide articulated control for attached payloads (Figure 3). The PPS will point payloads in order to acquire, reacquire and track targets as well as compensate for SSF motion. Payloads may be replaced on the PPS, as required, while

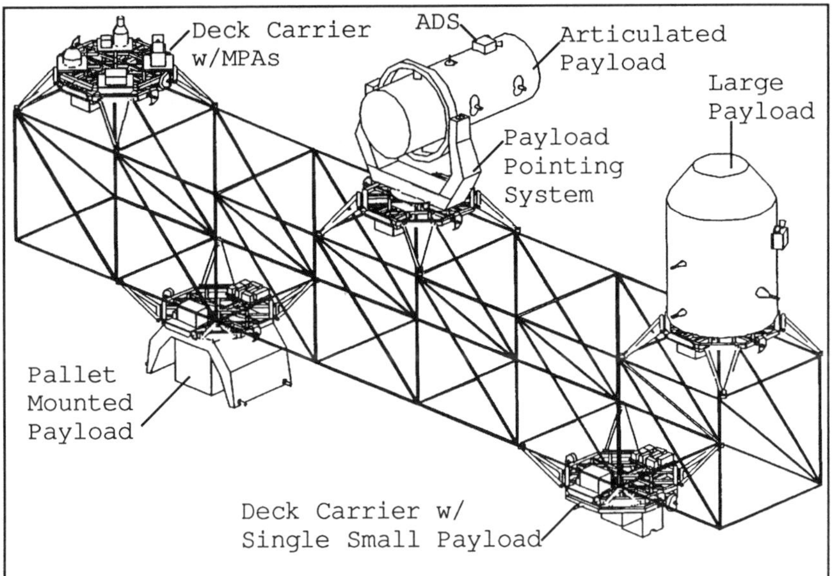

Figure 3. APAE Configurations

on-orbit. The PPS will provide a standard interface for payload attachment and the transfer of power and data utilities across the gimbal system.

The Attitude Determination System (ADS) contains attitude sensors and reference gyros which provide pointing knowledge and error correction signals to the PPS control loop or to self-pointed payloads. The ADS can also be used for non-articulated payloads to provide pointing knowledge if desired (See Figure 3).

The Contamination Monitoring Unit (CMU) will have the capability to monitor and measure molecular and particulate contamination levels in the vicinity of Attached Payloads. The CMU will provide status and warning information to alert users of changes in contamination levels that could degrade equipment, affect data accuracy, and/or require changes in operating procedures.

The SIA configuration features a hinged, deployable set of radiators for passive thermal control (Figure 2). These radiators provide thermal control for APAE. Payloads will be required to provide their own thermal control. The Space Station Program is presently considering introducing an active thermal control system as a system enhancement.

APAE INTERFACES

Pressurized Volumes: Nodes and Modules

The SSF has four resource nodes located at each end of the habitation and U.S. laboratory modules (modules are 4.3 m in diameter by 13.4 m long). The nodes are pressurized cylinders (4.3 m in diameter by 5.2 m long) that serve as: command and control centers; pressurized passageways to and from the various modules; and possible accommodation for test equipment racks. Certain nodes will contain berthing mechanisms for temporary attachment of either the Space Shuttle or the logistics modules. A Crew Support Station (CSS), will be located in one of the pressurized modules or in one of the nodes. The APAE has a significant interface with the CSS which provides the facilities for operation and monitoring of the payloads and the APAE by the crew. The CSS hosts payload unique software and hardware providing standard resource interfaces and access to Data Management System (DMS) services.

Mobile Servicing Center

The MSC, which will be transported on and operate primarily from the forward face of the central truss, is used for assembling, routine servicing, and maintaining the SSF and attached payloads. The MSC will remove cargo such as attached payloads and APAE from the Shuttle cargo bay, transport that cargo to the point of assembly, support assembly functions with crew or robot positioning devices, and provide for post-assembly inspection.

Flight Telerobotic Servicer

The FTS is a highly automated robotic device which works externally of the pressurized areas and is operated by astronauts from within the modules or nodes. The FTS will be used to improve efficiency and reduce risk during routine and hazardous tasks. Proposed functions include: installing truss members; installing fixtures such as the SIA on the truss; changing out of payloads; and performing inspection and maintenance tasks.

The FTS is designed to be operable from several different workstations on the SSF. Grapple fixtures provide the primary mechanical, electrical and data interfaces to the APAE and/or payload during FTS handling operations. The crew will control the FTS from an FTS workstation using direct manipulator control or by command of preprogrammed sequences. FTS hardware and software is structured in a modular fashion to ensure serviceability. The FTS configuration is flexible enough to accommodate capabilities upgrades.

APAE APPLICATIONS

APAE Users and User Sponsors

An APAE "User" is any group,individual or agency which has developed or is responsible for the development and operation of a payload that uses the APAE to access SSF resources. User interests and goals are expected to include: Scientific research,technology development and commercial ventures. In the United States Users interested in Space science research are typically sponsored by NASA's Office of Space Science and Applications (OSSA) or the National Oceanic and Atmospheric Administration (NOAA). Users developing Space technology are sponsored by the NASA office of Aeronautics and Space Technology (OAST). Finally, Users carrying out commercial ventures are the responsibility of NASA's Office of Commercial Programs (OCP). Similar agencies exist for foreign Users within ESA, CSA,and NASDA.

Space Science and Technology Research

The unique environment (e.g.,hard vacuum, microgravity, and unobscured view) surrounding Space Station Freedom make possible a plethora of heretofore impractical scientific and technological investigations. Research opportunities for APAE Users interested in Space science include: Astronomy and Astrophysics, solar system exploration, Solar and Terrestrial investigations and Material science. In addition, technology and engineering experiments in the following disciplines are being planned: Space environmental effects, automation and robotics, telecommunications, space structures dynamics and control, energy systems, propulsion systems, and thermal management. As with past space programs, the scientific and technological advances brought forth are eventually transferred to the private sector to improve the quality of life on earth.

USER INTEGRATION OFFICE

The goal of the APAE User Integration Office is to make the process of getting a User attached payload/mission onto the SSFMB as efficient and as convenient as possible. The User Integration Office plays an advisory role for the User community, including commercial payload developers. In this role, User Integration promotes User interests and describes their requirements at design reviews, utilization reviews, and resource allocation reviews where the user's interests may be impacted by decisions made during such meetings. In a similar manner, the User Integration Office will assist the User to insure that they are informed of program policies, requirements and constraints that may affect their mission objectives.

Early in the program the User Integration Office is collecting potential user and payload information to understand, assess, and correlate user requirements. These data are being provided to the APAE project to ensure that all appropriate requirements are considered in the APAE design.

The User Integration Office will provide the technical analyses to compare User mission requirements and the APAE, and make recommendations to facilitate the satisfactory integration of User objectives with APAE capabilities.

APAE COMMERCIALIZATION

The opportunities for attached payload commercialization fall into two major categories: manufacture of space-based commercial products and privatization of attached payload accommodations.

Commercial Products

The first class of opportunity, space-based commercial production, involves the use of Space Station Freedom resources and facilities by a commercially motivated agency, whether privately or publicly funded. The objective of such a venture would be to create a commercial product (e.g., alloys or crystalline materials) using processes or facilities that take advantage of the attributes of Space, specifically the micro-gravity environment and convenient access to a hard vacuum. Many such product investigations are initially planned, in a research mode, for the pressurized laboratories. The next step might be a pilot production project requiring greater volume or additional resources than are available in the laboratories. The APAE could be used to install and operate a prepackaged scaled-up test bed before the commercial agency would commit to a full scale, dedicated spacecraft/ facility. Such a test-bed on the APAE would allow more frequent quality control checks and troubleshooting through the use of the FTS, MSC or astronaut, if manned intervention were necessary.

Privatization

The second opportunity is the privatization of Space Station Freedom facilities. In the particular case of APAE, hardware might be purchased or leased by a commercial agency from an APAE hardware supplier just as commercial launch services can be purchased today. The private agency could then sub-let the hardware and other services to Users with scientific, technical, research or commercial interests.

The advantage to the commercial agency that establishes the facility is in cost and/or risk avoidance. The development cost of the hardware will have been paid by the Space Station Freedom Program and the basic designs will have been tested, verified, integrated and operated on orbit. The non-recurring costs normally associated with space flight hardware development would be avoided and the hardware would come with well developed documentation of its design, its capabilities and procedures for maintenance, operation and payload integration.

The advantage to potential Users of privatized APAE would be the enhanced potential for customized interfaces to meet the specific needs of unique payloads. The Space Station Freedom, as a whole, is being designed to meet the general needs of a large variety of applications and payloads. The program is typically not prepared to design and develop unique hardware or services for specific applications. Generally, unique payload interfaces or accommodations are the responsibility of the User or payload developer. With privatized APAE, the privatizing agency will have both the intent and capability to create and develop customized payload accommodations, providing the User with unique accommodations at a negotiated value-based price or standard accommodations to an established quote.

Incident Solar Flux at Fixed Orientations

Stephen Bilanow[1]

Abstract

Optimal orientations are computed to maximize or minimize the incident solar flux onto surfaces on spacecraft at fixed attitudes. This analysis was motivated by a potential need to optimize the power and thermal environment during the early assembly flights of Space Station Freedom. The time history of the incident solar flux is described using close analytical approximations.

Introduction

The early flights of the Space Station may have narrow constraints on the power and/or thermal environments due to the limited capabilities initially delivered on orbit. To analyze options available for meeting the constraints, examination was made of the optimal flight attitudes to either maximize solar energy on the solar panels, or minimize the solar energy on thermal radiators (Bilanow, 1989). Although motivated by Space Station analysis, the results presented apply to any spacecraft in a 28.5° inclination circular low Earth orbit, and can easily be extended to other orbits.

Coordinate Systems and Definitions

LVLH is a spacecraft centered coordinate system oriented relative to the instantaneous Earth direction as the spacecraft orbits: z is toward the center of the Earth, y is opposite the orbit normal direction, and to complete the orthogonal triad x is in the flight direction.

Solar α and β angles are used in this report to describe the Sun position in LVLH coordinates. The solar β angle is the elevation of the Sun from the orbit plane with positive angles being toward the $+y$ LVLH axis. The α

[1]Senior Analyst, Space Sciences Division, General Sciences Corporation, 6100 Chevy Chase Drive, Laurel, MD, 20707

angle is a rotation about +*y* LVLH from the <u>-*z* LVLH</u> direction to the Sun Direction, so that $\alpha=0$ occurs at "local noon," the orbit position when the Sun is "overhead." α and β are defined here in LVLH coordinates so they are independent of the station attitude.

The angles σ and ρ will define the direction of a surface normal in LVLH coordinates in a way analogous to α and β for the Sun direction. Thus σ is the "phase angle" of the surface normal in the *x*-*z* plane, with the -*z* zenith direction being the zero reference point. ρ is the elevation of the surface normal from the *x*-*z* plane toward the +*y* direction. We will refer to ρ as the "surface roll angle," because a roll attitude rotation moves a surface the same way from an initial zenith facing direction.

<u>Sun Motion in LVLH Coordinates</u>

For any given orbit the Sun stays approximately fixed in inertial space. For example during one orbit, the Sun will move less than 0.07 degrees along the ecliptic. During each orbit, LVLH coordinates rotate relative to inertial space about the orbit normal vector which aligns with the -*y* LVLH direction. Thus the direction of the Sun approximately prescribes a cone in LVLH coordinates. The Sun path in LVLH coordinates is illustrated in Figure 1.

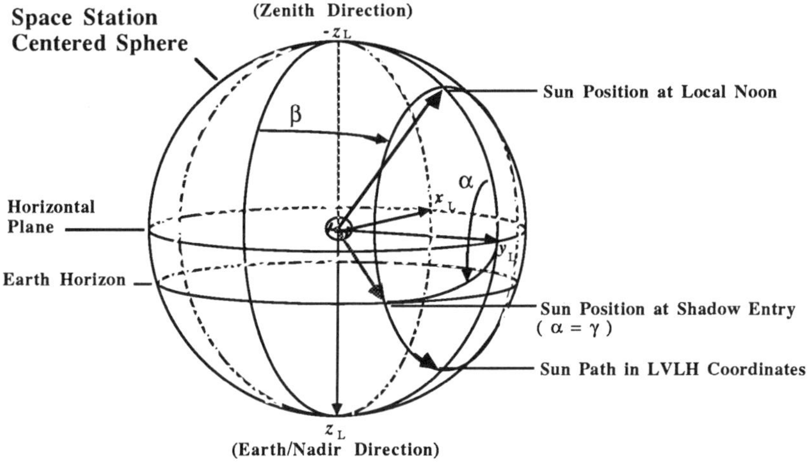

Figure 1. Sun Path in LVLH Coordinates over One Orbit.

While the α angle cycles 360° each orbit, the β angle does not change significantly in one orbit. The solar β angle changes gradually over the course of weeks due to the orbit precession and the seasonal Sun motion. The total range for the β angle is given by the combination:

Range in Orbit from Earth Equatorial Plane	$\approx \pm 28.5^\circ$
Range in Sun from Earth Equatorial Plane	$\approx \pm 23.5^\circ$
Total Range in β, Sun Elevation from Orbit Plane	$\approx \pm 52.0^\circ$

The $\pm 28.5^\circ$ β variation cycles about every 45 days and is superimposed on the ± 23.5 degree annual variation.

Sunlight Duration Each Orbit

The orbit angle at which the station enters the Earth's shadow can be derived from trigonometric relations assuming a spherical Earth of radius r and and a circular orbit at altitude h.

$$\gamma = \arccos\left(\frac{-\sqrt{2rh+h^2}}{(r+h)\cos\beta}\right) \quad (1)$$

Figure 2 shows the shadow entry angle as a function of solar β angle for orbits at 180 and 270 nautical mile (333 and 500 kilometer) altitudes, which span the expected operating altitude range for the station.

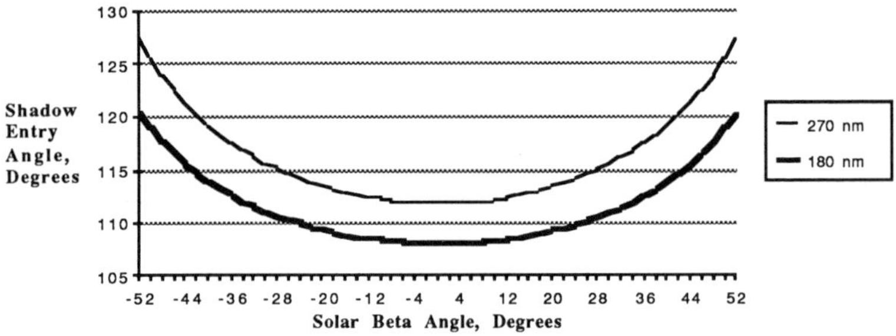

Figure 2. Shadow Entry Angle versus Solar β Angle.

Sun Projection versus Orbit Position

Given the Sun vector, **S**, as a function of position around the orbit and the surface normal vector, **N**, both in LVLH coordinates, the fraction of projected sunlight on the surface is given by the dot product of the unit vectors:

$$\hat{N} \cdot \hat{S} = \begin{bmatrix} -\cos\rho\sin\sigma \\ \sin\rho \\ -\cos\rho\cos\sigma \end{bmatrix} \cdot \begin{bmatrix} -\cos\beta\sin\alpha \\ \sin\beta \\ -\cos\beta\cos\alpha \end{bmatrix} \quad (2)$$

$$= \sin\rho\sin\beta + \cos\rho\cos\beta\cos(\alpha - \sigma)$$

Note that the panel phase angle, σ, just determines the orbit phasing of the Sun projection. A panel phase of $\sigma=0^\circ$ has the surface normal closest to the Sun at local noon.

Equation 2 is shown plotted in Figure 3 for σ=0°, various ρ angles and a solar β angle of 52°. Extreme condition occur for ±90° ρ where the projected area is constant. At lower β angles the constant projected flux at ρ = ±90° is less, going as the sine of β. At β = 0°, the projected flux versus orbit angle is just a cosine function whose amplitude is given by the cosine of the ρ angle.

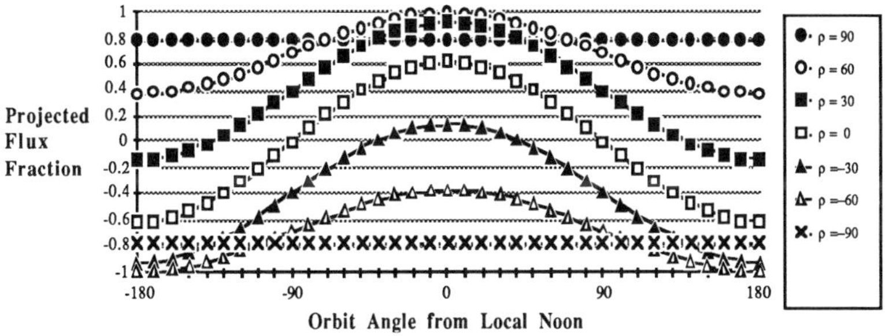

Figure 3. Projected Flux versus Orbit Angle with β = 52°.

<u>Conditions for Surface Facing Sun</u>

The plot above shows negative values for times when the Sun would be shining on the back of the surface. For many applications, one is only interested in flux on the front surface. The positions where the Sun is shining edge-on to the surface can be considered as "sunrise" and "sunset" <u>for the surfaces</u>. The Sun is edge-on when the dot product given by equation 2 is equal to zero. Solving for the α rotation angle where the "surface sunrise and sunset" occur we get:

$$\alpha_{\text{"sunrise/sunset"}} = \sigma \pm \eta \quad \text{where } \eta = \arccos(-\tan\beta \tan\rho) \tag{3}$$

η is the rotation angle from the position where the Sun is highest on the surface (local noon when σ=0) around to "surface sunset." The combinations of β and ρ where η is undefined correspond to regions which continually face toward or away from the Sun. The region where the surface normal stays within 90° of the Sun is a cone centered about the LVLH + or - y axis, whichever is toward the Sun. It is defined by

$$|\rho + \beta| > 90° \tag{4}$$

The region where the surface normal stays more than 90° away from the Sun is also a cone centered about the + or - y axis, but whichever is away from the Sun. It is defined by

$$|\rho - \beta| > 90° \tag{5}$$

Sunrise/set occur when

$$|\rho| + |\beta| < 90° \tag{6}$$

Conditions for Facing Sun throughout Daylight

Equation 3 gives the "surface sunrise/sunset" arc irrespective of whether the station stays in the daylight portion of the orbit. The surface is toward the Sun in daylight when

$$\sigma - \eta < -\gamma \quad \text{and} \quad \sigma + \eta > \gamma \tag{7}$$

The period during which the surface faces the Sun is centered about the daylight period when the surface phase angle, σ, is zero (surface normal closest to the Sun at local noon). Assuming $\sigma=0$, the surface stays toward the Sun during the daylight portion when $\eta > \gamma$. Substituting equations 1 and 3, assuming β is positive and solving for ρ, we get the following condition for continual sunlight:

$$\rho_{sunlit,\beta>0,\sigma=0} > \arctan\left(\frac{\sqrt{2rh+h^2}}{(r+h)\sin\beta}\right) \tag{8}$$

Figure 4 shows a plot of this result. With β negative, the inequality works the other way and ρ must be less than the equivalent negative values to keep the surface facing sunlight during daylight.

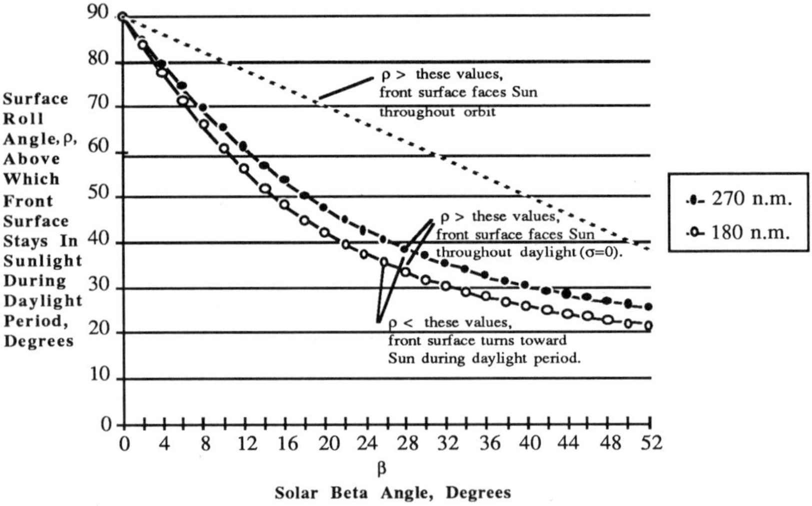

Figure 4. Minimum ρ to Face Sun During Daylight.

Integrated Flux on Front Surface at Large ρ and β Angles

At $\sigma=0$ and large positive values (or large negative values) for both ρ and β, the Sun will shine continually on the front surface during the daylight portion of the orbit. Thus the integrated flux is just the integral of **N•S** over the daylight period. We will be interested in the maximum integrated flux, and it can be seen from the characteristics of N•S in Figure 3 that $\sigma=0$ generally provides a maximum at any β. In order to normalize the integrated flux to a time-independent parameter, we will compute the integrated flux relative to the full flux that would occur with the surface normal pointed at the Sun and no shadow period in the orbit. We also assume a circular orbit so that the integral of the angle around the orbit is the same as a time integral over the orbit period. Thus

$$\overline{\mathbf{F}}_{\text{Large }\beta,\rho} = \frac{\int_{\text{daylight}} \widehat{\mathbf{N}}\cdot\widehat{\mathbf{S}}\,d\alpha}{\int_{\text{orbit}} d\alpha} = \frac{\gamma \sin\rho \sin\beta + \sin\gamma \cos\rho \cos\beta}{\pi} \qquad (9)$$

The surface ρ angles where maximum integrated flux occurs can be evaluated by taking the derivative of equation 9 with respect to ρ and finding where this derivative is equal to zero,

$$\pi\frac{\partial\overline{\mathbf{F}}_{\text{Large }\beta,\rho}}{\partial\rho} = \gamma\cos\rho\sin\beta - \sin\gamma\sin\rho\cos\beta = 0 \qquad (10)$$

Solving for ρ gives

$$\rho_{\text{max int. flux}} = \arctan\left(\frac{\gamma\tan\beta}{\sin\gamma}\right) \qquad (11)$$

Integrated Flux on Front Surface at Small ρ and β Angles

At small ρ angles, the surface will turn away from the Sun before the end of the daylight period and thereby define a different domain over which to calculate the integrated flux. Thus for these cases, integrating the flux over the period that the Sun is on the front surface, we obtain an equivalent result to equation 9, but with η providing the limits of integration instead of γ.

$$\overline{\mathbf{F}}_{\text{Small }\beta,\rho} = \frac{\eta\sin\rho\sin\beta + \sin\eta\cos\rho\cos\beta}{\pi} \qquad (12)$$

The derivative of this gives a result like equation 10:

$$\pi\frac{\partial\overline{\mathbf{F}}_{\text{Small }\beta,\rho}}{\partial\rho} = \eta\cos\rho\sin\beta - \sin\eta\sin\rho\cos\beta \qquad (13)$$

Since η is a function of ρ, this must be solved iteratively to find ρ for the maximum integrated flux.

ρ Angles for Maximum Integrated Flux

Figure 5 shows the ρ angle for maximum integrated flux on the front surface over the full range of positive β angles. Corresponding negative values of ρ give the maximum integrated flux for negative β angles. The altitude dependence occurs only at the higher β angles where the surface is already facing the Sun at emergence from Earth shadow.

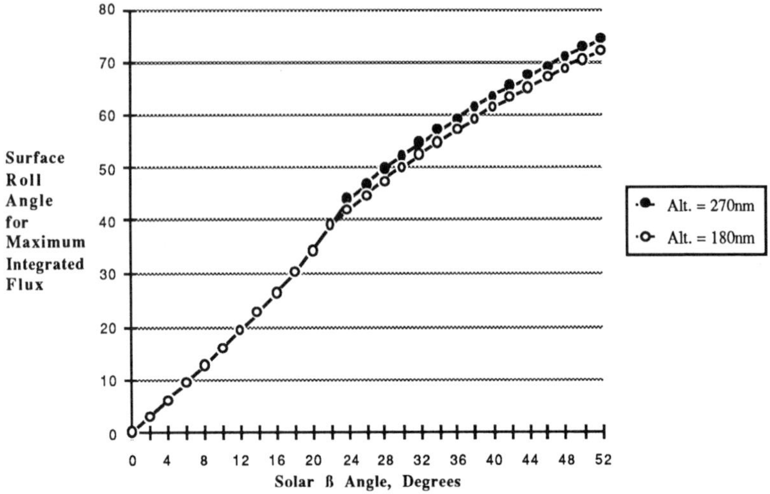

Figure 5. ρ for Maximum Integrated Flux vs. β.

The ρ angle for the maximum is generally larger than the β angle. A ρ equal to β would give the maximum flux at orbit noon, however a larger ρ gives significantly more flux around sunrise and sunset even though it gives slightly less flux at orbit noon.

Conditions for Zero Integrated Flux on Front Surface

Fixed orientations can always be found that will have the minimum integrated solar flux of zero. The required condition is that the surface normal remain at least 90 degrees away from the Sun during the daylight part of the orbit. The surface normal will stay 90 degrees away from the Sun for the whole orbit whenever the condition of equation 5 is met. This condition is required if the surface normal phase angle σ is such that the surface normal is closest to the Sun at any time during the daylight arc. i.e. $| \sigma | < \gamma$. Otherwise, the front surface may still avoid illumination as long as surface "sunrise" to "sunset" occurs during the shadow period. This requires that

$$\sigma - \eta > \gamma \text{ and } \sigma + \eta < 360° - \gamma \quad (14)$$

The optimal phase angle to have surface "sunrise" to "sunset" occur during shadow is $\sigma = 180°$. At $\sigma = 180°$, we can get a similar result to equation 8, differing only in the sign and the direction of the inequality. Thus for β positive and $\sigma = 180°$, if ρ is less than the negative of the values shown plotted in Figure 4, then the front surface stays in shadow. Likewise, the surface stays in shadow if β is negative and the roll is greater than the values plotted in Figure 4.

Total Integrated Flux on Front and Back Surfaces

For the thermal radiator panels we are interested in minimizing the integrated flux on both the front and back surfaces. Examination of the behavior of the flux around the orbit indicates that this minimum can be expected with a surface phase angle σ=0. Therefore we will analyze the integrated flux with σ=0 as a function of ρ angle and β angle. A formula for the total flux on the front and back surfaces can be derived by separating the domain of integration over each side and integrating with the proper sign for positive flux. Note that for σ=0 the dot product is positive over the domain from $-\eta$ to η. Thus the integral on the front and back can be computed as follows:

$$\begin{aligned} 2\pi\overline{\mathbf{F}}_{f/b} &= \int_{-\gamma}^{-\eta} -\widehat{\mathbf{N}}\bullet\widehat{\mathbf{S}}d\alpha + \int_{-\eta}^{\eta} \widehat{\mathbf{N}}\bullet\widehat{\mathbf{S}}d\alpha + \int_{\eta}^{\gamma} -\widehat{\mathbf{N}}\bullet\widehat{\mathbf{S}}d\alpha \\ &= 2\int_{-\eta}^{\eta} \widehat{\mathbf{N}}\bullet\widehat{\mathbf{S}}d\alpha - \int_{-\gamma}^{\gamma} \widehat{\mathbf{N}}\bullet\widehat{\mathbf{S}}d\alpha \end{aligned} \tag{15}$$

and we can combine the results from equations 9 and 12;

$$\begin{aligned} 2\pi\overline{\mathbf{F}}_{f/b} = {} & 2\eta\sin\rho\sin\beta + 2\sin\eta\cos\rho\cos\beta \\ & -\gamma\sin\rho\sin\beta - \sin\gamma\cos\rho\cos\beta \end{aligned} \tag{16}$$

Note that this formula is not valid when $\eta > \gamma$ (Sun on front surface only), and in that case equation 9 applies. When η=0 (Sun on back surface only), this equation also reduces to equation 9 except with the sign change. Applying these checks on the domain for the formulas, Figure 6 shows the integrated flux on both surfaces versus surface ρ angle for selected positive β angles. It is found that all of the minima occur in the domain where equation 16 is valid. The maxima all occur in the domain for equation 9 where the Sun shines on the front surface only, and thus are equivalent to the maximum flux ρ angles already discussed. At β = 0 a minimum total flux of zero occurs for ρ of either +90 or -90 degrees. Note that lower minima (as well as lower maxima) for the integrated flux are found at the lower β angles.

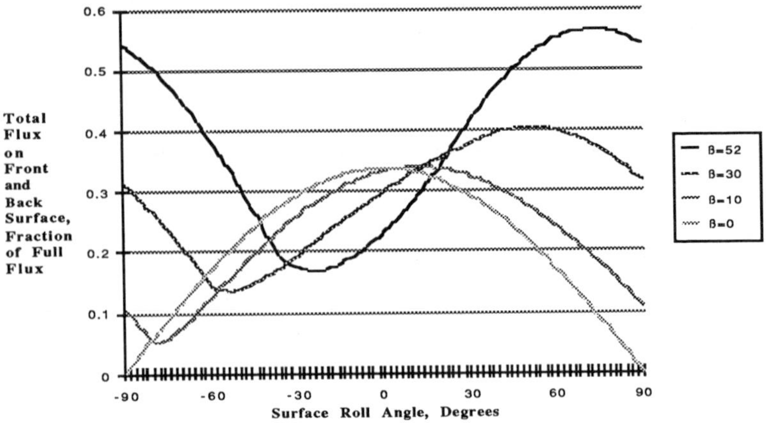

Figure 6. Integrated Flux on Front and Back versus ρ.

The ρ angles where the minima occur can be evaluated by setting the derivative of equation 16 to zero. This derivative is just a combination of equations 10 and 13:

$$\pi\frac{\partial \overline{\mathbf{F}}_{f/b}}{\partial\rho} = 2\,\eta\cos\rho\sin\beta - 2\sin\eta\sin\rho\cos\beta - \gamma\cos\rho\sin\beta + \sin\gamma\sin\rho\cos\beta \tag{17}$$

Like equation 13, this can be solved iteratively to find the ρ angles for the minima. Results are shown in Figure 7. This shows the optimal ρ angles to minimize the total flux on the front and back surfaces for an operating altitude of 220 nautical miles (407 kilometers).

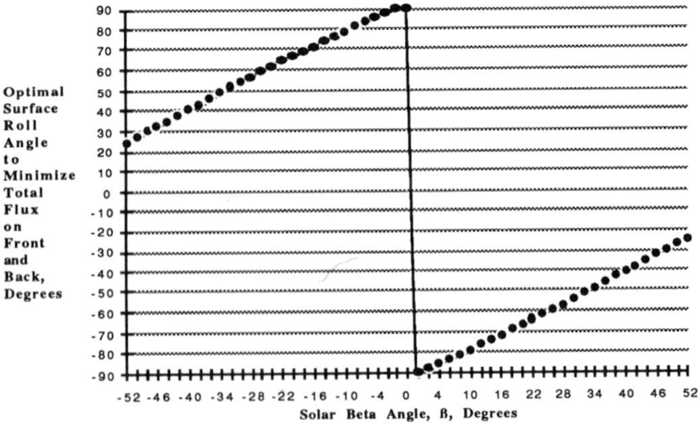

Figure 7. ρ Angles to Minimize Flux on Front and Back.

With ρ 90 degrees from the β angle, the flux becomes zero (edge to the Sun) at local noon in the orbit. However the ρ angles for minimum flux are somewhat less than 90 degrees from the β angle (except at the limiting case with β = 0). This extra ρ adds flux on the front surface at local noon, but removes more total flux from the back side over the rest of the orbit.

Additional Considerations and Conclusions

Additional effects need to be taken into account for detailed evaluation of the power and thermal environments on the Space Station. For the power evaluation, these additional effects include reflected Earthlight, possible shadowing of Sunlight on the arrays, battery charging limitations, and array temperature sensitivity. For the thermal system evaluation, additional effects include Earth thermal radiation, shadowing of Sunlight on the radiators, and nearby surface radiation. Nevertheless, it is felt that the optimal attitudes indicated under the simplified assumptions of this report represent a good first order approximation to the preferred flight attitudes if power or thermal constraints are crucial.

Other flight mechanics considerations also need to be taken into account in the selection of a flight attitude and orientations for PV array and radiator surfaces. In particular a Torque Equilibrium Attitude (TEA), which depends on the mass properties and surface projected areas, may be required to minimize the attitude control requirements. Also, some surface orientations will cause more orbital decay by putting more frontal area in the flight path. This analysis provides a starting point for the power and thermal inputs for the study of tradeoffs.

This report provides a first order estimate of the optimal orientations from the standpoint of the power and thermal environments. These results may by applied for any missions in low Earth orbit at a 28.5 degree inclination.

Acknowledgements

Thanks are extended to Jalal Mapar and Mike Mathews at SSEIC, and Doug Merkle at ARA, Inc. for useful comments.

Reference

Bilanow, S., Analysis of Incident Flux versus Time and Flight Attitude, Space Station Engineering and Integration Contractor SE&I Report, PSH-341-RP89-005, September, 1989.

Subsection 3.2: SPACE DEBRIS

SPACE DEBRIS - A DESIGN CONSIDERATION*

L. R. Utreja
BDM International, Inc.
950 Explorer Boulevard
Huntsville, Alabama 35806, USA

ABSTRACT

With the increased use of space by manned or unmanned missions, it has become necessary to include yet another parameter in the design of spacecraft, namely, the effect of a probable impact by the man-made space debris. Information about the number and size of small debris at the space station altitude is extremely limited and is mostly predicted by models. Protection of the space station against space debris is vital for its safe operation. The paper presents a review and discussion of the current space debris environment at the space station altitude, space station protection design analysis methods, and hypervelocity impact testing and analysis approaches.

INTRODUCTION

All long-term space-based vehicles (for example, Space Station Freedom, Space Shuttle, Orbital Maneuvering Vehicles) are susceptible to impacts by not only the natural meteoroid environment but also the man-made space debris. The space debris constitutes hazard to man's habitation of space; extra-vehicular activity involving inspection, maintenance, and assembly of large-scale structures; and construction in space. Space debris impacting a space vehicle may cause injury and loss of life, damage to equipment, and significant time loss in operations and experiments, and it may contaminate the space environment. It is important to know the space debris environment in terms of debris size, mass, velocity, and orientation so protective measures can be incorporated into the design and operating conditions of a particular spacecraft.

* A version of this paper was presented at the 40th IAF Conference in Malaga, Spain, October 1989.

Protection of a spacecraft from impacting debris can add significant weight to its structure. A structure must be designed considering collision probability for the spacecraft orbital altitude. Design optimization techniques are needed that will provide tradeoff studies for various impact models and probabilistic debris environment.

A typical encounter between a long-term spacecraft and space debris will occur at oblique angles and at hypervelocities. Computational and experimental approaches provide damage estimates under limited probable impact conditions. The existing hypervelocity gun facilities are limited in their velocity capabilities. Correspondingly, it is not possible to benchmark computational techniques for velocities above the test capability. Innovative techniques using material modeling are being investigated to predict damage at representative space debris impact velocities.

The purpose of this paper is to review the current manmade space debris environment with reference to the space station. Space station design analysis approaches, hypersonic test facilities, and oblique impact testing are also reviewed.

THE CURRENT ENVIRONMENT

An estimated 3,000,000 kg of manmade space debris exists within 2,000 km of Earth's surface. All catalogued objects, 6772 as of October 1989, are tracked by U.S. Space Command and are 10 cm in diameter or larger. Fig. 1 shows true earth satellite population and cumulative launches since 1957. Recent computer simulations and empirical data sources indicate the presence of approximately 17,000 objects 1 to 10 cm in diameter and 3,500,000 objects between 0.1 and 1 cm. Most are in high inclination orbits where they pass each other at an average relative velocity of 10 km/s.

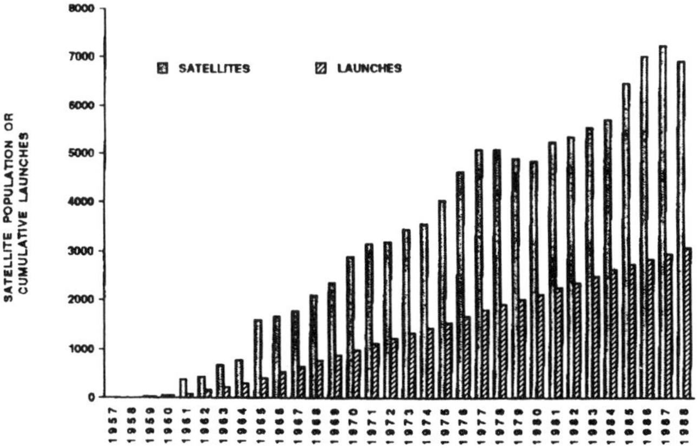

Fig. 1 True Earth Satellite Population. (Ref. 1)

Manmade space debris includes spent rocket stages, inactive and a few active payloads, intentionally discarded operational debris, fragmentation of satellites from explosions (accidental or induced), and other uncontrolled space vehicles. The debris growth rate fluctuates with solar cycles, launch activity, and collision frequency. Kessler, et al.[2] describe the current and future debris trends based on the most recent models and analyses from the Johnson Space Center (JSC). Predicted debris flux, as a function of debris size for the space station orbital altitude of 500 km and the year 1995, is shown on a log-log plot in Fig. 2.

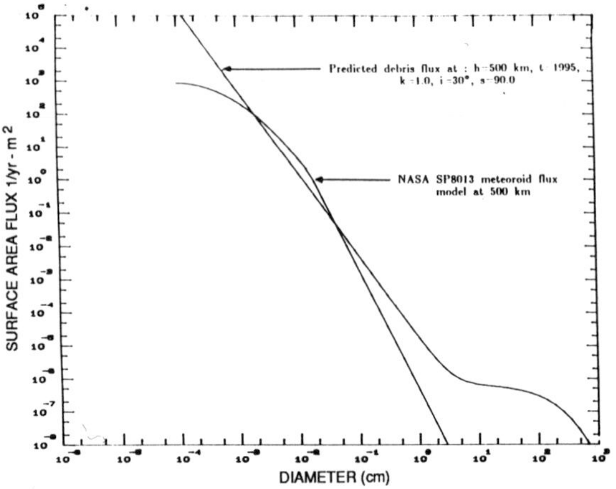

Fig. 2 Orbital Debris Flux at 500 km. (Ref. 2)

Uncertainties associated with the current debris flux model exist. The primary uncertainty is due to shortcomings in the data base. (This includes limited amount of data for 2 cm flux using ground telescope, measurement of debris size smaller than 0.5 mm only at 500 km altitude, and linear interpolation of data in between.) Other uncertainties are associated with future events (for example, solar activity, launch rate, and on-orbit fragmentation.) Prediction of solar activity and launch rate level beyond one year is not certain. Another probable consideration is encounters with objects in non-circular orbits that may cross the Space Station path at 20° from the direction of travel and with velocities less than 2 km/s. An average growth rate of 5 percent per year over the last year for the rest of the world, and a constant U.S. rate, suggests a projected growth rate of 5 percent per year. But an upper limit of 10 percent does not seem unrealistic.

According to another point of view[3], the tractable debris growth is in error. The reason quoted is that the data failed to account for the origin of newly catalogued pieces as well as recounting of objects that left Earth orbit on deep space missions. Based on these findings, a yearly linear growth of 240 objects since 1960 may be a good representation of the true debris growth. This rate will double the catalogued debris population in 31 years[2] compared to 14 years given by 5 percent growth rate.

The cumulative flux (F) of space debris of diameter (d) and larger depends on the spacecraft altitude (h), inclination (i), and the year (t) when the solar activity (S) was one year prior to t

$$F = F(d, h, i, t, S)$$

The flux is defined such that the average number of impacts on a spacecraft surface area (A) exposed to space debris for time (T) is given by

$$N = F\ A\ T$$

Figure 3 depicts debris flux angular distribution for an orbit of inclination of 30 degrees as seen in the spacecraft co-moving frame. The lines of constant debris flux are concentric circles. The corresponding normalized values averaged over altitude and over argument of latitude are shown on the ordinate. The curve shows that the debris flux exists within ± 90 degrees from the direction of travel, with an exclusion zone centered on the direction of travel. The direction of the relative velocity vector, with α as the angle (between the spacecraft velocity vector and the impact velocity vector) at the spacecraft orbital altitude of 500 km is given by,

$$\cos \alpha = -v/15.4$$

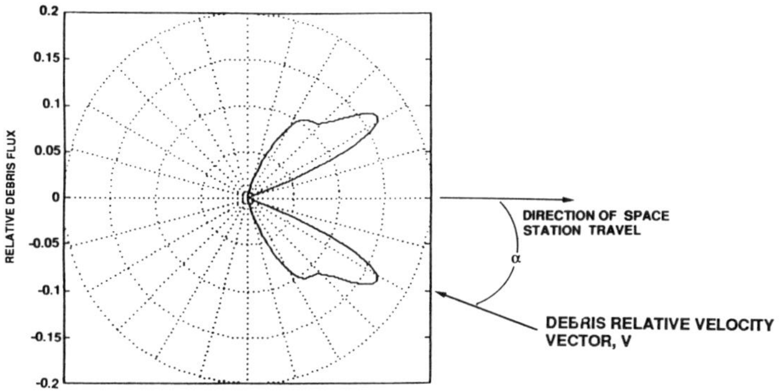

Fig. 3 Debris Distribution Relative to Space Vehicle.

PROTECTION STRUCTURE DESIGN

One of the considerations in the design of spacecraft is the protective shielding. The protective shielding must be designed to minimize damage caused by probable space debris impacts during the spacecraft mission life. Passive protection for the Space Station is provided by a debris shield, and a gap including multi-layer insulation surrounding the pressure wall.

Horn, et al.,[4] has developed a protective shield design methodology based on the probability of no penetration (PNP) for a given structure. The probability of no penetration refers to the probability of no impact and is a function of the critical debris flux (F), spacecraft exposed areas (A), and the spacecraft mission life time (T). The probability of no impact is derived from the Poisson distribution equation for zero occurrence and is given by

$$P_0 = e^{-FAT}$$

The mission life time is usually determined by program requirements and is defined in years for space debris analysis. Since space debris is directional, the effective area is either determined by the use of flux factors[2] (Fig. 4) if the spacecraft has fixed orientation or is the total surface area if the spacecraft has a random orientation. The critical debris flux is defined as the number of impacts per square meter per year of a critical particle mass and all larger particles. At 500 km altitude, 30° inclination, and for diameter, $d \leq 1$ cm, the debris flux is

$$\log F = -2.52 \log d - 5.82$$

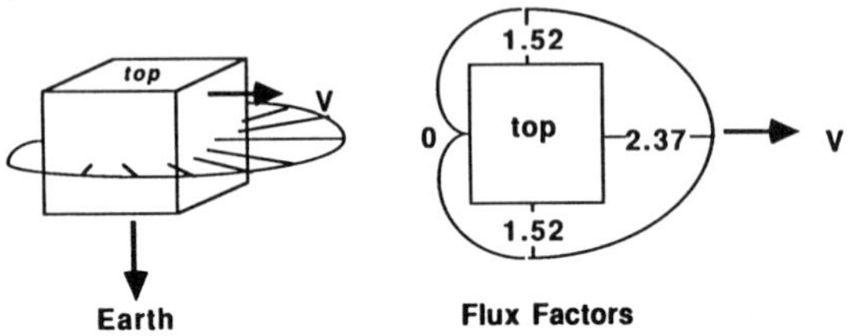

Fig. 4 Flux Factors for Spacecraft with Fixed Orientation.

The critical particle mass is then determined from the "Ballistic Limit" (a mass-velocity combination that will just penetrate the structure) of the structure. For a thin, ductile, metal plate of thickness t (cm) penetrated by a particle of mass, m (g) density, ρ (g/cc), and velocity, v (km/s) is given by

$$t = K\ (M)^{.352}\ (\rho)^{.167}\ (V)^{.875}$$

where K is material constant developed from tests (0.54 - 0.57 for Al). Equivalent diameter backed out from the particle mass is used to calculate the debris flux. The debris flux is then used to calculate PNP. With PNP as the design requirement, the protection system is designed with the aid of any one of the empirical, hydrodynamic, or optimization codes.

The collision probabilities for the Space Station habitation module (~700 m^2) and the entire exposed surface (~5000 m^2) for 10-year and 30-year mission durations is tabulated in Table 1.

PERIOD	COLLISION PROBABILITY WITH PARTICLES GREATER THAN			
	1 cm		10 cm	
	700 m^2	5000 m^2	700 m^2	5000 m^2
10 YEARS	.13	.63	.005	.034
30 YEARS	.34	.95	.015	.100

Table 1. Collision Probabilities.

The current design requirement for tolerable risk for the manned modules of Space Station for impact is 0.005. As is evident from Table 1, this value is reached for 10 cm size debris even for a 10-year mission. Objects less than 10 cm, however, have a greater collision probability. Protective shielding, therefore, must be used to protect the Space Station habitation module against all particles smaller than 10 cm to achieve tolerable risk.

A design is typically followed by optimization based on mission requirements. The protective system design optimization includes several tradeoffs that include program risk, mission life, shielding mass versus pressure wall mass, shielding material and debris vaporization versus damage to pressure wall, and spacecraft size versus design weight. Mog[5] describes several methods for optimization. The selected method can be used to minimize protection shield weight for selected materials and thickness. The techniques also provide optimum shield/pressure wall separation for minimum shield weight. Some of the results from optimization analysis follow. Out of the several aluminum alloys in the 2000, 5000, and 600 series studied, 2011-T8 is the best overall alloy for the shield as well as the pressure wall. It provides minimum weight shield and pressure wall for a 5-25 cm separation between the two. Graphite-aluminum composite shows further promise in reducing shielding weight. An increase in shield/wall separation from 10 to 15 cm results in a 25 percent weight savings. An increase in space debris generation above 6 to 8 percent results in a dramatic increase in structural weight. There is a large sensitivity to weight below 1000 km altitude. Increasing mission risk from 0.05 to .07 results in about 18 percent reduction in minimum structural weight.

HYPERVELOCITY IMPACT TESTING AND ANALYSIS

Using a 500 km, near circular orbit at 28.5° inclination, the probability of space debris impact with the Space Station occurs at oblique angles. The average velocity of the space debris impact is 10.5 km/s with a maximum of 15.4 km/s for a head-on collision. The dynamic response of full-scale structural configuration under these conditions is unknown. Two methods of approaching the problem are available: hydrocode or semi-analytical computations, and hypervelocity testing. Each of these methods can be used to develop a better understanding of hypervelocity interaction physics. However, individually each of these methods has limitations, as shown in Fig. 5.

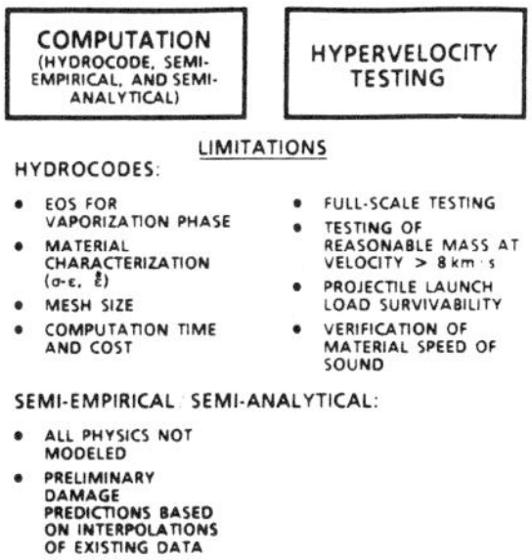

Fig. 5 Available Hypervelocity Impact Analysis Methods and their Limitations

The use of hydrodynamic codes has been well established in the prediction of mechanical damage. Many of the available codes provide excellent results for cases that are within the bounds of established accuracy for known equations of state and required material property information. A key issue is the validity of these codes beyond the melting and vaporization phases. The phase change conditions for many structural materials occur for impact velocities between 8 and 15 km/s. Present day test facilities have limited velocity capabilities for the anticipated debris mass as shown in Table 2.

TYPE FACILITY	BORE	MASS (g)	MAXIMUM VELOCITY	CAPABILITY GROWTH
• LIGHT GAS GUNS				
– AEDC	0.1–2.5 in	1–200	2.5–8	
– AEDC COUNTERFIRE		200/0.1" D	13–14	3-6 MONTHS+
– LANL	3 in		4.0	
– NMT-TERA	2.5 in		6.0	
– NRL	85 mm	250	6.5	
– SANDIA	20 mm	50	8.0	
	29 mm	50	7.0	
	8-12 mm	1-50	3–9	
– ERNST-MACH (FRG)	8.5 mm	<1	9.6	
• EML AND ET				
– GT DEVICES		3	10	*
– LANL	10.8–12.7	1	10.8–12.7	LGG INJECTED*
– LLNL	9 mm		12	LGG INJECTED*
– SANDIA		3	15	LGG INJECTED*
– WESTINGHOUSE	20	6	10	*
	56	100	10	*
• ELECTRIC GUN				
– LLNL		0.1	15	
		<0.1	30	
		FLYER PLATES		
• EXPLOSIVE GUN				
– LANL		10–15?	8–10?	DRAG LAUNCHER**
– PI		2–3	12.2	DEMONSTRATED IN 1960s
• SHAPED CHARGE ACCELERATOR		0.15	16.5	DEMONSTRATED IN 1965
• SLED-ON-SLED				
– HAFB			NET 13–15	DRAG LAUNCHER+
• ELECTROSTATIC ACCELERATOR				
– FRG		10–14	50	
– LANL		10–9	100	

R1-9642G-1

* PROJECTED CAPABILITY + EXPERIMENTAL
**PROOF OF PRINCIPLE

Table 2. Ground Test Facilities.

The data on hypervelocity impact at oblique angles are scarce, especially the damage to external structures caused by ricochet debris fragments. Shonberg's[6] work is noteworthy in this area. He investigated the problem of hypervelocity impact on multi-sheet specimens shown in Fig. 6. His findings revealed the existence of a critical oblique angle, between 60 and 65 degrees, beyond which the ricochet debris caused damage to the ricochet witness plate (Fig. 7). The ricochet ejecta contained fragments with varying velocities with small particles exceeding even the original impact velocity. In general, the ricochet damage increased with increasing impact obliquity, impact velocity, and the size of the impacting debris.

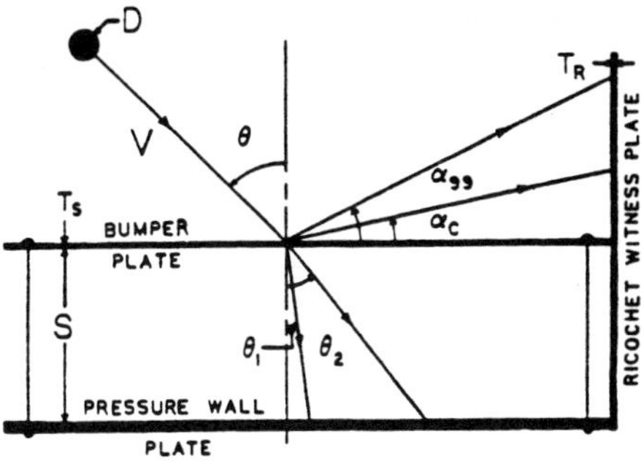

Fig. 6 Test Configuration and Definitions. (Ref. 6)

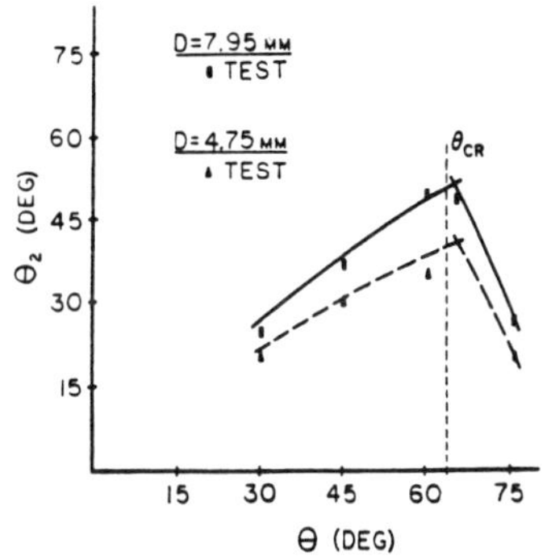

Fig. 7 Projectile Fragments Trajectory - Test Data Compared with Regression Equation Predictions. (Ref. 6)

The feasibility of scaled models and material models to predict damage effects at higher velocities has been investigated. Fig. 8 shows HULL code computations of aluminum-on-aluminum at 16 km/s compared with lead-on-lead at 4 km/s. Lead debris exhibits the same damage characteristics (note size and fragment distribution at the shield and the damage to the pressure wall) at 1/4 the velocity of aluminum debris. In general, however, because of large number of Pi terms that need to be matched in material modeling, there are some unresolved issues (Fig. 9). At hypervelocity impacts, several different mechanical and thermal processes take place, such as plastic flow, projectile fragmentation, melting, vaporization, and structure material spallation. The data on vaporizing materials do not exist to validate the existing scaling laws. In Fig.10, Hopkins et al.[8] show that in a double-plate impact situation when the impacting projectile velocity increases, the possibility of the second plate being penetrated depends significantly on melting and vaporizing phenomena. Since the Space Station with protective shields is a multiple plate structure that will be potentially struck at oblique angles, melting and vaporization must be examined closely for scaling at higher velocities. The effects of these phenomena must be extrapolated carefully to achieve confidence at velocities higher than present capabilities allow.

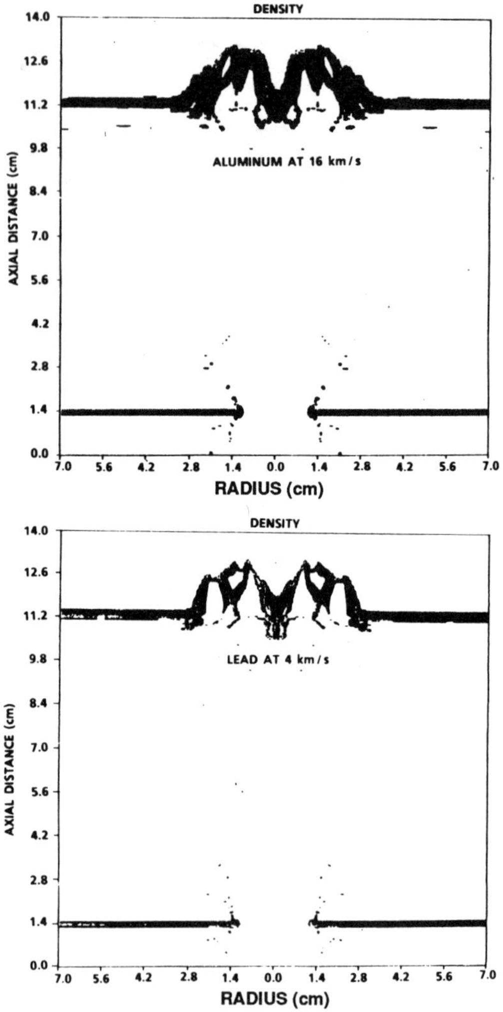

Fig. 8 Hull Code Computations on Material Modeling. (Ref. 7)

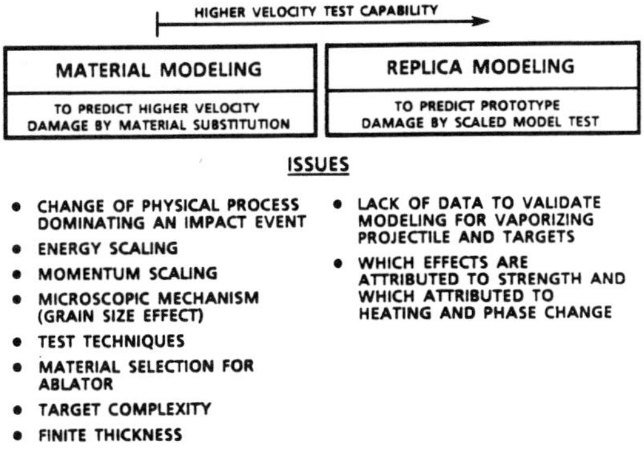

Fig. 9 Scale Modeling Issues.

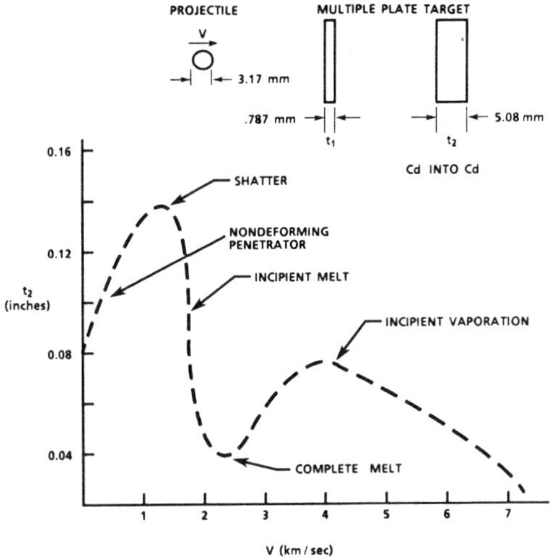

Fig. 10 Threshold Velocity for Just Penetrating Various Thicknesses of Shield Targets. (Ref. 8)

Several organizations have a requirement for hypervelocity test capability above 10 km/s. Some of them are close to their goal. For example, Ernst-Mach's light gas gun in the FRG has successfully achieved 9.6 km/s for fraction of a gram. Arnold Engineering Development Center's light gas gun can achieve 13-14 km/s in a counterfire mode, but this will require a preparation time of approximately one year after funding becomes available. Before the test facilities are available for verification of impact damage above 8 km/s; material and replica scaling can provide useful damage predictions right now.

CONCLUSIONS

Systematic improvements in space debris measurement technology are needed to improve data bases and correct knowledge of launch rates. Uncertainty in debris prediction will still persist as a result of on-orbit fragmentation.

Hydrocodes are available to design a protective shield for the Space Station, pending the availability of test facilities to achieve representative debris velocity and size environment. Design guidelines require that the habitation module be protected to withstand the impact of particles up to 10 cm. Optimal shield material among aluminum alloys is 2011-TB aluminum. Graphite aluminum metal-matrix shows promise in minimizing core module configuration weight.

Present-day hypervelocity test facilities are limited in their velocity capability to less than 8 km/s for 2 to 10 cm projectiles representative of the Space Station debris threat environment. Material modeling can provide useful damage predictions in the interim.

REFERENCES

1. "History of an Orbit Satellite Fragmentation," 4th Edition, Teledyne Brown Engineering Report CS 90-TR-JSC-002, January 1990.
2. Kessler, D. J., Reynolds, R. C., and Anz-Meader, P. D., "Orbital Debris Environment for Spacecraft Designed to Operate in Low Earth Orbit," NASA TM 100471, Johnson Space Center, Houston, Texas, April 1989.
3. McKnight, D. S. et.al., "Understanding the True Earth Satellite Population," IAA-89-627, 40th Congress of the International Astronautical Federation, Malaga, Spain, October 7-13, 1989.
4. Horn, J. R., et.al., "Analysis Procedures for Meteoroid/Debris Protection of Space Structures," TABES 89, Huntsville, AL, May 17, 1989.
5. Mog, R. A., "Spacecraft Protective Structures Design Optimization," TABES 89, Huntsville, AL, May 17, 1989.
6. Schonberg, W. P., "Hypervelocity Impact Testing of Dual Wall Structures," TABES 89, Huntsville, AL, May 1989.
7. Williamsen, J. E., "Space Station Freedom Meteoroid/Debris (M/D) Working Group Meeting, NASA/MSFC, January-February 1989.
8. Hopkins, A. K., Swift, H. F., and Lee, T. W., "Material Phase Transformation Effects Upon Performance of Spaced Bumper Systems," Journal of Spacecraft and Rockets, 9,5, May 1972

MAN-MADE DEBRIS ENVIRONMENT CHARACTERIZATION
FOR SPACECRAFT VULNERABILITY ASSESSMENT

BY ROBERT C. REYNOLDS†

Abstract

An issue of growing importance in the study of orbital debris is the need to understand the consequence of impact on a space structure. In this paper, a technique is developed to analyze the flux directionality of debris for spacecraft surfaces having different orientations, using catalog element set data.

Introduction

Man-made objects in low Earth orbit (debris) can be viewed as generating a flux field at each point in space accessible to that debris. This flux field, determined by the velocity and amount of time spent by each object crossing near the point, when convolved with the size and orbit of a spacecraft, can be directly related to a probability per unit time of collision occuring between that spacecraft and one or more debris objects. When debris size is accounted for, collision probability as a function of debris size can also be determined. In this paper, both the directionality and the magnitude of the debris threat will be investigated for several orientations of spacecraft surfaces.

The flux field will be defined in a spacecraft-fixed local-vertical, local-horizontal (LVLH) reference frame, in which the +Z axis is pointed toward the local zenith and the +X axis is pointed in-plane in the down-range direction; it follows that the +Y axis is directed along the orbital angular momentum vector and that the XY plane is the local horizontal plane. In this coordinate system, the direction from which the debris appears to be coming can be characterized by a yaw angle, ϕ, measured in the XY plane with the 0 point the +X axis and the +Y axis at 90°, and a pitch angle, γ, the angle out of the XY plane with a positive pitch angle indicating approach from above. The geometry of this coordinate system is shown in Figure 1.

The frequency at which objects generating this debris flux field impact a given surface area element on a spacecraft depends on the orientation of the element relative to the flux field as well as more complicated issues such as shadowing by other spacecraft structures and secondary impacts that depend on the overall spacecraft surface geometry.

In this paper, a simple box-shaped spacecraft geometry will be assumed. Two orientations will be examined. In the first case the long axis of the box will be aligned with the spacecraft velocity vector with one pair of long sides having their area normal vector in the local horizontal plane and the other pair having their area normal vectors aligned with the Z axis (Figure 2a). This is a good representation of the attitude for the habitability modules on the Space Station. The debris threat for this case will primarily

†Orbital Debris Project Manager, Lockheed Engineering & Sciences Co., Houston, Texas 77058

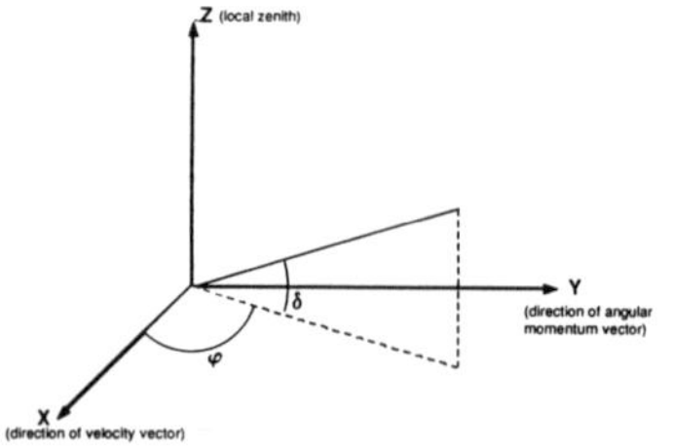

Figure 1. Definition of Yaw Angle (ϕ) and Pitch Angle (γ) in the Local Horizontal Local Vertical Reference Frame

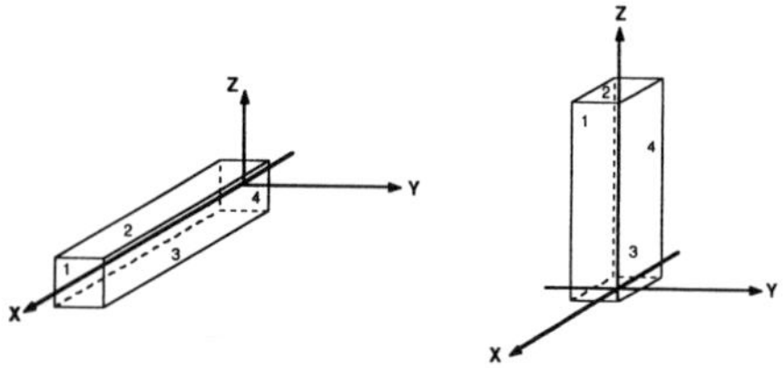

Figure 2. Two Cases of Spacecraft Attitude in Local Vertical Local Horizontal Reference Frame

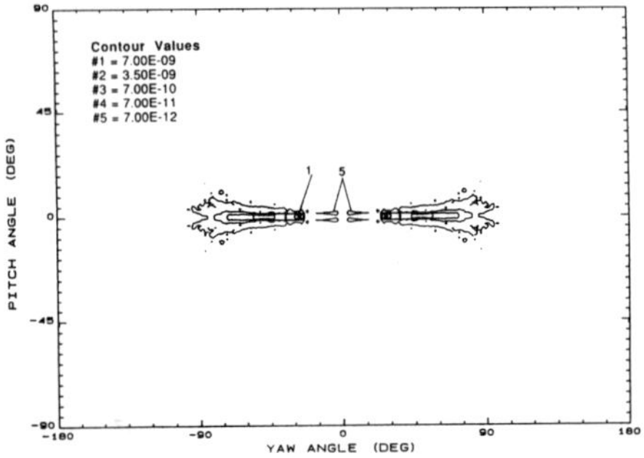

Figure 3. Flux Density Contours for All Relative Velocities

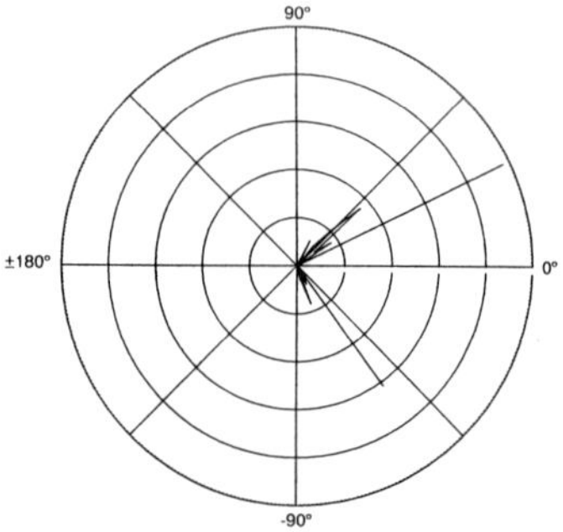

Figure 4. A Polar Plot in Yaw Angle of $\underline{G}(\phi)$ at the Ascending Node of the Orbit

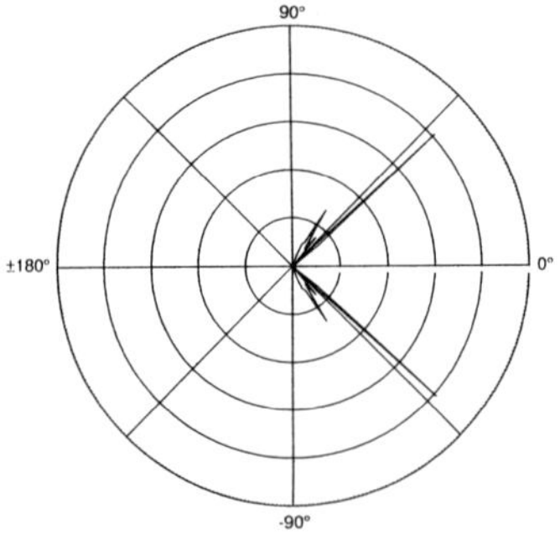

Figure 5. A Polar Plot in Yaw Angle of $\underline{G}(\phi)$ at Argument of Latitude 90°

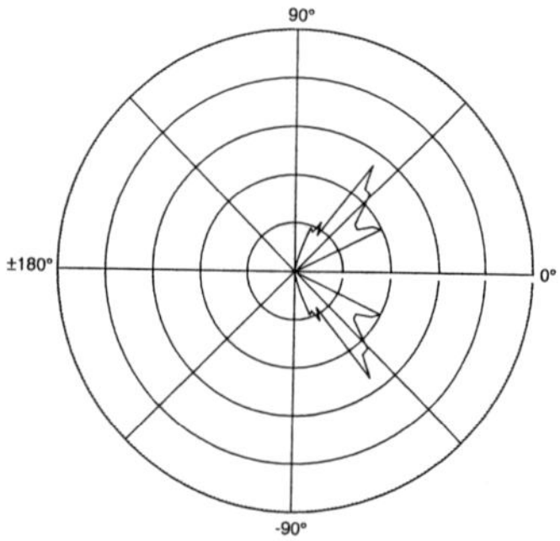

Figure 6. A Polar Plot in Yaw Angle of $\underline{G}(\phi)$ Averaged Over a Complete Orbit

appear as a level of hazard to the smaller leading surface (1), with a secondary threat level coming from the sides of the box having their area normal vectors co-aligned with the Y axis (2), i.e. the two sides exposed to debris moving nearly in the local horizontal plane, and much less threat coming from the sides of the box having their area normal vectors aligned with the Z axis (3) and minimal threat to the trailing end surface (4). Second, the box will be oriented so its long axis points toward the zenith (Figure 2b); this is the attitude such a spacecraft would assume if it was gravity gradient stabilized. Table 1 gives specific size information for each of the surfaces in these 2 attitudes.

TABLE 1. CHARACTERISTICS OF 4 SURFACE AREA ELEMENTS

AREA	DESCRIPTION	ORIENTATION (ϕ_S, γ_S)	SIZE (CASE 1) (m^2)	SIZE (CASE 2) (m^2)
1	leading face	(0,0)	16 (4x4)	60 (15x4)
2	facing zenith	(0,90)	60 (15x4)	16 (4x4)
3	facing horizontal	(90,0)	60 (15x4)	60 (15x4)
4	trailing face	(180,0)	16 (4x4)	60 (15x4)

Discussion

Flux Field

A flux field as sensed by the spacecraft can be defined from the orbital characteristics of the objects in orbit and from the orbit of the spacecraft in question. This flux will be directly related to collision probability or alternatively the collision rate.

Consider the calculation of collision probability for an area exposed to a uniform distribution of particles, all moving at $\vec{v}_o$ relative to the surface. This motion relative to the surface generates a flux field, $\vec{F}$, which is

$$\vec{F} = n\vec{v}_o$$

where n is the spatial density of particles. The units on flux are number/area/time. The probability of one or more collisions occuring with the area in time τ is

$$P_c(\tau) = 1 - e^{-F\sigma\tau} \tag{1}$$

where σ is the collision cross-section, which will in general depend on the size of the particles, the size of the surface, and the orientation of the surface relative to $\vec{v}_o$. A mean time between collisions may be defined as

$$t_m = \frac{1}{F\sigma} \tag{2}$$

For the case in which particle motion relative to a surface cannot be characterized by simple uni-directional flow, the flux vector becomes a flux density, $\underline{F}$, having directional dependence. The units on $\underline{F}$ are number/area/time/solid angle. In a spherical coordinate system with azimuthal angle ϕ (yaw) and polar angle γ (pitch), the flux from a small solid angle of $d\phi$ in direction ϕ_o and $d\gamma$ in direction γ_o, $dF(\phi_o,\gamma_o)$, will be given by

$$dF(\phi_o,\gamma_o) = \underline{F}(\phi_o,\gamma_o)C\gamma_o d\phi d\gamma \tag{3}$$

where

$$C_{\gamma_o} \equiv \mathrm{Cos}(\gamma_o)$$

The flux density distribution at a given point s is defined as

$$\underline{F}_s(\phi,\gamma) = \sum_i n_i(s) v_i(s) \delta(\phi-\phi_i)\delta(\gamma-\gamma_i) \quad (4)$$

where the summation is over all objects which can pass through the point in question, and

n_i = spatial density for each particle contribution to the flux field at the point in question (from /1/)

v_i = velocity at which the particle passes through the point

γ_i = - pitch angle of $\vec{v}_i$

ϕ_i = 180° + yaw angle of $\vec{v}_i$

δ = Kronecker delta function

To average over all or part of an orbit,

$$\underline{F}(\phi,\gamma) = \frac{\int_S n_s \underline{F}_s(\phi,\gamma) ds}{\int_S n_s ds} \quad (5)$$

where n_s = spatial density of the spacecraft at each point in the orbit

To relate the flux density distribution to collision probability or collision frequency, a collision strength density distribution per unit time as a function of direction may be defined as

$$\underline{F}(\phi,\gamma;d_o)\sigma(\phi,\gamma;d_o)C_\gamma d\gamma d\phi \quad (6)$$

where d_o is the size of the particles.

In the following calculations, the spacecraft will be taken to be in a 500 km circular orbit of inclination 28.5° and fixed in the LVLH coordinate system. In this coordinate system, the flux density for 10 cm and larger objects averaged over a complete orbit is given in Figure 3. The line of ϕ=0 in this figure is the local horizontal plane, the point ϕ=0°, γ=0° is the direction of motion of the spacecraft. The interval from ϕ=-90° to +90° is in front of the spacecraft. It can be seen from this figure that essentially all of the debris is encountering the spacecraft near the local horizontal plane. Furthermore, there is a strong focussing of debris in the forward direction. There is a broad maximum from 45° to 50° in yaw, and a secondary maximum around 30°. Because of this concentration of flux near the horizontal plane, a summed distribution over pitch angle, $\underline{G}_S(\phi)$, where

$$\underline{G}_s(\phi) = \int_{-\pi/2}^{\pi/2} \underline{F}_s(\phi,\gamma) C_\gamma d_\gamma \quad (7)$$

provides an accurate and more graphical picture of the directionality of debris impact on the spacecraft. It is in units of impacts/m^2/yr/deg, and can be plotted as a debris heading plot.

Examples of the **G** distribution for a spacecraft in 500 km circular orbit of inclination 28.5° at the ascending node is provided in Figure 4 and at argument of latitude 90° in Figure 5. The flux field averaged over a complete revolution is shown in Figure 6. The direction of motion for all cases is in the ϕ=0 direction. Figure 6 is replotted in a rectangular format, now showing yaw angle as the independent variable, in Figure 7 to show the similarity of this distribution to the inclination distribution at 500 km (Figure 8) for the objects creating the flux field. The yaw angle plot is symmetric about the 0° line, and is a mapping of the inclination distribution with an additional spreading function. Debris of inclination i will be spread non-linearly over a yaw range of 75.75°- i/2 to 75.75°+ i/2.

The fact that the yaw angle distribution is closely related to the inclination distribution is significant, for it means that if the inclination distribution for the small, unobserved objects is similar to that of the large, observed objects, as expected, then the flux density functions can be scaled down to small sizes by a simple scaling by the ratios of the numbers of objects at these two sizes. There is good reason to believe that the inclination distributions should be similar, since the processes that create small debris should do so with small plane change angle, and therefore by necessity small inclination changes.

The characteristic velocity on impact with the spacecraft surface for this flux distribution discussed to this point will be large. Shielding against these high speed impacts is quite different from low speed impacts. When the flux density for low speed impacts (<1 km/sec) is calculated, a much different picture emerges for the flux density, as seen in Figure 9. This figure shows that the significant low speed flux encounters the spacecraft from the rear, and can be explained as coming from objects in highly eccentric orbits, such as geosynchronous transfer orbits, near perigee. Such objects will have a larger horizontal velocity than the spacecraft in circular orbit. If the maximum impact velocity were to be reduced still further, a more isotropic distribution would emerge, arising from objects in low eccentricity orbits passing through 500 km altitude; some of these objects would be catching up to the spacecraft and some would be moving slower. The fluxes for this case would be lower still.

Interaction with Spacecraft Surface

In considering the interaction between the flux field and the spacecraft surface, if the surface is a simple convex figure all points on the surface will be exposed to 2π steradians of space. Many simple solids - spheres, cubes, cylinders - are examples. For such surfaces, the collision strength density of a unit surface element with a unit normal vector pointing to (ϕ_S, γ_S) with debris from a given direction (ϕ_O, γ_O) will be

$$\underline{\mathbf{H}}(\phi_o,\gamma_o;\phi_s,\gamma_s,d_o) = \underline{\mathbf{F}}(\phi_o,\gamma_o;d_o)\left(C_{\gamma_s} C_{\gamma_o} C_{\phi_s-\phi_o} + S_{\gamma_s} S_{\gamma_o} \right) \tag{8}$$

where

$$S_{\gamma_O} \equiv \mathrm{Sin}(\gamma_O)$$

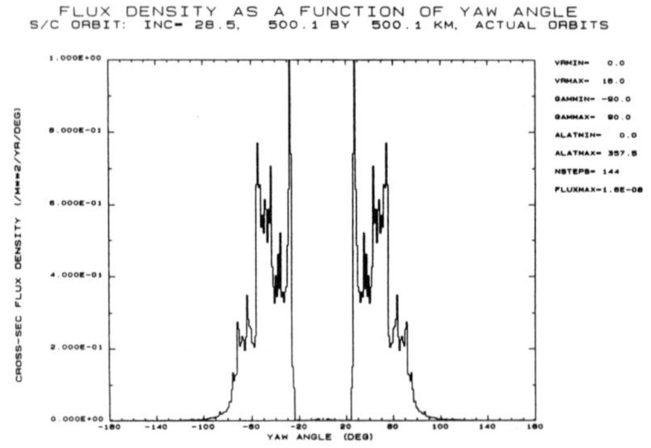

Figure 7. Rectangular Plot of $\underline{G}(\phi)$ vs. ϕ Averaged Over a Complete Orbit

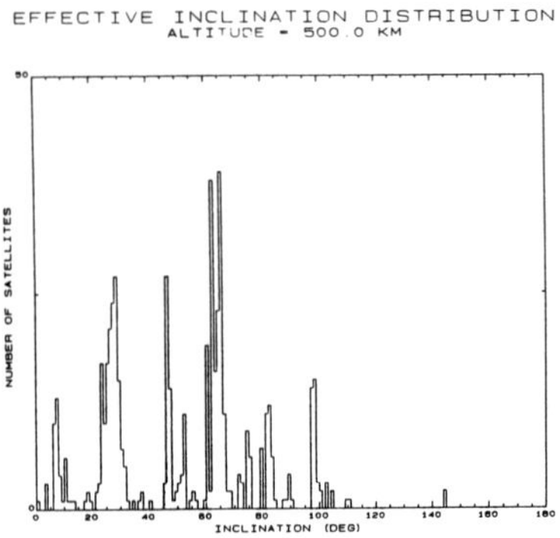

Figure 8. Inclination Distribution for Satellite Crossing Through 500 km Altitude

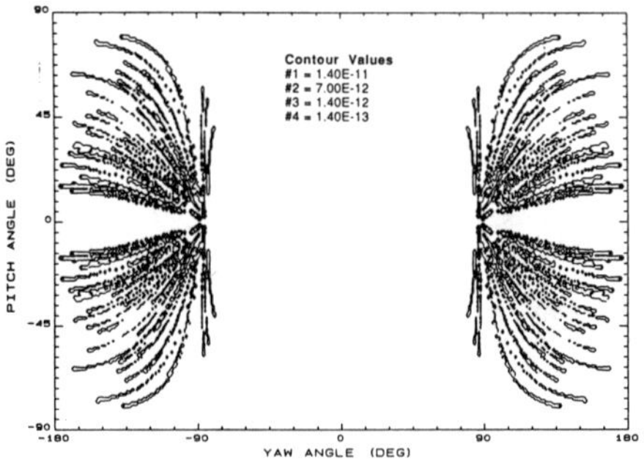

Figure 9. Flux Density Contours Constrained for Object Encountering the Spacecraft at Velocities <1 km/sec.

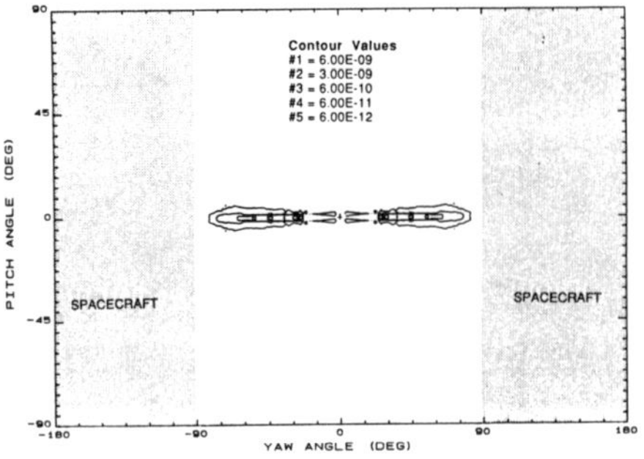

Figure 10. $\underline{B}$ Contours for a Surface Area Element Facing into the Direction of Motion

and

$$f = \begin{cases} 0 & \text{if } f \le 0 \\ f & \text{is } f > 0 \end{cases}$$

The integrated collision strength density function, $\underline{C}$, will then be defined as

$$\underline{C}(\phi_s,\gamma_s;d_o) = \int_{4\pi} \underline{H}(\phi,\gamma;\phi_s,\gamma_s,d_o)\sigma(\phi,\gamma;\phi_s,\gamma_s,d_o)C\gamma d\gamma d\phi \tag{9}$$

The generalized versions of Equations 1 and 2 become, for the probability of collision with the surface in time τ

$$P_c(\tau) = 1 - e^{-\underline{C}\tau} \tag{10}$$

and for the mean time between collisions

$$t_m = \frac{1}{\underline{C}} \tag{11}$$

For the simple convex spacecraft surfaces, the shadowing of a surface element by the rest of the spacecraft structure is that line on the (ϕ,γ) plane for which $\underline{H} = 0$. For geometrically more complex spacecraft surfaces, the outline of the spacecraft structure as seen at the surface element will be complex, and could lead to complex gamma limits as a function of phi in the integral.

Consider the case of a box oriented with its long axis aligned with the velocity vector and fixed in an LVLH coordinate system (Case 1, Table 1). This is the aspect ratio for the habitability modules in the Space Station. Also, consider the four surface area elements on this box as shown in Figure 2, and further defined in Table 1. The $\underline{H}$ function for these four cases is shown in Figure 10-13, and the the peak value of the function is given in Table 2.

TABLE 2. DATA FOR ALL ENCOUNTER SPEEDS; d_o = 10 cm

AREA	PEAK $\underline{H}$	H (EQN 12)	H (REF 2)	$\underline{C}$ Case 1	$\underline{C}$ Case 2
1	6.89E-9	4.40E-7	8.42E-7	7.05E-6	2.64E-5
2	6.78E-11	5.09E-9	0	3.06E-7	8.15E-8
3	4.95E-9	2.39E-7	4.40E-7	1.43E-5	1.43E-5
4	6.22E-12	2.20E-9	0	3.51E-8	1.32E-7

For surface 1 (Figure 10), the spacecraft occupies the shaded region from yaw angles -180° to -90° and from 90° to 180°. This surface is exposed to all pitch angles over the other yaw angles. The orbit normal vector is pointed in the direction marked with the "+". The plot looks very similar to Figure 3, but it can be seen that there is some reduction in the contribution from the lowest flux levels at the yaw angle farthest away from the area normal vector. Essentially all of the debris encountering this surface will come from a band of 10° width in pitch, centered on the horizontal plane, and running from 25° to 55° in yaw.

For surface 2 (Figure 11), pointed at the zenith, the spacecraft occupies the shaded region on the lower half-plane; the surface is exposed to debris at all yaw angles but only for positive pitch angles. The area normal vector is

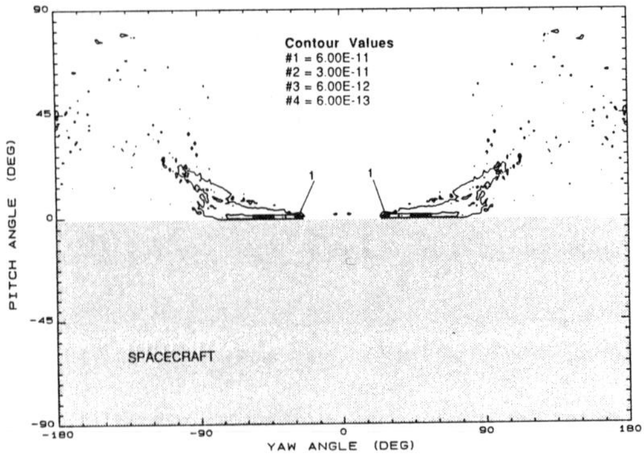

Figure 11. H Contours for a Surface Area Element Facing the Local Zenith

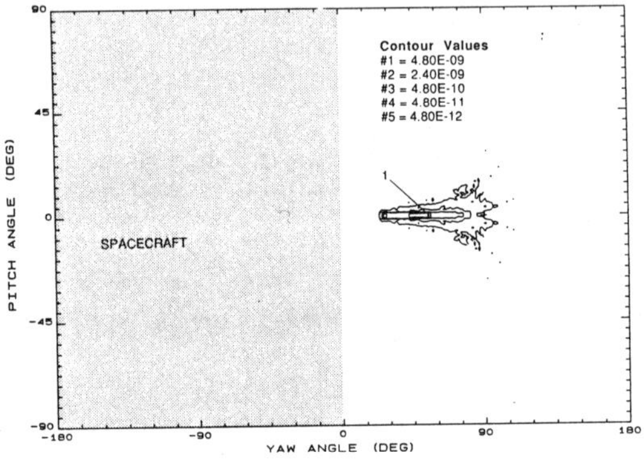

Figure 12. H Contours for a Surface Area Element Facing Yaw Angle 90° in the Horizontal Plane

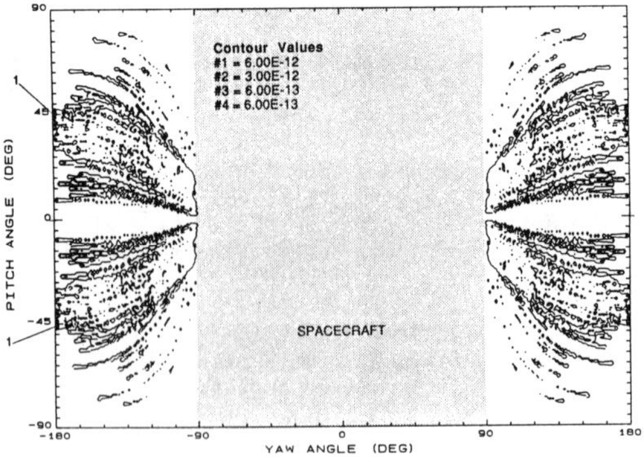

Figure 13. H Contours for a Surface Area Element Facing Directly Away from the Velocity Vector

pointed to a pitch of 90°, which, because of the latitude/longitude mapping used for the plot, is the entire line along the top of the figure. The orientation of this surface is such that the large flux contributions in the forward direction are so near to the horizontal plane that they are nearly perpendicular to the area normal vector, and therefor provide only a small contribution. As a consequence, the much lower flux contributions at large pitch angle, that do not appear in Figure 3 because they are so small, do provide a significant contribution to the perceived flux on this surface. The flux levels are small, as seen in Table 2.

The debris flux of consequence for this surface will come from objects encountering the surface at more than 85° from the perpendicular. These very shallow impact angles will mean that the impacting object will most likely glance off the surface and spray ejecta along the surface, rather than penetrate into the interior.

For surface 3 (Figure 12), pointed into the horizontal plane at a yaw angle of +90°, the spacecraft occupies the shaded area on the left half-plane. The area normal vector points in the direction denoted by "+". As in Figure 10, this surface is exposed to the large fluxes shown in Figure 3. There is some enhancement of the contours near yaw angle 90° compared to Figure 3 since this region is near the area normal.

The peak flux levels for the side surfaces are nearly as large as for the front surface, as might be expected in viewing Figure 6. This effect is strongly dependent on the inclination of the orbit of the spacecraft; had the spacecraft been in an 85° orbit the peaks in flux would have occured near 5° in yaw, and the exposure to the sides of the spacecraft significantly reduced.

For surface 4 (Figure 13), the trailing surface, the spacecraft occupies the middle half of the plane; the surface is exposed to all pitch angles for only those yaw angle directed behind the spacecraft. The area normal vector points in the direction denoted by "+", where the 2 locations so marked are really the same point. The contours in this figure are very much like those of Figure 9, as might be expected, and the flux levels are correspondingly low, as noted in Table 2. The debris encountering this surface will do so at low velocity.

For the case when the surface area, A, being exposed to the debris flux is considerably greater than the size of the particles forming the flux field, the collision cross-section in Equation 9 simplifies to being the surface area itself. In computing collision probabilities for debris objects <= 10 cm in diameter for the spacecraft surfaces used in this paper, this case holds and Equation 9 becomes

$$\underline{C} = A \int_{4\pi} \underline{H}C_{\gamma}\, d_{\gamma}\, d_{\phi} \equiv AH \qquad (12)$$

In this equation, H may be regarded as the perceived flux.

The values for H and $\underline{C}$ for each of the four surfaces for the two cases are provided in Table 2, where the simplification for $\underline{C}$ from Equation 12 has been used. For an object in a 28.5° orbit, the flux as perceived by a side looking surface element is comparable to that perceived by a forward looking element. This is an expected result, when the $\underline{G}$ distribution is plotted as in Figure 6.

If the spacecraft had been assumed to be in higher inclination orbit, the peak flux points would have been moved toward the velocity vector, and the perceived flux for side looking surface elements would reduce and forward looking elements increase. The second conclusion from this table is that the perceived flux for a zenith or nadir looking element is comparable to that of a trailing surface.

The equivalent calculation for H from Reference 2 is also given in Table 2 for a 10 cm 1990 population, growth rate of 5%, and solar flux at 10.7 cm of 150 x 10^4 Jansky. The values for H in this paper compare favorable with Reference 2 for surface 1 and 3, and the flux ratios for these 2 surfaces are nearly identical. Reference 2 does over-estimate the flux on the horizontal surfaces.

The $\underline{C}$ values for the 4 surface areas for the 2 cases is also provided in Table 2, where the areas are defined in Table 1. For Case 1, the greatest probability of collision comes from impact with the side of the spacecraft, since its surface area is nearly 4 times that of the front surface, and impact with the top or bottom of the spacecraft is a factor of 20 less likely than for impact with the front but a factor of 10 greater than for impact with the trailing surface. For the Case 2 attitude, impact with the front surface is greater than for any other single surface, but impact on the trailing surface is about twice as likely as impact with the small zenith or nadir pointing surfaces.

A net value for $\underline{C}$ for the entire spacecraft, $\underline{C}_{sc}$, may be calculated from these four $\underline{C}$ values using the simple relationship that

$$\underline{C}_{sc} = \underline{C}_1 + 2\underline{C}_2 + 2\underline{C}_3 + \underline{C}_4$$

where C_i refers to collision strength on surface i.

A value for $\underline{C}_{sc}$ and the corresponding t_m are provided in Table 3.

TABLE 3. COLLISION PROBABILITIES AND MEAN TIME BETWEEN COLLISIONS FOR THE 2 REFERENCE CASES. d_o = 10 cm

	COLLISION PROBABILITY (1/yr)	MEAN TIME BETWEEN COLLISIONS (yr)
Case 1	3.62E-5	2.76E4
Case 2	5.52E-5	1.81E4

Conclusions

A method has been presented for analyzing directionality of debris flux incident on a spacecraft surface. This information is carried primarily in the $\underline{H}$ function defined in Equation 8, a function of both the characteristics of the flux field and the orientation of the surface. It may be used to calculate collision probability or expected mean time between collisions. Knowing the directional composition for the sensed flux as a function of surface orientation could be important input for shielding considerations, and could affect the design of scientific experiments which are conducted in orbit.

The H function, by virtue of the definition of the F function, represents a mass or number flux for debris, as appropriate for collision rate calculations. Alternative forms for F and H types of functions could be defined to represent the flux of energy or some other penetrability parameter to more satisfactorily address consequences of impact. This will be developed in future work.

For the two cases studied, a box-shaped spacecraft having the size and aspect of the Space Station habitability modules and fixed in an LVLH coordinate system, the sides of the spacecraft experience a flux comparable to the leading surface. This is a consequence of the 28.5° inclination assumed for the spacecraft. In the first case, in which the long axis of the spacecraft is aligned with the velocity vector, collisions with the sides of the spacecraft will occur 4 times as often as with the front surface. For the gravity gradient stabilitzed attitude of the second case, collisions with the sides of the spacecraft will occur as often as with the front surface.

Although the results derived in this paper come from orbital element data from the large objects seen with ground-based sensors, it was shown that the inclination distribution of the objects in orbit is the important parameter in determining the flux density distribution. It is expected that the more numerous centimeter and millimeter debris will have roughly the same inclination distribution as the large objects, and so the results in this paper can be scaled to account for the differences in numbers of small objects compared to the large objects.

Acknowledgments

I wish to thank Eric Christiansen and Don Kessler for review comments, Phillip Anz-Meador for many discussions and help in preparing the figures, and to Cindi Karpiuk, Alene Simmons, and Anita Dodson for preparing the text and figures for publication.

REFERENCES

1. Kessler, D. J. and B.G. Cour-Palais 1978.

2. Kessler, Reynolds, and Anz-Meador 1989. NASA TM 100 471

CLEARING SPACE DEBRIS USING ENERGETIC MATERIALS

Laddie Marin, Jr., Major, USAF,[1] Charles A. Mulligan, Capt, USAF, and Joseph J. Secary, 1Lt, USAF

ABSTRACT

Space debris is growing at an alarming rate. It consists not only of old, unusable payloads but also aluminum oxide combustion products of various space delivery systems and small fragments from payload breakup. Although the thousands of large, trackable objects pose a collision hazard for future US space vehicles, the smaller, untrackable particles are an even larger threat due to their high velocities and nondetectability prior to impact. Even though efforts are under way to reduce the growth of this pollution problem, an effective debris clearing method is needed to ensure safe launch and deployment of expensive national space assets. This paper discusses the space debris problem, reviews some previously proposed solutions, and presents a new possible solution involving use of energetic material charges to cause reentry or escape of debris objects.

INTRODUCTION

Space exploration began with the 1957 Sputnik and 1958 Explorer I launches. The space exploration that followed resulted in deployment of civilian and military space vehicles, all of which contributed to the increase of space debris that was of little concern for many years. Unfortunately, the amount of space debris has reached such a magnitude that space pollution is now a safety concern for space operations. Space use is being driven by more than our thirst for knowledge and adventure. Today, space operations serve as a key part of the solution for many military and civilian information gathering, environmental research,

---

[1] Weapons Environment and Support Branch, Weapons Laboratory (AFSC), Kirtland AFB, NM 87117-6008.

communications, and meteorological needs. If space were unavailable, many of these missions would be degraded. (Cochran et al, 1985).

As many engineering and political issues of operating in space are being solved, a new issue is surfacing. The space environment has always been considered harsh, but it is becoming a literal "shooting gallery" due to the growth of man-made debris. Space has been affectionately termed the final frontier. Will space also become man's final junkyard?

BACKGROUND

Space Debris - What is it?

From the literature, an accepted definition for space debris is any non-operational or non-functional, uncontrollable object in space (Johnson and McKnight, 1987). In this paper, only man-made debris will be considered.

Where is it?

As might be expected, debris build-up patterns follow our exploration of space. (Interagency Group (Space) For National Security Council, 1989). As documented in the literature, there is more in low earth orbit (LEO) as this was the first region explored and later used operationally. Additionally, all subsequent exploratory and operational missions go through this region, regardless of their final destination, leaving additional debris deposits. This, combined with the natural slow gravitational decay process which brings all the near-earth debris "home", makes LEO the most debris littered region.

A second region of high interest is geosynchronous altitude. A strategically placed satellite at this altitude will remain relatively fixed over a given position on earth. With minimal station keeping efforts and the need for only a single ground control station, a single satellite can provide constant coverage for almost half of the globe. This makes geosynchronous altitude a prime parking site for communications and surveillance systems. Unfortunately, as these systems age and are retired, they become debris.

Early Concerns Over Space Debris

Early concerns regarding space debris were centered around natural space debris such as meteoroids (Brown,

1987). Efforts studied the possible effects of natural debris impacts and how to protect early exploration missions from them. Due to the hypervelocities of meteoroids (up to 72 km/sec), protection is difficult. Fortunately, due to the low density of such debris and the limited mission durations, the probability of impact was extremely low. As the amount of man-made debris increased, it quickly became more dense than natural debris. As mission durations went from days to years, the probability of impact increased.

Is Space Debris a Problem?

Is space debris significant enough to negatively impact our use of space? Minor damage from debris impact is well documented, such as the example of pitted Shuttle windows. Degraded mission performance due to inadequate power from damaged solar panels has also been cited. Additionally, there are over fifty unexplained instances of satellites becoming inoperative in space (Keesler and Su, 1985). A speculated cause is debris strikes. This type of damage will become more common as larger, long-term space structures, such as space stations or space telescopes, are fielded.

Debris Trends

Recognizing the size and range limitations of available debris data, reviewing NORAD's Satellite Situation Reports still provides some interesting results (NASA, various years). Although Australia, China (PRC), France, India, Japan, Mexico, United Kingdom, and West Germany are all space debris contributors, the major contributors are the USA, the USSR, and the European Space Agency (ESA). Collectively, they are responsible for 85.5 percent of the cataloged debris.

As shown in Figure 1, the growth rate of debris is larger than that of satellites. This should be expected considering that each satellite is responsible for several debris objects. What this data does not show, however, is the growth in small debris which is expected to be larger still. This is due not only to a larger number per launch, but also the fragmentation process as objects break up or hit one another and break into smaller objects.

Another interesting comparison is presented in Figure 2, which shows the step increase of ESA's debris due to a payload breakup. Imagine the debris cloud which would have occurred had the USA's Challenger accident occurred in space rather than in the atmosphere.

Figure 1.
Orbiting
Space Objects

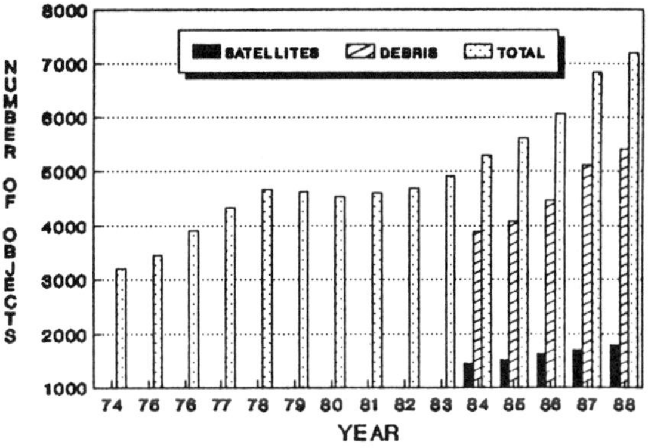

Figure 2.
ESA Space
Objects

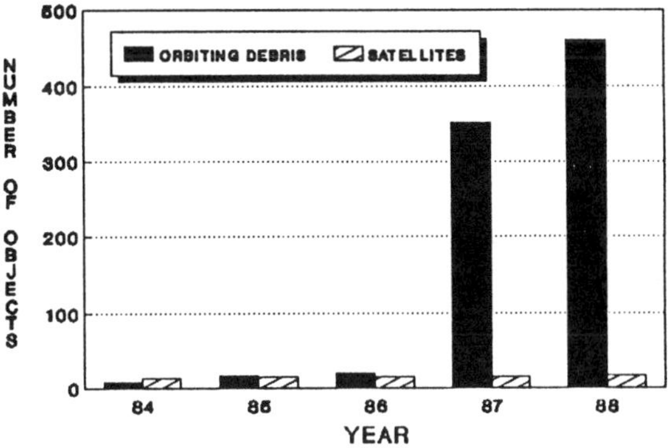

Figure 3.
Debris Growth
Predictions

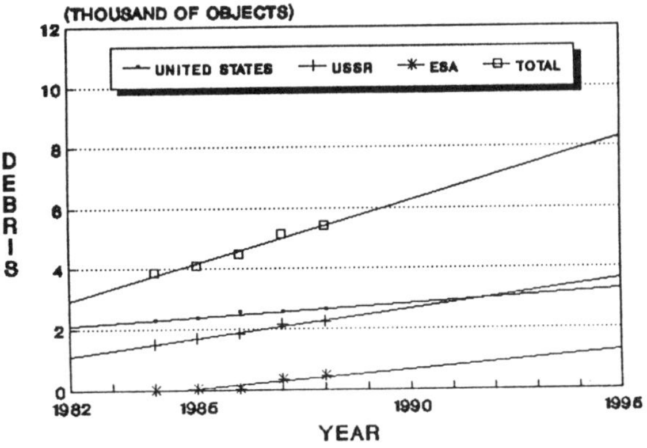

Four years of available debris data are presented in Figure 3, along with projections made using simple linear regression. Of importance to note is that although the USA is the leading contributor to reported debris, the USSR will soon surpass the USA. Even the ESA's slope is greater than the USA's. The difference in debris growth can be attributed to two factors: how different space users solve their individual mission requirements and their experience in space. For example, although the USA and the USSR have similar reconnaissance needs, the USA tends toward longer lived satellites to fulfill this requirement, while the USSR uses short lived systems (Johnson, 1987). As the life expires on a system, another must be launched to replace it. This results in added debris.

Experience in space operations is also a factor. When the ESA's first space efforts were undertaken, their payloads piggybacked on others countries' launch systems. As the ESA became more experienced, it not only developed its own delivery systems, but it delivered payloads into deeper space. As seen in the quantum jump of the ESA's space debris, explosive failures can be expected. All these factors led to increased debris.

The point is not who's winning the space debris race, but rather that major debris contributors are outside the control of the USA. Solutions such as legislation to limit debris may have some effect on the growth rate but will be ineffective overall.

A second point is that as other nations begin to explore and use space, they'll also be heavy contributors to the space debris problem. Even if they only contribute to the LEO debris problem, USA missions must fly through this region and will need effective protection from the debris.

POSSIBLE SOLUTIONS

From the literature, the following possible debris solutions are being either proposed or developed: hardening, avoidance, debris retrieval, earth reentry, transfer to dump regions, treaty/agreement, and collectors.

Hardening has been a traditional solution for preventing penetration damage. This solution, however, has two shortcomings. First, space debris travels at hypervelocities (as high as 15 km/sec), well outside traditional hardening regimes. Technology for developing hardened structures does not exist at these velocities.

Secondly, any hardening approach will most likely increase the system's total weight. Putting more weight into orbit would increase launch costs and might, in some cases, be beyond the capacity of current launch systems.

Avoidance schemes are being proposed: orbiting satellites would use thrusters to dodge debris on impact trajectories. These schemes, however, project tremendous fuel consumption, which for current satellites is unacceptable. Dodging debris might also require the development of an onboard debris detection and tracking system.

The debris retrieval concept has already been shown to be successful, at least in limited applications. A Shuttle retrieved an inactive satellite which was then repaired and returned to service. The drawback with this concept is the expense of retrieval and difficulty of the operation. Only high valued space debris would qualify for such a debris clearing solution.

Earth reentry has two aspects. The first aspect is to intentionally cause spent satellites to reenter by using the last remaining fuel to project the satellite into a decay orbit. One drawback with this concept is that energy needed to put a satellite into a decay orbit is significant, in many cases more than current satellites have on board. Even if future satellites were somehow required to carry a decay orbit propulsion reserve, this solution is limited to the satellite category of debris and does little to mitigate other categories of debris. Another drawback to this concept is the reentering satellite impact location. As seen with the USSR's satellite reentry into Canada, this can have undesired political and safety outcomes.

A second aspect to the reentry solution is the natural reentry process due to gradual orbital decay. This is nature's way of cleansing the environment. It unfortunately is a long process and, in the case of all but LEO debris, is much too slow.

Akin to the reentry concept is the transfer to a dump orbit. Recognizing the burden placed on the propulsion system to cause reentry, especially for satellites in other that LEOs, the concept is to move geosynchronous satellites from their strategic position when they become of little use. This concept also solves the earth-impacting satellite problem of the reentry solution, yet does little for other categories of debris and raises the question of determining suitable junkyard regions.

Legal aspects and efforts to legislate away the space debris problem are documented in the literature. To date, there has been little success. Currently, some LEO regions have been set aside for space developing nations to use. Also, discussions of restricting solid rocket propulsion systems, eliminating blow away clamps, and similar ideas are being considered. Although these potential solutions may prove to have some impact in the long term, litter laws have limited success. Debris will still exist.

Overall, the existing and proposed debris solutions appear inadequate. None appear able to ensure safe, debris-free missions.

DEBRIS CLEARING DEVICE

Could energetic materials be used to create debris free paths and parking spaces for future space missions? The energetic material would be castable, eliminating the need for a case which would add to the debris problem. It could be carried into LEO orbit with either a reusable or expendable launch platform and bussed or possibly catapulted into final position. The energetic material could then be initiated using lasers, eliminating the need for debris creating firing systems. The expanding detonation products would impact nearby debris and push it away. The castable charge can be tailored in both size and shape to achieve optimal effects. It can also be oxygen balanced to ensure that the reaction produces only gaseous products.

Figure 4 presents a qualitative picture of the possible effective limits of the energetic material debris clearing concept. Debris at LEOs would require small velocity changes and therefore a small charge mass to cause reentry. As the altitude increases, the charge mass needed to force a given debris particle to reenter would increase. On the other hand, debris at very high altitudes would only require small velocity increases to achieve escape velocities. At lower debris altitudes more energy would be needed to escape earth's gravity, requiring increased charge mass. These two curves provide a breakeven intersection. Based on payload limitations of the launch vehicle delivering the energetic material, one might have to choose between reentry and escape options for debris at a given altitude.

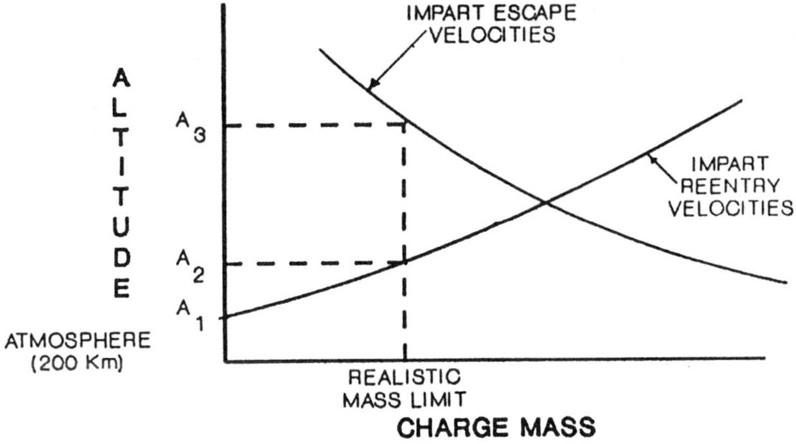

Figure 4. Realistic Altitudes for Debris Clearing

ANALYTICAL EFFORTS

Some preliminary calculations were performed to determine the feasibility of using energetic materials to clear debris. Neglecting atmospheric effects, the impulse needed to cause a 1 gram debris object in a circular orbit to hit the ground is 0.1094 N-sec from 370 km altitude and 0.2357 N-sec from 850 km. From geosynchronous altitude, about 35,863 km, the impulse needed to cause a 1 gram object to reach escape velocity is 1.2771 N-sec. From the literature (Falcovitz, 1986), the equation governing impulse from energetic material explosions in space is:

$$I = \frac{w\ (2Q_o)^{0.5}}{4\ \pi\ R^2}$$

where: I = impulse/square meter target projected area
w = charge mass
Q_o= charge energy (assumed to be 4000 KJ/kg)
R = range between charge and debris particle

From this equation, it can be seen that the impulse is inversely proportional to the square of the range. A more subtle point arises from the fact that "I" in the equation is in impulse per unit projected area of debris. The point is that, at a given range, the change in velocity imparted to a debris object is inversely proportional to the radius of the object, assuming spherical objects of identical density but different sizes. The reason for this is that larger objects have a greater mass to projected area ratio than smaller ones. These factors greatly limit the practical range over which energetic materials can be used to clear debris from orbit, since both range and size affect the change in velocity of a given debris object. Different changes in velocity result in different orbit changes. Therefore, a single charge of energetic material might cause some debris objects to reenter, others to achieve escape velocity, and still others to have their orbits changed without either escape or reentry. The problem becomes even more complicated when one considers non-circular initial debris orbits or the fact that debris might be located in all directions around a single energetic material charge.

Despite the above practical limitations, it is useful to examine some simplified cases. If one neglects atmospheric effects and assumes initially circular debris orbits, velocity changes parallel to the initial debris velocity vectors, and spherical particles of aluminium oxide debris with density of 3900 kg per cubic meter, then the following table can be generated for 1 gram debris objects. Keep in mind that many debris objects can be affected by a single charge.

	Reentry from 370 km Altitude	Reentry from 850 km Altitude	Escape from Geosynchronous Altitude
Range	Charge Mass(kg)	Charge Mass(kg)	Charge Mass(kg)
1 m	10	21	116
10 m	1000	2100	11600
100 m	100000	210000	1160000

Table 1. Energetic Material Mass versus Range

FUTURE EFFORTS

Assuming ideal rigid body momentum transfer, complete energetic material combustion, and optimum impact angles, necessary charge mass is a function of the debris' altitude, mass, initial orbit, and exposed surface, as well as the distance between the debris and the charge at initiation, and the detonation velocity of the energetic material. More work needs to be done to determine the orbital mechanics effects of non-optimum impact angles on a three dimensional distribution of different-sized orbiting debris objects. The effect of energetic materials on the ozone layer must also be considered. Energetic materials used to clear space debris could be tailored to enhance, or at least not deplete, the ozone layer. Vacuum experiments may be proposed to validate the analytical results.

CONCLUSIONS

Unfortunately, where humans go debris follows. Space debris will most likely always be with us. To date, proposed space debris solutions do not appear to be able to provide near-term answers allowing safe space operations. Using energetic materials to provide clear paths for space travel and operations may prove to be the most promising solution. Additional research is needed to understand the critical parameters to such a solution and optimization techniques.

ACKNOWLEDGEMENTS

Support for this project was provided by Captain William K. DeVault, US Air Force, of WL/NTED, who assisted with the literature search, and by Ms Cathi L. Wickesberg, WL/NTED summer hire from June to August, 1989, who provided assistance during the literature search and initial data reduction.

REFERENCES

Brown, D.C., NASA Activities, "Space Debris", Feb-Mar 87, pp 11 - 13.

Cochran, Curtis D., et al, Space Handbook, Air University Press, Jan 85.

Falcovitz, Joseph, Impulsive Loading From a Bare Explosive Charge in Space, NPS 72-86-004CR, Dec 1986.

Interagency Group (Space) for National Security Council, Report on Orbital Debris, Washington D.C., Feb 1989.

Johnson, Nicholas L., The Soviet Year in Space 1986, Teleydyne Brown Engineering, 1987.

Johnson, Nicholas L., Darren S. McKnight, Artificial Space Debris, Orbit Book Company, 1987.

Keesler, D.J., Shin-Yi Su, "Orbital Debris", NASA Conference Publication 2360, 1985.

NASA, Goddard Spaceflight Center, Satellite Situation Reports, Catalogued by NORAD.

Subsection 3.3: ORBITAL SUPPORT FOR EXTRATERRESTRIAL MISSIONS

TECHNOLOGY NEEDS DEVELOPMENT AND ORBITAL SUPPORT REQUIREMENTS FOR MANNED LUNAR AND MARS MISSIONS

Charles P. Llewellyn[1]
Karen D.Brender[2]

ABSTRACT

This paper will present an overview of the critical technology needs and the Space Station Freedom(SSF) focused support requirements for the Office of Exploration's(OEXP) manned lunar and Mars missions. Major emphasis will be directed at the technology needs associated with the low Earth orbit(LEO) transportation node assembly and vehicle processing functions required by the lunar and Mars mission flight elements. The key technology areas identified as crucial to support the LEO node function to be discussed in this paper include in-space assembly & construction, in-space vehicle processing & refurbishment, space storable cryogenics and autonomous rendezvous & docking.

INTRODUCTION

In early 1987, NASA Headquarters requested that the Langley Research Center's Space Station Freedom Office perform studies to assess the impact of manned lunar and Mars missions on the baseline Space Station Freedom(SSF). Agency-wide teams were formed to investigate the Station support necessary to accommodate such missions with emphasis on the precursor research, overall mission support in LEO, concurrent science requirements and impacts, technology needs and demonstration requirements and resource demands on station crew, power, volume and facilities. The results of these studies are published in references 1 and 2.

---

1 Analytical Mechanics Associates, Inc., Senior Engineer

2 NASA Langley Research Center, Deputy Manager, Evolutionary Definition Office, Space Station Freedom Office

Impetus was given to the exploration program in a directive issued by President Bush in July of 1989 to have a human presence on the Moon and Mars in the 21st century. The President announced a pathway to a Mars outpost by way of the Moon and a study process has been developed and implemented to accomplish this goal. (See references 3 and 4.)

CASE STUDIES OVERVIEW

For the Fiscal Year 1989(FY89) the Case Studies set by the Office of Exploration(OEXP) included two Mars and one Lunar scenario. A top level summary of these scenarios is shown in table 1.

TABLE 1. SUMMARY OF CASE STUDIES FOR FISCAL YEAR 1989

	Lunar Evolution (CS 4.1)	Mars Evolution (CS 5.0)	Mars Exped (CS 2.1)
Crew Size	8	5, 7	3
Staging node	SSF	dedicated new LEO node	none
Reusability of piloted vehicles	yes	yes	no
Aerobrake			
Use	at Earth	Mars, Earth	Mars only
Type	low L/D	low L/D	L/D = 1.0
Orbit at target body	LLO	Phobos (Gateway)	LMO
Trajectories			
Cargo	from SSF	Conjunction	Conjunction
Piloted	from SSF	Opposition & Conjunction	Sprint
ETO pipeline constraints			
Wet	570 t/yr	570 t/yr	570/yr
Dry	90 t/yr	90 t/yr	-
Propulsion			
Cargo	Chemical NEP, 2014	Chemical NEP, 2014	Chemical -
Piloted	Chemical	Chemical NTR, 2014	Chemical
In situ Resources Utilization	LLOX	Phobos LOX	none

In all of the FY89 Case Studies scenarios, major mission cargo is delivered to LEO on unmanned launches that occur at distinct intervals from the manned or piloted launches. A detailed description on each case may be found in reference 4.

In the Mars scenarios, the HLLV's payload and volume constraints drove the requirement for in-space assembly and construction capability at the node. In the lunar cases, where the flight rates were high, the lunar transfer vehicles were reusable. The need to process and service this reusable flight hardware on-orbit drove the requirement for an in-space vehicle processing/refurbishment capability.

TECHNOLOGY REQUIREMENTS/NEEDS

The top level technology requirements/areas needed to support manned lunar and Mars exploration missions are shown in figure 1. Although generic in nature, these requirements are relevant to both the lunar and Mars case studies. The technologies indicated on the figure are applicable to three key technical study areas which are: transportation node systems; transfer vehicle systems; and, extraterrestrial systems. The technologies were not time-phased nor prioritized but served as a point of departure in the studies for determining areas where additional emphasis was required. The remainder of this paper will concentrate on those technologies that are relative to the LEO transportation node function.

To provide better visibility and traceability to the technology needs evolving from the case study activity, the Exploration Technology Working Group (ExTWG) was formed with representatives from the various NASA centers, the Office of Exploration and the Office of Aeronautics & Space Technology. These representatives were designated as Technology Integration Agents(TIA's).

Analysis of the OEXP FY89 Case Studies by the ExTWG identified some fifty-six technology needs within the three key technical study areas mentioned earlier. All of the technologies were then ranked by the ExTWG TIA's and their recommendations were presented to the OEXP Technology Manager and the OEXP Technology Director. The results of this process are beyond the scope of this paper, however, the details of the analysis can be found in reference 4.

In the process of developing the technology needs for the three key technical study areas i.e., transportation node systems, transfer vehicle systems and planetary surface systems, it was found that there were specific technologies that were common to all of these key technical areas. These common or "cross-cutting" technologies were Automation, Robotics, Maintainability, Operability and FDIR (fault detection, isolation and recovery). Table 2 depicts a top level overview of the crosscutting technologies associated with the transportation node technology areas. As can be seen in the table the crosscutting technologies are common to all of the major technology areas identified for the node systems.

- o ADVANCED ENVIRONMENTAL CONTROL AND LIFE SUPPORT SYSTEM (ECLSS)
  - AIR, WATER, WASTE MANAGEMENT, FOOD PROCESSING
- o CREW SYSTEMS
  - ADVANCED EVA SUITS
  - HABITABILITY CONSIDERATIONS
  - HEALTH CARE AND MAINTENANCE CONSIDERATIONS
- o SURFACE TRANSPORTATION
  - ROVERS (UNMANNED, MANNED)
- o AUTOMATION AND ROBOTICS
  - CARGO HANDLING
  - ASSEMBLY
  - REMOTE SITE EXPLORATION
- o PROPULSION
  - CHEMICAL
  - NUCLEAR
  - SOLAR
- o STRUCTURES
  - AEROBRAKE/AEROSHELL
  - ASSEMBLY AND HANDLING
- o POWER/THERMAL
  - SOLAR
  - NUCLEAR
  - CHEMICAL
- o LONG-TERM PASSIVE STORAGE MISSION SYSTEMS/SUBSYSTEMS
  - RADIATION/TEMPERATURE EFFECTS
  - PROPELLANT STORAGE
  - MAINTENANCE/ACTIVATION

Figure 1. Top Level Exploration Missions Technology Areas

TABLE 2. CROSSCUTTING TECHNOLOGIES TRANSPORTATION NODE

Technology	Automation	Robotics	Maintainability	Operability	FDIR
In-Space Vehicle Processing/Servicing	X	X	X	X	X
Cryo Fluid Storage/Management	X	X	X	X	X
In-Space Assembly - Vehicle Level	X	X	X	X	X
Autonomous Rendezvous/Docking	X	X	X	X	X
In-Space Assembly - Element Level	X	X	X	X	X

The major technology needs associated with the transportation node systems definition, and the attendant infrastructure requirements, including the space station, needed to support the exploration missions are discussed in the following paragraphs. However, from the transportation node viewpoint, it was found that the technology requirements identified were relatively insensitive to the particular mission, or case study, under analysis. Generally, the only major differences between the requirements for the lunar and Mars case studies were the specific technology need dates. That is, the technology readiness level requirements were keyed to a particular case study's program and milestone schedule.

The top level technology areas unique to the LEO transportation node function are summarized in figure 2.

- **IN-SPACE ASSEMBLY**
  - **ASSEMBLY OF LARGE AEROSHELLS**
  - **ASSEMBLY OF LARGE SPACE TRANSFER VEHICLES**
  - **JOINING OF LARGE STRUCTURAL ELEMENTS (HANGARS, PROPELLENT STORAGE FACILITIES, ETC.)**
  - **AUTOMATION/TELEROBOTIC PRINCIPLES (PRECISION POSITIONING/ HANDLING)**
- **IN-SPACE VEHICLE PROCESSING/REFURBISHMENT**
  - **A&R/TELEROBOTIC TECHNIQUES & AIDS**
  - **AUTOMATED SYSTEMS TEST & CHECK-OUT**
  - **FAULT TOLERANT SYSTEMS**
  - **IN-SPACE SERVICING/DESERVICING & CHECK-OUT OF "WET" SYSTEMS (HYPERGOLS)**
- **CRYOGENIC FLUID MANAGEMENT & TRANSFER**

Figure 2. LEO Transportation Node Top Level Technology Areas

To accommodate the in-space assembly & construction needs associated with the LEO node, the capability to assemble, handle, and mate/demate very large, very massive and complex space vehicle and vehicle elements will be required. A high degree of confidence and reliability must be demonstrated and assembly/ construction operations must be conducted with minimum risks and minimum crew involvement, especially EVA. For the exploration space vehicles/elements (aeroshells, spacecraft, propulsion systems, interface structures, etc.), the on-orbit technology program must address handling, assembly and mating techniques using large capacity, highly articulated manipulators and robotic/telerobotic aids. The success of providing this capability depends upon major advancements in the discipline technology areas such as automation & robotics, telerobotics, development of advanced processes for joining/mating space vehicle elements & components(welding, bonding, snap connectors, etc.); and the associated controls-structures interactive systems necessary to maintain the close tolerances required while minimizing disturbances to the structure(s) and vehicle(s).

The in-space vehicle processing/refurbishment function, will require many of the same attributes needed by the in-space assembly/construction function, i.e., handling, mate/demate, manipulating large and massive mission elements with great precision. In addition to the integration and checkout functions, the capability to service/deservice, maintain, repair and refurbish all reusable flight hardware elements must be developed and demonstrated in the space environment. Technology issues associated with this on-orbit function include advancements in such areas as automation and robotics/telerobotics, automated systems test & checkout(fault-tolerant systems), on-orbit test, service/deservice checkout equipments/ hardware. Crew roles and interfaces must be an integral element in the design, development, test and engineering(DDT&E) process. In addition, when nuclear vehicles become part of the transportation inventory, research and technology programs will be needed to support remote/autonomous inspection, maintenance, servicing and checkout of the departing and returning spacecraft.

To accomplish on orbit what has traditionally been done using ground-based facilities will require a whole new set of in-space operational philosophies, procedures and orbital support equipments especially where manned systems are involved.

From a key technology standpoint, the capability to deliver and maintain large quantities of cryogenic propellants in space for long periods of time must be developed and demonstrated in the actual space environment before any of the proposed missions can seriously be considered. The major space cryogenic technology issues are the fluid storage, transfer, handling and management. Solutions to these issues are keyed to advancements in the supporting technology areas of automation and robotics and autonomous rendezvous and docking. From a "safe systems operations" standpoint, the propellant storage will probably be on a coorbiting facility and hence, the transfer, delivery and handling of these propellants will have to be conducted remotely.

Additionally, liquid slosh dynamics and control during docking and/or deployment of spacecraft with the propellent facility must be understood. In addition, methods for providing safe cryogenic tankage/storage prior to EVA proximity operations or emergency propellent dumps must be developed and fuel, oxidizer mixing must be avoided. System design criteria must also be established in the development of the cryo storage systems in order to minimize systems weight, understand systems integration(thermal/structural), provide high reliability, determine repeatable in-space fabrication techniques and assess potential material contaminants.

Autonomous Rendezvous and Docking(AR&D) is another key technology driver in implementing the exploration missions, particularly the planetary expeditions. For LEO node operations, this capability will be needed as traffic in the Command & Control Zone (CCZ) of the space station increases in order to accommodate the OEXP missions. For the lunar and Mars orbital operations suggested by the case study scenarios, the ability to perform this function totally autonomously (without any human intervention) is truly enabling primarily because the round-trip-light-time(RTLT) delays in deep space communications preclude Earth-based support.

THE ROLE OF SPACE STATION FREEDOM

A major Space Station Freedom (SSF) program goal is to design a facility capable of growth that will support future mission requirements and long-term national goals such as the human exploration initiatives.

The following discussion will concentrate primarily on the Lunar Evolution Case Study and how the Station may evolve to support this one particular case study.

Figure 3 shows the baseline Space Station Freedom "Assembly Complete"(AC) configuration scheduled to be operational in mid-1998. The station at this stage will have somewhat limited resources but will have the capability for growth as user needs and national goals mature. Briefly, it will support a crew complement of eight, provide up to 75kw of power and heat rejection with appropriate scars for addition of Solar Dynamic arrays and provisions for habitat/laboratory growth and utilities.

The transportation node resource requirements developed for the Lunar Evolution Case Study are shown in figure 4. Figure 5 depicts the time-phased station "growth deltas", or additions to the baseline station, necessary to meet the requirements shown in the previous figure.

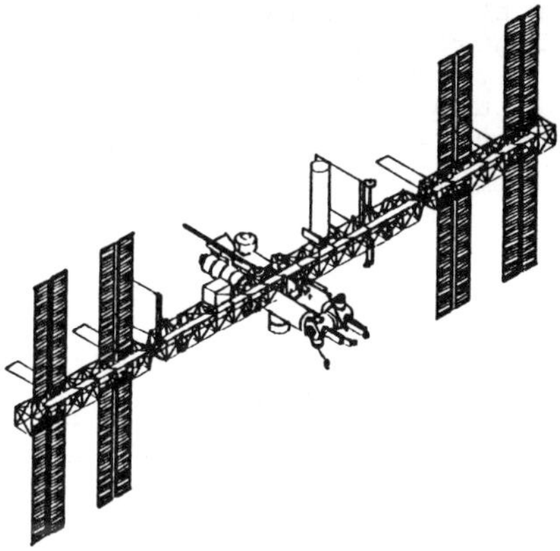

Figure 3. Space Station Freedom Assembly Complete Configuration.

Station Configuration	Recommendation
Power	
Average	175kw
Peak	(NA)
Crew	18
Pressurized Modules	
US Habitation	2
US Laboratory	1
ESA Laboratory	1
JEM Laboratory	1
Pocket Laboratory	1
Resource Node	1
Dual Keel	Scar for distributed systems
Servicing Facility	Scar for all possible locations
Power Modules	SD array growth on extensions of boom ends
Attached Payload	(TBD)
Accom. Equip.	
Tether Payloads	2

Figure 4. Transportation Node Resource Requirements

Δ1 (2 - 25kw) Solar Dynamic Modules; (2) 25 meter Transverse Boom Extensions; Space Based OMV & Space Based OMV accommodations

Δ2 Upper/Lower Keels & Booms; Utility Trays

Δ3 (1) Habitat Module; (2) Resource Nodes

Δ4 (2 - 25 kw) Solar Dynamic Modules; Servicing Facility Phase I

Δ5 Servicing Facility Phase 2

Δ6 Phase 1 STV Processing Facility; Phase 3 Servicing Facility (Completed CSF); (1) Mobile Servicing Center (MSC)

Δ7 Additional Lower Keel and Boom

Δ8 Part 2 LTV Processing Facility (LTV Processing Facility complete)

Figure 5. Time-Phased Station "Growth Deltas"

Figure 6 shows the Lunar Evolution Space Station reference configuration. The vehicle assembly, servicing and processing facilities are accommodated in an enclosed hangar attached to the lower keels and booms of the station. The station/node is capable of processing/storing two lunar transfer vehicles simultaneously.

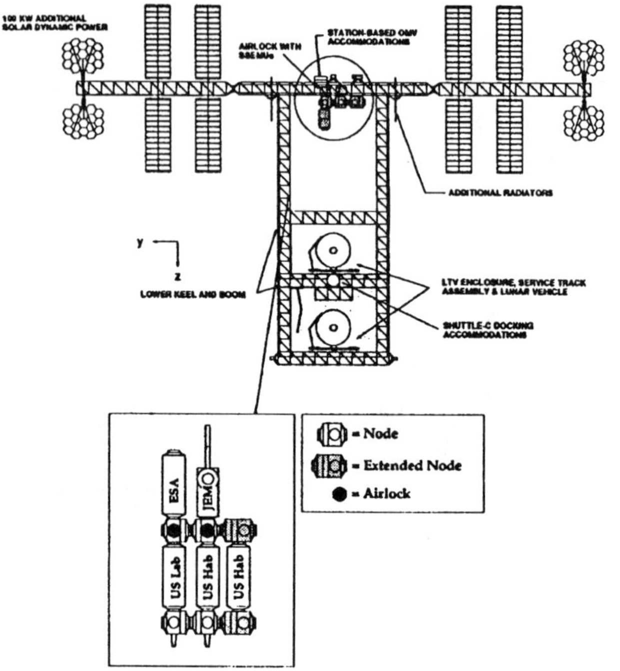

Figure 6. Lunar Evolution Space Station Reference Configuration.

Figure 7 shows the Space Station Freedom baseline and the evolution schedules and milestones for the exploration initiatives.

As the figure indicates, the evolution planning is currently in progress. Emphasis at this time is on design requirements leading to an evolution development decision for Phase A/B studies in late CY '90 or early CY '91. This decision point is key since the evolution Phase C/D activities must begin by mid-CY '94 in order to have the growth station operational for a manned lunar mission in the year 2004.

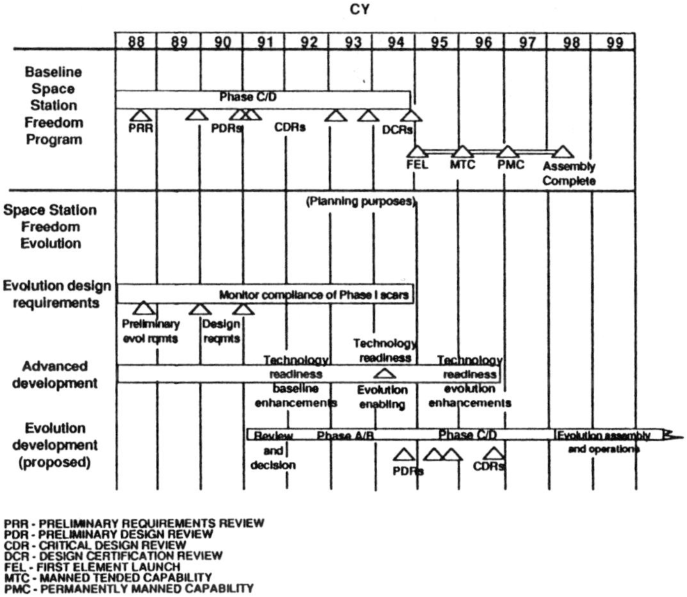

Figure 7. Space Station Freedom Evolution Schedule, Milestones

In addition to the technology development and operational support required by the station,the importance of demonstrating and validating the "serviceability" feature of each of the flight elements cannot be overemphasized. The ability to service and process the vehicle on-orbit in a timely and efficient manner with the high degree of confidence required for safe operations and mission success, is going to be a real challenge to both designers and operators alike.

End-to-end testing and all-up mission simulations with the totally integrated lunar vehicle configuration will also be required.

SUMMARY

Top level technology needs have been identified from a review of material presented in studies related to the Office of Exploration's (OEXP) transportation node activities and selected supporting study work from the Office of Space Station (OSS). The major technology needs associated with the transportation node system definition and the attendant infrastructure requirements needed to support the exploration mission have been developed.

Space Station Freedom activities required to support the Lunar Evolution Case Study have been addressed in some detail. However, it should be noted that the lunar initiative is essential to the pathway to Mars announced by the President and the Station evolution planning discussed in the paper reflects this direction.

The major role for the Station in the human exploration initiatives program will be the on-orbit technology development, testing and verification of the flight hardware and the in-space assembly and vehicle servicing/processing function development. The operational phases of the programs will require significant Station support for the assembly, processing, maintenance and refurbishment of the lunar and/or Mars mission hardware.

A commitment to provide the extensive LEO node support capability just discussed will require considerable study and major management decisions. In addition to the technological and engineering challenges mentioned above, two important factors will undoubtedly influence the decision process. These are the specific mission designs and the ability to demonstrate and subsequently conduct the many in-space operations required to implement and effectively sustain the proposed exploration missions.

References

Reference 1: Weidman, D. J., Llewellyn, C. P., et al: Space Station Accommodations for Lunar Base Elements - A Study: NASA TM 100501, October 1987

Reference 2: Cirillo, W. M., Llewellyn, C. P., et al: Manned Mars Mission Accommodation-Sprint Mission: NASA TM 100598, April, 1988

Reference 3: Exploration Studies Tecnical Report, FY 1988 Status, Volume 1: Technical Summary, Office of Exploration, NASA TM 4075, December 1988

Reference 4: Exploration Studies Technical Report, FY 1989 Status, Volumes I through V, Office of Exploration, NASA TM (TBD)

Low Earth Orbit Infrastructure to Accommodate Manned Lunar Missions

William M. Cirillo* , Eric L. Dahlstrom+ ,
Patrick A. Troutman** , J. Kirk Ayers+ ,
Laura M. Waters++ , Karen D. Brender+++

Abstract

In order to establish bases on the lunar and Mars surfaces, a significant amount of orbital infrastructure including assembly platforms, cryogenic fluids depots, and Space Station Freedom, will be required in Low Earth Orbit (LEO). These facilities will be required to perform a myriad of functions ranging from orbital demonstration of advanced technology systems, establishment of life science capabilities, to servicing and refurbishment of reusable Lunar Transfer Vehicles (LTVs). This paper addresses the requirements levied on these facilities and provides an overview of potential LEO infrastructure elements that satisfy various advanced manned missions. Finally, of key importance to the success of the manned lunar mission are (1) the evolutionary growth of Space Station Freedom to serve as a Transportation Node, and (2) the development of a Shuttle Derived Launch Vehicle (SDLV) to deliver mission elements to LEO.

---

* NASA Langley Research Center, Aerospace Technologist, Space Station Freedom Office's Evolutionary Definition Office, also Node Integration Agent for NASA Office of Exploration.

+ Lockheed Engineering and Sciences Company, Senior Engineer.

** NASA Langley Research Center, Aerospace Technologist, Space Station Freedom Office's Evolutionary Definition Office.

++ Analytical Mechanics Associates, Inc., Project Engineer.

+++ NASA Langley Research Center, Deputy Manager, Space Station Freedom Office's Evolutionary Definition Office.

Introduction

In 1987, the National Aeronautics and Space Administration (NASA) created the Office of Exploration (OEXP) to define and develop the requirements and elements necessary to establish a manned presence in the solar system. To date, OEXP has focused its analysis on establishing permanent bases on the near side of the moon and on the surface of Mars.

This paper will provide an overview of the Lunar Evolution case study analyzed by OEXP in 1989 while focusing on the orbital infrastructure and operations required in LEO to meet the given mission requirements.

Lunar Evolution Case Study

In 1989, the OEXP analyzed three different exploration missions referred to as Case Studies. These were the Lunar Evolution, Mars Expedition, and Mars Evolution case studies. Only the Lunar Evolution case study is discussed in this paper. A more thorough discussion of all three of the FY89 OEXP case studies can be found in the OEXP Study Requirements Document (SRD). (Craig, 3/1989)

The Lunar Evolution case study examined the development of a permanent lunar base using a long term evolutionary strategy. In this scenario, the lunar base progresses from an initial human-tended outpost (*emplacement phase*) to a small permanent base in the *consolidation phase*, and finally to an expanded self-sufficient facility in the *utilization phase*. The lunar base would support scientific research and exploration beginning with the initial human-tended outpost and would add research facilities as the lunar base matures.

The complete architecture for the lunar mission is illustrated in figure 1. The Earth-to-Orbit (ETO) delivery system includes both the Space Shuttle and an unmanned Shuttle Derived Launch Vehicle (SDLV) system. The SDLV is necessary to meet the large ETO payload requirements for this case study and to provide a launch system capable of delivering the Lunar Transfer Vehicles (LTVs) to orbit mostly assembled and integrated. The LTVs are used to deliver both crew and payload from LEO to Low Lunar Orbit (LLO). Once the payload and crew reach LLO they are transported to and from the lunar surface using the Lunar based Lunar Excursion Vehicles (LEVs).

The case study emphasized the use of oxygen produced from lunar raw materials, first to support closure of the lunar base life support system, and later to reduce the need to transport all of the LEV propellant from Earth. Key features of the case study included: (a) first lunar mission in 2004; (b) significant science research capability; (c) initial crew size of 4 growing to 30 at the lunar base; (d) reusable space transportation vehicles; (e) trans-lunar vehicles assembled and serviced at Space Station Freedom (SSF); (f) aerobraking in LEO for return vehicles; (g) chemical propulsion; and (h) lunar propellant production.

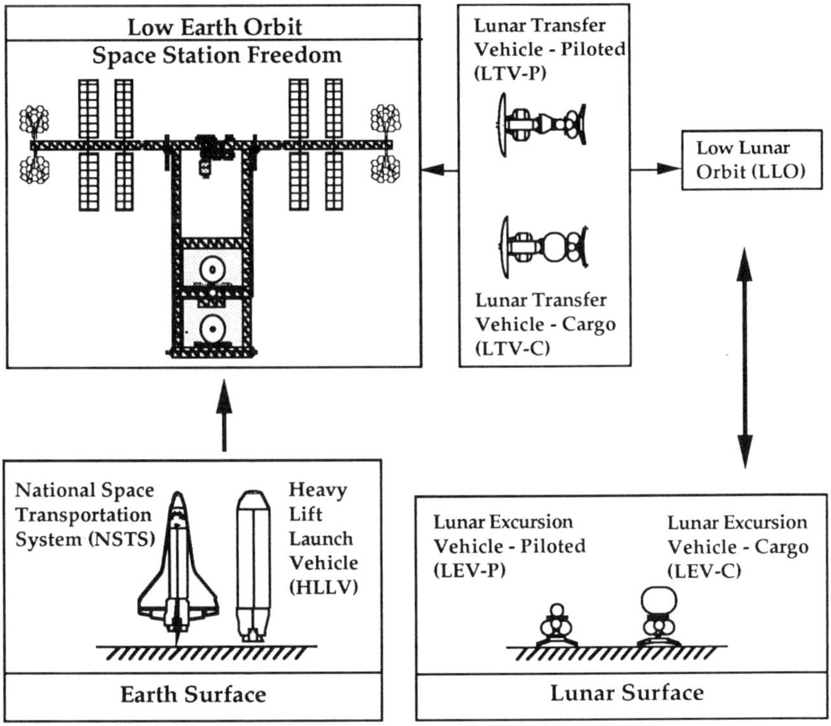

Figure 1. Lunar Evolution Case Study Mission Architecture

Case Study Space Station Requirements

The SRD specifies a major utilization of the Space Station's functional capabilities and resources in the Lunar Evolution case study. These capabilities include

Lunar Transfer Vehicle and payload storage and processing, lunar base and support crew housing, and propellant storage and transfer. A detailed list of requirements is given in reference 1. (Craig, 3/1989)

A detailed assessment of various growth concepts to support both initial Lunar Evolution mission objectives and advanced mission requirements is provided in the following paragraphs.

Time-Phased Space Station Freedom Growth

Accommodation of the Lunar mission requirements in LEO can be accomplished by a step-by-step growth of the Space Station configuration as shown in table 1. Space Station evolution was divided into two major phases that correspond with the evolutionary nature of the Lunar case study. The first phase, which supports vehicle flight testing and is capable of accommodating the first several expendable LTV flights to the moon is represented by deltas 1 through 6 in table 1. These deltas reflect major incremental increases in various resources including power, habitat volume, truss attach points, etc. The second growth phase requires additional increases in functional capabilities and resources to store and process both the LTVs and their payloads.

Table 1. - Time-phased SSF growth deltas for the lunar evolution configuration.

Phase 1

Δ1	(2) 25 kW Solar Dynamic (SD) Modules; (2) 25 meter Transverse Boom Extensions; Space Based Orbital Maneuvering Vehicle (OMV) and accommodations
Δ2	Upper/Lower Keels and Booms; Utility Trays
Δ3	(1) Habitat Module; (2) Resource Nodes
Δ4	(2) 25 kW SD Modules; Servicing Facility Part 1
Δ5	Servicing Facility Part 2
Δ6	Part 1 LTV Processing Facility; Part 3 Servicing Facility (Completed CSF); (1) Mobile Servicing Center (MSC); SDLV Docking Adapter

Phase 2

Δ7	Additional Lower Keel and Boom
Δ8	Part 2 LTV Processing Facility (LTV Processing Facility complete)

The programmatic schedule for Freedom evolution and other potential infrastructure is provided in figure 2. It is important to note that the Space Station Evolution

Phase B studies (if required) must start in 1992 in order for Freedom to support a late 1990's LTV demonstration flight.

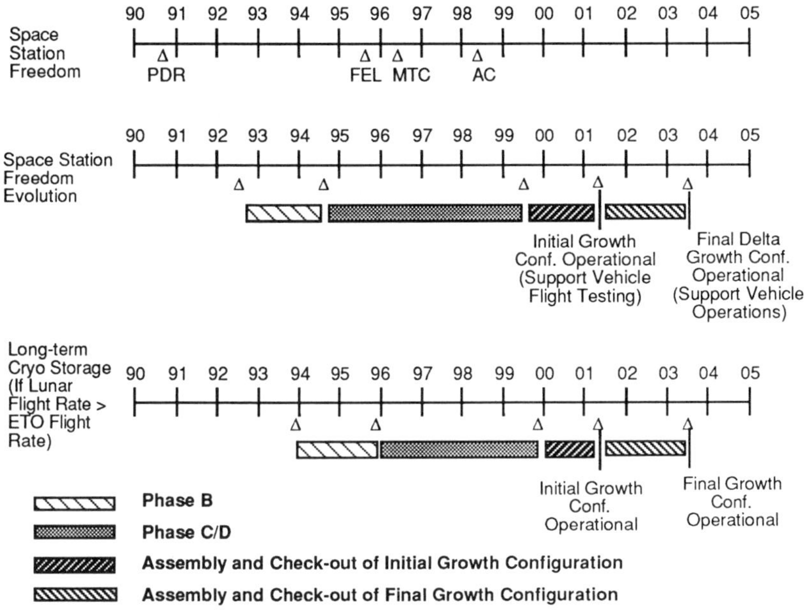

Figure 2. Lunar Evolution Case Study Programmatic Schedule

Growth Configuration Analysis

Several candidate evolution configuration options were analyzed using the ten discriminators listed in table 2. The discriminators with the greatest weighting factor were the controllability, operability, and the static micro-gravity environment for each option. It is desirable to have an evolution Space Station that would serve primarily as a transportation node, and yet provide a quiescent research and development environment when no major vehicle processing activities are occurring. To this end, configurations were sought which would provide all of the functional capabilities of a transportation node (LTV processing, propellant storage and transfer, payload accommodation, etc.), minimize operational impacts on Freedom (controllability, reboost propellant, crew time, etc.) and still provide a suitable micro-gravity environment.

Table 2. - Space Station growth discriminators

Discriminator	Measures of Effectiveness
Space Station Control	TEA, CMG momentum storage requirements Roll magnitude to dump secular momentum
Structural Capability	Truss member structural integrity under reboost loads Controls-Structures Interaction acceptability
Operational Capability	Ability to perform all vehicle processing requirements
Microgravity Environment	Highest microgravity level in the US Lab module TEA (affects microgravity gradient along the Space Station x-axis)
Viewing	Stellar viewing from the JEM exposed facilities: - Percent viewing blockage - Greatest field of view - Average swath angle
Main Radiator Interference	Clearance between radiator wing and growth elements
Module Closure/Dual Egress	Number and length of time with modules having less than half closure (racetrack) or lack of dual egress (two ways out)
Structure/Ease of Assembly	Number of modules along the transverse boom Number of modules within MSC reach
Ability to Accommodate Growth	Number of free and usable berthing ports on the final configuration
Growth Mass	Total weight of growth elements (approximate measure of cost)

Based on the ten discriminators listed in table 2 the configuration illustrated in figure 3 was deemed most suitable for accommodating the Lunar Evolution case study requirements. The transportation node configuration has a single hanger for performing all LTV and payload processing activities.

Two options were analyzed for this configuration. Option A had the processing hangers sized to accommodate an LTV with a deployed aerobrake, while option B had the LTV's aerobrake either removed or retracted for storage.

In order to determine which configuration(s) would be best suited for each phase of evolution the following analyses were conducted on each configuration using a computer-aided engineering analysis tool:

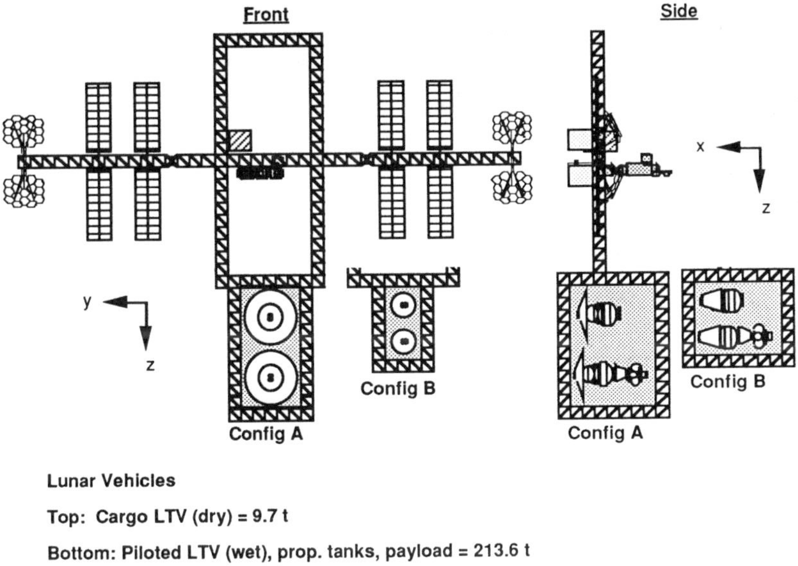

Figure 3. Space Station configuration for lunar vehicle accommodation.

a. Mass properties determination including center of mass and moments of inertia
b. Open-loop control system sizing including secular and cyclic momentum determination and torque equilibrium angle (TEA) determination
c. Closed-loop control system sizing including secular and cyclic momentum determination and TEA determination
d. Micro-gravity envelope calculation
e. Orbital decay and propellant reboost assessment
f. Space Station growth element and LTV/payload clearance analysis
g. Module cluster field of view determination
h. Electrical power and thermal rejection determination
i. Structural dynamic characterization including system modes and forced response (only on selected configurations)

The growth hardware elements for the transportation node configuration are listed in table 3. All of the analyses conducted have used scenarios which either stored part or all of the mission propellant directly on the LTV's themselves, or have assumed storage on a co-orbiting propellant facility.

Table 3. - Additional SSF Lunar Evolution elements

Components	Configuration 2A Number	Configuration 2A Mass (t)	Configuration 3A Number	Configuration 3A Mass (t)
Habitat Module	1	23.5	1	23.5
Resource Node	2	14.9	2	14.9
Truss Bays	220	16.5	164	12.6
Utility Trays	132	10.7	132	10.7
Meteoroid Shields	12	26.6	11	21.7
Solar Dynamic Units	4	27.6	4	27.6
Servicing Facility	1	11.5	1	11.5
Thermal Radiators	2	6.3	2	6.3
Docking Mast	1	1.4	1	1.4
(Payloads Removed)	all	-32.0	lower	-6.2
Total mass added		107.0		124.0

The analyses conducted to date indicate that a gravity-gradient stabilized configuration such as that shown in figure 3 is desirable from a controllability perspective.

Meteoroid and Orbital Debris Shielding for Hangers

The on-board LTV processing facility consists of an unpressurized hanger attached to the Space Station keel. The protection of the LTVs and payloads from meteoroids and orbital debris while at the Space Station involves balancing the benefits of additional protection against the costs of adding significant mass to the hanger walls. A hanger wall thickness was selected to provide some additional protection to EVA astronauts while operating in the unpressurized hanger.

Lunar Space Transfer Vehicle Accommodation

In addition to the micrometeoroid protection provided, a hanger also provides a suitable operational environment for EVA operations and teleoperated vehicle processing. The hanger is equipped with sufficient lighting, robotic access, orbital support equipment, and power, to perform all processing activities. A station arm is mounted on a mobile track assembly like that shown in figure 4. This arm will have the capability of reaching all of the LTV's orbital replaceable units (ORU) for ease of teleoperated vehicle servicing.

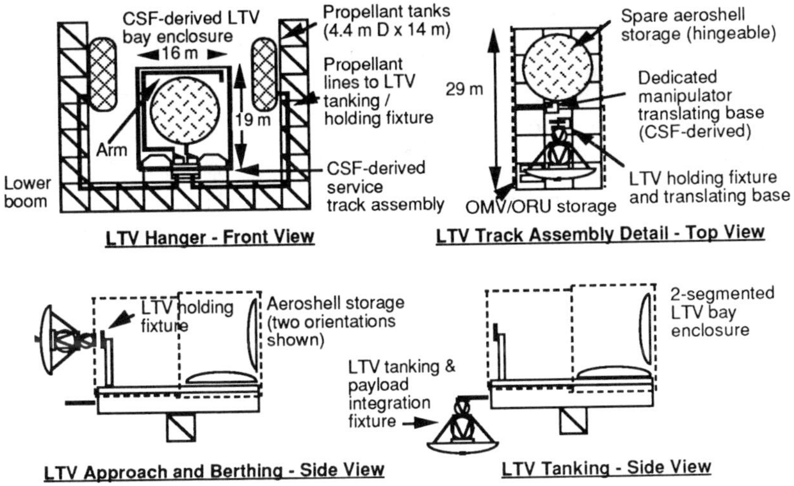

Figure 4. - Lunar Transfer Vehicle accommodation facility.

Lunar Vehicle Servicing and Refurbishment

The current analysis performed for the Node Integration Agent (Vargo, 1989), indicates that a vehicle processing crew of four will require approximately 123.5 shifts (of 8 hour work periods) to completely process a reusable LTV on-orbit assuming the current level of vehicle system designs. Figure 5 shows the required LTV processing activities for three major operational periods and provides a top-level breakdown of the required number of shifts to perform the various tasks.

Approximately twenty-five percent (180 hours/LTV flight) of the 123.5 shifts will consist of a crew of two performing extra-vehicular activities (EVA). The majority of the processing will be done using either (1) intra-vehicular activity (IVA) or, (2) Freedom based telerobotics.

Summary and Issues

The support of an evolving lunar base requires large vehicle and payload masses to be delivered, assembled and processed through Low Earth Orbit. Due to the nature of the transportation vehicles involved, a variety of orbital infrastructure elements is required in LEO to accomplish the exploration class missions being studied by the NASA Office of Exploration.

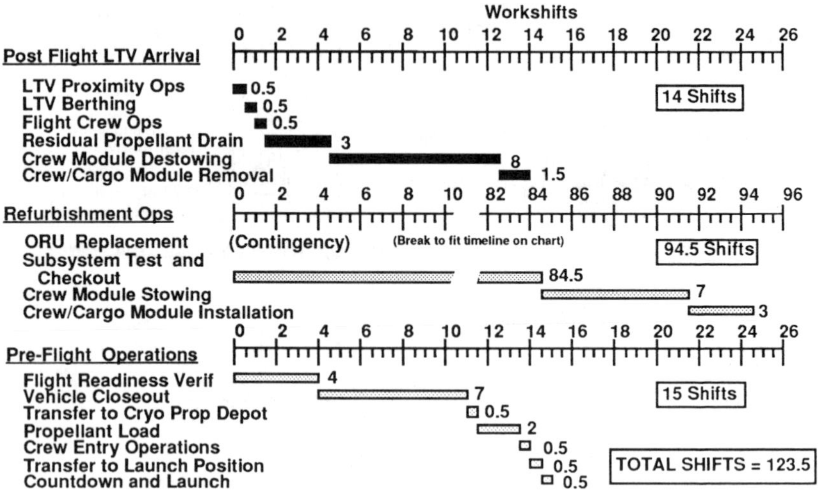

Figure 5. - Lunar vehicle processing operations.

The major open issues that will affect the accommodation of an evolving lunar base are (1) the optimum technique for on-orbit propellant storage and transfer, (2) a further definition of an acceptable micrometeoroid/orbital debris protection level, and (3) a refinement of the requirements for processing lunar transfer vehicles at Space Station Freedom (e.g. the use of advanced automation could significantly impact processing requirements). In addition, the choice of an ETO launch system capability will significantly impact the ability to emplace the infrastructure in a timely manner to support the Lunar mission.

References

Craig, Mark K. and Lovelace, Uriel M., Study Requirements Document, FY 1989 Studies, NASA Office of Exploration,Document No. Z-2.1-002 March 3, 1989.

Vargo, Rick, et. al., On-Orbit Assembly/Servicing Task Definition Study, GFY 1989 Summary Report, McDonnell Douglas Space Systems Company, Kennedy Space Center, #NAS10-11400, November 1989.

SPACE TRANSPORTATION NODE: THE ATRIUM FACILITY

Kriss J. Kennedy*

ABSTRACT

During the next thirty years the infrastructure of space transportation is expected to expand to a system of advanced vehicles. A transportation node is an important link, as a vehicle assembly platform, between Earth and our planetary neighbors. This paper presents a conceptual design of a Space Transportation Node used primarily as an assembly platform supporting the lunar transportation system. Transportation nodes are traditionally an extension of Space Station Freedom, which is a logical progression. However, this concept was initiated to study node architecture, not as a growth of Space Station Freedom but rather as an uninhibited concept synthesizing form and function.

This transportation node concept, The Atrium Facility, is designed to support lunar base activities before, during and after the lunar base build-up phase. It has a central assembly area surrounded by hangars and workstation platforms. This facility can support six permanent crewmembers, a four to six transient lunar crew and the shuttle crew. It has a dry mass of 318,710 kg, which does not include the cryogenic propellant stowage or the trans-lunar vehicle. This paper will discuss a few of the key element designs and rationale for those concepts.

INTRODUCTION

A space transportation node is a vital part of the infrastructure of space development and interplanetary travel. It plays an important role in the development of a lunar outpost and subsequent lunar base build up. The scientific work projected for Space Station Freedom is vital as the first step in the use of the space environment, but a transportation node will be the step towards inhabiting other celestial bodies.

The Atrium Facility is a product of NASA's continuing effort to investigate and analyze Transportation Nodes (TN). The ground rules and assumptions were predetermined by the Lunar Base Systems Study (LBSS) of the Advanced Programs Office (Eagle Engineering, 1988).

* Space Designer (M.Arch.), Advanced Programs Office, NASA/Johnson Space Center. Houston, TX

A few of the ground rules that shaped this concept are: 1) it is for flight dynamic analysis (Stecklein, 1988); 2) it is an investigation of node architecture; 3) it is not limited to Space Station Freedom technology; 4) it is for comparative purposes; and 5) it is not necessarily a baseline for transportation nodes.

Some of the given assumptions about this facility are: 1) it will have on-board propellant stowage; 2) it will provide habitation and command facilities for six permanent crew-members; 3) it is designed specifically for the lunar transportation system defined by the LBSS; 4) it will house and service orbital maneuvering vehicles (OMV); and 5) it will not accommodate scientific activities.

The Atrium Facility is designed to support the LBSS scenario of lunar development (Alred, 1988), which has a mission manifest of 3-4 flights per year. This scenario translates into several key design drivers the Atrium Facility had to meet.

TRANSPORTATION NODE ARCHITECTURE

The Atrium Facility is a design of form and function. This facility had several key design drivers that shaped it. The following design drivers are not inclusive to these few and could have an entire paper written on each. Thus they are briefly stated to give a greater comprehension of the Atrium's design.

Key design drivers in TN designs are: 1) an on-board propellant stowage; 2) orbital debris and micrometeoroid protection; 3) trans-lunar vehicle assembly and number of flights per year; 4) how payloads are moved and stowed; and 5) habitation and command facilities. Each of these drivers are discussed briefly as individual drivers and through the resultant solution, the Atrium Facility.

The TN has three primary requirements to satisfy. The first and major requirement is to support activities before, during and after the build-up phase of the lunar base. Secondly, the node shall act as a construction and assembly platform for interplanetary vehicles (manned and unmanned). Its third requirement is to act as a staging operations center for lunar crewmembers going to and from the lunar base.

Other requirements include operations such as maintenance and refueling of OMV's and Orbital Transfer Vehicles (OTV), tracking of Earth to LEO and LEO to lunar surface vehicles, communications with these missions, and support of orbital payload operations such as Lunar Oxygen (LOX) transfer.

ON-BOARD PROPELLANT STOWAGE

One of the given assumptions stated propellant is stowed on-board. There are many problems to be solved about propellant stowage, types of propellants used and on-orbit propellant transfers. This paper will not attempt to discuss all the ramifications of the various systems, but instead give the parameters this facility was designed to.

Propellant stowage requirements on the TN shall be sufficient to fuel two lunar departure stacks, one additional OTV and a 10% boiloff loss (Eagle, 1988). A total of 160 MT of LO_2 and 24 MT of LH_2 is required. This is the an equivalent delivery capacity of two heavy lift launch vehicle (HLLV) propellant tankers.

ORBITAL DEBRIS/MICROMETEOROID PROTECTION

Orbital debris and micrometeoroids can be damaging to an astronaut, space station, or spacecraft and some form of protection should be considered. Table 1 indicates an

Table 1. Orbital Debris Environment (Kessler, 1988)

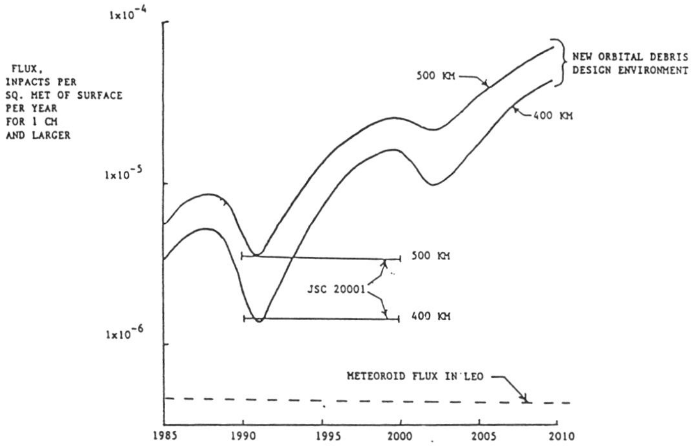

increased flux in the orbital debris environment as space activities increase toward the end of this century. This is of concern to any and all facilities placed in Low-Earth Orbit (LEO) and appropriate solutions should be implemented for protection of key elements.

TRANS-LUNAR VEHICLE ARCHITECTURE

Trans-lunar vehicle (TLV) designs vary in size and mass. These designs, in part, vary due to the method of capture for the return to Earth orbit. Figure one uses one aeroshell for the return of its OTV to Earth orbit. This design places its payload stack in front of the OTV. A reusable lander is maintained at the base and will dock with an OTV in lunar orbit for payload transfer to the surface. When fully assembled for a crew rotation, this configuration is 16m x 20m x 30m and has a mass of 115,000 kg (Petro, 1988).

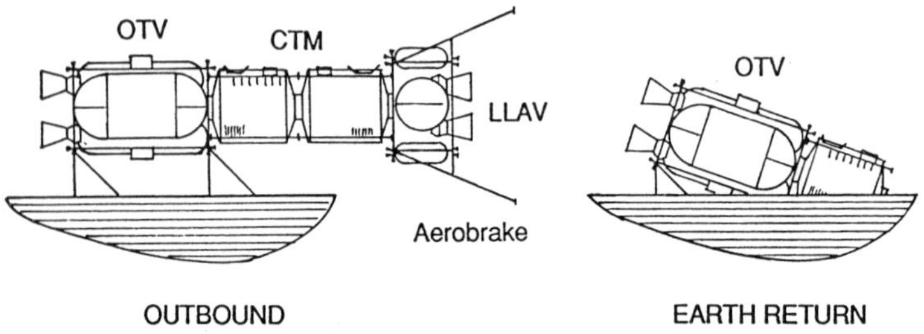

Figure 1. Trans-Lunar Vehicle

With this in mind, the common elements of a TLV are:
1) OTV (4 m x 8 m x 10 m)
2) Aerobrake (18.3 m dia. x 4 m)
3) Lunar Lander/Ascent Vehicle (LLAV) (14 m^2 x 7.2 m)
4) Crew Transfer Module (CTM): 4.45 m dia. x 6.2 m.
5) Lunar Surface Cargo (LSC): 4.45 m dia. x 9 m.

ATRIUM FACILITY

The Atrium Facility is a transportation node designed specifically for the assembly and maintenance of TLVs in support of lunar base activities. Numerous concepts were reviewed for assessing TN commonality, (Kaszubowski, 1988; Keaton, 1985, Ketchum, 1988). Figure two is an isometric view showing the overall relationship between the habitat facilities, assembly area, hangars and the fuel depot. The dominant feature of this concept is the use of hangars for stowage of TLV components.

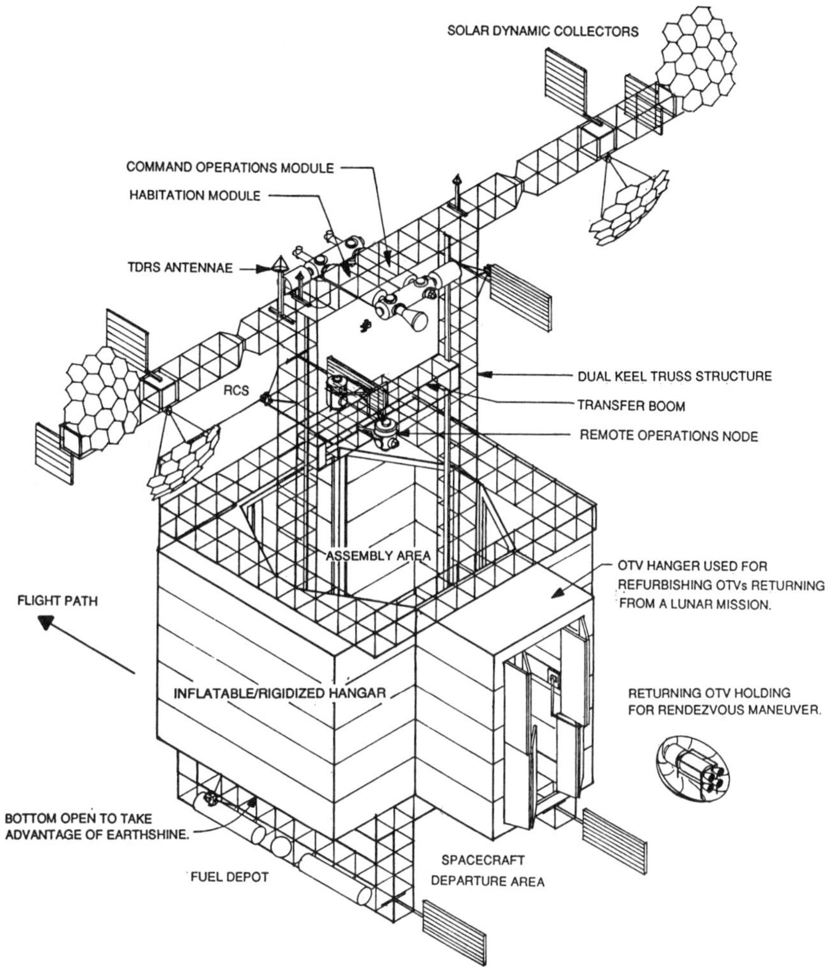

Figure 2. The Atrium Facility - Isometric

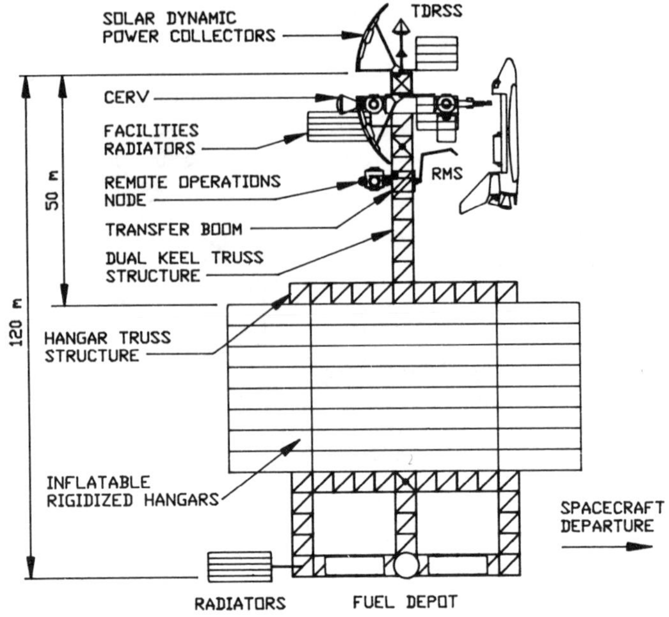

Figure 3. The Atrium Facility - Side View

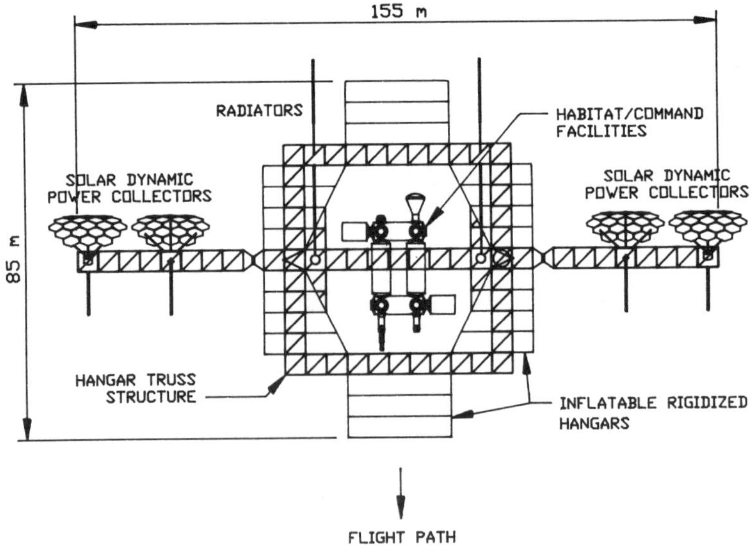

Figure 4. The Atrium Facility - Top View

Figures three and four are top and side views which depicts the relationship between the hangars and the others elements. This concept utilizes a series of hangars to encompass the assembly area. As the TLV components are needed for vehicle assembly, they are moved from the hangar into the assembly area. When the TLV is assembled, it is checked and then lowered for fueling in the fuel depot area. After fueling, the lunar crew boards the spacecraft for it's trans-lunar flight.

Inflatable/Rigidized Space Hangars

Figure five is a plan view of the assembly area and hangars showing the relationship of a transfer boom lowered into the assembly area. This illustration demonstrates the capable reach to all areas by the remote manipulator systems (RMS). The hangar skin will be on five sides, leaving the bottom open to utilize the natural light provided from Earth shine.

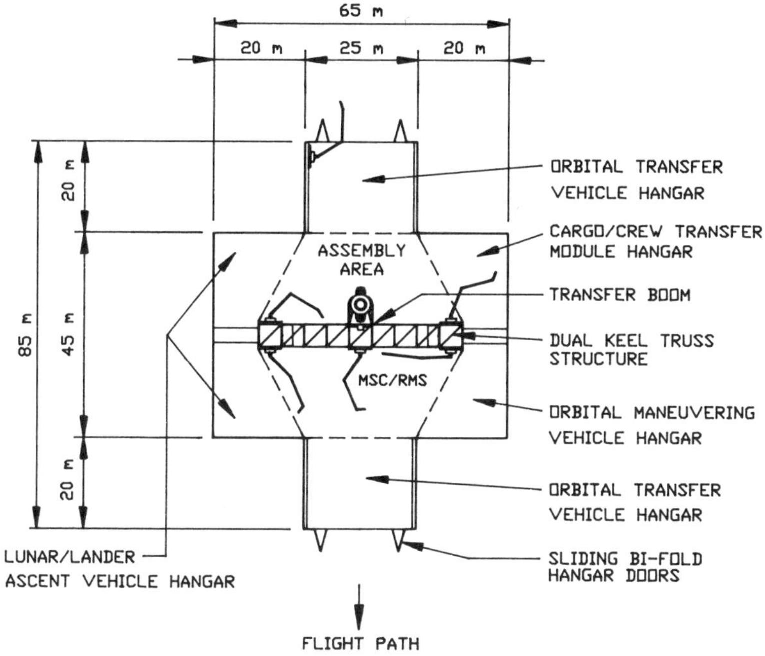

Figure 5. Assembly and Hangars - Plan View

The material used for an inflatable rigidized space hangar is composed of a sandwich construction. Goodyear Aerospace Corporation (GAC) developed an inflatable/rigidized structure of 2-1/2 inch foam with fiberglass sheathing or skins (Stimler, 1984). Inflation is used for control of deployment and a gelatin system is used for rigidization. As a micrometeoroid barrier this system was concluded to be equivalent to 0.53 cm (3/16") thick aluminum sheet with respect to penetration resistance. Table two is a preliminary graph showing the mass of various materials with similar penetration resistance to orbital debris. Where as this graph indicates an inflatable/rigidized system is the lightest, further research is required and this selection is preliminary.

Table 2. Hangar Material Mass Comparison

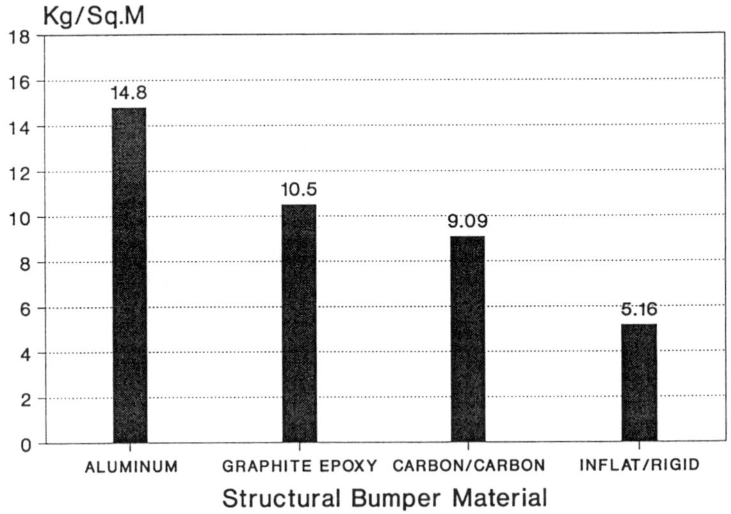

Transfer Boom

A movable transfer boom was designed to facilitate the need to move a payload from the orbiter to its proper hangar, figure six. Transfer of the payload will remain along the radial axis from Earth to the TN, thus minimizing disturbing variances with the center of gravity. Gravity gradient stability can be maximized by keeping the center of gravity at a near constant location in the lower third of the Atrium Facility.

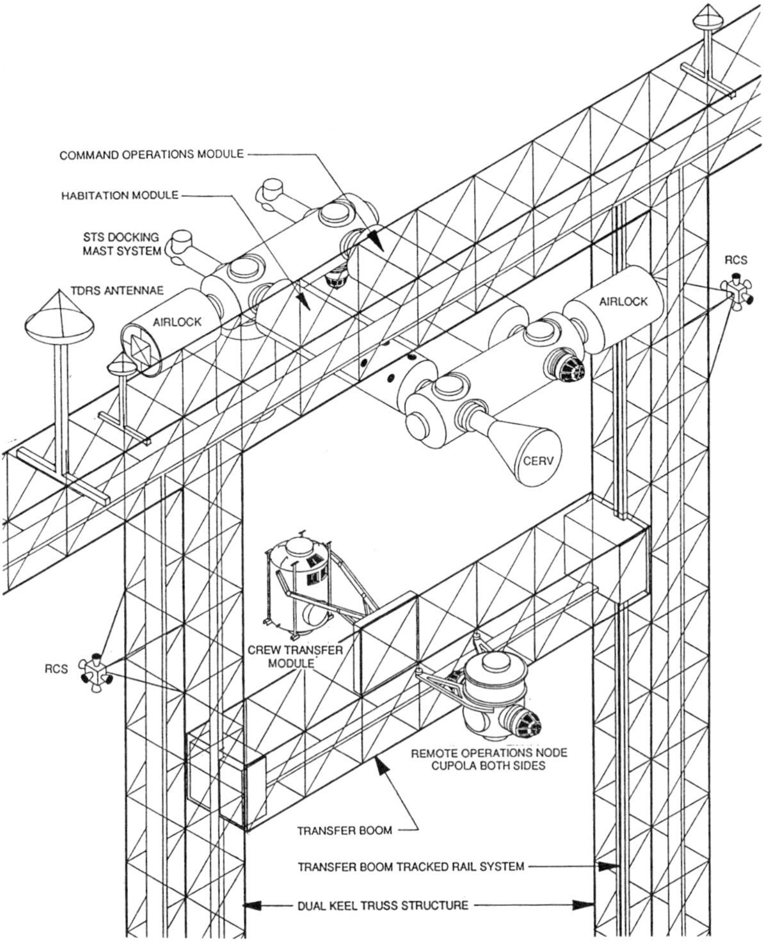

Figure 6. Transfer Boom - Isometric

The transfer boom is structurally composed of a five meter truss structure, similar to the main truss structure. It has a remote operations node (RON) system clamping attachment on the aft side and a RMS on the forward side. When the RON is attached to a module the node clamping device is detached and each arm is swung back to the boom truss. This allows usage of the transfer boom when the RON is not in use. The transfer boom's RMS is used to assist in payload operations, such as loading and unloading of the orbiter. Local RMSs are located in the assembly hangar for handling of lunar payload and to assist assembly operations.

Habitation and Command Facilities

Figure six shows the architecture of the habitat and command facilities. This facility has a habitation module, a command operations module, two interconnect modules, two airlocks, two cupolas, a crew emergency return vehicle and a logistics module.

Habitation and command operations facilities for the Atrium will support a permanent crew of six plus a temporary crew of nine. The temporary crew consists of three shuttle crewmembers and six transient lunar crewmembers. Thus a total of 15 crewmembers will need accommodations in the habitation module during a crew rotation cycle.

Space and volume requirements from the Space Station Freedom configuration indicate no need for two habitation modules. Ample space can be provided for the permanent and temporary crewmembers in the habitation (HAB) module when the Health Maintenance Facility (HMF) is relocated to an Interconnect Module. The HAB module houses the crew quarters (permanent and temporary), waste management systems, ECLSS systems, personal hygiene, shower, housekeeping, and stowage.

The permanent crew quarters (PCQ) will be the same as Space Station Freedom. Therefore a PCQ will be outfitted the same as, and will be interchangeable with, space station hardware. The temporary crew quarters (TCQ) will be one rack in size and offer its inhabitants a temporary place to sleep and some limited privacy.

The command operations facilities has several key functions that support the mission objectives. These functions include 1) a command post; 2) work stations for telerobotic operations; 3) communications; 4) traffic control; 5) logistics resupply; 6) data management and relay; and 6) orbital maneuvering operations. By grouping these activities in the same module efficient use of the space will be achieved and the daily operations can continue without interruption of the habitat environment.

MASS SUMMARY

Table three reveals that the Atrium Facility has a total mass of 615,710 kg. This is the total gross mass, which includes the stored cryogenic propellant and the trans-lunar vehicle. The hangar skin has a mass of 63,500 kg, the largest single element; but the habitation and command facilities have a total mass of 101,000 kg.

Table 3. Mass Summary Statement

Element	Kg.
Habitation Module	19,000
Command Operations Module	20,000
Interconnect Modules (2) (Replaces Nodes)	44,716
Logistics Module	7,657
Air Locks (2)	7,488
Cupolas (2)	2,724
Truss	7,014
Hangar Truss	5,055
Hangar Doors (3)	8,431
Hangar Skin (Inflatable/Rigidized Sandwich)	63,571
Transfer Boom (et al)	22,230
Power System (4 Solar Dynamic Units)	8,709
Remote Manipulator System (2)	3,273
Propellant Tanks (4)	55,751
Propellant Transfer Lines (et al)	16,000
Thermal Control	3,678
GN & C Pallets (2)	4,385
RCS Tank Pallets (2)	4,055
Utility Trays	13,550
Antennas	613
RCMs (6)	810
Total Mass (dry)	**318,710 kg**
Stored Cryogenic Propellant	182,000
Total Mass (wet)	**500,710 kg**
Loaded Lunar Vehicle Stack (1)	115,000
Total Gross Mass	**615,710 kg**

RECOMMENDATIONS

1) Space transportation hangars should be unpressurized due to the large work areas required. Pressurized hangar space would not be feasible for such an operation, because of the logistics and special hangar construction required for pressurized containers.

2) Design for maximum gravity gradient stability. A gravity gradient space station has favorable flight dynamics and control stabilization.

In conclusion, the Atrium Facility is a transportation node designed in an attempt to optimize a given set of requirements. This paper is not advocating this facility as **the** TN for a Lunar/Mars Initiative, but rather as a learning exercise directed towards a greater comprehension of the total architecture and the design drivers that influence its form and function.

REFERENCES

Alred, J., et al, (1988) "Development of a Lunar Outpost: Year 2000-2005", Paper submitted to the Second Symposium on Lunar Bases and Space Activities of the 21st Century in 1988, Lunar and Planetary Institute, Houston, Tx.

Eagle Engineering, Inc., (1988). Space Transportation Nodes Assumptions and Requirements, Eagle Engineering Report No. 87-174, NASA Contract No. NAS 9-17878.

Kaszubowski, Martin J., Ayers, J., (1988). The Transportation Depot - An Orbiting Vehicle Support Facility. PRC Systems Services, Aerospace Division, Hampton, VA. Paper in Lunar Bases II and Space Activities in the 21st Century.W.W. Mendell, Editor. Lunar and Planetary Institute, Houston, TX.

Keaton, Paul W., (1985). A Moon Base/Mars Base Transportation Depot. Los Alamos National Laboratory, Los Alamos, NM. Paper in Lunar Bases and Space Activities of the 21st Century. W.W. Mendell, Editor. Lunar and Planetary Institute, Houston, TX.

Kessler, D.J. and Reynolds, R. (1988), Orbital Debris Environment for Spacecraft Designed to Operate in Low Earth Orbit. Draft paper, April 1988. NASA Johnson Space Center, Houston, Tx.

Ketchum, W.J., Maloney, J.W., Pena, L.R., (1988). Servicing and Maintaining a Lunar Transportation System in Low Earth Orbit (LEO). General Dynamics Space Systems Division, San Diego, CA. Paper in Lunar Bases II and Space Activities in the 21st Century.W.W. Mendell, Editor. Lunar and Planetary Institute, Houston, TX.

Petro, Andrew J., (1988) Lunar Transportation System: Conceptual Design of Transport Spacecraft. JSC-23347, NASA Johnson Space Center, Houston, Tx.

Stecklein, Jonette M., (1988). The Flight Dynamics of a Transportation Node. NASA Johnson Space Center, Houston, Tx.

Stimler, Frederick J., (1984) Systems Definition Study of Deployable, Non-Metallic Space Structures. Goodyear Aerospace Corporation, Akron, Ohio 44315. NASA Contract No. NAS 8-35498.

Post-MECO Propulsion System Requirements for
Placement of an STS External Tank in Low Earth Orbit

William C. Stone [1]
Geraldine S. Cheok [1]

Abstract

This paper discusses the findings of a series of computer simulations carried out at the National Institute of Standards and Technology to investigate the requirements for powered flight of the external tank following separation from the Shuttle orbiter at main engine cutoff (MECO). The object of the investigation was to determine the minimum thrust and fuel requirements for an autonomous exterior propulsion package attached to the external tank to avoid re-entry on the critical first orbit, and to place the tank in a short term stable orbit from which customary orbit maintenance procedures may be carried out. It is shown that a 1000 lbf (4450 N) hydrogen/oxygen thruster is capable of boosting the tank to a transfer orbit with an apogee of 206 km and a perigee of 159 km by means of expending a 1100 lbm (500 kg) propellant at the initial orbit apogee following MECO. The subsequent orbit will have a decay time of 46 hours (32 orbits) to re-entry under solar maximum conditions. Estimates for propellant mass required for subsequent boost to a 500 km altitude storage orbit are presented.

Introduction

Significant interest has developed in the past two years for the on-orbit utilization of the external tank for the Space Shuttle, that is, the U.S. Space Transportation System (STS). The external tank is currently the only non-reusable component of the STS. On a nominal "standard insertion" launch these tanks, which carry cryogenic oxygen and hydrogen to fuel the three main shuttle engines, reach approximately 98% of orbital velocity at an altitude of about 105 kilometers, after which they separate from the orbiter and are left to re-enter the earth's atmosphere. Each tank measures 8.4 m in diameter by 46.5 m in length with an enclosed volume of 2069 cubic meters which is structurally capable of handling internal pressures necessary for human habitation. Potential commercial uses of these tanks in space include, among others, low-cost manned orbital workshops and man-tended manufacturing platforms [Sophron 1984]; fuel storage depots [Arnold, 1983]; and as building blocks for low-cost lunar spacecraft [King, 1989]. Although it is possible on most missions for the Space Shuttle to take the external tank into orbit [NASA, 1988], this has not yet been attempted for several reasons. First, the shuttle cargo bay payload capacity or the maximum mission altitude would generally have to be reduced to accommodate the increased amount of propellant needed to boost the external tank to orbit. Secondly, after achieving orbit, safety issues arise relating to the control of the tank in the vicinity of a manned orbital vehicle (i.e. the Shuttle orbiter), and in particular to the question of uncontrolled random re-entry ("Skylab Syndrome"). It is, therefore, desirable to consider the development of an economical means for autonomous placement of the tanks in long term stable orbits following the current separation sequence of the tank from the Shuttle orbiter.

Because of the potentially high payoff, in terms of enhanced on-orbit capability at vastly reduced cost, to U.S. companies seeking to conduct business in space, the National Institute of Standards and Technology has undertaken a research program to resolve the engineering questions relating to the placement, stabilization, and pressurization of Space Shuttle external tanks in low earth orbit. In this paper we discuss the requirements for powered flight of the external tank through the thermosphere following separation from the Shuttle orbiter at main engine cutoff (MECO).

---

1 National Institute of Standards and Technology, Washington, D.C.

Shuttle Operations

For the reader who is not familiar with shuttle operations, reference to Figure 1 will be of use. This shows a typical time sequence from launch to orbit for the STS for what is known as a "Nominal (or Standard) Insertion" mission, which is by far the most difficult case to solve in terms of propulsion requirements for an external tank. The methods which will be subsequently described apply equally well to "Direct Insertion" missions [see NASA, 1988]. In both cases solid rocket booster staging occurs approximately 2 minutes into launch, after which the orbiter and external tank continue to a predefined altitude of approximately 105 kilometers. There main engine cutoff (MECO) occurs and the external tank is jettisoned. On a Nominal Insertion launch the external tank will then trace an elliptical orbit with an apogee altitude of approximately 159 kilometers and a perigee altitude of about 7 kilometers. It is generally recognized that a spacecraft which descends into the atmosphere below an altitude of approximately 70 kilometers will shortly re-enter and burn up, unless specific steps are taken to increase its tangential velocity. For the external tank this point is reached approximately 45 minutes beyond MECO, after which atmospheric drag increases exponentially, leading to the destruction of the tank. Of particular interest, then, is the determination of the minimum thrust and fuel required to avoid re-entry of the external tank during the critical first orbit.

Orbital Dynamics

For the purposes of solving the problem described above, it is sufficient to consider the problem of orbital mechanics in two dimensions, as shown in figure 2. Perturbations due to the asphericity of the earth are not considered. Given an initial set of conditions such as the altitude of the external tank at MECO, the altitude at apogee, and the altitude at perigee [as provided in NASA, 1988], it is possible to determine both the initial location $[X_0, Y_0]$ and initial velocity components $[V_{x0}, V_{y0}]$ of the tank with respect to the cartesian coordinate system defined in figure 2 from orbital mechanics as:

$$X_0 = a[\cos(E) - e] \quad ; \quad Y_0 = a\sqrt{(1-e^2)}\ \sin(E)$$

$$V_{x0} = \frac{\sin(E)\sqrt{\mu a^3}}{r_0^2}[e(\cos(E) - 1] \quad ; \quad V_{y0} = \frac{\sqrt{\mu a^3(1-e^2)}}{r_0^2}[e(\sin^2(E) + \cos(E) - 1]$$

Where μ = 398600 $Km^3/second^2$; e = the orbit eccentricity; a = the semi-major axis length; E = the eccentric anomaly; and r_0 = the initial radius of the tank orbit.

Perturbing Accelerations

During the critical post-MECO period, the external tank will be acted upon by two disturbing forces which in turn produce perturbing accelerations which must be accounted for in the differential equations describing the motion of the tank. These disturbing forces are aerodynamic drag and the thrust produced by the propulsion package. Disturbing forces typically included in long duration orbital lifetime calculations, such as solar radiation pressure, represent second order effects for spacecraft at altitudes below 500 kilometers and are therefore not considered in this analysis. The aerodynamic deceleration per unit mass is given by:

$$T_a = -\frac{C_d\, A_d\, \rho\, V^2}{2m}$$

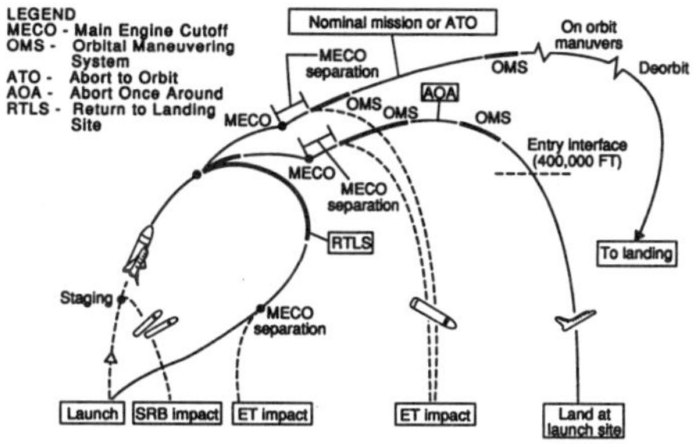

Figure 1: Standard Space Shuttle Launch Scenarios

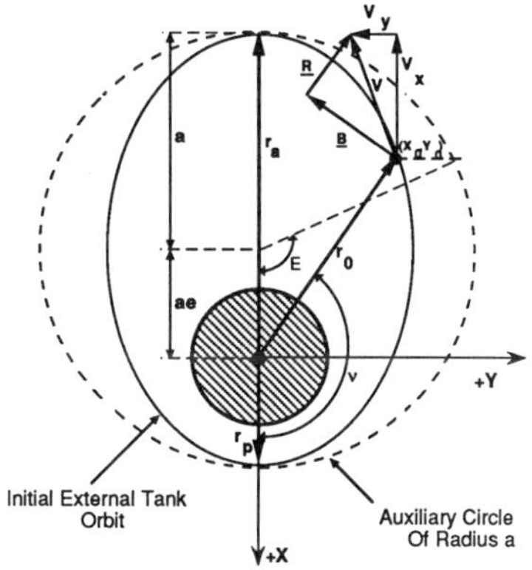

Figure 2: Orbital Mechanics Variables For External Tank Motion

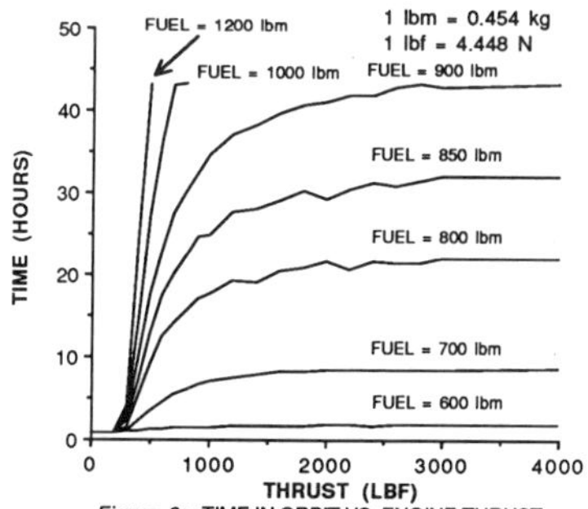

Figure 3: TIME IN ORBIT VS. ENGINE THRUST
Vehicle Mass = 74,000 lbm
ISP = 400 s, Burn at apogee
Solar minimum, Angle of attack = 0°

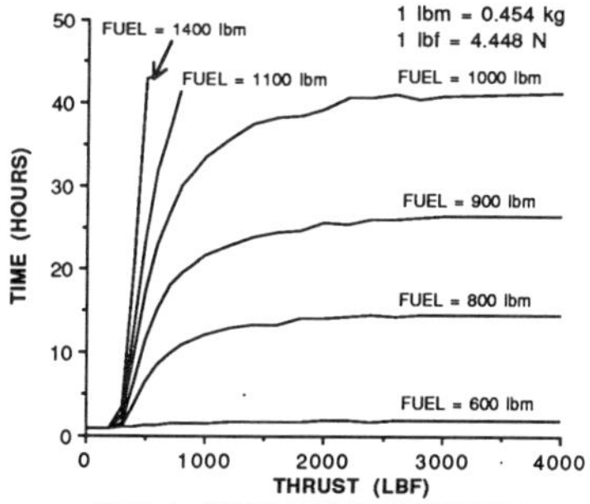

Figure 4: TIME IN ORBIT VS. ENGINE THRUST
Vehicle Mass = 74,000 lbm
ISP = 400 s, Burn at apogee
Solar maximum, Angle of attack = 0°

where:

C_d = drag coefficient, taken as 2.0 for the external tank.
A_d = projected drag area perpendicular to the velocity vector, in square meters.
ρ = atmospheric density in kg/m^3.
m = mass of the empty external tank plus any residual hydrogen and oxygen following MECO, plus the weight of any external propellants, storage vessels, and propulsion hardware, in kg.

A few comments are in order regarding the above. Accurate closed form solutions are now available for external tank drag area for any given angle of attack [Stone and Witzgall, 1989]. The minimum and maximum drag areas are 55.51 and 358.12 square meters for angles of attack of 0 and 90 degrees, respectively. The atmospheric density, ρ, used in this study is described by the National Standard Atmosphere [NOAA, 1976] for altitudes between 60 and 85 kilometers and by [Tobiska, 1989] for altitudes above 85 kilometers. Both the solar maximum and solar minimum conditions were considered in the analyses, since the atmospheric density may vary by as much as an order of magnitude during the 11 year solar cycle. Presently Solar Maximum is predicted to occur sometime in 1990 with Solar Minimum in 1997. These can represent real launch constraints if the mass of the external tank propulsion package is to be minimized.

It is assumed that a suitable attitude control system is provided which is capable of maintaining the longitudinal axis of the tank parallel to the velocity vector. This is essential to any practical application involving the tanks below an altitude of approximately 500 kilometers, since the drag area for a 90 degree angle of attack is nearly six and one half times that for the head-on configuration and would lead to premature re-entry. The thrust may be either positive or negative. The latter case applies specifically to the problem of de-orbiting the external tank, which is of considerable importance if the tank must be made to safely re-enter and land in a specific, uninhabited location on the earth's surface. The thrust acceleration (or deceleration) is given by Newton's law as $T_f = F/m$, where F is the thrust in Newtons of the propulsion system and m has units of kilograms. It should be noted that during an engine burn m is constantly changing since the propellant mass is decreasing. The overall spacecraft mass is thus given by:

$$m = M_{tank} + M_{fuel} - \frac{Ft}{I_{sp} g_0}$$

where:

M_{tank} = the structural mass of the external tank plus any residual cryogens (hydrogen and oxygen) not used for the propulsive maneuver, plus the structural weight of the propulsion module.
M_{fuel} = the mass of fuel <u>initially</u> available for use by the propulsion module.
t = the cumulative time of operation of the propulsion module at full rated thrust.
I_{sp} = Specific Impulse of the fuel used in the propulsion system.
g_0 = acceleration due to gravity at the earth's surface.

In the analysis the value of T_f is set to zero when the fuel is exhausted. This is distinctly different from Hohmann Transfer theory [Kaplan, 1981], which assumes all velocity change to occur instantaneously. For a system having a very low thrust level and a large quantity of available fuel, the burn arc, or arc angle of an orbit through which the motor is firing, can be significant. Numerical integration, therefore, is the only means of accurately tracing the resulting motion of the tank. It should be noted that the location of burn initiation will also significantly affect the path of the tank, and the available time in orbit.
The sum of equations 15 and 16 gives the cumulative perturbing acceleration acting upon the external tank: $T = T_f - T_a$.

Differential Equations of Motion

The motion of the external tank in two dimensions can be completely described by four differential equations, two containing derivatives of the velocity components V_x and V_y and two containing derivatives of the position vectors X and Y, as follows:

$$\frac{dV_x}{dt} = -\frac{\mu}{r^3} X + \tau_x T \quad ; \quad \frac{dV_y}{dt} = -\frac{\mu}{r^3} Y + \tau_y T$$

$$\frac{dX}{dt} = V_x \quad ; \quad \frac{dY}{dt} = V_y$$

Where $\tau_x = V_x/V$, $\tau_y = V_y/V$, and V is the velocity magnitude. The initial conditions for the above system are given by $[X_0, Y_0]$ and $[V_{x0}, V_{y0}]$ as defined above. The solution was carried out using single precision arithmetic on a Convex C-120 computer using the **SDRIV** numerical integration package [Kahaner, 1979] as a subroutine in the program, **ET_ORBIT**, which consisted of 12,000 lines of Fortran 77 code. Single precision accuracy was determined through simulation to be adequate for the results reported below, which represent a relatively brief time on orbit. Off-the-shelf software, which operates in double precision, exists for long duration orbital lifetime calculations but these do not permit examination of powered flight.

Results

The objective of this study was to determine the smallest values for the propulsion system thrust and fuel mass required to avoid re-entry for a period sufficient to permit subsequent orbital optimization burns. For the sake of limiting computational time, the required orbital duration for a "successful" boost was set to 30 orbits, or approximately 2 days. During this time the propulsion system will be required to carry out an additional series of burns in order to boost the tank to a parking orbit with sufficient altitude to allow for storage periods of up to 30 years, depending upon the desired use. One recent study [NASA, 1988] has indicated that such a long term parking orbit would have an altitude in the vicinity of 500 kilometers.

To solve the differential equations, certain additional initial conditions must be specified such as the spacecraft mass (structural mass and residual fuel not available for propulsion), the ΔV burn initiation point, the aerodynamic drag model (which includes the specification of solar maximum or minimum and the angle of attack), and the specific impulse, I_{sp}, of the propellant. It may be appreciated that no single optimum design will exist for all users, owing to the availability of an appropriate launch window, safety considerations which may affect where a velocity change burn may be initiated, and budget constraints which may limit a smaller company to consideration of a lower performance propulsion package. The results presented below, therefore, cover a wide spectrum of variables which present a range of options to the user.

Apogee Burn

From an analytical point of view, a propulsive maneuver may be initiated at any time following separation from the shuttle (MECO). From a practical point of view, a minimum separation distance must be allowed between the shuttle orbiter and the external tank in order to maintain safety for the orbiter and its crew. Thus one insertion scenario for the external tank would consist of firing the exterior propulsion package as soon after MECO as safety permits. The alternative would be to allow the tank to coast to its initial apogee (approximately 159 km vs 105 km at MECO) before initiating the burn. This would be the location of choice for a conventional boost intended to raise the perigee altitude.

The apogee-boost case represents the most conservative opportunity for placement of an external tank in orbit at minimum cost. "Conservative" in this context refers to the lack of sensitivity of the resulting motion of the tank to errors in the thrust magnitude and direction. Other scenarios discussed below can be more efficient, but at the price of sensitivity to propulsion system thrust level and attitude vector alignment variations. Figures 3 and 4 show families of curves in which the X-axis represents ΔV motor thrust and the Y-axis represents the time until re-entry in hours. An ascending curve which has been terminated at 43 hours indicates that the tank is still in orbit, and has therefore undergone a successful boost. The mass of the vehicle has been assumed to be 74,000 lbm (33,596 kg), which includes a 5,000 lbm (2,270 kg) budget for residual cryogens in the liquid hydrogen (LH_2) and liquid oxygen (LO_2) tanks and support structures for the propulsion package. This can be considered representative of the vehicle mass during most missions although the residual cryogen mass can be substantially higher, depending on the particular mission. The effect of changing vehicle mass is discussed below. The vehicle mass does not include the fuel used by the propulsion system, which is accounted for separately in the calculations since it is a dynamic quantity. An I_{sp} of 400 seconds (oxygen/hydrogen) has been assumed; the performance for different values of specific impulse is subsequently discussed.

Both figures 3 and 4 indicate that a threshold value of approximately 400 lbf (1800 N) exists for the thrust level necessary to achieve orbit regardless of the amount of fuel available or the prevailing solar flux. This threshold can be viewed as the aerodynamic drag compensation thrust. From an engineering standpoint it will be necessary to have sufficient reserve thrust to ensure attainment of orbit despite any problems which might arise, for example in the pointing accuracy of the reaction control system or a decrease in engine performance. A reasonable value of thrust would appear to be approximately 1000 lbf (4450 N). Such a thrust level could be obtained from a variety of off-the-shelf components and is in the range of the larger reaction control units presently employed on the space shuttle orbiter. Figures 3 and 4 indicate that in order to achieve an initial 40 hour period on orbit (during which additional altitude boost maneuvers would be carried out) a propulsion fuel budget of 1100 lbm (454 kg) is needed under solar maximum conditions. For an oxygen/ hydrogen thruster running a mixture mass ratio of 6:1 (oxidizer: propellant), this amounts to a storage requirement of only 374 liters LO_2 and 1023 liters LH_2.

MECO Burn

The apogee burn scenario described above was termed "conservative", because the time available on-orbit increases monotonically for increasing levels of thrust and amount of propellant. This is not the case when the burn is initiated close to MECO, as shown in figures 5 and 6. Here it is seen that the threshold thrust level to achieve orbit is approximately 200 lbf (900 N). However, there is wide variation in the time in orbit, depending upon the chosen thrust level and fuel mass. Furthermore, thrust levels above 500 lbf (2200 N) lead to significantly decreased time in orbit. The performance depicted in figures 5 and 6 can be better appreciated when one considers the following rule of thumb in conjunction with figure 2, which shows the initial orbit parameters for the external tank at MECO: an instantaneous thrust impulse at any point in an orbit will have the effect of raising the orbital altitude at a point 180 degrees opposite the point where the impulse was initiated. When low levels of thrust are applied continuously, the effect is generally to increase the orbital altitude at a point approximately 180 degrees opposite the center of the "burn arc". For the MECO burn scenario just described, high levels of thrust, combined with the relatively small quantities of fuel shown, represent nearly impulsive loading, which raises the initial orbital perigee only a small amount, compared with that for a smaller engine which burns over a much longer arc, indeed almost to initial apogee, before expending the same mass of fuel.

Despite the benefits indicated by figures 5 and 6, if one could achieve the peak performance shown, the MECO burn scenario represents a risky proposition, because minor variations in the level of thrust or drag would result in vastly reduced time in orbit.

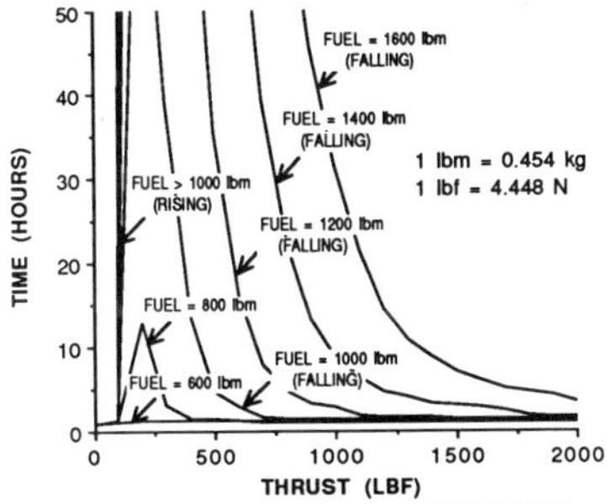

Figure 5: TIME IN ORBIT VS. ENGINE THRUST
Vehicle Mass = 74,000 lbm
ISP = 400 s, Burn at MECO
Solar minimum, Angle of attack = 0°

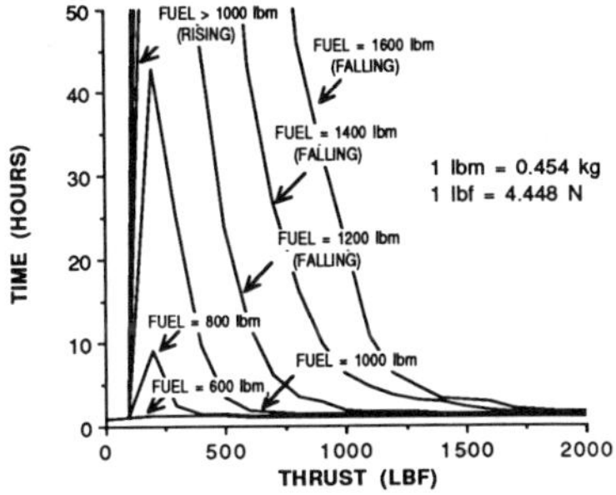

Figure 6: TIME IN ORBIT VS. ENGINE THRUST
Vehicle Mass = 74,000 lbm
ISP = 400 s, Burn at MECO
Solar maximum, Angle of attack = 0°

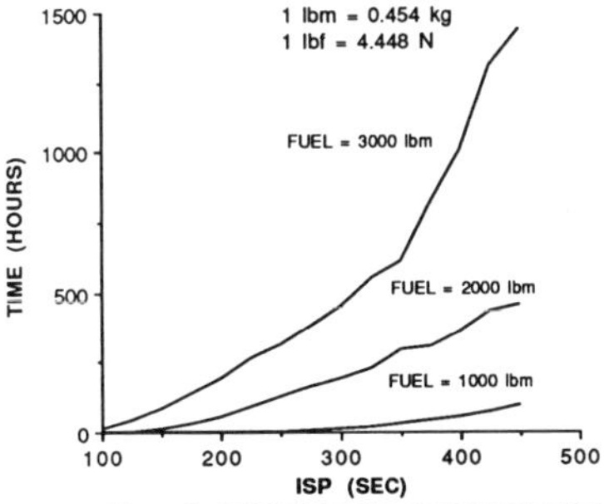

Figure 7: TIME IN ORBIT VS. SPECIFIC IMPULSE
Vehicle Mass = 74,000 lbm
Thrust = 1000 lbf, Burn at apogee
Solar minimum, Angle of attack = 0°

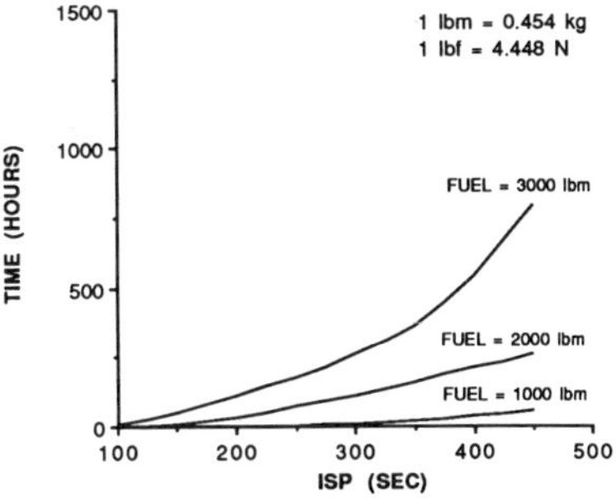

Figure 8: TIME IN ORBIT VS. SPECIFIC IMPULSE
Vehicle Mass = 74,000 lbm
Thrust = 1000 lbf, Burn at apogee
Solar maximum, Angle of attack = 0°

Effect of Specific Impulse

So far, it has been assumed that oxygen/hydrogen thrusters (I_{sp}=400s) would be used to boost the external tank. Other propellants could be used with varying degrees of reduced performance. Bipropellants, such as Nitrogen Tetroxide (N_2O_4) and Mono Methyl Hydrazine (MMH) are non-cryogenic and have an I_{sp} of about 300 s. Solid rocket motors are available in a wide variety of thrust and duration levels and have I_{sp} values between 200-250s. Figures 7 and 8 show the effect of specific impulse on time in orbit for varying quantities of fuel mass. The thrust level, using the apogee burn scenario, has been set at 1000 lbf (4448 N). Figures 7 and 8 indicate that time in orbit is approximately proportional to I_{sp}. A solid rocket motor with twice the mass of the liquid propellant for the oxygen/ hydrogen engine would be needed to achieve the same time in orbit.

Effect of Spacecraft Mass

Figures 9 and 10 show the time in orbit for an external tank equipped with a 1000 lbf (4448 N) propulsion package as a function of overall vehicle mass (propellant mass for the indicated burn excluded). The important aspect of these figures is that while all curves show decreasing performance for increasing vehicle mass, the penalty for additional mass, in terms of time in orbit following the initial boost, is small. This indicates that additional payload mass could be taken to orbit by the external tank -- perhaps inside the hollow Intertank structure -- with no significant re-scaling of the exterior propulsion package, provided that the overall shuttle system payload capacity was under-manifested.

Overall Propellant Budget for Orbital Storage of External Tanks

It has been recommended that a suitable long term circular orbit storage altitude would be in the vicinity of 500 km [NASA, 1988]. If, as previously suggested, a 1100 lbm (500 kg) apogee burn is carried out by the tank propulsion system using a 1000 lbf (4450 N) hydrogen/oxygen engine, then the resulting orbit is as described in figure 11a. This orbit has an apogee altitude of 206 km and a perigee altitude of approximately 159 km. If no further action is taken to boost the tank to a higher orbit, or if no other orbit maintenance burns are carried out, then the orbit will progressively decay due to atmospheric drag, as shown in figure 11b, ultimately ending with the re-entry of the tank 46 hours (32 orbits) after the initial apogee burn. Clearly, it is advantageous to act early to boost the tank, as the apogee altitude decreases by a greater amount with each successive orbit. An initial estimate of the additional fuel required to achieve a 500 km altitude circular orbit may be obtained using Hohmann transfer theory [Kaplan, 1981]. At least three additional burns are required for this maneuver. The first, a circularization burn carried out at the apogee altitude of 206 km, involves a velocity change of 14 m/s. The next burn, which may take place anywhere along the circular orbit, places the tank into a transfer ellipse with an apogee altitude of 500 km and a perigee altitude of 206 km. The third burn takes place at apogee of the transfer ellipse and results in the final 500 km altitude circular storage orbit. The combined velocity change for these two burns is 168m/s. This, added to the initial circularization burn amounts to 182 m/s. The propellant mass required to achieve this change may be determined from the standard rocket equation [Kaplan, 1981] to be 3516 lbm (1,596 kg) for a spacecraft mass of 74,000 lbm (33,596 kg). Thus, the total post-MECO fuel budget comes to 4616 lbm (2096 kg) or, in terms of volumetric storage requirements, 55 cu.ft. (1569 liters) LO_2 and 152 cu.ft. (4292 liters) LH_2. This is, coincidentally, quite close to the estimates [NASA, 1988] of the minimum residual cryogens suitable for propulsion which remain in the tank following MECO. The calculations thus far have counted this as deadweight and have assumed that all propellants would be carried in exterior tanks which comprise the autonomous propulsion package. Effective utilization of residual cryogens would eliminate any launch constraints which might exist due to Space Shuttle cargo bay manifesting. This paper has dealt solely with the subject of the initial boost requirements for placing external tanks in long term storage orbits. Other key facets of the problem include the determination of orbit maintenance propellant mass and the assessment of an appropriate deboost system should the tank need to be brought down in a controlled fashion. These subjects will be discussed in future papers.

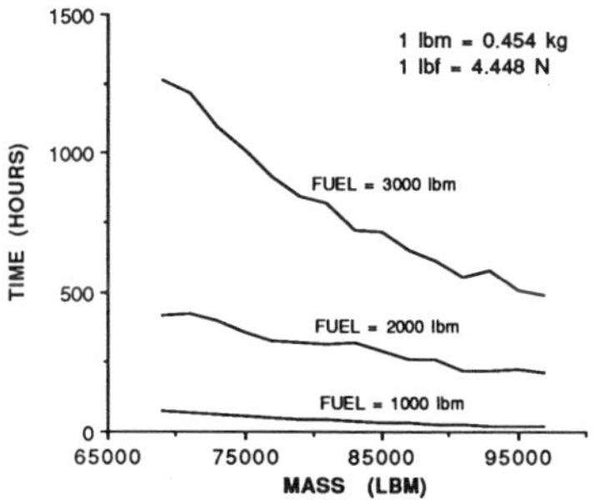

Figure 9: TIME IN ORBIT VS. VEHICLE MASS
Thrust = 1000 lbf
ISP = 400 s, Burn at apogee
Solar minimum, Angle of attack = 0°

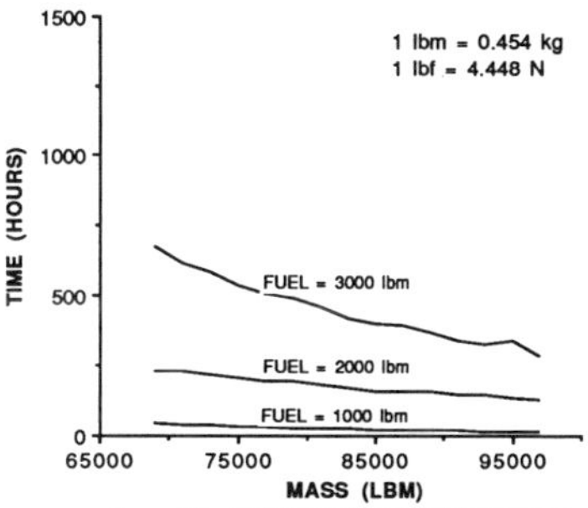

Figure 10: TIME IN ORBIT VS. VEHICLE MASS
Thrust = 1000 lbf
ISP = 400 s, Burn at apogee
Solar maximum, Angle of attack = 0°

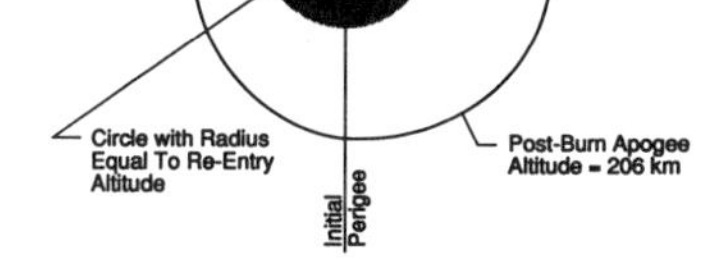

Figure 11a: First Orbit Following Apogee Burn

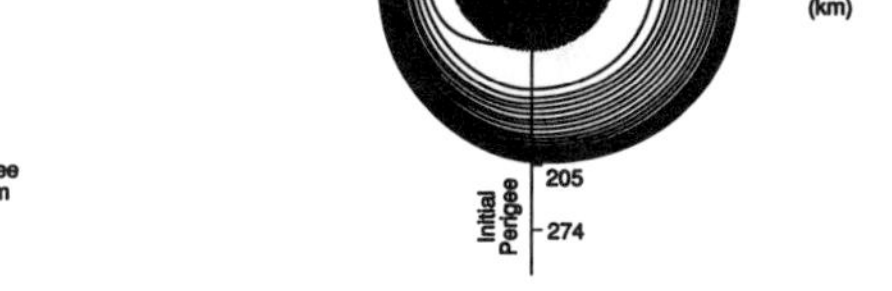

Figure 11b: Complete Decay Trace For External Tank Following Single Apogee Burn

Comparison of Propellant Budgets for Impulsive (Ideal) vs Liquid (Real) Thrusters

Figure 11 provides some insight into why numerical integration was a useful approach to investigating the initial boost of the external tank. In addition to allowing an accurate decay trace to be monitored, information is also available on the rotation of the semi-major axis of the orbit due to the finite burn arc (about 20°) which is of importance for rendezvous at a storage depot. It is interesting to note, however, that for this particular case (thrust = 1000 lbf [4450 N]) the propellant budget for an idealized Hohmann transfer also comes to, within two significant figures, 1100 lbm (500 kg), the same as for the numerically integrated solution. The Hohmann solution can be determined in two steps: a velocity change burn is made to circularize the initial orbit at the initial apogee altitude of 159 km, followed by a second burn to inject the tank into a transfer ellipse with a perigee altitude of 159 km and an apogee altitude of 206 km, as determined from the numerical integration. The delta-V values for these two impulsive maneuvers came to 46 and 13.6 m/s respectively leading to a propellant budget (hydrogen/oxygen) of 1100 lbm (500 kg). The point to be made is that beyond the threshold thrust required to overcome atmospheric drag, and for velocity change maneuvers in which the burn arc is small, the differences between ideal theory and real engine performance are negligible. This assumes that the burn is made at apogee; alternate thrust initiation points, and velocity change burns effected with low-thrust engines will lead to significant deviations in required propellant budgets from those that would be calculated using ideal theory.

Conclusions:

On the basis of extensive numerical simulations it was calculated that the space shuttle external tank can be boosted to a short term stable orbit following standard MECO separation from the shuttle orbiter, and without any direct interaction nor detriment to orbiter performance. An exterior propulsion package for the external tank equipped with a minimum thrust capacity of 1000 lbf (4448 N), a propellant mass of 1100 lbm (500 kg), and an I_{sp} of 400s appears sufficient to achieve an initial time in orbit of nearly two days under solar maximum conditions, provided the burn is made at initial apogee and the angle of attack is maintained near zero degrees by an onboard attitude control system. It is assumed that additional velocity change burns will take place following the initial burn which will place the tank in a circular orbit between 400-500 km altitude for long term storage. Initial estimates of the total fuel required to achieve a 500 km circular storage orbit come to 4616 lbm (2096 kg) based upon Hohmann transfer theory following the initial apogee burn. All calculations assumed 5000 lbm (2270 kg) of residual cryogens in the external tank following MECO as deadweight. Recovery and use of these propellants by the exterior propulsion package would lead to a dramatic increase in the time in orbit above the values reported in this paper.

References:

1. "The Feasibility of Using the Shuttle External Tank as a low Cost Warehouse-Workshop-Residence in Low Earth Orbit," Final Report, The Sophron Foundation, McLean, VA, August, 1984
2. "Utilization of the External Tanks of the Space Transportation System," Arnold, J.R., ed., California Space Institute, Scripps Institution of Oceanography, La Jolla, CA, April 1983.
3. King, C.B., et.al., "A Lunar Habitat Configuration Derived from a National Space Transportation System External Tank," NASA Langley Research Center Technical Paper, June, 1989.
5. "Preliminary Data for On-orbit Utilization of the External Tank," Special Publication of the Program Development Office, NASA George C. Marshall Space Flight Center, Huntsville, Al., April 1988.
6. Stone W.C., and Witzgall, C., "Evaluation of Aerodynamic Drag and Torque for External Tanks in Low Earth Orbit," Proceedings, 9th Conference on Space Manufacturing, Princeton, NJ, May 1989.
7. U.S. Standard Atmosphere, 1976, NOAA, NASA, USAF joint document, U.S. Government Printing Office, Washington, D.C., October, 1976
8. Tobiska,K., Ph.D. Thesis, University of Colorado, Boulder, 1989
9. Kaplan, M.H., "Modern Spacecraft Dynamics and Control," John Wiley & Sons, New York, NY, 1981
10. Kahaner, Moler, and Nash, "Numerical Methods and Software," Prentice Hall, Englewood-Cliffs, NJ., 1989, p295.

A SPACE STATION TRASH RECOVERY AND RECYCLING SYSTEM

Thomas C. Taylor[1], David Nixon[2] and Jan Kaplicky[3]

ABSTRACT

The quantity of trash produced on the Space Station during its lifetime will be substantial and will include useful waste materials. Given the high cost of transportation to orbit, it can be retained there for eventual recycling rather than returned to Earth. A trash management system enables crews to discard trash by type and store it in modular packages which are periodically removed by EVA to an external adjacent holding point. The holding point is visited annually by OMV and trash packages are removed to remote long-term storage which utilizes a Shuttle External Tank as a depository platform. The External Tank is modified to support eventual recycling hardware to process the stored waste material into useful components on-orbit. A system of this type has three potential benefits: it can liberate valuable Space Station internal stowage volume from dedicated trash storage; it can initiate long-term stockpiling of useful raw materials on-orbit for future use; and it can encourage commercial initiatives in the field of materials processing in space.

1 APPROACH

Broad design requirements for the Space Station Trash Management System have now been established (NASA, 1987). Trash collection, management and disposal will be a major housekeeping task on the Space Station for successive crews. The approach to the Space Station trash problem identified at Space Station definition stage is to collect it, compact it, and pack it into vacant food freezer racks in the Logistics Module for temporary storage. There it will await removal and shipment back to ground by a visiting Orbiter. The quantity of trash which will be generated by crews on the Space Station during its anticipated lifetime will be considerable. Each crewmember is expected to produce about 1.25 kg. of trash per day, amounting to a total of 3,650 kg. for one year alone for an 8-person crew complement, or 109,500 kg. over a 30-year period.

[1] President, Global Outpost Inc., 6836 Deer Run Drive, Alexandria, VA 22306, US.
[2] Partner, Future Systems, 1103 South Hudson Ave., Los Angeles, CA 90019, US.
[3] Partner, Future Systems, 19 Store St., London WC1 7DH, UK.

Since trash composition will include useful materials such as metal, paper and certain plastics, the concept of retaining and storing the trash on-orbit for eventual in-space recycling is attractive, given the cost and investment of delivering the cumulative weight of it in virgin form to orbit over successive Orbiter resupply missions.

A Space Station trash management system can be developed to enable trash products to be sorted at the point at which the trash item is discarded by crewmembers. Presorting of useful trash by type into metal, paper or plastic will utilize specially-designed receptacle bags. The bags are modular to enable them to be packaged together on a cumulative basis into portable 'trash-sacks' sized for efficient inter- and intra-module crew translation. A convenient maximum trash-sack volume is 0.96 cu.m., or approximately half the volume of a standard rack. This has outer dimensions of 1.07 m. diameter x 1.07 m. length, enabling crewmembers to negotiate successive hatches and changes in direction easily on routine IVA trash collection duty. This volume of trash will be produced by an 8-person crew every 22 days if a 2:1 trash compaction ratio is selected. This is a conservative trash compaction ratio chosen for preliminary estimating purposes. Actual Space Station trash compactor performance has yet to be finally established.

Proposed Space Station trash management policy requires the allocation of up to 2 refrigerated racks in the Logistics Module to store the trash between Orbiter visits. In view of the very high value placed on optimizing rack volume throughout Space Station, it makes sense to store the trash externally rather than internally if it can be deposited in a long-term holding facility on-orbit. This will reduce the trash storage capacity requirement of 2 racks to 0.5 of a rack (for the 22 days), resulting in the liberation of net volume equivalent to 1.5 x single standard rack volume which can be used for other purposes. 1.5 x single standard rack volume is equivalent to one crew sleeping compartment and represents a significant saving and a valuable resource for reallocation.

The regular 22-day trash-sacks will be designed for removal by EVA. Trash-sacks can be developed as sealable self-venting fabric shrouds to encapsulate the trash for removal through airlocks. Once outside, the trash-sacks are moved by EVA (with RMS-assist if required) to local storage at an interim holding facility located within, or attached, to the outside of the main beam truss structure. This facility will be positioned as close as possible to an appropriate airlock to minimize translation distances. The facility configuration will be designed to optimize interim trash-sack storage efficiency and EVA, RMS and OMV accessibility. Assuming that the interim holding facility is visited annually and the trash-sacks removed to a remote long-term depository, a holding capacity for 17 trash-sacks will be required. This can be easily accommodated over the surface of two standard structural bays of the Space Station truss beam, or in a volume of 3.35 m. x 3.05 m. x 1.83 m. within the beam framework itself.

Annual transportation of the trash-sacks from the interim holding facility on Space Station to long-term depository will be by OMV. It is assumed the depository is in a similar orbit to Space Station but maintained in a remote location beyond critical Space Station operational control zones.

The availability to the industrial and space community of Space Shuttle External Tanks in orbit has been made possible by President Reagan's Commercial Space Initiative announced in 1988 and reaffirmed recently by the Bush Administration. Global Outpost, Inc., proposes to place an ET in orbit and outfit it as an experiment platform from 1993 onwards (Mellors et al., 1989 and Taylor et al., 1989). Early use of ET as a commercial platform will provide data and experience leading to more advanced operational applications. An External Tank is ideally suited for the role of an orbiting trash depository because of its unique shape and size. The outer surface is smooth and generally free of projections enabling it to be utilized as a versatile restraint surface for long-term storage. The large size of the tank provides extensive curved surface area for the deposit and retention of accumulated waste material over time. Modifications required to convert an ET to a trash depository are minimal and can be carried out on-orbit. A possible technique utilizes a lightweight fiber net of orthogrid or isogrid pattern which is deployed around the cylindrical section of the ET and post-tensioned to form a shroud. The net incorporates microfastener pads at the nodes which are positioned to align with compatible pads on the outer surface of the trash-sacks. Simple visual recognition of, and substantial dimensional tolerance in the alignment of pads eliminates the need for accuracy in placing trash-sacks in position as it assumed this will be performed by teleoperation from an OMV controlled from Space Station or ground.

Debris ejection and outgassing from trash-packs caused by micrometeoroid impact is a potential issue that will require careful study. The operational acceptability of External Tanks in orbit requires that the outer surface is encapsulated in a protective shroud to prevent debris ejection from the ET thermal protection layer. This shroud can be integrated with the orthogrid or isogrid trash-pack retention net. It is considered that the fabric selected for the trash-packs, together with its method of lamination and fabrication into the bag configuration, will enable the system to comply with relevant test standards and safety performance requirements. Each deposited trash-sack will require a substantial free area allowance around its circumference to ensure safety and simplicity of initial emplacement and subsequent removal (i.e. sufficient operational distance from adjacent trash-sacks). This constraint determines that the exterior surface of the ET is capable of housing approximately 282 trash-sacks. With a 2:1 Space Station trash compaction ratio, this quantity would be achieved after 18 years of operation, requiring that the recycling and recovery system would need to be in place by this time. With a 4:1 Space Station trash compaction ratio, the ET depository operational lead time would double to 36 years.

The development of in-space processing technology to recycle trash material into reusable products indicates potentially valuable opportunities for new forms of commercial space endeavor. For example, if 30% of trash material is metal alloy from food containers, 10,900 kg. of it will be generated by Space Station during the first 10 years alone. The in-space processing of this material into components such as extrusions will yield products with considerable market value that will find many constructional applications in future space missions requiring on-orbit assembly of basic components. The ET itself provides an ideal platform for material processing operations and can be configured and developed as an in-space recycling 'factory' as well as a raw material depository.

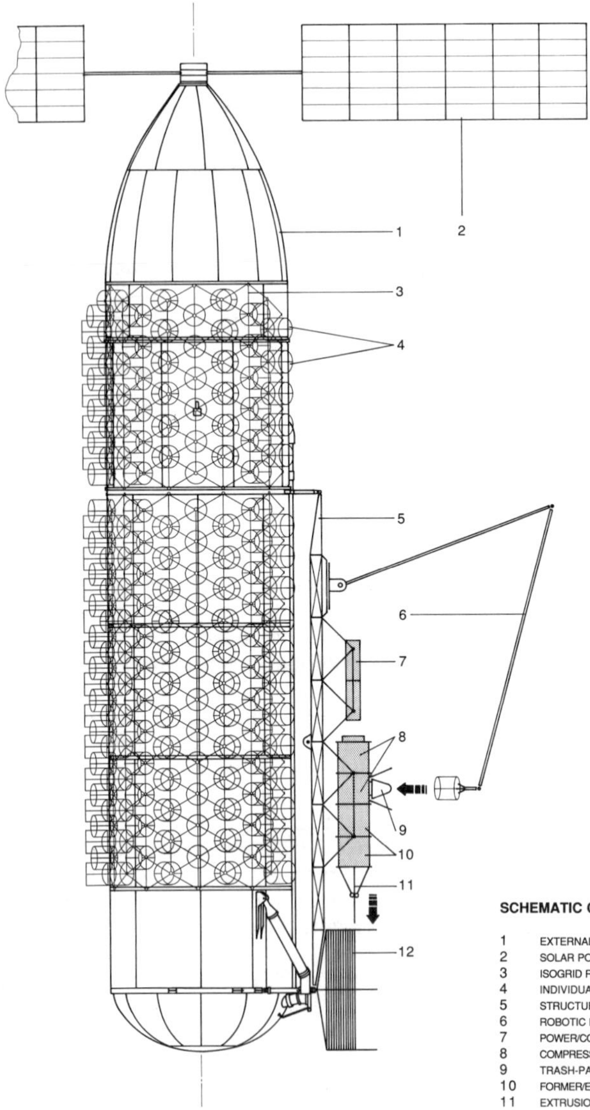

ET TRASH DEPOSITORY AND RECYCLING FACILITY

SCHEMATIC CONFIGURATION

1 EXTERNAL TANK PROFILE
2 SOLAR POWER PANELS
3 ISOGRID RETENTION NET
4 INDIVIDUAL TRASH-PACKS
5 STRUCTURAL PLATFORM
6 ROBOTIC MANIPULATOR/PICKER
7 POWER/CONTROL/COMMS. MODULES
8 COMPRESSOR/FURNACE MODULES
9 TRASH-PACK INPUT HOPPER
10 FORMER/EXTRUSION MODULES
11 EXTRUSION ALIGNMENT/SUPPORT
12 COMPONENT STORAGE CARTRIDGE

2 CONCLUSIONS

A Space Station Trash Recycling and Recovery System of the type described in this paper can be developed to provide the following significant benefits:

- *Liberation of valuable Space Station internal pressurized volume from long-term trash storage. Saved volume is approximately equivalent to 1.5 x single standard rack volume and can be allocated for other purposes.*

- *Stockpiling of substantial quantities of useful raw material for future in-space recycling. Eventual processing and availability of metal alloy, paper and composite products on-orbit by commercial initiative.*

- *Ideal application for Shuttle External Tank in Low Earth Orbit requiring minimal on-orbit modifications. The large External Tank exterior surface area is used to optimize trash long-term storage access and flexibility.*

3 SUPPORTING DATA

3.1 Trash Management System Requirements

Information on the requirements for the Space Station Trash Management System is derived from NASA Man-Systems Integration Standards, NASA-STD-3000, Volumes I, III and IV.

Background data used to calculate figures used in the synopsis are drawn from Space Station definition phase working documents and reviews. These include: Space Station Phase B Contractor Definition and Preliminary Design documents (1986); NASA-JSC Man-Systems Division studies (1988); and Space Station Phase C/D Contractor Reviews (1988).

3.2 Trash Generation Guidelines

- Expected trash generation per person on Space Station = 1.25 kg. (based on figures from NSTS missions which vary from 1.32 kg. to 1.73 kg.)
- 50% of trash comprises food/drink containers.
- 50% of trash comprises paper, plastic products, wipes, towels, bags.
- Some HMF/medical trash and misc. housekeeping trash is expected.
- Some Phase B studies recommend 1- or 2-day trash sorties to Logistics Module in preference to substantial trash storage in Galley.
- Combined wet and dry trash storage is expected (with water recovery).
- Trash excludes washables (e.g. discarded/soiled crew clothing).
- Trash excludes human waste products (dealt with separately).
- 75% of trash is generated in Galley.
- Standard Space Station rack volume = 1.93 cu.m.

3.3 Trash Generation Estimates

Table 1 summarizes 8-person Space Station Trash Generation for different durations.

TABLE 1	Total for 90 days	Total for 1 year	Total for 10 years	Total for 30 years
Weight in kg.	908	3,632	36,320	108,960
Uncompacted Volume in cu.m. (112.9 kg./m^3.)	8.50	33.98	339.81	1,019.43
Compacted Volume@ 2:1 ratio (225.8 kg./m^3.)	4.25	16.99	169.91	509.72
Compacted Volume@ 4:1 ratio (451.6 kg./m^3.)	2.13	8.49	84.95	254.86

3.4 Trash Material Breakdown

Table 2 lists a typical Space Station trash breakdown by material and physical condition and notional rate of generation. Regular generation implies several times per day. Occasional generation implies every few days, or longer.

TABLE 2	Wet Condition	Dry Condition	Regular Generation	Occasional Generation
Galley/paper	x	x	x	
Wardroom/paper		x		x
Laboratories/paper	x	x		x
Hygiene Compts./paper	x		x	
Sleep quarters/paper		x		x
Galley/plastics	x	x	x	
Wardroom/plastics		x		x
Laboratories/plastics	x	x		x
Galley/metal	x	x	x	
Laboratories/metal	x	x		x
HMF/paper	x	x		x
HMF/plastic	x	x		x
HMF/metal	x	x		x

Based on Table 2, a suitable recycling collection methodology will involve one of two approaches:

- Individual collection receptacles for paper, plastics and metal trash products with wet and dry trash combined (i.e. three receptacle variables).
- Individual collection receptacles for paper, plastics and metal trash products with wet and dry trash separated (i.e. 6 receptacle variables).

In each case, water recovery is performed on the trash while it is in the individual receptacles and before it is encapsulated in trash-sacks. The various receptacles are located throughout the Space Station at strategic points, within dedicated activity compartments, or at specific workstations.

3.5 Trash Storage Capacity

Trash storage volume required for a 90-day mission for 8 crewmembers is dependent on proven degree of trash compaction. This is unknown at this time but likely to fall between 2:1 and 4:1.

With a 2:1 compaction ratio, 90-day trash volume = 4.25 cu.m. = 2.2 racks.
With a 4:1 compaction ratio, 90-day trash volume = 2.13 cu.m. = 1.1 racks.

Since these estimates do not allow for skip-cycle capacity, they are probably on the low side, indicating that allocation of at least 2 trash storage racks in the Logistics Module may be required.

3.6 Proposed Initial 90-day Mission Food Stowage

At the beginning of each 90-day mission cycle, stockpiled food/drink in the Space Station will be at full capacity, having been resupplied by a visiting Orbiter. Since the removal of 1.5 x single rack trash storage volume in the Logistics Module also effectively removes the same volume of necessary food/drink storage capacity, a method is required to store 1.5 x single rack volume of food/drink at the beginning of each 90-day mission until it is diminished by depletion. Food/drink storables at ambient temperature in liquid, moist, dessicated or other condition can be easily packaged and stored in portable shallow soft stowage packs (as proposed by Boeing). The flexible stowage packs will be between 150 mm. and 330 mm. deep and attached to the faces of regular food storage racks in the Logistics Module without obstructing crew translation or access. These stowage packs will be fully depleted by about 42 days into each 90-day mission cycle.

3.7 Proposed Station Exterior Trash Holding Facility

If 1 trash-sack is generated every 22 days, total number produced in 1 year = 365/22 = 17. These can be configured in a modular grid 6 increments long x 3 increments high if they are stored only 1 increment (0.91 m.) deep. Therefore, total storage surface area required for 1 year is nominally 6.4 m. long (1.07 m. sack dia. x 6) x 3.2 m. high (1.07 m. sack length x 3). This can be housed within two standard truss bays with sufficient allowance for dimensional tolerances and misalignments. Alternatively, if the trash-sacks are stored 2 increments (1.07 m. x 2 = 2.14 m.) deep, all 17 can be stored across a single truss bay, or inside it. The holding facility will be located close to the pressurized module cluster and fully accessible by a truss-mounted RMS and FTS. The effect of the holding facility mass upon overall Space Station mass distribution is an issue that needs to be addressed. The RMS/FTS will be used to:

- Retrieve the sealed individual trash-sacks from crewmembers who remove them through an airlock on EVA every 22 days.
- Insert them in position in the interim holding facility on the exterior truss structure, using a specially-devised retention method.
- Retrieve all stored trash-sacks every 365 days and load them onto the OMV using a specially-devised payload attachment.

The interim holding facility will be designed as a lightweight tensile structure stretched over or between the truss structural nodes and longerons and attached at appropriate points. Retention method will be similar to the microfastener pad system proposed for the ET exterior surface with the possible addition of safety tethers or net shrouds as a precaution against accidental separation or jettisoning.

3.8 Proposed External Tank Trash Depository Capacity

The number of trash-sacks generated by Space Station for various durations is given in Table 3 as approximate numbers for two values of trash compaction ratio.

TABLE 3	Total for 90 days	Total for 1 year	Total for 10 years	Total for 30 years
Number of trash sacks @ 2:1 compaction ratio	4	16	160	480
Number of trash sacks @ 4:1 compaction ratio	2	8	80	240

If trash-sacks are stored on the outer surface of the ET, they will each occupy a surface area of 0.93 m^2. Each deposited trash-sack will require a substantial free area around its perimeter to ensure safety and simplicity of emplacement and removal. This would be defined as a minimum operational distance from adjacent trash-sacks. For preliminary estimating purposes, a factor of 2.5 is used as a multiplier. Thus, allocated surface area per trash-sack = 0.93 x 2.5 = 2.32 m^2.

Usable exterior surface area of ET for long-term trash depository is provisionally defined as follows: The longitudinal dimension commences at the LO_2 tank true cylinder forward circumferential line, continues across the intertank skirt and along the LH_2 tank, and terminates at a circumferential line defined by the forward end of the twin aft thrust struts. This dimension is approximately 28.65 m.

The circumferential dimension is represented by the outer circumference of the ET along the main length of the cylinder with a deduction for a segment obstructed by the recycling equipment platform which would be attached to the ET/Orbiter interface struts. This dimension is approximately 22.86 m. Approximate usable surface area = 28.65 x 22.86 = 655 m^2.

The maximum depository capacity of the ET is found by dividing the usable surface area by the allocated trash-pack area =

$$\frac{635}{2.32} = 282 \text{ trash-sacks.}$$

This quantity of trash-sacks would be produced by the Space Station after approximately 18 years of operation at a 2:1 compaction ratio, or 36 years of operation at a 4:1 trash compaction ratio.

4 REFERENCES

National Aeronautics and Space Administration: Man-Systems Integration Standards; Vol. I, March 1987; Vol. III, January 1988; Vol. IV, December 1986, National Aeronautics and Space Administration, NASA-STD-3000, March 1987.

Taylor, T.; Hodge, J.; and Good, W.: The Outpost Concept - A Transportation and Service Platform in Low Earth Orbit, Global Outpost Inc., Symposium on Space Commercialization: Roles of Developing Countries, Nashville, TN, USA, March 1989.

Mellors, W.; Taylor, T.; and Wilson, J.: The Global Outpost Program - Commercial Services in Low-Earth Orbit, Global Outpost Inc., 40th Congress of the International Astronautical Federation, IAF/IAA-89-711, Malaga, Spain, October 1989.

MALEO: MODULAR ASSEMBLY IN LOW EARTH ORBIT: AN ALTERNATE STRATEGY FOR LUNAR BASE DEPLOYMENT.

Madhu Thangavelu *

Abstract

Modular Assembly in Low Earth Orbit (MALEO) is a new strategy for building an initial operational capability lunar habitation base. In this strategy, the lunar base components are brought up to Low Earth Orbit by the Space Transportation System (STS)/Heavy Lift Launch Vehicle (HLLV) fleet, and assembled there to form the complete lunar base. Modular propulsion systems which are also assembled from components brought up by the STS/HLLV, are then used to transport the MALEO lunar base, complete and intact, all the way to the moon. Upon touchdown on the lunar surface, the MALEO lunar habitation base is operational. The strategy is unlike conventional concepts which have suggested that the components of the lunar base be launched separately from the Earth and landed one at a time on the moon, where they are assembled by robots and astronauts in extravehicular activity. A typical MALEO scenario is described. Major operational advantages of this strategy over conventional ones are discussed. The the need for further study in structures, modular propulsion, guidance, navigation and control of such a large spacecraft are pointed out. It is concluded that the MALEO strategy holds promise for international collaboration, and for a rapid lunar base deployment about the year 2000.

Introduction

In reaffirming our commitment to lead humanity in manned space exploration, on the 20th anniversary of the Apollo 11 mission, President Bush charged NASA with developing plans for a permanently manned lunar base, and then extending the human domain on to Mars(1). NASA had been working on lunar return concepts all along(2,3,4) and the recent reports from the Office of Exploration suggest several case studies for both manned lunar and planetary missions under consideration(5,6,7). Assembly in Low Earth Orbit has been suggested for the missions to Phobos and Mars(8).

A phase 1 lunar habitation base (LHB-1) is conceived as the first permanently manned facility on the lunar surface. It will provide

* Institute of Aerospace System Architecture and Technology
University of Southern California, Los Angeles,CA 90089-1191

a test bed for extended habitation, exploration, and scientific investigation. LHB-1 will form the nucleus, if needed, for further expansion and experimentation, during the evolution of what would become the first fully self-sustained lunar colony. This evolved lunar colony will then support spacecraft operations in cislunar and interplanetary space, by providing propellants and other material manufactured on the lunar surface(9).

The first priority in a primary phase 1 extended duration mission of this nature, is the provision of an assured safe haven for the astronaut crew, which will alleviate astronaut anxiety associated with build-up operations, followed by a comfortable environment within the facility, which will enhance crew productivity.

Development of the MALEO Strategy

Conventional lunar surface assembly strategies suggest launching space station derived modules from the Earth, landing them one at a time on the lunar surface, and assembling them there, using robots and astronauts, in extra-vehicular activity(EVA)(10,11,12). Extensive precursor missions are employed to land crew and assembly equipment. If Earth based robotic teleoperation is employed for assembly operations, a rather complex cislunar satellite tele-communication network is required(13).(Figure 1)

At the 1988 inaugural summer session of the International Space University (ISU), held at MIT in Cambridge, Massachusetts, M.Thangavelu, a graduate student from the University of Southern California, proposed the idea of assembling the components of the lunar base in Low Earth Orbit. Using modular propulsion systems, the entire assembly could then be directly transported to the lunar surface, for immediate occupation by the astronaut crew. The cost and risk associated with primary manned extravehicular activity on the lunar surface, which had been proposed in several earlier concepts(13,14,15), are thereby circumvented. M.Thangavelu along with another ISU student, G.E. Dorrington from Cambridge University, presented the idea in a paper entitled" MALEO: Module Assembly in Low Earth Orbit. Strategy for Lunar Base Build-up" at the 39th Congress of the International Astronautical Federation, in October 1988, held in Bangalore, India(14).(Figure 2)

The architectural drivers for the MALEO concept are the following:
1. The need to provide an assured safe haven / comfortable working environment for the astronaut crew, who are on a primary extended duration mission, as safely and quickly as possible, minimizing the initially risky EVA on the lunar surface,
2. The maximum exploitation of the evolutionary benefits derived from the assembly and operation of space station Freedom.
3. Every attempt is made to design the MALEO systems using established or state-of-the-art technology, thereby minimizing the design, development, testing and engineering(DDTE) costs associated with realizing new systems.

The strategy is architected to make full use of the U.S. Space program infrastructure that would be in existence by the year 2000. The support elements of the MALEO strategy include the Space Transportation System (STS), Space Station Freedom (SSF-1), and the Heavy Lift Launch Vehicle (HLLV). A cryogenic propellant depot is also assumed to be operational in Low Earth Orbit (LEO).

Several ETO launches	Several ETO launches
Several component wise TLIs	Assembly in Low Earth orbit
Several " LOIs	One or two TLIs
Several "lunar landings	One or two LOIs
Assembly on lunar surface	One lunar landing
Substantial Precursor Missions	Minimum Precursor Missions

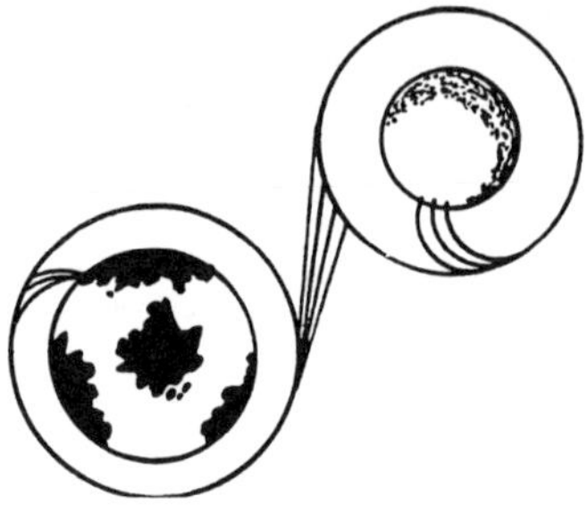

Figure 1.Lunar Surface Assembly

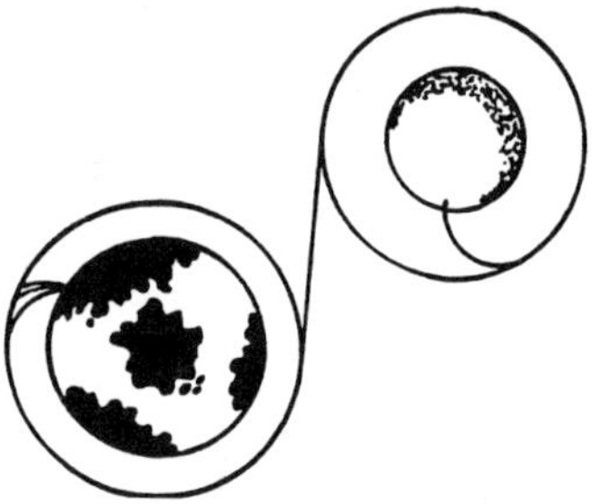

Figure 2.MALEO Strategy

ETO - Earth To Orbit, TLI - Translunar Injection, LOI - Lunar Orbit Insertion

<u>Configuration of the Lunar Habitation Base-1 (LHB-1)</u>

Several configurations are possible for the design of LHB-1. In the past, both horizontal and vertical configurations have been proposed for lunar habitation bases(10,12,15). In the MALEO strategy, the horizontal space station-like configuration was preferred to the vertical one for the following reasons :

1. Commonality with space station Freedom enhances design and engineering economy.
2. Commonality with space station design enhances crew adaptation and productivity.
3. Better work spaces for extended duration missions
4. Wider footprint for better landing stability.
5. Ease of expansion during evolution by attaching additional horizontal modules.

Issues of fire control and cabin contamination dictated a dual egress design. A three month crew rotation period enabled a design that did not require added regolith radiation protection since the nominal dosage during this period on the lunar surface is well within the NASA astronaut career limits(16). A separate solar storm shelter(S3) will have to be constructed and maintained for protecting the astronaut crew against a solar particle event.

<u>The components of the Lunar Habitation Base-1 (LHB-1)</u>

The major components of LHB-1 are the modules and the nodes. The three modules and three nodes which constitute the manned core of LHB-1 are described below.(Figure 3)

* The habitation module will house the four astronauts / mission specialists, and will contain crew sleeping quarters, a gymnasium / recreation facility, and a galley.
* The sanitation / hygiene node is conveniently located adjacent to the habitation module and besides serving those functions, will contain water recycling and waste management equipment, which are linked with the ECLS system of the LHB-1.
* The laboratory module is also adjacent to the sanitation / hygiene node. This module is equipped with interchangeable racks which are partially outfitted, so that modifications and new setups are possible, as the base evolves.
* The primary EVA node is the airlock which is used for all of the extra-base activity. It contains the EVA suits, and equipment to prevent regolith back tracking and contamination.
* The power and logistics module contains the power generation, storage, and regulation equipment. The module contains solar power panels which are deployed externally, can accomodate a nuclear power source that could be deployed externally.
* The ECLSS/Command Center/Expansion node is designed so that the base may be extended by attaching additional modules, if necessary. Delivered initially as logistics / consumables modules, when depleted, they would find other uses in the evolving base. The strategy is similar to the logistics module proposed for space station Freedom(17), except that the depleted modules are not brought back to Earth, but used for the expansion of the lunar base. The components which integrate the modules together as a complete spacecraft are the following :

1. The truss superstructure
2. The landing gear / airbag deployment system
3. The attitude control system pallet
4. External storage pockets for EVA equipment / solar arrays
5. Storage pockets for lunar rovers.

1. The truss superstructure has three functions. It supports the thrust structure of the modular orbital transfer vehicle (mOTV) and distributes the forces transmitted from the mOTV uniformly through the entire LHB-1 during TLI, LOI and lunar landing. It also offers attachment points for the attitude control system pallet(ACSP), the landing gear/airbags and storage pockets.
2. The landing gear / airbag deployment system is designed to absorb the shock on impact and may be conventional lunar excursion module type absorbers or controlled gas escape airbags or both.
3. The three attitude control system pallets(ACSP) are assembled and fuelled on Earth, brought up to LEO by the STS, and attached to the three corners of the LHB-1, after the truss superstructure has been built around the modules. They stabilize the attitude of MALEO LHB-1 during transit and landing operations.
4 & 5. The storage racks are placed symmetrically about the MALEO truss superstructure in order to maintain the thrust structure symmetry. They contain the rovers, the solar panels, and other EVA equipment, which are essential for the effective IOC of LHB-1.

<u>The Modular Orbital Transfer Vehicle (mOTV)</u>

The mOTV is used to transport the MALEO lunar base, complete and intact, to the prescribed lunar parking orbit. The mOTV provides the required delta V for TLI, LOI, and midcourse impulse and correction manuevers. Two current or near term options are available for the mOTV propulsion system. They are:

1. Chemical cryogenic propulsion
2. Electric propulsion

1. In the first option, LH2 / LO2 engines are employed to make up the mOTV cluster. Engine redundancy and gimballing are employed for handling engine-out emergencies. Advanced space shuttle main engines (ASSMEs) may be used as the propulsion system for the mOTVs and the advanced space engine (ASE) is also a candidate for this purpose. The maximum accelerations occur at TLI and is estimated at about 2gs using existing SSMEs. If the lower limit of trottleabiltiy could be extended for the ASSME, lower thrusting acceleration would minimize the stresses induced on the MALEO LHB-1 during TLI. If in-space restart capability is incorporated in the ASSME, the same set of engines could be employed for TLI as well as LOI.(and landing ?)

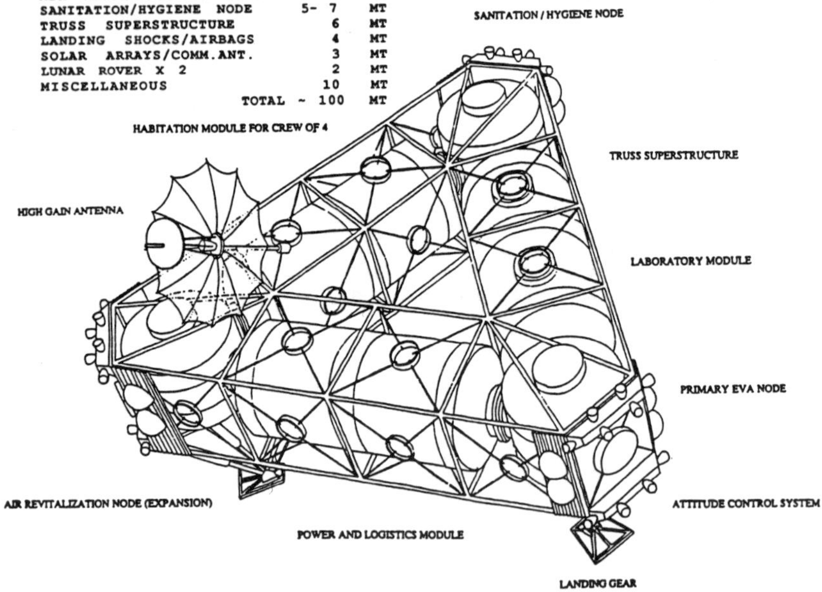

MALEO LHB-1 MASS SUMMARY

HABITATION MODULE	15-17.5	MT
LABORATORY MODULE	15-17.5	MT
POWER/LOGISTICS MODULE	15-17.5	MT
PRIMARY EVA NODE	5- 7	MT
AIR REVITALIZATION NODE	5- 7	MT
SANITATION/HYGIENE NODE	5- 7	MT
TRUSS SUPERSTRUCTURE	6	MT
LANDING SHOCKS/AIRBAGS	4	MT
SOLAR ARRAYS/COMM.ANT.	3	MT
LUNAR ROVER X 2	2	MT
MISCELLANEOUS	10	MT
TOTAL	~ 100	MT

Figure 3. Components of a 3 Module MALEO Lunar Habitation Base and Mass Summary

2. In this option, the MALEO LHB-1 is gradually spiralled out of LEO all the way to a suitable lunar parking orbit using the low but continous thrust of the ion engines. This option drastically reduces the total mass of the assembly (30-50% lower in mass than the conventional chemical mOTVs)(6) as well as the thrusting forces encountered by LHB-1 during TLI and LOI and may be employed if rapid deployment of the order of days is not a priority. The recent advances in SP-100 nuclear space reactor power system using an ammonia arcjet system (SRPS) holds promise not only for mOTV

applications, but also for power generation at the base, after touchdown(18).

The Lunar Landing System (LLS)

The LLS is used to de-orbit and softland the MALEO LHB-1 on the lunar surface. Chemical cryogenic propulsion is favored for the descent and landing manuever. Advanced RL-10 rocket engine technology would be applicable for the development of the LLS. The LLS is assembled in LEO from components brought up by the STS / HLLV fleet and integrated with the MALEO LHB-1 and the whole assembly is transported to the moon by the mOTV cluster.

The MALEO assembly and deployment of LHB-1

At least four options exist for the MALEO assembly of LHB-1. They are :

1. Free Space Assembly using the STS as the assembly platform.
2. Free Space Assembly using STS as the primary assembly platform with assistance from Freedom and her crew.
3. MALEO assembly attached to space station Freedom.(SSF-1)
4. MALEO assembly connected to the manned core of Freedom.

The fourth option utilizes the space station infrastructure most effectively by providing the following benefits:

a. The inertia and structure of Freedom provides the stable platform and scaffolding required for MALEO assembly of LHB-1.

b. Option 1 and 2 require a separate reaction control system (RCS) to maintain orbit and orientation of the MALEO LHB-1 assembly configuration during the several Earth to orbit(ETO) sorties by the STS, needed to bring up all the components for LHB-1 assembly. Option 3 and 4 utilize the SSF-1 RCS for this purpose.

c. Once connected with the manned core of Freedom, much of the EVA associated with outfitting the LHB-1 modules might be carried out as IVA, in a shirt sleeve environment.

d. The manned core of the LHB-1 would offer added utility and space for Freedom and her crew, until it departs for the moon.

Two options are possible for the deployment of MALEO LHB-1.

1. Single phase direct lunar landing option
2. Two phase Lunar Orbit Rendezvous (LOR) option

1. In the simplest and quickest direct lunar landing option, the LHB-1 is first assembled at Freedom. The lunar landing system(LLS) is then integrated with the LHB-1. Finally the mOTV units are clustered and docked with the MALEO LHB-1 + LLS spacecraft. After topping off the mOTV and the LLS, TLI is effected. After lunar orbit insertion(LOI), the mOTV is jettisoned. The LLS propulsion system is then employed to de-orbit, descend, and softland the MALEO LHB-1 on the lunar surface. The entire operation could be manned or unmanned. This option is particularly suited for cryogenic propulsion orbital transfer as the entire mission from TLI to lunar landing could be effected in a few days. Boil-off may not be a concern in this option.

2. The two phase option is as follows. In phase A, the MALEO LHB-1 is transported from LEO to the lunar parking orbit (LPO) by an mOTV. In phase B, the lunar landing system is transported by a similar mOTV and the LHB-1 docks with the LLS in Lunar Orbit

Rendezvous(LOR). The LHB-1 + LLS then descend and touchdown on the lunar surface as in the first option. The two phase option is architected in order to reduce the TLI mass of the MALEO operation which will facilitate easier propellant management in LEO. And this it does effectively, but at the cost and risk of EVA associated with the LOR operations. This option is suited for the slow nuclear electric orbital transfer.

The following sequence of illustrations (figures 4 - 6) help to visualize the MALEO strategy single phase direct lunar landing option for the deployment of the lunar habitation base LHB-1.

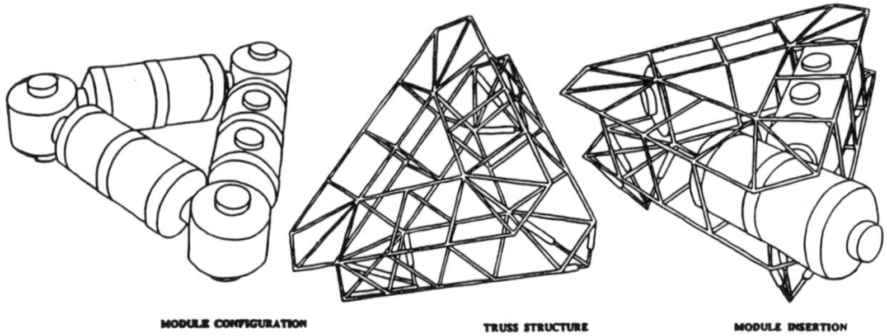

Figure 4. MALEO Assembly of LHB-1

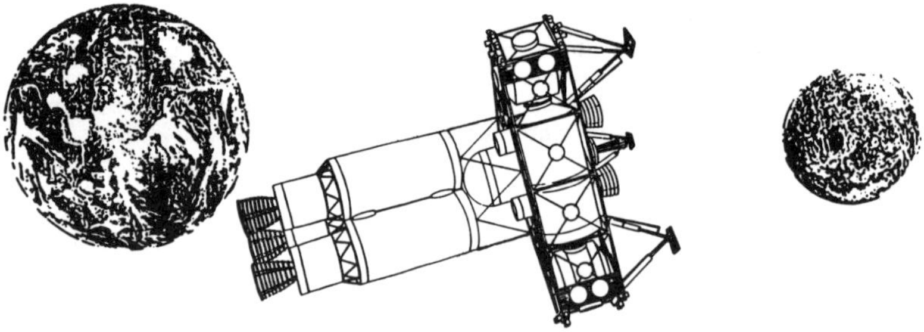

Figure 5. Cislunar transportation of MALEO LHB-1

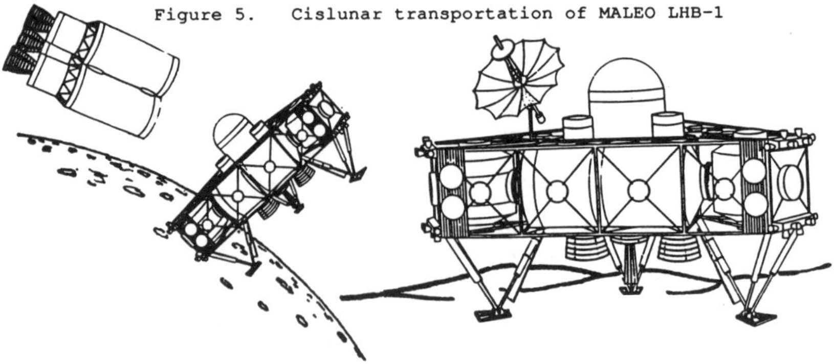

Figure 6. MALEO De-orbit, descent and touchdown

Advantages of the MALEO Strategy

If space station derived modules are to be employed in the construction of a phase 1 lunar base, then the MALEO strategy offers the following advantages:

1. The safer LEO radiation environment for EVA, is also less risky and more economical than EVA on the lunar surface(19).
2. Spares and replacements are easily and more economically flown to LEO or borrowed from the LEO SSF-1 infrastructure.
3. The clean LEO assembly environment avoids dealing with the lunar soil which has undesirable abrasive and cohesive properties. Regolith has interfered with manned EVA systems in the past(20). During phase 1 build-up, paucity of man and equipment demands high reliability, and MALEO avoids the initial lunar surface assembly operations entirely.
4. In the event of an assembly accident, a crew rescue is more feasible from LEO using the assured crew return capability(ACRC) of SSF-1, than from the lunar surface.
5. The inheritability of the module assembly and truss erection techniques acquired while building space station Freedom.
6. LEO offers SSF-1 assisted assembly and the possibility of Earth-based real-time telerobotic assembly operations.
7. Commonality with Space Station Freedom hardware enhances cost / unit economy as well as spares and replacement units for both the MALEO as well as the space station program.
8. Avoids the risk of the repeated Earth to Orbit (ETO) launchings, TLIs, LOIs and lunar landings associated with the typical componentwise sequential build-up prescribed by the assembly on lunar surface strategy.
9. The clustered engines which form the modular propulsion system in the MALEO strategy would be more reliable than small single engines suggested for component-wise launching and landing maneuvers.
10. Reduces precursor missions, equipment and associated lunar surface activity.
11. The tight power constraints encountered in the early phases of lunar surface base development is avoided (21).
12. The LHB-1 is safely configured for habitation upon touchdown. Conventional strategies involving lunar surface assembly cannot offer a comparably safe environment during base build-up without additional investment.
13. It may be possible to develop a MALEO based assured crew return capability(ACRC) system for the entire crew of LHB-1.
14. The MALEO strategy might be adopted for landing outposts on Phobos, or for lunar and Mars orbiting stations.
15. In this primary extended duration mission, the MALEO LHB-1 offers an assured and substantial safe haven at IOC which will help to diminish anxiety and enhance productivity among the astronaut crew.

The Challenges

The risk of losing the entire LHB-1 in the event of an accident during the transportation and landing of the base is a disadvantage of the MALEO strategy but should be evaluated carefully against the repeated component-wise Earth to Orbit (ETO) launches, TLIs, LOIs and lunar surface landings associated with the conventional strategies in which the operations are arranged more or less sequentially, and hence susceptible to delay and operational collapse in the event of a failed component launch or landing.

The MALEO strategy suggests a prefabricated phase 1 lunar base which deprives the assembly crew of the initial experience of learning to work on the lunar surface(22). It is important to realize that a substantial safe haven is the first priority for the astronaut crew who are on an extended mission, establishing a permanent base on the moon. Lunar surface EVA experience will be gained as exploration and experimentation continue and the assured safety of the MALEO base would also help to relieve astronaut crew anxiety while the base is operating in the IOC mode.

MALEO is a large manned spacecraft. All the MALEO systems, the LHB-1, the LLS, and the mOTV needs to be studied, their dynamic characteristics examined, and their limitiations confirmed. Though the TLI and LOI can be effectively controlled using an electric propulsion mOTV for the orbital transfer manuever, the vibration and resonance characteristics of such a large structure during lunar descent and a heavy impact or unsymmetric landing are being studied and will require detailed analysis.

Conclusion

One scenario for Modular Assembly in Low Earth Orbit has been discussed in this paper. The systems described herein need to be designed and built for assembly in space. The MALEO strategy minimizes the initial risky manned EVA associated with lunar surface assembly operations and provides a safely configured working environment for the astronaut crew upon touchdown. The LHB-1, thus deployed, will alleviate crew anxiety and enhance productivity. The MALEO strategy maximizes the commonality and inheritability from the LEO infrastructure that will be in operation by the year 2000. The experience gained while assembling space station Freedom will be invaluable for the MALEO assembly and deployment of LHB-1. More study is required to confirm the feasibility of clustering existing propulsion systems in space to build the modular orbital transfer vehicles. High power electric propulsion like the SRPS ammonia arcjet offers promise for a nuclear electric mOTV for use in the MALEO deployment of LHB-1.

MALEO LHB-1, is in essence, a large manned spacecraft and could be used as orbiting stations for the moon and Mars. The lack of aerodynamic contours and for reasons of stability, control, and propulsion requirements, the concept is particularly suited to landing on low gravity bodies without an atmosphere, and the Moon is an ideal first choice. MALEO bases could land on Phobos or could be used for prospecting the near-Earth Asteroids in the future.

The importance of highly reliable structural systems for the truss superstructure of the LHB-1, the LLS, and the mOTV, are clearly evident for the success of such a mission. In the event that a decision is made for the deployment of a national or an international lunar base about the year 2000, using space station-like modules, the MALEO strategy holds promise and needs detailed engineering analyses along with other strategies.

Acknowledgements

This paper is dedicated to my mother, Nanoo Saraswathy Thangavelu. I would like to thank P. Diamandis, T. Hawley, G.E. Dorrington, Dr.R.Shaefer and the whole crew at ISU for helping me to carry through the idea. I would like to thank Profs. R. Harris, G.G. Schierle, R.F. Brodsky, D. Vergun, E. Rechtin , R.K. Miller, E.P. Muntz, all of whom helped me to develop a framework in which to carry out a detailed study of the MALEO concept at the University of Southern California.

References

1. Bush, George. Presidential Address on the 20th Anniversary of the Apollo 11 Moon Landing. The White House 20 July 1989
2. Alred, J.W. et al. Lunar Outpost. Systems Definition Branch, Advanced Programs Office NASA JSC 1989.
3. Alred, Petro et al.,Development of a Lunar Outpost-Year 2000-2005 AIAA 2nd Symposium on Lunar Bases 1988.
4. Alred, J.W.& Bufkin A.L., Lunar Surface Construction & Assembly Equipment. Eagle Engineering Report NAS9-17878.
5. Office of Exploration.Exploration Studies Technical Report FY 1988 Status.Volume 1 & 2. NASA Technical Memorandum 4075
6. Office Of Exploration NASA,Beyond Earth'sBoundaries Nov1988
7. Lineberry E.,Lunar Outpost to Early Mars EvolutionO'EXP 4075
8. Bell J. et al., Low Earth Orbit Assembly Strategy MASE JSC O'EXP 4075 1988.
9. National Commission on Space.Pioneering the Space Frontier. Bantam Books 1986.
10. Duke, Wendell, Roberts. Strategies for a Permanent Lunar Base.Lunar Bases and Activities for the 21st Century.LPI1985
11. Duke, M.B. and Alred J.W., A Lunar Base Scenario Emphasizing Early Self Sufficiency, LBS-88-241, Symposium on Lunar Bases, 1988
12. Hoffman, Stephen J. and Neihoff, John C., Preliminary Design of A Permanently Manned Lunar Surface Research Base. Lunar Bases and Activities of the 21st Century. LPI 1985
13. Iwata, T. Unmanned Surface Development for Manned Surface Activities. LBS 88-7232. Symposium on Lunar Bases, 1988.
14. Thangavelu, Madhu and Dorrington Graham E., MALEO:Strategy for Lunar Base Build-Up. ST-88-15. IAF 1988.
15. Johnson, Stewart W. and Leonard, Ray S. Evolution of Concepts for Lunar Bases. Lunar Bases Conference LPI 1985
16. Kline R., McCaffrey R., Stein D. Potential Designs of Space Stations and Platforms. AIAA Vol.99 Ed. Bekey & Herman 1985
17. Natchwey, Stuart., Radiation Protection on the Lunar Surface.International Space University(ISU)Lecture July1988.
18. Deininger, William D.et al, Arcjet Propulsion System for an SP-100 Flight Experiment NASA JPL NASA Recon. 89A24247*#
19. Bufkin A.L. et al., EVA Concerns for a Future Lunar Base LBS-88-241, Lunar Base Symposium 1988.
20. Weaver, Leon B. and Laursen Eric F., Techniques for the Utilization of Extraterrestrial Resources IAF 1988.
21. Fordyce, J.Stuart., Space Power, ISU Lecture 20 July 1988.
22. Bekey Ivan, Correspondence with Rechtin,Eberhardt,University of Southern California, March 1988.

Automated Manufacturing in Platform-Based Orbital Systems (APROPOS)

Dipl.-Ing. Elmar Degenhart[1]
Dr.-Ing. Jürgen Geissinger[2]
Dipl.-Ing. Arwed H. Exner[3]

Abstract

As an example of an orbital manufacturing scenario, an automated orbital manufacturing system is designed for the production of turbine blades for high pressure turbines. The target here is to verify the technical feasibility of the application of automation and robotics (A & R) based on a highly standardized design suitable for automation as well as to develop a method for the design of orbital manufacturing systems.

By means of relevant constructional measures and standardization steps, an automation and maintenance-oriented design of the mission scenario can be achieved. Efforts are made to reach the highest possible degree of automation with minimum complexity of the orbital system. At the same time, the system is to possess high task flexibility.

Based on the elaborated results fundamental rules and principles for the design of automation manufacturing facilities in space are developed /1/.

Introduction

Automation and robotics are key technologies and will become an important element in future space activities /2, 3, 4, 5/. A sound concept for the automation of space systems, however, cannot be derived from preceding developments. Investigations on the subject A & R concentrate on the robotics portion.

----

[1] Head of group "Space Automation", Fraunhofer Institute for Manufacturing Engineering and Automation, Nobelstr. 12, 7000 Stuttgart 80, West Germany
[2] Head of department, Fraunhofer Institute for Manufacturing Engineering and Automation
[3] German Space Agency DARA, Department for Space Technology Development, 5300 Bonn 2, West Germany

In preceding investigations, robot systems were designed for tasks in known system environments which until then had been carried out manually. These robot systems are structured rather anthropoidly, but in spite of all development efforts so far can not replace the human capabilities. Up to now all solutions combined a very low degree of automation with elementary involvement of humans in the execution of tasks.

In automation attempts of the last few years (with ESA-projects in relation with the COLUMBUS-program) emphasis has been put on automation suitable design principles mainly in relation with the automated carrying out of experiments.

As far as the layout of orbital manufacturing systems is concerned, few approaches have been made because it has not been verified with the help of experiments yet which manufacturing scenarios are economical.

There can be no doubt, however, about the increasing significance of the commercial use of space in the future /6/. General Electric /7/ has made first conceptional investigations regarding the layout of an entirely automatical orbital manufacturing system for the production of VLSI-chips.

Procedure

After having completed a functional analysis of the processing system, material flow system, maintenance system, control and ground/orbit transport system in question, as well as of the construction of the manufacturing system, specifications profiles are construed and corresponding concepts for the automation-oriented system design are compiled. This requires a consideration of the overall system in order to incorporate all principal system interactions and their interfaces in the design.

For designing the production system a method is applied succesfully that derived from terrestrial planning procedures for complex automated systems. This has two advantages: the range of solution steps is not restricted from the start by limitations of technical equipment and a reliable manual is available for the technical equipment design which makes an objective-oriented system design according to prescribed criteria possible. The result of the automation-oriented planning does not necessarily lead to rigid automation structures (Hard

Automation). Rather, the objective of the system is to work out flexible solutions (Soft Automation) which can accomplish a prescribed variety of production tasks with no changes required to the system's basic concept.

Description of the Investigation Scenario in Relation to the Product

The stressability of gas turbine blades manufactured under terrestrial conditions can actually not be further increased. The mechanical and thermic features of turbine blades applied in high pressure turbines, however, should be improved because by this means the turbine's efficiency and usable life would be increased. It is expected that the features mentioned above can be significantly improved with the help of a thermic process carried out in space. The process mainly includes the controlled melting and solidification of turbine blade blanks which have already been prefabricated on earth.

The procedure applied here is called skin technology /8, 9/. This technology allows the controlled solidification of an eutectical high temperature alloy conserving the original shape of the blank in weightlessness. Figure 1 shows the 4 basic steps in the manufacturing of turbine blades.

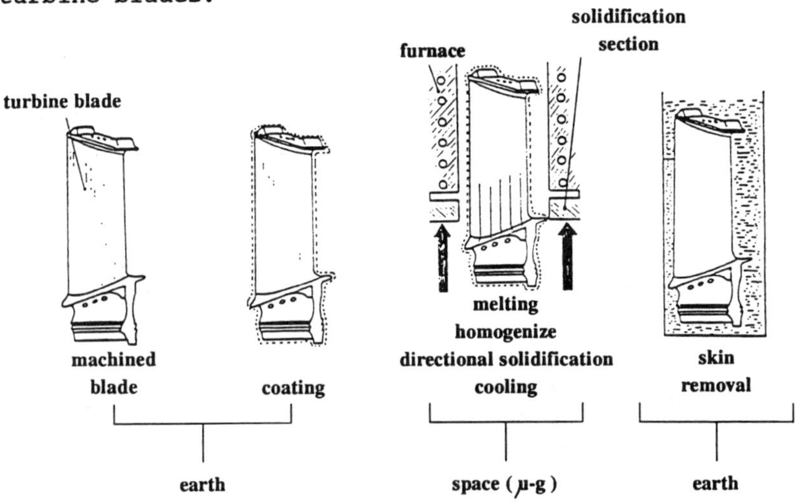

Figure 1. Steps in the manufacturing of turbine blades with the help of skin technology

The blanks on the platform have to be taken off the blank stock and put into the furnace where they are processed and afterwards brought from the furnace into the stock for processed blades. The isothermal stock melting is done at a temperature of approx. 1,600° C. After the melting the blades are homogenized for about half an hour. Therefore micro-gravitation conditions of approx. 10^{-6} g are required. The next step is the actual procedure of directional solidification of the melt down to a temperature of 800° C and the further cooling of the blades in separate cooling units.

Structure of the Orbital Manufacturing System

The structure of the orbital manufacturing system to be designed is shown in a reference model (see figure 2).

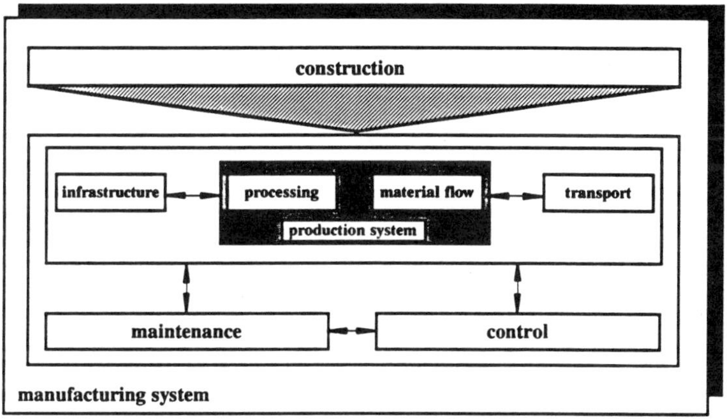

Figure 2. Reference model for the break down of the orbital manufacturing system into main systems

The selection of adequate concepts for the individual main system is based on the following systematics.

(1) Working out of alternative system concepts considering the respective requirements
(2) Drawing up of selection criteria and evaluation of alternative solutions
(3) "Trade Off" for solution comparison
(4) Selection of adequate solution alternatives
(5) Integration of the solution into the whole concept of the orbital manufacturing system

Design of the Manufacturing

Considering the technological aspects of production and space, the following design options are chosen with respect to the manufacturing system design.

Construction Principle:

Considering technological aspects of production and especially of space, small platforms are better suited for automated manufacturing in space than large ones.

In order to prevent orbital operational efforts and the application of handling devices for the construction, the manufacturing platform is brought into the orbit as a whole unit in one single launch. The infrastructure system is spatially separated from the production system, but can not be undocked!

The functional break down of the orbital manufacturing system with its main systems is shown in figure 3.

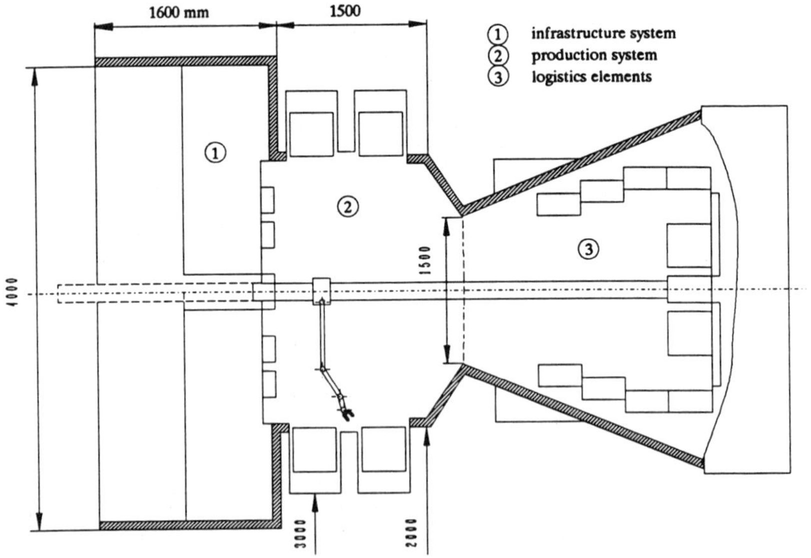

Figure 3. Functional break down of the orbital manufacturing system

Mating the platform with the logistics element in orbit is carried out by Rendezvous and Docking. The logistics element contains the active gadgets. It remains docked to the production system during the processing phase. As the logistics element is used as a stock during manufacturing time, no additional stocks for blanks and processed turbine blades on the platform are needed.

Arrangement:

Since the basic structure for the arrangement of the production and infrastructure systems consists of a central tube or of an object which is spatially closed and limited by flat lattice structure units, the modules for the material flow system are made accessible from inside. Thus especially the complexity of the handling and transport system is greatly reduced.

Subsystems of the infrastructure and production system which have limited lifetime and are prone to failure are arranged in such a way that they are accessible for both the platforms' maintenance system and Extra Vehicular Activities (EVA).

The process units of the production system are arranged in the area of the centroidal position of both the overall system and the logistics element which is passively docked during the manufacturing process.

Contamination of the production system and disturbance of the micro-g environment caused by the firing of the altitude and orbit control system is prevented by the arrangement of the nozzles on the outer skin of the cylindrical infrastructure system.

Construction:

An open and non-pressurized construction is preferred to the closed construction of a pressurized module because of its lower system complexity.

A central tube is chosen as a carrying structure. It allows the shielding of sensitive components against space radiation, contamination and meteorite debris and has advantages concerning the layout of the material flow system. If pressure discharge of individual subsystems is necessary, these subsystems are encapsulated.

Figure 4 shows the orbital manufacturing system in a 3-dimensional depiction.

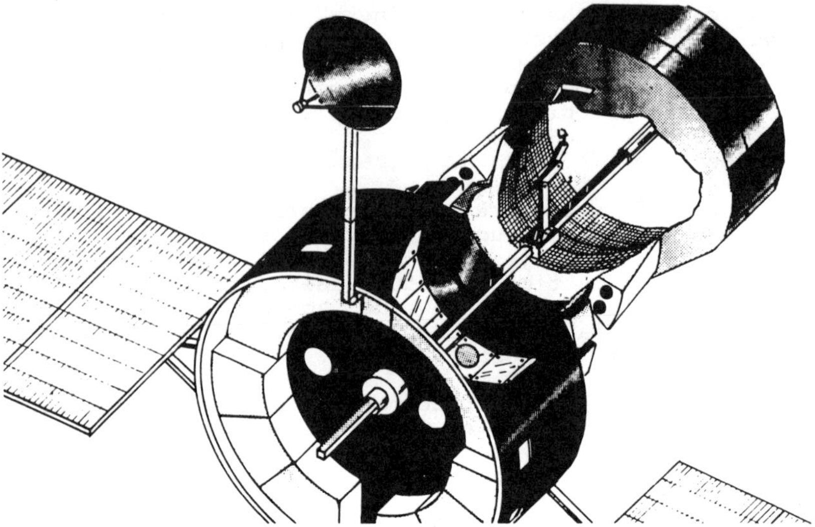

Figure 4. 3-dimensional depiction of the orbital manufacturing system

Extension:

In order to be flexible regarding new products and the required increase of production rates, the extension of the orbital manufacturing system is to be carried out by realizing further mini-platforms.

Servicing concept:

The servicing concept is a combination of a redundant layout and the exchange of defective subsystems. Overhaul will always mean an exchange. Repairs will not be performed!

Mission Strategy:

An earth based mission support for the manufacturing system is strived for in order to minimize interdependencies with other orbital systems.

<u>Processing System</u>

The starting point is the assumption that 2,000 turbine blades per 90 day mission are manufactured. This production rate is based on the actual need of turbine blades for the high pressure turbines of Airbus turbine engines.

On these suppositions six furnace units are planned in each of which one turbine blade is processed. Each unit is equipped with an independent cooling facility as well as with drives for the heating blocks. For the insertion of turbine blades and thermical elements, suitable insertion aids are planned at the front of each furnace unit. Each furnace unit is equipped with standardized gripping elements for the exchange of units. Five furnace units are always in operation, whereas the sixth unit is for stand by as backup. The process, stock and control modules have a square cross section of 400 x 400 mm. The length of the modules is 200 mm, 400 mm or 600 mm. In addition to the furnace units, 4 cooling units with 3 cooling places each and 2 calibration units for the calibration of the thermo couples are required as process modules.

Based on the choosen structure for the overall system (central tube) the components are arranged on two rings in the area of the production system.

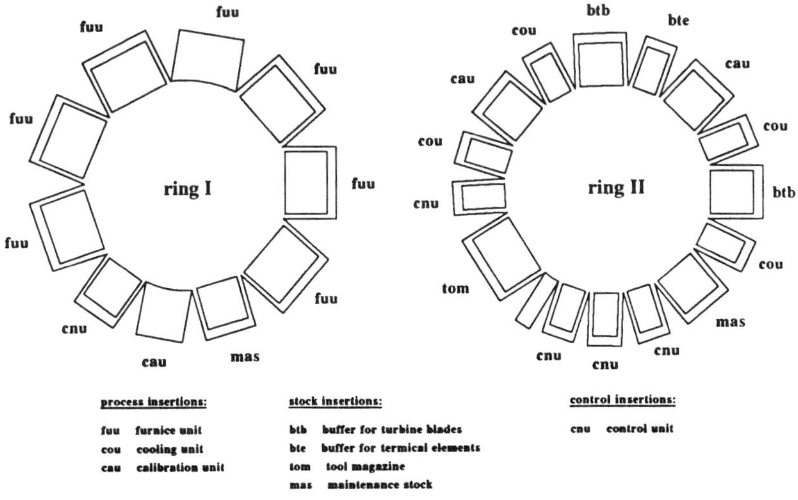

Figure 5. Arrangement of the insertion modules on both rings

In order to have enough space for handling operations the internal diameter of the rings is fixed on 2,000 mm. From this ensues the arrangement of the insertions on both rings as shown in figure 5.

During a production time of 85 days a handling phase and a process phase will run alternately. Loading and unloading of the furnace and cooling units with turbine blades and the exchange of the thermo couples are part of the handling phase. The next step is the process phase during which the thermo couples are also calibrated.

Material Flow System

The material flow system is a central system for the orbital manufacturing system. Besides the actual handling tasks for the maintenance of the manufacturing processes, the material flow system should carry out, if possible, additional tasks, e. g. in the area of maintenance. Considering the correlation with other main systems the material flow concept of a handling system traversable on a linear axis is choosen. This concept has the lowest number of necessary subsystems for the material flow tasks arising within the scenario in view. If a robot is used as a central handling and transport system it can carry out all handling tasks arising in the manufacturing system as well as maintenance tasks.

Figure 3 shows the layout of the material flow system.

The handling system is able to traverse with its slide on the linear axis, which means that it can move within the area of the manufacturing system as well as in the area of the logistics element which serves as the main storage system. After docking the logistics element the linear axis is inserted where it remains during the whole manufacturing phase. To be able to carry out all handling tasks the handling system must have 8 axes.

Altogether three exchangeable gripper types are intended for the handling of the turbine blades and the standardized modules as well as for the exchange of small subsystems. Figure 6 shows the kinematics of the handling system.

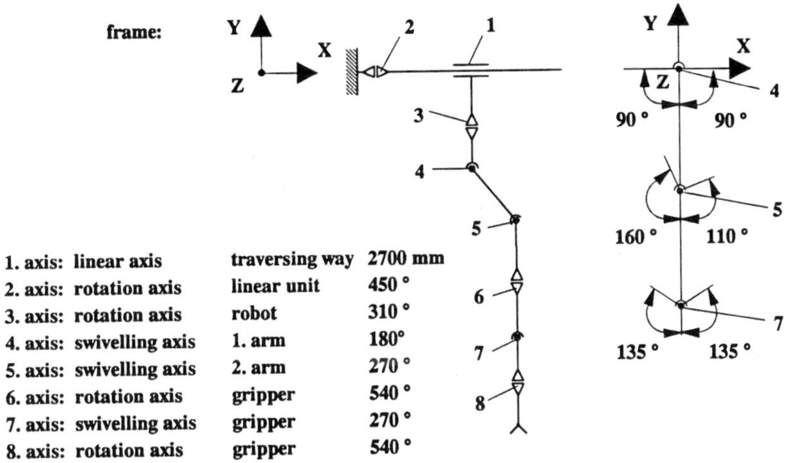

Figure 6. Kinematics of the handling system

Earth/Orbit Transport System

For the supply of the manufacturing platform an unmanned reusable capsule in APOLLO shape is applied. The capsule's docking adapter must have a minimum diameter. The capsule is designed in such a way that capsule subsystems (e. g. tanks) can be put in its front area. The retro-engines and the rescue equipment are placed in the capsule's aft-cone and are arranged outside around the docking port.

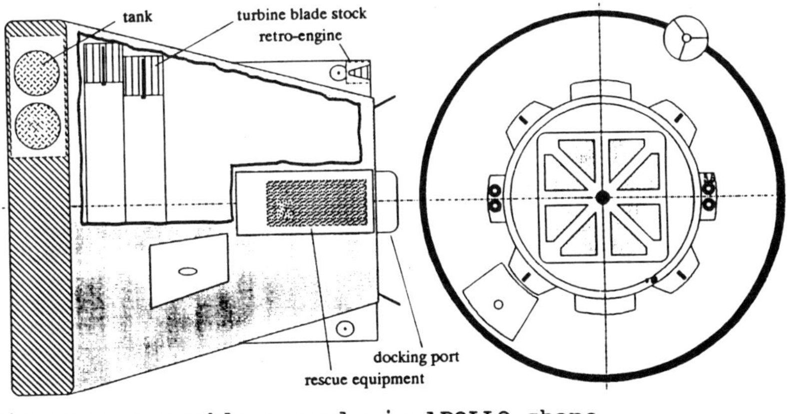

Figure 7. Reusable capsule in APOLLO shape

Maintenance System

The maintenance of the orbital manufacturing system is automated as far as possible in order to guarantee a high degree of availability. Manual overhaul is only done exceptionally, e. g. in case of defects in areas which are not accessible to the handling system. In order to minimize the necessary efforts for automated maintenance and to optimize the rate of utilization of the conceived flexible handling system, all necessary maintenance tasks are carried out by this system applying additional peripherical equipment. For that reason subsystems or components which have to be maintained are arranged in such a way that they are accessible to the handling system. All maintenance measures can be realized with just two grippers which have different sizes but operate identically. The smaller gripper is only applied in maintenance. The bigger gripper is the standard gripper applied in the handling of the different modules. All objects to be exchanged are provided with standardized gripping elements as well as with swiveling safety elements against unintentional changes of position. The real exchange which is carried out by the handling system is done by simple linear movements.

Control System

The manufacturing technological control system is divided into two parts: one on board and one on earth. The communication between the two spatially separated systems is effected by the mission specific control system.

The control system on board can be divided on the host level into a higher control system and on the component control level into a handling system control, process control and calibration control.

The control system on earth carries out all complex planning, programming, monitoring and diagnostic checking functions. In addition to that it provides all required facilities for monitoring and telemanipulation.

Execution planning contains the sequence planning of event oriented transactions, the movement planning for the central handling device and the generation of alternative control strategies for exceptions. At the end of planning the user programs are verified in a ground based test bed before they are transferred to the manufacturing control system and finally activated.

The system's autonomy which is indispensable with orbital manufacturing systems requires an efficient diagnostic checking system and failure management system for process monitoring, localization of faults and their removal. According to the layout philosophy drawn up in the requirements profile, fault diagnosis and failure removal on board are limited to parts which require real time in order to minimize the system's complexity on board. Wide-ranging fault diagnosis up to alternative planning of task execution is done on earth.

Guidelines for a Design Suitable for Automation

Some of the most important guidelines for the automation-suitable design of the orbital manufacturing system are listed below. The guidelines are based on the idea to reach a high degree of automation of the manufacturing process with minimal system complexity.

- Standardization of module dimensions
- Integration of mechanical interfaces and interfaces of the supply system in interface units
- Standardization of interface units
- Application of mechanical rapid locks
- Execution of handling functions, if possible, without sensor support
- Application of sensors only for control and inspection tasks
- Joining aids have to even out positioning and path faults of the handling device
- Minimizing of the material flow distances
- Application of standardized gripping elements
- Realization of linear movements of the handling system
- Usage of material flow systems for the execution of maintenance measures
- All diagnostics and failure management functions which do not require real time are executed on earth

Conclusion

In the described study a systematic automation concept for the application of A & R systems in orbital manufacturing devices is developed based on a representative manufacturing task and then verified by a real system layout.

In consideration of the limited transport capacity, the manufacturing platform is construed in the form of a production system with modular distribution possibilities and a functionally segregated infrastructure.

Earth/orbit transport here is effected by a logistic element, e. g. a capsule, which, in orbit, simultaneously serves as a storage system. The handling system takes the form of a central horizontal-track robot.

The consequent application of the above procedure for the automation of orbital manufacturing systems leads to the minimization of the system's complexity. This ensues a reduction of the construction and operation costs as well as an extensive reduction of the requirements for both the subsystems and the necessary intelligence of the production specific control systems on board. As a result, the operating reliability is increased because decision requirements depending on real time are restricted to a minimum. The basic technologies required for the realization of the system are already available in the manufacturing technology as well as in the field of space technology.

Acknowledgement

The study "Automated Manufacturing in Platform-Based Orbital Systems" was elaborated by the following institutions on behalf of the German Federal Ministry for Research and Technology:

Fraunhofer Institute IPK, Berlin
MBB-ERNO, Bremen
Fraunhofer Institute IPA, Stuttgart

References

/1/ Automated Manufacturing In Platform-Based Orbital Systems (APROPOS). Study for the German Federal Ministery of Research and Technology, March 1989

/2/ Automation Technologies for Space Travel. Support Concept of the German Federal Ministery of Research and Technology, March 1989

/3/ Chandler, P., et al.: Automation and Robotics in the Future German Space Program. German Aerospace Research Establishment DLR, November 1986

/4/ Advancing Automation and Robotics Technology for the Space Station and for the U.S. Economy. NASA TM-87 566, Volume I, II, III, March 1985

/5/ Report of Research Forum on Space Robotics and Automation. Japan Space Utilization Promotion Center, March 1989

/6/ Shahrokhi, F., et al.: Commercial Opportunities in Space. AIAA Inc., Washington DC, 1988

/7/ Hallett, H. R.; Kugath, D. A.: Automation Requirements Derived from Space Manufacturing. Proceedings of the Space Tech Conference, Anaheim, California, September 1985

/8/ Staub, F., et al.: An Alternative Single Crystal Manufacturing Process for Gas Turbine Blades. FOCUS Maschinen- und Turbinen-Union, Munich, 1/1989

/9/ Sprenger, H.: Skin Technology. MAN Technologie GmbH, München, August 1984

FORCE AND GEOMETRY CONSTRAINTS IN ROBOT EXCAVATION

Darcy M. Bullock, A.M.ASCE[1], Sudheer M. Apte[2], and
Irving J. Oppenheim,M.ASCE[3]

Abstract

Humans have the capability to match a limited work envelope to arbitrarily large purposeful excavation tasks, primarily by replanning when constraints are encountered in either the force or spatial domain. Machines for automated excavation must achieve a robust capability for completion of their excavation task subject to these same constraints. One interest in this research effort was to develop and test an automatic control approach to adjust machine motions to the force conditions encountered during excavation -- essentially constructing a robot which digs by *feel*. The problem is approached by *supervisory control*, consisting of discrete adjustments to the digging trajectory. A laboratory manipulator approximating a backhoe was configured with four actuated degrees of freedom, and an object-oriented programming environment was built to permit rapid prototyping of the supervisory control level. Strain data taken on the "boom" were used as the internal measure of forces developed during excavation. Simple rules for supervisory control were programmed and tested in laboratory studies of excavation in sand, and were effective in adjusting the digging actions to maintain forces within the target envelope. As a result the manipulator has a robust capability, effectively adjusting its "appetite" as different conditions (density, cohesion, friction, geometry) prevail. The object-oriented programming environment was also used to model spatial planning associated with excavation. The first approach applies to a Cartesian system and models the excavation volume (or its bounding box) as a rectangular prism, for decomposition into layers and further decomposition into unit cuboids forming an intended excavation plan. This is greatly facilitated by the programming environment, and is readily programmed as a dynamic (non-linear) planner/replanner. The work next addressed the mapping of an excavator in polar coordinates, typifying a backhoe at one fixed base position, to a prismatic excavation target.

---

[1]Intec Controls Corp., Walpole, MA, 02081; formerly graduate student, Department of Civil Engineering, Carnegie-Mellon University.

[2]Graduate student, Department of Civil Engineering, Carnegie-Mellon University.

[3]Associate Professor, Departments of Architecture and Civil Engineering, Carnegie-Mellon University, Pittsburgh, PA, 15213

Introduction

A man working with a spade possesses a work envelope exceedingly small in both the force and spatial domains, but he is capable of matching that limited envelope to arbitrarily large tasks. Force domain limitations include payload limits and overturning stability; spatial domain limitations include the unit size, the reach problem, and the coupling to the evolving site itself. Excavation is important in space as a unit process in sampling, mining, and construction. While actual systems for excavation in space will use engineering improvements, much can be gained studying the problem as phrased above. In accomplishing the excavation task the human demonstrates various planning and replanning capabilities at strategic and tactical levels of control, paraphrased here as processing constraints on force and geometry.

The act of digging appears simple only to the most casual observer. On closer inspection, it is extraordinarily complex and difficult. The subsurface conditions, which include soil density, cohesion, internal friction, inclusions and discontinuities, are clearly unknown, and most digging is accomplished without explicit knowledge or study of them. In fact, the influence of such conditions is generally evidenced only in the very act of irreversibly interfering with the medium while digging. A man digging with a spade adjusts his angle of attack, depth of penetration, and angle of removal in response to these varying characteristics; those adjustments are essentially done by feel. Moreover, the man with a spade also adjusts his digging mechanics to the force envelope within which his muscles, joints, and limbs can work. He generates a plan (at this tactical level) responding simultaneously to these different influences; as a result, he has extraordinary ability to match a very limited work envelope to arbitrarily large excavation tasks. Our laboratory manipulator is configured, like a human, with weak joints and limbs and with a need to maintain its own orientation and stability. In addition to acting on these force domain constraints, the man with a spade processes all the spatial implications of reach and stability. Moreover, at a higher level, he embraces the considerable range between an immediate (tactical) objective, which is the next spadeful of soil, and the overall (strategic) objective, which is the target geometry to be excavated. This study reports separate results on force-cognitive excavation and on spatial task planning, but their coupled nature should be emphasized.

On the first question of force domain constraints, it is instructive to compare the human (man and spade) excavator to a conventional terrestrial machine excavator. The machine is characterized by sufficient power and stiffness to perform a typical excavation action (one bucket in volume, common granular medium, normal reach) under position control with the geometric stability of the excavator deriving from system size and weight; such a matching of the work envelope to its (tactical level) task is implicit in the machine design. However, it would be mistaken to think the operation free from force constraint processing; the operator-in-the-loop performs the same type of tactical level replanning steps that he would if using a spade. The machine is fully reliant on its operator for stability in the force domain, and is equally reliant on its operator to choose its attack geometry, or it may find itself attempting to lift the mountainside rather than simply filling its bucket.

Our objective is to use cognitive system capabilities -- sensing and thinking -- to achieve a robust fit between the robot and its task which cannot be achieved mechanically. The reader is asked to accept a problem statement assigning to the manipulator certain excavation tasks which will challenge its inherent mechanical limits, forcing reliance on some operator to accomplish control in the force domain. The research effort then seeks an autonomous system to replace that operator. This capability is needed before any reasonable excavator could be developed for automated use. While a mechanically robust device will no doubt be used for that purpose, and while actual supervisory intervention will be less frequent than in this experimental study, it is argued here that the best mechanism for developing this capability is study of a system in which force-cognition is highly demanded. Moreover, when we consider extra-terrestrial applications it is certain that the excavator will be light and of limited strength, thereby requiring the same cognitive capabilities employed by the spade-user and by the manipulator in this research program.

In the proposed approach the manipulator is commanded under position control in its digging motions. Forces are sensed and processed at a higher level, where there exist computational objects which create the *supervisory control* and which alter the intended digging trajectory. The updating of trajectory plans at the higher level can be considered a *tactical planner* which is sensor driven and which interacts with the ongoing position control of the device. The same manipulator system is used to demonstrate results in spatial task planning applied to excavation. That constitutes a level of *strategic planning*, and with its ongoing dynamic replanning it descends to the level of *tactical planning*.

Description of Experimental System

Manipulator

An experimental manipulator of modular design was constructed in the Civil Engineering Department. Six actuated rotary joints were fabricated; each joint is a DC motor controlled through a power amplifier, driving a harmonic gearing, and turning a potentiometer as a rotation sensor. End plates and fittings permit rapid link changeout and reassembly into different experimental configurations. Figure 1 depicts the manipulator as the 4-DOF system used in this research effort. It corresponds to a backhoe at one particular chassis position. The link denoted L2 corresponds to the "boom" and the link L3 corresponds to the "stick" of the backhoe. In the laboratory configuration the length L3 is 16 inches. Position control was implemented for the four joint rotations using digital control programmed on a Motorola VMEbus real time development system. On a real backhoe excavator candidate locations for force sensing would include pressure in hydraulic cylinders, axle loads, boom and stick strains, and so on. In this research effort force sensing was accomplished by strain measured on the manipulator links and processed by a MicroMeasurements 2100 strain conditioning device. A sandbox was assembled, and a sand blasting medium was placed as the excavation material; it is a poorly graded sand, with 100% passing a U.S. #10 sieve, 65% passing a U.S. #20 sieve, and 1.1% passing a U.S. #40 sieve.

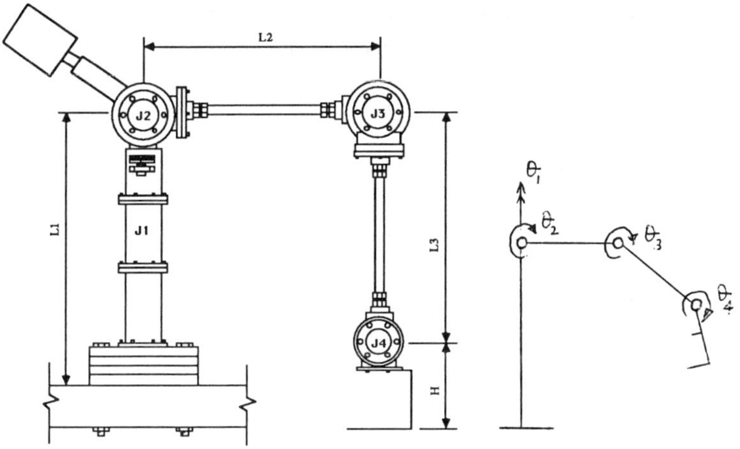

Figure 1: Excavator

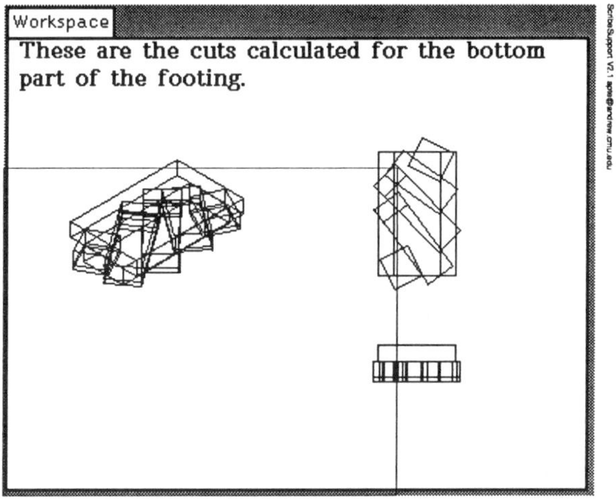

Figure 2: Radial Excavation, Mapped to a Rectangular Footing Geometry

A Smalltalk "Robot Programming Language"

Most commercial or laboratory robots accomplish the equivalent of a "Robot Programming Language" by establishing a command set (with its set of arguments) such that the description of a task is accomplished at a high level. A typical command might be MOVETO with arguments {tool position and orientation in Cartesian space}; the "language" is a procedure which executes the transforms that return the necessary joint rotations. Most investigators acknowledge the utility of highly capable RPL's, and in this effort we specifically sought an RPL useful for rapid prototyping at the symbolic level, and compatible with representations that include geometric objects. Therefore we undertook the construction of a Smalltalk-80 environment for manipulator modelling, control, and task planning. While object-oriented environments have been recognized (LaLonde, 1987) for their promise, they have not been well suited for interfacing with real-time control, and this constituted a portion of our research task. In our earlier effort a Smalltalk environment was implemented as a concurrent process on the VME system, and in this research effort a Smalltalk environment on a Sun-3 was extended to achieve serial line or Unix-socket communications with the VME system. A small set of primitive commands, essentially in joint-space, was established on the real-time system. All manipulator and path modelling was then performed in Smalltalk. Supervisory control approaches were programmed as appropriate classes, objects, and methods within the Smalltalk environment. Symbolic programming was sought, in which the full sequence of excavation is characterized as *swing-sweep-scoop-raise-swing-dump*, with appropriate arguments (ie, bucket positions) for each motion in the sequence. For instance, the *swing* motion includes the rotation in θ_1 about the vertical axis; the first *swing* motion in the sequence would take the bucket from the dump position (truck location or spoil pile) back to the point in space at which the next digging cycle is to begin. Actual digging occurs within the *sweep-scoop-raise* portion of the sequence.

Research Results; Processing Spatial Constraints

Background; Spatial Task Planning

Planning the geometry of the excavation activities was the other major role described in the Introduction. Stated simply, a human decomposes the spatial description of the goal state, such as a trench geometry, into a set of subgoals down to the level of tactical control, typically one spadeful or one bucketful of soil. Various spatial precedence relationships must be enforced, as well as spatial constraints. Replanning occurs constantly, processing changes in machine location or sensor observations of domain changes (perhaps the sand has slid back) or of the failure to achieve the last subgoal. Task planning for these purposes entails operations on spatial properties. Ideally it requires first a spatial representation, and then a proper set of spatial operators and a constraint language to permit search, planning, and replanning. This is a major research frontier in robotics and artificial intelligence; it is noteworthy that the simple topic of robotic excavation demands it and moreover couples it to dynamic planning in the force domain.

An Object-Oriented Programming Approach

In these research activities some useful first results (Apte, 1989) were demonstrated using the inherent capabilities of object-oriented programming. Those results can be traced as follows:

- An object class, *cuboid*, was created to represent a rectangular trench. Numerous useful geometric operators were thereby automatically imparted to it. As a geometric object it could be further developed for compatibility with sensing operations, although this was not undertaken.
- An object in the *trench* or *footing* subclass of *cuboid* is a goal state object. These objects have localized knowledge in which they can decompose themselves into *unit-of-excavation cuboids*, ordered vertically and horizontally into an excavation plan. If the system were interfaced with sensor updating, the program could replan anew at each step. The *unit-of-excavation cuboid* is a rectangular extrusion in Cartesian space.
- The solution is then extended to include *unit-of-excavation cuboids* in polar co-ordinates. A fan array of cuboids corresponds to the excavation which can be accomplished from a backhoe at fixed base position; moving the machine estalishes a new origin and a new fan.
- Spatial constraints can be processed to map such a fan to a trench or footing geometry, generating the task plan for that machine geometry. An example is shown in Figure 2.

Research Results; Processing Force Constraints

Background; Robot Control Approaches

At the signal level, which is the lowest level of control, manipulators are most typically under position control, while force control, hybrid control, and impedance control exist in laboratory examples. Perfect position control applies when no resisting forces exist; corrupted position control describes the attempt to assert position control despite forces from the environment, and it may perform acceptably if the forces are small compared to manipulator capacity, if the error in position is acceptable, and if stability is achieved. Force control (Whitney, 1985), hybrid control, and impedance control (Hogan, 1987) constitute important contributions, but are not in themselves a solution to the force cognitive excavation problem. In common with all signal level control approaches they are expressed at the level of manipulator state variables, and must represent task objectives at that level. In contrast, force cognitive manipulation as demonstrated by humans and as sought in the laboratory represents task planning at a higher level which nonetheless descends to manipulator control at the lowest level. Our research attempts to achieve this by supervisory control, resorting to an act-sense-think-plan cycle which is performed incrementally and discretely. Although crude in comparison to responses obtained at signal levels of control, it is more powerful in its embrace of representation, models, abstractions, goal knowledge, action knowledge, and so on.

Experiments without supervisory control on strain

The *intended* digging trajectory is the *sweep-scoop-raise* sequence. The full experimental *sweep* is a 30° rotation in θ_3 about a center of rotation which produces an arc approximately 12 inches long with about 2 inches of penetration into the medium. Figure 3 records the motion of a single point on the blade; the bucket itself and the manipulator links are not shown, and an approximate sand level is shown. Digging proceeds from right to left. The individual joint motions were recorded (not shown) and reveal the *scoop* motion (a 60° rotation in θ_4) to commence near bottom dead center of the trajectory, and the *raise* motion (a -2° rotation in θ_2) to complete the path. This trajectory can be considered an *intended unit of excavation*; given an excavator sufficiently strong, stiff, and properly positioned, this trajectory could be completed under position control, and the supervisory control exerted by the operator would be limited to the proper geometric sequencing of the motions. However, these experiments were chosen to demand significant supervisory control on internal forces. In principle the system could operate on several concurrent channels of force sensing (joint torques, link strains, axle/outrigger pressures, etc.); three channels were examined in this laboratory study, and the strain in link 3, the backhoe "stick," was observed to be an effective overall surrogate. The link 3 strain signal as a function of blade position is recorded in Figure 4. The desired strain range is +/- 1048 units; in this experiment the signal violated the strain limit shortly into the *sweep*, after only 5 inches of blade motion and went fully out of range of the strain conditioning for a portion of the path. In this experiment the system is "stiff" with respect to blade position and the target strain range is "tight," creating a severe test for force-cognitive excavation.

Experiments with supervisory control on strain

A series of experiments (Bullock, 1988) were conducted in which various force-cognitive tactics were programmed at the supervisory level of control. The most effective was supervisory control on both *sweep* and *scoop* motions, producing experimental results shown in Figures 5 and 6. The trajectory begins as a *sweep* but is overriden at the supervisory control level after approximately 6 inches because of excessive link 3 strain. The *scoop* motion commences but supervisory control then shortly demands a partial withdrawal of the bucket (approximately 1.5 inches of *reverse sweep*) during the *scooping*, again because of strain observations. The remainder of the motion is the completion of the *scoop* and the *raise*. Figure 6 records the strain signal and demonstrates that force-cognitive tactics have accomplished a complete excavation cycle while maintaining the strain signal within its desired range.

Supervisory control on *sweep* and *scoop* actions constitutes a robust demonstration of force-cognitive excavation. The laboratory manipulator will make appropriate adjustments to its digging trajectory to accomplish excavation while maintaining force (strain) conditions within a desired range. Experiments demonstrated that capability over many trials, and the experiment recorded in Figures 5 and 6 displays a rather complex behavior, ably reproducing the role of an equipment operator. For example, note that the stiffness of the excavator-soil system at the start of digging is much higher in Figure 6 than in Figure 4, as evidenced by the slopes at the start of each record. This would be caused by a

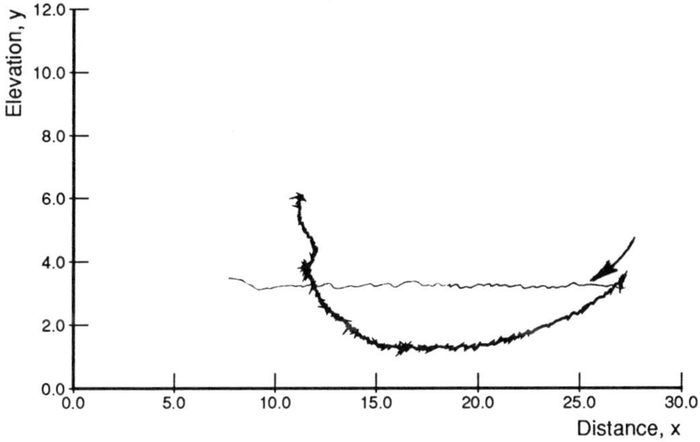

Figure 3: Excavation Path, No Supervisory Control on Strain

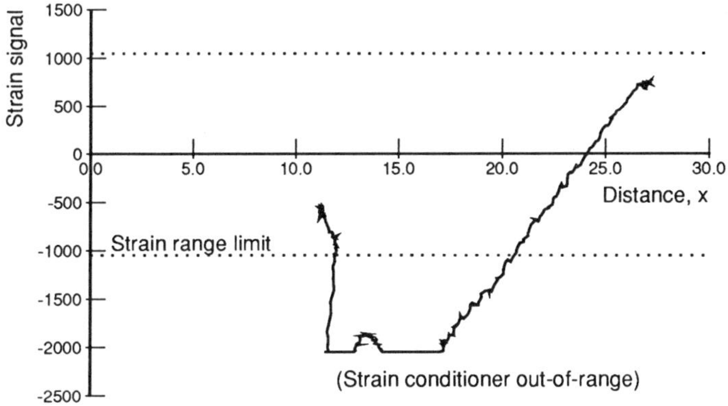

Figure 4: Strain vs. Blade Position, No Supervisory Control on Strain

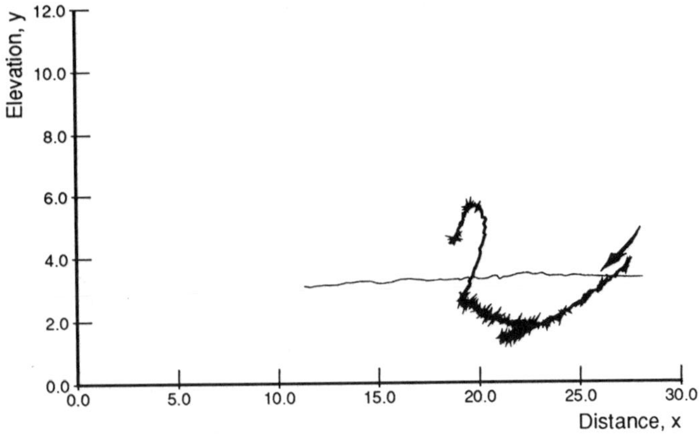

Figure 5: Excavation Path, Supervision on Sweep and Scoop Actions

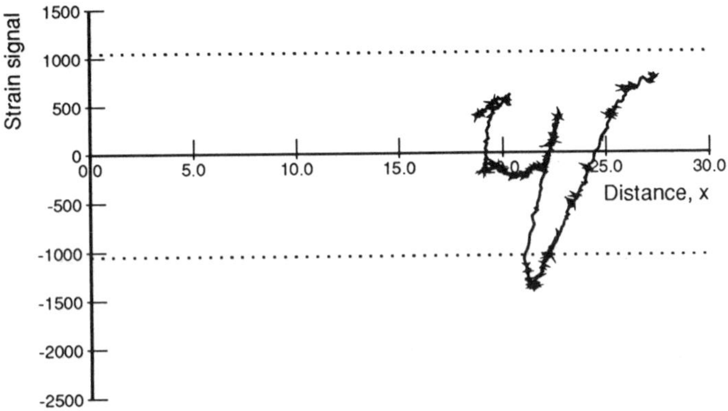

Figure 6: Strain vs. Blade Position, Supervision on Sweep and Scoop Actions

sharper angle of attack, by an upslope to the soil surface, by compaction, and so on; it is noteworthy that the force-cognitive control is insensitive to the origin of the condition.

Conclusions

These conclusions are offered as pilot study observations:

1. The results described can be taken as demonstration of the robustness which can be obtained using solutions built upon sensing and processing functions, as compared to that which might be achieved through mechanical design.
2. The Smalltalk-80 programming environment proved its compatibility with various objectives such as rapid prototyping, modular program construction, and rich representational ability. The program for the laboratory digging sequence is only 18 lines long, and the difference between versions with and without control on strain is restricted to minor changes in two of those lines.
3. Supervisory control was demonstrated to support the laboratory implementation of force-cognitive excavation.
4. A facility in task planning was demonstrated using the object-oriented programming approach to create a meaningful representation and a localized knowledge of spatial constraints and spatial operators.

Acknolwedgements

This work constituted one portion of research funded by the National Science Foundation under Grant DMC-86-15072 for *Research on Cognitive Excavation Automation.* The authors also wish to thank Professor I. Shimoyama of the University of Tokyo who contributed to our development of these capabilities, and to other research sponsors (Department of Interior, Bureau of Mines, and the Department of Energy) for support which provided portions of the experimental and computational environments. Some portions of this paper have previously appeared in a non-copyrighted publication (Bullock, 1989).

Appendix: References

Apte, S. M. Object-Oriented Implementation of a Domain-Specific Planner for a Robotic Excavator. Master's thesis, Carnegie-Mellon University, 1989.

Bullock, D. M. Supervisory Control for Cognitive Excavation. Master's thesis, Carnegie-Mellon University, 1988.

Bullock, D. M., and Oppenheim, I. J. *A Laboratory Study of Force Cognitive Excavation.* Proceedings Sixth International Symposium on Automation and Robotics in Construction, June, 1989.

Hogan, N. *Stable Execution of Contact Tasks using Impedance Control.* Proceedings IEEE International Conference on Robotics and Automation, April, 1987.

LaLonde, W.R., Thomas, D.A., and Johnson, K. *Smalltalk as a Programming Language for Robotics?.* Proceedings IEEE International Conference on Robotics and Automation, April, 1987.

Whitney, D. E. *Historical Perspective on the State of the Art in Robot Force Control* . Proceedings IEEE International Conference on Robotics and Automation, April, 1985.

Application of Automation and Robotics to Lunar Surface Human Exploration Operations

by
Gordon R Woodcock,[1] Brent Sherwood,[2] Patricia A Buddington,[3] Leona C Bares,[4] Rolfe Folsom,[5] Jack Lousma,[6] Robert Mah,[7] and Dr. William "Red" Whittaker[8]

Abstract

Major results of a study (RLSO, 89) applying automation and robotics to lunar surface base buildup and operations concepts are reported. The study developed a reference base scenario with specific goals, equipment concepts, robot concepts, activity schedules and buildup manifests. It examined crew roles, contingency cases and system reliability, and proposed a set of technologies appropriate and necessary for effective lunar operations. This paper refers readers to four companion papers for quantitative details where appropriate.

Introduction and Technical Approach

This study was initiated to explore possible high-leverage uses of automation and robotics (A & R) for the resumption of human exploration of the Moon. The Apollo missions twenty years ago accomplished initial human exploration, but present ideas for further lunar activities center on permanent human presence. Potential efforts include performing in-depth lunar planetology, emplacing or constructing astrophysics observatories, developing the technology and technical systems maturity needed for human missions to Mars, and developing lunar resources for various uses (including Mars exploration).

1,2,3 Boeing Aerospace & Electronics, JX-23, Hunstville AL 35824
4,8 RedZone Robotics, 401 Bingham St, Pittsburgh PA 15203
5 Boeing Aerospace Operations, 244-2A, Moffett Field CA 94035
6 Former Skylab and STS astronaut, consulting engineer
7 NASA Ames Research Center, 244-7, Moffett Field CA 94035

Permanent human presence on the Moon is, to a degree, a chicken-and-egg proposition. We need facilities on the Moon to support the people and we need people there to construct the facilities. While it is certainly possible to devise operations scenarios that resolve this dilemma (basically, crews could live in their transport vehicle until at least minimal facilities were erected), difficulties with these scenarios make the use of A & R to construct the initial facilities quite attractive. If that can be accomplished, it is only a small extrapolation in thought to ask the A & R system also to emplace an initial oxygen production facility. Then routine transportation operations to support permanent presence can gain the advantage of lunar oxygen (LLOX) propellant production from the outset. Consequently, the two driving tasks for our lunar robotic concept were: emplacing a shielded crew habitat system, and initiating LLOX production. Many other required tasks spun off from those two.

Our study philosophy was guided by two principles. First, our intent was to produce feasible concepts for applying A & R to the very earliest stages of lunar development. Early application of A & R techniques would be required for a lunar base program to begin flights just after the turn of the century. We aimed for solutions within the near-term state of the art. However, we were aggressive in thinking about what the near-term state-of-the-art is, and in some instances adopted technical solutions that might be considered novel. We relied mainly on well-understood technologies and those assured of being available. For example, we did not presume availability of nuclear reactor surface power systems; our base systems use solar sources, mostly powering down during lunar nights. We considered robotics technologies now in use or in advanced development for Earth-based applications to be "available".

Second, we developed concepts for the base elements (the robotic workpieces) and the robots (the workers) together, iteratively. This resulted in a more coherent scenario than could be achieved by simple sequential treatments (defining the base elements and only then defining the necessary A & R). Our element configurations, production flows and operation scenarios explicitly incorporate our proposed robotic capabilities. Further, the robot types we developed, albeit versatile beyond our scenario, grew in turn directly out of specific, identified requirements. The value of our scenario resides not so much in details of the many elements as in the system interdependencies among them.

We discovered early that a complete end-to-end systems analysis and operations scenario for emplacing an initial lunar base had never been published; thus part of our job was to create a strawman concept and scenario. Although the schema we developed included A & R embedded from the beginning, it is useful for general lunar base studies. The requirements and functions we identified must be accommodated in some fashion, even if not in the particular ways we selected.

Functional and Operational Requirements

Our prospective site is on the southern edge of the Sea of Tranquility (on the equator, about 27 deg east longitude, between Moltke and Maskelyne). Apollo 11 sampled the regional geology: mature, flat, deep regolith, rich in ilmenite. The Nearside equatorial site simplifies Earth communications; equatorial sites in general simplify trajectory design. The Tranquility site appears promising for future base growth, being flight mechanics-wise suitable for a ballistic payload launcher to send lunar resources from the surface to the L2 libration point.

The site plan, shown in this volume in (Sherwood, 90b Figure 4), is clustered around an open workyard. Functions are zoned as indicated. Areas requiring routine vehicle travel are paved minimally with compacted gravel generated by mining. Building the base produces a 3 yr supply of industrial feedstock. Eventually, growth of the industrial area northward sustains material resource needs. The site plan is pre-adapted for indefinite base expansion, including element upgrades. Its development and contraints are detailed in (Sherwood, 90b).

All identified base elements are listed concisely in this volume in (Sherwood, 90b Figure 1), and described in detail in (Sherwood, 90a). For easy reference, the primary and mobile elements are illustrated here in Figure 1. Systems are provided for landing equipment on the Moon, moving it around, and installing, operating and maintaining it; for sustaining and shielding human crews; for providing power; for extracting and storing liquid oxygen for propellant. Here we note concisely the overriding design constraints for the entire set.

Throughout the concept, simplicity and certain achievability were given higher priority than strict efficiency. During the high-risk early phase, greater requirements for power and mass-throughput purchase greater reliability. The equipment is sized for an oxygen production rate of 100 t/yr (metric tonnes per year). Since this requires excavating of order 1 kg/s of mare regolith, simple and low-power methods can be used. Furthermore, since mare regolith is well-comminuted, careful surface and subsurface surveying permits using small excavation tools, by avoiding areas with large inclusions. By only running the excavation

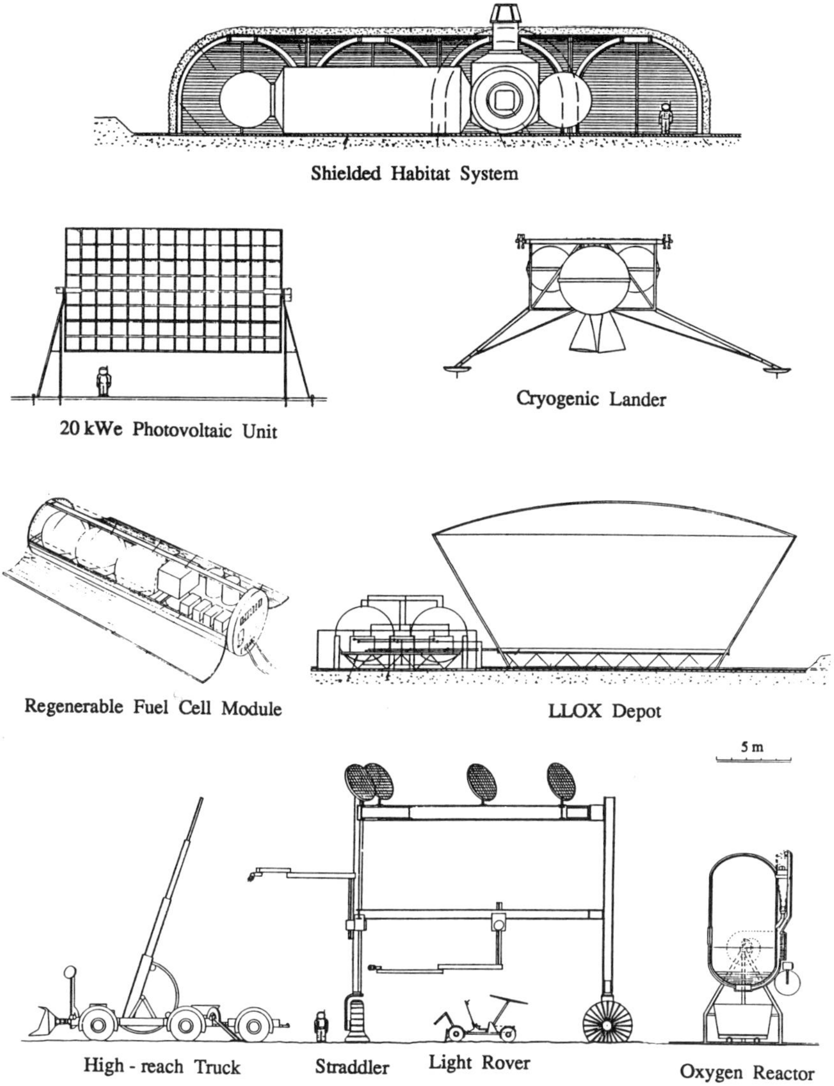

Figure 1 *Primary and mobile element concepts.*

and oxygen processes during the lunar day, the base can function with simple, reliable solar photovoltaic power production. This allows credible cost, schedule and maintenance estimates. The high (1 t/kWe specific mass) cost for storing power through the 2 wk lunar night is reserved for life-support and critical equipment. All equipment concepts are optimized both for their main function and for robotic deployment, connection, operation, inspection and maintenance. Most equipment arrives intact and ready for use. Only the passive radiation shelter is "constructed" on the Moon.

Robotic Capabilities

Selecting the Approach - The emphasis on robotics in this study was directed to minimizing the need for on-site, hands-on, EVA human activities. It made little difference to our end results whether robotic activities employed "true" (highly autonomous) robots or human-directed machines as long as the human director did not have to be on the Moon in an EVA suit. Considerations of the signal time delay from the Earth to the Moon and back (several seconds including switching delays) led us to closely-supervised modes rather than "joystick" modes. We determined that the primary need for human involvement was in task planning, sequencing, and problem/workaround recognition. It is straight-forward enough to program a robot for task primitives: install a bolt, move to a predetermined location, or pick up/put down a load. The need for "intelligence" is in deciding which primitive to invoke next, and why, and how best to coordinate the activities of several machines to accomplish complex overall objectives.

We assigned tasks and functions to the robotic systems to survey the base site above and below the surface, and construct the human habitation area and the base propellant production facilities. Human control is implemented as symbolic commands from base control center, a direct extension of Earth. The robotic systems act in a task query - response mode to the central command station. The supervisory control scheme can accommodate the several seconds of signal delay from Earth's surface to the Moon, as well as command-loop timing discrepancies when both local and Earth-based control are used together.

Robotics System Configuration - We developed the system configuration by defining requirements, analyzing them for commonality and design drivers, and then involving all of the authors in iterative, structured brainstorming. The expertise at hand included systems engineering, aerospace engineering, A & R hardware and software development, construction engineering, EVA operations research, astronaut flight experience, reliability analysis, and configuration design. Here we will not present an anecdotal recounting of our discussions, but describe and rationalize our results as concisely as possible.

Requirements are summarized in Figure 2. These came from a functional analysis of operations, whose sequences are detailed in (Buddington, 90). We found that the requirements could be satisfied by a set of three robotic vehicle classes: a light mobile transport (rover), a median-mass, mobile transport with "high-reach" boom, reconfigurable end effectors and towed trailers (truck), and a heavy mobile gantry able to straddle work areas, and carRy, lift and place massive objects (straddler). The vehicle concepts are depicted at the bottom of Figure 1.

Much of our discussion centered on the straddler concept; it seemed off the beaten track of thinking about lunar surface construction and operations equipment. Three basic considerations drove us to this concept: (1) in our scenario, the first trip to the Moon is a cargo trip which must unload itself with no assistance from people or any local equipment; (2) we could combine capabilities for the inescapable functions of offloading heavy, bulky payloads from elevated locations (such as the lunar lander), transporting them across the surface, unloading them from the transporter and emplacing them precisely, in one machine; (3) we found other cases, such as mining regolith, loading oxygen production reactors or moving a crippled lander, where the heavy mobile crane served the need very well. Once we accepted a machine concept like this, it seemed to solve all our open operational requirements problems.

There are Earth analogs to the heavy mobile crane. Yacht basins use a machine called a "comporter" to lift boats (especially sailboats) out of the water and place them in drydock, and then to refloat them. Lumberyards use a similar machine to move and load piles of lumber, and railyards use captive gantries for reloading trailers among freight cars. These analogs give us confidence in the mechanical advantages of selecting the straddler as a reasonable solution.

<u>Robotics Design Considerations</u> - Careful attention to tribology is enabling for lunar mechanisms. In addition to the familiar space lubrication problems of vacuum (metals cold-weld, greases turn to glue, liquids evaporate and intercalation fails) and temperature extremes (ranging from -170 C at night to +110 C during the day), the Moon introduces potentially severe abrasive wear. 50 % of the regolith is finer than the human eye can resolve (about 70 microns), and this highly abrasive

Functions \ Tasks	Site survey	First cargo down	Cargo unloading	Landing pad prep	Road building	Transfer line deploy	PV deployment	Habitat emplacement	Regolith shelter setup	Radiator / sunshade setup	Comm system inst'l	RFC emplacement	Oxygen reactor setup	LLOX depot setup	Ilmenite mining	Ore transport	Gangue deposition	Ox reactor loading	Ox reactor dump	Ox rctr clng & servc'g	Moving disabled lander	Servicing lander	Servicing PV unit	Servicing RFC	Servicing utilities	Servicing robots
Self - Unload	●	●																								
Light mobility, materials & equipment transport	●			●		●	●		●	●	●			●						●		●	●	●	●	●
Heavy mobility, materials & equipment transport		●	●	●	●			●	●			●	●	●	●	●	●	●	●		●					
Light low - lift / positioning			●			●	●	●	●	●				●								●	●	●	●	●
Light high - lift / positioning			●				●		●	●	●					●				●		●	●		●	●
Heavy low - lift / positioning		●	●	●	●			●	●			●	●	●	●		●		●					●		●
Heavy high - lift / positioning		●	●						●				●			●		●			●					●
Light materials placement	●			●		●	●		●	●	●			●						●		●	●	●	●	●
Heavy materials placement		●	●	●	●			●	●			●	●				●	●	●		●	●		●		●
Manipulation / tool use	●	●	●	●		●	●	●	●	●	●	●	●	●				●	●	●	●	●	●	●	●	●
Excavation / grading				●	●	●	●	●					●	●	●		●									

Figure 2 *Lunar surface robotic tasks/functions matrix.*

dust sticks electrostatically to virtually everything it touches. We propose overlapping countermeasures, both prophylactic and compensatory. First, we try to keep dust off. There is no atmosphere to suspend lofted particles, and the creeping motions of most robotic activity will not "kick up" much dust. Human activity and lander exhaust plumes are thus the dominant sources. Keeping critical components at least 1 m up off the ground minimizes the former, and intercepting ballistic blast ejecta at the landing pad with high barriers can virtually eliminate the latter.

Some components, like electronics, many sensors, and traction-drive motors, can be hermetically sealed against dust. (Unit closure is independent of thermal control in vacuum, since conduction to other hardware and ultimately radiation to space are the only means available for rejecting waste heat. Equipment units optimized for lunar use must be able to run somewhat hot.) Subcomponent maintenance would be accomplished by robotically replacing the entire unit and removing it to the pressurized workshop, where crew members can clean, open and repair it. Sensor lenses, solar arrays and radiator surfaces can periodically be cleaned electrostatically; and optical polymeric films that are peeled off in layers show promise for protecting larger windows.

Many gears and joints just cannot be kept dust-free, because they are either part of dust-handling systems, located under dust-shedding mechanisms or low to the ground, or even operate in the regolith. Our approach here is to configure such mechanisms openly enough to let all but the inevitable dust film fall through, size them robustly, and use hardened metal inserts for the bearing surfaces most subject to abrasive wear. Plasma deposition of diamond-like carbon (DLC) and real diamond films on hard alloys shows promise for these devices.

Molybdenum disulfide is still the best dry vacuum lubricant available. Relatively long-lived surfaces can be made using applied coatings impregnated with such chalcogenide materials. The essential problem with dry lubricants is that relubrication means replacing the part. To minimize transported mass and yet simplify robotic changeout, critical mechanism bearing surfaces should be designed as separable, modular parts, and as interchangeably common as possible.

Hydraulic mechanisms, typically used for large machines on Earth, are undesirable in the lunar environment due to leakage and their strict requirements for tight, moving seals. We specify rack-and-pinion (R & P) drives for robotic mechanisms with a large range of motion, such as extension arms, swing arms and actuated pivots. The R & P approach follows the design philosophy outlined above, allowing relatively open, contamination-tolerant joints with replaceable, hardened bearing-surface inserts.

Robot components for remote use benefit greatly from commonality and modularity. Common fittings, protocols, testing and repair procedures maximize the efficiency of logistics. And modularity can translate the problem of fitting tools to a particular job from the hardware domain to the control, or software, domain. A laboratory development called the Reconfigurable Arm consists of a kit of common joints, simple inter-joint arm lengths, and a sophisticated control system. Each part contains an embedded processor which contributes knowledge of its own dynamical behavior to a simple serial bus threading the system. The controller then adjusts its data interpretation and signal generation based on the parts in use for any given job. Reconfigurable robotic devices will prove enhancing for lunar operations, where the range of dexterity required for maintenance and changeout operations will inevitably be quite wide.

Sensors for robotic systems continue to be developed rapidly, with some recent advances showing particular promise for lunar application. Light, reliable fiberoptic sensors embedded directly in mechanical components can provide required information on strain, position, acceleration, temperature, even magnetic fields. Small, CCD cameras with fisheye lenses can provide hemispherical video coverage which, albeit difficult for humans to interpret, is adequate for robotic operations. (The images can be deconvolved for more conventional presentation to human operators when desirable.) Scanning laser rangers yield distance-driven information extremely useful to machines and humans alike for navigation and manipulation, especially in high-contrast or completely unlit work areas such as will characterize the lunar environment.

Software Considerations - We avoided solutions that invoke "magic" (software technologies that might be invented in the future). We developed a modular and hierarchical approach, depicted in Figure 3. This scheme has operator entry points at several levels. The lowest level is device control primitives, such as "move a specified distance and direction"; "rotate the bolt wrench". The operator can either engage the entire primitive, or engage it by modules, such as "engage the bolt head", "apply a certain torque", etc. At this level, feedback loops are mainly internal to the robot (speeds, torques, angles), with processing located in the tool itself. External environment feedback is fundamental parameters, like "strain". Human control at this level is "joysticking", for which sensor data are "processed" by the operator.

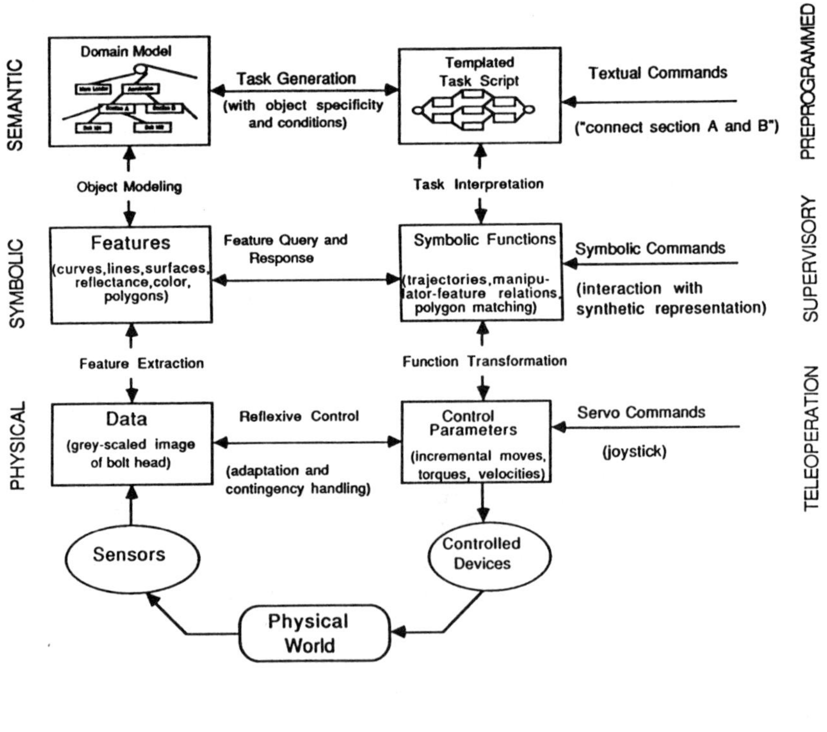

Figure 3 *Multi-tiered robotic control architecture.*

Alternatively and more efficiently, the operator can engage the next higher, "supervisory" level, where simple sequences of primitives constitute unambiguous operations: "locate bolt xyz, install it and confirm proper installation". At this level, the robot generates feedback through feature extraction of raw sensor data, so that for instance it recognizes a mis-oriented bolt head as such. It must determine its location relative to target objects, and it must perform positive part identification, for example by bar-codes. Internal processing at this level is somewhat slower, but can still best be located locally at the worksite. Injection of human supervisory commands and data analysis can occur offsite, buffering short delays like those encountered for telemetry between the Moon and Earth.

Software is deterministic up through this level. The next possible level involves automated task planning, as in the case where a command is to "unload the habitat and emplace it in the pre-planned location". At this level, several courses of action are possible, and the robot must identify feasible paths through the network of possible sequences that reach the end objective, selecting a good one. At this level, the robotic system could justifiably be said to understand substantially both a model and the reality of its work environment.

The modular and hierarchical approach outlined enables us to begin operations with deterministic software, and then effect a gradual transition into extensive artificial intelligence as the technology evolves. Eventual autonomous operation of a lunar base will introduce a new generation of performance goals capable of driving A & R technology to greater maturity than have undersea mining and the nuclear industry (two terrestrial applications which have already implemented the lower, deterministic levels of control quite successfully). A detailed explication of the control of lunar robots is contained in (Bares et al, 90).

Robotic and Manned Operations Analyses

Our operations analyses concentrated on a classical task/time/resource analysis of base and production facility buildup, tied to both the robotic equipment and flight schedule (Buddington, 90). An event logic network was created and analyzed to schedule the operations and uses of equipment. We found that the operations were mainly constrained by frequency of lunar transport flights rather than by capacities of the robotic equipment. Consequently, the schedule for building up the base and production operations is essentially entirely controlled by the transportation manifest. Only at the beginning of the buildup schedule, when activity is paced by site preparation milestones, are the vehicles used almost full-time. Later, as the base industrial equipment arrives, is set up, checked out, and used, extensive downtime results from the infrequency of lunar cargo delivery (an assumed fixed rate of four flights per year). This planned downtime permits machines to be overhauled, and used for other purposes like scientific investigations.

The relative density of undisturbed lunar regolith is extremely high due to eons of micrometeorite gardening. Excavating such a material in familiar ways (front shovels, dumptrucks and dozers) would require levels of mobile power that an early lunar operations scenario just could not provide. Most likely, high-power lunar equipment eventually will use fuel cell plants, or even tethered nuclear generators. But limiting ourselves to solar power constrains the rate and type of excavation that a mobile robot can reasonably achieve. Our modest excavation rate, and the rather homogeneous character of mare regolith, together make possible another approach. Consequently we developed our operations concept to use of order 10 kW per miner/straddler, with a tool bite 5 x 30 cm, moving continuously at 10 cm/s. Clearly this rate and method is more compatible with continuous robotic control than with human driving. The slow-moving straddler is envisioned to be entirely robotic. The truck has an on-board operator station for EVA use, and only the light rover is optimized as a manned transporter.

We also examined what could go wrong with our "best-laid plans", since complex operations as contemplated here never occur nominally. Our system design was purposely redundant, but redundancy alone does not provide robustness. Our analysis of likely problems resulted in levels of response ranging from preventative countermeasures and built-in redundancy (for debris damage or electronics failures), to robotic work-arounds (for "stuck" vehicles), to teleoperated work-arounds (for mechanical problems), to actual human intervention (to remedy design flaws or recalcitrant mechanical problems), to replacement (for catastrophic damage, such as from a lander crash). Since preparing and launching an unscheduled crew visit could take from two to five months, our analysis validated the philosophy of keeping the robotic tasks straightforward, designing the equipment simultaneously with the operations scheme, and providing the robots with a controllable, well-characterized work environment.

One thing that can surely go wrong is equipment failures. A robotics system design and operations plan that assumed no failures of equipment would be doomed. Accordingly, we made estimates of failure rates, based on historical failure rates of similar equipment in analogous environments, and estimates of the nature and complexity of components (especially for the robotic elements). The data were culled from a wide variety of sources, since operational data for actual lunar equipment involve only short-duration usage; our projections represent a first attempt to bound the problem. We can expect the light rovers, for example, to experience a failure (minor or major) every 573 hr. The entire base will experience a few failures every 24 hr, highlighting the need for an optimal strategy of robotic changeout and human repair. Our analysis identifies both specific areas where technology development would be highly beneficial (such as durable metal flex-wheels) and an appropriate spares manifest for the scenario (15 % of the active mass).

Initial human activity will be limited to remote, supervisory operations of the robotic systems. With the present manifest, manned missions are scheduled twice before full lunar liquid oxygen production begins, and base growth and human occupation become routine. Both of these visits last about one lunar diurnal cycle.

The first of these is the seventh flight (19th lunar cycle after initial landing). Primary goals are to verify and adjust the functioning of the habitat-workshop complex, and to perform investigations with fluid bed reactor elements and soil reactions. The crew will have the opportunity to observe the mining operations and review, eyeballs-on, the performance of robotic maintenance actions. They will be capable of intervening in these operations as necessary or performing additional checks and corrections, fine-tuning the robotic operations.

The second crew landing occurs as the twelfth flight (34th cycle after initial landing). At this point the oxygen liquefaction and storage plant, and the first fluid bed reactor, have been erected and put into operation. This flight verifies their functioning, and perform adjustments and repairs as required. Additional in situ resource utilization tests (recovering iron from reactor slag, sintering construction materials, producing glass fiber or performing other experiments) can be conducted at this time. Both human crews can expect to have a backlog of broken parts (which have been changed out with spares, robotically, in their absence) to attempt to repair. This kind of work, done IVA, is a good candidate for lunar nighttime activity during their stay, while the base machinery is shut down.

Lunar Base Buildup

Details of the buildup schedule and delivery manifest are contained in (Buddington, 90). The priority for base buildup is: spaceport, habitat area, industrial production site, base expansion. The work overlaps, due to site-wide infrastructure emplacement and each area's need for more paving gravel than it generates. The landing pad is done first (cleared, graveled, compacted, and critical blast deflectors erected), to provide a predictable surface and limit contamination of other base elements due to blast effects. The habitat and supporting facilities (arrays, RFC, modules and shelter) are erected on cleared and (where required) graveled and compacted surfaces. Then the first manned flight occurs. The industrial buildup takes another year, before initial plant startup (one oxygen reactor). Full production is achieved 3 flights later (after the manned mission and installation of the two remaining oxygen reactors). In all, 15 flights are required to build up the base. The 13 cargo flights each land 30 t.

Projecting our scenario beyond the first four years of buildup and operation, we would bring backup mining equipment, and complete two more landing pad complexes in the spaceport. That would allow simultaneous lander servicing and broaden contingency options. Next, the habitat system would be expanded to support (psychologically as well as logistically) continuous human presence as desired. The habitat power plant would grow commensurately. Finally, other industries (recovering volatiles from regolith, sintering paving and shielding blocks and recovering alloyed metals from oxygen reactor byproducts, making glass fibers from sorted, ilmenite-deficient leftover regolith) would be emplaced, and oxygen production increased. The base could begin to support more frequent lander flights, and to supply propellant for transfer vehicles in LLO. Nuclear power plants could augment the solar array field. Lightly shielded reactors could be placed arbitrarily far away to the west. Scientific and prospecting expeditions could be sent all over the Moon from such a base.

Supplying large amounts of lunar products to staging points in space such as L2 would require a mass-driver. It would be sited in the spaceport, firing eastward horizontally under the lander approach path. Regular, economical transportation to a Farside science base could be accomplished with a robotically-built surface rail system heading east around the limb. The Tranquility settlement would develop a cardinally-oriented, cross-axis service infrastructure connecting a human community to the south, a robotic industrial complex expanding to the north, power sources to the west, and a spaceport bordering the wilderness to the east. The modest initial site could thus evolve smoothly into a true lunar colony.

References

Bares, L.C., Kweon, I.S., Osborn, J.F., Wettergreen, D.S., Whittaker, W.L. (1990). "A Mobile Robot for Lunar Surface Operations" SPACE 90, ASCE.

Buddington, P.A. (1990). "Manifesting for a Lunar Robotic, Oxygen-Producing Base." SPACE 90, ASÇE.

RLSO - Robotic Lunar Surface Operations: Engineering Analysis for the Design, Emplacement, Operation and Performance of Robotic Lunar Surface Systems, (1990). The Boeing Company D615-11901.

Sherwood, B. (1990a). "Lunar Base Elements Designed for Robotic Operations." SPACE 90, ASCE.

Sherwood, B. (1990b). "Site Constraints for a Lunar Base." SPACE 90, ASCE.

Site Constraints For A Lunar Base

Brent Sherwood [1]

Abstract

A rigorous site plan is a prerequisite for quantitative analyses of lunar base operations. The development of such a site plan, done for a detailed lunar base study (RLSO, 89), is described and the result presented. Standard methods are used; the site plan accommodates the detailed and combined requirements of all base elements, as well as indefinite growth.

Introduction

A site plan is the fundamental tool for organizing architecture. By addressing the largest dimensions and relationships of a project -- the largest "scale" -- it relates that project most directly to the larger world. The site plan reflects and guides daily and developmental activities within the site, directing its maturation over time. A successful site plan reconciles conflicting requirements so as to avoid inefficiency, unsatisfactory accommodation, or haphazard growth which would lead to them. By establishing a framework for meshing the detailed considerations of specific architectural elements, the site plan insures that elements work together to serve the larger goals. As a dynamic document, the site plan must itself evolve in response to real conditions, both when first conceived and later, as built lessons are learned. The best site plans are resilient enough to accommodate changing requirements over long times, while still expressing their originally conceived integrity and order. Some urban plans on Earth have served, recognizably intact, for millennia.

Many artists' concepts of lunar bases have been published, and used to illustrate papers, proposals, presentations and program plans. By showing explicit relationships among base elements,

---

1 Senior Engineer, Advanced Civil Space Systems, Boeing Aerospace & Electronics, JX-23, Huntsville AL, 35824-6402

each such image implicitly records a site plan. Yet most intentionally feature some particular aspect of the base, and so evince consideration of only a few of the serious constraints we can really expect. Unfortunately, the site plan plays such a fundamental role that, if superficial, it limits any ability to develop credible detail in a scenario.

This paper reports methods and results of developing a lunar base site plan. We required a credibly detailed context for studying the application of automation and robotics (A & R) to lunar surface operations (Woodcock et al, 90). Specifically, our study goal was to evaluate the feasibility of emplacing, operating and maintaining an early, oxygen-producing lunar base before extensive or continuous human presence. While not proffered as universally valid, our methodically developed result outlines a representative array of issues relevant to lunar bases, and provides an example of how to produce a real site plan.

The Method

Developing a site plan is not a mysterious, nor a haphazard, activity. Straightforward, rigorous methods for designing successful site plans are available. Some may be unfamiliar to most people in the "space community", but they are practiced commonly by the architectural professions. We outline a simple method here and demonstrate its use with the example of an early lunar base.

The essential problem of the site plan consists of accommodating simultaneously the conflicting needs of the site elements. If all the multivariate relationships among elements were identified and quantified, and penalty functions accurately assessed, linear programming could be used confidently to arrive at an optimal site arrangement. Two general factors confound this approach for our purpose. One is that for sites including human use, those relationships are so extensive and complex that they defy simple numerical modeling: an experienced designer provides more useful results! In our case (laying out an early lunar base), this complicating human "factor" is much less prominent than the other, enabling (and quantifiable), engineering considerations.

The second confounding factor is that, during the very earliest study phases of such an unprecedented project as a lunar base, available element details are quite insufficient to enable any productive quantification of their interrelationships. Again, the synthesizing approach of an experienced designer is the only practical solution to working with incomplete data.

The four-step method explained below provides the human designer with a tool to record, organize and finally visualize the most critical relationships of the site. The first step is to define the site program. By this we mean the activities that the site will support, the goals it must achieve, and finally the aspirations it should embody. The second step is to define as completely as possible the site elements, those discrete pieces of the site architecture which introduce unique, intrinsic requirements. These may be hardware systems (like buildings), or siteworks (like roads), or more abstract elements (like a required activity space among buildings). We include location-specific, geographical, topographical and geological information of the chosen site as part of the element definition. The third step is to identify connections among the elements, which constrain how they may, or must, be arranged within the site. An effective way to record these is to list them as part of the element descriptions of each element they affect. The fourth step is to generate a proximity diagram which graphically represents the relative sizes of the elements, as well as the relative importance (and interference) of the element connections. Iterating to an optimal proximity diagram produces the skeleton of a workable site plan.

Our early, solar-powered lunar base fulfills a fourfold program: it must produce oxygen (for space transportation propellant) from lunar resources (LLOX); it must facilitate the in situ human investigation of lunar science and engineering; it must be able to grow in capacity and capability over time; its early, heavy, dangerous, and routine work must be performed robotically. Justification for this choice of program is reported in (Woodcock et al, 90).

The primary reason for past difficulties in developing realistic lunar base site plans is their lack of sufficient, consistent element detail. A useful site plan cannot be generated until everything large enough or important enough to affect it is defined to comparable levels of detail. That includes all buildings, spaces, vehicles, industrial plants and utilities. Our study of robotic lunar surface operations was charged with developing one, reasonable scenario down through several layers of engineering detail. The disadvantage of this approach is that the untraded underlying base concept lacks universal relevance. But a tremendous advantage is that it requires, and generates, a detailed program with consistent element definitions. That in turn exposes many of the real engineering issues to analysis. The base elements we used, and their major connection constraints, are detailed in (Sherwood, 90).

For convenient reference, Figure 1 lists the primary, mobile, utility, and sitework elements of our lunar base. To this list we can append topological provisions for the expansion necessary for long-range base occupation and growth: habitation, industry, mining, spaceport and wilderness. Soberingly, this extensive list comprises an approximately minimum element set for a lunar base capable of supporting a handful of people (intermittently) and producing 100 t of LLOX per year (the amount required for four round trips between the surface and lunar orbit). It takes 13 flights delivering 30 t each, over almost four years, to get all the Earth-supplied equipment to the site (Buddington, 90); thus even an early, modestly productive lunar base represents a substantial undertaking.

Primary Elements	Landers	(up to 3)	(cryogenic, 30 mt landed capacity, manned or unmanned)
	Habitat System		(habitat module, 2 airlocks, connecting node w/ cupola, pressurized workshop)
	Radiation Shelter		(regolith - filled, encloses hab system, provides unpressurized garage)
	Solar Arrays	(24)	(20 kWe, deployable rigid tracking)
	Regenerable Fuel Cell Modules	(2)	(nighttime power, 20 kWe output, 50 % overall efficiency)
	Oxygen Reactors	(3)	(batch fluid - bed, hydrogen reduction of ilmenite)
	LLOX Depot	(1 per pad)	(liquefaction, refrigeration, redundant storage, pumping)
Mobile Robots	Straddler	(2)	(robotic mobile gantry; lifting, moving, positioning, mining)
	Truck	(2)	(outrigger - stabilized, high - reach boom w/ utility suite, front loader, rear tow with utility trailer suite, robotic w/ onboard operator station)
	Rover	(2)	(light duty, site survey & crew transport, robotic or manned)
Utilities	Radiator Modules	(8 total)	(5 and 3 ganged together, with fixed, deployable sunshades)
	Debris Barriers	(12 / pad)	(deployable, intercept ejecta from lander exhaust to protect base elements)
	Hoppers	(22)	(with chutes, hold 27 t material each)
	Storage shed	(optional)	(made of habitat shelter vault sections)
	Guidance Beacons		(lander targeting and local site navigation)
	Communication Transceivers		(link all mobile elements, and site to LLO and LEO)
	LLOX terminal	(1 / pad)	(buried valve box and conditioning - umbilical connectors)
	Gas Lines		(conduct gaseous oxygen from reactor field to depot)
	Liquid / Vapor Lines		(connect depot to LLOX terminal under landing pad)
	Power Substation		(regulates industrial load and crossover distribution to hab power system)
	Power, Data & Grounding Cables		(throughout site, linking all fixed elements)
	Sensor Heads		(monitor critical views)
	Local Lights		(augment Earthlight during lunar night for critical areas)
	End Effectors & Tools		(for robot manipulators and crew)
Siteworks	Spaceport		(up to 3 landing pads w/ prepared surface)
	Foundations		(undisturbed, naturally consolidated regolith, 1 m overburden scraped off)
	Open Workyard		(paved area for equipment staging, disassembly and reconfiguring)
	Connecting Roads		(leveled, paved with compacted gravel for dust control)
	Deposition Sites		(receive rocks, gangue, spent oxygen - reactor solids)

Figure 1. Lunar site elements list.

A site at the southern edge of the Mare Tranquillitatis, near Apollo 11's Tranquility Base, is attractive for our base program for several reasons. Its geology is empirically known through in situ human observation and returned samples. Its mature mare regolith is comparatively rich in ilmenite, level, and well-comminuted by meteorite gardening. Thus a usable soil feedstock, low in rock content, can be scraped up with minimal mining infrastructure. The Nearside site allows simple, continual communication with Earth orbit. The specific location allows mining growth to the north, with highland regions to the south for exploration. Finally, its essentially equatorial location minimizes propulsive requirements for transportation operations to and from orbit, and pre-adapts the base for efficient mass-driver launches to libration point targets in an industrial future.

The Proximity Diagram

For surface sites, the proximity diagram should be two-dimensional. Figure 2 represents the plan areas of our major base elements simply, as circles at the same scale. Such a catalog compares concisely the relative element sizes as measured in the plane of their interactions. Figure 3 shows these element-circles arranged into a proximity diagram. Lines represent functional connections between pairs of elements; their thickness represents the relative importance of the connections they represent. Ranking the connections derives in this case from engineering judgments, based on estimating the complications introduced by relaxing certain proximities.

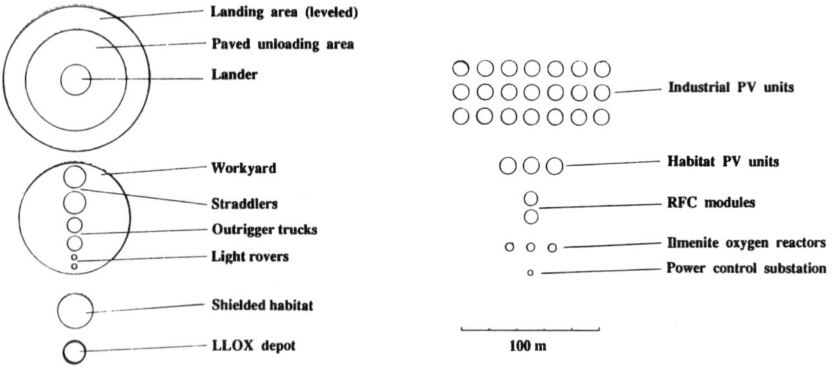

Figure 2. *Scaled element plan-area catalog.*

For example, grounding cables may be required to connect all base elements (on the Moon, only common "chassis" grounding appears possible because of the anhydrous regolith), to prevent potentially hazardous differential charging. The total cable length should be minimized. But such cables are cheaper than cryogenic fluid transfer lines, which would be heavier, require more maintenance attention, and introduce the hidden cost of additional refrigeration power per unit length. So keeping fluid lines short and accessible, at the expense of longer grounding cables, is probably a good trade.

The rules of a proximity diagram are simple: the connection lines should be kept as short as possible, and none should cross. Satisfying both of these rules strictly is hardly ever possible, but the distinction between a good solution and a poor one is usually obvious. Often, iteration between the proximity diagram and the element set produces better results. For example, without a "workyard" element, the lunar base diagram would show a tangled web of activity lines through its center, representing physical vehicular access among all the other elements. Explicitly requiring a workyard replaces that tangle with a definable element, within which the access interferences can be resolved by scheduling rather than by facility design.

When generating the proximity diagram, it is critical to recognize any peculiar requirements which might influence the ultimate physical element arrangement, if such constraints are so fundamental that they appear even in the diagram's abstract language. The solar power plant of our equatorial lunar base has such a requirement: none of the individual arrays may obstruct another's view of the sun from dawn to dusk. This leads most simply to a linear, north-south installation, which appears clearly in Figure 3. Our power utilization is based on a 300 hr lunar day. This is about 90 % of the full lunar day, thus discounting the sun when it is nominally within 9 deg of either horizon. Given the dimensions of our solar arrays, we could base them along the sun's track, on 55 m centers in flat terrain, without any further view penalty. Indeed, several parallel north-south lines would be the proper choice for a solar field a few times larger than ours, because it would minimize both DC transmission losses and driving distance to the remote arrays. The diagram immediately shows the importance of providing a single power plant bus, rather than separate long feeds from each array.

Another particular requirement is for the oxygen reactors to be located in a line, so that the hoppers receiving their spent solids can be removed along a simple, common rail system. The final, orientation-specific requirement is that the spaceport reflect astronautical constraints. Landings at this equatorial base would typically be from the east, along a terminal approach path elevated at only 15 deg. Launches would be practically vertical for a considerable distance. Short landings are the most probable failure mode during terminal approach. Thus the safest place for the landing port is east of the base. The necessary reliability of a man-rated vehicle, its ability to hover upon touchdown, and pinpoint navigation using beacons all allow the pad to be positioned adjacent to the rest of the base if debris barriers are employed. Close proximity shortens access roads, access time, and permanent utility lines for the initial base. Later, as more-frequent flights increase the absolute probability of a landing failure, base growth will compensate by increasing the separation of active pads from other base elements.

Figure 3 indicates practical zoning for base growth, to avoid topological interferences into its future. Power plants, supplying both industrial and human needs, would grow to the west, away from the polluting spaceport. Nuclear reactors would probably eventually supplant solar arrays for a large base, and could be sited to the west with minimal shielding. Industry can grow to the north, into the rich and accessible plains of the Sea of Tranquility. The pressurized habitation "village" can grow unimpeded to the south. If the base continues growing into a real lunar city, and if sited close enough to the highlands at the Sea's edge, we could envision the human quarter eventually spreading into the interesting topography of the foothills there.

The spaceport can grow as needed to the east, with the region surrounding and beyond it remaining "wilderness". Preserving untouched lunar terrain within sight of the settlement may well prove critical for psychological relief if the base grows into a densely populated city. In the enclosed, technological human environment of a lunar settlement, viewing barren wilderness would provide the only access to native order available to the inhabitants. Accommodating such a use may become a major constraint eventually, albeit one easily foreclosed by short-sighted planning. It makes sense to combine an exclusion zone with the spaceport, since the latter's function will keep it the most growth-choked base element indefinitely into the future.

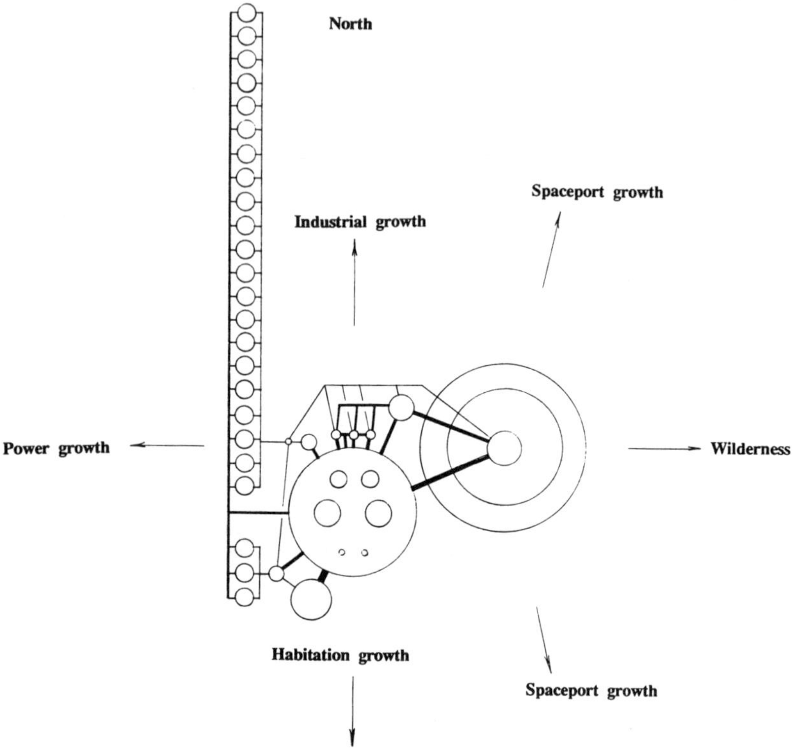

Figure 3. *Element proximity diagram.*

The Site Plan

We can now trust the proximity diagram as a guide in laying out an actual site plan, knowing that element relationships have been accommodated according to their importance. Essentially our task at this point is to introduce actual element details, and the physical infrastructure necessary to realize the element relationships throughout the site. Figure 4 shows the final site plan for our study. Power, grounding and fluid lines are not shown, but reference to Figure 3 verifies that their layout is sensible. Line lengths between connectors (which require inspection) are buried in 1 m deep trenches. The power plant bus line runs on the graded but unpaved surface beneath the solar arrays.

The maneuvering room required by the large robotic vehicles has led to merging the small workyard with minimum workzones around the oxygen reactors, power storage modules and power conditioning substation. This simplifies the grading, graveling and compacting activities necessary to build these paved areas. Paving itself is necessary to limit dust-contamination of critical systems and to provide a predictable surface for repeated travel. Roads to accommodate the straddler regularly are 25 m wide; truck access roads are 10 m wide. At reasonable baseline excavation rates (governed by the oxygen production goal), it takes 1 yr to prepare the substrate area shown (Buddington, 90). Most areas are scraped 20 cm deep; excavating foundations for the heavy habitat shelter, oxygen reactors and LLOX depot to 1 m allows reasonably small footings. Generating sufficient gravel and sand to pave most of this area to a depth of 5 cm requires almost an additional year. This 2 yr site preparation period amounts to only half the time required to bring all the base elements from Earth.

Most of the roads per se are required for the spaceport. Two north-south roads establish the pattern. The west service road allows maintenance access to the LLOX depots which will be associated with each landing pad developed; these need to be behind the debris barriers, and connected with power and gas lines to the rest of the base. The straddlers require easy access to landers, and so use the east road in front of the barriers. As more pads are developed, barriers can be erected between them to limit adjacent blast effects. An east-west artery links the spaceport with the rest of the base. A pair of north-south dust-control roads allows truck and straddler access to the solar power plants. The arrays themselves are staggered to permit both close packing and maintenance access to their critical tracking and control mechanisms.

The modest separation between the base center and the spaceport service road allows the gradual buildup there of an "earthwork" consisting of the gangue separated from the ilmenite reactor feedstock. Deposited and compacted incrementally, such a berm could reach 35 m in height, providing both a retrievable stockpile of these materials and a debris shield wall west of the spaceport. Five years of production at the baseline rate (100 mt oxygen per year) yields a sitework only 10 m high, 50 m wide and 110 m long; however, provision of a useful tailings location seems sensible. Developing more landing pads will fulfill mining requirements for a few years. After that, the habitation and power areas should grow; ultimately mining will lead industrial growth to the north.

Progressively folding in constraints and details deriving from a specific program and element set has enabled us to apply a simple method of developing site plans to the example of an early lunar base. The result is a document useful for quantifying

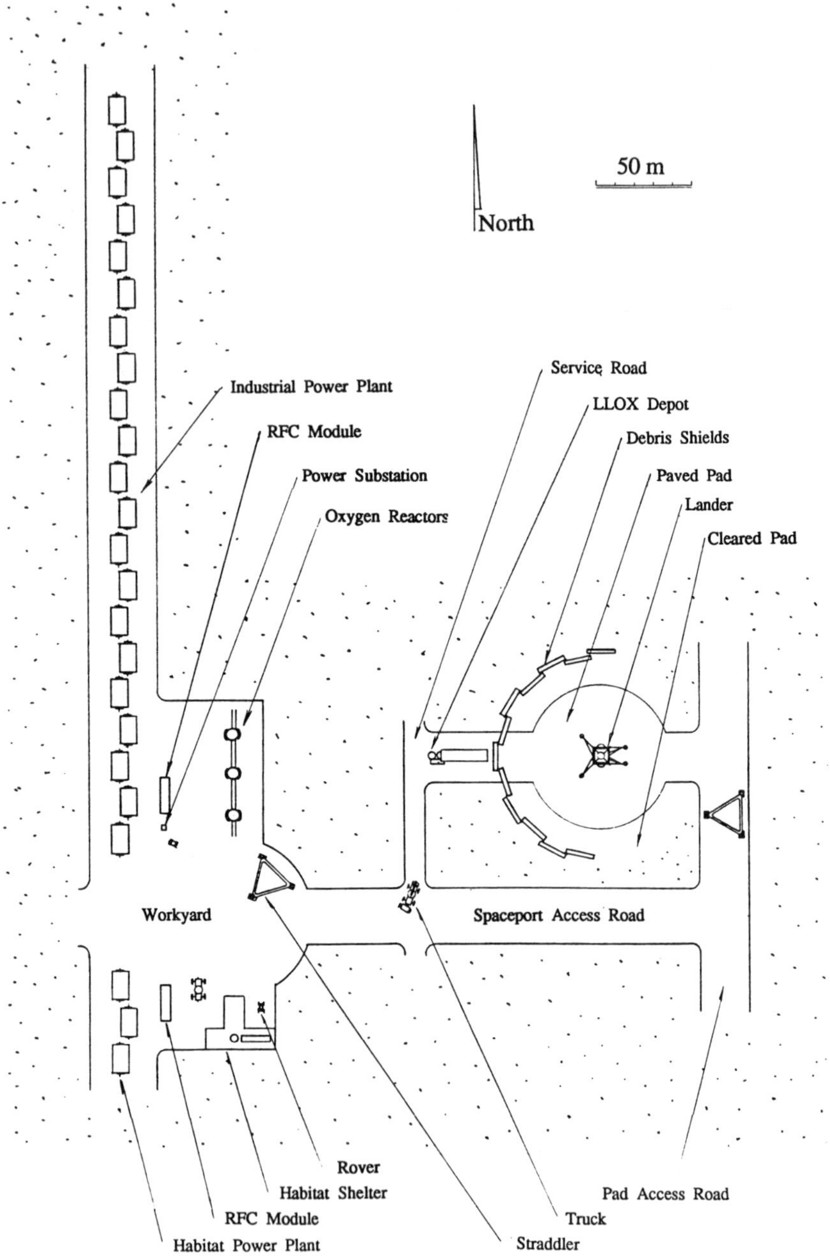

Figure 4. *Lunar base site plan.*

operational activities in and around the base. In particular, the phased buildup of the base, constrained by flight, diurnal and surface-vehicle schedules, can be credibly investigated (Buddington, 90).

If such a base, established for research, demonstration and production purposes, did continue to grow indefinitely, the zoning system we have outlined would lead naturally to a cruciform, cardinally-oriented transportation and utility service infrastructure. The city center would grow into an "activity hub" at the intersection of these cardinal spines, which could grow to immense scale, absorbing the initial base site but preserving the original regional separation. Evolved directly out of our site constraints, this orthogonal infrastructure is immediately recognizable to city planners as the decumanus, kardo and forum layout of all Roman cities, one of the most pervasive, persistent and successful designs in the history of human culture. The simplicity of such a sectored scheme appears able to serve well the changing requirements in scale, element type, and even program emphasis as a permanent Tranquility Base grows.

References

Buddington, P.A. (1990). "Manifesting for a Lunar Robotic, Oxygen-Producing Base." SPACE 90, ASCE.

RLSO - Robotic Lunar Surface Operations: Engineering Analysis for the Design, Emplacement, Operation and Performance of Robotic Lunar Surface Systems, (1990). The Boeing Company D615-11901.

Sherwood, B. (1990). "Lunar Base Elements Designed for Robotic Operations." SPACE 90, ASCE.

Woodcock, G.R., Sherwood, B., Buddington, P.A., Bares, L.C., Folsom, R., Lousma, J., Mah, R., Whittaker, W. (1990). "Application of Automation and Robotics to Lunar Surface Human Exploration Operations." SPACE 90, ASCE.

Lunar Base Elements Designed for Robotic Operations

Brent Sherwood [1]

Abstract

Equipment concepts developed for a study of robotic lunar surface operations are presented. The concepts were evolved together with a consistent robotic buildup and operations scenario; they reflect contemporary understanding of lunar base technologies, and specifically accommodate robotic constraints. The entire inventory of elements required for an initial base is discussed.

Introduction

The history of lunar base concepts is filled with images of astronauts busily building themselves a scientific outpost. Today's more realistic view recognizes that the ongoing construction, operations and maintenance work associated with a lunar base is far too extensive, heavy, hazardous and time-consuming to burden space-suited people with it sensibly. The base should serve the crew, not vice versa. A complementary mix of both people and machines is required to maximize base productivity. A recent study (RLSO, 89) investigated the high-leverage application of automation and robotics (A & R) to early lunar surface operations. The major conclusions are presented in a companion paper (Woodcock et al, 90), which also justifies the base program (an initial, solar-powered, robotically operated, oxygen-producing, habitable facility which largely predates onsite crew presence). This paper details the base elements developed in the course of that study.

The elements evolved along with specific robot concepts (also presented here), and a detailed operational scenario, together and iteratively. The robot machine types thus grew directly out of specifically identified tasks, incorporating mechanism and control concepts which are advanced but attainable.

1 Senior Engineer, Advanced Civil Space Systems,
Boeing Aerospace & Electronics, JX-23, Huntsville AL, 35824-6402

They reflect real lunar conditions, and are not merely borrowed from terrestrial concepts. Finally, they constitute a minimal set of versatile equipment, rather than an unaffordable catalog of unique surface systems.

Earlier studies (Sherwood et al, 89) have concluded that factoring specific advantages and limitations of A & R into space systems destined for remote operations, from the earliest stages of conceptual design, is a precondition for success. The A & R operations context has fundamentally affected the lunar base element concept designs we report here. Principles of predictability, modularity, commonality, minimal assembly, maintenance access, positive identification, and self-monitoring have guided the workpiece configurations. Our goal was to insure credible conclusions about performing lunar work remotely, by arranging a navigable, manipulable, understandable environment conducive to robotic activity.

Primary Elements

Figure 1 shows our single-stage, man-rated, cryogenic lander, considered a base element because of its need for servicing there. Flown robotically, its down-cargo capacity is 30 t. Initially, it would be supplied with propellants by the transfer vehicle from Earth. After the first 2 yr of base buildup, its liquid oxygen would be produced and tanked at the base. Flight design (including engine-out safety) has

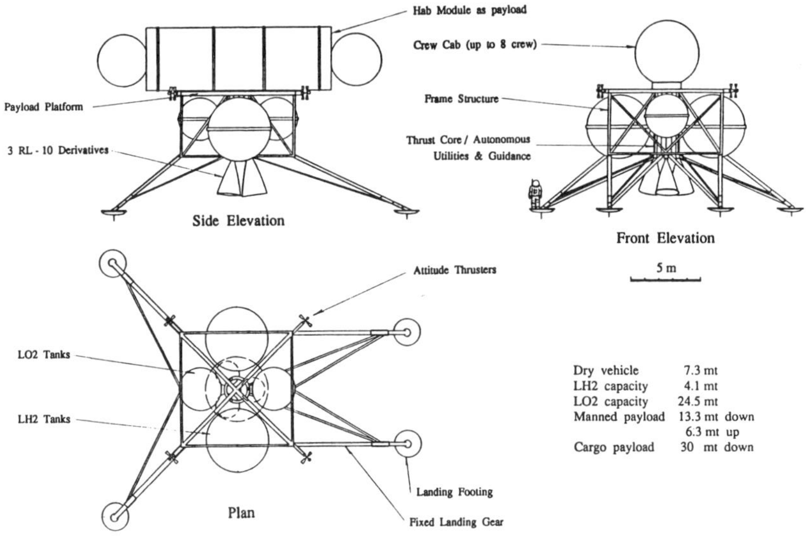

Figure 1. *30 t capacity lander.*

not been compromised at all for surface functions; there is no mass penalty for offloading equipment. The unique landing gear footprint accommodates the straddler described next.

The straddler (Figure 2) is a versatile, unmanned robot optimized for those jobs which clearly exceed human scale: offloading landers; moving empty landers; carrying heavy or bulky elements around the base; positioning elements requiring deployment or assembly; mining, moving and depositing lunar regolith. Two are required at the initial base. Assembled in orbit from a few large parts, each drives itself off the lander which delivers it. Solar powered, they operate only during the lunar day, and can act as mobile power utilities if needed.

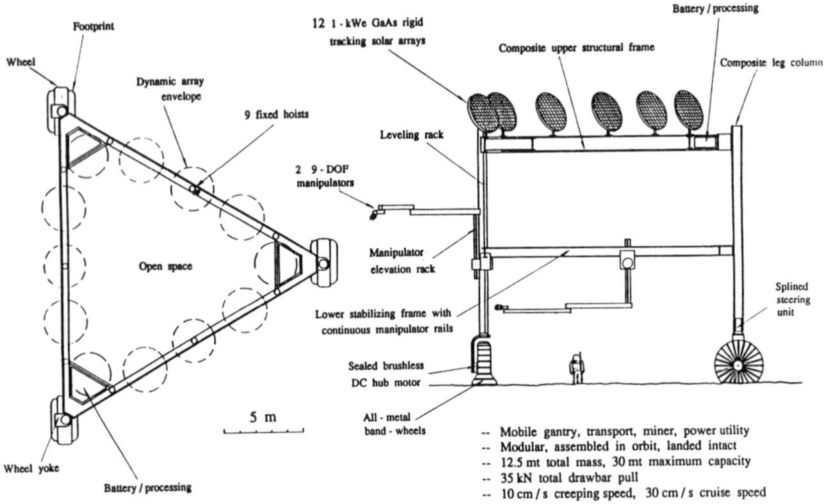

Figure 2. *Straddler mobile gantry.*

The straddler's two open triangular frames ride vertically on three columnar legs with captive rack-and-pinion drives. Independent leg motion allows the robot to be self-leveling on cratered ground. The top, "strong" frame carries 12 tracking, 1 kWe solar arrays, peaking batteries and onboard processors, communication equipment, lights and sensors, a manipulator end effector set, and 9 fixed cable hoists with which to lift, position and carry payloads massing up to 30 t. The lower, "light" frame stabilizes the vehicle laterally, and deploys jacks to the ground to allow leg maintenance. Two multi-jointed manipulator arms ride continuously around a rail on this frame. Each can reach beyond the middle of the straddler envelope from any perimeter point, below or above the top frame. Independent, brushless DC motors provide motive power in the wheel hubs, and steering at the leg bases. The metal wheels deform when loaded more heavily to maintain sufficient flotation even on unimproved lunar soil. Power and control lines from the top frame to the drive units are coiled within each leg. Structural members are made of graphite composites; leg racks, manipulator rails and wheel yokes are of titanium. Maximum speed is 30 cm/s; when mining or positioning large payloads, "creeping motion" of 10 cm/s or less is used to limit dynamic effects and maximize available torque.

The mobile miner/separator plant (Figure 3) is an attached payload for the straddler. It excavates regolith, mechanically separating useful fractions as it moves. The amount of material excavated to support 100 t/yr of LLOX production (and paving gravel generation) is extremely small (1582 t/300 hr lunar day) by terrestrial standards, and is achieved by taking 5 cm slices off the lunar surface. 1/6 of the miner's time is spent dumping materials at separate deposition sites. 1/4 of the remainder is lost to steering changes. 1/11 of that remainder is spent hoisting native material to the top of the beneficiation stack. A combination of scalpers, hoppers, seives, chutes, and magnetic separator divides native regolith into stones, gravel, sand, 55 % enriched ilmenite feedstock, and leftover gangue. The miner covers 1.8 m^2/min when excavating. Motors and actuators are positioned accessibly around the outside.

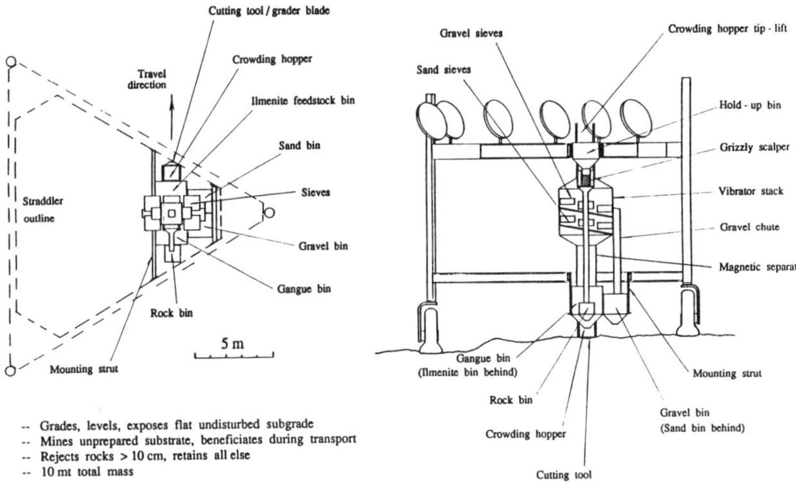

Figure 3. *Miner/separator plant attachment.*

Three ilmenite oxygen reactors (Figure 4) evolve oxygen gas from the enriched fines. Each is brought intact by one lander flight, and positioned by a straddler on a prepared foundation. Fluid-bed batch reactors, each consumes about 130 kWe while running during the lunar day. A straddler charges each reactor with two hoppersful of ilmenite feed (53.2 mt) at the beginning of the day. They are then sealed and pressurized to 10 atm with hydrogen. A pump circulates the gas through a plenum at the bottom of the vessel, to fluidize the bed and increase its reaction surface area. Heating the flowing gas heats the solids to 900 C over 150 hr. The pressure vessel is made of carbon/carbon composite, with a vapor-deposited alumina abrasion liner on the inside and flexible ceramic insulation (TABI) on the outside.

The gas (90 % H_2, 10 % H_2O reaction product) removed near the top of the vessel is stripped of 96 % of its suspended dust by two, staged cyclone-separators, before passing by the zirconia membranes of a gas-phase electrolyzer. The pure oxygen gas thus extracted is piped away as the plant runs at temperature for another 150 hr. Keeping the spent solids fluidized as they cool down into the lunar night helps prevent caking. Then the hydrogen is compressed into storage bottles, and the evacuated reactor is opened and tipped. The spent solids empty into hoppers positioned by a cart on rails beneath the three reactors and carried to the deposition site by a straddler.

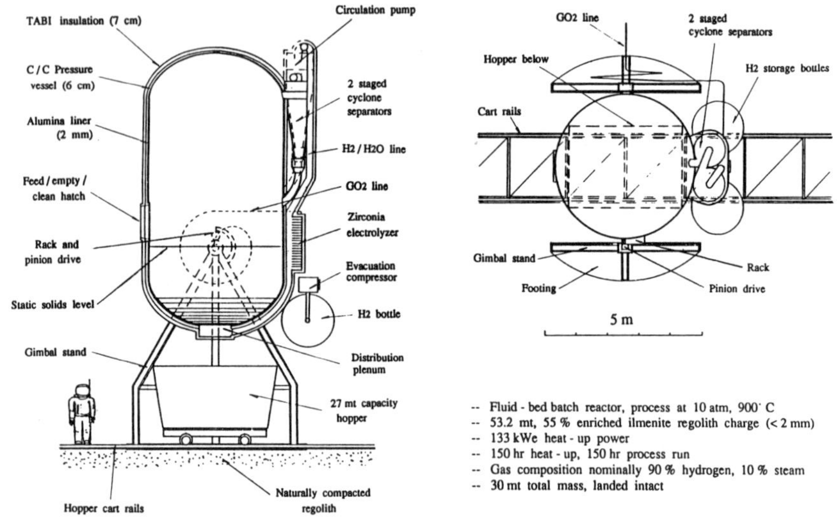

Figure 4. *Lunar ilmenite oxygen reactor.*

Oxygen gas flowing from the reactors is collected into one line which runs underground, cooling passively, to the LLOX storage/liquefaction depot (Figure 5), which condenses and stores it cryogenically until needed to fill a lander's LOX tanks. The depot, closely associated with a single landing pad, stores enough LLOX, in two tanks, for two entire lander loads. Three refrigeration units are each capable of half the cooling load, allowing offline maintenance. The machinery and radiator are both stood off the graveled ground to minimize casual dust contamination from crew activity. A fixed lightweight sunshade treated with selective coatings is mounted over the radiator to prevent it from viewing the sun directly most of the time. Its geometry is optimized for an east-west orientation so as not to limit its space view factor excessively. The storage tanks, the pallet of liquefaction machinery, and the radiator modules arrive intact but separate. Required assembly amounts to deploying the trussed platforms, emplacing the units, deploying the fixed radiator sunshade, and connecting power and fluid lines.

Our daytime power plant consists of 24 freestanding 20 kWe photovoltaic units (Figure 6). Designed to be transported intact, they require only deployment, and connection to a utility bus. Each 8 x 12 m rigid, composite waffle-plate carries a GaAs array sized for the ambient lunar radiation and temperature environment. The plate folds for transport, protecting its active surface. Hoisted by a straddler into position, it is deployed by the straddler's manipulators. The panel unfolds, the tripod-legs are extended and locked open, and auger pins are screwed down into consolidated regolith to anchor them in place. When aligned with its pivot axis north-south, the unit tracks the sun throughout the lunar day. Redundant tracking motors allow online motor changeout.

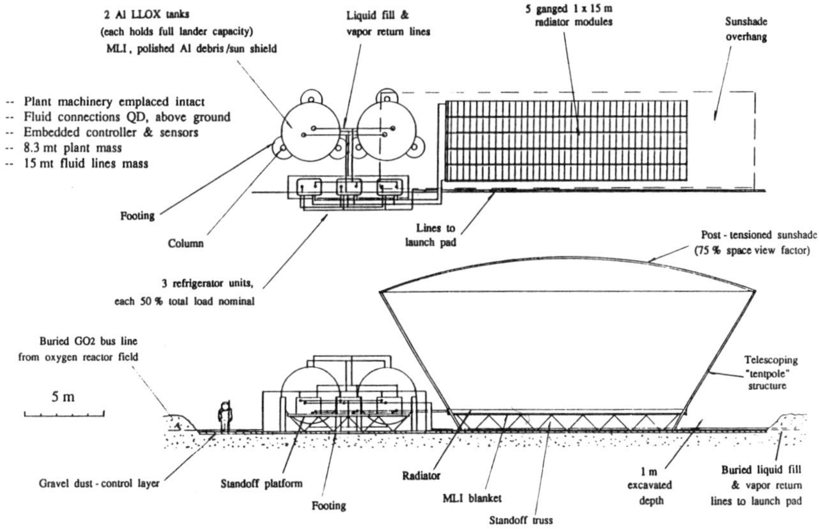

Figure 5. *LLOX liquefaction/storage depot.*

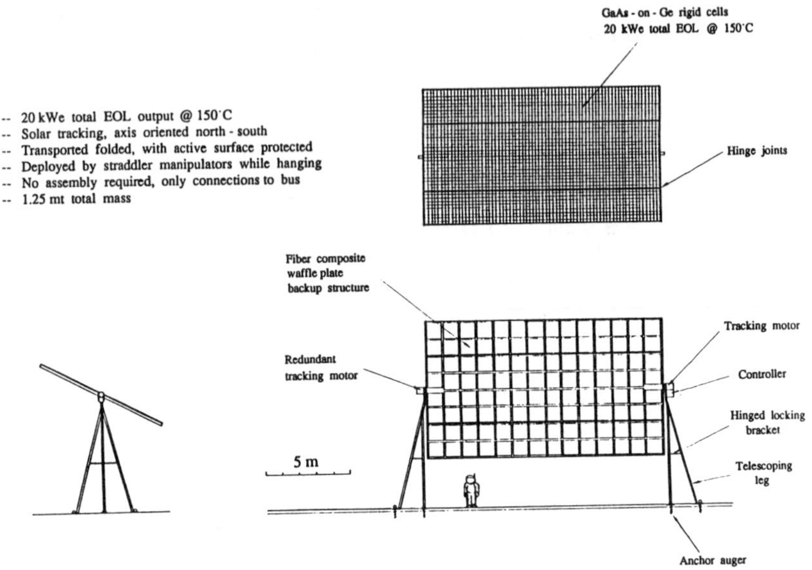

Figure 6. *Freestanding 20 kWe photovoltaic unit.*

Our nighttime power plant consists of two cross-bussed regenerable fuel cell modules (RFCs, Figure 7), one dedicated to the habitat system and the other to the industrial elements. These operate at 50 % overall efficiency, splitting water when being charged and recombining the gaseous hydrogen and oxygen to provide a modest 40 kWe total power during the 336 hr lunar night. This sustains vital functions such as life-support, and keeps critical equipment from cooling to -160 C. They are landed intact, ready to be towed or carried into position and plugged in. Maintainable components are located accessibly around the perimeter.

An initial habitat system of pressurized modules derived from Space Station Freedom hardware (Figure 8) would consist of a large living/science module, two simple airlocks with dust-shower facilities, a connecting node with lookout cupola, and a pressurized workshop featuring a large end-hatch, through which equipment needing dextrous repair can be placed. A separately constructed radiation shelter is required to mitigate galactic and solar radiations (Figure 9). Building the system robotically before people ever arrive allows even early crews to take full advantage of safe accommodations.

Our shelter uses lunar regolith as the shield material, and is also expandable. A 0.5 m thickness of unconsolidated (average density = 1.5 t/m^3) regolith without macroscopic voids will limit annual radiation exposure to about 20 rem/yr (well below the recommended astronaut allowance) even pessimistically assuming one large solar flare (SPE) per year. The basic shelter components are vault sections of corrugated aluminum with inner structural ribs and outer tension spacers. Transported nested, they are emplaced on metal footing pads by a straddler and riveted together robotically. The surface around the footing pads is graveled for dust control. Corrugated panels are riveted to the tension spacers as an outer skin, and the void created is filled with regolith as it is built up from

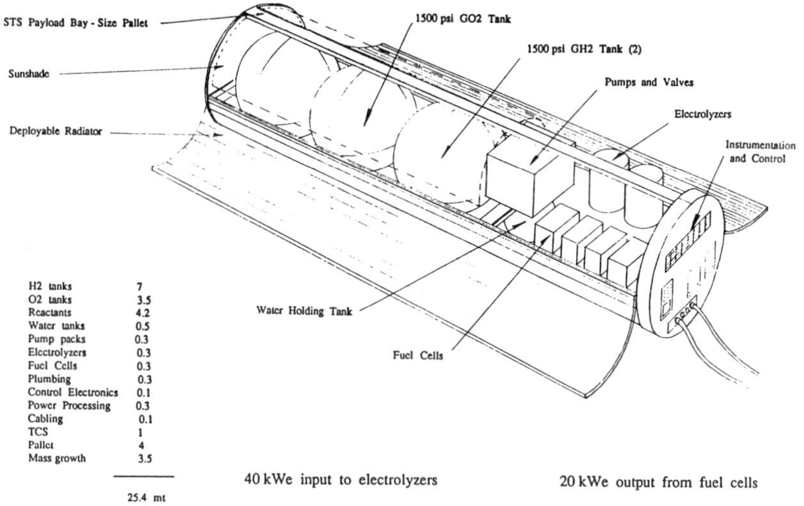

Figure 7. *Regenerable fuel cell module.*

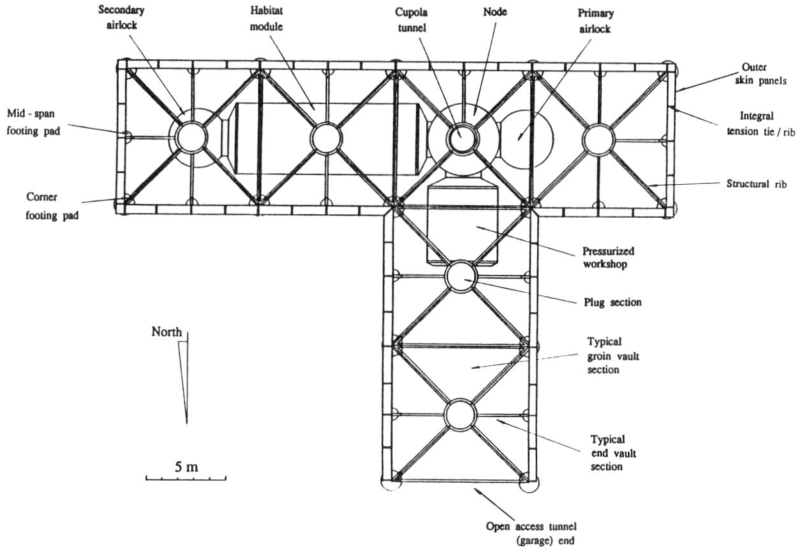

Figure 8. *SSF-derived habitat system plan.*

ground level. The tunnel end beyond the workshop hatch is left open, providing a largely shielded, unpressurized garage volume. For SPEs, a straddler lifts a separate regolith-sandwich wall section into place over the open tunnel end. Utility connections for rooftop radiators and communications equipment occur through the circular plug sections in the vault crowns. The required east-west orientation of the rooftop radiator establishes the long tunnel's site orientation.

- 0.5 m regolith (average density = 1.5 mt / m3) yields 7 rem max SPE dose, 13 rem / yr GCR dose (50 rem / yr astronaut allowable)
- Straddler holds 0.3 m regolith sandwich panel over open tunnel entrance for SPE
- Inner vault sections transported nested
- Foundation pads positioned & cabled together, then gravel layer deposited
- Construction starts from tunnel intersection around cupola, inner structure built first
- Outer skin panels attached from bottom up, cavity filled with regolith as built
- External connections for power supply, thermal rejection, communications
- 3 ganged 1 x 15 m rooftop radiator modules with post - tensioned sunshade
- 14 mt total shelter hardware mass

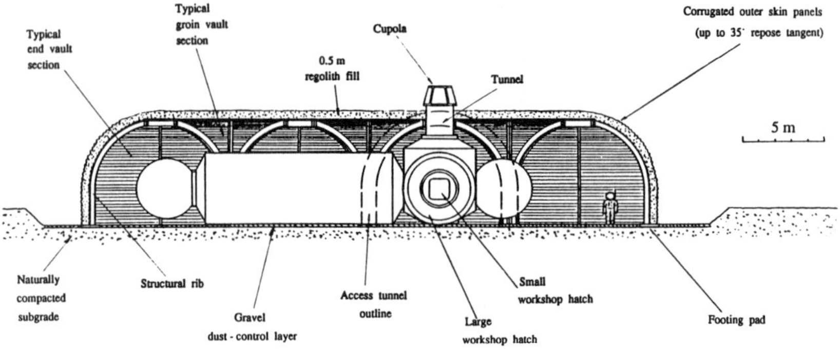

Figure 9. *Habitat system radiation shelter.*

A mobile robot is required for intermediate jobs: filling cavity walls with regolith; compacting roads; moving small piles of soil; scraping out the oxygen reactors; cleaning solar arrays and radiators; positioning crew, specialized manipulators or sensors high up on base equipment; driving into the garage to place components in the workshop; excavating trenches for buried fluid and power lines; and lander maintenance including connecting LLOX-fill and conditioning umbilicals. For all this we use two high-reach trucks (Figure 10). Stabilized by deployable outriggers, this vehicle can extend its turntable-mounted, telescoping boom throughout a range from the top of the tallest base element to below grade. It can tow trailered accessories or move modest amounts of regolith. The truck masses 6 mt including boom and basic attachments, and stores 60 kWhr of power, giving it up to 10 hr of operation on one charge. All large-travel actuators are robust, open rack-and-pinion drives. Nominal driving speed is 10 km/hr, and the machine can be handled by an onboard operator when desirable.

The final mobile robot we need is a light <u>rover</u>, akin to the Apollo LRV but heftier and solar-powered for longer-lived operation. Unpressurized, this is used primarily for crew transport around the base, and short excursions beyond. A 1 kWe tracking solar array also provides a convenient mobile power source. It would seat two, with a top speed of be about 15 km/hr on a paved surface. Autonomous navigation increases its usefulness, particularly for rescue. Before the first 30 mt cargo flight ever arrives at the base site, two of these rovers would be flown there on an Atlas-Centaur flight. They would survey the prospective base location, using primarily video sensors above ground and ground-probe radar (GPR) below ground, to map the site's character in three dimensions down to 10 cm-resolution. (A complete geometric model of the site and mining volume can thus be developed, and the site plan finalized, before landing any heavy equipment there.)

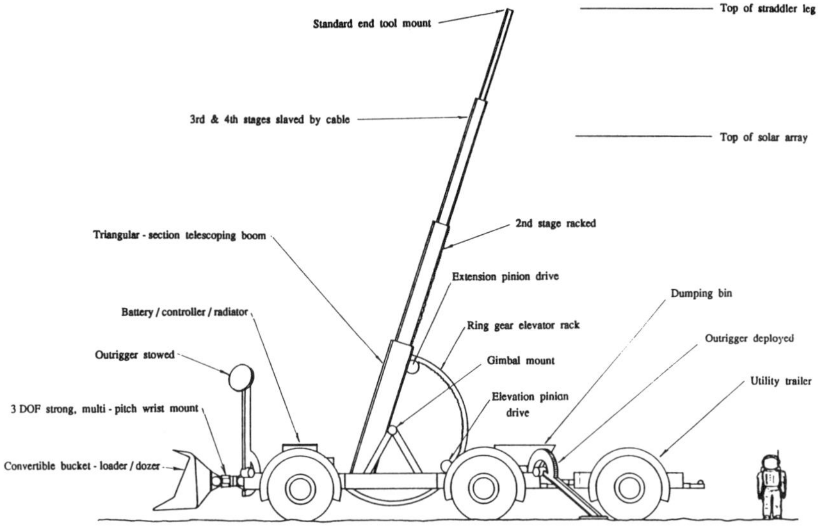

Figure 10. *Outrigger high-reach truck.*

<u>Utilities, Siteworks and Conclusion</u>

"Utilities" are defined here as ancillary elements that also must be brought from Earth. <u>Debris barriers</u> are 10 x 14 m corrugated composite panels deployed (even more simply than the solar arrays) between landing pads and the base, to limit abrasion damage from soil particles ejected by the lander exhaust plume. Specialized <u>end effectors</u> and accessory tools are required for robotic manipulators, human crews, and particularly for the truck. <u>Guidance beacons</u> permit accurate landings, and local versions assist robot navigation around the site.

Communication transceivers maintain datalink with Earth orbit and among mobile base elements. Power, data and grounding lines link all fixed base elements. A power substation regulates the bus loads of the industrial solar arrays, reactors, depot and RFC modules, and controls contingency switching between industrial and habitat loads. Fluid lines connect the oxygen reactor field with each storage depot, and each depot to a buried LLOX depot in its associated landing pad. Small local lights, both freestanding and attached to the shelter, augment Earthlight for nighttime human activity in critical areas. And deployable tripod-mounted sensor heads transmit critical views and data to the base control consoles.

"Siteworks" are base infrastructure elements composed entirely of indigenous material. Chief among these is the spaceport, consisting of a cleared, leveled and paved (5 cm-deep compacted gravel) pad 100 m across in which a hovering lander can safely touch down, and be offloaded and maintained. Extremely heavy base elements (habitat shelter, oxygen reactors, and storage tanks) require foundations to prevent excessive settling; these are prepared by removing 1 m of native overburden, exposing naturally consolidated regolith. A leveled and paved workyard is necessary for limiting dust contamination of reconfigurable mobile elements. Paved roads connecting fixed elements are required to limit dust kickup at higher driving speeds. Deposition sites receive unused fractions separated from the mined material, and spent oxygen reactor solids, which may be recovered later for other industrial processes.

The extensive set of equipment described here comprises a representative, arguably minimal, catalog of the items really needed to establish a modest, crew-supporting, oxygen-producing lunar base. It includes all the mobile robotic equipment required to keep the base operational, from the first precursor landings all the way through full production and growth. These robot concepts have grown strictly out of the tasks required by operating on each other and on the fixed facilities. All of these elements are matched to the capabilities and limitations of such a small robot suite.

References

RLSO - Robotic Lunar Surface Operations: Engineering Analysis for the Design, Emplacement, Operation and Performance of Robotic Lunar Surface Systems, (1990). The Boeing Company D615-11901.

Sherwood, B. (1990). "Site Constraints for a Lunar Base." SPACE 90, ASCE.

Sherwood, B., Buddington, P.A., Whittaker, W.L. (1989). "Earth Orbital Operations Supporting Manned Interplanetary Missions." AAS 89-184, Proceedings of the 1989 Goddard/AAS Symposium on Orbital Mechanics and Mission Design. Univelt.

Woodcock, G.R., Sherwood, B., Buddington, P.A., Bares, L.C., Folsom, R., Lousma, J., Mah, R., Whittaker, W. (1990). "Application of Automation and Robotics to Lunar Surface Human Exploration Operations." SPACE 90, ASCE.

MANIFESTING FOR A LUNAR ROBOTIC, OXYGEN-PRODUCING BASE

Patricia Buddington[1]

Glossary

Cycle ....The 28-day diurnal cycle, consisting of one lunar daytime and one lunar night.
Interval ....The period between lander arrivals (3 cycles)
Lunar day ....The sunlit, daytime fortnight of a lunar cycle.
Lunar night ....The dark, nighttime fortnight of a lunar cycle.
Cislunar Space......The area between and including the orbits of Earth and the Moon
Supervised ....A robotic control mode in which the machine performs detailed task planing and execution, and processes raw data, in response to and in support of intermittent, more abstract human commands.
Tele-operation .....Strictly, the direct control of a robot by real-time human operation.
Telepresence ....The virtual participation of humans in remote robotic activity, through sensors communications links, and manipulators.
Tele-robotics ......The use of supervised autonomy, backed up by tele-operations when required.

Abstract

Boeing has done a study (Woodcock et al,90), for NASA Ames, on using lunar robotic operations for surface base development. Our study objective was to define automation and robotics systems that could enable two high-leverage accomplishments: 1) the early emplacement of a radiation shielded habitat, and 2) an early start on lunar oxygen production. The study identified where such systems may be employed, with human missions, to develop a lunar liquid oxygen (LLOX) production plant system. The goal of this LLOX plant is 100 metric tonnes per year at full production with 3 oxygen reactors to support any lunar lander vehicle oxygen needs. As the base grows, it is possible that LLOX will be available to transport into cislunar space. The ground rules limited the number of flights to four per year on 90 Earth day launch centers (3 lunar cycles), with delivered payloads of 30 metric tonnes per flight. This included the human missions of 6 people staying for one lunar cycle, which occurs twice in the base buildup. The level of robotics was considered human supervisory and below, being directed from the Earth and/or by the manned missions. The levels of human and robotic involvement this entails is shown in the robotic software architecture diagram of a companion paper (Woodcock et al, 90). Further investigations indicated that under the groundrule conditions a site development plan using only manned missions could not perform the necessary major construction in the

---

[1]Specialist Engineer, Boeing Aerospace and Electronics, The Boeing Company, 499 Boeing Blvd. M/S JX-23, P.O. Box 240002, Huntsville Al 35824-6402

available sortie time and that the resultant extended schedule of base buildup would not allow a production start within five years. Five years is considered a goal for a reasonable return on development investment (Woodcock 90). The use of telepresent and tele-operated robotic equipment would allow a practical buildup schedule. This schedule will allow a production start within five years of initial cargo landing, particularly if the site has been preselected and surveyed before a major operations commitment has begun (Barnes, et al, 90). Enough contingency time is by nature built into such a schedule, with the 90 day launch centers and the rate of required paving material generation, to be able to expand the base and begin production with stockpiled ore before regular human presence is established. This paper addresses the manifests and schedules for this type of development.

Introduction

The flight sequence over the initial five year period contains mostly cargo manifests to support the robotics operations needed for base construction. The construction would proceed using two each of the three types of vehicles: 1) a large straddler to unload cargo from the lander, do mining operations, and dump separated regolith where it is required and erect large structures; 2) a "truck" with a "high reach" extended arm with repair function attachments and bulldozer/trailing excavation equipment accessories that make this a utility vehicle, and 3) a "rover" that will first function as a robotic vehicle to survey and deploy beacons on a precursor mission and will later be converted to manned transport vehicles. These will vehicles be covered in more detail in a companion paper (Barns, et al, 90). Only two manned missions are scheduled in the first 15 flights. These are to support the site development, and are not to be involved in the construction operations in and of themselves. It is only after the base has been roboticly constructed and the oxygen plants are in full production do the manned flights dominate the transport schedule, until then the flights are mostly material and equipment deliveries.

The surface development dictates that ilmenite mining will be a consequence of site preparation until LLOX production begins. This surface development along with the site plan requirements will have internal priorities for the timelines that affect the surface buildup operations and supporting manifests.

Manifests Development

Selection of the site and the establishment of a site plan are explained in a companion paper(Sherwood,90b). The base is to be established at Mare Tranquillitatis. The site plan is shown, for convenience, to be divided into four major areas: the spaceport, the base center, the industrial area and the expansion area. The Spaceport consists of the central pad and its unpaved border, the oxygen refrigeration and storage area (LLOX storage area and Depot), the straddler roads, and the truck access road. The base center consists of the habitat area, the paved and unpaved habitat power area with the habitat regenerable fuel cell area and the workyard. The industrial plant area consists of the oxygen reactor area with the industrial regenerable fuel cell area, the paved and power area roads and the unpaved area under the industrial arrays. Expansion areas are any portion of the base not shown in the site plan and would include additional landing pads, extension of the habitat area and industrial area (including power arrays). An overall timeline schedule showing the arrival and operation of each major piece of equipment to be set up the base is given in Figure 1.

Sections of these major areas will have priority over other parts of the same area, as an example: the paved and unpaved landing area will have clearing and paving priority over the straddler roads due to the need to control dust spread by the lander activities. The straddler can move over unimproved terrain, therefor it does not need the road for mobility and at a maximum of 30 cm /sec it does not disturb much dust. The roads are there to ease straddler mobility as a convenience rather than a necessity. Both of these areas are within the spaceport but have widely different priorities.

Priorities for clearing and paving the sections are based on dust control

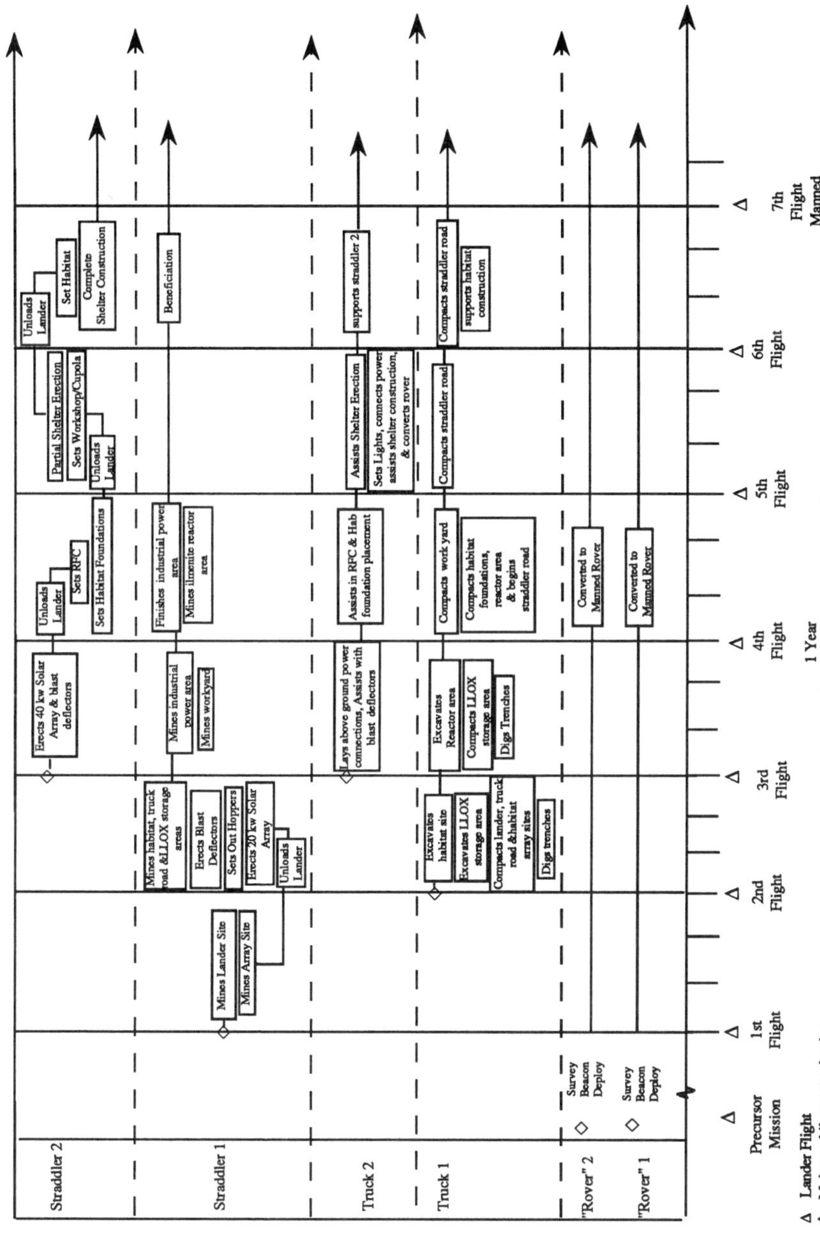

Figure 1. Robotics Operations Sequence

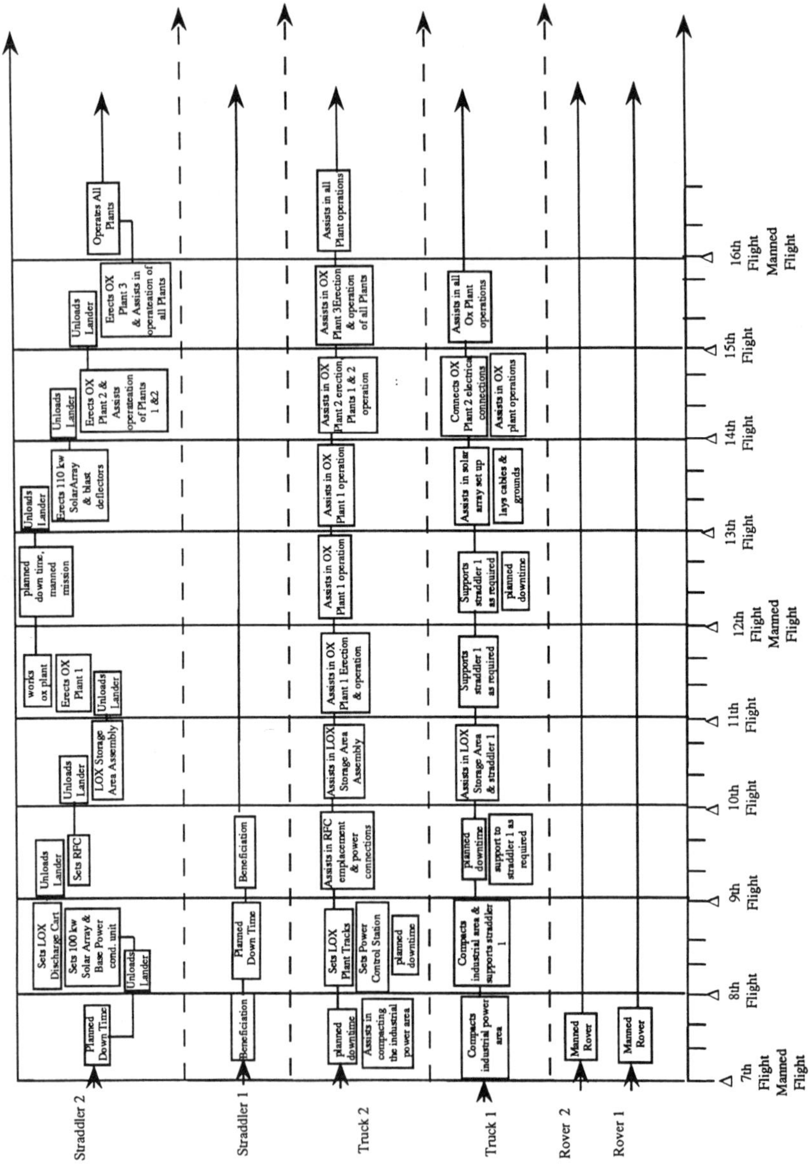

Figure 1. Robotics Operations Sequence(Con't)

over critical areas such as the solar arrays, habitat and work areas. Critically timed construction areas which require excavation to 1 meter depth and /or have early use have high priority. However, even with this hierarchy all areas in the original site plan have been cleared by the time of the first manned mission, which occurs on the seventh flight (19th lunar cycle after the initial cargo landing). All graveling and compaction of predesignated areas are completed the second lunar cycle after the eighth flight.

Table 1 shows the clearing (excavation) and paving as they occur on the lunar cycle schedule between lander flights. The rate of land clearing and beneficiation yields the amount of gravel produced per cycle with one hopperful equal to 27.14 metric tonnes (152 metric tonnes per lunar cycle). This is for an area cleared and processed by the straddler of 1013 cubic meters per lunar cycle (67 sq.m/hr. at 5 cm depth, 5/6 duty cycle for to account for dumping trips and 3/4 duty cycle to account for steering loses for 300 hr. per cycle). The straddler with the mining/beneficiation box does the processing of materials into ore,sand and gravel, rocks and gangue (leftover fines) and is the final level clearing apparatus. The truck is used as an auxiliary excavator around sites that are deeply emplaced (Habitat, LLOX storage, and oxygen reactor area working at 54 sq.m /hr. at 5 cm depth with steering losses and a 2/3 duty cycle for recharging accounted for in a 300 hr. cycle). The details of the equipment are found in a companion paper (Sherwood, 90a).

Beyond this, the human occupation flights begin with cargos supporting these missions and delivering additional equipment for the base, such as an additional intact miner box (self contained mining and beneficiation equipment used with the straddler), additional shelter and habitat hardware, additional power systems and additional processing systems (not necessarily LLOX production and storage equipment).

Throughout the first 15 flights there are two critical surface buildup phases where the surface construction is driven by the need for site preparation by a given date. The first phase extends from the first cargo landing to the first manned flight. A representative section of this schedule is shown in Figure 2 . The second phase from the arrival of equipment for construction of the LLOX storage area (10th flight, 28th lunar cycle) to the first oxygen reactor (11th flight, 31st lunar cycle). All other surface buildup is driven by the delivery schedule of cargo on the preset launch centers of 90 Earth days.

A third phase, in the overlap between initial production (one reactor on-line) and the end of buildup (three reactors on-line), have some critical actions conducted in daylight that bring the oxygen production to full capacity. Once oxygen production begins, slag dumping and reactor filling take place in pre-dawn hours. The equipment will be run off the industrial RFC as a regular part of the production cycle

First Critical Buildup Phase:

This phase is driven by the need to 1) clear then gravel the landing pad area, 2) establish the habitat area (including the workyard) for human occupation as soon as possible, 3) finish the deep excavations (habitat, LLOX storage and reactor areas) and 4) emplace the underground cables and piping that must be finished before construction can begin. The first manned mission will be used to checkout not only the habitat, but to conduct tests on the beneficiated regolith that may affect the design of the the fluidized bed oxygen reactors and observe the robotic operations, modifying them for efficiency as required. All of these human interfaces are done early enough for any impact to the schedule to be absorbed in either lunar contingency time, or in Earth based processing.

Second Critical Buildup Phase:

This phase is driven by the need to assemble the oxygen liquefaction / storage plant, and check it out in time for the first oxygen reactor to be emplaced and functionally checked. This activity lasts for two lunar cycles. It is time spent constructing the LLOX storage area following the arrival of flight 10. The LLOX plant functional checks must be performed in advance of the first oxygen reactor plant arrival on the next flight for a reactor and system functional check, so any occurring problems may be isolated to the additional equipment. It is likely that some of the contingency time will be used here in the

Lunar Cycle	Lander Flight	Area Cleared	Area Paved	Gravel deposited (hoppersful)	Gravel stored (hoppersful)
1	1	Lander Area (central pad, unpaved border & access road)	Lander Central Pad	5.6	----
2		Lander Area	Lander Central Pad	5.6	----
3		Lander Area	Lander Area / Depot	5.0 / 0.6	
4	2	Lander Area	Truck Road	1.9	3.7
5		Habitat Array	Habitat Array	8.7 *	0.6
6		Habitat Area/ Depot	Habitat Array / Lander Area / Depot Truck Road	5.0	1.2
7	3	Workyard	LOX Storage	1.8	5.0
8		Workyard / Industrial Power	Workyard	5.6	5.0
9		Industrial Power	Workyard	5.6	5.0
10	4	Industrial Power/ Reactor Area	Workyard	5.6	5.0
11		Reactor Area	Workyard/ Reactor Area Habitat Area	3.1/ 2.5 3.8 *	1.2
12		Base Extension	Reactor Area/ Straddler Roads	4.1/ 1.5	1.2
13	5	Base Extension	Straddler Roads	5.6	1.2
14		Base Extension	Straddler Roads	5.6	1.2
15		Base Extension	Straddler Roads	5.6	1.2
16	6	Base Extension	Straddler Roads	5.6	1.2
17		Base Extension	Straddler Roads/ Industrial Power Area Access Roads	1.1/ 4.5	1.2
18		Planned Downtime			
19	7	Base Extension	Industrial Area Access Roads	5.6	1.2
20		Base Extension	Industrial Area Access Roads	5.6	1.2
21		Base Extension	Industrial Area Access Roads	5.6	1.2
22	8	Base Extension	Industrial Area Access Roads	5.6	1.2
23		Base Extension	Industrial Area Access Roads	1.1	5.7

* Deposition tasks requiring < 5.6 hoppersful result in gravel storage for later use.

Deposition tasks requiring > 5.6 hoppersful tap the stored surplus

Table 1: Lunar Excavation / Paving Schedule

plumbing connection checkout, with the usual complications of a checkout program. All other flights carry equipment that should take no more than on lunar cycle to erect and checkout.

Production / Construction Phase:

This is the time between the on-line operation of reactor #1 and the full operation involving all three reactors. It begins the second lunar cycle after

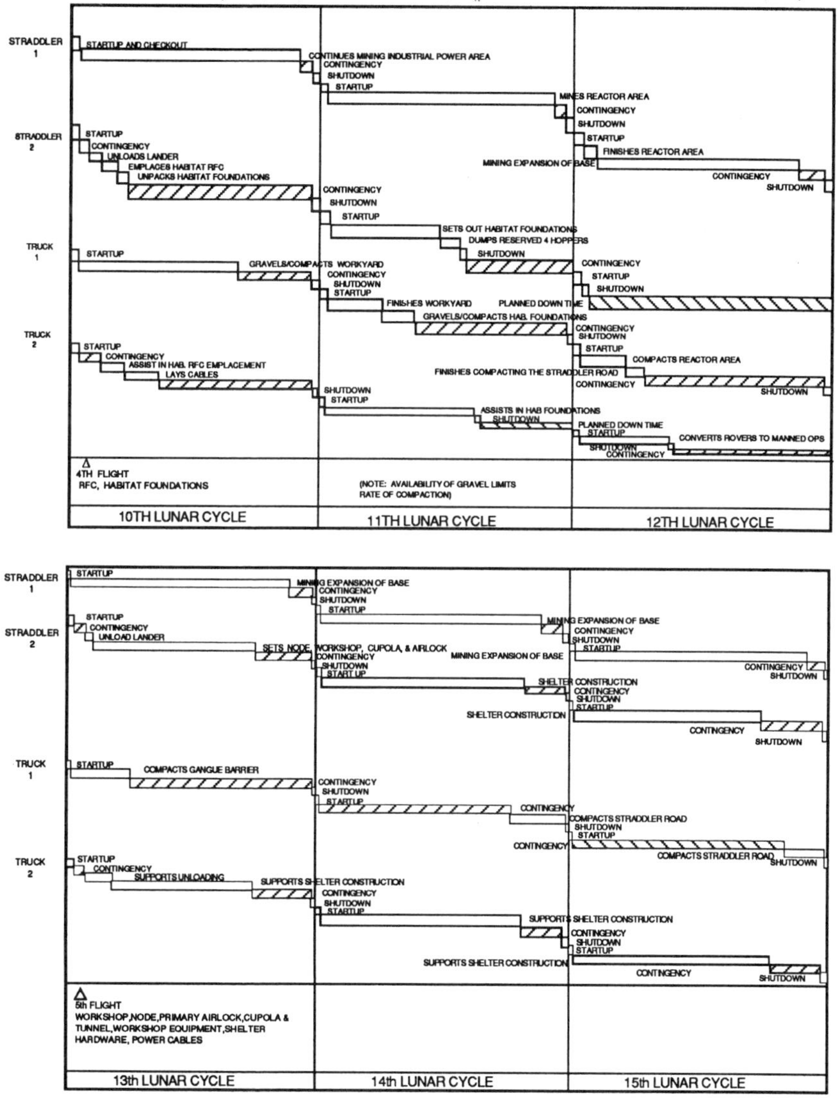

Figure 2: Selected Critical Surface Buildup Schedules

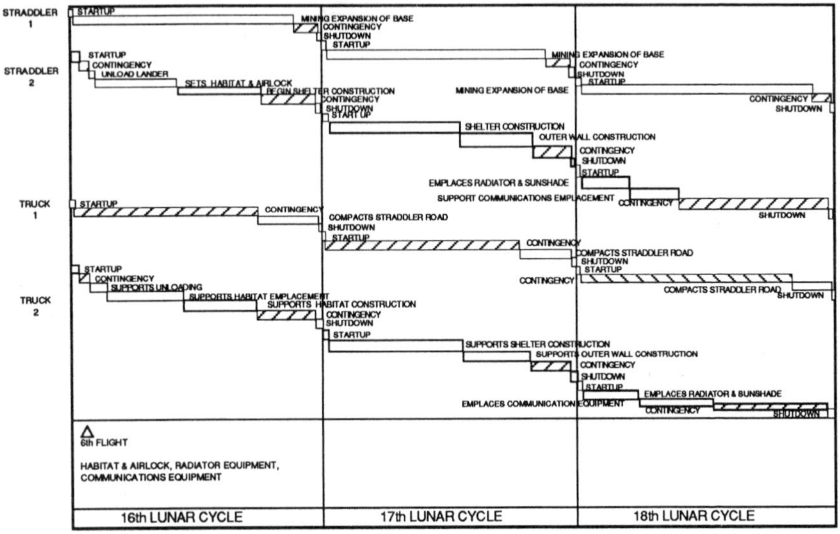

Figure 2: Selected Critical Surface Buildup Schedules (Con't)

flight 11 and continues through the second lunar cycle after flight 15. During this time, additional support power and operations equipment are landed and made operational besides the two reactors needed to provide a full production complement. The reactors will operate during the daylight, cooling down overnight. They will dump slag in the pre-dawn hours using the dedicated industrial Regenerable Fuel Cell (RFC) power. The trucks will clear away the discharge carts, clean and inspect the reactors in the pre-dawn by working off batteries and recharging off the industrial RFC. The straddler will be used in the pre-dawn to lift the full slag hoppers out of the way, dump and return the hoppers on industrial RFC power. The reactors will be loaded in the early daylight by a straddler with a charge of 53.2 metric tonnes of 55% enriched ilmenite ore. In this fashion the reactors have most of the lunar cycle daylight to work and the vehicles remain available for other base activities, such as maintenance, or exploratory activities.

Manifests are payload equipment lists for each flight and are dictated by the needs of the surface buildup schedules, the launch rate and landed load capacity. The schedules in turn are the results of the base buildup priorities.

Figure 3 shows the manifests required over a five year period to construct the base as outlined above and further described in companion papers. On close inspection, some critical items do not seem to have redundancy; however there is "hidden redundancy" in the manifests lists and in how the robotics are used. An example of this is the miner box that does the excavation. it appears to have no backup, but the system allows for modularity of parts and use of robotic remove and replace operations. Hence, redundancies are buried in the spares inventory and designed commonalty of parts. Early manifests will have large spares weight allowances for critical parts to cover a lack of redundancy in fully configured large/complicated equipment items.

The first flight brings the beneficiation / mining box on the straddler. the hoppers, which are scattered throughout the manifest table, will be used to store ore, gravel and gangue for later use. Rocks will be dumped and sand can be stockpiled for possible later use and the gangue used to begin building an in-situ barrier between the habitat / industrial areas and the landing pad. This is in addition to the blast deflectors brought on latter flights. the second flight brings the first truck and the underground plumbing and cabling that has to be positioned and buried as the foundations are made before they are graveled and

Flight 0 (direct Atlas-Centaur mission)

- Rovers
- Site Survey equipment

Flight 1

• Straddler	12.5
• Miner	10
• 5 hoppers	6
• Packaging	1
TOTAL	29.5 t

Flight 2

• Truck #1	6
• Truck boom tools	4
• Fliud & power lines for burying	1.9
• LLOX terminal	0.1
• 3 hoppers	3.6
• (1) 20 kWe PV unit	1.25
• Vibrating compactor	1
• Bucket-wheel excavator trailer	0.8
• 4 debris shields	3.2
• Spares	6.5
• Packaging	1.5
TOTAL	29.85 t

Flight 3

• Straddler #2	12.5
• Truck #2	6
• Truck boom tools	4
• 2 PV units	2.5
• Utility trailer	0.8
• 4 debris sheilds	3.2
• Packaging	1.0
TOTAL	30.0 t

Flight 4

• RFC module (for habitat system)	25.4
• Habitat shelter foundation materials	4
• Packaging	0.6
TOTAL	30.0 t

Flight 5

• Workshop module, node & primary airlock	17.3
• Cupola & tunnel	2.2
• Shelter structure	7.5
• Power cables	2.0
• Packaging	1.0
TOTAL	30.0 t

Flight 6

• Main habitat module & secondary airlock	20.0
• Shelter structure	3.0
• Communications equipment	0.5
• Radiator and shade	0.7
• Local lights	1.5
• Spares and stores	3.0
• Packaging	1.3
TOTAL	30.0 t

Flight 7 (crew - carrying mission)

- Stay time 30 days
- Verifies habitability
- Monitors robots
- Qualifies oxygen process

Flight 8

• 10 PV units	12.5
• Slag-hopper cart and rails	3.5
• 5 hoppers	6.0
• Power and ground cables	1.0
• Power switching substation	0.3
• 2 debris shields	1.6
• Spares	3.0
• Packaging	1.8
TOTAL	29.7

Flight 9

• RFC module (for industrial plant	25.4
• Slag-hopper cart	1.0
• 3 hoppers	3.6
• Packaging	1.0
TOTAL	30.0 t

Flight 10

• LLOX depot tanks	1.0
• Depot support structure	5.5
• Depot refrigeration units	0.5
• Depot radiator, sunshades & platform	1.3
• Depot plumbing	10.0
• 3 hoppers	3.6
• 2 debris shields	1.6
• 2 LLOX terminals (future growth)	0.2
• Hydrogen make-up gas reserve	0.5
• Spares	3.0
• Packaging	2.0
TOTAL	29.2 t

Flight 11

• Oxygen plant #1	30.0 t

Flight 12 (crew-carrying mission)

- Stay time 30 days
- Verify LLOX production
- Investigate other ISRU
- Detailed repairs as needed

Flight 13

• 11 PV units	13.8
• Lander conditioning trailer (LLOX transfer)	1.0
• Power and grounding cables	0.5
• 3 hoppers	3.6
• 1 spare slag-hopper cart	1.0
• Spares	9.0
• Packaging	1.0
TOTAL	29.9 t

Flight 14

• Oxygen reactor #2	30.0 t

Flight 15

• Oxygen reactor #3	30.0 t

Total equipment mass	**388.2 t**
Spares provisioned	**24.5 t**

Figure 3: Lunar Base Manifests

compacted over. It also brings the tools to do these operations as well as the robotic fine manipulator repair set. The third flight brings the second straddler and truck, freeing the first straddler/ truck set to clear, beneficiate, excavate and pave the site on a continuous basis. With the first straddler doing the beneficiation continuously, stockpiles of ore and other separated materials will be available for use in construction or contingency reserves Flights 4,5,and 6 bring the material to construct the habitat and workshop. In recognizing that an early manned visit should be used to verify the oxygen reaction process, it was also recognized that items break and need replacement and some human evaluations were going to be necessary. While replacement and some repair may be done by robots, fine work must be done by people. The evaluations of conditions under which things break and the development of steps to modify and prevent further damage would also be part of the early human "hands on" venue. Therefore a workshop is provided for human shirt-sleeve repair operations after the 7th flight. Flights 8 through 11 establish the industrial portion of the base and bring the first oxygen reactor into operation. This includes landing and installing a base power and conditioning unit to coordinate the demand priorities and switching of all power systems in operational or emergency conditions. Setting these control priorities will be the function of the Earth or habitat control stations. The impact of the 10th flight is described above. Flight 12 is the second manned mission in the buildup schedule. It is used to verify operations of the LLOX production and find uses for the slag and other stockpiled beneficiated materials. Flights 13 through 15 land the items needed to complete the additional power requirements and supply the remaining reactor complement. Flights beyond the 15th are generally manned flights with cargo supply taking a support role to human missions or industrial spares.

As the schedule is now seen, there is enough beneficiated ore stockpiled, as a result of site preparation, at the beginning of full LLOX production to last for 2 to 3 years of continuous production at 100 metric tonnes per year. This could make some of the mining vehicles available for alternative tasks during the hiatus between cargo flights, if an increased flight schedule is not followed.

Conclusions

Using robotic systems and human supervisory control, a lunar base reasonably be established to achieve full production within five years. Intermittent human presence works in coordination with teleoperation and telepresence within a robotic dominated buildup scheme to accomplish this schedule with considerable planned down time and contingency time to cover almost every circumstance.

Our analysis suggests that this five year buildup could be better accomplished if the schedule beyond the seventh flight could be compressed by launching on centers closed than 90 Earth days (3 lunar cycles). Before the seventh flight the main driver for mining is not ilmenite production but site preparation. Once the site preparation is complete, the schedule is driven by the availability of landed cargo.

References

Bares,L.C, Kweon,I., Osborn,J.F., Wettergreen,D.S., Whittaker,W.L, "A Mobile Robot for Lunar Surface Operations." SPACE 90., ASCE.

Sherwood,B., (1990a)"Lunar Base Elements Designed for Robotic Operations." SPACE 90. ASCE.

Sherwood,B., (1990b)"Site Constraints for a Lunar Base." SPACE 90. ASCE.

Woodcock,G.R., (1990) "Economical Space Exploration Systems Architectures." SPACE 90. ASCE.

Woodcock,G.R., Sherwood,B., Buddington,P.A., Bares,L.C., Folsom,R., Lousma,J., Mah,R., Whittaker,W.L. (1990) "Application of Automation and Robotics to Lunar Surface Human Exploration Operations." SPACE 90. ASCE

A ROBOTIC RETURN TO THE MOON

by

A. Steve Johnson[1], student member AIAA and AAS
William E. Faller[1],
Alex Hoehn[1],
and Marvin W. Luttges[2], member AIAA

ABSTRACT

An autonomous Lunar precursor mission is proposed as the first step to an effective, low cost return of man to the Lunar surface. The mission would use two self similar robotic vehicles and three experimental bio-volumes. The volumes are derived directly from the three lunar landers required to fulfill mission hardware delivery needs. The autonomous lunar vehicles are scheduled to establish "ground truth", carry out resource exploration, develop sites and roads, and to provide resource recovery and processing support. Operationally, the two vehicles would conduct these activities employing the "*buddy system*". The operations, thus, should be very reliable. By exchange of data, each vehicle would proceed to learn Lunar maps and pathways from the other. These vehicles would support the experimental bio-volumes by returning resources for use in physico-chemical or biological processing. Delivered resources coupled with *in situ* resources would provide the reconfigured bio-volumes with the capability to carry out biological growth and waste processing demonstrations autonomously. Success in growth and processing would lead to the stockpiling of life support consumables for future manned missions. The overall scenario would enhance the value of manned Lunar activities. The implementation of the Lunar Oasis by Automation and Robotics (*LOAR*) has immediate returns to the U.S. technology base in the crucial areas of automation and robotics. Since the activity, initially, is not manned it should be readily carried out with substantial assistance from industry and academia ... at substantial reductions in overall mission cost. The *LOAR* could be the first step in an enduring human presence on a surface other than Earth's.

---

[1]Graduate Student, [2]Professor, Department of Aerospace Engineering Sciences, Campus Box 429, University of Colorado, Boulder, Colorado 80309-0429.

INTRODUCTION

One of the major goals envisioned for the United States' space program is that of returning to the moon. Missions such as the Lunar Polar Orbiter (*LPO*) would provide information vital to assessing both science and technology objectives critical to the return of humans to the Lunar surface. To enhance the effectiveness of astronauts on the Moon, an autonomous precursor mission is proposed. This mission uses multiples of two major hardware elements: an autonomous lunar rover and an autonomous bio-volume.

The integrated operation of two vehicles and three volumes leads to a Lunar Oasis produced by Automation and Robotics (*LOAR*), a precursor to a manned Lunar base. The robotic vehicles will provide on-site verification of resources, route improvements, and site improvements. The Experimental Bio-Volume (*EBV*) will establish a biological processing capability, as well as, provide an initial habitat and life support consumable cache. Additionally, the robotic vehicles can introduce lunar resources, such as carbonaceous meteorite materials, for bioprocessing and enhanced consumable production.

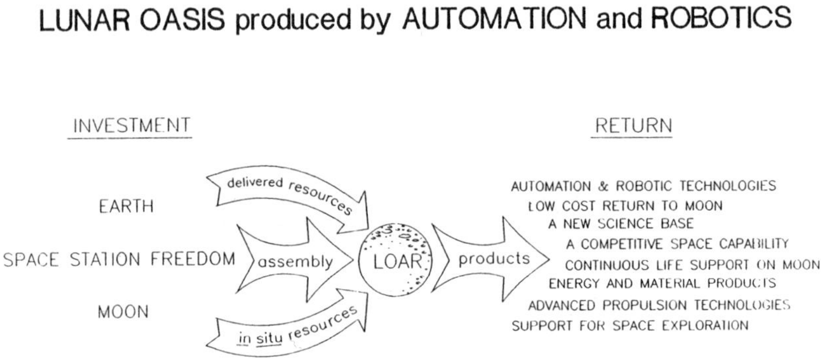

Figure 1 - *LOAR* Investment and Return

RATIONALE

Returning to the Moon has been proposed in several studies (NCOS, 1986 and Ride, 1987) as well as part of the space initiative outlined by president Bush. However, effective exploitation of Lunar environment and resources will occur only through the establishment of a low cost but effective supporting infrastructure. Such an infrastructure would provide the basic facilities, equipment and service necessary to support life, to conduct meaningful developmental operations and to enable innovative science on the Lunar surface. Flexibility, cost efficiency and timely availability would be benchmark characteristics of this well-established infrastructure. The mission proposed here provides the mechanism to establish the beginnings of a Lunar based infrastructure as a prelude to humans returning to the Lunar surface and to human presence on Mars.

The investment and anticipated returns from the *LOAR* program are depicted in Figure 1. In order to initiate the foundation of the Lunar infrastructure a modest investment in Earth-originated resources is delivered to the Lunar surface. Potential *in situ* Lunar resources for future exploitation would be rapidly identified through robotic activities. In addition the capacity to generate and process biomaterials and to store life support consumables is staged. These activities will enhance the options and performance of future manned activities.

The envisioned space activities could be the apex of needed industrial, technological and commercial growth destined to stimulate the educational and economic future of the United States. Automation and robotics technology would enable a low cost return to the Lunar surface without the safety and high reliability concerns or the expense associated with manned missions. At the same time, the development of this technology is applicable to a wide range of industrial needs on Earth. It would also support space applications and would serve as a cornerstone for maintaining a competitive U.S. space capability. Investments in such technology holds enormous spin-off potential for Earth-based applications. Both industry and academia are likely, accordingly, to eagerly support automation and robotics developments.

Both *in situ* resource recovery and processing techniques, as well as, biological production data bases obtained autonomously can be used in the design and certification of requirements necessary for man-rated missions. Waste processing and low power food production technologies will be of direct commercial interest on Earth. Research in controlled environments may provide insight and understanding of the mechanisms underlying the dynamic environmental alterations that appear to be occurring on Earth. The proposed mission, thus, has immediate technical implications for the continuing prominence of the U.S. in the world economic and environmental areas.

Once these envisioned technologies have been proven in the Lunar environment, they will become the foundation for other space missions. The continuous progression of autonomous space programs could lead to lower cost space exploration, utilization and habitation. More sophisticated manned space missions could then be carried out. Although the envisioned *LOAR* mission is a powerful tool as an unmanned space mission, the investment provides significant returns by complementing and enhancing the manned activities that would occur on the Lunar surface and, eventually on the Martian surface.

MISSION SCENARIO AND IMPLEMENTATION

The primary goal of this mission is to facilitate effective but low cost exploration and development of the Lunar surface. The Lunar Polar Orbiter (*LPO*) will provide a topographic and resource concentration map of the Lunar surface. However, these maps can not include surveys of sub-surface or inaccessible area resources. The Lunar Robotic Vehicles (*LRV*) proposed here, will provide on-site verification and calibration for *LPO*-identified Lunar resources. In addition, lava tubes and subsurface resources will be surveyed. All surveys will identify the location and concentration of Lunar mineral and volatile resources that may be of value to future space missions. Such comprehensive surveys also have profound science implications. Polar ice, ilmenite and carbonaceous meteor materials are examples of both some potentially valuable mission resources as well as informative science resources. Prototype processing equipment would be staged and used to demonstrate the effectiveness of using various Lunar resources for products such as oxygen for propellants and Lunar fiberglass for habitation and tankage volumes. Processing successes underwrite a continued series of science opportunities.

The robotic vehicles would identify Lunar sites deserving or requiring manned exploration or evaluation. Route identification or improvement would enable both astronaut effectiveness and safety on the Lunar surface by reducing both travel time and risk. The robotic vehicles would transport some Lunar materials back to the base (*EBV* locations) either for direct storage or for use in the Experimental Bio-Volume.

Two robotic vehicles are required in order to enable the "*buddy system*" that permits additive or even synergistic Lunar capabilities. Typically, the "*buddy system*" is one where two individuals operate independently but may combine information or efforts during critical situations. An otherwise dangerous or disabling situation may be avoided or mitigated due to the intervention and/or support from the second team member. An example of this capability might be towing assistance provided by one vehicle to rescue the second. Another would be the sharing of path or roadway information that would allow more direct and safe transport. Although the vehicles are envisioned to operate nominally in autonomous fashion, a teleoperational capability will be provided to permit human intervention during failures, to allow human input to "unrecognized" circumstances or to aid in the performance of complicated operations.

The "*buddy system*" provides an effective means of performing the mission objectives while assuring hardware and operational reliability. Operationally, the two *LRV*'s would exchange information via teleoperational links. Ranging from a distress call to route information this capability would provide redundant stored information as well as enabling each machine to "know" twice as much Lunar environment information than it has actually experienced. The "*buddy system*" enables the performance of modular change-outs and repairs. For example, if a wheel were broken on one vehicle, the second vehicle could retrieve a replacement wheel and aid in replacing the broken one.

The initial design, development and evaluation of vehicle automation capability will be performed during intensive Earth-based field testing (Mel, 1988). This field testing will determine the effectiveness of *LRV* hardware, *LRV* autonomous routines, emergency recognition routines and manned intervention guidelines. Manned teleoperated change-outs and repairs would be experimentally demonstrated.

The delivery system for the *LRV*'s is envisioned to be a liquid oxygen/hydrogen propelled lander. The propellant tankage, similar to a shuttle external-tank in configuration (much smaller), would be designed to be reconfigured into a closed bio-volume. To relax storage requirements, any residual propellants would be combined and stored as water. On-board robotic capabilities would be used to reconfigure the internal volume to support various levels of microorganism and plant growth, as well as, waste processing and soil composting.

Before humans return to the Lunar surface, the *EBV* would have demonstrated plant growth, biological waste processing and small scale animal growth. The robotic vehicles will have supported this activity through the return of Lunar resources to the *EBV* such as water (Lunar ice) and materials to support plant growth (Lunar regolith and carbonaceous meteorite debris). Thus, this precursor mission helps delineate the Lunar resource availability and the Lunar biological impacts to be dealt with in follow-on manned activities.

Having demonstrated autonomous resource recovery and processing techniques, later, short-term manned visits may be able to utilize life support consumables produced and/or stored in the Experimental Bio-Volumes. Waste generated during the manned mission, a valuable resource, would be left for batch processing in the *EBV* to support the arrival of the next manned mission. The amount of life support consumables would expand with every additional mission and with the "leveraged" addition of Lunar resources.

As manned visits increase, the mission emphasis of the Lunar Robotic Vehicles would be changed to support manned mission objectives. Site improvements, ranging from roads to *EBV* bulk radiation shields for safe havens, would become priority activities aiding long term Lunar uses. New missions may be enabled, such as the robotic construction of a Lunar far-side observatory or deployment of complex scientific payloads. A robotic vehicle, for example, utilizing a sophisticated modular soil analyzer, would later provide transport for a portable processing unit. This latter unit allows the robotic vehicle to travel to identified concentrations of resources and then return partially processed products to the base. For example, trace amounts of hydrogen in the regolith may be scavenged and added to the bio-volume. A parallel between extended human presence and enhanced robotic capabilities of the modular *LRV*'s will inevitably exist.

VEHICLE DESIGN

The mechanical and locomotion design of these vehicles will be based on existing technologies with proven success rates and reliability. Also, modular simplicity will be maintained to the maximum extent practicable. However advanced robotic and automated systems emphasizing sensor and data handling capabilities will be necessary to enable these systems to operate autonomously. They must be capable of maneuvering in and manipulation of the surrounding environment.

The overall size of the *LRV* is based on the Apollo lunar rover. The weight is estimated to be 1500 to 3000 kg, two to four times the weight of the Apollo rovers. The vehicles are anticipated to be approximately 4.2 m long by 2.5 m wide six-wheeled, powered platforms (each wheel with an independent motor). The platform will be designed to accept modular units that perform various tasks. A conceptual view of the *LRV* is illustrated in Figure 2.

Figure 2 - Conceptual Rover Transporting Lunar Resources

Proposed power for the *LRV*'s is a 5 kW electric Dynamic Isotope Power System (*DIPS*). This type of power system uses a radioisotope unit to produce heat, which is transferred to a working fluid and used to dynamically generate power. Since a working fluid is being used for power generation in a closed cycle, a higher efficiency (20%) is obtained compared to that achievable (8%) through the use of conventional Radioisotope Thermoelectric Generators (*RTG*) (Angelo and Burden, 1985). For a given design power, the *DIPS* translates into a smaller radioisotope generator and, just as important, lower heat rejection requirements.

It should be noted that there are some launch safety concerns associated with this type of power supply. Public concern was expressed in the recent Galileo. Additional cost may be incurred for flight qualification of such a system. However, this type of power unit would provide the greatest operational capacity for the vehicle. Alternatives, such as solar power, would provide power only during Lunar daylight, with only minimal maintenance power available for the 2 week lunar night. The associated weight and volume penalty for full power night operations is prohibitive. Additionally, as solar array size increases, the arrays must be mounted higher, resulting in elevating the center of gravity and eventually causing vehicle stability concerns. Modest power storage would allow only short-term normal operations in encountered shadowed areas that would be frequently encountered in mountainous terrain.

Fuel cells also were considered. In this configuration water would be electrolyzed using power available at the base and the resultant gases would feed the fuel cells of the vehicles. Water from vehicular power generation would have to be transported back to the base for regeneration. Obviously, large volume commitments for fuel storage are necessary to enable long duration missions away from the base and a large storage weight penalty is exacted throughout each vehicular mission. The dependency on the base for fuel would severely limit the range of the vehicle.

The *DIPS* enables the greatest mobility for the vehicle. The *DIPS* would be mounted on the chassis, which aids in maintaining a low center of gravity. Day, night and shadowed operations are possible. Solar independence is especially important, since the Lunar poles are potentially one of the most interesting areas to explore, but are areas of long shadows and long nights due to the low sun angle. Finally, there is no dependence on the base, allowing the rovers to range as far as necessary to explore the lunar surface. This freedom of travel could allow the retrieval of the Apollo Lunar Rovers as a prelude to astronauts returning to the Moon. Minor refurbishment and battery replacement may be all that is necessary to reuse these rovers. This potential reuse is an example of how the *LRV*'s might gather resources to lay the foundation of a Lunar infrastructure. The *LRV* range and endurance required in this example simply would not be possible with currently available alternative power supplies.

A wide range of specialized equipment will be necessary to enable the *LRV* mission objectives to be met. The capability of sensors, such as, pulsed laser ranging and infrared proximity sensors will be needed for manipulative operations (Cheatham et al., 1988). Semi-dexterous manipulators will be needed for sample handling as well as many other mission tasks.

Although conventional computers will suffice for normal "expert system" control functions, navigational capacity may be best served through a computer using a neural network architecture. Semi-autonomous Navigation may be facilitated by stereoscopic near-infrared and visual light cameras (Wilcox, 1988) allowing up to 7 km per day excursions (Ruoff, 1989). Greater distances may be covered during a teleoperational mode of operation.

A significant volume of the Lunar Robotic Vehicle will be dedicated to the sample analysis module. This module will contain a variety of analyzers and equipment. Samples will be collected through the use of drills (core samples), manipulator arms and the teleoperated Lunar Remote Access Technology (LRAT). The latter is a small expendable, teleoperated, tethered robot that can be driven into areas either inaccessible to the rover or too dangerous to risk the rover, in order to retrieve a small sample which is returned to the *LRV* analyzer. Once in the analyzer module, the samples can be prepared for various tests. Internal modules would determine characteristics such as, density, hardness, magnetic properties, solubility, temperature and thermal characteristics. Some samples may be polished for evaluation using teleoperated optical microscopes. The analyzers could include gas chromatographs and x-ray fluorescent spectrometers for composition analysis. Geophysical data may be determined through the use of magnetometers, gravimeters and seismometers. Underground composition may be analyzed using radio-synchronized computer seismophones, deployed by the *LRV*.

As the lunar exploring phase of the mission ends, the analyzer module will be replaced by prototype processing units. The *LRV*'s will then be used to test different processes in the Lunar environment or commence collecting easily isolated resources. Iron particulates may be magnetically collected. Trace hydrogen may be collected by "cooking" the regolith using waste heat, electric heaters, solar concentrators or a combination of these to achieve the necessary temperatures to liberate the hydrogen. Key resources isolated by the *LRV*'s will be returned to the base for stockpiling.

Other equipment also will be needed. The analyzer module will need servicing and replenishing. Construction tools will be necessary for site improvements at the base or to improve access to some areas otherwise inaccessible, such as crater bottoms. Specialized tools will come later with special payloads. Such tools might include those items necessary for the deployment of a Lunar farside observatory.

OTHER MISSION ELEMENTS

Although the focus of this paper has been on the lunar robotic rover possibilities, the Experimental Bio-Volume offers several novel opportunities. One key feature of the *EBV* is the fact that the lunar lander is convertible, teleoperationally, into an operational bio-volume. Much of the lander will be scavenged for future uses. Residual cryogenic fuel will be combined to form water. Reaction control nitrogen gas will be used to initiate the inert portion of the atmosphere. Exhausted reaction control nitrogen tanks may be used to store excess gas, water or other fluids as necessary. The usefulness of every pound of payload delivered to the lunar surface will be maximized by careful planning that includes conversion possibilities. This ensures maximum utility of the equipment in establishing the foundation of the lunar surface infrastructure.

The deployment of the envisioned mission can be achieved through three Shuttle-C launches. The first two launches will transport the two rovers and their respective landers to low Earth orbit. The third launch will transport a third lander and a payload package of spare parts, extra tools, and additional consumables for the analyzer and bio-volumes. Each of the three landers will be configured into Experimental Bio-Volumes.

COST

Relaxed reliability, acceptance of non-critical failures and the use of simple redundant systems could be the cost reducing benchmarks of the envisioned, unmanned mission. Small entrepreneurial companies having committed academic partners would develop germane technologies simultaneously supportive of the mission and U.S. advancements in technology. To encourage investments by the private sector, such entities could retain technology exclusivity. Both business and academia would benefit while supporting a low cost Lunar mission. The cooperative, low cost development is enabled by the unmanned precursor nature of the mission.

The total investment for such a mission may be as low as 5 billion dollars. Based on current launch costs ($8000/kg), launching the three elements to the lunar surface would be approximately 1 billion dollars. An estimated 2.5 billion dollars may be invested for research and development activities leading to production of the hardware elements, which may run 1 billion dollars more. Utilizing a university based operations center such as the *SME* Explorer satellite operations center, missions operations may be kept well below 500 million dollars.

Low continuing costs would be related to private industry having vested interests in the Lunar *in situ* resources (Burt, 1989). The "Moonsteading" partners would recover returns on investment through the use of the Lunar resources identified and utilized for infrastructure support as well as through the technologies developed in supporting a successful mission.

CONCLUSIONS

The initial foundation of a Lunar surface infrastructure can be enabled through the use of Automation and Robotic technologies. Autonomous Lunar rovers can be used for on-site resource verification as well as the identification of future sites for manned investigation or Lunar bases. The effectiveness of the Lunar roving vehicles is significantly enhanced through the use of the "*buddy system*". Vehicle flexibility enables a wide variety of operations. Dynamic isotope power supplies would provide the greatest degree of mission operations. As a precursor to manned Lunar missions, the *LRV*'s can identify important resources, routes, potential base locations and preliminary site improvements. When astronauts return to the Lunar surface, they find that their activities will be complemented by the robotic capabilities of the rovers. Perhaps they will arrive to a site with used Apollo rovers! Additional missions, such as the deployment of a farside radio-observatory or other special science payloads may be deployed by the *LRV*'s inexpensively. And they may find that a variety of life forms had already thrived on the Lunar surface.

Very little is to be discarded. Operational bio-volumes will be established with the main lander propellant tanks. A habitational volume with stored life support consumables and stockpiled Lunar resources will have been accomplished by the *LRV*'s to provide an operational Lunar Oasis (*LOAR*), the foundation of a Lunar surface infrastructure.

This autonomous mission is to be accomplished at low cost would be achieved by accepting lower reliability and some degree of non-critical failure. Novel consortiums between academia and private industry would benefit technology while promoting low cost. In such a "Moonsteading" partnership, participants are vested in technology development as well as Lunar resource recovery and utilization accomplished by this mission.

This mission could substantially increase the effectiveness of a manned return to the Moon, it will stimulate research and development in automation and robotic technology for Earth applications, it would demonstrate the feasibility of complex life systems on a non-terrestrial surface and it should excite the imagination of a new generation of scientists and engineers.

ACKNOWLEDGMENTS

Some of the efforts of this report have been the result of the NASA Advanced Design Program (Faller et al, 1989).

APPENDIX-REFERENCES

Angelo, Joseph A. and David Buden, "Space Nuclear Power", Orbit Book Company, Inc., Malabar Florida, 1985.

Burt, Donald M., "*Mining the Moon*", American Scientist, Vol. 77, pp. 574-579, Nov/Dec 1989.

Cheatham, J.B., C.K. Wu, P.L. Weiland and T.F. Cleghorn, "*A Multi-sensor System For Robotics Proximity Operations*", Second Annual Workshop on Space Operations Automation and Robotics (SOAR '88), NASA Conference Publication 3019, July 1988.

Cleghorn, Timothy F., Paul T. Baffes and Lui Wang, "*Robot Path Planning Using A Genetic Algorithm*", SOAR '88, NASA Conference Publication 3019, July 1988.

Faller, William et al., "*Cis-Lunar Space Infrastructure: Lunar Technologies*", NASA/USRA Advanced Mission Design Summer Conference, 1989.

Mel, Barlett W., "*Murphy: A Robot That Learns By Doing*", Report No. UIUCDCS-R-88-1397, Center For Complex Systems Research, University of Illinois, Urbana, Illinois, Jan. 1988.

National Commission on Space (NCOS), "*Pioneering the Space Frontier*", The Report of the National Commission on Space (NCOS), Bantam Books, Inc., New York, New York, 1986.

Ride, S. K., "*LEADERSHIP, and America's Future in Space*", August, 1987.

Ruoff, Carl F., "*Space Robotics In the 90's*", Aerospace America, August 1989.

Wilcox, Brian H., "*Machine Vision for Space Telerobotics and Planetary Rovers*", SOAR '88, NASA Conference Publication 3019, July 1988.

A Robotics Testbed for In-Space Assembly Tasks

John Blanco[1]
Jack Faber[2]
Nick Wilde[3]

Abstract

The operations testbed of the Center for Space Construction at the University of Colorado is described. This testbed serves as a centralized testing environment for interdisciplinary research in the problems of construction and assembly in space. The three main components are described: NoPumpG — a flexible user interface system; OASIS (Operations and Science Instrument Support) — a teleoperations package; and a SCORBOT robot with its associated assembly mock-ups. Finally, integration efforts involving all three components and some on-going robotics assembly tasks using the robotics testbed are described.

Introduction

The University of Colorado Center for Space Construction's Operations Testbed is designed to be a central testing facility for the Center's on-going research program on manual, automated, and telerobotic in-space assembly.

The Center for Space Construction emphasizes inter-disciplinary research. Following this theme, the testbed is designed to allow research in, and testing of, the broad spectrum of issues involved in space construction. It incorporates ideas from several different areas of on-going research at the center: human-computer interaction, systems level design of assembly tasks, and robotics control for in-space assembly.

The testbed consists of 3 main modules: NoPumpG — an interactive

----

1 Graduate Student, Electrical Engineering, Campus Box 425, University of Colorado, Boulder, CO 80309.
2 Engineer, Laboratory for Atmospheric and Space Physics, Campus Box 392, University of Colorado, Boulder, CO 80309.
3 Graduate Student, Computer Science, Campus Box 430, University of Colorado, Boulder, CO 80309.

user interface system; OASIS — a teleoperations package designed at the University of Colorado; and a SCORBOT robot. Each of these modules is loosely coupled to the others, allowing interdependence of function and possible usage of similar modules for testing or special purposes.

NoPumpG is an interactive user interface system built on some of the fundamental principles of spreadsheets. Since it is built on a spreadsheet model, it allows for rapid modification of the interface by the user.

OASIS is a telecommunications package produced by the University of Colorado's LASP (Laboratory for Atmospheric and Space Physics). The software tools provided by OASIS make up a complete teleoperations package. They allow for a local/remote operator to control, monitor, and communicate with robots and other machines. Within OASIS, the system to be controlled and monitored is described generically by several database tables. By serving as the central switchboard of the testbed, OASIS allows a variety of physical and simulated devices to be attached to either end of the control chain (Figure 1).

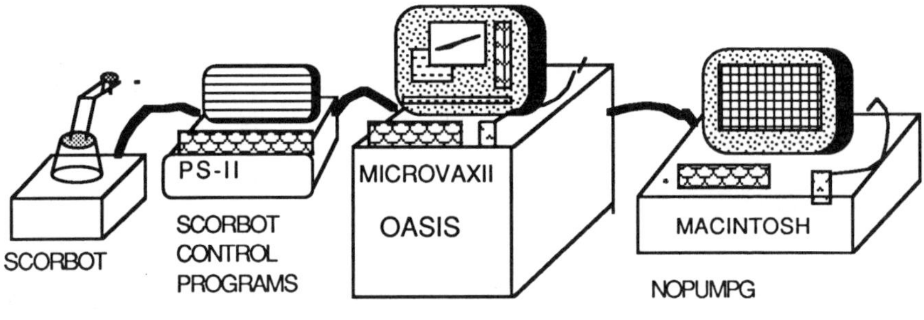

Figure 1 - Center for Space Construction Testbed

The SCORBOT robot is but one of a number of robotic devices that could be used at the end of the testbed control chain. This robot's purpose is to provide physical manipulation through the NoPumpG-OASIS control loop, and to allow demonstration of possible scenarios for in-space assembly tasks. In conjunction with the SCORBOT robot, a number of elementary assembly tasks have been designed. Hardware mock-ups for these tasks have been constructed and are available for experimentation, such as a standard Orbital Replacement Unit (ORU) replacement task.

The three components of the current testbed cooperate with, but do not depend directly upon, one another. Using OASIS as a central switchboard, any number of user interface components can be tied into the current testbed. In the same manner, any number of physical or simulated devices could be used to replace the current

SCORBOT robot. In fact, work is currently underway to tie in a robotic simulation package.
In the following sections, we will describe the data flow through the current testbed and provide a description of the individual components of the testbed. We will conclude by describing some of the integration efforts and test scenarios that have been applied to the testbed.

Data Flow Through the Testbed

Data flow through the robotics testbed bi-directionally (Figure 1). Thus, when commands are issued through a particular interface to the robotic device, the user can be kept apprised of the device's status through telemetry feedback.

Command data flow from the interface (in this case NoPumpG) to OASIS. OASIS formats the given high-level commands into raw commands for a particular physical device. The formatted and interpreted command stream then flows from OASIS to the currently selected physical device, where it is acted upon.

In addition, telemetry is sent from the given physical device back to OASIS, which again formats the data into something the interface can understand. OASIS then relays this information back to the user. In this manner, the user is kept continuously updated of the state of the robot, without having to directly query the physical robotic device.

The NoPumpG User Interface

NoPumpG is a user interface with a twist. Rather than being the standard point and click graphical interface we are all familiar with, NoPumpG combines spreadsheet technology with interactive graphics to produce a fully programmable, yet easily definable, interface.

The idea behind NoPumpG, and behind the name, is that programming in a conventional sense is difficult. It is difficult, in part, because it is so hard to get information into and out of a normal computer. A barrier exists between the user and the computer. This barrier manifests itself physically as the screen. Information on one side of this barrier must be pumped over to the other side via input and output program code. Without input and output code, the most sophisticated computer program is useless. NoPumpG attempts to bridge this gap between user and computer by supplying a pumping engine that manages the I/O process.

NoPumpG supplies this pumping engine in the form of a spreadsheet. The spreadsheet allows for ready and easy access to all important computational details in progress, thus allowing for ready modification by the user. All partial results are available in spreadsheet cells. If a user wishes to modify part of the current computation, s/he may do so simply by editing a cell formula or value.

NoPumpG attempts to bring the power of the spreadsheet interface to other facets of computer usage by combining the spreadsheet with interactive graphics (the G in NoPumpG stands for graphics). The combination is accomplished in the following manner: the user creates graphical primitives (lines, small bitmaps, pieces of text, etc.) in a manner similar to that of a typical mouse-driven draw program. Once these primitives are created, the primitives are then available for modification or inspection as the values of dedicated spreadsheet cells.

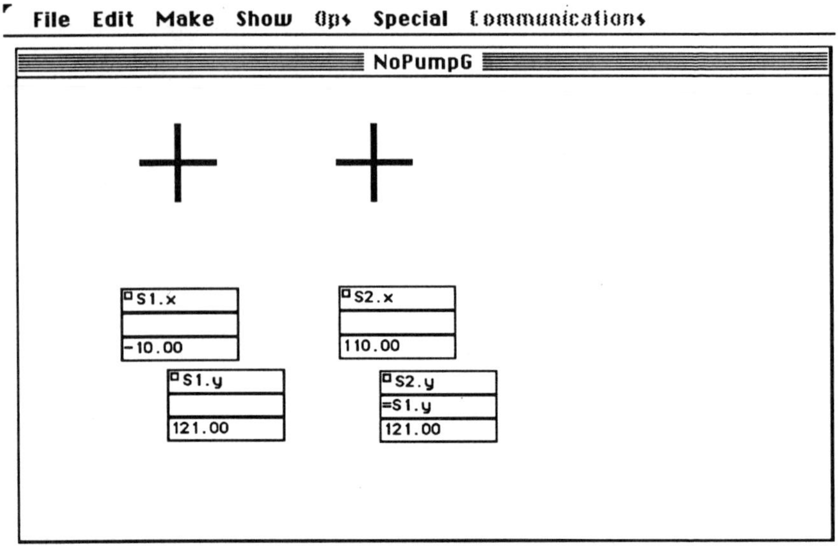

Figure 2 - NoPumpG Screen

As an example, Figure 2 shows a simple NoPumpG screen. Two bitmapped sketches have been created. These sketches have their X and Y coordinates controlled by a pair of spreadsheet cells. (Our cells are not fixed in a grid as a normal spreadsheet, but are liberated and allowed to move around the screen — as well as hidden). Thus, when the user drags one of the sketches around the screen, current X and Y coordinates are still available for the given cells.

Cells not only report behavior, they control it. Thus, if a value is typed directly into one of the controlling cells, the drawing changes to reflect the new value. Formulae can be input into these cells, as in conventional spreadsheets. Thus, a primitive's behavior can be made dependent upon other primitives. Thus, putting a formula into a Y coordinate cell to constrain the value to be equal to another value also constrains the two sketches to be of the same height. Formula in the cells controlling and reporting behavior back to the same or other cells allow for relatively complex interactions to be programmed fairly easily.

In addition to X and Y coordinate cells, NoPumpG contains cells for controlling other aspects of the given primitives. These include visibility and pen cells (a trail of ink left behind when the cell is moved). A more complete description of the NoPumpG system is available in Lewis (1987).

All information is tied into the operating testbed, in general, and OASIS, in particular, by specifying send and receive operations. Thus, the X coordinate of a specific graphical primitive, for example, can be sent to OASIS through any one of several virtual channels. OASIS receives the information, and uses the virtual channel to determine which aspect of the physical device to command. The current state of the physical device is then sent back through the same virtual channel.

In this manner, complete flexibility is achieved as to exactly which aspect of NoPumpG controls what aspect of the physical device. The mapping of a given NoPumpG application to a device can be changed at several different levels: the NoPumpG application level itself (i.e. a particular graphical primitive is sent and a particular graphical primitive is changed upon receiving incoming data); the initial input/output channel specified to OASIS; and the OASIS level (i.e. what OASIS does with a particular input/output channel in terms of device commands).

OASIS

OASIS software tools executing on either the Microvax-II or the SUN 3/60 make up a complete control and monitoring system. The heart of the system is a software program written in Ada and its complementary databases. In the VAX implementation, the Graphics Kernel System (GKS) subroutines and DEC's UIS windowing environment are used. In the SUN application, X windows and the Transportable Applications Environment (TAE) developed by Goddard Space Flight Center (GSFC) are also available.

OASIS is a flexible system that allows a user to change an application simply by modifying database tables with an editor or a newly developed TAE database builder. These tables define everything from the protocols to be used by the external communication software to the shape and color of displays and icons on the screen.

When data arrives from an external source, the stream of bits is decomposed into individual data items which are then converted to the appropriate numerical or state variables. The system then activates a variety of tasks as defined by this data. These tasks include updating displays, storing formatted or unformatted data in a file, and triggering commands to be issued from one or more external devices.

A command can be requested in a variety of ways. A hexadecimal pattern can be explicitly called for, or an English-like command can be entered from the keyboard. In addition, commands can be called from a menu – with options for filling in required values. OASIS also includes an English-like control language which allows users to code repetitive procedures for future use.

OASIS has a query system for displaying or capturing the current contents of the database. A privileged user can modify or add these records – an important feature in database development. There is an on-line help system whereby each addressable object can be associated with a help file. Help can be executed from the keyboard or a mouse. The help facility is easily maintained and developed in a series of ASCII files.

OASIS provides an interface that allows for a complete control and monitor system to be developed and maintained by scientists and engineers rather than by software programmers. The system is designed to evolve with the hardware and user requirements, and designed to be used throughout the lifecycle of a project. The CSC testbed, with its OASIS switchboard, uses two external protocols: one talks to the robot controller and one talks to the NoPumpG application.

Without NoPumpG, commands can be generated by OASIS at the request of the user, then transmitted to the SCORBOT controller in message blocks. These blocks are checked first for transmission errors. If such an error is detected, the block is discarded and an error message is returned. If no transmission errors are detected, the commands in the block are checked for reasonableness, such as whether the commanded movement is within the actual range of the robot. If a command fails the reasonableness check, further command execution is halted and the failure of the command is transmitted to OASIS.

Once a command block has been executed, telemetry packets are sent back on any item that has been changed by the command block. These packets inform OASIS that new commands can be sent. Commands can take the form of requests for more information, movement of a series of individual joints, movement to a position known by the controller, control of auxiliary devices, or initialization of the robot. In general, any function can be executed from any component of the testbed allowing transfer between control software packages.

There are several options for controlling the robot from NoPumpG through OASIS. The simplest option is for NoPumpG to send pre-arranged binary commands containing parameters for robot movement. These commands are then deciphered into the correct format for the particular robot being controlled. This frees NoPumpG to specify the motor, position, and whether the command is relative or absolute. Therefore, with a robot upgrade or change, NoPumpG does not need to be modified.

A more complex option for controlling the robot is for NoPumpG to generate a robot-specific command and send it directly to the OASIS language processor. More of a burden is put upon NoPumpG to be aware of robot changes. OASIS grants access to NoPumpG by switching on one stream. Another stream is switched on for the robot. Commands from NoPumpG will then trigger commands to the robot, which then sends data back to NoPumpG.

SCORBOT and Current Assembly Tasks

The robot arm used in the Center for Space Construction (CSC) testbed is a SCORBOT-ER III Educational Robot, built by Esched Robotec Ltd. This robot arm has five degrees of freedom plus a gripper. It has a lift capacity of 2.2 pounds. The robot hardware consists of a robot arm, a motor controller, and an IBM PS2 personal computer.

The motor controller acts as the interface between the PC and the mechanical arm. It receives ASCII-formatted commands from the PC through a serial connection, then controls the stepper motors in the robot arm accordingly. The motor controller also checks that commands from the PC are executed properly. It returns status information about the position and condition of the motors.

Software supplied with the SCORBOT allows for various levels of control of the robot arm. A utility package provides user access to most of the functions available in the controller, such as moving individual motors, setting motor speeds, etc. It also allows some higher level operations, such as moving the robot to previously defined positions or poses. This utility package allows users to access robot functions from high-level and assembly programming languages. A menu-driven program is also supplied, allowing users the ability to move the robot as well as define robot positions from the computer keyboard. The user can also write programs in an interpreted, robot-specific language.

Two main manipulation and assembly tasks have been the focus of the work in the CSC operations testbed. The first task, common in the field of basic robotic research, is the moving and stacking of blocks. The second task, representative of the usage of robots in space construction, is more complex. It consists of the removal and replacement of an ORU (Orbital Replaceable Unit). Each of these experiments are described in the following section.

Integration Efforts

A number of different methods of remotely controlling the robotic arm are under investigation at the CSC. The simplest of these methods is a menu that allows the user to control each of the individual degrees of freedom of the robot. This menu is run on the OASIS system and consists of a list of the various degrees of freedom or axes, their current position, an increment that tells how far and what direction the axis is to move, and the speed of movement. The user has control of each of these fields. By clicking on an axis name, s/he sends a command message to the SCORBOT system with the corresponding axis, direction, and speed.

Only one axis can be moved at any one time, making it difficult for tasks of any complexity.

An extension of this interface gives the users graphical control of each degree of freedom or axis. This involves using NoPumpG to graphically represent each axis as a line with a cursor on it. As the cursor is moved, information about its location is sent to OASIS. This information is translated into commands that are sent to the PC controlling the SCORBOT, causing the corresponding axis to be moved by the appropriate amount.

The next experiment involves interacting with the robot arm on a higher level. Instead of controlling individual degrees of freedom, commands are sent to the robot for moving blocks from one location to another. The NoPumpG system is used to create a 3 x 3 grid on the screen. Two numbered icons represent two blocks (Figure 3). The user is able to pick up either of the two blocks by moving the mouse to the block's location on the screen, then clicking the mouse. The block can then be moved to a new location in the grid on the screen. To do this, commands are sent from the NoPumpG program to the OASIS program, then to the SCORBOT system.

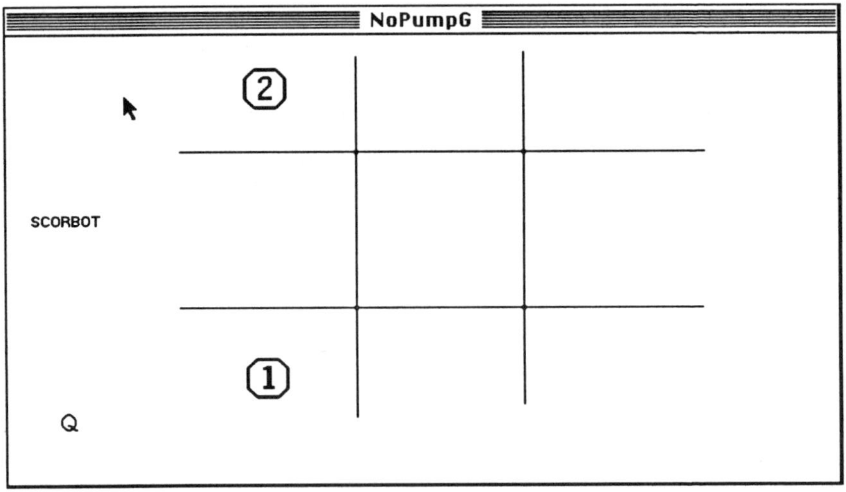

Figure 3 - NoPumpG interface for block movement experiment

Positions are predefined in the PC attached to the SCORBOT. A C program translates the commands from NoPumpG (via OASIS) into a series of positions for SCORBOT movement. When these commands are received at the SCORBOT robot arm, it then moves blocks on the grid.

The last experiment removes and replaces a simplified ORU (Figure 4). Removal of the ORU involves decoupling a fluid connector and moving it aside, removing an RS-232 connector, removing four bolts, and pulling the unit itself from its location. Replacement is the reversal of these steps. Currently, a script is stored on the OASIS system that contains an ordered list of the positions and motions that the robot must go through in order to effect the removal of the ORU. This script is sent to a program running on the SCORBOT PC. The program interprets the script, issues commands to the robot arm, thus causing motions necessary to accomplish removal of the ORU.

Figure 4. Remove and replace a simplified ORU

In the experiments described above, little feedback is returned to the program from whence the commands were issued. In other words, a user isolated from the SCORBOT system has little information on the status of the robot arm, nor the success (or failure) of the commanded operation. For this reason, planned future experiments will incorporate this feedback into the testbed.

Another problem that often occurs in the execution of the more complex and higher level instructions is the repeatability of finer motions. For this reason, we plan to incorporate more sensing capabilities, allowing for more dynamic behavior of the robot arm. A vision system will be the first such sensor system added to the testbed. All other sensors, such as tactile and range-finding devices, are under investigation.

References

R. Davis and E. Hansen, "OASIS Teleoperations Package Makes Its Debut", March 1986, Information Systems Newsletter, NASA OSSA Publication.

A. Jouchoux, J. Cowley, R. Davis, E. Hansen, M. Klemp, S. Lasater, D. Mullens, T. Sparn, G. Tate, "Developing a Spacecraft Monitor and Control System in ADA, Joint Conference on ADA Technology", March 1987.

Lewis, C.L., "NoPumpG: Creating Interactive Graphics with Spreadsheet Machinery", Technical Report CVCS-372-87, Department of Computer Science, University of Colorado - Boulder, Colorado

Subsection 3.5: MATERIALS APPLICATIONS

EXPLOSIVE METAL WORKING FOR SPACE

A. E. DOHERTY[1]

Abstract

As the United States looks towards larger and larger lift vehicles for the transportation of the space station and other large payloads, the application of explosive metalworking becomes a more attractive technology for the production of the components that make up the launch vehicles and space station. This man-rated manufacturing method, which had its start in the 1800's, is providing engineers with a technique to increase strength, reduce weight, and use material combinations for improved performance that are not possible with other manufacturing methods.

Introduction

The use of explosives to shape metals is not a new space age technology. In 1888, C. E. Munroe described a method for transferring a design to iron plates by the use of explosives[2]. The earliest patent on explosive forming was an English patent (No. 21,840) issued in 1898 for the joining of bicycle frames[2]. The first patent issued in the United States was No. 939,702. This was issued to Isaac Newton Jones in 1909. This technology lay dormant until about 1954, when it was rediscovered and applied to the manufacture of jet engine and missile components. Explosive metalworking was a subject that was taught in many of the engineering schools and practiced by almost every major aerospace manufacturing company. With the decline of missile and space construction, which occurred in the `60's, there was no longer sufficient production for the many facilities that existed, and they slowly began to close. This resulted

----

[1] Vice President & General Manager Formed Products Explosive Fabricators, Inc., 1301 Courtesy Rd., Louisville, CO 80027

[2] High-Velocity Forming of Metals ASTME, 1968

in a loss of interest at the schools, and the technology quickly became unknown to the new generations of engineers.

Today with the desire to reduce launch vehicle costs and the interest in larger launch vehicles, attention to this technology has blossomed anew.

Explosive Forming-Fundamentals

Explosive Forming, as previously stated, has been around for a long time, but the basic process has not changed dramatically since its inception. Explosive forming uses a single die -- normally female; thus, the tooling requirements are half that of conventional press or drop hammer forming. Parts are formed from flat plates or from preforms.

Preforms can be simple rolled and welded cones or tubes, or a part that has been partially formed by a conventional process. Preforms are generally used when the desired shape requires the material to move beyond the mechanical property limits.

The material to be formed is loaded into the die and clamped with a fixed hold-down pressure, as shown in Figure 1. The air is removed from the space between the blank and the die. A calculated amount of explosive in a given shape is positioned in the predetermined location(s). Normally, the entire system is then submerged in water. The water acts as the "punch" when the explosive is detonated, and this action takes place in microseconds. The pressure generated at the explosive/water interface can approach 3 million psi, depending upon the explosive used. The detonation velocity may be 24,935.6 feet per second. These factors result in the high strain rate (100 inch/inch/second to 1,000 inch/inch/second) forming conditions associated with this process.

Because the entire blank surface is instantaneously being moved, the stresses in the formed part are generally more uniform than those found in conventionally formed parts. The high strain rate will, with some materials, produce results that exceed the mechanical properties given in the texts. However, it is best to design the forming around the text book properties.

Explosive Forming-Applications

The explosive forming process has been used to fabricate parts of all sizes and shapes ranging from

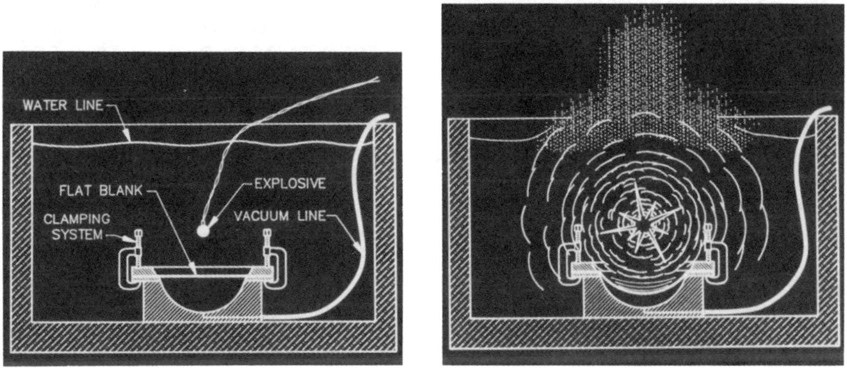

FIGURE 1: EXPLOSIVE FORMING PROCESS

FIGURE 2: 120–INCH DIAMETER ALUMINUM DOMES.

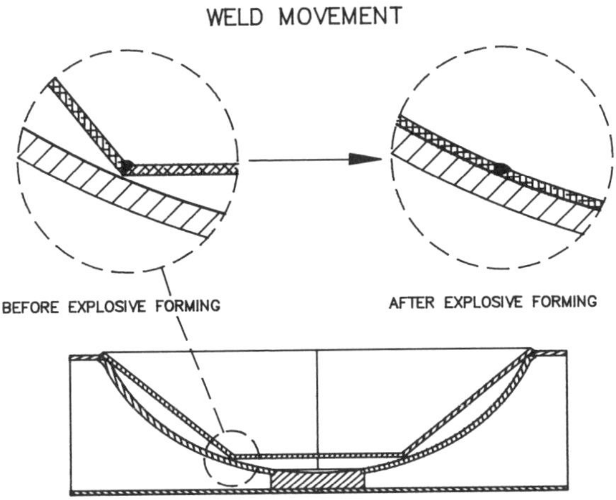

FIGURE 3: PREFORM IN DIE

0.5-inch diameter tubes to segments of 260-inch diameter domes. Figure 2 shows 120-inch diameter aluminum domes. The process has been used to form parts made from thin materials (0.010-inch) and thick materials (4-inch), and from almost all available materials. Tungsten plate has been hot-formed using explosives. Many parts that have been formed by conventional methods have been explosively sized in order to achieve the required contour and tolerances.

This process has been considered by many to be usable only when there are a few pieces to make. This is an incorrect conclusion. Production runs of 1,000 or more pieces have used this technique with complete success. A typical example is the early Polaris, which used explosively formed or sized end closures.

As previously stated, the current interest is directed toward present and future launch vehicles, such as Atlas, Titan, and ALS. This interest is driven by the fact that this technology offers the opportunity to produce parts from fewer pieces, to closer tolerances, and from materials that create problems for other fabrication processes. In addition, the industry is now faced with foreign competition, which is driving companies to look at other fabrication processes to reduce vehicle costs.

Let us consider the domes or end closures of launch vehicles from 120-inch diameter (Atlas, Titan) to 30-foot diameter (ALS). The current methods of manufacture are welded gore segments and some spinning. The use of explosive forming will greatly reduce, if not eliminate the welding associated with the gore segment fabrication technique, which will, in turn, reduce costs. It will also, with some materials (i.e., aluminum), reduce weight because of the thickened weld areas.

If the vehicle is to be fabricated from a material like 2219-T87 a new set of problems arise. Conventional techniques, such as spinning, require that the material be in the "O" condition. After spinning, the dome must be heat treated and quenched in 15 seconds. Quenching a 120-inch diameter dome causes some serious problems, but quenching a 30-foot diameter dome causes very serious facility and production problems. There is also the problem of distortion after heat treatment, which will require re-sizing.

Explosive forming could be used in the following ways to produce these same domes with fewer problems:

Method A: The technology could be used to produce gore segments of a larger size, thus reducing weld length. At the same time, the gore segments could have a waffle or grid pattern machined in the flat blank prior to forming. Most people have forgotten that the Saturn launch vehicle, which placed men on the moon, used explosively formed end closures pre-machined in the flat. The aluminum parts could also be formed in the T-351 condition, thus, the finished dome would require only a low temperature age to obtain the desired strength.

Method B: The technology can be used to form domes from a flat plate (as was shown in Figure 2). The limiting factor is the material's mechanical properties. Explosive forming has no size limitations, therefore, the shape or size of the part is not a restrictive consideration.

Method C: A preform is fabricated with conical sides and a flat or preformed bottom. This preform is produced from heat treated material that will only require a low temperature age after forming. The construction of the sides will be controlled by material size availability. The fabricated preform will then be placed in the die and formed (Figure 3). Figure 4 illustrates how the use of a preform allows us to stay within the mechanical property limits of the material. An alternate approach to reduce the bending of the weld joining the cone to the cap, is to form the cap before attaching it to the cone, as illustrated in Figure 5. The explosive forming method and tooling (Figure 6), allows the user to return the formed dome to the die and re-size it if there is any change during the aging cycle.

Thus, explosive forming gives the manufacturer more flexibility, and eliminates many problems associated with the conventional methods.

Facility Costs

It has been estimated that a conventional spinning and heat treating facility for the ALS domes will cost approximately $75 million dollars. The explosive forming facility is estimated to cost $750,000 dollars. The construction time for these two facilities is also significantly different -- the explosive forming facility takes far less time to construct. Since we use water, a location near an ocean or a river would be fine, thus

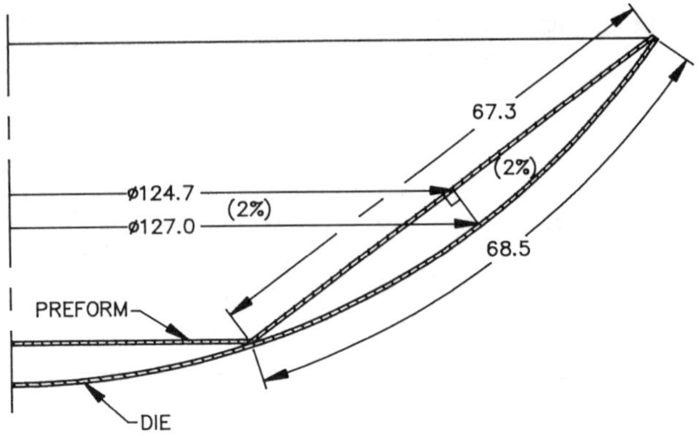

FIGURE 4: ESTIMATED ELONGATION USING PREFORM.

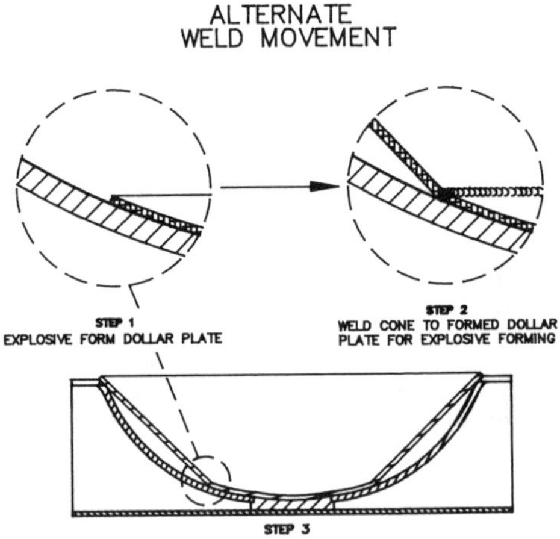

FIGURE 5: ALTERNATE PREFORM CONSTRUCTION

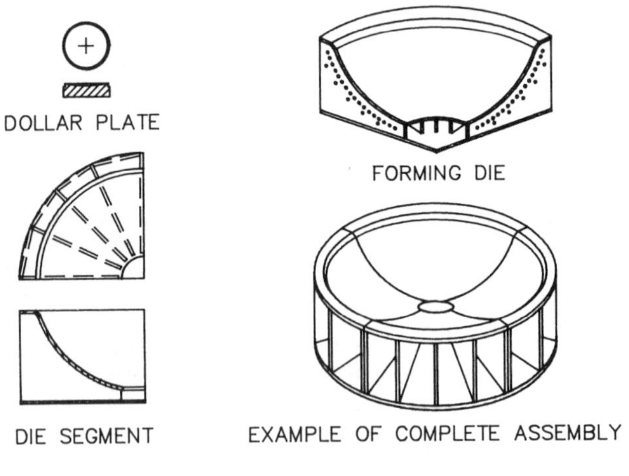

FIGURE 6: EXPLOSIVE FORMING DIE FOR LARGE DOMES

easing the shipping problem. The simple facility requirements of this technology allow it to be easily located near the vehicle manufacturing site.

Explosive Bonding-Fundamentals

Explosive bonding (also referred to as explosive welding) is a technology which began to draw interest in the late `50's. The initial evidence of bonding was observed with projectile impacts on plate, and occasionally on explosively formed parts. Explosive bonding occurs when plates impact each other under the proper conditions. Almost any combination of metals can be joined by this process. The bond is a true diffusion bond, as shown in Figure 7. The bond strength is equal to that of the weakest metal in the combination, and the tensile failure will occur in that metal and not in the joint. Another advantage of this joining system is that it is a cold bond, therefore, the properties of the metals are not changed. Also there is no heat effect zone with this joining technique. In the basic assembly setup necessary to achieve bonding, there is a space between the backer and the cladder. This is referred to as the stand-off distance. This variable, along with the explosive velocity and quantity, will be different with each material combination that is bonded.

Figure 8 illustrates the dynamics of this system. As the explosive detonation front moves across the driver plate, that plate is impacted against the backer. At the point of impact, an extremely high pressure exists which causes the metals to behave like a fluid. Additionally, a thin layer of both metals is removed by the jet and expelled out, thus providing an oxide free condition at the bonding point.

Explosive bonding is primarily used to produce a billet or plate from which the finished part is taken. Because of the pressures associated with the process, it is rarely used to join finished hardware.

Explosive Bonding-Applications

The primary space applications of this technology to date have been transition joints. These are generally used in the fuel systems where the tankage and plumbing are of different materials.

Probably the most famous transition was used on the Lunar Module, and, thus, the process is man-rated. Figure 9 shows the Titanium/Stainless Steel transition joints used on the fuel system for that vehicle.

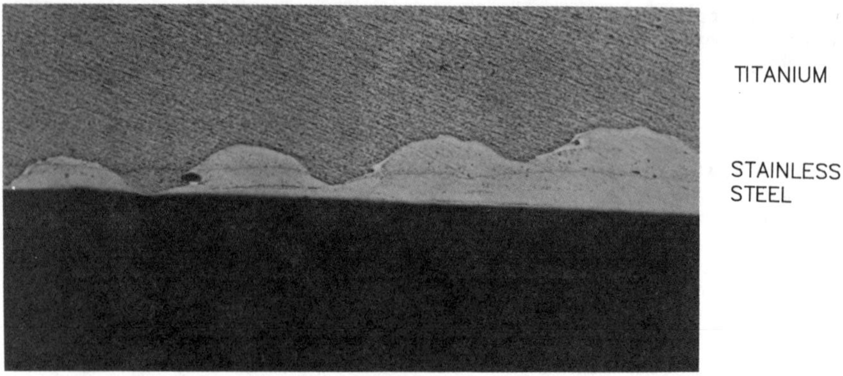

FIGURE 7: 6A1–4V TITANIUM TO 304 S.S.

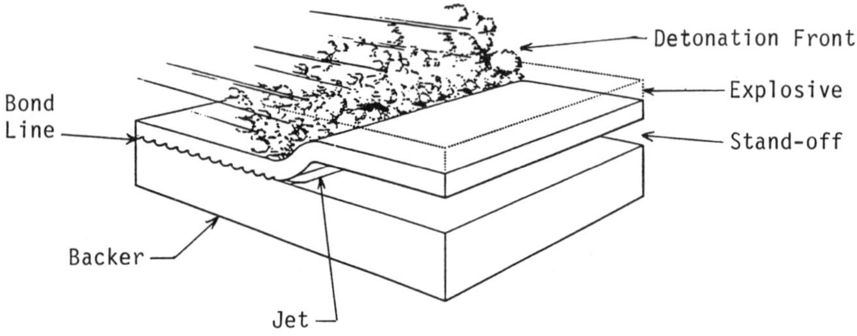

FIGURE 8: EXPLOSIVE BONDING SYSTEM

FIGURE 9(A): LUNAR MODULE TRANSITION JOINTS

FIGURE 9(B): LUNAR MODULE

FIGURE 10: EXPLOSIVELY BONDED
TRANSITION JOINTS

These products have passed extensive testing -- both destructive and non-destructive. Typical tests are Helium Leak, Burst, Ultrasonic, and Tensile testing.

As we look ahead for application on the vehicles we are talking about, we see the following possibilities.

Flat Plate Joining

The aluminum plate and sheet material produced by the mills is not of adequate size (width) for the larger domes. When the material is joined together by conventional welding methods, it is necessary to thicken the cross-section in that area due to the low strength weld. If the sheets were joined explosively, the joint would have parent metal strength, thus, weight savings and strength improvement.

It may also be possible to join the materials in the heat treated condition for further cost reduction when combined with explosive forming.

Dissimilar Metals

The ability to join dissimilar metals by this technique opens the door for the engineer to combine metals in areas where additional strength or thermal resistance are needed. This can apply to the propulsion systems, as well as the vehicle. We envision applications on nozzles, injectors, and structural transitions. Figure 10 illustrates a variety of transition joints currently in use.

Tubular Transitions

As previously mentioned, dissimilar metal transition joints have many applications in fuel systems, ducts, and manholes. These offer the engineer the opportunity to eliminate bolted joints, for example, and replace them with a leak-tight lightweight joint.

Space Applications

As we look forward to fabrication and assembly in space, explosive packages can offer a method of assembling structures. The framework and plumbing of space stations could be welded together with pre-packaged explosive charges.

Prefabrication of Ceramic Segments by Compaction and Microwave Fusion Technology

Chao-Lung Hwang[1]

Ren-Yih Lin[2], Suh-Lan Shieh[3], and Long-Cherng Lee[3]

Abstract

Material source and environmental impacts are two major concerns in the application of conventional concrete technology in lunar base construction. In this study, the simulated lunar soil which is similar to lunar material in composition was adopted for small ceramic segment production. Compaction, resistance heating, and microwave heating technologies were applied to process the test specimens; furthermore, the produced test specimens were analyzed by scanning electron microscope and mercury intrusion porosimetry technologies. The test results reveal that under controlled heating temperature and heating time by resistance heating, the compressive strength of the test specimens with little or no water content has satisfactory outcomes. In addition, in considering the processing efficiency, the utility of the microwave heating process with regard to some constraints on the moon was also preliminarily compared with a traditional resistance heating process.

Introduction

Construction of a manned lunar base or even the supplement of structural segments from the moon to the space center is no longer a far-fetched dream. However, because of the interdisciplinary nature of the lunar base project, both technological soundness and economical feasibility are equally challenging to scientists and engineers. Currently , one of the major concerns in constructing a lunar base is to overcome the environmental impacts on the moon, the gravitational effect and the extreme temperature gradient in particular. Furthermore, the molecular behavior of water in the concrete is irregular under

----

[1]Member, ACI Committee 125, Lunar Concrete, Professor, National Taiwan Institute of Technology (NTIT), PO Box 90-130, Taipei, Taiwan 10772, ROC.

[2]Ph.D.Candidate, NTIT.

[3]Research Assistant, NTIT.

extreme temperature; this phenomenon implies that conventional concrete technology is inapplicable under the special environment on the moon.

In this study, the feasibility of the ceramic technologies in the production of small segments using simulated lunar soil was preliminarily analyzed. Material that is in composition similar to lunar soil was selected and formed in a confined model to produce test specimens; subsequently, either resistance or microwave heating technology was applied respectively to process the test specimens. The direct application of lunar soil for structural segment production would be a great breakthrough in the study of lunar base construction.

Simulated Lunar Material

According to the analysis of Apollo lunar soils and rocks, lunar materials mainly consist of silicate, alumina, calcium oxide, iron oxide and other minor components (Lin 1987). From the phase diagram shown in Fig. 1 (Lea 1980), the similarity of lunar materials and fly ash products collected from local heat plants can be noted. Because no lunar materials were available in the local area, Type F fly ash was selected as the test material throughout this study .

The Application of Compaction Technology

To offset the gravitational effects on the moon and in space, extrusion or compaction technology is proposed to replace the traditional mixing methods of concrete practice. Compaction technology uses little water to produce the specimens. This is a great advantage because water is not available on the moon and the water transportation from the earth would be economically infeasible. In this study, specimens made from either pure fly ash products or fly ash with cement were each produced by a confined mold as shown in Fig. 2. A 200-ton universal testing machine was adopted to apply compression pressure. The use of compaction pressure is valid for further fusion reactions because the applied force can shorten the distance among particles of the specimens.

The Selection of Sintering and Fusion Technology

Compaction and fusion technologies are typical material processes in the conventional ceramic industry. The rich content of silicate and alumina in lunar materials is practically useful for constructing the lunar base because these components can be fused at about 1200°C, forming silica-alumina products which can withstand the extreme temperatures. In addition, the inexhaustible solar energy on the moon facilitates the heating applications. On the other hand, conventional Resistance Heating (RH) technology in ceramic industry needs huge containers. When compared with MicroWave Heating (MWH), RH seems to be awkward (Hwang 1989). MWH uses a wave guide to conduct energy for local fusion reactions and enhanced heating rates. In this study, along with RH, MWH was

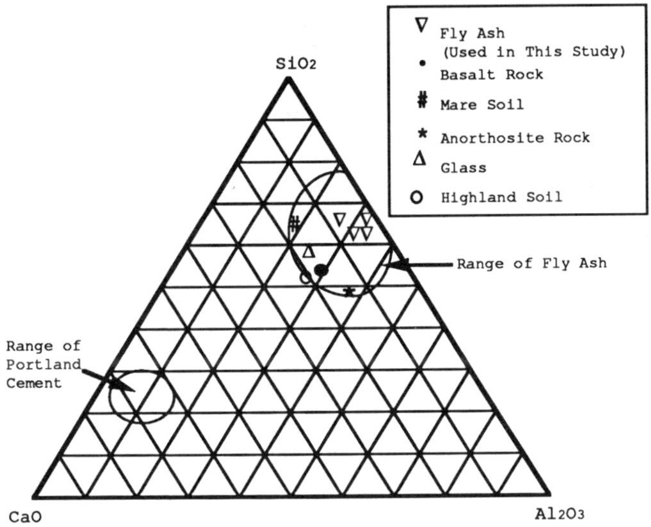

Figure 1 The correspondence among lunar soil, portland cement, and fly ash in the SiO_2-Al_2O_3-CaO phase diagram.

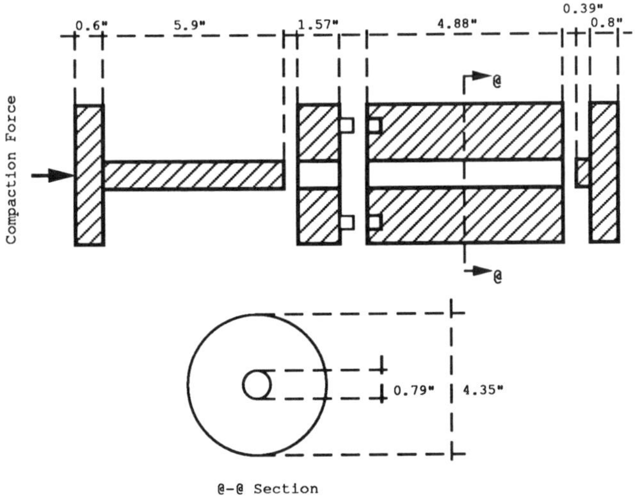

Figure 2 The dimensions of the casting mold.

applied to the control and test of specimens. A MWH system (model MAS-300) with 1200°C heating capacity was adopted for the assessment.

Specimen Casting and the Adopted Fusion Technology

In the ceramic production process, the higher the applied compression pressure, the better the formability of the segments. In this study, the compressive strength of specimens was proportional to the applied pressure, heating time, and heating temperature. RH and MWH were applied respectively to induce fusion reaction. Table 1 and Fig. 3 are the test results from RH; Fig. 3 indicates clearly that right after the specimen was fused, the specimen strength increased rapidly. And, the compressive strength of the specimens with cement is higher than that of specimens with pure fly ash products. Fig. 4 reveals the appearances of test specimens produced by different heating temperature and heating time. The dimension and the weight of the specimens were reduced along with the sintering process; moreover, the fused specimens had the largest weight loss and its color was changed from light-gray to light-yellow.

Microstructure

The properties of fused specimens were governed by the heating time and heating temperature as mentioned above. Glass substance in the original fly ash specimens had partially recrystallized and subsequently several crystal peaks indicated by the arrows in X-ray diffraction patterns (see Figure 5), had appeared after the controlled heating process. The presence of diffraction peaks implies the rapid strength development as shown in Fig. 3. The micrographs from Scanning Electron Microscope (SEM) show that fusion reaction around the contact area of particles had occurred after the compaction/heating process. At 1000°C, the particles had not been totally fused; at 1200°C, however, the fused particles formed a continuum and the recrystallization occurred (see Fig. 6, 7). This fusion process explains the mechanism of rapid strength development. Fig. 8 reveals that when the heating temperature is fixed, the longer the heating time, the larger the reacted area. This phenomenon means that heating time is an important factor controlling the physical properties of the products. Additionally, the pore structure of the test specimens, analyzed by means of Mercury Intrusion Porosimetry (MIP), was also altered. Fig. 9 portrays the MIP pore size distribution curves, indicating that the pore volume will be reduced with respect to the increased heating temperature. Furthermore, after thirty-minute heating time which is quite shorter than that of RH, MWH can significantly reduce the internal pore volumes of the segments.

Table 1 The compressive strength of the fly ash specimens under different heating temperature and heating time.

Compaction Pressure (psi)	Heating Temperature (°C)	Heating Time (hour)	Water/Cement Ratio*	Compressive Strength (psi)
53470	1200	4	0.10	22017
53470	1200	2	0.10	20780
53470	28	-	0.10	5680

* By weight

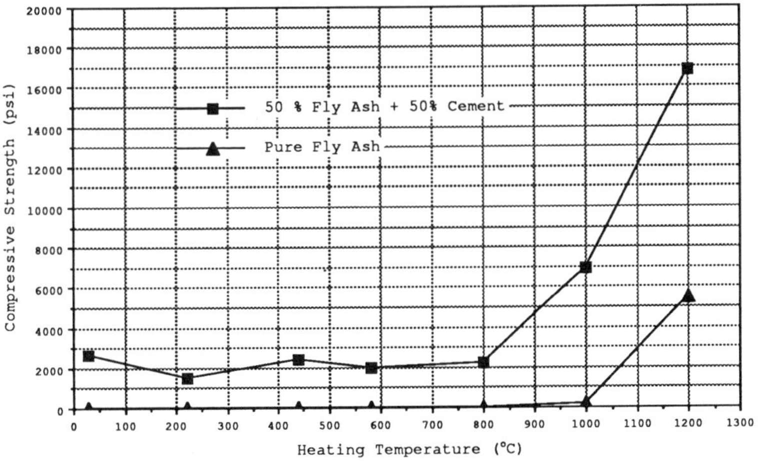

Figure 3 Compressive strength development with respect to different heating temperatures.

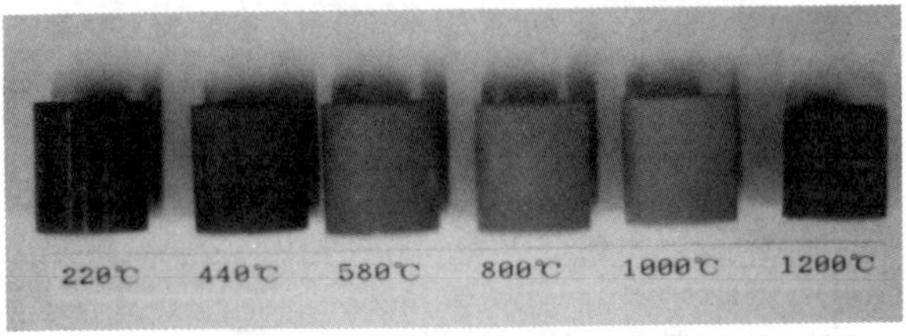

(a) 50% Fly Ash + 50% Cement

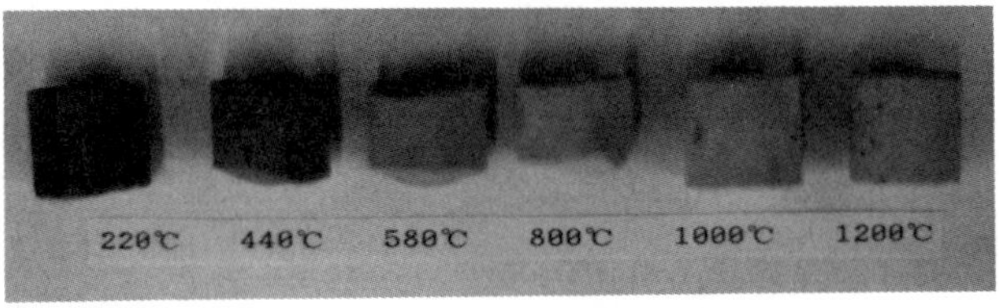

(b) Pure Fly Ash

Figure 4 The test specimens under varied heating temperature.

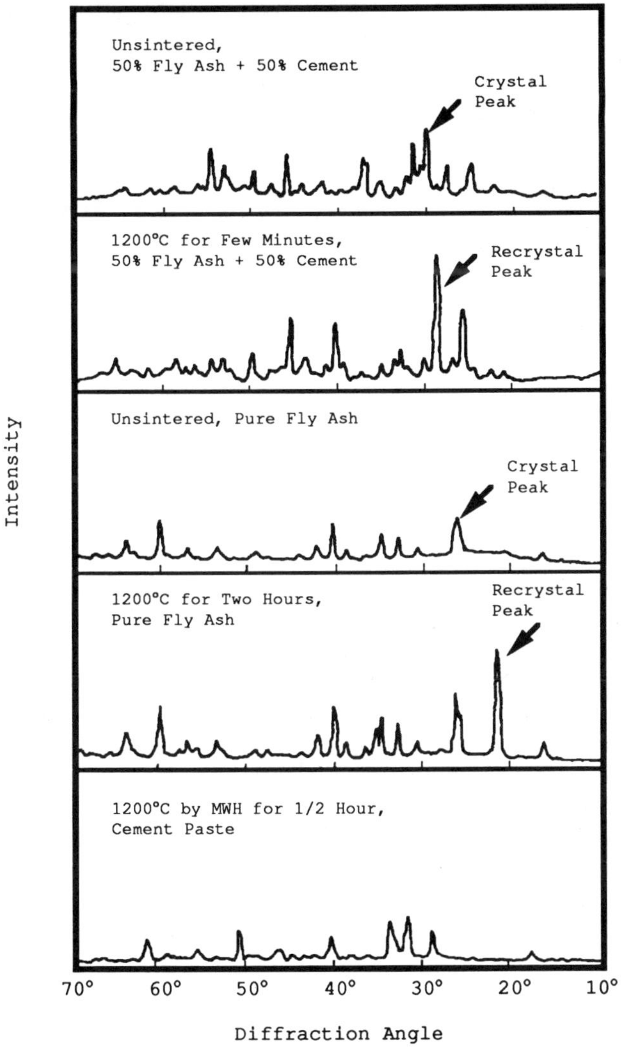

Figure 5 X-ray diffraction patterns.

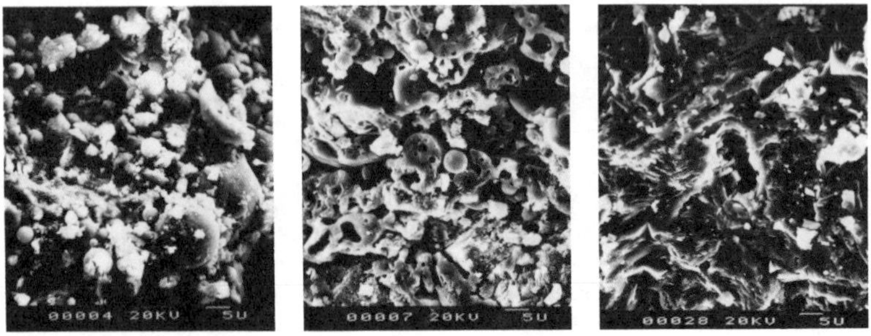

(a) Unsintered (b) 1000°C for Few Minutes (c) 1200°C for Few Minutes

Figure 6 The crystal patterns of specimen with 50% fly ash under varied heating temperature from scanning electron microscope.

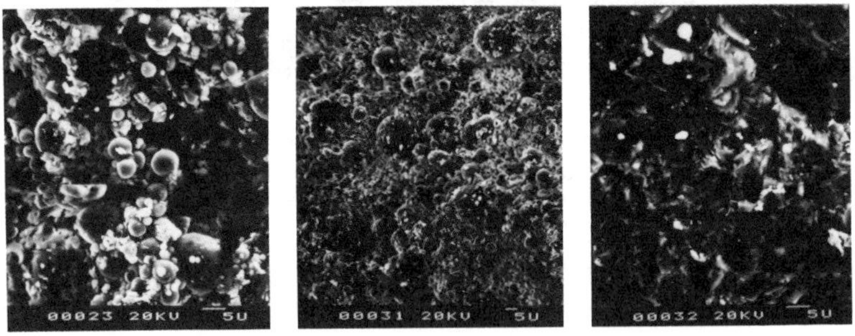

(a) Unsintered (b) 1000°C for Few Minutes (c) 1200°C for Few Minutes

Figure 7 The crystal patterns of specimen with pure fly ash under varied heating temperature from scanning electron microscope.

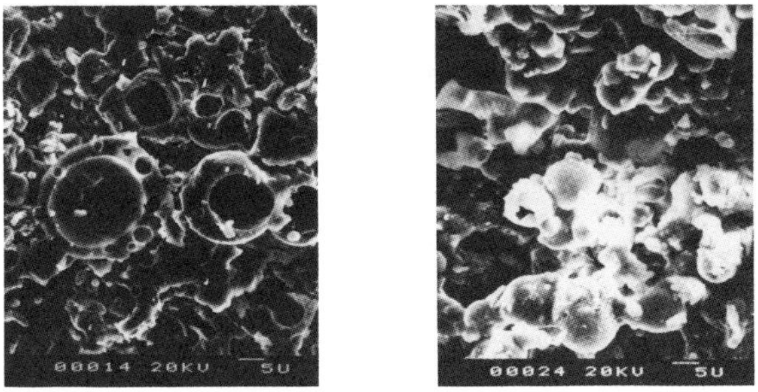

(a) 1 Hour (b) 4 Hours

Figure 8 The crystal patterns of specimen with 95% fly ash under varied heating time from scanning electron microscope.

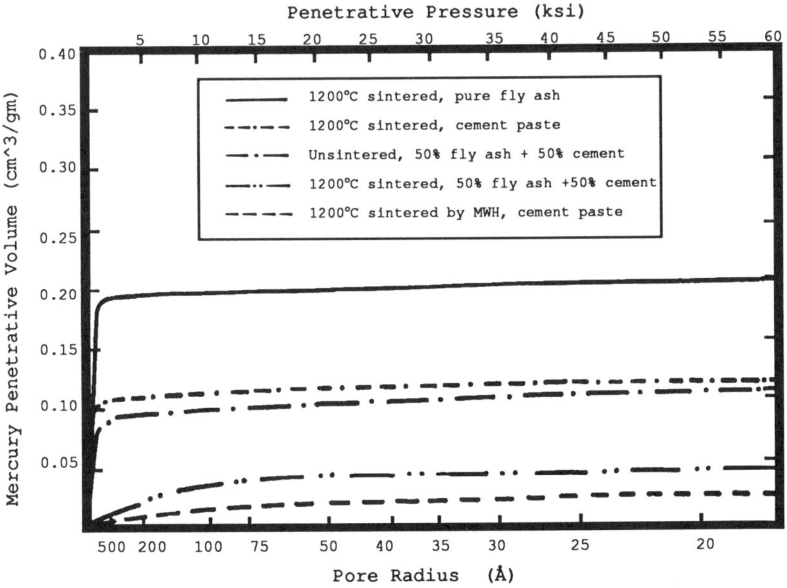

Figure 9 Pore size distribution from MIP test.

Conclusion

In this test program, the compressive strength of the partially fused specimens reached over 20 ksi. This primary result indicates that high temperature and high compaction pressure can be used to produce small segments using simulated lunar materials. In addition, if more sophisticated manufacturing process were available, the production of large segments could be expected. And large segment products can then be assembled to construct structural members. The assembling process could be quite similar to that proposed by Lin (1987). Furthermore, MWH technology seems to be a more practical way for the ceramic segment process on the moon in contrast with traditional RH process used in ceramic industry. However, the efficiency of microwave heating in processing large structural segments is still awaiting further study.

References

Lin, T. D. (1987)."Concrete for Lunar Base Construction." Lunar Bases and Space Activities of the 21st Century, Concrete International, 9(7), 48.

Lea, F. M.(1980)."The Chemistry of Cement and Concrete." Chemical Publishing, New York, 497.

Hwang, C. L., and Lin, C. Y. (1989)."IR and MW Heating for Concrete Materials." Report of National Science Council, Taipei, Taiwan.

UNUSUAL SPACECRAFT MATERIALS

Jonathan V. Post [1]

0.0 Abstract

For particularly innovative space exploration missions, unusual requirements are levied on the structural components of the spacecraft. In many cases, the preferred solution is the utilization of unusual materials. This trend is forecast to continue. Several hypothetical examples are discussed in this paper.

1.0 Sails, Sungrazers, and Iceships

For particularly innovative space exploration missions, unusual requirements are levied on the structural components of the spacecraft. Titanium and carbon-carbon have become materials of choice, the Lunar Excursion Module's thermal protection blanket was literally gold-plated, and so on. But this trend is likely to continue, straining the ingenuity of space engineering, construction, and operations capabilities.

To list three examples: (1) solar sail missions require extremely lightweight, reflective, flexible deployment of very large surface areas [Clarke, 1990]; (2) Starprobe is a mission being considered at NASA-JPL which would send a spacecraft close to the sun, and future follow-on Sungrazer missions may go even further: into the corona of the sun, requiring structural stability at extreme temperatures; (3) unmanned interstellar probes [Jaffe, 1968] powered by nuclear fusion require a minimum deadweight ratio (fraction of non-payload mass remaining after all fuel is expended) and minimum molecular weight of exhaust material.

----

[1] Mailcode 841-FB72, Space Transportation Systems Division, Rockwell International Corporation, 12214 Lakewood Blvd., Downey, California 90241

In many cases, the preferred solution is the utilization of unusual materials. In the three examples above, solutions include: (1) monatomic layer ion sputtered deposition of metal on a substrate subsequently etched or sublimed away as described by K. Eric Drexler [Drexler, 1983]; (2) tungsten carbide, rhenium, and other extreme refractory materials for the structural components of Starprobe successors; (3) The fuel and the structural components of a fusion probe being one and the same. Among contenders, beryllium (molecular weight 8) is rather difficult to vaporize, leaving the prime choices as either water ice (molecular weight 18) or lithium (molecular weight 6) stiffened by boron (molecular weight 11) or carbon (molecular weight 12) fibers, being vaporized, ionized, and expelled as reaction mass until no structural components remain besides the payload and now-useless engine.

In the last case, recent analysis by J. B. Stephens et. al. at NASA-JPL suggests that even hydrogen ice (molecular weight 2) can be stiffened by admixture of fibrous or particulate material far beyond its normal pliability -- about the same as butter. Hydrogen ice can also be adequately protected against sublimation by very modest insulation. A one meter radius sphere of Hydrogen ice, insulated by a centimeter of low-density Hydrogen ice fluff and one centimeter of layered reflector, can last ten years in Earth orbit.

Each of these cases, all feasible within the next 30 years, raises further material-related concerns.

(1) What is the best metal for solar sail deposition in strength/weight tradeoff of ultra-thin films, and to what extent can weight be further reduced by perforation with holes smaller than the wavelength of light? What is the ideal disposable substrate? How can enormous numbers of microscopic holes best be engineered? Aluminum has been recommended, but additional analysis is required.

Dr. Robert L. Forward assumed aluminum could be used for a laser-propelled round-trip interstellar lightsail [Forward,1984]. But, as Geoffrey A. Landis pointed out [Landis, 1989], achievable thrust is limited by the maximum acceptable temperature that the sail attains by inadvertant absorption of light.

Landis ignores reflectivity and high-temperature tensile strength for sail metals, using a figure of merit of melt temperature divided by density, normalized for aluminum (T = 940°K, density = 2.7 grams/cm^{3}) as Al=1. Other metals have superior figures of merit, including boron (3.6), beryllium (2.8), scandium (2.1), and titanium (1.5). Richard Feynman [Feynman, 1986] pointed out that scandium was hitherto unique among elements in having NO known useful applications. Since beryllium (T = 1550°K, density = 1.8 grams/cm^{3}) is also very reflective, Landis considers it the best metallic sail material.

Provocatively, Landis compares transparent nonmetallic dielectric materials, which can be deposited as quarter-wavelength thin-film reflectors. Forward suggested an 8-layer stack alternating diamond [T > 3500°K] with vacuum [Forward, 1986]. Materials with good figures of merit include diamond (2.41), silicon carbide (2.65), zinc sulfide (2.35), tantalum pentoxide (2.5), zirconium dioxide (2.15), and silicon dioxide. Furthermore, if the incident laser light is infrared or ultraviolet instead of visible, then silicon (3.5 at 1200 nm) and lithium fluoride (1.45 at 130 nm) become viable candidates. silicon carbide is judged to be best, but any of these materials offers accelerations "orders of magnitude higher than those achievable with metal films":

Diamond, SiC, ZnS, Ta_2O_5, ZrO_2, SiO_2, Si, LiF

Diamond thin-film can be made by glow-discharge decomposition of methane; tantalum pentoxide, zirconium dioxide, and zinc sulfide are commonly deposited as thin-film for optical coatings by electron beam evaporation; silicon carbide is currently grown by epitaxial deposition on silicon, with the silicon substrate capable of being "etched away to leave a free-standing film ... eminently suitable for a small-scale demonstration" [Landis, 1989]. It isn't clear if LiF would make a suitably strong free-standing film.

The solar sail concept represents a spacecraft which needs no fuel [Clarke, 1990], and has historical/ political significance [Uphoff, 1990]. Refinements to the concept give it an almost poetic elegance [Bradbury, 1990].

(2) What are the optimal refractory materials for Starprobe successors, and how can they best be combined

and engineered? Some particular materials can be easily manufactured from in situ regolith on the lunar surface:

Al_2O_3, CaO, MgO, TiO_2, SiO_2, Ti_5Si_3, Cr_2O_3, K_2TiO_3

Problems remain to be solved. What are the optimal mixtures of these materials (and others, such as the spinels)? Is the moon an appropriate site for materials extraction and processing for near-sun spacecraft components? How may passive refractory insulation and dynamic cooling best be combined? What is the optimum refractory geometry? Robert Waldron [Waldron, 1989] suggests a sharply tapered wedge pointing toward the sun for optimal combination of reflection angle and radiator surface area.

(3) What are the structural limits of fiber-stiffened water ice, hydrogen ice, and lithium? Is it preferable to filter out the fibers, adding them to the deadweight, or to add engine weight to allow them to also be ionized and added to the exhaust? And, in the case of a water-ice fusion spacecraft, can we legitimately refer to this as the ultimate steamship?

2.0 Hydrogen Iceship: Detailed Examination

This section of the paper concentrates on the hydrogen ice spacecraft concept, and consists of an introduction, thermal analysis, experimental results, conclusions, and suggestions for future research.

Much of this analysis was contributed by James Salvail, of SETS, Inc. Honolulu.

Rather than having astronauts perform extravehiclular activity to chip off chunks of hydrogen ice for fuel, we visualize a spacecraft made of a structural cluster of hydrogen ice spheres which can be robotically detached, one at a time as needed, and melted or slushified [Post, 1990] in a conventional fuel tank.

2.1 Hydrogen Iceship: Introduction

A hydrogen ice spaceship can be modelled as a cluster of concentric metal spheres each of which has hydrogen ice filling the spaces between its spherical surfaces. Most of the mass of the spacecraft is composed of these spheres, with the mass of the engine, avionics, payload, and so on comprising a much smaller fraction. The spheres are designed to provide some structural capability and to maintain fuel in solid form until needed (then they are liquefied for propellant usage).

Concentric spheres are connected to each other by at least two rods made of a material that has very low thermal conductivity, such as hard rubber. This is necessary so that the spheres above the instantaneous level of the subliming ice surface do not move relative to each other. The outer shells are made of a highly reflective material, such as aluminized (more easily ionized in propulsion: lithium-ized) mylar, thick enough to provide reasonable structural integrity. The inner spheres are made of the same materials, but much thinner (<<0.1 cm), as they are merely radiation shields.

The radiation shields and outer hulls must contain enough sufficiently sized holes or pores so that sublimed hydrogen molecules are quickly lost into space. The evacuated spaces between the slowly receding ice surface and the outer hulls thus have negligable gaseous heat conduction because the gas is very rarified. Gas flux is small enough (barring close flyby of the sun, nuclear explosions, or laser heating) that heat convection is also negligible. Under the listed abnormal operating conditions, gaseous conduction/convection would still be much smaller than radiative heat transfer. The effects of varying sphere radii, radiation shield number/spacing, outer hull albedo, and external environment are investigated to optimize the design for cost, size, and lifetime.

2.2 Hydrogen Iceship: Thermal Analysis

Thermal analysis consists of temperature calculations for outer hull, radiation shields, fixed radii from centers, and (crucially) at ice surfaces. Hull and shields are sufficiently thin and heat conductive as to be effectively isothermal through their thicknesses.

The energy balance at the outer hull consists of incoming solar radiation, emitted radiation from both sides, incoming radiation from the adjacent lower surface, and downward gaseous heat conduction. The two sides of the outer hull have, in general, different albedos and emissivities (i.e. painted black externally for visual stealth, as the low temperature already offer IR-stealth). These effects are described by:

$$E_m Sb T_{i,j}^4 + E_b Sb T_{i,j}^4 - f_b (1-A_m) E Sb T_{i-1,j}^4 - K \frac{dT}{dr} = \frac{(1-Ab)Sc}{4R^2}$$

Where E_m is the emissivity of the natural metallic surface, E_b is the emissivity of the outer surface (possibly black), unsubscripted E is the emissivity of the adjacent lower surface. If the lower adjacent surface is a radiation shield, then $E=E_m$. If the lower adjacent surface is the ice surface, then $E=E_h$, the emissivity of the ice. A_m is the albedo of the natural metallic surface. T is temperature, with subscripts i and j denoting depth and time. f_b is a geometric factor accounting for the smaller area of the adjacent inner surface. Sb is the Stefan-Boltzmann constant. Sc is the solar constant at 1 AU (Astronomical Unit). K_g is the gaseous thermal conductivity of hydrogen. r is the radial coordinate (from the center of the ice sphere). R is the heliocentric distance in AUs. The factor 4 in the right hand solar insolation term reflects the assumption of rapid rotation. Constant values are given in section 2.3. This equation, and a related one for the moving ice surface (with an energy balance including upward radiation from the ice surface, downward radiation from the adjacent metal surface, heat conduction into the ice, and the latent heat energy due to sublimation) are the basis for the results of section 2.3.

2.3 Hydrogen Iceship: Computer Simulation Results

Computer simulation based on the preceding analysis calculated temperature at outer surface, radiation shields, surfaces and interior of the hydrogen ice. Ice surface temperature allowed derivation of hydrogen gas flux and radial position of the receding ice surface as a function of time, and thereby deriving hydrogen ice lifetime. Various runs determined the effects of radiation shields, outer albedo, and outer hull radius, in normal conditions and in a simulated nuclear blast. Kirchoff's law was assumed for metallic and ice surfaces. Parameter values are as listed below:

Property/Parameter		Value
A	Vapor pressure constant	6.17×10^{9} dynes/cm^2
B	Vapor pressure constant	149.44 °K
A_m	Albedo of metal	0.95
A_h	Albedo of hydrogen ice	0.65
E_m	Emissivity of metal	0.05
E_h	Emissivity of hydrogen ice	0.35
C_h	Specific heat of hydrogen ice	2.7×10^{7} erg/g-°K
D_h	Density of hydrogen ice	7.06×10^{-2} g/cm^3
K_h	Thermal conductivity of ice	1×10^{7} ergs/cm-s-°K
H_s	Latent heat of hydrogen ice	6.2×10^{9} ergs/g
S_c	Solar constant at 1 AU	1.3928×10^{6} ergs/cm^2-s
T_i	Initial temperature of ice	5°K

M	Molecular weight of hydrogen	2.0 g/g-mole
Sb	Stephan-Boltzmann constant	5.6×10^{-5} ergs/cm^2-s-$^\circ$K
R_u	Universal gas constant	8.315×10^7 ergs/mole-$^\circ$K

First, a 1 meter radius sphere with 50 radiation shields spaced 2 centimeters apart and natural metallic surface was simulated. The ice remained nearly isothermal at the initial temperature of 5°K, with a negligible temperature gradient and a near-constant mass flux of 17.8 nanograms/cm^2-sec. The outer hull remained at a temperature of 236°K at 1 AU from the sun. After a simulated 10 years, the sphere had shrunk to 21 cm in radius, and the total lifetime was roughly 12 years.

Reducing the radiation shields from 50 to 10 had no effect. Painting the outer surface black (albedo = 0 for stealth) gave a tripled mass flux of 53.8 nanograms/cm^2-sec, a surface temperature of 5.2°K, and a reduced lifetime of 4.2 years.

At 0.1 AU from the sun a 50-shield 1 meter shiny sphere stays at 5.81°K, with a mass flux of 1.06 micrograms/cm^2-sec, and a lifetime of 75 days. With 10 shields, a 1 meter shiny sphere stays at 6.39°K, with a mass flux of 10.5 micrograms/cm^2-sec, and a lifetime of 35 days. Hence radiation shields are important for larger thermal loads, such as would occur if the hydrogen iceship mission began with a gravity assist swingby close to the sun.

The thermal effects of nearby nuclear detonation were simulated as a temporary change in heliocentric distance from 1.0 AU to 0.01 AU (where the radiative equilibrium temperature for a black body is 2808°K) for 20 seconds. If the outer metallic coat doesn't melt at the maximum temperature attained (2361°K), then the hydrogen ice adjacent to the outer surface peaks at 8.73°K, with a gas flux of 4.7 milligrams/cm^2-sec, decreasing after 10 minutes to 5.85°K (ten shields) or 5.79°K (fifty shields), at which time the ice has receded 2.5 cm.

All other things being equal, the lifetime of a hydrogen ice sphere was found to be directly proportional to the first power of the the initial radius. Thus, a 2 meter radius sphere has a 24 year lifetime at 1 AU, and 2 years at 0.1 AU. For deep space missions, loss becomes rather small for spheres several meters in radius.

Similar thermal analyses has been performed for slab and

cylindrical geometries [Hustvedt, 1986].

2.4 Hydrogen Iceship: Summary of Concepts

The greatest advantage for a hydrogen ice spacecraft is obtained if the craft is an unmanned monolithic composite solid cryogen with embedded insulation and superconducting avionics. As disclosed by J. Stephens at JPL (who generated the original concepts in 1984 and 1985, while in communication with this author), the primary intellectual properties for patent purposes are: (1) Ice embedded insulation, (2) Vapor cooled insulation, (3) Isomer conversion catalyst integral with insulation (activated carbon), (4) IR photon reflective and vapor conductive insulation (variable mesh cloth multi-layer insulation), (5) Vapor cast crystalline hydrogen ice using nuclear magnetic resonance heating of non-crystalline ice, (6) Self-forming filamentary insulation from dispersed particles in the ice that cohere due to ice cleaning.

The attributes of the primary intellectual properties are: (1) Unitized design; ice is the cryogen, structure, propellant, shielding, absorber, power source, window, and insulation support during launch, (2) Superconducting temperature cryostat (<5°K for Hydrogen), (3) Self-insulating solid cryogen, (4) Long lifetime in earth orbit, (5) Low cost materials (<$10/lb), (6) Low cost fabrication (casting process), (7) Low launch cost (withstand high acceleration forces), (8) Low cost operation (efficient superconducting solid-state system), (9) Acoustically quiet (no moving parts) so good for very accurate optics or interferometry, (10) Thermally stable (large thermal capacity well insulated), (11) High density ice vapor cast and used at same temperature avoiding shrink stresses in insulation and components embedded in ice.

The ancillary intellectual properties enabled by the primary concepts are: (1) Vapor cooled refractory insulation, (2) Neutron absorbing cryogen (Hydrogen), (3) Microwave absorbing ice/insulation, (4) Microwave reflecting ice/insulation.

The advantages of the ancillary concepts are: (1) Laser tough shielding, (2) Neutron tough shielding, (3) Neutral and charged particle beam tough shielding, (4) Radar stealth, (5) Superconducting phased array radar.

Concepts enabled by Cryostat primary and secondary concepts are: (1) Propulsion and electric power systems: (a) Solar powered ion rocket and superconducting magnet power generator and storage system, (b) Magneto-hydrodynamic detonation wave ion rocket and superconducting magnet power generator and storage system (detonate layers of solid oxygen alternated with solid hydrogen); (2) Guidance and control: (a) Superconducting computer, (b) Superconducting gyroscope, (c) Superconducting magnet attitude control, (d) Superconducting radio and antenna; (3) Launch forces resistant structure: (a) fiber reinforced composite ice; (4) Remote sensing synersensory systems in phased arrays in Cryostats orbiting in formation: (a) Synthetic aperture superconducting phased array radar, (b) Synthetic aperture superconducting phased SQUID array Magnetic Anomaly Detection, (c) Synthetic aperture superconducting phased SQUID array Gravimeter, (d) Synthetic aperture superconducting phased array Altimeter, (e) Blue-Green synthetic aperture superconducting phased array lidar, (f) Thermal IR telescope spectrometer.

This astonishingly rich set of concepts of Jim Stephens is only moderately challenged by demands of the interplanetary or interstellar regime, as opposed to the near-Earth applications originally envisioned.

Individual hydrogen ice spheres can be orbited by small boosters, and later assembled into a large spacecraft. Solid hydrogen is inherently safer than liquid hydrogen. The spheres can have embedded avionics, providing distributed redundant capability for the spacecraft at superconducting temperatures. Once assembled, the low accelerations typical of an ion, fission, or fusion propulsion would not endanger the inherently low tensile strength of hydrogen ice as a structural material. The hydrogen ice spheres would be between the crew (or payload) and the nuclear propulsion, providing neutron-absorbant shielding at no extra cost.

Space exploration applications include: (1) Sungrazer, (2) Outer planet explorer, (3) 1000 AU mission (TAU), (4) Subterranean radar mapping of planets, (5) Manned Mars Mission, (6) Propellant transfer and storage for Space Station refueling depot [Post, 1990].

2.5 Hydrogen Iceship: Future Research

Future areas of hydrogen iceship analysis include: (1) Comparisons of fiber-stiffened water ice, carbon dioxide, lithium, or other alternatives to hydrogen ice; (2) Experimental determination of strength, stiffness, etc. for hydrogen ice with various fiber compositions (boron, carbon); (3) Structural design optimization for various types of propulsion system; (4) Exploration of the concept of detonation wave propulsion/attitude control with alternating layers of hydrogen ice and oxygen ice; (5) Sensor capabilities of phased arrays of embedded cryogenic detectors in a fleet of coorbiting iceships; (6) Relativistic kinematics of multi-staged interstellar iceships.

References

Bradbury, R., and Post, J.V., "To Sail Beyond the Sun", in Clarke, A.C., ed., "Project Solar Sail", ROC Books, New American Library, April 1990

Clarke, A.C., ed., "Project Solar Sail", ROC Books, New American Library, April 1990

Drexler, K. Eric, U.S. Patent, personal communication, 1983

Jaffe et. al., JPL Document 32-233: "Feasibility of Interstellar Travel", ca. 1968

Feynman, R., personal communication, 1986

Forward, R.L., "Roundtrip Interstellar Travel Using Laser-Pushed Lightsails," AIAA J. Spacecraft and Rockets, Vol.21:2, 187-195, 1984

Forward, R.L., "Laser Weapon Target Practice with Gee-Whiz Targets", Laser Propulsion Workshop, Lawrence Livermore National Laboratories, 7-18 July 1986

D. Hustvedt, personal communication, 17 January 1986

Landis, G.A., "Optics and Materials Considerations for a Laser-Propelled Lightsail", Paper No. IAA-89-664, Proc. 40th Congress of the International Astronautical Federation, Malaga, Spain, 7-12 Oct 1989

Post, J.V., "Hydrogen Ice Spacecraft", to appear in Proc. AIAA Space Programs and Technologies Conference, Huntsville, AL, 25-28 September 1990

Uphoff, C., and Post, J.V., "A Rebel Technology Comes of Age", in Clarke, A.C., ed., "Project Solar Sail", ROC Books, New American Library, April 1990

Waldron, R., Rockwell International, personal communication, 1989

Subsection 3.6: STRUCTURES AND STRUCTURAL RESPONSE

MOTION CONTROL OF SPACE STRUCTURES

Larry Silverberg* and William Dunn**
Member AIAA

Abstract

A new area of civil engineering is emerging as we begin to establish a permanent presence in space. The new area of civil engineering is the motion control of space structures. This paper describes why the motion control of space structures problem is fundamentally a civil engineering problem.

Introduction

Civil Engineering on Earth has made great strides since the days of early man. We have progressed from mud huts to skyscrapers and from logs across creeks to towering bridges spanning rivers. And from the mud hut to the skyscraper, the principles have remained unchanged. We use building material able to support itself and the weight of the structure above it, and we build upon a firm, steady foundation capable of supporting the entire structure. However a new area of civil engineering is emerging as we begin to establish a permanent presence in space. For the first time the presence of a zero-g environment means that there is no structural weight to support and the absence of ground means that there is nothing to which a foundation can be anchored. It would seem that the most widely used principles of civil engineering on Earth would be different than those used in space. Yet a deeper look exposes similarities that will enable us to draw upon the thousands of years of experience on Earth to design systems for the motion control of civil engineering structures in space.

The space structures of the near future will be constructed and then function in a weightless environment. In a sense, weightlessness is an attractive feature since the space structures must be strong enough only to withstand the relatively small disturbances encountered in space. For instance, construction in space, docking with other vehicles, atmospheric drag, and crew

* Assistant Professor, Mars Mission Research Center, North Carolina State University, Raleigh, North Carolina 27695-7910.
** Graduate Research Assistant, Mars Mission Research Center, North Carolina State University, Raleigh, North Carolina 27695-7910.

movement will all produce relatively small levels of force throughout the space structures.

On the other hand, the absence of an anchoring point in space has led to the demand for an artificial foundation designed to control the space structure's motion. This new artificial foundation, referred to as the control system, is comprised of sensing devices measuring the space structure's position, actuation devices exerting forces, and a brain dictating commands to the actuators to exert forces corresponding to measurements received from the sensing devices. Together the actuators, sensors and brain produce an artificial foundation designed, in part, to replace the ground. Moreover, the design of this artificial foundation is intimately related to the most fundamental concepts in Earth-based civil engineering.

The earliest control systems oriented artificial satellites to point to ground stations on Earth. Such sensing devices as gyroscopes, earth horizon sensors, or star trackers, measured orientation errors. The actuation devices were either internal torque wheels exploiting the principle of conservation of angular momentum, or small rocket thrusters providing an external torque. The brain simply commanded actuators to return the satellite to its desired orientation by producing a torque in proportion to the indicated error and in the opposite direction. Indeed, relatively primitive control systems were used for these early satellites in much the same way as their earthbound counterparts, mud huts, were built on simple foundations.

In contrast, with the coming of larger and more flexible space structures the next generation of control systems will be much more sophisticated. They will be required to orient space structures, implement controlled maneuvers, and suppress structural vibration. NASA is currently planning over 20 projects that can be classified as Large Space Structures. The structures of the near future are the size of a football field, and those planned for the more distant future will be ten times that size. Just as civil engineers "scraped the sky" with their enormous towering buildings, so too will space civil engineers soon "scrape space" with the coming spacescrapers.

Natural Motions of Space Structures

Accompanying the demand for larger size space structures designed to carry out more ambitious missions is the equally pressing demand to reduce the cost of launching these structures into orbit by lowering their mass. The combination of large size and low mass forces the space structures to be highly flexible.

The implication is that every point on the structure can move relative to the next. Since there are hundreds of thousands of points (theoretically an infinite number), the prospect of supplying an artificial foundation to control the motion of these highly flexible space structures would seem to be insurmountable. Fortunately scientists have known for some time that there is another simpler way to view structural motion. Rather than look at hundreds of thousands of individual points on the structure each contributing to the overall motion, structural motion is traditionally viewed as the sum of several dozen or more independent motions called natural motions. The problem of controlling the motion of these large and flexible space structures can be treated by controlling their natural motions.

Associated with each natural motion are three parameters. The first is called the natural mode shape, or mode. The mode is the unique spatial shape of a particular natural motion. The second parameter is called the natural frequency. The natural frequency expresses the rate in time at which the natural motion oscillates. The third parameter is a measure of the time required for the natural motion to decay and is called the natural decay rate.

The space structure's natural motions correspond to the three directions in which the structure can translate, and the three axes about which the structure can rotate, while the remaining correspond to bending, twisting and stretching motions. Each natural motion participates in the overall motion depending upon the degree to which it is excited by the external forces. Generally, the natural motions participate inversely proportional to their corresponding natural frequency: the natural motions with the larger frequencies participate less in the overall response. There are two reasons for this effect. First, the natural motions with higher frequencies are excited less by the forces typically found in space. Secondly, the internal friction found in space structures acts to damp out the natural motions with higher frequencies more rapidly than the lower ones. So the overall structural motion is viewed as the sum of three translations, three rotations and several dozen or more low frequency flexural motions.

To date the space structures launched into orbit have been predominantly rigid. Their overall motion consisted of at most six natural motions: translations in three directions, and rotations about three axes. Typically, the three translations were constrained by orbital mechanics. Restated, the space structures followed a predetermined orbit or trajectory without need of a control system. This left at most three rotations in need of control. As stated earlier, a relatively simple artificial foundation controlled these rotations. In contrast, the artificial foundations anticipated will soon require the control of dozens of complicated flexible natural motions in addition to the translational and rotational natural motions.

The Control System

As mentioned earlier, the control system brain senses certain physical quantities, then multiplies them by control gains, and finally commands actuators to exert the appropriate forces to return the structure to its undisturbed state. The undisturbed state can correspond to the structure's fixed position and orientation (like foundations on Earth) or the undisturbed state can correspond to a moving position or orientation such as the case when we wish to maneuver the structure. In either event, the physical quantities sensed are displacements and velocities relative to the undisturbed state. Scientists have long known that producing control forces in proportion to these quantities yields distinctly different results. Control forces proportional to the displacement and in the opposite direction return the structure back to its undisturbed state, and in doing so, they represent an artificial form of stiffness. Control forces proportional to the velocity and in the opposite direction resist structural motion and cause the motion to damp out in time, and so they represent an artificial form of damping.

Scientists in the 1970's understood that measurements of the displacement and velocity of each participating natural motion were needed in order to produce the proper control forces. Dr. Haim Baruh, now at Rutgers University, developed a method to synthesize the displacement and velocity of each participating natural motion from displacement and velocity measurements over the space structure (Baruh and Choe, 1989). Using the synthesized displacements and velocities of the natural motions appropriate control forces could be produced to individually damp out each participating natural motion.

The motion control problem seemed to be solved, because control gains could be selected to produce forces to control each natural motion in any way desired. Certain natural motions could be targeted for control while leaving others uncontrolled, or certain natural motions could be damped out more rapidly than others --- all depending upon the choice of the control gains.

However, the question remained to what degree each natural motion should be controlled. Some scientists pursued elaborate schemes designating the control of each natural motion at different rates depending on the disturbances encountered. Others sought to minimize the number of gain calculations in order to reduce the computational effort and to reduce the number of communication lines and the software complexity. Still others sought to minimize the fuel or power required to produce the control forces.

It was acknowledged that all of these aspects were important, and it seemed apparent that optimizing only one aspect of the control system would compromise other aspects. It was not anticipated that the most desirable characteristics could all occur simultaneously. Yet this is precisely what happened when scientists who were trying to minimize fuel, looked at the control system in a new and unique way. Moreover, they unexpectedly uncovered an intimate relationship between the motion control of space structures and the static support of civil engineering structures on Earth.

The Controlled Structure

Recall that the structure's overall motion has traditionally been viewed as the sum of natural motions each acted upon by the control system. Recently, many scientists have adopted the view in which they treat the structure plus the control system together as one new system. The corresponding motion of the controlled space structure was then viewed as the sum of new "unnatural" motions. The control system with the gains turned on was viewed as changing the very nature of the uncontrolled structure. The old, uncontrolled structure had a set of mode shapes, natural frequencies, and decay rates. Now, with the control system turned on, a new set of unnatural motions was obtained, each with a new mode shape, frequency and decay rate. The designer individually specified the new mode shapes and their corresponding frequencies and decay rates merely by appropriately selecting the control gains.

What benefit does this new viewpoint yield? The second perspective allows us to compare the two systems; to compare the natural motions comprising the uncontrolled system's motion with the "unnatural" motions that form the controlled system's motions. Specifically, the two systems' decay rates, mode shapes and frequencies were compared.

The choice of the controlled system's mode shapes was first investigated by Leonard Meirovitch, a Distinguished Professor at Virginia Polytechnic Institute who specified the new controlled mode shapes and the natural mode shapes to be identical (Meirovitch and Silverberg, 1983). With this selection, the total effort to produce the forces (power or fuel) was shown to be minimized.

The question remained how to specify the frequency and decay rate for each of the controlled natural motions? The author followed Meirovitch's lead in his selection of the controlled mode shapes and found that it requires increasing amounts of effort to produce forces to change the natural frequencies (see figure 1). In other words, a minimal level of effort is required when the controlled frequencies are specified to be identical to the natural frequencies (Silverberg and Morton). Next, considering the choice of the decay rate for each natural motion, recall that the decay rate expresses an exponential function continuously decreasing in time that envelopes the natural motion. The author argued that artificially specifying a higher decay rate requires increasing amounts of effort in the form of fuel or power, that is the more rapid a natural motion is suppressed.

But how should the modal decay rate for one natural motion relate to the modal decay rate for another natural motion? When different decay rates are chosen for the natural motions, the natural motion with the lowest decay rate lingers on to dominate the overall motion after the others have subsided. Since a higher decay rate requires more effort, an unnecessary price in terms of fuel or power would have been paid to quickly damp out some natural motions while allowing the overall motion to continue due to the presence of one slowly decaying natural motion. Therefore, one overall decay rate for the natural motions was selected (Silverberg, 1986). The selection of the overall decay rate was then left up to the control system designer, and depended upon how much overall effort the designer was willing to expend to produce the control forces.

In view of these developments, scientists now recognize that changing the structure's natural characteristics requires an expenditure of effort and that changing the mode shapes and frequencies is an unproductive use of this effort. It became efficient to expend effort only for the purpose of increasing the natural decay rates to one uniform value to uniformly suppress the overall structural motion. But when the associated control gains were determined, scientists were startled with the result.

Having determined the control gains, the control forces became functions of the displacements and velocities of the structure itself rather than of the displacements and velocities associated with the natural motions. It became unnecessary to synthesize and separate out each of the natural motions. Furthermore, only the displacement and velocity measurements in the region containing the actuator determined the control forces, because the control gains became zero except for the two in the region of the actuator. In other words, the entire control system became decentralized. The structure could now be divided into regions each controlled independently of the others, without need of synthesizing the natural motions. The whole control system together formed a distribution of sensor/actuator packages over the structure each independently responsible for the motion control of its region.

The control gains turned out to be simple algebraic expressions depending on the overall decay rate and the masses of the regions. The control gain multiplying the velocity measurement was 2×(mass)×(decay rate). Associated with the displacement measurement, the control gain was (mass)×(decay rate)^2. As the reader can see, the gains are proportional to mass and independent of the structure's stiffness.

Although the dynamic characteristics (mode shapes and frequencies) depend on both the mass and stiffness of the structure, the control system only requires knowledge of the mass, the time desired for the overall motion to cease, and the measurements of displacements and velocities. Indeed, the control system uniformly damps out the natural motions without knowing or explicitly sensing them.

The Link Between Civil Engineering and Motion Control

Can the experience of thousands of years of civil engineering on Earth help us in space? To answer this question let us compare the support system used on a representative civil engineering structure such as a bridge with the representative space structure illustrated in figure 2 (a brief description of the control system design process is found after figure 2).

The ideal method of supporting a long bridge over a river (or a long log across a creek for that matter) is a continuous support system. For instance, the cables of a suspension bridge provide continuous support over the length of the bridge. Often, this solution is too expensive, so the bridge is divided into regions each supported by independent, discrete supports. In a decentralized fashion, each support is designed strong enough to support only the structural mass and the loads expected in its region. Likewise we discovered earlier that the ideal method of supplying artificial support to a large space structure is a continuous distribution of control forces, and this solution as well can be too expensive. Then, the large space structure is divided into regions each supported by discrete supports --- our sensor/actuator packages. The support strength of each actuator is determined by the mass of its region, likewise in a decentralized manner.

Given a certain number of bridge supports, a good civil engineer would place the supports so that the shear forces and bending moments at any point along the bridge are reduced. Analogously we can envision our space structure as being "dynamically supported" by the control forces. We could conceive of supporting our space structure by hanging it (computationally or even physically) using vertical cords from the ceiling of a hanger. As a rule of thumb, we could use the locations of the cord attachments as good approximations for the locations of the actuators. In other words, support our space structure with an actuator just as a civil engineer would most efficiently support the structure on Earth with a cable. Then we can use the percent of the total weight each cable supports to compute the percent of the total mass that the corresponding actuator dynamically supports. Indeed, designing support systems for both the static bridge and the dynamic space structure involve the same steps.

Perhaps in the future as we more clearly understand the similarities between the disciplines of Earth and space based civil engineering, more ambitious designs will be implemented in space. We will also more clearly understand how the mass of future space structures can be reduced to extremely low levels never before envisioned. Space may some day be populated with enormous, wafer thin sheets of solar panels and ultra thin antenna arrays with large numbers of small micro sensor/actuator packages that together efficiently maintain the shape, orientation and even maneuver the structures. But whatever the future holds, we can feel confident that the benefit of thousands of years of experience on Earth will enable us to design simple, reliable control systems for the generations of space structures to come.

Suggested Reading

NASA to Improve Design of 21st Century Satellites, *Aviation Week*, Nov. 4, 1988.

"Mission to Mars" by Michael Collins, *National Geographic*, pp. 733-764, Nov. 1988.

References

Baruh, H. and K. Choe, 1989, "Reliability Issues in Structural Control," Vol. XXXI of the Annual International Series on *Advances in Control and Dynamic systems*, C. T. Leondes, Editor, Academic Press.

Meirovitch, L. and L. Silverberg, 1983, "Globally Optimal Control of Self-Adjoint Distributed Systems," *Journal of Optimal Control, Applications and Methods*, Vol. 4, pp. 365-386.

Silverberg, L. and M. H. Morton, "On the Nature of Natural Control," *Journal of Vibration, Acoustics, Stress and Reliability in Design*, to appear.

Silverberg, L., 1986, "Uniform Damping Control of Spacecraft," *Journal of Guidance, Control and Dynamics*, Vol. 9, No. 2, pp. 221-227.

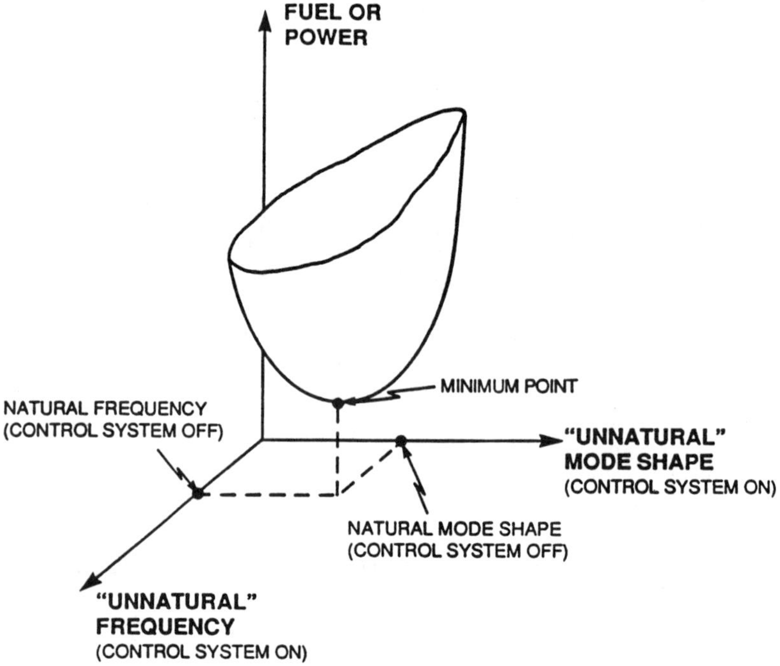

Figure 1. Fuel or Power to Exert Control Forces

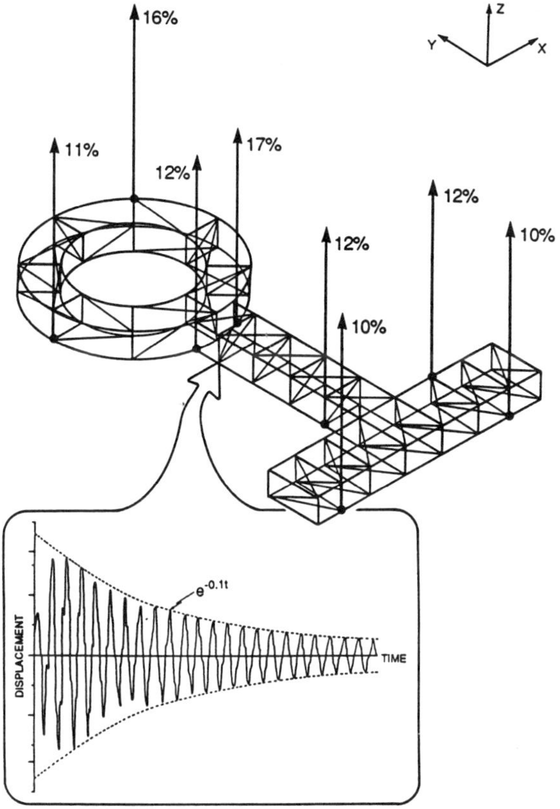

Figure 2. Design Example

Caption for Figure 2: Design Example

Suppose we desire to install a control system of our own on the structure shown. The following describes the four steps in the process.

STEP 1 DETERMINE THE RATE OF DECAY

A decay envelope is selected to dampen 90% of the motion in 25 seconds. Based on our knowledge of the exponential function we obtain a decay rate of 0.1 rad/sec.

STEP 2 DETERMINE THE NUMBER OF EXCITED MODES

The mode shapes and frequencies of our structure are shown in figure 2. Let us assume that disturbances are caused by high frequency motors attached to the structure, crew movement, and shuttle docking. The high frequency motors will excite only the higher modes and we may let internal friction or a vibration isolation pad take care of the higher modes. Crew movement will excite the first few modes, and suppose we have determined experimentally that, in all, the lowest 25 natural motions are excited by these disturbances. Recognizing that the fewer actuators used, the more critical their locations become, suppose we select eight sensor/actuator packages.

STEP 3 DETERMINE THE PLACEMENT OF SENSOR/ACTUATOR PACKAGES

Drawing upon our experience with Earth-based civil engineering, we hang the structure using eight cable supports. As shown, the amount of tension in each cable support is proportional to the percent of the total weight carried by the cables. We then might move the supports around until each support has approximately the same tension so that the structure is as uniformly supported as possible. The sensor/actuator packages are located at the same places as the support cables would be with their mass region equal to the same percentage of the total mass that the cable would have supported. If the actuator size that would be required to exert a control force for any of the regions is too large, we simply choose more packages, (hang the structure with more supports) decreasing the amount of mass for which each package is responsible. We end up with a control system whose actuators are all approximately the same size. With the control forces as evenly distributed as possible, the effect of any one actuator's location is minimized.

STEP 4 DETERMINE THE CONTROL GAINS

The gains are determined by two simple algebraic expressions. The formula for the control gain multiplying the displacement measurement is mass $\times$ decay rate^2, where the decay rate is the previously determined decay rate for the overall motion and the mass is associated with the local region the package is responsible for controlling. The control gain multiplying the velocity measurement is 2 $\times$ mass $\times$ decay rate. For example suppose the mass of a region is 10 kg and the decay rate for the structure has been chosen to be 0.1 rad/sec. Associated with the displacement, the control gain becomes 0.1 Newton/meter and associated with the velocity, the control gain becomes 2 Newton/(m/s). Then, if the sensor in region 1 senses an instantaneous displacement of 0.1 m and an instantaneous velocity of 1 m/s, the control force becomes 0.01N + 2N = 2.01N.

Having completed our design, the gains now accurately translate the measured displacements and velocities into actuator forces over each region using a minimum amount of fuel or power and with the overall structural motion decaying uniformly as shown. As illustrated here, the control system artificially supports the space structure using forces proportional to the mass of each region. This same principle has been applied to Earth based foundations for thousands of years in which structural components are designed to support the weight of large civil engineering structures.

EFFECTS OF INITIAL CONDITIONS ON LARGE SPACE STRUCTURES

by Ramesh B. Malla,[1] Associate Member, ASCE

ABSTRACT

Effects of initial conditions of axial deformation (length) and attitude (pitch) angle have been investigated on the subsequent orbital motion, attitude (librational) motion, and axial deformation (length) of a large space structure under the action of the earth's gravitational force. The space structure is assumed to be axially flexible and is considered to be executing a planar motion around the earth. It is observed that the initial values of the parameters can have appreciable effects on the structure's axial deformation and its attitude motion, whereas they have negligible effects on the structure's orbital motion (orbital radius and angular velocity).

INTRODUCTION

The main focus of scientific and engineering community interested in space related activities at present has directed toward design, construction, and operations of facilities and bases in orbit and on the moon and nearby planets which will be continuously occupied by human beings (Johnson and Wetzel 1988). Future space structures are conceived to be very large in size and minimum in weight, hence extremely flexible, and complex in geometry as well as function.

It is necessary to eliminate or at least minimize structural dynamic responses of these space systems to space environmental disturbances, such as, docking, deployment, maneuvers, meteoroid impact, and thermal changes. The problems of devising a properly stable control scheme, modeling the structure, and making in-

---

[1]Visiting Assistant Professor, Department of Civil Engineering, University of Connecticut, Storrs, CT06269.

flight response measurements of the systems have received a great deal of attention in recent years (Meirovitch 1982, Pinson 1989). Prior to achieving any of the above functions for engineering and construction of space structures, the behavior and responses of the structures under a particular environment must be known as accurately as possible in advance.

This paper presents the results of an investigation delineating the effects of initial values of axial deformation (change in length) and the attitude (pitch) angle of a large space structure on its subsequent orbital motion (orbit radius and angular velocity), attitude (pitch librational) motion, and axial deformation simultaneously. The structure is considered to be axially flexible and is assumed to execute a planar orbital motion under the action of earth's gravitational force.

EQUATIONS OF MOTION

For a space structure represented by a dumbbell-shaped geometry consisting of two end masses m_1 and m_2 connected by an axially flexible link structure which moves in a general non-circular orbit in an orbital plane around the spherical earth of radius R (Fig. 1), the writer has previously derived a set of four equations of motion using Lagrange's principles

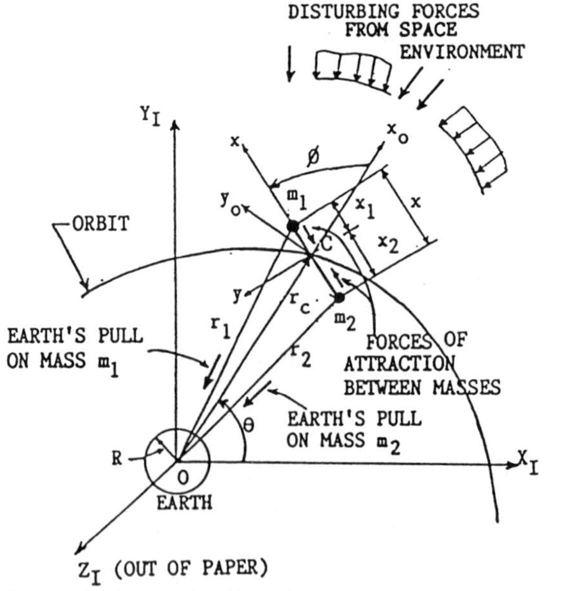

Figure 1. Dumbbell-Shaped Large Space Structure in Orbit.

(Malla et al. 1989).

The equations of motion are derived in terms of the four generalized coordinates of the problem: (1) the radial distance r_c between the center of the earth and the center of mass of the space structure, (2) the angle θ describing the sweep of the radial line r_c in the plane of motion, (3) the distance x between the two dumbbell masses m_1 and m_2, and (4) the angle ϕ (attitude or pitch angle) between the local vertical line OC and the line joining the two masses. Unfortunately, there are few misprints in the equations of motion reported in the reference Malla et al. (1989) mentioned above. The correct form of the equations of motion are rewritten below for the continuity of the material, or they can also be found in Malla et al. (1988).

$$m\ddot{r}_c - m r_c \dot{\theta}^2 + \mu\left[\frac{m_1(r_c + x_1\cos\phi)}{r_1^3} + \frac{m_2(r_c - x_2\cos\phi)}{r_2^3}\right] = Q_r \tag{1a}$$

$$m r_c(r_c\ddot{\theta} + 2\dot{\theta}\dot{r}_c) - \bar{m}\mu\left[\frac{(-1)}{r_1^3} + \frac{1}{r_2^3}\right] x r_c \sin\phi = (Q_\theta - Q_\phi) \tag{1b}$$

$$\bar{m}\ddot{x} - \bar{m}x(\dot{\theta} + \dot{\phi})^2 + \bar{m}\mu\left[\frac{(x_1 + r_c\cos\phi)}{r_1^3} + \frac{(x_2 - r_c\cos\phi)}{r_2^3}\right] + \frac{Gm_1m_2}{x^2} + \frac{\partial U_e}{\partial x} = Q_x \tag{1c}$$

$$\bar{m}[x^2(\ddot{\theta} + \ddot{\phi}) + 2x\dot{x}(\dot{\theta} + \dot{\phi})] + \bar{m}\mu\left[\frac{(-1)}{r_1^3} + \frac{1}{r_2^3}\right] x r_c \sin\phi = Q_\phi \tag{1d}$$

Where r_1, r_2 = distances from the center of the earth to masses m_1 and m_2, respectively; x_1, x_2 = distances from the center of the mass of the structure to the dumbbell masses m_1 and m_2, respectively; $m = m_1 + m_2$; $\bar{m} = m_1 m_2/m$; $\mu = gR^2$, in which g = acceleration due to gravity at the earth's surface; G = universal gravitational constant; U_e = strain energy; and Q_r, Q_θ, Q_x, Q_ϕ = external forcing functions corresponding to the generalized coordinates, r_c, θ, x, and ϕ, respectively. Dot (.) on top of a character represents time derivative of the corresponding quantity.

One of the reasons behind choosing the dumbbell model to represent a space structure is that the link joining the two dumbbell masses can be considered to be a truss-like structure. It has been envisioned that a strong candidate for a low-mass and high-stiffness structure is an open-truss configuration.

The space structure is considered to be subjected to earth's gravity alone. In such a situation, Q_r, Q_θ, Q_x, and Q_ϕ are all zero. For an axially flexible structure with elastic deformation, the strain energy corresponding to axial force can be determined from $dU_e/dx = K_a (x-x_s)$, where K_a is the force required to stretch the structure by a unit distance. The axial stiffness K_a is equal to AE/x_s, in which, A is the area of the cross section of the structure, and E is the modulus of elasticity of the material of the structure. x_s is the static (reference) length of the dumbbell structure.

APPLICATION

Presented below are the physical parameters and initial conditions selected to obtain a solution of the complete equations of motion (Eqns. 1). The response of the space structure is given by variations of r_c, θ, x, and $\emptyset$ with time, t.

Selection of Physical Parameters

The link strtucture between the two dumbbell masses may be a continuum model of an open truss configuration. There are several approaches developed to determine equivalent mechanical properties for such a continuum model (Noor et al. 1978; Nayfeh and Hefzy 1981). For the present example, the space structure is considered to have axial rigidity (EA) equal to 1.63859×10^8 N. The total mass of the system is chosen to be 3600 metric tons equally distributed between the two ends (i.e., $m_1 = m_2 = 5.0 \times 10^5$ N-min^2/km). The link structure is assumed to be massless. The static length (x_s) of the structure selected is one kilometer. The general terrestrial constants used are: $G = 8.64432 \times 10^{-16}$ km^4/N-min^4; g= 35.316 km/min^2; and R= 6374.33 km.

Selection of Initial Conditions

A total of eight initial conditions, one each on r_c, θ, x, $\emptyset$, $\dot{r}_c$, $\dot{\theta}$, $\dot{x}$, and $\dot{\emptyset}$, are required to solve the four second order differential equations of motion (Eqns. 1). They are respectively denoted as: r_0, θ_0, x_0, $\emptyset_0$, $\dot{r}_0$, $\dot{\theta}_0$, $\dot{x}_0$, and $\dot{\emptyset}_0$, where the subscript "0" indicates the value of a variable in the beginning of the motion of the structure in an orbit (i.e., at time equal to zero). In general, any set of values may be assigned to the initial conditions. For the present example, following initial values were selected.

The value of initial orbital radius (r_0) selected is 6574.33 km. This represents the low earth orbit (LEO) at an altitude of 200 km from the surface of the earth. The first set of values of the initial length x_0 equal to 1.000023 km and initial angular velocity $\dot{\theta}_0$ equal to 0.0710629204l142 radians/minute are determined from the equilibrium state of the space structure in a circular orbit in LEO (i.e., r_0 = 6574.33 km) with $\theta_0 = 0$, $\emptyset_0 = 0$,

$\dot{r}_0 = \dot{x}_0 = \dot{\varnothing}_0 = 0$, $x_s = 1$ km, and $m_1/m_2 = 1.0$. Other value of x_0 considered is $1.05x_s$. Similarly, additional values of $\dot{\theta}_0$ considered are 0.1 and $\pi/2$ radians. The initial values θ_0, $\dot{r}_0$, $\dot{x}_0$ and $\dot{\varnothing}_0$ are arbitrarily set equal to zero for all cases.

Computer Solution Method

The equations of motion are solved using a fifth and sixth order Runga-Kutta-Verner integration method on a Cyber 175 digital computer. The step size for the integration is 0.2 minutes. A tolerance limit is maintained at 1.0×10^{-6}.

RESULTS AND DISCUSSION

As mentioned in the previous section, the first set of initial conditions were determined by satisfying the steady state equilibrium configuration of the space structure in a circular orbit at an altitude of 200 km from the surface of the earth (i.e., $r_c = 6574.33$ km). The computer results obtained with these initial conditions verify that the orbit remains perfectly circular at the altitude of 200 km for the 1200-minute duration of motion. The period of the orbit is found to be approximately 88.4 minutes.

It has been observed that the axial length (x) and the pitch angle ($\varnothing$) oscillate about their respective initial values with negligibly small amplitudes (for example, see Fig. 2 for $\varnothing_0 = 0$ radians). During the 1200-minute duration, the maximum amplitude of oscillation in x is approximately 0.0006 percent of its initial value and that in $\varnothing$ is approximately equal to 1.63910×10^{-6} radians.

Effects of Initial Attitude Angle

In order to study the effects of the initial attitude (pitch) angle ($\varnothing_0$) on the motion and deformation of the space structure, two additional values of $\varnothing_0$ (equal to 0.1 and $\pi/2$ radians) were selected. The initial value of x considered in these examples is 1.000023 km. The results show that the smaller value of $\varnothing_0$ equal to 0.1 radians does not perturb the orbit at all. The larger value of $\varnothing_0$ equal to $\pi/2$ radians tends to perturb the orbit by a negligible amount. The change in r_c is between -0.0003 km and 0.0005 km during the 1200-minute duration. The orbital period remains unchanged at 88.4 minutes.

Fig. 2 shows the plot of x/x_s versus time for $\varnothing_0 = 0$, 0.1 and $\pi/2$ radians for the 1200-minute duration. The oscillation in x for $\varnothing_0 = 0.1$ radians is seen to be sinusoidal about the initial value of x with period 1.1 minutes. Amplitudes of oscillation increase

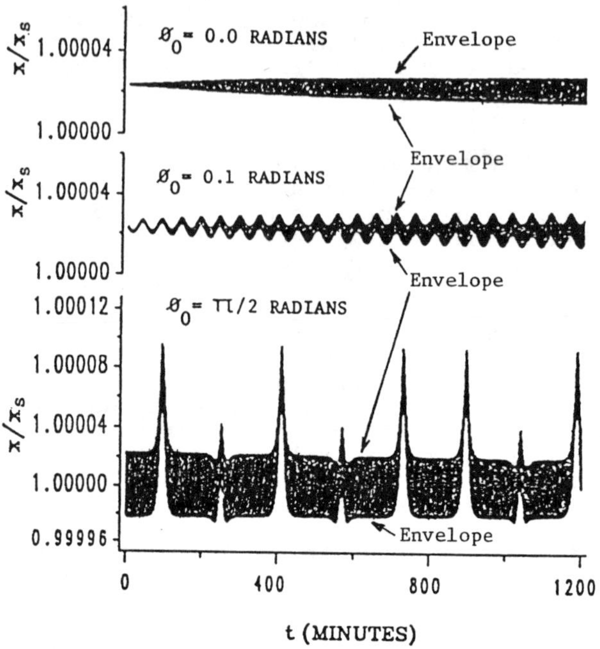

Figure 2. Effects of Initial Attitude (Pitch) Angle ($\emptyset_0$) on Axial Deformation (Length, x)

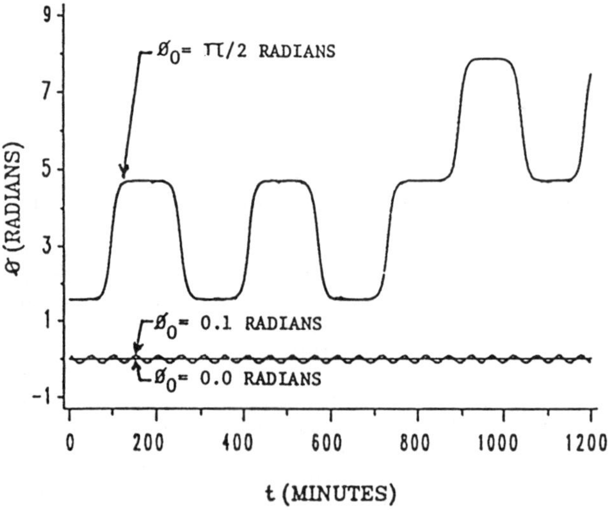

Figure 3. Effects of Initial Attitude (Pitch) Angle ($\emptyset_0$) on Attitude (Librational) Motion ($\emptyset$)

with time, the increase being less distinct near the end of the 1200-minute duration. The maximum amplitude of oscillation in x is approximately 0.0009 percent of the initial length x_0. This type of fine oscillation in x was present even for $\phi_0 = 0.0$ radians. In addition, it can be observed that the envelope curves (curves passing through the peaks of the small-amplitude oscillation in x) are sine-wave type periodic functions with period approximately 51.3 minutes. This period of oscillation in x is found to be the same as that of oscillation in ϕ for $\phi_0 = 0.1$ radians (Fig. 3).

For $\phi_0 = \pi/2$ radians, there are again two types of oscillations in x (Fig. 2). The first one is the small-amplitude oscillation with period approximately 1.1 minutes, which, as pointed out earlier, exists even for $\phi_0 = 0.0$ radians. However, except at certain regular intervals, this oscillation in x now occurs about the reference value (x_s = 1 km) with maximum amplitude approximately 0.0023 percent of its initial value x_0. The other oscillation in x, which is more prominent, is represented by envelope curves passing through the peaks of the above described oscillation. The envelope curves have the following features:

First, they indicate a "beat-type" phenomenon. Second, the envelope curves are periodic in either side of t approximately equal to 800 minutes with period equal to approximately 314.5 minutes. This period of oscillation in x is again found to be the same as that of oscillation in ϕ for $\phi_0 = \pi/2$ radians (Fig. 3). Third, studying the variations of x and ϕ for $\phi_0 = \pi/2$ radians in Figs. 2 and 3 together, it can be clearly observed that building of the larger magnitude peaks in envelope curves of the x-oscillation correspond to the period during which ϕ is increasing. On the other hand the smaller magnitude peaks in the envelope curves correspond to the period during which ϕ is decreasing. The maximum change in the structural length for this case during the motion is approximately 0.0073 percent of the initial value of x.

Fig. 3 shows that the initial pitch angle (ϕ_0) has significant effects on the attitude motion in the obital plane of the space structure. For $\phi_0 = 0.1$ radians, ϕ oscillates about the r_c vector (local vertical) sinusoidally with a constant period of 51.3 minutes and amplitude equal to its initial value (0.1 radians). The ratio of the orbital period (88.4 minutes) to the period of libration (51.3 minutes) is approximately equal to 1.7232. The ratio of orbital to libration periods is close to $\sqrt{3}$ as predicted for small initial pitch libration by Klemperer and Baker (1957) for an ideal rigid dumbbell.

For the large initial pitch angle (ϕ_0) equal to

$\pi/2$ radians, the pitch angle ϕ during the 1200-minute duration has rectangular-type oscillation with constant period of 314.5 minutes (Fig. 3). ϕ oscillates between the maximum value of $3\pi/2$ radians and the minimum value of $\pi/2$ radians during the first part of the motion and between the maximum value of $5\pi/2$ radians and the minimum value of $3\pi/2$ radians during the second part of the motion. The transition from one position to the other takes place in a relatively short duration of approximately 40 minutes. It can also be inferred from Fig. 3 that for the large initial pitch angle, the flexible space structure tumbles more and more with time.

Effects of Initial Axial Deformation

The simultaneous equations of motion are solved for one additional initial value of x given by x_0/x_s = 1.05. The initial value of ϕ considered in these example is zero radians. The computer results show that the orbit remains unperturbed in all respects for the value of initial axial deformation of the structure (that is, the orbit remains circular at 200 km altitude with a orbital period approximately 88.4 minutes).

The initial conditions on axial deformation (length) are quite significant in perturbing the axial length of the space structure during the motion. Fig. 4 shows the plots of x/x_s versus time for x_0/x_s = 1.000023 and 1.05. The plot for x_0/x_s = 1.05 clearly shows that x oscillates sinusoidally about its static reference value x_s with a constant period of approximately 1.1 minutes, and the amplitudes of oscillation vary between 0.05 km and 0.041 km. As it is seen from the equations of motion, the axial motion is coupled to both the attitude (pitch) and orbital motion. The fact that the amplitudes of the oscillation in x are other than constant indicate the effect of the coupling, even though the axial modal motion appears to dominate.

Fig. 5 shows that the initial condition on the axial deformation also affects the attitude motion (ϕ vs t) greatly. For the case where x_0/x_s = 1.05, the dumb-bell oscillates sinusoidally about the radial vector r_c with nearly constant amplitudes of approximately 0.0617 radians and periods of approximately 50.8 minutes.

CONCLUSIONS

Effects of initial structural conditions on the responses of the space structure were investigated. Some important findings are stated below:

(1) Increasing the value of the initial attitude (pitch) angle causes an increase in the libration of the large space structure about its center of mass. An

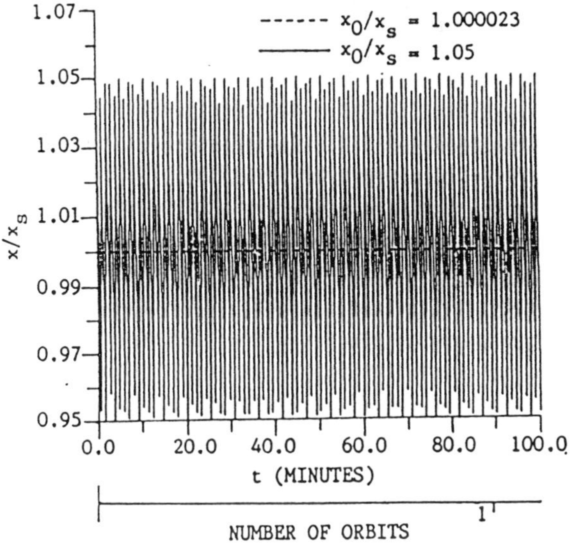

Figure 4. Effects of Initial Axial Deformation (Length, x_0) on Axial Deformation (Length, x)

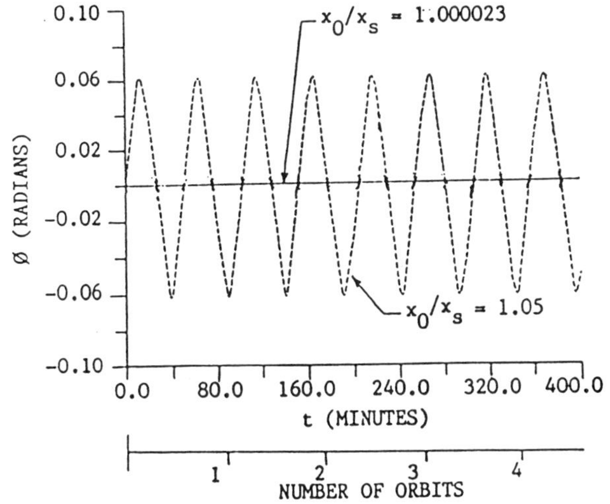

Figure 5. Effects of Initial Axial Deformation (Length, x_0) on Attitude (Librational) Motion (Ø)

initially imposed attitude misalignment is also seen to produce increased axial deformation. However, it does not produce notable changes in the orbit of the structure. (2) Initially induced axial deformation of a large space structure has considerable effect in changing the length and attitude (pitch) angle of the structure, but has no effect on perturbing the parameters of the orbit of the structure.

ACKNOWLEDGEMENT

The work reported herein is taken in part from the writer's Ph.D. thesis submitted to the University of Massachusetts, Amherst, MA, September 1986. The writer extends his sincere thanks to Dr. William A. Nash and Dr. Thomas J. Lardner for their guidance and suggestions. The project was in part supported by Air Force Office of Scientific Research, Washington, D.C. under Grant No. AFSOR - 83 - 0025. The writer also acknowledges facilities and partial financial support provided by the Department of Civil Engineering, University of Massachusetts at Amherst.

APPENDIX I - REFERENCES

Johnson, S. and Wetzel, J., eds. (1988). Engineering, Construction, and Operations in Space. ASCE, New York, N.Y.

Klemperer, W.B. and Baker, R.M., Jr. (1957). "Satellite Librations," Astronautica Acta, Vol. 3, 16-27.

Malla, R., Nash, W., and Lardner, T. (1988). "Thermal Effects on Very Large Space Structures," J. of Aerospace Engineering, 1(3), 171-190.

Malla, R., Nash, W., and Lardner, T. (1989). "Motion and Deformation of Very Large Space Structures," AIAA Journal, 27(3), 374-376.

Meirovitch, L. (1982). "Control of Self-Adjoint Distributed Parameter Systems," J. of Guidance and Control, Vol. 5, Jan.-Feb., 60-66.

Nayfeh, A. H. and Hefzy, M.S. (1981). "Continuum Modeling of the Mechanical and Thermal Behavior of Discrete Large Space Structures," AIAA Journal, 19(6), 766-773.

Noor, A.K., Anderson, M.S., and Greene, W.H. (1978). "Continuum Models for Beam- and Platelike Lattice Structures," AIAA Journal, Vol.16, Dec., 1219-1228.

Pinson, Larry D. (1989). "Recent Advances in Structural Dynamics of Large Space Structures," Acta Astronautica, 19(2), 161-170.

Actively Controlled Optimum Structural Design

N. S. Khot*
D. E. Veley*
R. V. Grandhi†

Abstract

This paper deals with the investigation of the number and location of actuators and sensors on the optimum structure and control design of space structures. The optimization problem is posed with the objective of minimizing the weight of the structure and satisfying the constraints on the closed-loop eigenvalue distribution and damping parameters. The allocation of the actuators for a specific disturbance is based on the work done by the individual actuators.

Introduction

Recently there has been considerable interest in the structure and control interaction of flexible space structures (Ryan and Scofield, 1986). These structures require an active control system in order to reduce the structural response due to disturbances. A number of investigators have considered an integrated approach to the optimum design of structure and control systems instead of treating the two disciplines separately. Hale et al (1983) and Salama et al (1986) have considered minimization of the sum of the mass of the structure and quadratic control cost function with constraints on structural and control design parameters. Haftka, Martinonic and Hallauer (1985) have investigated the sensitivity of control system to structural modification. Junkins, et al (1984) have presented an eigenspace optimization approach to design feedback controllers for flexible structures. The integrated approach with weight as the objective function and constraints on the damping and frequency distribution of the closed-loop system was presented by Khot (1988). Khot, et al (1988) considered the simultaneous optimization with constraints on the closed-loop eigenvalues and the control gain norm. The selection of actuator locations to control the shape of the space structure was discussed by Haftka and Adelman (1985) and by Lindberg and Longman (1984).

---

* Aerospace Engineer, Flight Dynamics Laboratory (WRDC/FIBRA) Wright-Patterson AFB OH 45433-6553

† Associate Professor, Department of Mechanical and Materials Engineering, Wright State University, Dayton OH 45435

This study focuses on the effect of the number and location of actuators on the minimum weight of the structure and on the total work done by the actuators for specified constraints and disturbance. The controls approach used is a linear quadratic regulator theory with constant feedback. At the beginning colocated actuators and sensors are provided in all the elements. The actuator doing the least work is removed and the structure is reoptimized for the specified constraints on the closed-loop eigenvalues and the damping parameters. The procedure of eliminating an actuator is continued until an acceptable design satisfying the constraints is obtained.

Problem Formulation

The equations describing the dynamic behavior of a structure are as follows:

$$\{\dot{x}\} = [\mathrm{A}]\{\mathrm{x}\} + [\mathrm{B}]\{\mathrm{f}\} \tag{1}$$

$$\{y\} = [\ \mathrm{C}\ ]\{x\} \tag{2}$$

where $\{x\}$ is the state vector, $\{f\}$ is the control vector and $\{y\}$ is the output vector. In Eq. 1, $[A]$ and $[B]$ are the plant and input matrices given by

$$[\mathrm{A}] = \left[\begin{array}{c|c} 0 & \mathrm{I} \\ \hline -\omega^2 & 0 \end{array}\right] \tag{3}$$

$$[\mathrm{B}] = \left[\begin{array}{c} 0 \\ \hline \phi^{\mathrm{T}}\mathrm{D} \end{array}\right] \tag{4}$$

where I is the identity matrix, ω^2 is the diagonal matrix of square of structural frequencies and $[\phi]$ is the modal matrix. In Eq. 2, $[C] = [B]^T$ if the sensors and actuators are colocated. The optimal control $\{f\}$ is selected to minimize the following performance index integral:

$$\mathrm{PI} = \int_0^t \left(\{x\}^{\mathrm{T}}[\mathrm{Q}]\{x\} + \{f\}^{\mathrm{T}}[\mathrm{R}]\{\mathrm{f}\}\right)\,\mathrm{dt} \tag{5}$$

where $[Q]$ and $[R]$ are a positive state weighting matrix and positive definite control weighting matrix, respectively. Optimum feedback control law is given by

$$\{f\} = -[\tilde{\mathrm{G}}]\{x\} \tag{6}$$

where $[\tilde{G}]$ is the optimum gain matrix given by

$$[\tilde{\mathrm{G}}] = [\mathrm{R}]^{-1}[\mathrm{B}]^{\mathrm{T}}[\tilde{\mathrm{P}}] \tag{7}$$

In Eq. 7, $[\tilde{P}]$ is the Riccati matrix obtained by the solution of the algebraic Riccati equation.

$$[\mathrm{A}]^{\mathrm{T}}[\tilde{\mathrm{P}}] - [\tilde{\mathrm{P}}][\mathrm{B}][\mathrm{R}]^{-1}[\mathrm{B}]^{\mathrm{T}}[\tilde{\mathrm{P}}] + [\tilde{\mathrm{P}}][\mathrm{A}] + [\mathrm{Q}] = 0 \tag{8}$$

Using Eqs. 1 and 6 the optimum closed-loop system can be written as

$$\{\dot{x}\} = [\overline{\mathrm{A}}]\{x\} \tag{9}$$

where

$$[\overline{\mathrm{A}}] = [\mathrm{A}] - [\mathrm{B}][\tilde{\mathrm{G}}] \tag{10}$$

The complex eigenvalues of $[\overline{\mathrm{A}}]$ can be written as

$$\lambda_i = \tilde{\sigma}_i \pm j\tilde{\omega}_i \tag{11}$$

The damping factor of the closed-loop system is given by

$$\xi_i = -\frac{\tilde{\sigma}_i}{(\tilde{\sigma}_i^2 + \tilde{\omega}_i^2)^{1/2}} \tag{12}$$

For the initial given condition $\{x(o)\}$ the solution to Eq. 9 is given by

$$x(t) = \mathrm{e}^{[\overline{\mathrm{A}}]t} x(0) \tag{13}$$

The work done by the ith controller is given by

$$\mathrm{W}_i = \int_0^t f_i\, \dot{U}_i\, dt \tag{14}$$

where $\dot{U}_i$ is the velocity component associated with the ith actuator. The total work done by the actuator can be calculated by taking the product of control force, velocity component and time increment at each time interval and summing it over the simulated time.

The optimization problem for the integrated design of structure and control system can be stated as: Determine design variables A_i that would minimize the structural weight,

$$\mathrm{W} = \sum_{i=1}^{m} \rho_i A_i \ell_i \tag{15}$$

such that

$$g_j(\tilde{\omega}_i) \leq 0 \tag{16}$$

$$g_j(\xi_i) \geq 0 \tag{17}$$

$$g_j(A_i) \geq 0 \tag{18}$$

where

$$g_j(\omega_i) = \tilde{\omega}_i - \overline{\tilde{\omega}}_i \tag{19}$$

$$g_j(\xi_i) = \xi_i - \overline{\xi}_i \tag{20}$$

$$g_j(A_i) = A_i - \overline{A}_{i(min)} \tag{21}$$

In Eq. 15 ρ_i is the density, and ℓ_i is the length of the element. In Eqs. 16 and 17, $g_j(\tilde{\omega}_i)$ and $g_j(\xi_i)$ are the constraints on the imaginary part of the closed-loop eigenvalues and on the damping parameters. In Eqs. 19 and 20 $\overline{\tilde{\omega}}_i$ and $\overline{\xi}_i$ are the limiting values of the eigenvalues and damping parameters. In Eq. 21, $\overline{A}_i$ is the minimum allowable cross-sectional areas of the element. The optimization problem was solved by using the NEWSUMT-A (Thareja and Haftka, 1985) optimization program based on the extended interior penalty function method with Newton's method of unconstrained minimization.

The gradients of the objective function are obtained by differentiating Eq. 15 with respect to the design variables A_l. The gradients of the closed-loop eigenvalues λ_j with respect to the structural design variable A_l are given by (Khot, 1988)

$$\frac{\partial \lambda_j}{\partial \mathrm{A}_l} = \{\beta\}_j^{\mathrm{T}}[\overline{\mathrm{A}}]_{,l}\{\alpha\}_j \tag{22}$$

where $\{\beta\}_j$ and $\{\alpha\}_j$ are the left-hand and right-hand eigenvectors of the closed-loop matrix $[\overline{\mathrm{A}}]$.

Illustrative Example

The structural model of ACOSS-FOUR selected for this study is shown in Fig. 1. This structure models the feed tower in a generic class of large antennas. The dimensions of the structure are specified in consistent unspecified units. The coordinates of the node points are given in Table 1. The structure is pinned at nodes 1, 2, 3 and 4 and thus has twelve degrees of freedom. Four masses of 2 units each are attached at nodes 1 through 4. Youngs modulus of the member is 1.0 and the density of the structural material is assumed to be 0.001. The apex represents the antenna feed and its motion in $X - Y$ plane defines the line-of-sight error denoted by LOS. If the motion of the apex in the X and Y directions is denoted by $LOS - X$ and $LOS - Y$, respectively, then $LOS^2 = (LOS - X)^2 + (LOS - Y)^2$. The transient response was simulated for all the designs with a unit displacement at node 2 in X direction at $t = 0$, and the work done by the actuators was calculated for the total period of 25 seconds with Δt being equal to 0.05 seconds. All the actuators and sensors are colocated in the elements of the structure. Fig. 1 also shows the actuators located in each element.

The cross-sectional areas of the nominal design were equal to those used (Strunce, et al, 1980) for the design of the control system for ACOSS-FOUR studies. The constraints imposed on the optimum designs are $\bar{\tilde{\omega}}_1 = 1.341, \bar{\tilde{\omega}}_2 \geq 1.6$ and $\bar{\xi}_1 = 0.2574$. For the nominal design $\tilde{\omega}_1 = 1.341, \tilde{\omega}_2 = 1.66$ and $\xi_1 = 0.054$. Thus the optimum design would have the same $\tilde{\omega}_1$ as the nominal design, and the damping parameter ξ_1 associated with the first closed-loop frequency of the optimum design would be nearly four times that of the nominal design. The constraint on $\tilde{\omega}_2$ was imposed in order to prevent coalescence of the first two structural frequencies during optimization. The weighting matrices $[Q]$ and $[R]$ in the definition of the performance index in Eq. 5 are diagonal matrices. In the $[Q]$ matrix elements associated with the displacements and velocities were equal to the square of the structural frequencies and unity respectively. The $[R]$ matrix was the identity matrix.

The optimum design with twelve actuators was first obtained and the work done by each actuator was calculated by simulating the transient response for the initial disturbance. The actuators were ranked based on the amount of work done. The first number was assigned to the actuator doing maximum amount of work and the last number to the actuator doing minimum work. The actuator doing least work was removed and the structure was reoptimized. This process was continued until a design satisfying all the constraints could be obtained.

Table 2 gives the rank of the work done by each actuator for different designs. The first row gives the actuator number as defined in Fig. 1. The first column on the left gives the number of actuators present in each design. The first design with 12 actuators is the nominal design which is nonoptimum. All other designs are minimum weight designs satisfying the specified constraints. Comparing nonoptimum and optimum designs with twelve constraints it is seen that the ranking of the work done by the actuators is not the same. In the nominal design rank 2 is associated with actuator number 9 while in optimum designs actuator number 1 is ranked second. In the optimum design with twelve actuators the least work is done by actuator six.

Hence, actuator six was removed and the structure was reoptimized with eleven actuators. In order to obtain a feasible solution at least five actuators were required. The bottom row in Table 2 shows the ranking of the work done by the five actuators. For designs with more than six actuators, actuator seven did the maximum work. For the remaining two cases with six and five actuators, actuator one did the maximum work.

Table 3 summarizes the weight, the total work done by all the actuators and the magnitudes of the performance index for all the designs. A minimum weight design with minimum total work done is obtained with eleven actuators. There is not much difference in the weights and the total work done between the designs with ten or eleven actuators. Reducing the number of actuators below ten started increasing the optimum weight and the total work done by the controllers.

The cross-sectional areas of the members for all the designs are given in Table 4. Table 5 gives the square of the structural frequencies for all the designs. The band of frequencies for a design with ten actuators is the smallest. As the number of actuators is reduced this band widens. The frequencies associated with first and second modes are nearly equal for all the designs. This is due to the constraints imposed on the closed-loop frequencies.

Summary

The optimum number and location of actuators and sensors were selected based on the amount of work done by the controller at that location. The actuator doing the least work was eliminated and an optimum structural design was obtained. This procedure was continued as long as the integrated system constraints were met. This work concludes that there is a minimum number of actuators needed to control the system without resulting in a heavier structural mass.

References

1. Haftka, R.T. and H.M. Adelman, "Selection of Actuator Locations for Static Shape Control of Large Space Structures bt Heuistic Integer Programming," Computers and Structure, Vol. 20, 1985.

2. Haftka, R.T., Z. Martinonic and W.L. Hallauer, Jr., "Enhanced Vibration Controlability by Minor Structural Modification," AIAA Journal, Vol. 23, No. 8, 1985.

3. Hale, R.J. Lisowski and W.E. Dahl, "Optimizing Both the Structure and the Control of Maneuvering Flexible Structures," Proceedings of the AAS/AIAA Astrodynamics Conference, Lake Placid, New York, August 22-24, 1983.

4. Junkins, J.L., D.S. Bodden and J.D. Tuner, "A Unified Approach to Structure and Control System Design," A paper presented at the Fourth International Conference on Applied Numerical Modelling, Tainan, Taiwan, December 1984.

5. Khot, N.S., "Structure/Control Optimization to Improve the Dynamic Response of Space Structures," Computational Mechanics, Vol. 3, 1988.

6. Khot, N.S., Öz, H., Grandhi, R.V., Eastep, F.E. and V.B. Venkayya, "Optimal Structural Design with Control Gain Norm Constraint," AIAA Journal, Vol. 26, No. 5, 1988.

7. Lindberg Jr., R.E. and R.W. Longman, "On the Number and Placement of Actuator for Independent Modal Space Control," Journal of Guidance and Control, Vol. 7, 1984.

8. Structural Dynamics and Control Interaction of Flexible Structures, NASA Conference, George C. Marshall Space Flight Center, Huntsville, Alabama, edited by R. S. Ryan and H. N. Scofield, April 22-24, 1986.

9. Salama, M., M. Hamidi and L. Densetz, "Optimization of Controlled Structures," Proceedings of the JPL Workshop on Identification and Control of Flexible Space Structures, San Diego, California, June 4-6, 1986.

10. Strunce, R.R. eta al., "ACOSS FOUR (Active Control of Space Structures) Theory Appendix," RADC-TR-80-78, Vol. II, 1980.

11. Thareja, R. and Haftka, R.T., "NEWSUMT-A, A Modified Version of NEWSUMT for Inequality and Equality Constraints," VPI Report 148, Aerospace Engineering Department, Virginia Tech, Blacksburg VA, 1985.

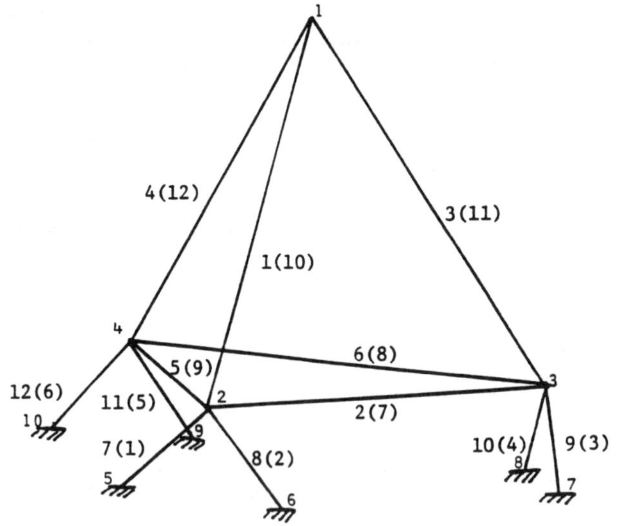

FIGURE 1: Structural model of the feed tower in a generic class of large antennas. Actuator numbers are in parentheses.

TABLE 1: NODE POINT COORDINATES OF ACOSS-FOUR

Node	X	Y	Z
1	0.0	0.0	10.165
2	-5.0	-2.887	2.00
3	5.0	-2.887	2.00
4	0.0	5.7735	2.00
5	-6.0	-1.1547	0.0
6	-4.0	-4.6188	0.0
7	4.0	-4.6188	0.0
8	6.0	-1.1547	0.0
9	-2.0	5.7735	0.0
10	2.0	5.7735	0.0

TABLE 2: RANK ORDER OF ACTUATOR INPUT

NUMBER OF ACTUATOR	1	2	3	4	5	6	7	8	9	10	11	12
12*	6	5	8	7	11	12	1	3	2	4	9	10
12	2	3	5	6	4	12	1	8	7	9	10	11
11	2	3	4	9	11		1	7	5	6	8	10
10	2	3	4	9			1	7	5	6	8	10
9	2	4	9	3			1	5	6	8	7	
8	2	4		3			1	5	6	8	7	
7	2	3		4			1	5	6		7	
6	1	3		6			2	4	5			
5	1	4					2	3	5			

*INITIAL DESIGN

TABLE 3: PERFORMANCE INDEX, TOTAL WORK AND WEIGHT

NUMBER OF ACTUATORS	PI_1	PI_2	PI	TOTAL WORK	WEIGHT
12*	159.3	159.8	319.1	79.44	43.69
12	41.15	40.28	81.43	21.92	14.52
11	39.76	38.86	78.62	19.82	14.39
10	40.72	40.11	80.83	19.82	14.40
9	48.56	47.93	88.49	24.08	14.43
8	52.02	49.10	101.12	24.07	14.43
7	64.29	64.63	128.92	28.18	15.22
6	77.27	80.52	157.79	35.71	21.50
5	91.56	96.01	187.66	35.60	21.55

*INITIAL DESIGN

TABLE 4: AREAS OF MEMBERS

ELE	12*	12**	11**	10**	9**	8**	7**	6**	5**
1	1000	238	205	204	123	123	262	379	373
2	1000	192	137	136	227	227	276	342	342
3	100	125	217	218	214	214	130	487	493
4	100	184	133	134	212	212	257	276	276
5	1000	144	239	240	237	237	133	92	93
6	1000	264	219	218	140	140	148	112	111
7	100	142	213	214	191	190	199	387	384
8	100	198	121	121	179	178	227	186	186
9	100	149	115	114	203	203	241	378	387
10	100	218	205	203	120	120	121	394	419
11	100	130	187	192	212	212	160	182	173
12	100	233	171	168	110	110	161	89	90
WT	43.69	14.52	14.39	14.40	14.43	14.43	15.22	21.50	21.55

*INITIAL DESIGN
**NUMBER OF ACTUATORS

TABLE 5: STRUCTURAL FREQUENCIES (ω_j^2)

12*	12**	11**	10**	9**	8**	7**	6**	5**
1.80	1.80	1.79	1.79	1.77	1.77	1.79	1.79	1.79
2.77	2.56	2.56	2.56	2.58	2.59	2.57	2.56	2.56
8.35	7.63	7.69	7.70	7.48	7.48	7.46	10.34	10.34
8.74	9.31	9.21	9.19	9.20	9.20	10.55	13.17	13.08
11.55	13.19	12.72	12.70	12.72	12.72	14.09	17.81	17.98
17.68	26.41	24.70	24.61	24.48	24.48	24.24	24.30	24.11
21.73	27.78	26.65	26.59	26.64	26.64	30.45	36.23	35.97
22.61	34.33	35.03	34.89	35.82	35.78	33.10	44.00	43.67
72.92	40.32	40.46	40.47	40.58	40.58	42.40	49.52	49.24
85.57	44.70	41.59	41.91	41.78	41.77	45.61	74.11	76.27
105.8	46.32	44.16	44.17	44.20	44.21	50.66	83.69	83.75
166.5	50.10	48.11	48.08	48.08	48.11	54.14	97.87	100.6

*INITIAL DESIGN
**NUMBER OF ACTUATORS

WAVE PROPAGATION IN HYPER-STRUCTURES

M. Ettouney[1] H. Benaroya[2] J. Wright[3]

ABSTRACT

The concept of hyperelement was introduced by (Kausel and Roesset, 1977) in connection with the derivation of the dynamic stiffness matrix for layered continuum problems involving two widely separated boundaries. Their approach is extended here and the term hyperstructure is used in an analogous way for structural systems with a large number of periodically spaced, identical structures. For these structures, it is shown that the steady state responses can be calculated more accurately and with fewer operations when compared to more general methods. In order to demonstrate the method, two planar systems are analyzed.

INTRODUCTION

Structures with a large number of identical bays are utilized, and will be utilized, to a great extent in space-related ventures. These types of structures in space are subject to different types of dynamic forces and some of these dynamic forces inherently excite higher structural modes either because of their spatial distribution or frequency content. This calls for models and analysis methods which are accurate for the higher structural modes.

Two types of analysis methods are available for this class of problems. The direct stiffness approach (sometimes referred to as the finite element approach) yields the "exact" solution of the problem at hand. (Of course, we imply here that the mathematical model is exactly representative of the physics, which it never is.) However, for large space structures, such solutions may be costly both in time and computational effort. In addition, if one is considering the coupled dynamics and control of

[1]Applied Science Division, WEIDLINGER Associates, 333 Seventh Avenue, New York, NY 10001

[2]Department of Mechanical and Aerospace Engineering, RUTGERS University, P.O. Box 909, Piscataway, New Jersey 08855–0909

[3]Applied Science Division, WEIDLINGER Associates, 333 Seventh Avenue, New York, NY 10001

large space structures, then the finite element approach proves to be unsatisfactory in predicting higher order modes. The sources of the inaccuracies are apparently due to joint modeling, and to a lesser extent numerical accuracy. Joint modeling is extremely important; joints will dominate model error issues.

Another approach is to approximate trussed structures as continua (Noor and Mikulas, 1988) then evaluate the solution using simple elasticity. This approach is economical and may be accurate for lower structural modes in some cases. The accuracy of this approach for higher modes is at least questionable.

The present study, which introduces a new class of solutions for the problem at hand, is based on the semi–infinite transmitting boundary method (Ettouney, M., Benaroya, H., Wright, J., 1988). This method is applicable for finite length structures, constructed of repeated identical substructures, in steady state behavior.

In the present study we will demonstrate the applicability of the approach to discrete large (finite length) space structures. The approach is based on evaluating all wave numbers and wave mode shapes for all the waves which propagate in the finite length periodic structure for each forcing frequency. It will be shown that the number of such waves is finite for each forcing frequency. The wave numbers and the wave mode shapes can be evaluated by solving a quadratic eigenvalue problem of very small order for each forcing frequency. The final solution is obtained by superposing all the resultant wave mode shapes.

The order of the quadratic eigenvalue problem is independent of the size of the truss under consideration. This means that the solution is extremely economical computationaly, and since all the possible waves are considered in the solution, it will result in the exact solution. In addition, added insight is possible using this approach. Furthermore, the proposed method offers possibilities for accurate modeling of the joints. All these factors make the proposed method ideal for the class of problems being addressed, especially for application to large space structure dynamics and control.

This method of analysis has another benefit in that it accounts, in a natural way, for the acoustical properties of the structure, which may be a prerequisite for predicting accurate results for the problem on hand. As a case in point, it has been argued that an acoustic structural model may be needed for accurate analysis of control problems of large space structures. The method of analysis which will be presented in this study does couple both structural dynamic and structural acoustic techniques, resulting in an efficient, accurate and simple analytical tool for this class of engineering problems. In order to demonstrate the analytical method presented, two planar systems will be considered.

THEORETICAL DEVELOPMENT

Waves in Semi–infinite Structures

Figure 1 shows a structure which is semi–infinite in the x direction. This structure is composed of a set of repetitive, identical cells. For a wave propagating in the x direction in the structure, the displacement field $u(x,t)$ can be expressed as:

$$u(x,t) = u_0 e^{-i(\omega t + kx)}, \tag{1}$$

where $u_0 = u(0,0)$, ω is the forcing frequency and k is the wave number, which is a complex number in general.

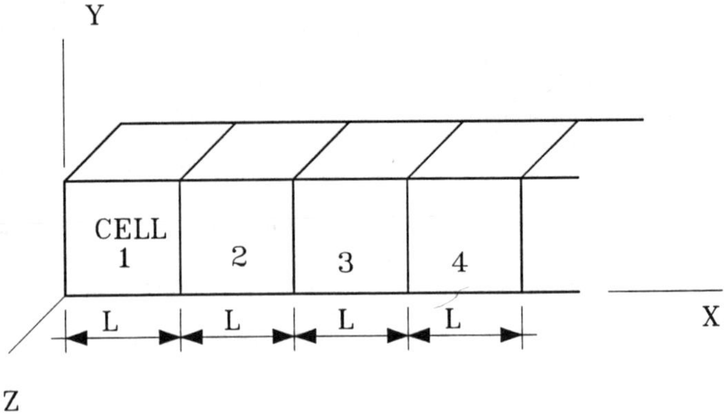

FIG 1–SEMI–INFINITE SYSTEM

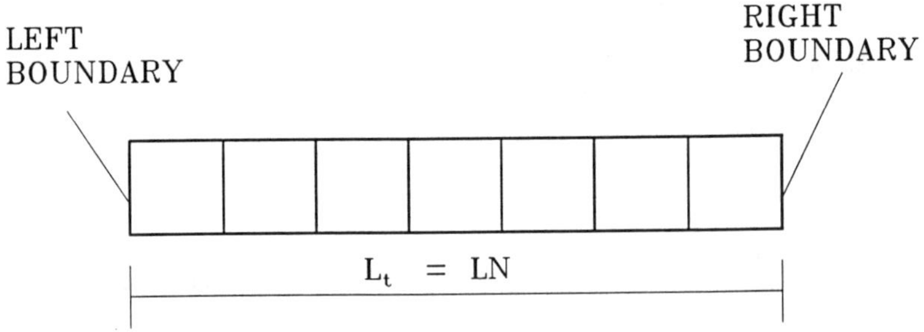

FIG 1A–HYPER SYSTEM

Quadratic Eigenvalue Problem

Consider an idealized semi–infinite structure composed of identical substructures that are periodic in the x–direction. The equations of motion are of the form

$$\mathbf{M}\ddot{\mathbf{u}} + \mathbf{C}\dot{\mathbf{u}} + \mathbf{K}\mathbf{u} = \mathbf{f}(t), \tag{2}$$

where $\mathbf{u}$ is a vector of nodal displacements, $\mathbf{f}$ is a vector of nodal applied forces, and $\mathbf{M}, \mathbf{C}$ and $\mathbf{K}$ are the mass, damping and stiffness matrices, respectively, with $\mathbf{C}$ and $\mathbf{K}$ symmetric, positive semi–definite, and $\mathbf{M}$ symmetric, positive definite. Due to the periodicity, it is convenient to order the equations and partition the matrices in terms of $\mathbf{u}_i$, the displacements of the ith set of nodes which lie between substructure i and substructure $i+1$. Solutions are sought based on Fourier transform from the time into the frequency domain which leads to modal analysis.

Consider the equilibrium of the panel between $\mathbf{u}_i$ and $\mathbf{u}_{i+1}$, where mass and damping matrices are incorporated and there exists harmonic time dependence:

$$\begin{bmatrix} \mathbf{K}_{11} & \mathbf{K}_{12} \\ \mathbf{K}_{21} & \mathbf{K}_{22} \end{bmatrix} \begin{bmatrix} \mathbf{u}_i \\ \mathbf{u}_{i+1} \end{bmatrix} = \begin{bmatrix} -\mathbf{P}_i \\ \mathbf{P}_{i+1} \end{bmatrix}. \tag{3}$$

where $\mathbf{K}_{jl} \equiv \mathbf{K}_{jl}(\omega) = \mathbf{K}_{jl} - i\omega\mathbf{C}_{jl} - \omega^2\mathbf{M}_{jl}$, and $\mathbf{u}_i \equiv \mathbf{u}_i(\omega)$. Subsequent expressions do not explicitly show this frequency dependence.

It was shown (Ettouney, Benaroya and Write, 1988) that the equations of motion of the semi–infinite system can be reduced to:

$$\left\{\kappa^2\mathbf{K}_{12} + \kappa\left[\mathbf{K}_{11} + \mathbf{K}_{22}\right] + \mathbf{K}_{21}\right\}\mathbf{u}_B = 0. \tag{4}$$

Equation (4) is a quadratic eigenvalue problem with κ as the eigenvalue. It is of order M, where M is the number of degrees of freedom associated with each set of boundary displacements, $\mathbf{u}_i$. For a specified frequency, the solution of the problem results in $2M$ pairs of eigenvalues and corresponding eigenvectors. It can be shown (Waas, 1972 and Kausel, 1974) that the $2M$ solutions have certain relations among them. One half of the eigensolutions, denoted by κ_s , $\mathbf{V}_s$ for $s = 1, 2, \ldots, M$ are characterized by $|\kappa_s| \leq 1$, which corresponds to solutions that decrease in amplitude as the distance , x, increases (outgoing waves). The other half of the eigensolutions, denoted by $\tilde{\kappa}_s$, $\tilde{\mathbf{V}}_s$ for $s = 1, 2, \ldots, M$ are characterized by $\tilde{\kappa}_s = 1/\kappa_s$ so that $|\kappa_s| \geq 1$, which corresponds to solutions that increase as x increases (incoming waves). Special cases, corresponding to $|\kappa_s| = |\tilde{\kappa}_s| = 1$, may arise (e.g. rigid body motions with $\omega = 0$ and zero damping). These cases can be accounted for through a special quadratic eigenvalue computer coding. Note that we can relate the eigenvalues,κ_s, to traditional wave propagation concepts (Crandall and Achenbach) by relating the wave number k_s to each of the resulting eigenvalues κ_s through $\kappa_s = \exp(-ik_sL)$, where L is the typical bay length. Doing this will facilitate studying each wave mode in the x–direction. It will also make it easier to generalize the concepts of semi–infinte structures to structures with finite length.

In order to completely define the properties of structures with finite lengths (hyper structures) we will need both sets of solutions, since the motions within the truss are a linear combination of both incoming and outgoing waves. Kausel and Roesset, 1977, used this approach for layered soil media. They called the resulting continuum element "the hyper element" and we will refer to the resulting structural element as the hyper structure element. Following their lead, consider now the following relation:

$$\tilde{\mathbf{V}}_s = \mathbf{T}\mathbf{V}_s. \tag{5}$$

Note that $\mathbf{T}$ is a simple, diagonal transformation matrix which was discussed by (Waas, 1972) for plane strain/plane stress problems, and (Kausel, 1974) for three dimensional continua. In the present study, $\mathbf{T}$ will be slightly different from those references if the system on hand has bending (rotational) degrees of freedom; otherwise, it will be similar to them. The solutions $\mathbf{V_s}$ and $\tilde{\mathbf{V}}_\mathbf{s}$ will be used to form the dynamic stiffness matrix of the hyper truss element.

Dynamic Stiffness Matrix

Consider the finite length truss of figure 1a. It consists of N identical panels. The truss has a total length $L_t = NL$. The dynamic stiffness matrix relating the forces and displacements at the nodes at $x = 0$ and $x = L_t$ can now be derived using the eigensolutions of the quadratic eigenvalue problem of the previous section. The displacement vectors, $\mathbf{U}_{B1}$ and $\mathbf{U}_{B2}$, at $x = 0$ and $x = L_t$, respectively, can be expressed in terms of the eigenvalues and eigenvectors as

$$\begin{bmatrix} \mathbf{U}_{B1} \\ \mathbf{U}_{B2} \end{bmatrix} = \begin{bmatrix} \mathbf{V} & \tilde{\mathbf{V}}\vec{\mathbf{G}} \\ \mathbf{V}\vec{\mathbf{G}} & \tilde{\mathbf{V}} \end{bmatrix} \begin{bmatrix} \Gamma_1 \\ \Gamma_2 \end{bmatrix} \tag{6}$$

with

$$\begin{aligned} \mathbf{V} &= [\mathbf{V}_1\mathbf{V}_2\ldots\mathbf{V}_M] \\ \tilde{\mathbf{V}} &= \left[\tilde{\mathbf{V}}_1\tilde{\mathbf{V}}_2\ldots\tilde{\mathbf{V}}_M\right] \\ \vec{\mathbf{G}} &= \operatorname{diag}\left[\kappa_1{}^N\ \kappa_2{}^N\ \ldots\ \right] \\ \Gamma_j &= [\Gamma_{j1}\Gamma_{j2}\ldots\Gamma_{jM}]^T. \end{aligned}$$

where Γ_{1s} = modal participation factor of outgoing waves, and Γ_{2s} = modal participation factor of incoming waves for the $s-th$ mode with $s = 1,2\ldots M$ and $j = 1,2$.

Equation (7) can be written as:

$$\mathbf{U}_B = \mathbf{D}_0\Gamma \tag{7}$$

$$\Gamma = [\Gamma_1\ \Gamma_2]^T. \tag{8}$$

Using equation (3), the boundary force vectors, $\mathbf{P}_{B1}$ and $\mathbf{P}_{B2}$, at $x = 0$ and $x = L_t$, respectively, can be expressed as

$$\begin{bmatrix} \mathbf{P}_{B1} \\ \mathbf{P}_{B2} \end{bmatrix} = \begin{bmatrix} \mathbf{K}_{11}\mathbf{V} + \mathbf{K}_{12}\mathbf{V}\vec{\kappa} & -\left(\mathbf{K}_{22}\tilde{\mathbf{V}} + \mathbf{K}_{21}\tilde{\mathbf{V}}\vec{\kappa}\right)\vec{\mathbf{G}} \\ -\left(\mathbf{K}_{11}\mathbf{V} + \mathbf{K}_{12}\mathbf{V}\vec{\kappa}\right)\vec{\mathbf{G}} & \mathbf{K}_{22}\tilde{\mathbf{V}} + \mathbf{K}_{21}\tilde{\mathbf{V}}\vec{\kappa} \end{bmatrix} \begin{bmatrix} \Gamma_1 \\ \Gamma_2 \end{bmatrix} \tag{9}$$

where

$$\vec{\kappa} = \operatorname{diag}[\kappa_1\ \kappa_2\ \ldots\ ]$$

thus, the boundary force vector $\mathbf{P_B}$ can be expressed in terms of the wave modal participation factors as:

$$\mathbf{P_B} = \mathbf{D}_1\Gamma \tag{10}$$

Finally, from equations (8) and (11)

$$\mathbf{P_B} = \mathbf{D}\mathbf{U_B} \tag{11}$$

$$\mathbf{D} = [\mathbf{D}_1][\mathbf{D}_0]^{-1}, \tag{12}$$

where $\mathbf{D}$ is the required dynamic stiffness matrix of the hyper structure. This matrix is of order $2M$, it is in general complex and symmetrical.

EFFICIENCY AND ACCURACY

For a structure that has N identical panels and M degrees of freedom per set of boundary displacements, $\mathbf{u}_i$, the total number of equations in the system stiffness matrix is $n = (N+1)M$ and its bandwidth is $b = 2M$. Hence, for large N and M, the number of operations required to factor the system matrix is dominated by $nb^2 = (N+1)M^3$. Therefore, the amount of computation required to find all wave numbers and corresponding mode shapes for a given forcing frequency by means of standard iterative methods (e.g. inverse power) is dominated by a constant times $(N+1)M^3$. On the other hand, solution of the quadratic eigenvalue problem (4) involves factorization of a fully populated matrix of order $2M$, which means that the number of operations is dominated by a constant times M^3, regardless of the number of panels. Thus, by taking advantage of the special form of the system matrix, the reduction in computational costs can be extremely large. Furthermore, this drastic reduction in the number of operations for large N will reduce the magnitude of the accumulated errors due to roundoff, thereby producing more accurate solutions.

EXAMPLES

Axial Motion of Beams

Consider the beam of figure 2. It is fixed at one end, and subjected to an axial steady state force at the free end. The material and geometric properties of the beam are shown in the figure. The axial steady state motion of the free end is obtained using regular finite element analysis by subdividing the beam into 10 elements, figure 2a (10 elements, 11 nodal points and 33 degrees of freedom). The same motion is also obtained using a single hyper element using the method and equations of this study, figure 2b (two nodal points and six degrees of freedom). The amplitudes of the axial displacements are plotted against the forcing frequencies in figure 3 for the two methods. The agreement between the two solutions is so close that it is difficult to differentiate between the two curves.

Truss With cross Bracing

Now consider the more interesting case of a cross braced plane truss of figure 4. The truss is built from ten identical cells. It is fixed at one end. The other free end is subjected to a lateral steady state force with unit amplitude. The material and

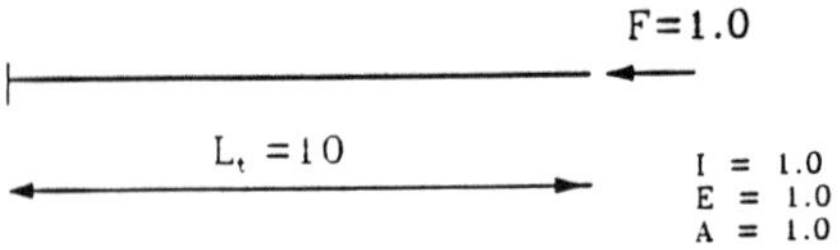

FIG 2-BEAM EXAMPLE

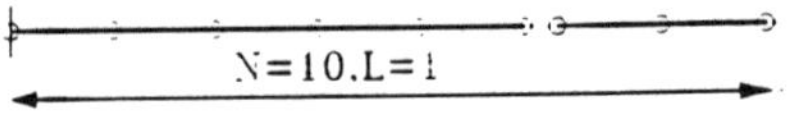

FIG 2a-FINITE ELEMENT MODEL

FIG 2b-HYPER STRUCTURE MODEL

Fig. 3 - Axial Displacements of a Beam
(Displacement Amplitudes)

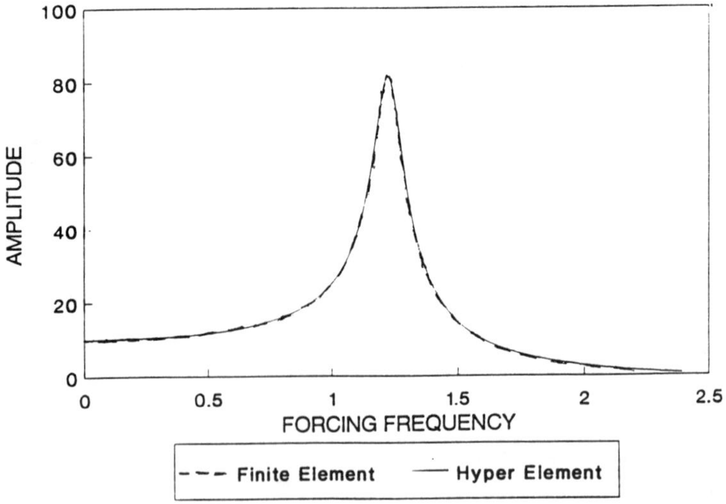

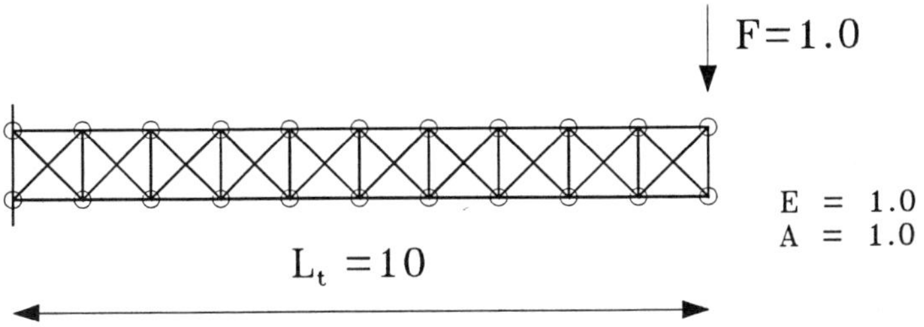

FIG 4-CROSS BRACED SYSTEM

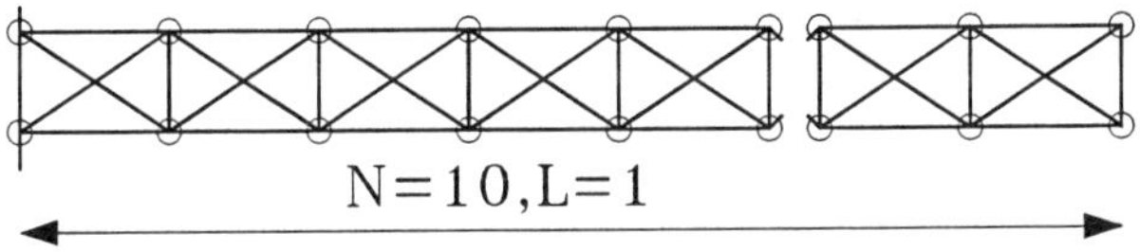

FIG 4a-FINITE ELEMENT MODEL

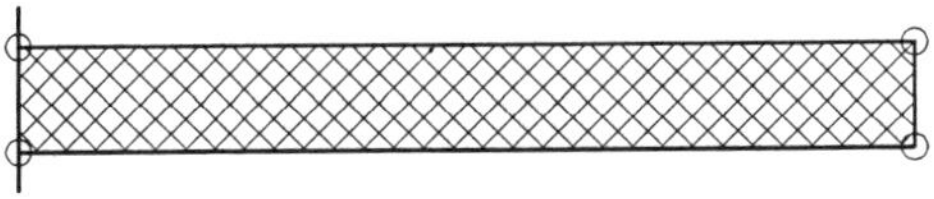

FIG 4b-HYPER STRUCTURE MODEL

Fig. 5 - Cross-Braced Truss
(Amplitude of Lateral Displacements)

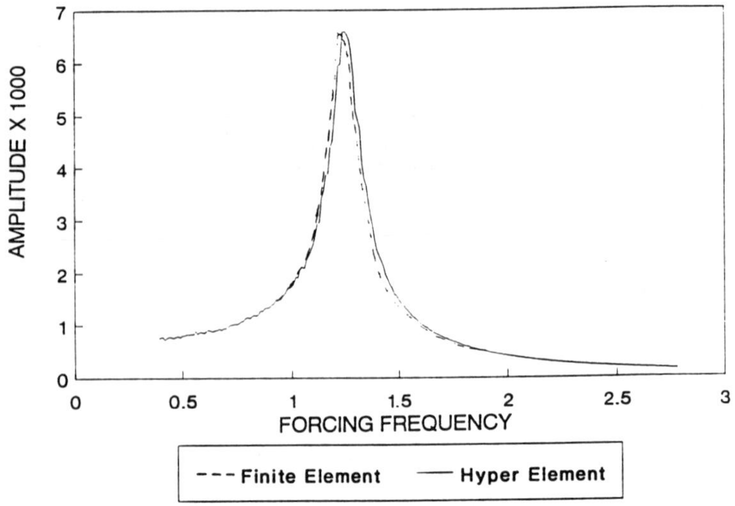

Fig. 6- Cross-Braced Truss
(Axial Displacement Amplitude)

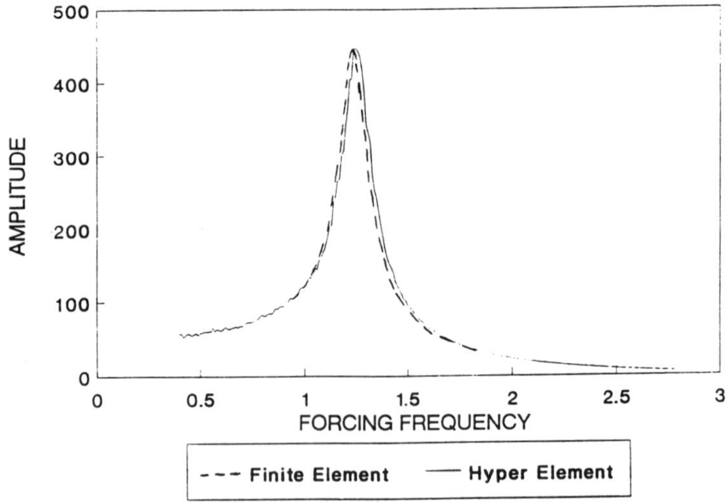

geometric properties of the truss are shown in figure 4. The steady state lateral and axial motions of the free end of the truss are obtained using the conventional finite element method using the model of figure 4a (50 elements, 22 nodal points and 44 degrees of freedom). They were also obtained using a single hyper element, figure 4b (4 nodal points and 8 degrees of freedom). Figure 5 shows the frequency dependent lateral displacements of the free end of the truss as were obtained using the two methods. Figure 6 shows similar comparison for the axial displacements of the free end due to the same loading conditions. Again the two methods produce remarkably similar results.

CONCLUDING REMARKS

A method of analyzing structures which are built from a set of repeated panels (cells) is presented. The method reduces the structural equations to a quadratic eigenvalue problem where the eigensolutions represent the wave numbers and the wave mode shapes. The dynamic stiffness matrix of the system can then be established, thus the whole system is reduced to a single element which was named in this work as the hyper element. This approach presents a major savings in calculational efforts. It also presents a more accurate representation of the systems under consideration, especially when compared to the well known continuum approach. Conversely, utilizing the accuracy and efficiency of this method, the basic panel can be modeled with greater detail, without a major increase in the numerical effort that will result if the same amount of modeling detail is required during the traditional analysis methods.

APPENDIX I - REFERENCES

Achenbach, J.D., **Wave Propagation in Elastic Solids**, North–Holland, 1973.

Crandall, S. **Wave Propagation**, Academic Press.

Ettouney, M., Benaroya, H., Wright, J., "Linear Dynamic Behavior of Semi–Infinite Trusses," presented at the Seventh VPI&SU Symposium on Dynamics and Control of Large Structures, and submitted to the **AIAA Journal**.

Kausel, E., "Forced Vibrations of Circular Foundations on Layered Media", Sc. D. Thesis, Civil Engineering Dept., MIT, Cambridge, Mass., January, 1974.

Kausel, E., and Roesset, J. M., "Semianalytic Hyperelement For Layered Strata", Journal of Engineering Mechanic Div., **ASCE**, Vol. 103, No. EM4, August 1977.

Noor, A.K., Mikulas, M.M., "Continuum Modeling of Large Lattice Structures: Status and Projections", in **Large Space Structures: Dynamics and Control**, Atluri S.N. and Amos A.K., Eds., Springer-Verlag, 1988.

Waas, G., "Linear two dimensional analysis of soil dynamics problems in semi infinite layered media", Ph. D. Thesis, U. of California at Berkeley, Berkeley, Ca. 1972.

Space Construction; A CSI Perspective

John R. Sesak[1]
J. Mel Waldman[2]

Abstract

This paper outlines the control structure interaction (CSI) problem for space construction engineers. An overview of the Control Astrophysics and Structures Experiment in Space (CASES) is provided. This overview discusses pertinent CSI issues.

Introduction

As Space platforms grow larger, they will require construction in space. Because of launch constraints these structures will have low mass, have little intrinsic structural damping, and the resulting vibration modes will have frequencies that are fractions of a Hertz. The free decay settling of such structures will be on the order of hours. This time penalty will prove too costly for the timely completion of many programs. Thus some means of vibration damping will be required during the construction phase of these platforms to ensure reasonable settling times. e.g, a space construction damper (SCD). The design and application of these devices requires some knowledge of CSI technology.

Overview of the CASES Program

The CASES mast, a 14.4 inch boom, when fully deployed to its 30 meter length, as shown in Figure 1, is required to point and maintain the pointing direction and stability of an x-ray telescope to 0.4 arc sec. This is accomplished with a set of bi-linear thrusters (BLTs) at the boom tip which continuously positions the orbiter so that the detector package sitting in the orbiter bay will point accurately at its target. The stiffening of the boom and the alignment of the mask at the tip are dynamically controlled by three sets of angular momentum exchange devices (AMEDs) located along the boom.

---

1 Senior Staff Engineer

2 Program Manager, Lockheed Missiles & Space Company, Inc., 1111 Lockheed Way, Sunnyvale< CA 94089

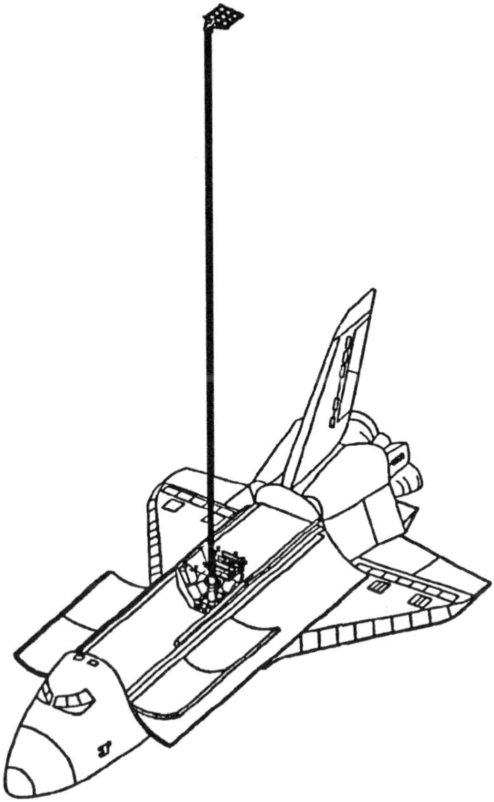

Figure 1. CASES Deployed in Orbiter

In order to successfully accomplish this pointing function, application of Controls, Structures Interaction (CSI) technology is required. This has classically been accomplished by designing the structure stiff enough so that the structure modal frequencies are above the control system frequencies. CASES provides a structure that doesn't meet these conditions and therefore the control system, in order to meet the pointing requirements, must operate in a frequency region which contains the structural modal frequencies. CSI technology provides this capability, but the technology needs demonstration and further development.

Future NASA missions, especially those associated with Solar/Astrophysics, will involve larger, more flexible structures with varied mass distributions and with greater requirements for disturbance suppression, surface stability, and pointing precision. Because of their flexibility, interactions of the components with both the spacecraft and the experiment control systems will more likely occur, reducing performance or restricting operations. CASES, is developing the technology to support the design and development of large space structures where these interactions are controlled and the system optimized.

The active control of space structures requires the solution of a multi-input/multi-output control problem as depicted in Figure 2. The operation of the closed-loop system is as follows: Disturbance excites the system. Sensors respond to the disturbance-induced vibration. A state estimator (systems identification function) reduces the sensed motion to modal coordinates for feedback control. The control law develops signals that command the actuator response to damp the disturbances and produce the "quiet" response shown. The guest investigator program provides a primary input to both the control law and systems identification function.

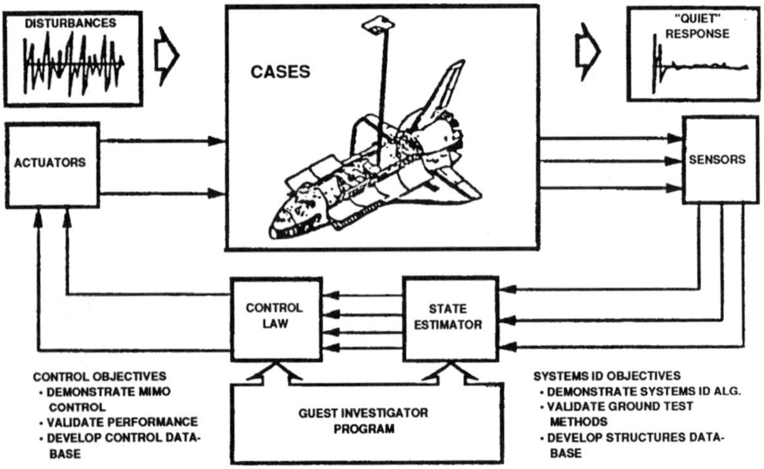

Figure 2. Active Control of Space Structures

The CASES program provides a unique opportunity to integrate slewing, pointing, vibration and shape control in a single control subsystem. The integration of these major subsystems with overlapping bandwidths, as shown in Figure 3, represents a major CSI step beyond mere vibration control. The parameter modification system figures in the quasistatic shape control subsystem. The integration of these major subsystems stems directly from realistic requirements imposed by the astrophysics experiment.

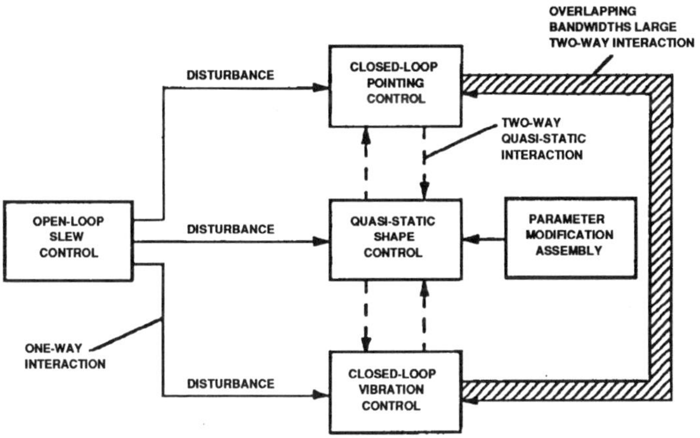

Figure 3. CASES Integrates Major Control Subsystems

Figure 4 shows the fifteen control system trades conducted in order to produce a viable CSI testbed. First, one must consider the algorithms to be used as a baseline. Linear Quadratic Gaussian (LQG) and collocated velocity feedback were the algorithms of choice. The selection was based primarily on their maturity (LQG) and robustness properties (collocated velocity feedback). In order to reduce computer requirements, off-line capability was desired for systems identification algorithms. With regard to the test article properties, it was decided that ten modes would be adequate, damping levels should be low, two-dimensioned modal coupling would be sufficient, and low-frequency modal behavior (below 10 Hz) was a must; in addition to low frequencies, the modes should be coupled and occur in clusters. A multi-input/multi-output system is required. To achieve this MIMO control system capability, the number of sensors should exceed 25, and the number of actuators should be 11 (2BLTs and a three-axis AMED at the boom tip and two more identical three-axis AMEDs symmetrically placed along the boom). The sensor complement, in addition to gyros, accelerometers, and optical systems, should include load cells, strain gauges, and thermistors. For technology transfer, the structure should be a combination of a continuous longeron and a joint-dominated structure. A parameter modification system with capability for large structural variations and enhanced modal coupling would be a stiff arm section with variable weights attached at the boom tip. It is felt that this combination of structure and instrumentation provides sufficient testbed capability to demonstrate sophisticated CSI algorithms.

CSI TRADES	SELECTED OPTIONS
CSI CONTROL ALGORITHMS	LQG, COLLOCATED VEL FB
SYSTEM I.D. ALGORITHMS	OFF-LINE
NO. OF MODES CONTROLLED	TEN MODES
DAMPING LEVEL	LOW
COUPLING COMPLEXITY	TWO DIMENSIONAL
FREQUENCY REGIME	<<10 Hz
MODAL CHARACTERISTICS	COUPLED
CSI CAPABILITIES	MIMO
DYNAMICS	CLUSTERED MODES
TYPES OF SENSORS	FORCE, STRAIN, TEMP, ACC RATE, POSITION
NUMBER OF SENSORS	>25
TYPES OF ACTUATORS	GAS JET, REACTION WHEEL
NUMBER OF ACTUATORS	11
STRUCTURE	COMBINATION
PARAMETER MOD SYSTEM	STIFF ARM

Figure 4. CSI Configuration Summary

The CASES baseline as developed by Lockheed and shown in Figure 5 consists of a configuration packaged into five modules:

- Astrophysics Module
- Propulsion Module
- Boom Module
- Astrophysics Electronics Module
- Control and Data Handling (C&DH) Electronics Module

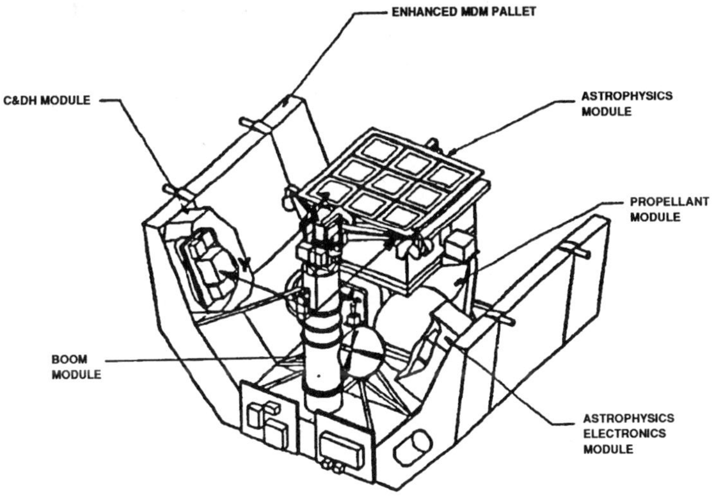

Figure 5. CASES Baseline Design

The five modules are designed to facilitate testing and integration at Lockheed and to simplify installation into the Enhanced MDM Pallet (EMP) and Shuttle at KSC.

The EMP was selected after several design trade-offs because of the services it provides to the experiment payload. It contains the required signal interface to the orbiter through its dual wide smart flexible MDM and high data rate subsystem. The EMP also provides the power distribution and the active cooling and telemetry interfaces for the experiment.

The Astrophysics Module contains 8 proportional counter X-ray deflector assemblies and their structure together with 4 star trackers. The module also consists of the mask and the mask assembly restraint system which clamps the mask to the detector assembly during ascent and orbiter re-entry. The packaging of both the mask and detectors into a single module facilitates integration and testing at both Lockheed and at KSC.

The Propulsion Module houses three spherical propellant tanks and provides the structure support for the Astrophysics Module.

The Boom Module consists of the experiment boom and its canister housing, the tip, mid-boom and boom position assemblies, the remote alignment measurement system (RAMS), IMU, propellant valves/regulator, and the jettison assembly (for orbiter safety should the boom not retract).

The electronics are housed in two modules: the Astrophysics Electronics Module which contains the Astrophysics Controller Buffer Unit and the High Voltage Power Supply, and the C&DH Electronics Module which contains the experiment computer, amplifier units, power supplies, the data acquisition unit and the orbiter I/O interface.

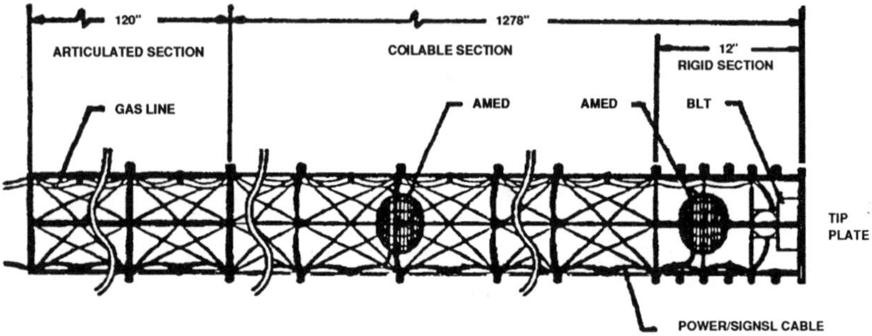

Figure 6. CASES Mast Schematic

Figure 6 shows the CASES mast schematic that has evolved from the CASES study. Integration of the Angular Momentum Exchange Devices (AMEDs) and Bi-Linear Thrusters (BLTs), plus securing cables and gas lines, as shown, have been bread-boarded. Also bread-boarded are the derotator design for the electrical cables and the pneumatic swivel for the BLT gas lines. Lockheed has evolved a concept of including a joint dominated articulated mast segment to allow a greater technology transfer capability in the CSI experiments and a stiff arm segment with variable weights in the rigid section to enhance the system identification experiments.

Advancing technological capabilities in the area of large space structures is clearly a boon for future astronomical instrumentation; the evolution of CASES to a permanent Pinhole/Occulter Facility on the Space Station is evident. However there are an impressive amount of "breakthrough observations" which CASES will already be able to obtain during a ten-day mission. The solar observations should verify or refute the concept that coronal heating is a stochastic flare-related process. The presence of a massive black hole at the center of our galaxy should be established if that is indeed the case.

The CSI technology developed on CASES will provide a practical demonsration of the enabling technology that will be required in the future to make space construction of large structures feasible.

THE TRUSS-BEAM AND SPACE TECHNOLOGY

by

ELIAS G. ABU-SABA[1],
WILLIAM M. MCGINLEY[2],
RAYMOND C. MONTGOMERY[3],

ABSTRACT

A method is presented herein to determine the dynamic characteristics of truss-beam systems for both planar and space structures. Results from this approach are compared with those of the finite element method and test results. The truss-beam model is based on the assumptions of negligible web contribution, i.e., very large flexible structures. Future experimental evaluation of the plane truss-system is planned at North Carolina Agricultural and Technical State University and this experimental program is described.

INTRODUCTION

As space explorations continue at present and in the future, the need for space stations fully equipped with machines used for instrumentation and life support systems will increase dramatically. Previous experience shows the truss-beam element to be an integral part of the state-of-the-art in space construction. Dynamic control of space vehicles and stations is thus significantly dependent upon our understanding of the dynamic behavior of the truss-beam

---

[1]Member ASCE, Associate Professor of Architecural Engineering, North Carolina A & T State University, Greensboro, NC, 27411
[2]Associate Member ASCE, Assistant Professor of Architecural Engineering, North Carolina A & T State University, Greensboro, NC, 27411
[3]Aerospace Technologist, NASA Langley Research Center/ 161 Hampton, VA, 23665

element. Analytical models have been developed by many researchers and used subsequently to make predictions concerning mode shapes and natural frequencies of such systems.

The truss-beam model developed by Elias G. Abu-Saba (Abu-Saba, 1988) was used in the dynamic analysis of a plane truss shown in Figure 1. The same model was also applied to the mini-mast structure which has been tested at NASA-Langley (Abu-Saba, 1989) and to the box truss system analyzed by Noor and Andersen (Noor et. al., 1979). The analytical predictions for the plane truss as well as the space truss systems (Mini-mast and Box) will be compared with those obtained from the other analysis techniques, finite element method and the Noor-Anderson model. Some experimental results for the mini-mast are available and will be included in the comparison.

In addition, the authors plan an experimental study at North Carolina Agricultural and Technical State University in the coming few months to experimentally verify and refine the truss-beam model. This proposed experimental program will be outlined.

While the present study does not include joint participation, follow-up research will introduce the effect of the joints on the behavior of the system.

ANALYTICAL PROCEDURE

The truss-beam model was applied to the plane truss system shown in Figure 1. The development of the truss-beam model is based on the following assumptions:

1. The material is linearly elastic.
2. Deflections are small.
3. Coupling between vibratory modes, both torsional and bending, is negligible.
4. Contributions of the web members to the lateral stiffness of the plane truss is insignificant for structures where the planar span to depth ratio is greater than 80.

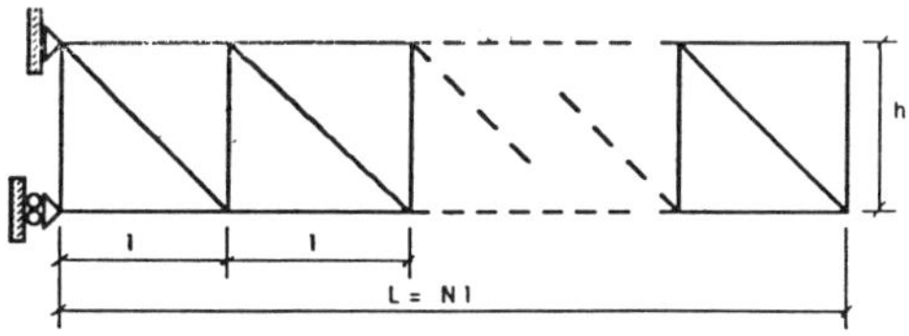

Figure 1 Truss Beam

5. All masses are lumped at the center of the panel length.
6. All panels are identical.

Based on the above assumptions the following set of algorithms was developed and used to determine the elements of a cantilever beam flexibility matrix for bending:

$$A_{ii} = \frac{1^3}{24EI}(2i - 1)^3, \; i = 1, N \quad [1]$$

$$R_i = \frac{1^3}{24EI}(2i - 1)^2, \; i = 1, N$$

$$A_{ij} = A_{ij-1} + R_i, \; j=i + 1, N$$

$$A_{ij} = A_{ji}$$

where
- N = Number of panels
- l = Length of the panels
- E = Elastic modulus
- I = Moment of inertia about bending axis, for chord area only

Based on the previously stated assumptions, the flexibility matrix for the twisting mode of the space truss system shown in Figure 2 is generated by the following algorithm:

$$A\theta_{ij} = (M-j)\frac{1}{G\ J} \qquad (i=1, N) \quad [2]$$
$$(j=i, N)$$

where
- $A\theta_{ij}$ = Torsional flexibility matrix
- M = N + 1

The general equation of motion for a linear undamped structural system is expressed in the following form:

$$\{Y\} = \omega^2 [A] [M] \{Y\} \qquad [3]$$

where
- N = number of panels
- [A] = flexibility matrix (bending or torsion)
- [M] = Mass matrix
- {Y} = displacement matrix
- ω = natural frequency of the system.

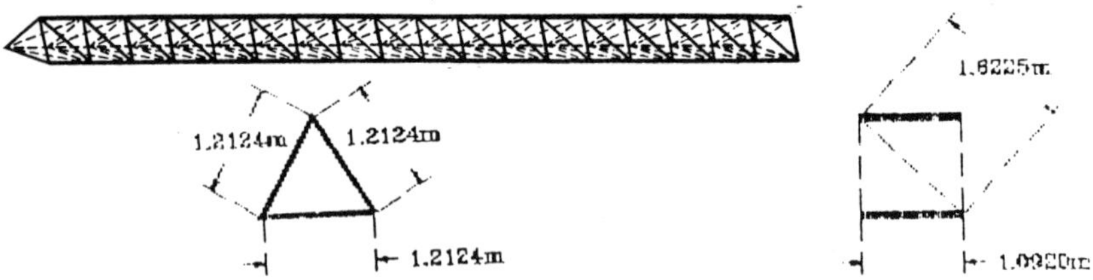

Figure 2 Mini-Mast Truss

Once the flexibility matrix has been determined from Equation 1, an interactive computer program was developed to provide an iterative solution to Equation 3. Because of the assumptions mentioned earlier, the model provides only bending modes for the plane truss. Using the torsional flexibility matrix developed by Equation 2, the same procedures were used to predict the torsional behavior of the space truss system shown in Figure 2 (Mini-mast). It should be noted that this torsional analysis is valid as long as the space truss system meets the same assumptions as listed for the plane truss system.

It should be further noted that the vibratory responses measured by Montgomery et al (Montgomery, 1989) show little difference between X and Y axis bending mode frequencies. These results appear to confirm negligible coupling of the X and Y axis bending and torsional vibratory modes for long flexible structures, thus validating the earlier simplifying assumption used for the development of the truss-beam model.

APPLICATION OF THE METHOD

Example 1

A plane truss with panels 20 in (508 mm) and 20 in (508 mm) in depth was analyzed using the truss-beam model and the procedures outlined in the previous section (Abu-Saba, 1988). The cross-sectional area of the truss chords was 0.25 in^2 (161.3 mm^2). The values of E and the mass of each

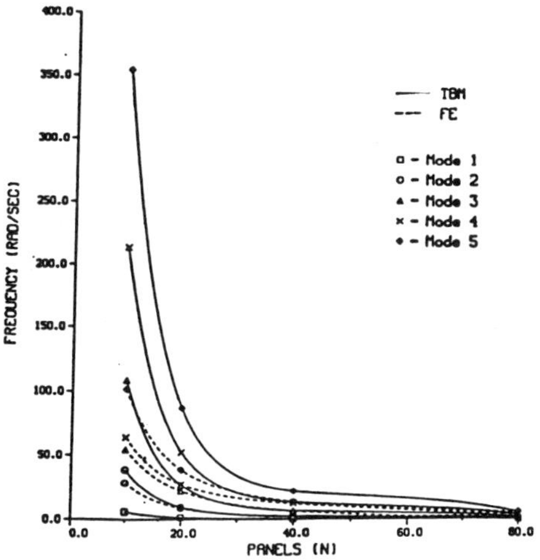

Figure 3 Variation of Frequencies with Panel Number

panel were 3 x 10^7 psi (207,000 MPa) and 0.282 pci (7.8 Mg/m^3). A linear elastic finite element model was also used to analyze the plane truss described above and the two sets of results were compared (Abu-Saba, 1989). The curves shown in Figure 3 show the frequencies thus determined for panel numbers, N, ranging from 10 to 80.

Example 2

The truss-beam model and finite element analysis techniques were applied to the Mini-mast truss designed and manufactured by Astro Aerospace Corporation of Carpinteria, California. This system was used as the first ground testbed at NASA Langley for the Controls-Structures Interaction (CSI) program (Anon, 1988).

The total length of the Mini-mast is 66.1 ft (20.16 m) and it contains eighteen bays at 3.67 feet, with every other bay repeating. All major members are tubular and are constructed of graphite/epoxy composite with the following geometrical and material properties:

	Outside/Inside Diameters	
Longerons	0.795/0.587 in	(20.2/14.9 mm)
Battens	0.594/0.469 in	(15.1/11.9 mm)
Diagonals	0.594/0.441 in	(15.1/11.1 mm)

Modulus of elasticity 17.9 x 10^6 psi (1.24 x 10^{11} N/m^2)

Table 1 presents the results of the truss-beam analysis for the first six bending mode frequencies. Table 1 also shows the results of a finite element analysis and experimental measurements for the same vibratory modes (Montgomery, 1989).

TABLE 1 NATURAL FREQUENCIES (HZ)

MODE	MODE TYPE	FE	TB	MEASURED
1	1B	0.688	0.849	0.856
2	1T	4.417	14.00	4.299
3	2B	6.225	7.458	6.167
4	3B	30.94	32.839	
5	4B	43.124	58.805	
6	5B	70.252	108.830	

FE - Finite Element Model ; TB - Truss Beam Model

Example 3

The third example of an application of the truss-beam model is the box type truss previously analyzed by Noor and Andersen (Noor et. al., 1979) (see Figure 4). The truss contains ten bays that are alike. The length of each panel is 24.60 ft (7.5 m) and the width and height are both 16.4 ft (5.0 m). The geometrical and material properties of this truss are listed below.

	Cross-sectional Area
Longitudinal Bars	0.062 in^2 (40×10^{-6} m^2)
Battens	0.093 in^2 (60×10^{-6} m^2)
Diagonal Bars	0.062 in^2 (40×10^{-6} m^2)

Modulus of Elasticity 10.4×10^6 psi (71.7×10^9 N/m^2)
Density 0.355 pci (2.768 Mg/m^3)

The first three bending mode frequencies are presented in Table 2 showing the comparison between the method used by Noor and Anderson (Noor et. al., 1979) and the truss-beam method.

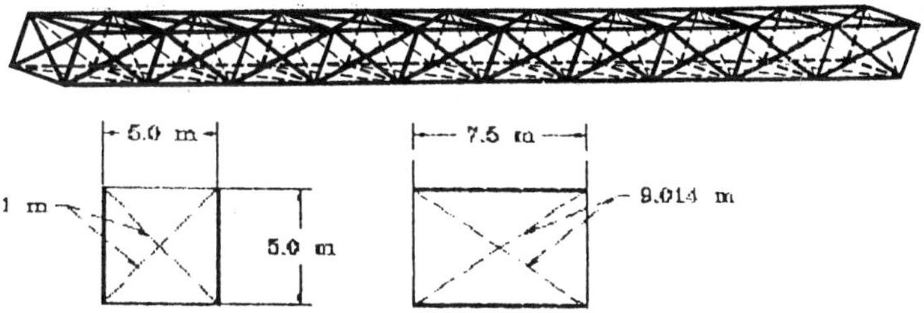

Figure 4 Truss Beam (Noor et. al., 1979)

TABLE 2 NATURAL FREQUENCIES (HZ)

MODE	MODE TYPE	Noor, Andersen	Truss-Beam
1	1 Bending	0.8335	0.7718
2	2 Bending	4.5723	4.8639
3	3 Bending	10.9937	13.6870

COMPARISON OF RESULTS

By confining the analytical model of the truss-beam to a cantilever form, a plane system yields only bending moment mode shapes and frequencies. However, a space truss will have torsional modes in addition to bending. With the truss-beam model coupling between the bending and torsional modes was assumed to be small and thus was neglected.

In Example 1, Figure 3 shows that for the first five bending modes of the plane truss the modal frequencies predicted by the finite element method and the truss-beam model approach the same values, as the number of panels in the truss increases. Further, it is noticed in Example 2 (Table 1) the first bending mode predicted by the truss-beam model is very close to the value obtained experimentally by the NASA researchers (Montgomery, 1989). For this first bending mode, the accuracy of the prediction of the truss-beam model exceeds that of the finite element method (when compared to the experimental results). At higher bending modes the accuracy of the truss-beam model decreases. This decrease in accuracy is likely due to more significant contribution of the web members at the higher vibratory modes and round-off inaccuracies in the iteration of the higher vibratory frequency calculations. Some error is also introduced by the unknown play and slip in the joints in the experimental specimen which was not accounted for in the truss-beam model.

Examining the results in Table 1 shows that the truss-beam model does not accurately predict the first torsional mode of the Mini-mast. Further refinement of the torsional flexibility algorithm will likely produce much better results and is presently underway.

In example three, the first three bending mode frequencies predicted by the method used by Noor et al (Noor et. al., 1979) and by the truss-beam method also compare well. It appears that the simple approach presented by the truss-beam model predicts, with reasonable accuracy, the dynamic behavior of plane truss systems, especially for the bending modes of long flexible systems.

EXPERIMENTAL PROGRAM

To further confirm the accuracy of and refine the truss-beam model an experimental program has been proposed. This experimental program is needed to define the behavior of these systems without the unknown contribution of joint play and slip. As was discussed in the previous discussion of the analytical model, the behavior of the joint dominated beam may be dependent on the flexibility of the truss panels, the mass of the panel and the flexibility of the truss joints. All these quantities must be accurately measured if a reliable evaluation of the analytical method can be made. Thus, the proposed experimental program will consist of two stages. The first stage will involve the measurement of the material properties of the truss panels and joints, and the second stage will investigate the dynamic response of the truss assembles.

STAGE 1

During the first stage of the experimental program, the dimensions, mass and flexibilities of the truss panels, joints and chord tubing will be measured. Figure 5 shows a typical truss panel specimen. Each panel is 30" long and 18" deep and constructed primarily of 1" x 1" x 0.065" tubing. It should be noted that the joint connections are designed to have approximately the same axial stiffness as the chord members. The joints are also designed to have no slip while approximating a pinned joint. The proposed experimental investigation will test a total of six truss panels, individually and in multi-panel assembles.

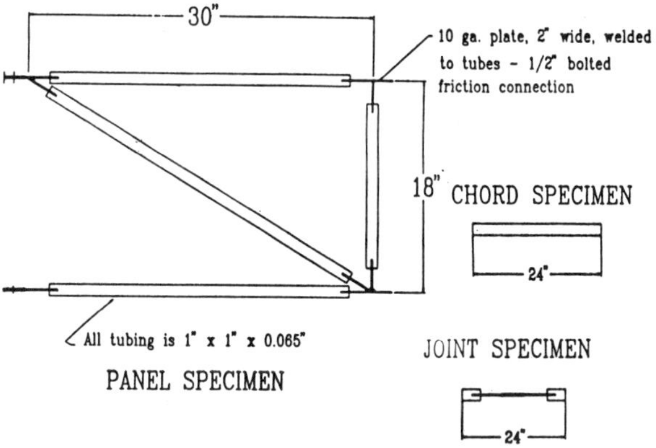

Figure 5 Proposed Truss Specimen

Joint and Chord tests

A total of five chord specimens and five joint specimens will be fabricated and tested (see Figure 5). After the dimensions of the specimens are determined, each specimen will be subjected to a number of static axial tension loads using a Forney material tester. By measuring the elongation of the specimen under load, the axial flexibility of each component type can be determined. This elongation will be measured by LVDT's on the joint specimens and by strain gauges on the chord specimens.

Truss Panel Tests

Each of the six truss panels will be fastened to the supporting frame shown in Figure 6 and subjected to a static point load at the free end. Deflection of the panel free end and chord strains will be measured for a number of load levels. Detailed measurements of the cross-sectional properties of the specimen will be made and all tests will be repeated three times.

Upon completion of the individual tests, the truss panels will be joined in two, three, four, five and six panel assemblies and the static test described above will be repeated for each assembly.

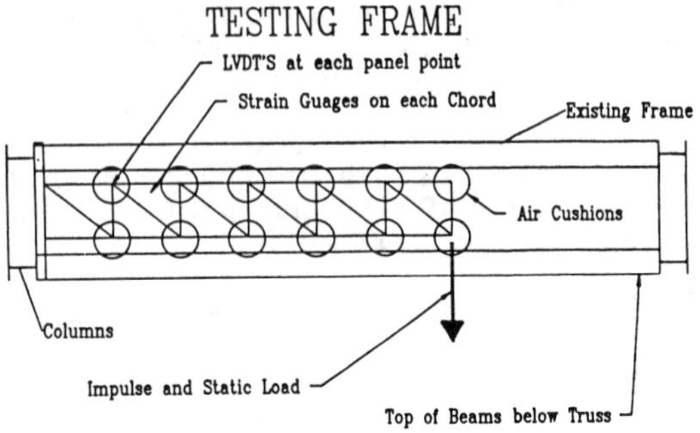

Figure 6 Proposed Testing Apparatus

STAGE 2

The primary goal of the second stage of the experimental program is to determine the natural frequencies and mode shapes (in planar bending) of truss specimens fabricated from an increasing number of panels. After each specimen is attached to the support and the air cushions adjusted for free planar movement, the panel will be excited by a pendulum and the chord strains and panel deflections will be measured for approximately twenty seconds. This test will be repeated three times for each panel. The fundamental frequency and mode shape for each panel specimen will be determined from this data.

After completion of the individual panel tests a, two, three, four, five and six panel specimen will be constructed using the individual panels and tested as above. In addition, the pendulum will be moved to each joint on the multi-panel specimens and used to apply an impulse load at these locations. By measuring chord strains and nodal deflections after excitation of each joint location, the significant higher vibratory modes of the multi-panel specimens can be determined. These tests will also be repeated three times.

By measuring the variation of both the chord strain and the nodal deflections, the frequencies of each bending vibration mode can be determined from two independent sets of readings. This provides redundancy of measurements and provides a higher level of confidence in the experimental results.

CONCLUSIONS

A mathematical model without joint participation was developed for plane trusses and space trusses. Joint participation will be included in a future research effort. An interactive computer program has been developed to predict the undamped bending and torsional vibrations using the truss-beam model and requiring only the elastic and geometric properties of the chord members.

Based on the previous discussion the following conclusions can be made.

1. The truss-beam model yields increasingly accurate results for structures with larger length to depth ratios (ratios>80).
2. The bending mode results of the truss-beam model for the Mini-mast compare favorably with those of the finite element analysis and the experimental values found by Montgomery et al (Montgomery, 1989), especially for the lower modes.
3. The truss-beam model is simple to use and requires less computational effort than a finite element model. The order of the flexibility matrix is N for a plane truss using the truss beam model. The finite element flexibility matrix has an order of 4N + 1 and 6 N + 1 for a plane and space truss, respectively.
4. The experimental procedures outlined in this paper will be used to further refine the truss-beam model for the prediction of the vibratory behavior of plane truss systems.

REFERENCES

1. Abu-Saba, E. G., Sherbourne, A.N., and Montgomery, R.C.;"Dynamic Analysis of the Joint Dominated Beam", Engineering, Construction and Operations in Space, ASCE proceedings, August 29-31, 1988, P. 495-505.
2. Abu-Saba, E. G., McGinley, W. M., and Montgomery, R.C.; "Mini-Mast Dynamic Analysis Using the Truss-Beam Model", 3rd Annual Conference on Aerospace Computational Control, Oxnard California, August 28-30, 1989.
3. Anon, " New Technology Program for Spacecraft Design: Control-Structures Interaction Focus for Future Design", NASA Activities, December, 1988, pp. 5-6.
4. Montgomery, R. C., Gosh, D., and Sulla, J, "Mini-mast Testbed, NASA Langley Research Center, Hampton, VA, March 1989.
5. Noor, Ahmed K., and Andersen, C. M., "Analysis of Beam-Like Lattice Trusses", Computer Methods in Applied Mechanics and Engineering, North-Holland Publishing Company, 1979, pp 53-70.

Dynamics of Composite Tubes

S.K. Datta[1], T. Kohl[1], A.H. Shah[2], and R. Rattanawangcharoen[2]

Abstract

A displacement-based stiffness method is used to study the characteristics of dispersive wave propagation in laminated composite cylindrical tubes. It is assumed that the tube is composed of several bonded, uniaxial or angle-ply composite laminates. The objective is to model modes of propagation in composite tubular struts that are to be used for truss elements of space structures. Numerical results are presented showing the dependence of dispersion characteristics on the fiber orientations in the laminae.

Introduction

Fiber reinforced laminated composite tubes are currently being studied as candidate struts for trusses to be used in space structures (see, for example, Dorsey and Mikulus, 1989). These have requisite strength-to-weight ratios and dimensional stability for space environments. In this paper, we present a model study of wave propagation in long cylindrical composite laminated tubes.

Axially symmetric vibrations and waves in an infinitely long, isotropic solid circular cylinder are governed by a frequency equation formulated by Pochhammer (1876). The frequency equation contains Bessel functions and implicitly relates the frequency to the wave number. Although the equation was introduced as early as 1876, only in recent years were numerical results over a wide range of frequencies and complex wave numbers presented by Onoe et al. (1962). The studies reported for isotropic materials since then, which are relevant to our work, are the vibrations of hollow circular cylinders by Gazis (1959) and of composite cylinders by McNiven et al. (1963) and Armenakas (1965, 1971). Within the framework of three-dimensional linear elasticity, Mirsky (1964) and Armenakas and Reitz (1973), employing Frobenius-series, formulated the frequency equation for homogeneous orthotropic solid circular cylinder.

Since the dispersion relation for anisotropic laminated tubes is intractable to analytical procedures, many alternate approximate formulations have been proposed. The most common ones are shell theories. There are numerous shell theories for homogeneous or laminated media consisting of isotropic or anisotropic

1. Cooperative Institute for Research in Environmental Sciences, Center for Space Construction and Department of Mechanical Engineering, University of Colorado, Boulder, CO 80309-0427.
2. Department of Civil Engineering, University of Manitoba, Winnipeg, Canada, R3T 2N2.

materials. No attempt is made here to give any reference to various shell theories.

Although in infinite tubes only propagating modes corresponding to real wave numbers are physically realizable, for finite tubes both propagating and non-propagating modes are needed to satisfy the end conditions. But even for isotropic homogeneous cylinders, complex wave number solutions of the transcendental frequency equations are difficult to obtain. To circumvent the difficulties Mindlin and McNiven (1960), McNiven (1961), McNiven and Perry (1962), and McNiven and Shah (1967) proposed a three-mode theory to solve for the frequencies of vibration of finite and semi-infinite cylinders. For laminated anisotropic cylindrical shells Tasi (1971) and Tasi and Roy (1971) presented a two-mode shell theory.

Since the number of propagating and non-propagating modes needed to describe the frequency spectrum of a finite tube can be large depending on the frequency range of interest, it is necessary to consider alternate numerical procedures. In this paper we have used a stiffness method that was discussed previously by Huang and Dong (1984) and Dong and Huang (1985). Recently, Datta et al. (1988) have presented a higher order stiffness method that gives better accuracy at high frequencies. Because the displacement continuity based stiffness method presented by Huang and Dong (1984) is simpler we have used that to present dispersion spectra for laminated uniaxial and angle-ply fiber reinforced tubes with and without aluminum cladding. Aluminum cladding has been proposed to protect the tubes from environmental damage.

Governing Equations

We consider time harmonic waves in a circular cylinder composed of m laminae. Discretization occurs in the radial direction. Each lamina is divided into several sublayers. Each of these sublayers may possess distinct constitutive elastic properties and is geometrically defined by an inner radius r_b and outer radius r_f. In this paper, the constitutive elastic properties of a lamina are assumed to be cylindrically anisotropic. With reference to cylindrical coordinates (r,θ,z), the radial, circumferential, and axial displacements (u,v,w) are assumed here to be independent of θ. If the tube is composed of laminae possessing cylindrically orthotropic constitutive elastic properties, the wave propagation problem reduces to two uncoupled ones: torsionless axisymmetric (longitudinal) in which the displacement components are (u,0,w) and torsional with displacement components (0,v,0). In the more general case of cylindrical anisotropy, equations governing the displacement components are coupled. We treat the latter which encompasses the former as a special case.

The strain-displacement relations are given by

$$\varepsilon_r = \frac{\partial u}{\partial r}, \quad \varepsilon_\theta = \frac{u}{r}, \quad \varepsilon_z = \frac{\partial w}{\partial z}, \quad \gamma_{\theta z} = 2\varepsilon_{\theta z} = \frac{\partial v}{\partial z}$$

$$\gamma_{rz} = 2\varepsilon_{rz} = \frac{\partial u}{\partial z} + \frac{\partial w}{\partial r}, \quad \gamma_{r\theta} = 2\varepsilon_{r\theta} = \frac{\partial v}{\partial r} - \frac{v}{r} \tag{1}$$

Consider the kth sublayer bounded by $r=r_b$ and $r=r_f$. The stress-strain relation in this sublayer is given by

$$\{\sigma\} = [C^{(k)}]\{\varepsilon\} \tag{2}$$

where σ is the stress tensor written as

$$\{\sigma\}^T = [\sigma_{rr}, \sigma_{\theta\theta}, \sigma_{zz}, \sigma_{\theta z}, \sigma_{rz}, \sigma_{r\theta}] \tag{3a}$$

and the strain tensor ε is

$$\{\varepsilon\}^T = [\varepsilon_{rr}, \varepsilon_{\theta\theta}, \varepsilon_{zz}, \gamma_{\theta z}, \gamma_{rz}, \gamma_{r\theta}] \tag{3b}$$

$C^{(k)}$ is the stiffness tensor with superscript (k) denoting kth sublayer. Using the interpolation polynomials in the radial direction, the displacement components are approximated as

$$\{U\} = [N]\{q\} \tag{4}$$

where

$$\{U(r,z,t)\}^T = [u(r,z,t), v(r,z,t), w(r,z,t)] \tag{5a}$$

$$\{q(z,t,)\}^T = [u_b, v_b, w_b, u_m, v_m, w_m, u_f, v_f, w_f] \tag{5b}$$

and

$$[N] = \begin{bmatrix} n_1 & 0 & 0 & n_2 & 0 & 0 & n_3 & 0 & 0 \\ 0 & n_1 & 0 & 0 & n_2 & 0 & 0 & n_3 & 0 \\ 0 & 0 & n_1 & 0 & 0 & n_2 & 0 & 0 & n_3 \end{bmatrix} \tag{5c}$$

In Eq. (5) the generalized displacement $u_b(z,t)$, $v_b(z,t)$, $w_b(z,t)$, $u_m(z,t)$, $v_m(z,t)$, $w_m(z,t)$, $u_f(z,t)$, $v_f(z,t)$, and $w_f(z,t)$ are taken at the back, middle and front nodal surfaces of the sublayer. The interpolation polynomials n_i are quadratic functions given by

$$n_1 = 1-3\hat{r}+2\hat{r}^2, \quad n_2 = 4\hat{r}-4\hat{r}^2, \quad n_3 = -\hat{r}+2\hat{r}^2 \tag{6}$$

with normalized r defined by

$$\hat{r} = \frac{(r-r_b)}{h}, \quad h = r_f - r_b \tag{7}$$

Inserting Eq. (4) into the strain-displacement relations (1) we get

$$\{\varepsilon\} = [b]\{q'\} + [a]\{q\} \tag{8}$$

where prime denotes differentiation with respect to z. For the sake of brevity the strain displacement transformation matrices [a] and [b] are not reproduced here.

The equation governing the nodal displacement (u_j, v_j, w_j) will be obtained using Hamilton's principle. We calculate the Lagrangian of each sublayer as

$$L^{(k)} = \frac{1}{2}\int_t\int_z\int_r\left\{\int_\theta [\rho^{(k)}\{\dot{\bar{U}}\}^T\{\dot{U}\} - \{\bar{\varepsilon}\}^T[C^{(k)}]\{\varepsilon\}]d\theta\right\} rdrdzdt \tag{9}$$

where ρ is the mass density, overbar and overdot denote complex conjugate and time differentiation, respectively, and the integral over r is from r_b to r_f. Substituting Eqs. (1), (4) and (8) into Eq. (9) and integrating over the circumferential and radial coordinate yields

$$L^{(k)} = \pi \iint_{tz} [\{\dot{\bar{q}}\}^T[m]\{\dot{q}\} - \{\bar{q}'\}^T[k_1]\{q'\} - \{\bar{q}'\}^T[k_2]\{q\} - \{\bar{q}\}^T[k_2]^T\{q'\} - \{\bar{q}\}^T[k_3]\{q\}]\, dzdt \tag{10}$$

where $[k_1]$, $[k_2]$ and $[k_3]$ are three stiffness-related matrices and $[m]$ is the consistent mass matrix given by

$$[m] = \int_r \rho^{(k)}[N]^T[N]rdr \qquad [k_2] = \int_r [b]^T[C^{(k)}][a]rdr \tag{11}$$

$$[k_1] = \int_r [b]^T[C^{(k)}][b]rdr \qquad [k_3] = \int_r [a]^T[C^{(k)}][a]rdr$$

The integrations specified in Eq. (11) may be computed exactly given the elastic moduli, density, and limits of integration r_b and r_f for the kth sublayer. Observe that $[k_1]$, $[k_3]$ and $[m]$ are symmetric matrices, while $[k_2]$ is not.

The Lagrangian for the complete cylinder is found by summing the contributions from all the sublayers. Setting the first variation of the Lagrangian to zero and integration by parts yields a system of second order partial differentiation equations

$$[K_1]\{Q''\} + ([K_2] - [K_2]^T)\{Q'\} - [K_3]\{Q\} = [M]\{\ddot{Q}\} \tag{12}$$

Matrices $[K_1]$, $[K_2]$, $[K_3]$ and $\{\dot{M}\}$ are obtained by assembling the corresponding elemental matrices.

A solution to Eq. (12) can be assumed in the form

$$\{Q(z,t)\} = \{Q_0\}e^{-i\omega t}e^{-\gamma z} \tag{13}$$

where $\{Q_0\}$ represents the nodal amplitude vector, ω the circular frequency and $i\gamma$ the complex wave number. Solutions of the form

$$\{Q_1\}e^{-i\omega t}e^{\gamma z}$$

are discarded since all waves are assumed to propagate in the direction of increasing z.

Substitution of Eq. (13) into Eq. (12) results in a two-parameter algebraic eigenvalue problem (EVP), with γ or ω serving as eigenvalue parameter. If ω is given, then EVP-I is obtained as

$$[\gamma^2[K_1] - \gamma([K_2] - [K_2]^T) - ([K_3] - \omega^2[M])]\{Q_0\} = 0 \tag{14}$$

If γ is specified, then EVP-II is obtained as

$$[[K(\gamma)] - \omega^2[M]]\{Q_0\} = 0 \tag{15}$$

where

$$[K(\gamma)] = -\gamma^2[K_1] + \gamma([K_2] - [K_2]^T) + [K_3] \tag{16}$$

EVP-I is of particular interest because of the physical significance of the solution. The specified frequency ω is real and the solution for the wave number $i\gamma$ that results is of the form

$$i\gamma = i\eta + \zeta \tag{17}$$

Only solutions with $\eta \geq 0$ are allowed from the point of view of stability of the system. If $\eta=0$ we have a propagating mode. Otherwise the mode is nonpropagating as it decays with distance in an exponential or oscillatory exponential form for $\zeta=0$ and $\zeta \neq 0$, respectively. Such modes are not allowable from a physical standpoint in an infinite tube, but are significant in problems involving tubes of finite length.

Solution of EVP-I is obtained by introducing

$$\{Q_1\} = -\gamma\{Q_0\} \, . \tag{18}$$

The the quadratic eigenvalue problem in Eq. (14) is reduced to a generalized eigenvalue problem:

$$[A]\{Q\} = \gamma[B]\{Q\} \, . \tag{19}$$

In Eq. (19)

$$\{Q\} = \begin{Bmatrix} \{Q_0\} \\ \{Q_1\} \end{Bmatrix} \tag{20a}$$

$$[A] = \begin{bmatrix} [0] & -[I] \\ (\omega^2[M]-[K_3]) & ([K_2]-[K_2]^T) \end{bmatrix} \quad [B] = \begin{bmatrix} [I] & [0] \\ [0] & [K_1] \end{bmatrix} \tag{20b}$$

and [0] and [I] are null and identity matrices, respectively.

Numerical Results and Discussion

The method outlined above was implemented in a computer program to obtain the eigenvalues of problems I and II (Eqs. (14) and (15)). A general purpose routine was constructed to deal with any number of cylindrically anisotropic laminae.

The correctness of the program was verified by comparing results with available exact solutions. Excellent agreement was found to exist for the results of this approach when compared with those of Armenakas (1971) for a laminated isotropic hollow cylinder and with Armenakas and Reitz (1973) for an orthotropic hollow cylinder.

For illustration purposes we present here results for four cases:

1. A graphite/epoxy tube with 0° fibers (Fig. 1).
2. A graphite/epoxy tube with a [15°/-15°/15°/-15°] ply lay-up (Fig. 2).
3. An aluminum cladded graphite/epoxy tube with 0° fibers (Fig. 3).
4. An aluminum cladded graphite/epoxy tube with a [15°/-15°/15°/-15°] ply lay-up (Fig. 4).

In these examples dispersive spectra are plotted for dimensionless frequency Ω and dimensionless phase velocity c. The phase velocity is related to the frequency and wavenumber in dimensionless form by

$$c = \frac{\Omega}{\xi}$$

If γ is a root, then its complex conjugate is also a root. Only the phase velocities of propagating modes with $\zeta>0$ and $\eta=0$ are shown below. The numerical convergence of the method can be verified by increasing the number of sublayers used. It was found that approximately twenty sublayers were sufficient.

In these examples, the elastic properties of graphite/epoxy used are:

$$E_L = 20 \times 10^6 \text{ psi.} \quad G_{TT} = .5 \times 10^6 \text{ psi.} \quad \nu_{LT} = .25$$
$$E_T = 10^6 \text{ psi.} \quad G_{LT} = .6 \times 10^6 \text{ psi.}$$

where L signifies the direction parallel to the fibers and T is the transverse direction. The elastic moduli are computed according to the relations given by Vinson and Sierakowski (1986). Properties of aluminum are:

$$E = 10.3 \times 10^6 \text{ psi.} \qquad G = 3.8 \times 10^6 \text{ psi.}$$

and the material is assumed to be isotropic. The elastic moduli and density of the aluminum and graphite/epoxy are normalized by E_L and ρ of the graphite/epoxy. The density ratio of aluminum to graphite/epoxy was taken as 1.62. The normalized frequency, wavenumber, and phase velocity are given by

$$\Omega = \frac{\omega H}{V_L}, \quad \xi = \frac{\zeta H}{\pi}, \quad c = \frac{\omega\pi}{\zeta V_L}$$

where

$$V_L = \left[\frac{E_L}{\rho}\right]^{\frac{1}{2}}_{\text{graphite/epoxy}}$$

The ratio of the total thickness of the tube to the radius of the middle surface, H/R, was 1/3. In the examples with cladded tubes, a 1/20 ratio of cladding layer thickness to total tube thickness was used and the tube was cladded on both inner and outer walls.

In comparing the results for graphite/epoxy with 0° fibers and [15°/-15°/15°/-15°] ply lay-ups, shown in Figs. 1 and 2, respectively, the effect of the displacement mode coupling due to off-axis fiber orientation can be seen. In Fig. 1 the dashed lines are torsional modes and the solid lines are longitudinal modes. Graphite/epoxy with 0° fibers is orthotropic and the torsional and longitudinal displacement modes are therefore uncoupled. This allows for the crossing of the dispersive spectra mode lines. Referring to Fig. 2 for a [15°/-15°/15°/-15°] ply lay-up graphite/epoxy, because of the deviation of the fiber directions in a lamina from the cylinder axis, the longitudinal and torsional displacements are coupled. The coupling is indicated by the mode lines not crossing, since modes of the same family may not have the same phase velocity for a given frequency. The same conclusion may be drawn in comparing the orthotropic (0° fiber) versus anisotropic (off-axis fiber) materials in Figs. 3 and 4.

The effect of the cladding on the dispersive behavior of the graphite/epoxy can be seen in comparing results shown in Figs. 1 and 3 (uncladded and cladded 0°

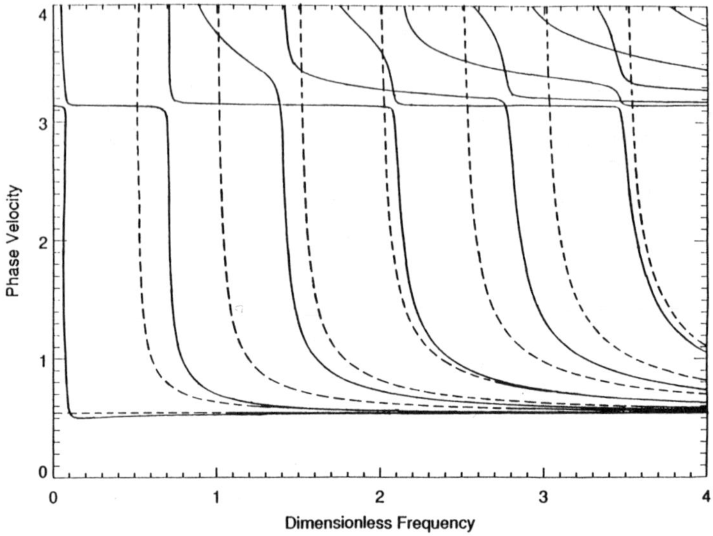

Figure 1. Dispersion curves (phase velocity vs. frequency) for a uniaxial graphite/epoxy tube.

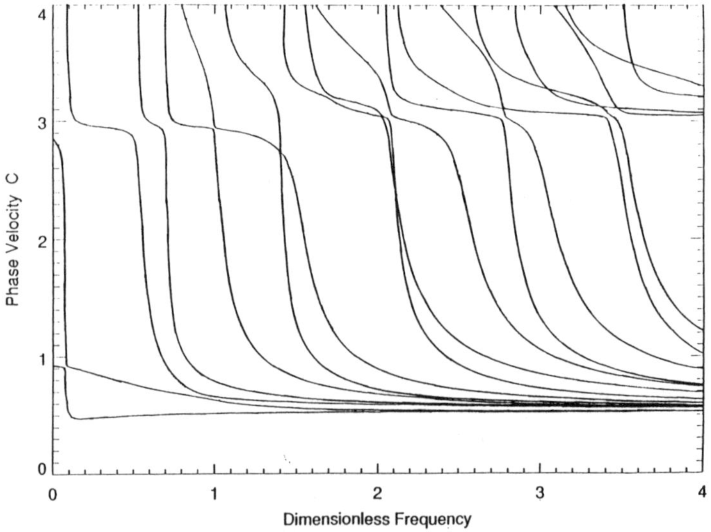

Figure 2. Dispersion curves for a 15°/-15°/15°/-15° graphite/epoxy tube.

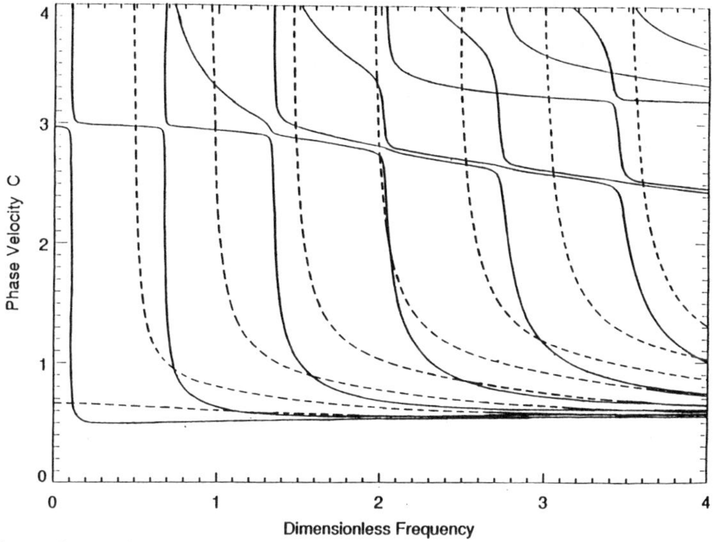

Figure 3. Dispersion curves for an aluminum cladded uniaxial graphite/epoxy tube.

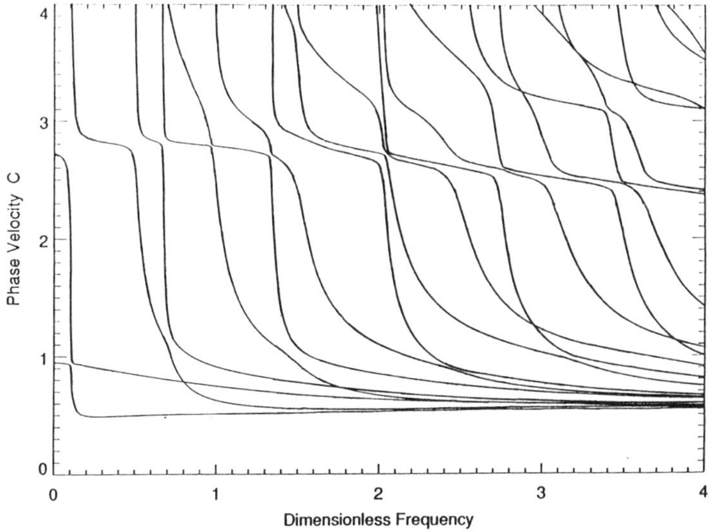

Figure 4. Dispersion curves for an aluminum cladded 15°/-15°/15°/-15° graphite/epoxy tube.

fiber orientation). The plateaus that occur in the longitudinal modes have lower phase velocities and have some additional slope for the cladded graphite/epoxy. This may be seen as indicative of being intermediate between isotropic and fiber reinforced materials. In the purely fiber reinforced material as in Fig. 1, there exists a pronounced non-sloping plateau region in the dispersion spectra curves of the longitudinal modes. In a purely isotropic material, longitudinal and torsional modes will have very similar dispersion spectra curves with no plateau region, as in the torsional modes of Fig. 1.

Referring to Figs. 2 and 4, this observation regarding a plateau region in the longitudinal modes of fiber reinforced materials offers additional evidence of the coupling of displacement modes for off-axis fibers. As can be seen in Figs. 2 and 4, all modes have sloping plateau regions, indicating lessening of anisotropy effects.

Conclusion

A displacement-based stiffness method using through-thickness interpolation functions is used to study guided waves in a laminated composite tube. Such tubes have important applications in space structures due to their high strength-to-mass ratio and their ability to be manufactured in configurations with zero coefficient of thermal expansion.

Results are presented for uniaxial and off-axis fiber orientations with and without aluminum cladding. The coupling of longitudinal and torsional modes for composite tubes made of laminae with off-axis fiber orientations is evidenced in dispersion spectra by the nonintersecting mode curves and in the occurrence of sloping plateaus of non-constant phase velocities for all modes. Conversely, the uncoupled behavior of modes in composites with uniaxial fibers resulted in the intersection of torsional and longitudinal modes, and the occurrence of plateaus of constant phase velocity only in longitudinal modes. The addition of cladding shifted the occurrence of plateaus to lower phase velocities and to non-zero slopes which may be interpreted as dispersive behavior intermediate between fiber reinforced and isotropic materials.

The results derived here are also useful for solving boundary-value problems involving tubes of finite lengths in order to satisfy end conditions. For instance, in a dynamic problem when a steady-state incident wave impinges on the free end of a tube, a reflected wave alone will not be sufficient to satisfy stress-free edge conditions. It will be necessary to include waves with real, imaginary and complex wave numbers obtained through EVP-I. In a subsequent paper we wish to investigate the case of finite tubes.

References

Armenakas, A.E. (1965). "Torsional waves in composite rods". *J. Acoust. Soc. Am., 38*, 439-446.

Armenakas, A.E. (1971). "Propagation of harmonic waves in composite circular cylindrical shells. Part II: Numerical analysis." *AIAA Journal, 9*, 599-605.

Armenakas, A.E. and Reitz, E.S. (1973). "Propagation of harmonic waves in orthotropic circular cylindrical shells." *J. Appl. Mech., 40*, 168-174.

Datta, S.K., Shah, A.H., Bratton, R.L., and Chakraborty, T. (1988). "Wave propagation in laminated composite plates." *J. Acoust. Soc. Am., 83*, 2020-2026.

Dong, S.B. and Huang, K.H. (1985). "Edge vibrations in laminated composite plates." *J. Appl. Mech., 52*, 433-438.

Dorsey, J.T. and Mikulus, M. (1989). "Preliminary design of a large tetrahedral truss/hexagonal heatshield panel aerobrake." NASA TM 101612, NASA Langley Research Center, Hampton, VA.

Gazis, D.C. (1959). "Three-dimensional investigation of the propagation of waves in hollow circular cylinders - I. Analytical foundation, II. Numerical results." *J. Acoust. Soc. Am., 31*, 568-578.

Huang, K.H. and Dong, S.B. (1984). "Propagating waves and edge vibrations in anisotropic composite cylinders." *J. Sound and Vibr., 96*, 363-379.

Mindlin, R.D. and McNiven, H.D. (1960). "Axially symmetric waves in elastic rods." *J. Appl. Mech., 27*, 145-151.

McNiven, H.D. (1961). "Extensional waves in a semi-finite elastic rod." *J. Acoust. Soc. Am., 33*, 23-27.

McNiven, H.D. and Perry, D.C. (1962). "Axially symmetric waves in finite elastic rods." *J. Acoust. Soc. Am., 34, 433-437.*

McNiven, H.D., Sackman, J.L. and Shah, A.H. (1963). "Dispersion of axially symmetric waves in composite, elastic rods." *J. Acoust. Soc. Am., 35*, 1602-1609.

McNiven, H.D. and Shah, A.H. (1967). "The influence of the end mode on the resonant frequencies of finite, hollow elastic rods." *J. Sound and Vibr., 6*, 8-19.

Mirsky, I. (1964). "Axisymmetric vibrations of orthotropic cylinders." *J. Acoust. Soc. Am., 36*, 2106-2122.

Onoe, M., McNiven, H.D. and Mindlin, R.D. (1962). "Dispersion of axially symmetric waves in elastic rods." *J. Appl. Mech., 29*, 729-734.

Pochhammer, L. (1876). "Ueber die fortpfanzungsgeshwindigkeiten schwingungen in einem unbregrenzien isotropen kriescylinder." *Zeitschrift für Mathematik, 81*, 324-336.

Tasi, J. (1971). "Effect of heterogeneity on the axisymmetric vibrations of cylindrical shells." *J. Sound and Vibr., 14*, 325-338.

Tasi, J. and Roy, B.N. (1971). "Axisymmetric vibrations of finite heterogeneous cylindrical shells." *J. Sound and Vibr., 17*, 83-94.

Vinson, J.R. and Sierakowski, R.L. (1986). *The Behavior of Structures Composed of Composite Materials*, Martinus Nijhoff Publishers, Dordrecht, Netherlands, p. 46.

An Introduction to Pressurized Volumes in Space

Michael Roberts*

ABSTRACT

Space habitation designers come from diverse backgrounds. As it is human nature to emphasize that which is familiar, habitat design is strongly influenced by the designer's professional background. The purpose of this paper is to provide a uniform approach to space habitation design, by offering an even-handed treatment of the many design issues involved. The nomenclature outlined in this paper should also prove useful as a basis for discussing the merits and deficiencies of particular designs.

INTRODUCTION

Pressurized volume beyond the confines of the Earth is a valuable resource. In order for humans to live and work in space, their space habitat must provide them with a pressurized, thermally-controlled, radiation-free environment. There must be enough space to accomplish mission objectives, and to maintain good mental and emotional health.

The space habitation designer must determine how best to provide this resource: what the best configuration is; how much radiation shielding is needed; at what level the internal pressure should be maintained. The answers are not always clear, often because the designer is not asking the proper questions. Space habitation design is an optimization problem with many variables. The "best" habitat for a space station in 1995 may not be the "best" habitat for a lunar base in 2005.

The purpose of this paper is to guide the space habitation designer through the design process. By understanding the entire problem from the beginning, the

*Engineer, Advanced Programs Office, NASA/Johnson Space Center, Mail Code ED2, Houston TX 77058.

designer can work more efficiently, and avoid wasting time on a detailed design that is fatally flawed when viewed from a broader perspective.

DATA GATHERING

The first step in the design process is to gather information. Before starting in on the design, the habitat designer should have in hand a working description of the local environment, and a general statement of the habitat's purpose and requirements.

This step involves obtaining data on such topics as the radiation environment, the thermal profile of the site, and the meteoroid flux. It also involves answering the question, "What is the habitat there for?" The answer need not be a minute-by-minute mission timeline: a general statement of the purpose of the facility, with some examples of activities, is sufficient to get started.

DECIDING ON THE BASIC CONFIGURATION

In any optimization problem, one must define the parameters to be optimized. Once the requirements have been decided upon (e.g., "The habitat will support 12 people for 70-day tours on the surface of the Moon over a period of 20 years..."), the designer need only optimize two parameters: cost and safety. The habitat must meet the requirements while minimizing the overall cost and assuring the maximum practical safety. Unfortunately, the relationship of cost and safety to the habitat design may not be readily apparent. These two parameters must be broken down into their component parts, so that the relationships can be more easily seen.

The suitability of a space habitat can be measured by five parameters:

1) Failure Resilience

2) Reliability

3) Habitability

4) Transportability

5) Constructibility

Each of these principles can be applied in the design of a space habitat; each relates in some way to the cost or safety of the habitat. As defined below, each of these parameters suggests some ideal configuration. Looking at each criterion individually, one can imagine a concept that best satisfies that criterion. The conceptualizations can then be compared to identify those criteria that are in conflict.

1) Failure Resilience. The habitat should "fail safe." Once a failure occurs, quick action should be possible to limit the impact of the failure.

One failure of particular concern in a space habitat is a catastrophic failure of the pressure vessel, so that a large amount of air escapes in a short period of time. This failure would be difficult to repair immediately, due to the unpredictable nature of the hole, the rapid loss of pressure, and the high-speed airflow in the vicinity. The best way to deal with such a blow-out is to seal off the affected area, allow the sealed volume to depressurize, and attempt a repair later. Redundant habitable volume is required for this procedure.

From the standpoint of resilience, then, a highly modular configuration would be optimum, so that a ruptured module could be isolated with a minimal effect on the rest of the habitat.

The connection to safety is obvious here. While the added safety may cost more, it acts as insurance against a mishap, and protects the hardware as well as the people.

2) Reliability. The failure rate should be low, and the chance of a catastrophic failure should be remote.

The reliability criterion is different from failure resilience. Designing for resilience means designing the structure to withstand failure; designing for high reliability means designing so that failure doesn't occur.

Absent the threat of puncture by meteoroid (meteoroid protection is discussed below), the habitat wall would most likely fail due to some defect, brought to a critical point by some event, perhaps very minor. This hypothesis is supported by the body of data on pressure vessel failures. These data suggest that most such failures are the result of defects in seams and welds (Nichols, 1971).

Seams and welds occur wherever discontinuous pieces of material are joined. The total length of such joints on a pressure vessel is proportional to the surface area (because many discrete pieces are required to form the surface) and to the number of attach points (such as hatches, windows, and other fittings). By minimizing the sum length of the seams and welds, the pressure vessel should be made more reliable.

A habitat optimized for reliability would have the minimum surface area for its contained volume: in other words, it would be spherical. It would also have as few hatches and windows as possible.

Safety is served whenever the risk of failure is reduced. The inherent simplicity of a reliable design may well reduce the immediate costs as well, by reducing the amount of quality control required. Even if, for some reason, the simpler design was the more expensive, the insurance argument noted above would also apply here.

3) Habitability. The habitat should provide sufficient usable volume in an open, flexible environment.

Habitability is a measure of the habitat's usefulness as a place to live and work. The first requirement for habitability, aside from providing shelter, air, water and food, is that there be sufficient free volume for the crew to do their work. Data from past programs and Earth analogs is drawn on heavily for this stage of the design. Several authors have made recommendations regarding the quantity of volume per person that is required for various missions. Perhaps the best known of these is the Celentano Curve, which maintains that the amount of volume required per person in a confined environment is proportional to the number of people and the length of their stay (Fraser, 1969).

Because of the somewhat unpredictable nature of the crew and the work to be done, the best way to ensure the continued usefulness of the habitat is to design it so that it may be reconfigured to meet new needs as they arise. Besides making the crew's work more efficient, it will benefit them psychologically by giving them control over their immediate environment. A tour of duty in space would be highly motivating for many scientists and researchers, no matter how poor the living conditions might be expected to be. But once they have been there for a few days, minor inconveniences may become major irritations, to the point of affecting performance (Stuster, 1986). Providing flexibility in the design, so that the crew can implement their own solutions to any problems of space or interior arrangement that arise, should be a primary goal in engineering a space habitat.

The concept of open volume is a key element in this philosophy of flexible design. The walls of the pressure vessel form the ultimate boundary that the crew must work within. They can arrange rooms inside, but they cannot rearrange the walls of the pressure vessel. To optimize for habitability, the pressure vessel should be unconfining: opposite sides of the habitat should be as far as from each other as possible. There should be enough space to rearrange the interior, and to expand the capabilities of the base without major exterior construction.

The impact of habitability on cost and safety is debatable. It is clear that there is a relationship between habitability and human performance (Stuster, 1986). That relationship has not yet been quantified. It is impossible to say that X number of mishaps will be prevented if Y cubic meters of volume are provided per crewperson. However, there is a link, and if there is to be an error, it should be on the side of safety rather than on the side of cost-effectiveness.

4) Transportability. The habitat should be partitionable into discrete payloads, and provide the most usable volume per unit mass.

The habitat must be compatible with the transportation system. It is unrealistic to expect that a mammoth launch vehicle will be designed solely to accommodate the large habitat. Vehicle designs should be optimized in view of their lifetime manifest, and not to carry rare, anomolously large payloads. Therefore, the habitat must be designed in tranportable pieces.

Here there are two choices. The pressure vessel can be partitioned into multiple modules, and thereby violate several of the other design criteria. Alternatively, the pressure vessel may be collapsible, with the interior fittings transported separately for installation on-site. Both approaches represent a necessary compromise to the constructibility criteria.

The mass of a pressure vessel depends upon its geometry and the interior pressure. A spherical pressure vessel has the highest volume/mass ratio of any geometry.

Transportability bears little relationship to safety, because theoretically all of the transportation could be unmanned. However, it does have a great bearing on cost. It is very expensive to transport mass into space, and it will probably remain expensive for a long time. The habitat designer must be keenly aware of the mass and volume of the facility he or she intends to send into space.

5) Constructibility. The habitat should require a minimum of time, manpower, and support equipment to construct or expand it.

The tasks required for these activities should be simple, few, and quick. Extra-vehicular activity should be minimized where possible, as should reliance on complex or bulky equipment. A habitat optimized for constructibility would be a single, self-contained unit.

The construction phase exposes the crewmembers to high risk. There may be EVA requirements, and while the habitat is being constructed, there may be reduced failure resilience due to lack of redundancy. Safety is enhanced by making this period as short as possible, and by reducing the requirement for high-risk activity such as EVA. Cost may also be reduced, by lowering the cost of training, and by bringing the habitat into the operational phase as soon as possible to start recouping the investment.

Three of the five criteria, then, favor a single large pressure vessel. One criterion, failure resilience, suggests a multi-module approach. The tranportability criterion is inconclusive: it may be satisfied by either a modular design, or a collapsible habitat that could be folded into a transportable package. The habitat engineer must determine the relative importance of these parameters for his or her particular application, and then make the appropriate compromises between conflicting criteria.

DETAILED DESIGN

Once the basic configuration is chosen, more specific design issues must be addressed. The design is influenced architecturally by the activities carried on within the habitat. It is also influenced by the physical forces imposed on the structure, and by the environment in which it operates. Finally, the transportation system to be used remains a key design factor. An examination of these design considerations follows.

Architectural Design Considerations:

As used here, the term architectural refers to those factors that arise from the interaction of humans with the habitat. The physical and psychological needs of humans in a confined environment must be balanced against the physical limitations of a pneumatic structure.

Function. What types of activities will be pursued in the habitat? If the primary function is to support EVA construction tasks, then the airlock and EVA equipment will figure prominently in the design. If the primary purpose is to perform long-term scientific research, then the laboratory will be the centerpiece. Of course, regardless of the habitat's purpose, there are some functions which must occur: sleeping, eating, and excretion. These must be taken into account as the habitat takes form.

Human factors. Those design features which enhance human performance, but which are not needed to establish basic habitability. These features include the use of esthetically pleasing colors, acoustic insulation, and the relative placement of facilities (placing noisy equipment away from crew quarters, for example).

Structural Design Considerations:

Structural design issues arise from the physical forces acting on the structure. Gravity loads must be taken into account, as well as loads applied due to accelerations during transport (e.g., launch loads). The internal pressure produces membrane stress and leakage.

Gravity. Whether or not the habitat sits on a planetary surface will strongly influence the design. Supporting structure, ladders and flooring may be needed in a planetary habitat where they were not required in orbit.

Transportation loads. The accelerations experienced during transport can greatly exceed gravitational loads. For example, the Space Shuttle imposes a maximum steady acceleration of 3 times Earth's gravity on its payload, and transient loads (vibration) may be much higher. (Shuttle Operational Data Book, 1988.)

Membrane stress. A pneumatic structure is stressed primarily by its internal pressure, overshadowing external forces such as gravity. The internal pressure creates stress in the wall of the structure called the membrane stress. Because the wall thickness in a pneumatic structure is much less than the local radius of curvature and therefore may be neglected, it is customary to express the membrane stress in N/m rather than N/m^2. The membrane stress depends on both the internal pressure and the local radius of curvature. In a sphere, the membrane stress is given by:

stress = (pressure * radius)/2

There are similar equations for the stress in a cylinder or torus. The membrane strength is the principle qualifying characteristic for the material to be used in the habitable pressure vessel.

The internal pressure to be maintained in the habitat depends on many factors, but will range between 3.6 psi (the minimum pressure at which a human can function, with 100% oxygen) and 14.7 psi (standard sea level pressure). (NASA-STD-3000, 1989.)

Leakage. Aside from resisting the membrane stress due to the internal pressure, the structure must also be sealed against leakage. The wall material must be impermeable to air, and care must be taken in the design of hardware interfaces (such as hatches or windows) to limit potential leak paths.

Environmental Design Considerations:

When designing a space habitat, the effects of the local environment over a long period of time must be considered. Radiation, temperature extremes, vacuum, and meteoroids all act to erode the integrity of the structure over time.

Radiation. The radiation hazard comes from three sources of varying character: galactic cosmic radiation, "background" radiation that permeates space at roughly constant levels; solar energetic particle events that produce extremely high levels of radiation for short periods of time; and radiation emanating from space nuclear power sources. Radiation can affect both crew and equipment, including the materials used in the habitat.

Temperature. Temperatures in space are generally more extreme than on Earth. In a vacuum, thermal radiation is the only means of heat transfer, and surfaces heat quickly in full sunlight and cool rapidly when viewing deep space. On a planetary surface, an atmosphere and even the mass of the planet itself can moderate the temperature. The habitat engineer should have a time history of the thermal environment in which the habitat is to operate.

Vacuum. Some materials give off gases in a vacuum, possibly affecting their physical properties over time. This must be taken into account when materials are selected for the habitat.

Meteoroids. Meteoroids and debris exist throughout the solar system, posing a threat to the pressurized habitat. The habitat engineer must obtain a good statistical model of the likelihood of an impact in the vicinity of the habitat, in order to specify the degree and type of shielding required. Fortunately, the radiation shielding and the debris shielding are generally complimentary: radiation shielding contributes to the meteoroid shielding and vice-versa.

In addition, some or all of these environmental factors may produce different effects when they act in combination than they do when present singly under laboratory conditions. These combined effects must be considered when materials are being selected and tested.

CONCLUSION

The habitat designer should follow a three step process in the preliminary design of a space habitat.

Step 1: Define the environment in which the habitat is to operate, and the functions to be performed at the habitat.

Step 2: Define the basic configuration of the habitat based on the mission requirements and the five parameters of failure resilience, reliability, habitability, transport-ability, and constructibility.

The designer should always keep in mind the true optimization criteria of cost and safety. The environmental data and associated risk assessment, together with the mission requirements and their associated risk assessment, should provide a guide to any compromises that must be made between cost and safety.

Step 3: Refine the basic concept by looking at the functional requirements and the environment.

Once the basic configuration is selected, the habitat designer can turn to the details of the design: how many hatches are required, and where are they to be placed? What size should the crew quarters be? Which type of thermal control system is best?

In the process of refinement, the designer may discover that the basic configuration is unworkable. Design is an iterative process. However, by following the rational procedure described above, the space habitat designer can minimize the iterations required by making the right decisions early in the process.

APPENDIX. References

Fraser, T.M. (1969). "Confinement and Free Volume Requirements." Space Life Sciences, 1(4), 428-466.

NASA-STD-3000, Man-System Integration Standards, Volume I, Revision A. (1989). National Aeronautics and Space Administration.

Nichols, R.W., ed. (1971). Pressure Vessel Engineering Technology. American Elsevier Publishing Co., New York, N.Y.

Shuttle Operational Data Book, JSC-08934, Volume 1. (1988). National Aeronautics and Space Administration.

Stuster, J.W. (1986). Space Station Habitability Recommendations Based on a Systematic Comparative Analysis of Analogous Conditions, NASA CR-3943. NASA Scientific and Technical Information Branch, Washington, D.C.

Effect of Gravity on Dynamic Behavior of Beams

Ralph J. Dornsife[1]

Introduction

Large space structures will be needed to provide structural framework for maintaining relative positions of individual components of systems in low earth orbit. Particularly because of the high cost of transporting payload into low earth orbit, large space structures must first be dynamically analyzed and tested on earth. The resulting ground based test behavior must be correlated to behavior in a microgravity environment. Past laboratory testing has, largely, examined dynamic behavior only in directions perpendicular to the gravity vector. This study examines the correlation between earth based natural frequencies in vibration directions parallel to the gravity vector and the theoretical natural frequencies encountered in a microgravity environment. The structure considered is a prismatic beam.

Beam on an Elastic Foundation

The natural frequencies of a structure depend on the physical properties of its components and the way in which it is supported. For example, a cantilever beam will exhibit different natural frequencies than will the same beam simply supported at both ends. Likewise, an unsupported beam in space microgravity will have different natural frequencies than the same beam resting on an elastic foundation within a gravity field. However, the unsupported beam cannot be tested as such on earth because some type of support constraint is required to counteract the gravity imposed dead load.

---

[1]Structural Engineer, U.S. Army Corps of Engineers, Construction Engineering Research Laboratory, P.O. Box 4005, Champaign, IL 61824-4005

Thus, first consider a prismatic beam of length L, resting on a uniformly distributed elastic foundation (Meirovitch, 1986). The governing differential equation of free vibration is

$$EI\frac{\partial^4 w(x,t)}{\partial x^4} + \kappa w(x,t) + m\frac{\partial^2 w(x,t)}{\partial t^2} = 0 \qquad (1)$$

where

E	=	modulus of elasticity of the beam
I	=	moment of inertia of beam about bending axis
w(x,t)	=	deflection of beam with respect to static configuration
κ	=	foundation stiffness per unit length
m	=	beam mass per unit length
t	=	variable time
x	=	distance measured along beam from far left end toward right end

Assuming a solution of the form

$$w(x,t) = w(x) \sin \omega t \qquad (2)$$

where

ω = circular frequency of vibration

and applying boundary conditions at the ends yields an eigenvalue problem. The roots are the circular natural frequencies. These are

$$\omega_i = \left(\frac{16EI\alpha_i^4}{L^4 m} + \frac{\kappa}{m}\right)^{1/2} \qquad (3)$$

where

$\alpha_1 = 0$
$\alpha_2 = 0$
$\alpha_3 = 2.36502035$
$\alpha_4 = 3.9266023$
$\alpha_5 = 5.49778715$
$\alpha_6 = 7.0685828$
$\alpha_i \approx \frac{(2i-3)\pi}{4} \quad i = 7,8,9,...$

In summary, the natural frequencies of a beam resting on an elastic foundation depend upon the stiffness of the foundation relative to the stiffness of the beam and the mass of that beam per unit length. The beam can experience free vibrations only at circular frequencies which are greater than $[\kappa/m]^{1/2}$.

Unsupported Beam In Microgravity

The theoretical solution for the unsupported beam in a microgravity environment is actually a special case of the beam on elastic foundation, where the foundation stiffness, κ, is zero. Now,

$$\omega_i = \alpha_i^2 \left(\frac{2}{L}\right)^2 \left(\frac{EI}{m}\right)^{1/2} \tag{4}$$

where α_i is as previously defined. The lowest two circular natural frequencies are obviously zero. Thus, the square of each circular natural frequency of the beam on an elastic foundation on earth will be κ/m greater than the square of each circular natural frequency of the same beam in a space microgravity environment. Or,

$$\omega_{sp\text{-}i} = \left[\omega^2_{ef\text{-}i} - \frac{\kappa}{m}\right]^{1/2} \tag{5}$$

where

$\omega_{sp\text{-}i}$ = i th circular natural frequency of the beam in a space microgravity environment

$\omega_{ef\text{-}i}$ = i th circular natural frequency of the beam on an elastic foundation in gravity

Beam Supported by Discrete Springs

Experimentally, it would be rather difficult to support such a prismatic beam on a continuous elastic foundation. A much more practical situation occurs when the beam is supported by a number of discrete springs equally spaced. This problem is considerably more complicated because the effect of each spring must be considered. Instead of one governing differential equation, separate equations describe the dynamic behavior between each pair of springs.

The piecewise governing differential equation of free vibration for a prismatic beam, including the spring masses, can be expressed as

$$EI\frac{\partial^4 w(x,t)}{\partial x^4} + m\frac{\partial^2 w(x,t)}{\partial t^2} + \frac{1}{3}\mu^*\frac{\partial^2 w(0,t)}{\partial t^2}\delta(x) + \frac{1}{3}\mu^*\frac{\partial^2 w(L,t)}{\partial t^2}\delta(x-L)$$

$$+\frac{1}{3}\mu\sum_{j=1}^{n-1}\frac{\partial^2 w(x,t)}{\partial t^2}\delta(x-x_j) = -k^*w(0,t)\delta(x)$$

$$-k^*w(L,t)\delta(x-L) - k\sum_{j=1}^{n-1}w(x,t)\delta(x-x_j) \tag{6}$$

where

μ = mass of each interior spring
μ^* = mass of each exterior spring
k = stiffness of each interior spring
k* = stiffness of each exterior spring
n = number of equal length segments into which beam is divided
j = 0 corresponds to left end of beam
j = n corresponds to right end of beam

$$\delta(x - x_j) = 0.0 \text{ for } x - x_j \neq 0 \tag{7a}$$

$$\delta(x - x_j) = 1.0 \text{ for } x - x_j = 0 \tag{7b}$$

The spring mass terms are the inertial effect of the springs due to their axial accelerations. The 1/3 factor is a well known approximation based upon the solution of the wave equation for axial vibration of a continuous bar fixed at one end and having a tip mass attached to the free end (Roark and Young, 1975).

Again, assuming a solution of the form

$$w(x,t) = w(x)\sin \omega t \tag{2}$$

and applying boundary conditions at each discrete spring and at the ends yields a rather complicated eigenvalue problem. Because of the complexity of the problem and because the mathematics is very well suited for automatic computation, a FORTRAN 77 program was developed to solve the problem. This code searches for intervals of ω where roots exist. Interval halving is then used to close in on a root. The beam must be divided into an odd number of segments, "n". The user has the option of including the inertial effect of the spring masses. After evaluating each circular natural frequency, the corresponding mode shape is calculated for the left half span by evaluating the relative deflection at each spring.

Example

Consider a 6 m long aluminum beam having a 2.5 cm x 2.5 cm cross section. The bending stiffness, mass per unit length, and foundation stiffness required for a dead load deflection of 15 cm are calculated as follows:

$$EI = \left(\frac{6.9 \times 10^{10} N}{m^2}\right)\left(\frac{1}{12}\right)(.025m)^4 = 2250 \text{ N} \cdot \text{m}^2$$

$$m = \left(\frac{2690 \text{ kg}}{m^3}\right)(.025m)^2 = 1.681 \frac{\text{kg}}{\text{m}}$$

$$\kappa = \frac{(\frac{1.681\ \text{kg}}{\text{m}})(\frac{9.80665\ \text{m}}{\text{s}^2})}{(.15\ \text{m})} = \frac{109.9\ \text{N}}{\text{m}^2}$$

The unsupported beam in a space microgravity environment has circular natural frequencies calculated from Equation 4.

$$w_{sp\text{-}i} = \alpha_i^2 \left(\frac{2}{L}\right)^2 \left[\frac{EI}{m}\right]^{1/2} \qquad \text{yields}$$

$$w_{sp\text{-}1} = 0.000\ \text{sec}^{-1} \qquad w_{sp\text{-}2} = 0.000\ \text{sec}^{-1}$$

$$w_{sp\text{-}3} = 22.7371\ \text{sec}^{-1} \qquad w_{sp\text{-}4} = 62.6756\ \text{sec}^{-1}\ \text{, etc.}$$

The same beam resting on a continuous elastic foundation on earth has circular natural frequencies calculated by rearranging Equation 5.

$$w_{ef\text{-}i} = \left[w_{sp\text{-}i}^2 + \frac{\kappa}{m}\right]^{1/2} = \left[\alpha_i^4 \left(\frac{2}{L}\right)^4 \frac{EI}{m} + \frac{\kappa}{m}\right]^{1/2} \quad \text{yields}$$

$$w_{ef\text{-}1} = 8.0857\ \text{sec}^{-1} \qquad w_{ef\text{-}2} = 8.0857\ \text{sec}^{-1}$$

$$w_{ef\text{-}3} = 24.1320\ \text{sec}^{-1} \qquad w_{ef\text{-}4} = 63.1950\ \text{sec}^{-1}\ \text{, etc.}$$

Consider the same beam on six discrete springs ($n = 5$). In order to maintain the same dead load deflection of 15 cm, each interior spring must have a stiffness of (109.9 N/m^2) (6 m/5) = 131.88 N/m. Each exterior spring has half this stiffness. The resulting circular natural frequencies evaluated by the program are listed in Table 1. The unsupported microgravity case, elastic foundation case, and twelve discrete spring case ($n = 11$) are also included in the table for comparison. The program was compiled and executed on a Harris 500 mini-computer.

Several observations can be made from the tabulated results. First, natural frequencies for the discrete spring cases are very close to those for the continuous elastic foundation case. The two zero frequency rigid body modes for the beam in a space microgravity environment correspond to two very closely spaced non-zero frequency rigid body modes for the supported beam. The more discrete springs, the closer together these two frequencies are, until they become equal for the beam on elastic foundation. Beyond the second mode, natural frequencies for the supported beam become increasingly closer to the natural frequencies of the beam in a space microgravity environment.

Conclusions

Because of the high cost of transporting payload into low earth orbit, techniques must be developed to predict from ground tests the dynamic behavior of large space structures in microgravity. For simplicity, a prismatic beam was chosen for examination. The dynamical vibration characteristics of a given prismatic beam depend upon the support conditions imposed. The natural frequencies for the beam supported on a uniformly distributed elastic foundation can be easily evaluated from the dynamical differential equation of motion. The beam in a microgravity environment is simply a specialized case of the beam on an elastic foundation where the foundation stiffness is zero. The foundation supported beam will always have higher natural frequencies than the one unsupported in a microgravity environment and cannot vibrate at circular frequencies lower than $[\kappa/m]^{1/2}$.

For experimental purposes, a beam supported by discrete springs at equally spaced intervals is a much more practical case for study. It is also a much more difficult problem computationally. Separate equations describe the dynamic behavior between each pair of springs. Fortunately, the entire procedure of evaluating natural frequencies is very well suited for automatic computation. A FORTRAN computer code was developed to evaluate the circular natural frequencies and mode shapes.

An example problem demonstrates the variation in natural frequencies of a prismatic beam having different support conditions imposed. Section properties were selected so as to be representative of the flexibilities most likely to be encountered in large space structures. The resulting natural frequencies for the beam supported on 1) an elastic foundation, 2) six discrete springs and 3) twelve discrete springs were compared to those of the same beam unsupported in a microgravity environment. The supported beam always has higher natural frequencies than the unsupported beam. The two zero frequency rigid body modes of the unsupported beam correspond to two closely spaced nonzero frequency rigid body modes for the supported beam. The variation in natural frequencies between the supported and unsupported beam is much less pronounced for the higher frequency modes.

ω_i	Unsupported in Microgravity	Elastic Foundation	6 Discrete Springs (n = 5)	12 Discrete Springs (n = 11)
ω_1	0.0000	8.0857	8.0823	8.0855
ω_2	0.0000	8.0857	8.3968	8.1519
ω_3	22.7371	24.1320	24.4646	24.2012
ω_4	62.6756	63.1950	63.4146	63.2398
ω_5	122.8686	123.1343	123.3029	123.1673
ω_6	203.1092	203.2701	203.4568	203.2953
ω_7	303.4101	303.5178	303.5146	303.5386
ω_8	423.7711	423.8483	423.8836	423.8662

Table 1 Lowest eight circular natural frequencies for the prismatic beam with various support conditions

APPENDIX I. References

Meirovitch, L. (1986). Elements of Vibration Analysis. McGraw-Hill, Inc., New York, NY, pp. 220-227.

Roark, R.J. and Young, W.C. (1975). Formulas for Stress and Strain. McGraw-Hill, Inc., New York, NY, p. 577.

Subsection 3.7: MISSION TO PLANET EARTH

GLOBAL PROBLEMS: IMPLEMENTING MISSION TO PLANET EARTH

Stewart W. Johnson,[1] Member, ASCE
John P. Wetzel[2]

Destiny is not a matter of chance - it is a matter of choice. It is not a thing to be waited for - it is a thing to be achieved.
William Jennings Bryan

ABSTRACT

Serious global environmental problems are facing our generation on Earth. The decrease in ozone concentrations and the greenhouse effect have implications for all life forms on the planet. Computer modeling of the global environment is important in understanding the difficult problems we face, but there are many uncertainties with predictions made using these models. Mission to Planet Earth is an important international effort to understand global problems of the environment. Eos and other satellite systems are being planned to provide the information so essential to the understanding of our Earth's environment as an integrated planetary system. Tremendous amounts of data generated by these satellites will require improved data storage, analysis, and retrieval. This paper discusses some of the Mission to Earth concepts and their implementation. Satellite and ground segments and human interfaces pose significant challenges for engineering, construction, and operation of the system.

INTRODUCTION

On July 20, 1989 United States President George Bush spoke on the 20th anniversary of the Apollo 11 Moon Landing:

> ...environmental destruction knows no borders. A major national and international initiative is needed to seek new solutions for ozone depletion, and global warming, and acid rain. And this initiative -- "Mission to Planet Earth" -- is a critical part of our space program. And it reminds us of what the astronauts remember as the most stirring sight of all. It wasn't

---

[1]Principal Engineer, Advanced Basing Systems. BDM International, Inc., 1801 Randolph Road S.E., Albuquerque, NM 87106.

[2]Associate Staff Member, Space Systems. BDM International, Inc.

> the Moon or the stars, as I remember. It was the Earth -- tiny, fragile, precious, blue orb -- rising above the arid desert of Tranquility Base.

The mental image of our home planet floating in the void of space conjures up a fragile image of Earth, making us realize how finite our little World actually is.

But what are the problems facing our planet and what are we doing about them? This paper draws upon a summary of critical global problems, methods to investigate and combat these problems, and discusses implications for engineering, construction, and operations.

THE GLOBAL PROBLEM

The Earth is under environmental stress from a continually increasing world population which approaches six billion people and may exceed 10 billion by A.D. 2100. Substantial increases in energy requirements suggest a need for burning vast coal reserves and, hence, liberating carbon dioxide and other products. Land needs motivate clearing more forests, particularly in the tropics. Industries produce chemicals which find their way into the environment (e.g., chlorofluorocarbons) which in turn lead to poorly understood, and potentially adverse, consequences. Figure 1 presents a summary of some of the issues and areas of concern facing our generation. Note the international nature of these issues; no country is isolated from these global problems.

Human Population Growing and Approaching 6 Billion	**Energy Needs and the Potential for Use of Large Coal Reserves**
Rain Forest Destruction at the Rate of Millions of Acres/Year	**Climate Warming and Climate Stability**
Ozone Layer Depletion and Consequences	**Desert Growth in Some Parts of the World**
Atmospheric CO_2 and other Gases Increasing	**The Earth's Environment and the Interdependence of Nations**
Acid Rain	**Ocean Dumping and Pollution**

Figure 1. Global issues and areas of concern.

Ozone

In the winter of 1987 concentrations of ozone over Antarctica dropped about 50% over an area as large as the United States (Hively, 1989). There is now general agreement that chemicals responsible for destroying ozone should be phased out of production as soon as possible. The chemicals most active against ozone are of the chlorofluorocarbon (CFC) family (Levi, 1988) and they are beginning to be phased out. CFCs are very useful and benign on Earth (in air conditioners, refrigerators, aerosol spray cans, plastic foams, and other sources). When liberated on Earth they drift into the stratosphere where solar radiation breaks them down so chlorine is released which ultimately destroys ozone. Ozone in the stratosphere, although present in less than a part per million, protects plants and animals on Earth from harmful ultraviolet radiation.

The NIMBUS-7 satellite launched in 1987 helped track global averages of ozone. Since that time both ground-based and satellite measurements have detected a decline in ozone every spring. It has recently been confirmed that ultraviolet irradiance at the Earth's surface is increased in the Antarctic in the presence of the thinning of the stratospheric ozone layer (Maggs, 1989). Between 1969 and 1986 stratospheric ozone over the Northern Hemisphere decreased by 2%. Over the United States, Europe, and the Soviet Union the average depletion was 3% (Hively, 1989). The amount of harmful ultraviolet radiation on Earth is expected to rise 2% for every 1% decrease in ozone concentration. A 2% decrease in ozone could bring a 4-6% rise in basal and squamous cell carcinoma (Hively, 1989) or three to 15 million more cases among Americans born before the year 2075. Fatalities could be 50,000 to 250,000. People in Argentina, New Zealand, and Chile may be more at risk. Some plants will be damaged and crop yields will be lowered. CFCs not only destroy ozone but are thought to account for 15-20% of the current global warming. CFCs are 10,000 times more efficient (per molecule) in trapping atmospheric heat than CO_2 (Carbon Dioxide).

The NIMBUS-7 satellite, which has a Total Ozone Mapping Spectrometer (TOMS) and Solar Backscatter Ultraviolet (SBUV) sensor, has played a significant role in providing maps of total ozone (Stolarski, 1988). Dobson spectrophotometers at 100 ground stations worldwide measure ozone in the overhead air column. Some have data since 1957. The best global coverage is by satellites; but satellite instruments need calibration which can be difficult. By flying an instrument on the Shuttle, calibration can occur before and after flights. This data can be used to calibrate NIMBUS-7 instruments also.

"Greenhouse Effect"

The Earth, the home planet for the human race, has been made habitable by the presence of certain gases. These gases trap long-wave radiation emitted from the Earth's surface. The result of the entrapment is a global mean temperature (Mitchell, 1989) of 15°C contrasted with -18°C in the absence of an atmosphere. This greenhouse effect results primarily from the presence of water vapor; but substantial contributions come from carbon dioxide and to a lesser degree from the presence of ozone, methane, and nitrous oxide.

It is known that concentrations of carbon dioxide, methane, and nitrous oxide are increasing and that other greenhouse gases such as CFCs, like $CFCl_3$ and CF_2Cl_2, are being added to the atmosphere. The consequences of these increases and additions in terms of climatic effects are difficult to deduce but may be of great importance. Since 1860 the global mean surface temperatures have increased by about 0.5°C. A warming of approximately 1.5°C over the next 40 years is possible (Mitchell, 1989). The temperature rise with associated changes in climate could profoundly impact societies and their economic activities (Revkin, 1988).

IMPORTANCE OF MODELS

General circulation models of the atmosphere coupled to simplified models of the ocean and sea ice have been used to study possible climatic effects. Results of calculations suggest the following (Mitchell, 1989):

1. The troposphere and Earth's surface show a warming trend. In the tropics the warming varies little with seasons. In winter the largest warming occurs over sea ice.
2. Sea ice will become less extensive.
3. Atmospheric water vapor content rises as temperature rises. The radiative heating of the Earth's surface increases and precipitation and evaporation rates increase, but not necessarily in the same areas of the world.

These general models play an important role in enhancing our understanding of the problem and its consequences. The models can predict far-reaching causes and effects of our actions today, assisting us in choosing the best means with which to combat this global problem.

PREDICTION UNCERTAINTIES

There are strong links between the radiation budget of the Earth, the Earth's climate, and the circulation of the atmosphere and oceans on our planet. Several uncertainties exist which reduce our confidence in the results of three-dimensional models which are used in studying long-term (decades or longer) greenhouse effects and global warming. Effects of cloud cover and ocean atmosphere interactions are major sources of uncertainty in our understanding of climate and the response of climate to human activities.

The net energy input to our climate system is a function of the balance between absorbed solar energy and energy emitted from the Earth system. The net energy input to the climate system is a function of the irradiance of the sun, the planetary albedo (amount of electromagnetic radiation reflected), and the way the surface-atmosphere system emits, which is a function of the planet's effective radiating temperature. The effective radiating temperature is observed to be 288K. Satellite measurements (Ramanathan, et al., 1989a,b) show that the Earth's surface emits 390 W/m^2 but that only 237 W/m^2 is radiated into space by the atmosphere. Hence the infrared radiation from the ground is trapped in the atmosphere by a "greenhouse effect" attributed to the previously mentioned gases.

The three-dimensional models need to be constrained by accurate long-term data from satellite and other observations. To do proper modelling, the absorption of solar radiation by oceans, sea ice, tropical forests, deserts, and other features as a function of zenith angle of the sun must be known. The clear-sky albedo of the planet must be known as well as how clouds modulate solar absorption by geographical location. Since the mid 1980's ERBE (Earth Radiation Budget Experiment) instruments carried on the Earth Radiation Budget Satellite (ERBS), NOAA-9, and NOAA-10, have been collecting these data which are so essential to defining how climate change is a function of human activities. They have provided some of the necessary information to establish parameters relating to cloud-climate feedback (Ramanathan, et al., 1989a,b). Observations are needed over a time period of several decades.

Another uncertainty (Waldrop, 1989) in predicting the global greenhouse is the sun which exhibits variability in energy output. Solar energy output has been correlated to the sunspot cycle which has been observed to be eleven years. The sun has other cycles such as in the 17th and early 18th centuries (the Maunder minimum) when sunspots were rare and Europe experienced unusually cold weather.

In an ongoing project, 1000 nearby stars are being monitored (Waldrop, 1989) to obtain a better understanding of sunspot-like activity on other stars. Only 22 years of data are available so far but results show that the eleven-year periodicity of our own sun is probably not common for other stars. Maunder minimums may be rather common. Extending our observational capability to use telescopes in space and on the Moon such as the proposed Lunar Optical-Ultraviolet-IR Synthesis Array (LOUISA), could yield a needed understanding of how the activity of stars (and our sun) vary with time (Burns, 1990).

There are many issues related to a better understanding of climate trends and the associated role of human activities. There is renewed interest in supersonic commercial airplanes (SST). A multi-year program is scheduled to develop engine technology and study SST engine exhaust effects on ozone depletion with the provision that the engines are not to be built if their emissions (eg., NO_x) in the stratosphere, at 18 to 24 km altitude where ozone is densest, would seriously perturb the ozone layer (SST Emissions, 1989).

THE APPROACH FOR MISSION TO PLANET EARTH

Eos (Earth Observing System) will be built on the foundation established by precursor satellites such as the Upper Atmospheric Research Satellite (UARS), the Ocean Topography Experiment (TOPEX), and the Geopotential Research Explorer Mission. "We must address the problems of understanding the Earth as a planetary body if we are going to learn to predict this changing system. The primary goal of Eos is directed toward understanding the Earth on a global scale; the data, not the hardware are the hallmark of Eos" (Soffen, 1989). The Eos science effort is led by 28 interdisciplinary investigator teams.

Eos requires a new satellite bus. The synthetic aperture radar (SAR), just one payload of Eos, has operating power requirements (5700 watts average and 13,500 watts peak) in excess of those of entire suites of instruments on recently launched satellites. Large payloads and extended duration missions are involved in Eos (figure 2).

INTRUMENT	MEASUREMENT
Advanced Data Collection & Location System (ADCLS)	Data and command relay and location of remotely sited measurement devices
SISP-Surface Imaging and Sounding Package	
Moderate-Resolution Imaging Spectrometer (MODIS)	Surface and cloud imaging in the visible and infrared 0.4 to 2.2 nm, 3 to 5 μm, 8 to 14 μm resolution varying from 10 nm to 0.5 μm
High-Resolution Imaging Spectrometer (HIRIS)	Surface Imaging 0.4 to 2.2 nm; 10 to 20 nm spectral resolution
High-Resolution Multifrequency Microwave Radiometer (HMMR)	1 to 94 GHz passive microwave images in several bands
Lidar Atmospheric Sounder and Altimeter (LASA)	Visible and near-infrared laser backscattering to measure atmospheric water vapor, surface topography, atmospheric scattering properties
SAM-Sensing with Active Microwaves	
Synthetic Aperture Radar (SAR)	L-, C-, and X-band radar images of land, ocean, and ice surfaces at multiple incidence angles and polarizations
Radar Altimeter	Surface Topography of oceans and ice, significant wave height
Scatterometer	Sea surface wind stress to 1 m/s 10° in direction Ku band radar
APACM-Atmospheric Physical and Chemical Monitor	
Doppler Lidar	Tropospheric winds to 1 m/s doppler shift in laser backscatter
Upper Atmosphere Wind Interferometers	Upper atmospheric winds to 5 m/s doppler shift in O2 thermal emissions
Tropospheric Composition Monitors	Trace chemical constituents of the troposphere
Upper Atmosphere Composition Monitors	Trace chemical composition passive emission detectors at wavelengths from UV to microwave
Energy and Particle Monitors	Solar emissions from 150 to 400 nm, 1 nm spectral resolution, Earth radiation budget, total solar irradiance, particles and fields environment

Figure 2. Instrumentation concepts for Eos from Butler et al. (1987, p. 10).

IMPLEMENTATION

International Nature

We want a planet as healthy and as biologically diverse as ever for those who follow us. Each generation of humankind is responsible for passing on to the next and succeeding generations an acceptable environmental legacy. What is required to achieve that end is not only the technical means as exemplified by Mission to Planet Earth but also the political means. The international nature of the Mission to Planet Earth, with its space and ground segments, may be a great aid in generating the trust and credibility of results to facilitate international agreements to keep a healthy planet for humankind.

There is not only competition but also potential constructive interaction between nations to serve the global satellite remote-sensing needs of humankind to support Mission to Planet Earth. The burgeoning capabilities in existence include those offered by ECOSAT (USA), Spot Image (France), and Soyuz Carta (USSR). Spot Image and Soyuz Carta remote sensing/imagery have both been marketed in the U.S. and each firm offers important capabilities. Soyuz Carta offers five-meter resolution and stereoscopic coverage with 60% overlap from the Soviet KFA-1000 camera.

The environmental challenges which are the focus of Mission to Planet Earth are not readily addressed with the current international legal structures. The United Nations Environment Program (UNEP), which is concerned with climate change, has representation from 58 of a possible 170 UN delegates. A "trustees of the Earth" council has been suggested to deal with change (Henson and Hively, 1989). Decisions must be made on what sort of environment is wanted and how to achieve and retain it.

Proposed Satellite Constellation

Figure 3, which has been developed from data presented by Covault (1989a,b) and Soffen (1990), lists some major spacecraft elements for the projected satellite constellation. The polar platforms would be replaced at five year intervals during a fifteen year program. The use of robotic servicing will extend vehicle life but replacement spacecraft may be required not only by NASA but also by ESA and Japan. Figure 4 shows a scaled drawing of an Eos satellite concept compared to current generation spacecraft.

The first NASA polar platform will be a 40 foot by 15 foot vehicle and will include a Japanese microwave scanner. Some spacecraft concepts developed are large (60 feet long) and will include microwave dishes to obtain temperature and moisture data. Several smaller satellites are scheduled for Earth observation missions (Marshall, 1989a). The Upper Atmospheric Research Satellite (UARS) is scheduled for 1991 and the joint NASA-French Ocean Observer is to be launched in 1992 or 1993.

ELEMENT	INTRUMENTS	LAUNCH	NOTES
First NASA Polar Platform 15 tons (GSFC)	19 Earth-monitoring Eos instruments	Late 1996 or 1997 by Titan IV	Sun-synchronous orbit at 705 km
Second NASA Polar Platform (JPL)	3 instr.; includes high-res. imaging radar by JPL	~1998 by Titan IV	
European Polar Platform (two)	4000 lb Earth viewing	by Ariane 4	700 to 850 km
Japanese Polar Platform	includes laser atm. wind sounder & other instr.	1998 on H-2 booster	In phase A studies
Space Station elements	tropical rain-fall & winds; equatorial rad. Earth photos	1996 or later	
Geostationary-orbit platforms	continuous hemis-pheric data	first launch at end of century	Four or five (3 US; ESA & Japan)
Earth probes	eg: TOMS; others: tropical rainfall	~1991 USSR ~1995 H-1	Soviet Meteor 3 spacecraft

Figure 3. Projected spacecraft elements.

<u>Data Analysis and Reduction</u>

There is a strong need to build up the human resources in the related scientific disciplines to support Eos. The Earth-orbiting instruments of Eos will produce a tremendous amount of digital data. Existing satellites have already overloaded the information system. There is also need for educating scientists and engineers to interpret and apply the data from existing Earth-monitoring systems. These systems have created archives of data that could be used in climate research. The existing space network for Earth observing

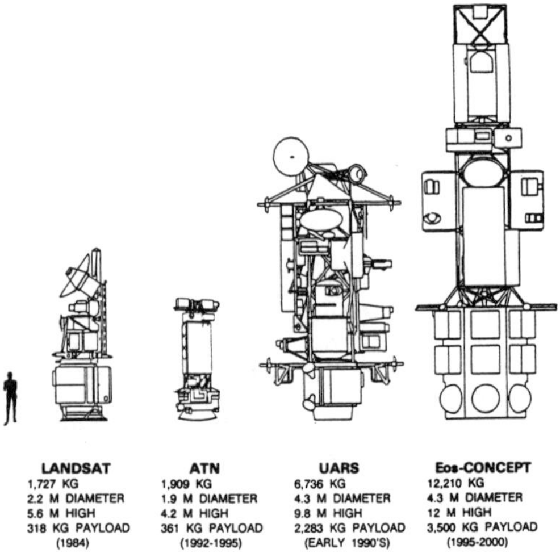

Figure 4. Scale drawing comparing an Eos payload concept with current generation spacecraft (Butler, 1987, p. 20).

includes, for example, Landsat 4 and 5 (which scan the Earth's surface) and GOES (which monitors weather and tracks hurricanes).

Satellite data management for the present is a challenge and for the future will be much more demanding. The Director of the Earth Remote Observation System data center in Sioux Falls, South Dakota estimates that the 1990's Eos satellites could provide as much data in two weeks as Landsat has created in 17 years (Marshall, 1989a,b). Another challenge is to enhance viable working relationships between government agencies in using and sharing data.

SUMMARY: Implications for Engineering, Construction, and Operations

Many significant challenges are ahead for Eos instrumentation. A prime example is the Synthetic Aperture Radar (SAR) which is anticipated to have an antenna measuring 20.2 meters by 4.1 meters. The concept (Carver, et al., 1987) encompasses electronic beam steering and mechanical roll, pitch, and yaw for range and azimuth steering. Data rates from SAR alone could exceed 200 Mbps and ground processors to process raw data with speeds of 1 GFlop/s are required. These tremendously large rates and needs for analysis and archives are some of the most difficult challenges Mission to Planet Earth will face.

The Eos platform design and instrument accommodation require engineering studies relating to cooled focal planes, power consumption, electromagnetic and mechanical interference, and data handling. The Eos platforms are large and require precise orbit determination and offer significant challenges for engineering, construction, and operations as relates to large and complex systems in Earth orbit. Meeting scientific goals within engineering constraints will require multidisciplinary guidance and careful management to achieve the goals of the program not only with respect to the space segment, but also for the ground segment, and the human interfaces. All elements must function so that humankind can gain a predictive understanding of the interactive physical, geological, chemical, biological, and social processes that regulate the total Earth system.

ACKNOWLEDGEMENTS

The constructive comments from various people are appreciated in the preparation of this manuscript, especially reviews from Phil Richter, Ron Eck, and Wes Huntress.

APPENDIX I. REFERENCES

Burns, J. O., et.al. (1990). "LOUISA: A Lunar Optical-Ultraviolet-Infrared Synthesis Array," in Space 90, Engineering, Construction, and Operations in Space, ASCE, New York.

Butler, Dixon M. et al. (1987). EOS Science Steering Committee Report, Vol. II - Eos Reports, "From Pattern to Process: The Strategy of the Earth Observing System," NASA, Washington, D.C., 140 pages.

Carver, Keith, et al. (1987). Synthetic Aperture Radar Instrument Panel for Earth Observing System, Vol. IIf - Eos Reports, NASA, Washington, D.C., 233 pages.

Covault, Craig (1989a). "Major Space Effort Mobilized to Blunt Environmental Threat," Aviation Week & Space Technology, Vol. 130, No. 11, pages 36-44.

Covault, Craig (1989b). "New Earth Observing Platforms to Study Global Water, Biology" Aviation Week & Space Technology, Vol. 130, No. 11, pages 46-50.

Hively, William (1989). "How Bleak is the Outlook for Ozone?," American Scientist, Vol. 77, No. 3, pages 219-224.

Henson, Robert, and William Hively (1989). "Toward an International Law of the Atmosphere," American Scientist, Vol. 77, No. 4, pages 324-325.

Levi, Barbara Goss (1988). "Ozone Depletion at the Poles: The Hole Story Emerges," Physics Today, Vol. 41, No. 7, pages 17-21.

Maggs, William Ward (1989). "Ozones Effects on UV Documented," Eos Transactions, American Geophysical Union, Vol. 70, No. 34, page 791.

Marshall, Eliot (1989a). "Bringing NASA Down to Earth," Science, Vol. 244, No. 4910, pages 1248-1251.

Marshall, Eliot (1989b). "Early Data: Losing Our Memory," Science, Vol. 244, No. 4910, page 1250.

Mitchell, John F. B. (1989). "The 'Greenhouse' Effect and Climate Change," Reviews of Geophysics, Vol. 27, No. 1, pages 115-139.

Ramanathan, V., Bruce R. Barkstrom, and Edwin F. Harrison (1989a). "Climate and the Earth's Radiation Budget," Physics Today, Vol. 42, No. 5, pages 22-32.

Ramanathan, V., et al. (1989b). "Cloud-Radiative Forcing and Climate: Results from the Earth Radiation Budget Experiment," Science, Vol. 243, No. 4887, pages 57-63.

Ride, Sally K. (1987). "Leadership and America's Future in Space," NASA, Washington, D.C., 63 pages.

Revkin, Andrew C. (1988). "Endless Summer: Living with the Greenhouse Effect," Discover, Vol. 9, No. 10, pages 50-61.

Soffen, Gerald A. (1989). Eos Meeting, Science, Vol. 245, No. 4919, pages 691-692.

Soffen, Gerald A. (1990). "NASA's Mission to Planet Earth", NASA Tech Briefs, January 1990, pages 12-13,81.

SST Emissions (1989). EOS, Transactions AGU, Vol. 70, No. 33, page 779.

Stolarski, Richard S. (1988). "Watching a Disappearing Shield," Aerospace America, Vol. 26, No. 10, pages 32-35.

Waldrop, M. Mitchell (1989). "Our Future in the Stars," Science, Vol. 243, No. 4893, pages 890-891.

Space-Based Control of the Climate.

Michael Mautner[1] and Kelly Parks[2]

Abstract.

The expected global greenhouse warming of 2 ^{o}C may be reversed by a space-based screen that intercepts 3% of the incident solar radiation. Space-based devices may also mitigate the ozone/UV problem and extreme weather phenomena such as cyclones. Against greenhouse warming, a fleet of screen units can be deployed in Earth orbit, at the sun-Earth L1 point, or in radiation-levitated solar orbit. For the latter, the required areal density at 0.1 au from the Earth is 5.5 g m^{-2}, and the total required mass is $1.7x10^{10}$ kg, equivalent to a hill or asteroid with a radius of 120 m. The required material can be obtained from lunar sources by mass drivers. Active orbital control can be achieved using solar sailing methods. In the long run, similar devices may also modify the climates of other planets for terraforming. Space-based climate control is possible through the use of large energy leverage, ie., solar screens and reflectors manipulate 10^6 times the energy required for deployment. Therefore in the space scenario projected for several decades, substantial intervention with the climate will be facile, and this suggests serious studies.

1. Background and General Concept.

Increased CO_2 levels and industrial trace gases in the atmosphere are expected to raise the global temperature by 2 - 5 ^{o}C by the middle of the next century (Jager, 1983; Dickinson, 1987). This may result in major global economic losses, including wide-spread flooding and desertification (Schneider, 1989). The damage is compounded by increased UV incidence at surface level, caused by ozone loss. It would be desirable to avoid these effects by environmental measures, but with increasing worldwide industrialization this may not be possible.

Could technology, and particularly space technology, mitigate these effects? Possible space-based solutions include the use of satellite solar power to replace the burning of fossil fuels, and ultimately the use of space colonization to reduce terrestrial populations (O'Neill, 1974). However, these solutions

---

[1]Department of Chemistry, University of Canterbury, Christchurch, New Zealand;[2]American Rocket Co.m Camarillo, California 93010.

require long-term foresight and are politically difficult, since governments respond more readily to crises than to vision.

Space-based climate engineering may provide a possible answer. The fundamental principle is the energy leverage by solar screens and reflectors. These devices manipulate energy many orders of magnitude larger than that invested in deployment. This is needed since climatic phenomena involve larger energies than the total global industrial production, which makes Earth-based climate control impractical.

We suggested recently (Mautner, 1989a) that the warming and UV effects may be reversed by a space-based screen that will reduce the solar radiation incident on the Earth. For reasons given below, 3% shielding is required to reverse 2° warming. The required screen must have a large area, but if it is constructed mostly of thin film, the mass can be relatively modest.

Considering the Sun as a point source at infinite distance and denoting the radius of the Earth as R_E, the effective area of the screen must be $0.03x(3.14R_E^2) = 3.8x10^{12}$ m^2. Using a film and wire mesh structure with a realistic overall areal density of $1g/m^2$ as the main component (Drexler,1979; Forward, 1986) the required mass is $3.8x10^9$ kg. This is comparable to a spherical boulder (of density 3 g/cm^3) with a radius of 67 m, which is the size of a small lunar hill or asteroid. The material can be launched from the moon in 1 - 10 years by a single mass driver with a capacity of 10^9 kg/year (O'Neill, 1974).

The greenhouse effect could become significant by about the middle of the next century. According to current US space goals, a permanent lunar base should exist by then. The additional cost of a space screen will be then smaller than the damage caused by climatic change. More imminent and fundable applications of space-based climate control can include the dispersal of low-pressure centers leading to hurricanes and typhoons; this was also suggested by Marwick (1989). Space-based devices can control the temperatures of other planets.

Space-based climate engineering may be within reach in a few decades. It is necessary to study the multifaceted aspects of this development, including astronautical, geophysical, climatic, materials, economical, and even political and ethical questions.

2. Deployment and Mass Requirements.

There are three basic approaches to deployment: in Earth orbit, near a libration point, or in levitated solar orbit. The screen may be made of dust particles or continuous surfaces. The stability and practicality of dust shields, which do not require processing, warrants further study.

As to a structured screen, a connected surface with an area of 10^{12} m^2 will be unstable. In the discussion below, the "screen" means a collection of many individual units, and the area and mass of the "screen" mean the total area or mass of the screen units. The screen in Earth orbit would be a belt of independent units, dispersed in various orbits to avoid collisions and to create a globally uniformly dispersed shading effect. The mutual re-radiation and re-absorption of energy by the Earth and an absorbent screen were considered (Mautner 1989b). For an overall screening of effect of 3%, with a totally absorbent film screen of 1 g/m^2, these factors lead to minimum mass requirement of 3.4×10^{10} kg at an altitude of 2,200 km.

Near-Earth screens would intefere with space activites, and may suffer from atmospheric drag. Screens deeper in space could be deployed near the Sun-Earth L1 libration point. Position control of a 10^4 kg screen unit could be achieved using an on-off control system with a 172 m^2 solar sail (Farquhar, 1970). Screens at L1 are discussed in more detail elsewhere (Mautner 1989c).

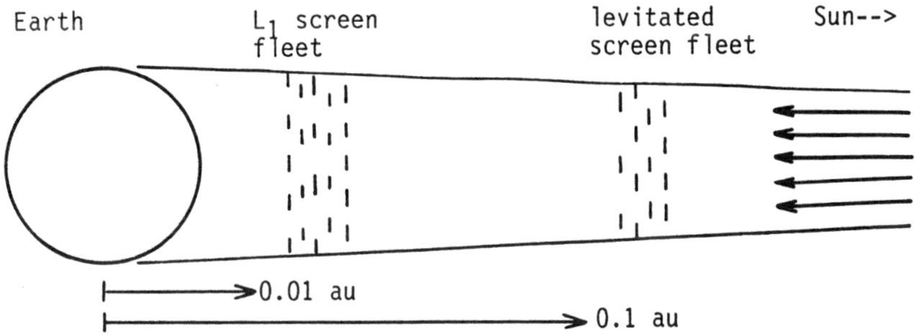

Figure 1. Schematic drawing (not to scale) of screen deployment.

We shall focus here on a third alternative, a fleet of film structures in levitated solar orbit. Levitated orbits for research and communication devices were proposed (Forward 1981, 1984, 1986). The principle of the levitated orbit is that the radiation pressure decouples the orbital period from the radius, allowing non-Keplerian orbits permanently between the planet and the Sun. The orbit is determined by the balance of forces comprising the gravity of the Sun inward, and the gravity of the planet, the centrifugal force and the radiation pressure outward. For a planar totally reflective sail perpendicular to the Sun-sail axis, the balance of forces is given by equation (1).

$$\frac{G\ M_S m_s}{R_{S,s}^{\ 2}} - \frac{G\ M_p m_s}{R_{p,s}^{\ 2}} - m_s w^2 R_{S,s} - \frac{2ASR^2}{c\ R_{S,s}^{\ 2}} = 0 \quad (1)$$

Here G is the gravitational constant and c is the velocity of light. M_S, M_p and m_s the masses of the sun, planet and of the sail, respectively; $R_{S,s}$, $R_{p,s}$ and R are the Sun to sail, planet to sail and Sun to planet distances, respectively. The angular velocity w of the sail in its solar orbit is equal here to the Keplerian angular velocity of the planet that it shadows. S is the solar constant at the planet; c the velocity of light.

The area A of the screen can be expressed as m_S/d where d is the overall average areal density of the sail plus payload. In this manner the area A cancels from equation (1). The distances $R_{S,s}$ and $R_{E,s}$ to the sail can be expressed as fractions of the Sun - planet distance as $R_{S,s}$ = fR and $R_{p,s}$ = (1-f)R. Equation (1) then reduces to (2).

$$d\left(\frac{GM_S}{f^2 R^2} - fw^2R - \frac{GM_p}{(1-f)^2R^2}\right) - \frac{2}{c}\frac{SR^2}{f^2R^2} = 0 \qquad (2)$$

For planetary motion, $w^2 = GM_S/R^3$. We also note that S is the solar constant at distance R; since S varies as R^{-2}, relation $4x3.14SR^2 = S_S$ applies, where S_S is the total radiation output of the Sun. At distances where the gravity of the planet is neglible (in the case of Earth, past 0.05 au), these relations lead to the general equation (3) and with the constants in SI units, equation (4), for a levitated solar sail orbiting at a fraction f of its Keplerian distance from the Sun.

$$d = \frac{S_S}{2x3.14GM_Sc(1-f^3)} \qquad (3)$$

$$d = 1.50/(1-f^3) \quad g\ m^{-2} \qquad (4)$$

Therefore the required areal density is a function only of f, ie., of the ratio of the radius of the orbit of the screen to the Keplerian radius with the same angular velocity. In particular, d is independent of the value of R itself. Therefore the same relations apply to sails shading any planet at a fraction f of the orbit of the planet.

Numerical solutions of equation (2) yield the relationship between the distance from the Earth and the required density. At 0.01 au from the Earth the object is at a libration point. In this case a screen with an arbitrarily large areal density can be stable. At 0.02, 0.03 and 0.05 au the required density d decreases to 29, 18, and 11 g m^{-2}, repectively. Past this distance the gravity of the Earth is negligible, and at 0.1, 0.2, and 0.5 au the required areal density decreases to 5.5, 3.1 and 1.7 g m^{-2}, respectively. The required density at 0.1 au from the Earth, possibly a practical distance, is 5.5 g/m^2. For this distance, even using a robust film of 2.3 g/m^2 still allows half of the total mass to be used for structural and steering components. For 3% shielding, the required area is $0.03x3.14xR_E^2$ = $3.82x10^{12}$ m^2, and the mass is then $2.10x^{10}$ kg. It is an interesting coincidence that general physical constants and factors such as the Sun's mass and luminuosity and the Sun - Earth distance, combine in equation (2) to require feasible film thicknesses.

3. Orbital Stability.

The mission of the screen units requires that they stay within the cone segment (to a good approximation, a cylindrical volume) defined by the Sun and the diameter of the Earth. Considering the Sun as a point source, this confines the screen units at 0.1 au from the Earth to stay within an area with a circular cross-section of radius 0.9 R_E = 5.7×10^6 m. Considering that the length of the screen's orbit about the sun is 2x3.14x 0.9 au = 8.42×10^{11} m, this means that the orbital velocity of the screen must be accurate to 1 part in 1.5×10^5 to stay within the required area for one year.

Suppose that the screen units require positional corrections equal to transiting the radius of the shading zone of 5.7×10^6 m within one year. Suppose that this is done by solar sailing, by accelerating and decelerating the sail during the rotation period of 3.15×10^7 sec., tangential to the orbital radius. This requires an acceleration of 4.6×10^{-8} ms^{-2}, ie, for a screen unit of 10^5 kg, a force of 4.6×10^{-3} N. If the acceleration is attained by tilting the entire sail in an angle to the Sun - sail axis, then the required angle is 0.00012 degrees. Alternatively, if a steering sail section is used, which is tilted 45° to the sun-sail axis, then the required area of this steering section is 812 m^2, ie., 8.1×10^{-6} times the total sail area.

The force for orbital corrections is easily provided therefore by solar radiation pressure. However, this also shows that the screen units can be displaced from the effective orbit by small forces, for example if tilted by more than 0.00012 degrees to the solar flux. Active attitude control may be necessary, for example by adjusting a fraction of the screen as a position and attitude control device. This will require sensors, computers and actuators aboard. Each screen unit will be therefore a high-technology device. A convex conical screen geometry may help with attitude control (Farquhar, 1970).

Some of these problems could be avoided by making the screen units spherical, with small maneuvering sails attached. As we saw, small sails, with areas of several hundred m^2, will be sufficient. Large film spheres may require a supporting grid and beam structure and possibly, a screen made of a cloud of small spherical reflective balloons could be considered.

A screen of flat surfaces must be composed of many independent units. Assuming that the upper limit for a stable size is 100 km^2 = 10^8 m^2, and the total required area for 3% shielding of the Earth at 0.1 au from the Earth is $0.03\times(0.9\times3.14R_E)^2 = 3.1\times10^{12}$ m^2, then the screen must be made of 3.1×10^4 independent units. Evidently, these units must be mass produced. Assuming that the screen is constructed in ten years and needs to be replaced at the rate of 10% per year, screen units must be produced at the rate of 1,000 per year. In this respect, the total complexity of the sensors, computers and steering mechanism of a screen unit may be comparable to a computerised automobile. In practice, factories commonly mass produce such commodities on the order of 1,000 units annually.

4. Material Requirements and Sources.

Solar sail materials in a NASA study were polymer films coated with a reflective layer of metal, for example, 2000 A° aluminum. It was estimated that 1.5×10^6 m^2 of about $2g/m^2$ Kapton film could be produced by a plant per year (Friedman, 1978). For the large screen masses, lunar or asteroid sources are more practical (O'Neill, 1974). However, the problem here is that even carbonaceous chondrites contain only 0.1 - 3.8% carbon. Therefore a carbon-based film could increase the raw material requirements by about a factor of 100.

On the other hand, lunar and asteroid materials are rich in silicon, aluminum, titanium, iron and nickel. Screens of metal films could be stretched on a wire mesh. With the above production rate, the requirement is 3.1×10^{11} m^2/year, using 1.7×10^9 kg of material per year. Space-based manufacturing of thin films was discussed (Drexler, 1979).

5. Filters Against UV Wavelengths.

UV wavelengths below 320 nm represent about 3% of the solar spectrum. This is about the amount of energy that needs to be eliminated against greenhouse warming. However, most of the solar UV is absorbed in the upper atmosphere and its contribution to the surface temperature may be smaller. A filter screen that absorbs or scatters the UV portion of the solar spectrum would solve the health and environmental problems of increased UV incidence and ameliorate greenhouse warming. An ordinary lime glass screen could be used.

However, eliminating the entire solar UV would also offset the chemistry of the upper atmosphere, including the formation of ozone itself, which results from UV induced radical photochemistry. It may help to filter out selectively the UV from 250 to 320 nm (although most of the UV energy is in this range, and chemical effects would be still significant). Narrow band rejection filters opaque exclusively to 250 - 320 nm are available commercially.

To reduce the mass, the thin rejection filter films must be deposited on a thin substrate transparent to vacuum UV, such as quartz or fused silica.

Filter screens could be deployed in Earth orbit, or at L1, where mass and thickness considerations are not critical. As to levitated orbits, the UV filter screen is not suitable as solar sail since it transmits most of the radiation. Such screens must be levitated on solar sails. Of course, the sail will then block a portion of the total spectrum. For example, the overall areal density required at 0.9 au is 5.5 g/m^2. If the sail density is 1 g/m^2, then 1 m^2 sail can carry 4.5 m^2 filter, also of 1 g/m^2. In this case, if the screen blocked out 3% of the total radiation, it would also block out 16.5 % of the incident UV. At the present the surface-incident UV is increased by about 2 - 5%, depending on season and latitude, over natural levels. Therefore the above combination could correct simoultaneously both the expected greenhouse and UV problems.

5. Climatic Considerations.

The shielding requirements depend on the magnitude of greenhouse warming, and possibly of ozone depletion for UV screens.

The rise in atmospheric CO_2 depends on expected energy sources and uses, and the changes in biota (deforestation, desertification, or conversely, possible biota increases due to increased CO_2 levels). Different scenarios project that the CO_2 levels will increase from pre-industrial levels of about 200 ppm to 300 - 500 ppm by the year 2030 (Jager, 1983; Schneider, 1989).

The doubling of atmospheric CO_2 causes increased heating of the lower atmosphere, oceans and land by a global average of 4 W/m^2, ie., a 2.8% increase over natural insolation (Jager, 1983). The resulting heating depends on feedback factors, quantified by the relation dT = Q/L where L is the feedback parameter. Considering feedback effects of increased humidity gives L = 2.0 W/m^2 K, ie., a 2 ^oC warming. Altogether, a 1979 NAS report estimates L = 1.7$\pm$0.8 W/m^2 K, and therefore a dT for doubled CO_2 of between 1.6 to 4.4 K (NAS, 1979).

As to the relation of CO_2 heating and insolation, historic records or General Circulation Models should be used. A formula for the radiation balance, for conditions of 50% cloudiness and a constant albedo of 0.3, yields a 1.5 ^oC change for 1% change in solar radiation (Budyko, 1971). This is similar to an estimate which gives 1.2 ^oC per 1% change in insolation (Manabe, 1980). The most pertinent finding is that, including the effects of cloudiness, the doubling of atmospheric CO_2 is similar to a 2% increase in incoming solar radiation (Wetherald, 1980).

Sensitivity to insolation is illustrated by the finding that a 1% decrease in the solar constant can cause glaciers to move to the present temperate zones of 50° latitude, and with albedo feedback effects, even to the equator (Budyko, 1969; Verbitskiy, 1988). Therefore, a solar screen would have to be designed prudently, such as not to cause over-cooling.

In summary, the complexity of the climatic system leads to significant uncertainties. Nevertheless, most models show that CO_2 levels will probably double, leading to a 2$\pm$1 ^oC warming, equivalent approximately to a 3% change in insolation, which can lead to substantial climatic changes.

These models are the basis to our present estimates of the needed screening factor. However, climatic interventions may have more unpredictable effects than the greenhouse effect itself

(Schneider, 1989). Models with increased CO_2 and decreased insolation simoultaneously are needed. Ulitmately, the necessary screening factor may be found empirically. These uncertainties suggest that the space screen should be easy to manipulate to vary the screening factor as needed.

6. Other Climate Control Applications: Severe Weather Phenomena on Earth, and Temperature Control of Other Planets.

In the preceding sections we discussed solar screens to correct against pollution-related climatic changes. However, space-based climate engineering may be applied more imminently to correct harmful weather effects on Earth, such as the development of cyclones (hurricanes and typhoons), and El Nino conditions. Such phenomena start with small atmospheric temperature and pressure differentials, which grow slowly due to positive feedback effects. For example, a tropical depression that may grow into a hurricane could be dissipated at an early stage (see also Marwick (1989)). To illustrate the power requirements, an air mass at constant volume, confined to a volume with a diameter of 50 km and 10 km height with a pressure of 980 mbars at 300 K is heated for a month to raise the pressure to 1000 mbars by raising the temperature by 6 K. The required power is equal to that supplied by seven satellite solar power stations (SSPS) of 10 GW each. The same energy could be supplied more simply to the depression from a space-based solar reflector of 7.2x7.2 km^2.

Similarly, directed energy from SSPS systems may be used to dissipate incipient conditions that ultimately amplify into El Nino effects. Directed SSPS energy may be also used to decrease the antarctic ozone hole by dissipating airborne ice crystals that catalyze the ozone loss. Prospects to eliminate natural disasters may promote SSPS development. Experiments to dissipate early hurricanes and to direct energy into critical air masses in a controlled way, in order to understand weather systems, could be part of the mission of an experimental SSPS energy system.

On the longer term, space screens may be applied to other planets. Screens in orbit about other planets would not interfere with astronautics, as they would about the Earth. Screens at the Sun-planet libration points, and levitated screens as discussed above are also possible.

A possible application is to reduce the temperature of Venus for terraforming. Although the present atmosphere of Venus causes a severe greenhouse effect, we assume that ultimately an Earth-like atmosphere will be produced. To maintain Earth-like temperatures, the insolation will have to be reduced to the level of the terrestrial solar flux. For this purpose it is then necessary to reduce the insolation of Venus by a factor $(R_{Sun,Venus}/R_{Sun,Earth})^2 = 0.52$, ie, to screen out 48% of the radiation. As we noted above, for levitated screens d is the same function of f for any planet. Therefore at f = 0.9, ie., at 0.1 $R_{Sun,Venus}$ from the planet, again d = 5.5 g m^{-2}. With the required screening factor, the required area is 4.5×10^{13} m^2, and the mass is 2.5×10^{11} kg.

In contrast, terraforming Mars will require warming the planet. For example, additional solar radiation could be reflected onto the cold night side. This requires reflectors placed behind the planet, ie., at distances larger than the Keplerian orbit. This cannot be done by levitated sails, since beyond the Keplerian radius, ie. for $f > 1$, the solution for d in equations (1 - 4) is negative. Reflectors can be positioned at the Sun - Mars L2 point or in orbit about Mars in a plane perpendicular to the ecliptic can be used.

7. Conclusions.

Climatic engineering, which is impractical with Earth-based technology, can be achieved by space-based methods. The key is a very large leverage, achieved by re-directing a large amount of energy at the cost of investing a much smaller amount of energy. For example, suppose that constructing and maintaining the screen requires removal of 10^{10} kg/year of material from the moon to re-direct 3% of the energy incident on the Earth. In this manner, with the expenditure of 2.8×10^{16} Joule/year it is possible to manipulate 1.7×10^{23} Joule/year. This amounts to a multiplier factor of 6×10^{6}. In comparison, the total global energy production is on the order of 10^{22} Joule/year (Jager, 1983). Therefore the effect of diverting 3% of the incident solar radiation is larger than the total global energy production.

Climate corrections for the Earth may become necessary in several decades, when large-scale space activities should be well advanced. Conversely, climate correction may motivate space development. Handling masses equal to a 100 m lunar hill or asteroid should then be facile, and technologies for large space structures should be well developed. The deployment and maintenance of 10,000 mass-produced 10x10 km^2 screen units may occupy 1,000 workers, which is 10% of the population of a Model 1 O'Neill type space colony. Assuming a cost of $100,000/man-year (O'Neill, 1974), the manpower will cost $100M/year. Therefore a space-based screen could prevent the estimated greenhouse damage of $1 trillion over 100 years at 1% of the cost.

The possibility to eliminate hurricanes, typhoons, El Nino damages, and UV/ozone health hazards can contribute to the incentive and political support for large-scale space development. This will in turn open the way to the ultimate solution of environmental and population problems by space colonization as envisaged by O'Neill (O'Neill, 1974).

Any intervention with the climate, including the greenhouse effect or the space-based solar screen, can have major effects on the life and welfare of many millions of people. The point of the present paper is that such intervention may be readily achieved within the space development of several decades. The serious study and debate of the subject is well justified.

References.

Budyko, M. I. (1969). The effect of solar radiation variations on the climate of the Earth. Tellus, 21, 611-619.
Budyko, M. I. (1971). Climate and Life. Hydrometeorolical Publishing House, Leningrad.
Dickinson, R. E., and Cicerone, R. J. (1987). Nature 319, 109.
Drexler, K. E. (1979). High Performance Solar Sails and Related Reflecting Devices. Forth Princeton/AIAA Conference on Space Manufacturing Facilities, Princeton, N. J., May 14-17, 1979, AIAA Paper No. 79-1418.
Farquhar, R. W. (1970). The control and use of libration-point satellites. NASA Technical Report TR R-346.
Forward, R. L. (1981). Light-levitated geostationary cylindrical orbits. J. Astronautical Sciences, 29, 73-80.
Forward, R. L. (1984). Light-levitated geostationary cylindrical orbits using perforated light sails. J. Astronautical Sciences, 32, 221-226.
Forward, R. L. Exotic propulsion in the 21st century. Aerospace Century XXI: Space Flight Technologies, Proc. 33d Ann. AAS International Conference, Boulder, Colorado 1986, AAS Paper 86-409, 717-744.
Friedman, L. et al. (1978). Solar sailing - the concept made realistic. AIAA 16th Aerospace Meeting, Hunstville, Alabama, 1978. AIAA Paper No. 78-82.
Jager, J. (1983). Climate and Energy Systems. Wiley, New York.
Manabe, S., and Wetherald, R. T. (1980). On the distribution of climate change resulting from an increase in the CO_2 content of the atmosphere. J. Atmosph. Sci., 37, 3-15.
Marwick, E. F. Space mirrors, sails and screens for better environments and more useful energy. IECEC-89 24-th Intersociety Energy Conversion Engineering Forum: "International Forum on Energy Engineering."
Mautner, M. (1989). a. Climatic Warming. Chemical and Engineering News, June 5, 33. b. Space Based Screens Against Climatic warming. J. British Interplaneatry Soc., Nov. 1989, in press. c. Deep-Space Screens Against the Greenhouse Effect. Proceeding of the Third AAS Pacific Intenational Symposium on Space Technology and Applications, Los Angeles, Nov. 4 - 6 1989, in press.
NAS, (1979). Carbon Dioxide and Climate: A Scientific Assesment. National Academy of Sciences, Washington, DC.
O'Neill, G. K. (1974). The colonisation of space. Physics Today, 27, September 1974, 32-42.
Schneider, S. H. (1989). The changing climate. Scientific American, 261, September 1989, 70-79.
Verbitskiy, M. Ya. (1988). Role of astronomical factors in the evolution of glacial periods. Izvestiya, atmospheric and Oceanic Phys., 24, 224-227.
Wetherald, R. T. and Manabe, S. (1980). Cloud cover and climate sensitivity. J. Atmosph. Sci., 37, 1485-1510.

Large Space Systems In Global Change Mitigation

Lyle M. Jenkins[1]

Abstract

Alternative courses of action are an important aspect of reacting to climate changes, including the greenhouse effect, and other environmental issues such as depletion of the ozone layer. Earth orbit represents a vantage point to overview the global environment. This capability is being used to develop a scientific understanding of Earth system processes and their interactions. As this understanding is incorporated into computer models with the proficiency to predict the course of global change, deliberate action may be needed to prevent or mitigate the effects of change. An activity is proposed that defines and evaluates very large space systems concepts for feasibility in actively reducing or averting the effect of environmental deviations. Such a study analyses alternative modes of intervention in critical cases of global change. This effort must build on the Earth system research objectives to be valid and to result in feasible concepts that are well understood and to show benefits that overwhelm risks in implementation. Some of the issues are addressed in this paper.

Introduction

The serious environmental problems of today invoke not dreams but nightmares of what they may become in the near future. "It is difficult to say what is impossible, for the dream of yesterday is the hope of today and reality tomorrow." This

[1]Project Management, Johnson Space Center, Houston, Texas 77058

expression by Dr. Robert Goddard opens consideration of alternative solutions to environmental issues. There must also be a dream and hope for compensating for these expected catastrophes. Technology is producing negative effects on Earth system processes. It should be possible to use technology as a positive influence on the environment. The Earth's present population precludes a return to a simpler, less technical era. Much of our technology resulted from the deficiencies of earlier eras. Unfortunately, technology implementation has often been flawed by lack of a total system approach. For example, environmental impacts have often been neglected for reasons of economics or ignorance. Nevertheless, within our lifetime, we have watched fantasies become visions and visions become reality.

Now many scientists and engineers are participating in the hope of understanding our Earth system by observing and modeling in the U.S. Global Change Research Program, the International Geosphere-Biosphere Program and other similar programs. As this vision approaches the ability to predict responses of the Earth's processes, the importance of alternative courses of action is clear. It is time to examine concepts that may have been fantasies in earlier times, to find those realistic visions for the future and to be ready with the concepts and tools to enable the human race to survive and prosper.

Earth orbit is an invaluable observation site to overview, sense, and understand Earth system processes including natural disaster sources. This vantage point may serve as a location for active intervention to mitigate the effects of civilization on the environment and the effects of the environment on civilization.

Objectives

The U.S. Global Change Research Program objectives are listed below:

1. Establish an integrated, comprehensive monitoring program for Earth system measurements on a global scale
2. Conduct a program of focused studies to improve our understanding of the physical, chemical, and biological

processes that influence Earth system changes and trends on global and regional scales

3. Develop integrated conceptual and predictive Earth system models

This paper proposes an activity to define concepts for large space systems that can intervene with Earth system processes for the prevention or alleviation of catastrophic environmental conditions. The proposed study goes beyond the global change research objectives to consider modes of reaction to global change. Definition and analysis of reaction modes that intervene in critical cases provide alternatives for the decision and policy-making process. This effort must build on the Earth system research objectives to be valid and to result in feasible concepts for space intervention systems.

Issues

The relationship of an intervention concepts study to planned and developing programs needs to be complementary rather than competitive. Consideration of international and domestic implications is a major factor in the development of study criteria and requirements. The scope of the study should address a broad range of intervention situations and should not limit definition of potential concepts. There are numerous examples of schemes that were 'crackpot' when described and are now an accepted part of our culture. The feasibility study will first focus on the realm of possibility. Then the study will address the most viable alternatives from a cost/ benefit perspective.

Intervention Candidates

Opportunities for beneficial intervention include natural events and events caused by human activities. Global warming, including the effects of greenhouse gases, is a significant concern. Other such situations are depletion of the ozone layer in the stratosphere, hurricanes, tornadoes, volcanic eruptions, drought, flooding, nuclear disasters, and similar pollution events. Currently, these conditions are handled through post-event reaction or through attempts to reduce effects by prohibiting offending actions. Intervention from space, as applied to these situations, should be evaluated even though many schemes may not be practical.

Other Earth system phenomena are candidates for space intervention concept feasibility evaluation. Ground-based reactive measures cope with floods, hurricanes, volcanoes, and pollution. Space systems ideas may also reduce the impact of natural disasters that are not global. The overview capability of orbiting spacecraft provides an expanded dimension in anticipating and reacting to catastrophic phenomena. These ideas include identifying, data gathering, coordinating, and controlling ground-based emergency responses. Space systems solutions that interact positively with man-made catastrophes or pollution cases may also exist.

Potential Concepts

The generation of power in space for use on Earth reduces the requirements for fossil fuel burning. The Solar Power Satellite (SPS) concept, a renewable energy source, has the potential to supply commercial power. For example, each satellite in the system could supply commercial power for a city the size of Houston, Texas. The concept has been analyzed as the result of intense study that occurred in the late 1970's. The studies of SPS included consideration of the environmental effects as they were perceived at that time. These studies of very large space systems support the feasibility analysis of concepts with other intervention objectives.

The critical ozone layer has become a sort of waste dump site from the migration of chlorofluorocarbon constituents. Active cleanup of this region in the stratosphere is a prime candidate for direct intervention. Feasible extenuation concepts will be difficult to create when dealing with such a remote region that has gases and plasmas in extremely low concentrations. Ground-based microwave systems have been conceived for counteracting the depletion of the ozone layer by converting neutral chlorine atoms into negative ions. Operations from space give direct access to the stratosphere and, when coupled with an immense source of orbital power, the microwave concept may be useful. Dispersal of materials to neutralize the offending chemicals may be an effective concept for space-based implementation. Other exotic concepts may involve ion generators on tethers, tuned lasers, or charged particle beams grazing the atmosphere.

Products

The principal product of the feasibility study will be space-based concepts for alternative solutions to global environmental problems. From the feasible space system concepts, requirements for data and modeling to support these alternative solutions can be defined. An additional product is the definition of technology needed for effective implementation of alternative solutions.

Summary

Space systems with the potential to influence the direction of global change should be assessed for feasibility to provide solutions to negative global environmental changes and severe sudden disasters. Technology has been blamed for exacerbating problems between man and the environment. This has largely been the result of ignorance and short-term economic decisions. The sensitivity to environmental issues is now addressed in the development and use of technology. The continued use of technology to help solve world problems appears to be the only reasonable course for improving living standards in the undeveloped countries and for maintaining standards in the developed nations.

The ability to overview a disaster situation on the Earth is a unique capability. In addition to observation, systems can be conceived that increase the effectiveness of ground-based operations through communication and coordination. There may also be space systems that could intervene in natural processes to alleviate or mitigate the effects of catastrophe. By searching for feasible space systems concepts, several near-term products are anticipated. Definition of data and modeling requirements are needed to support decisions on feasibility and application of these concepts. Sensor and system technology advances are needed to support development of beneficial intervention systems.

The U.S. Global Change Research Program Goal is "To establish the scientific basis for national and international policymaking related to natural and human-induced changes in the global Earth system." Initiation of the evaluation of space-based alternatives for dealing with the effects of global change and the impact on the international community is proposed as a component of the program. Implementation of feasible alternatives may be decades into the future. Other specific courses of action may represent little risk with immediate benefits from early action. All response strategies and approaches must build on reliable information, dependable predictions, and a thorough assessment of policies and actions affecting the Earth system. Understanding and documenting intervention concepts will aid in separating rational, viable schemes from crackpot ideas. Through interaction with observation and modeling programs, an interdisciplinary perspective aids in the definition of requirements and technology needs. Just as space systems are a substantial tool in monitoring the Earth system, space systems may also be a powerful tool in the mitigation of the effects of global change. Creative intervention in natural processes should not be overlooked if it is approached with respect and intelligence.

References

Budyko, M. I.: Climatic Changes. American Geophysical Union, 1977.

Committee on Earth Sciences: Our Changing Planet: A U. S. Strategy for Global Change Research. Office of Science and Technology Policy, 1989.

Committee on Global Change: Toward an Understanding of Global Change. National Academy Press, 1988.

Fowler, Wallace T.: A Feasibility Study of Methods for Stopping the Depletion of Ozone Over Antarctica. University of Texas, Austin, 1988.

Schneider, Stephen H: The Genesis Strategy. Plenum Press, New York and London, 1976.

Wong, A.Y.; Steinhauer, J.; Close, R.; Fukuchi, T.; Milikh, G.M.: Conservation of Ozone in the Upper Atmosphere by Selective Ion Removal. Comments Plasma Physics and Controlled Fusion (UK). Vol. 12, no. 5, 1989.

A Mission to Earth: Global Rural Electrification

Raymond S. Leonard[1]

Abstract

No nation can long remain morally strong or economically competitive without an inner purpose. There are many noble projects but few which will both kindle the individual's imagination and provide economic as well as spiritual benefits to a nation in its battle to remain economically competitive and financially secure. The human exploration of space has been proposed as one such project. However, unless human activity in space can provide tangible benefits to the man in the street such activity will not be funded. What is needed are missions to Earth which provide benefits that the factory worker and the suburban commuter or salaryman can understand and appreciate. Electric power from orbit as a gift from this century and the industrialized nations to future generations and the Third World is such a program. Providing electric power from orbit offers mankind a rationale for the exploration of space similar to the reasons that fueled the voyages of Magellan, Columbus and Cook during the first age of exploration.

Introduction

President Bush has asked that a new course be charted that will take man back to the Moon, on to Mars and beyond. Before a proper course can be charted the questions that must be asked are: what are the purposes of these new voyages and how will they benefit mankind, our Earth, America or any participating nation and in particular the average citizen. The answers will define missions as well as help prioritize those missions amongst the other demands for Federal funds.

The first answer and the one that NASA has unconsciously focused on is the human exploration of space for the purpose of nurturing and stimulating the spirit of mankind. This answer while recognizing some of the results of exploration ignores the motivating reasons for most of human exploration to date. Those reasons were primarily economic.

The age of exploration from 1400 to approximately 1800 was created by the desires of nations for a new shorter route to the orient, territory or furs and gold and silver from the New World. Exploration because of the challenge, the races to the poles, or science came much later when there was a healthy economic base, both private and public, to support such endeavors. The voyage of the Beagle in 1831 on which Darwin served as a naturalist is a prime example. Even then there were underlying reasons of national pride or acquisition of militarily useful data. Exploration alone, whether it is for the human spirit, for knowledge, or for geopolitical competitions, cannot justify the tremendous expense involved in sending a few representatives of mankind to Mars. Space must be more to mankind than a mere stimulus or occasional news story if the tremendous cost of exploration is to be justified.

[1] President, Ad Astra, Ltd., Rt. 1 Box 92 LL, Santa Fe, NM 87501

The second and equally important answer is to utilize the resources of the solar system for the benefit of the Earth's biosphere and mankind. This, in the writer's mind, means extending mankind's resource base out to at least to the orbit of Jupiter. Then mankind's resource base will encompass solar energy found in Earth's orbit, the mineral resources of the Moon and the asteroid belt, water from the moons of Mars and volatiles from the gas giants and their moons. How we construct this answer will determine whether mankind moves out of the cradle of Earth to the nursery of the solar system.

Background

Mankind is placing increasing stress on the Earth's biosphere. One can no longer speak of economic growth without considering the environmental costs. Neither can we attempt to maintain the status quo without considering the environmental and societal cost. Clearly we need to look at how we can use our technology to both minimize further impacts and reduce or reverse the damaging effects our current operations are having on the terrestrial biosphere. Moving electric power generation out of the Earth's biosphere would would allow us to avoid further air pollution from coal and oil fired plants. Placing new and replacement generation capacity in geosynchronous orbit eliminates the fly ash and scrubber sludge disposal problem for coal fired plants. The thermal pollution of water ways caused by all types of thermal plants would also be eliminated. Of equal importance is the fact that such a move would make available the hydrocarbons which would have been burned for long term use by other industries.

In solving the environmental problems of the degradation of the ozone layer and CO_2 buildup we must find a way to involve the Third World. China and India alone account for 40 percent of the total world's population. If China, with one third of the the World's proven coal reserves, fully exploits those reserves to fuel its economic growth, any efforts by the industrialized nations to slow global warming will be negated. Unless the more developed nations with their richer populations begin to understand that protecting the global environment is inextricably linked with eliminating poverty in the Third World we will continue to witness the burning of rain forests and the desertification of agricultural lands along with degradation of the ozone layer and climatic change. The elimination of poverty requires economic development which historically has been based on easily available cheap energy. Potentially, electric power from space offers mankind a way to eliminate a major pollution source, promote economic development in the Third World and stimulate education and technology and provide a rationale for the exploration and utilization of space.

Why Global Rural Electrification

Quoting from Dregne (1985) "Perhaps the single greatest contribution that could be made to environmental conservation would be the invention of a satisfactory fuel-wood substitute." He goes on to state that a major contribution to desertification is the lack of "an economical and socially acceptable substitute for fuel wood as an energy source. Massive tree and shrub cutting is now epidemic in developing countries. Apart from the damaging effect of excessive wood cutting on soil productivity and flooding, the increasing scarcity of fuelwood has raised its price and necessitated that more time be spent in collection"

Other reasons generally cited by the proponents of rural electrification as justification for introducing electricity to rural villages in developing countries are: (Barnes, 1988) gains in productivity, income, expanded irrigation and the creation of jobs as well as intangible quality of life improvements. Video education using television and satellite distribution of programing is yet another reason.

A reason to consider energy as a Foreign Aid and Foreign Policy initiative is that the countries with the lowest levels of Gross National Product and which can least afford rural electrification programs are the ones most in need of alternatives to fuel-wood and dung used for fuel rather than fertilizer.

An advantage of working with the developing nations to prove the concept of electric power from orbit is that they need and can use small, modular increments of power. The small size of each increment of power makes it affordable and feasible to launch as well as build the ground station with available resources.

Finally, given the long term advantages of opening the space frontier, to both the developed and developing nations, it behooves us to give electricity to the rural populations of developing nations. In doing so we also promote world peace through international cooperation and gain economical access to the solar system.

Costs & Technological Readiness

Simply put the technical component of the concept of global rural electrification involves beaming microwave energy from space to receiving stations and power conversion substations located next to villages. The technology needed to place a prototype unit in geosynchronous orbit consists of: solar cells in arrays similar to those in use on communication satellites and the Soviet space station, attitude control similar to that already in use, microwave generators in use for a variety of purposes here on Earth, and a diode receiving antenna. We could assemble a demonstration power satellite system including a ground station from off the shelf components. Davis, 1988, proposed, based on prior work done by NASA and DOE, a modest $40 million per year development program.

Most analytical evaluations of proposals for electrification projects are based on a cost per kilowatt-hour. However projects are often chosen for other reasons which include both economic and subjective policy factors. Some of those subjective factors are: 1) Cost per kilowatt installed; 2) Total cost per project, i.e. do I have to build a 100 megawatt plant to get the 1 megawatt I need today?; 3) Sources of capital; 4) Dependency on energy imports especially those which adversely affect balance of payments; 5) Donor assistance; 6) Cost of transmission system; 7) Environmental impacts and humanitarian benefits.

The kilowatt per hour (kw-hr) cost of power from space lies somewhere between the very optimistic DOE/NASA estimate of 6 cents a kilowatt hour and the very pessimistic National Research Council estimate of 45 cents per kilowatt hour. Similarly the capital cost per kilowatt installed ranges between 4 and 35 thousand 1988 dollars. The three main cost variables or drivers are launch costs, cost of solar cells, and the weight of the solar cells.

Table 1 lists data on costs and payload performance for various launch vehicles past and present. Using that data plus costs and weights for currently available solar cells we can come up with a ball-park figure for a demonstration power satellite.

Table 1 Performance and cost data for various launch vehicles

(References at end of paper)		Dollars per kilogram of payload to LEO (1986$)			
Type of Vehicle	Payload to LEO (kg)	Low	Median	High	Quoted
US Shuttle [1]	25,000.	1,965.	3,931.	5,055.	
Titan 4	15,700.				13,165. [2]
Saturn 1B	22,700.				no quotes
Ariane 4 (44 L)	4,200.				varies
Proton (SL 13)	19,500.				1,540. [3]
Energiya	100,000.				no quotes
NASA Ref. Sys.[4]	424,000.	44.	115.	175.	<--1980 $
escalated 1986 $ ----	------->	50.	131.	200.	

Costs for space qualified solar cells are a 1,000 dollars per watt with government documentation or 340 dollars per watt for a fixed price contract, no government red tape. Weight of the array is between 30 and 50 watts per pound or

7.58 kilograms per kilowatt which works out to 1.53 kilograms per square meter. For comparison purposes an off the shelf terrestrial array costs $691 per square meter. This works out to approximately $6.30 per watt. An 18″ x 40″ module for use on Earth weights about 13 pounds. This weight includes a protective glass cover, metal backing and frame. Table 2 summarizes the above data with the weight for the terrestrial array being reduced by one-half to account for being able to eliminate some of the structure needed for support in the terrestrial environment. If we assume the use of a Soviet Proton launcher as part of an international endeavor we could place in LEO either 200 kW_e of terrestrial solar cells or 1,280 kW_e of space qualified cells. Table 2 summarizes this data.

Table 2 Cost and Weight Data for Photovoltaic Cells (1988 $)

	Space Qualified	Terrestrial
Cost per watt	340.00	6.30
Cost per kilowatt	340,000.	6,300.
Watts per sq. meter @ 12 % eff. and 1.35 kw incident.	162.	162.
Weight per watt (kgs)	0.0152	0.0963
Weight per kilowatt (kgs)	15.2	96.3
Weight per sq. meter (kgs)	2.45	15.6

Based on the data shown in Table 3 it would appear that a demonstration power system could be deployed for between 100 and 450 million dollars which at the upper end is less than the cost of one B-2 stealth bomber. The question then is what would be the cost for a production system?

Previous work, (Leonard 1987, 1989), showed that if you can mass produce expendable Saturn I or PROTON type launchers for the same cost per pound that it costs Boeing to manufacture 747's you should be able to achieve a launch costs to LEO of 240 dollars per kilogram ($110/pound) for hardware and propellant, no operating costs included. Achieving this cost reduction will be hard unless the launchers are built in either Eastern Europe or a developing country such as India or South Korea which have both lower labor costs and the necessary skills or craftsmanship. American aerospace industry, if it wears its commercial hat as opposed to its defense or NASA hat, could probably meet the design to cost goals.

The cost of moving cargo from LEO to GEO using nuclear electric ion drives was calculated to be 210 dollars per kg if the launch cost from Earth was 240 dollars per kg (Leonard 1987). This cost included charges for propellant and capital recovery but not for operations. Interest was assumed to be 12 percent with a recovery period of 10 years.

The capital recovery factor equation used to calculate both the launch costs as well as the power cost is given below and was taken from the 1970 edition of Grant & Ireson's Engineering Economics, page 36. The capital recovery factor when multiplied by the present debt or present investment gives the uniform end-of-year payment necessary to repay the investment in *n* years at an interest rate of *i*. A 10 percent interest and a 30 year recovery results in a capital recovery factor of 0.1061.

Table 3 Cost for a Demonstration Power System (1988 $)

Cost Component	Space Qualified	Terrestrial
Launch to LEO 19,500 kgs @ $1,540/kg	30,030,000.	30,030,000.
LEO to GEO assumed	30,030,000.	30,030,000.
Photovoltaics 100 kW_e	340,000,000.	6,300,000.
Microwave Generation 100 one kW tubes @ 3k	300,000.	300,000.
Receiving Station $5,200 per kilowatt	520,000.	520,000.
Satellite Control System	TBD	TBD
System Integration 10 % of cost	40,880,000.	6,718,000.
Program Management 5 % of cost	20,440,000.	3,359,000.
Total	462,200,000.	**77,257,000.**

Dividing the year end payment by the number of kilowatts delivered (not-necessarily used) in a year gives an approximate cost per kilowatt-hour. The number of kilowatts delivered is equal to the plant factor or amount of time the plant is on line multiplied by 8,760 hours which is the number of hours in a year times the plant capacity, assumed to be 0.95. The result is the cost of power in mills (one thousandth of a dollar) per kilowatt-hour which when multiplied by 100 gives cents per kW_e-hr. The composite factor is 2.55 E-6 for those who wish to run their own quick checks of the numbers in Table 4,.which summarizes the possible cost of power for mass produced power systems.

If a fully reusable vehicle can be manufactured for the same cost per pound that the US taxpayer paid for the B-1 and which has the weight to payload ratio of the Saturn I, then a cost per kilogram of payload to LEO of about 30 dollars per kilogram ($12/pound) is possible. A $30/kg cost to LEO results in $60/kg cost from LEO to GEO. These costs would result in a power cost of 15.73 cents per kW_e-hr and an installed cost of $12,378 per kilowatt. Given recent and demonstrated advances in materials and computer systems a fully reusable unmanned cargo launcher should be possible. The Soviets have demonstrated the remote operation of their shuttle on its first test flight.

Table 4 Cost for a 500 kWe Power System in 2001 (1988 $)

Cost Component	Total Cost	$/$kW_e$ Installed	cents per kW_e-hr
Launch to LEO 19,500 kgs @ $240/kg	4,680,000.	9,360.	11.93
LEO to GEO 19,500 kgs @ $210/kg	4,095,000.	8,190.	10.44
Photovoltaics 500 kW_e @ 30 kg/kW	500,000.	1,000.	1.24.
Microwave Generation 500 one kW tubes @ 5 kgs	750,000.	1,500.	1.91
Receiving Station $2,600 per kilowatt	1,300,000.	2,600.	3.31
Satellite Control System assumed	750,000.	1,500.	1.91
System Integration Charge 1 % of cost	189,000.	378.	0.48
Program Management Charge 5 % of power cost	945,000.	1,890.	2.41
Total	13,209,000.	26,418.	33.63

We can ask how the above costs, 34 and 16 cents per kW_e-hr, compare to power prices and costs in the Third World and the US. Table 5 is a summary of electric power costs for both domestic and foreign users. The price deflators data is taken from Table No. 732 of the 1988 Statistical Abstract of the United States of 67.3 for 1977 to 1982 and 114.1 for 1982 to 1986 with 1982 as the base the DOE/NASA estimated cost of power becomes 7.31 cents for 1980 (4.92/0.673) and 8.34 cents for 1986 (7.31 times 1.141). Foreign data was taken from Electric Power for Rural Growth, Barnes 1988. My conclusion is that electric power from orbit on economic terms alone could be competitive.

One of the features of Global Rural Electrification is that it proposes to add power in small increments. Granted this might not be as efficient as building one gigantic power satellite but the approach is affordable and relatively opened ended.

Impacts and Paths

How do Global Rural Electrification and electric power from orbit fit with the Human Exploration of the Solar System? So far my costs have been based only on bringing all material needed to build the power satellites from Earth. From both a cost and environmental viewpoint this is not an acceptable approach even if all the launch vehicles use only oxygen and hydrogen for fuel. Bringing building materials from the Moon along with oxygen for propellant and eventually volatiles from the moons of Mars and the asteroid belt offer even greater savings as well as a rationale to explore space.

Table 5 Electric Power Costs (cents/kW_e-hr)

Domestic	Low	Average	High
Residential (1986)	7.8	8.5	15.35
Hawaii(all fuel imported, 1986)	10.03		14.61
Portable Power Supplies (1988)	Capital Cost	Oper. Cost	Total
Gasoline (5 kW)	25.89	16.40	42.29
Gasoline (25 kW)	12.82	14.56	27.38
Diesel (50 kW)	7.36	7.00	14.36
Diesel (100 kW)	5.86	8.10	14.79
Rural India (Price not cost)	Low	Average	High
Punjab (1980)	4.8		
Grid supplied power (1980)	12.6		18.0
Andhra Pradesh (1980)			55.0
Northern & Equatorial Africa		25.0	

Consequently, the first prototype power satellite creates both the need and rationale for exploring the Moon and the remote sampling of Phobos and Diemos. Successful tests then lead to a manned base on the Moon to develop processes to produce the construction materials. Having building materials available on the Moon makes building a 16 meter telescope on the Moon to search for human habitable planets around other stars economically feasible. Should remote probes confirm that Phobos and Diemos do in reality contain volatiles including water, then economic necessity pulls mankind outward to Mars. The need for specialty metals will lead mankind further out to the asteroids while increasing space industrialization will compel him to tap the atmospheres of Jupiter and Saturn. Consequently the mission to provide power from space to the rural poor of the Third World will require that we learn to utilize the resources of the solar system and that need will drive us to explore. The two reasons, utilization and exploration, will together provide the stimulus to the human spirit that President Bush was seeking when he asked for a new course to be charted to move outward into the solar system.

It is doubtful that any government agency will initiate or propose a program to provide the Third World with electric power from space. Furthermore, as pointed out in a letter to me from the Indian Space Research Organization, a program to provide electric power from orbit to developing nations must be carried out under the auspices of an impartial international organization. Having been held hostage once by high energy prices the nations are justifiably leery of being held hostage again as political parties and policies change with elections. Consequently an international foundation is being established to promote and deploy at least the first power satellite. It is thought that once a successful demonstration is made an international consortium similar to Intelsat can be established which would receive foreign aid funds from participating nations as well as loans from the World Bank. The consortium would operate similar to ESA in awarding hardware contracts for the space based systems and would work with the receiving nations to create in country capability to build the ground stations.

Summary

No nation can long remain morally strong or economically competitive without an inner purpose. That purpose may be unstated, as it was during America's westward expansion,or it can be explicitly stated, as it was for World War II, the Marshall plan and Apollo. Kennedy, perhaps unconsciously, recognized this fact when he presented America with the challenges of Apollo and the Peace Corps. President Bush appears to be seeking a project with a similar purpose to those listed above. There are many noble projects and needs to choose from but few which will both kindle the individual's imagination and provide economic as well as spiritual benefits to a nation in its battle to remain economically competitive and financially secure. Perhaps we need a new breed of merchant adventurers, such as John Cabot in the early 1500's, who will explore the solar system looking for resources for the benefit of mankind. We should remember that the merchant adventures who went looking for spices found instead land, furs, gold and silver and new thoughts about government and the basic rights of man.

Many nations and a few U.S. politicians have already seen that economic power based on scientific and technological achievements rather than military power has been the dominant force for change during the last decade. If the current superpowers are not to slip into the status of Second World nations they must pick distant vistas which will motivate and challenge both their populations and their politicians to strive for excellence. Global rural electrification with the ultimate goal of providing non-polluting energy from orbit to all of mankind could be one of those vistas.

Exploration should not, at least in the 21st century, be limited to poking around new real estate. There are areas in both space science and space engineering which will open up new vistas for mankind. Vistas which include new real estate as well as new paths for the spirit of mankind to walk.

Acknowledgements

I wish to thank Bill Stanbro of Los Alamos National Laboratory for clarifying the exploitation versus exploration initiative, his comments on competition for fuels, scrubber sludges and the energy needs of the Haber-Bosch process. Phil Newman reminded me of E. F.Schumacher's work and provided helpful comments. Steve Howe of Los Alamos provided the concept of mankind on two worlds by the end of the millennium. Finally my thanks to Stewart Johnson and John Wetzel, editors of the proceedings for their patience and careful scrutiny.

While funding for this paper came from internal research funds some of the supporting material and referenced work was done in collaboration with staff members at Los Alamos National Laboratory and their support as well as the Laboratory's is gratefully acknowledged.

References

Barnes, D. F., 1988, Electric Power for Rural Growth: How electricity affects rural life in developing countries, (Rural Studies), Westview

Berman, M., 1989, Despite bleak Predications, Careful Planning Can Alter Future, p. B3, Albuquerque Journal, Sunday, December 31, 1989.

Davis, H.P., 1988, Solar Power Satellite: A Re-Visit, pages 866-869, Space 88: Engineering, Construction and Operations in Space, edited by Stewart W. Johnson and John P. Wetzel, American Society of Civil Engineers, New York, NY.

Dregne, H. E., Aridity and Land Degradation, pp. 16-33, Environment, Vol. 27, No. 8, October 1985.

Leonard, R. S., Blacic, J. D, and Vaniman, D. T., 1987, The Economics of Mining The Martian Moons, Space Manufacturing VI, May 1987, Princeton, NJ.

Leonard, R. S., Global Rural Electrification: The Role of Developing Countries, paper presented at the Symposium on Space Commercialization: Roles of Developing Countries, March 1989, Nashville, Tennessee, University of Tennessee.

Reference Notes - Table 1

1 Roland, A., The Shuttle: Triumph or Turkey, pp 29-49, Discover, November 1985.

2 Jane's Spaceflight Directory 1987-1988, page 462.

3 Jane's Spaceflight Directory 1987-1988, page 465, "In Feb 1988 at the Space Commerce Congress in Montreux Glavkosmos president Alexander Dunayev also hinted that a promotional launch might be arranged.

4 Table 3.6, page 41, Electric Power From Orbit, National Research Council, 1981

Biosphere II, located near Oracle, Arizona: Longitudinal Section showing wilderness biomes (from left to right): desert, marsh, marine, savannah and rainforest. (from Dempster, this volume)

SECTION 4

SUPPORT INFRASTRUCTURE

The development of the capability to go into space and provide the basis for a sustained presence there requires a substantial support infrastructure. In this section that support infrastructure is highlighted including terrestrial analogs, life support, waste management, and power and heat rejection. The substantial support also encompasses attention to the education of our people, our cultural diversity, and our experiences with project Apollo. Seven high school student papers are included at the end of this section. The papers in this section are subdivided into 9 sections:

EARTH ANALOGS AND TERRESTRIAL TESTING: Proposed Earth-based testing facilities are examined.

LIFE SUPPORT SYSTEMS: Closed Ecological Life Support Systems (CELSS), waste management, and water recycling processes are discussed in this section.

LAUNCH FACILITIES AND SUPPORT: Launch vehicles and facilities necessary to support a strong, on-going space initiative are presented including the National Aero-Space Plane and commercial facilities.

SPACE POWER: Power requirements, power systems, and waste heat rejection to support extraterrestrial and orbital construction and operations are examined.

HUMAN FACTORS: Topics in this section include the health hazards due to radiation exposure, and psychological and physiological symptoms of long-duration space travel.

ARTIFICIAL GRAVITY: The possible use of artificially induced gravity on long-duration space missions is discussed, as well as some facility concepts.

EDUCATION: Authors discuss education challenges, programs, and curricula dedicated to education for the future in space.

RESOURCE MANAGEMENT, ECONOMICS, AND LOGISTICS: Management and planning tools are examined for use in construction and operations of space facilities.

STUDENT PAPERS: Student papers cover various engineering, construction, and operations topics.

Antarctic Testbed for Extraterrestrial Operation and Technology

Larry Bell[1] and Deborah J. Neubek[2]

Abstract

A remote and harsh environment, Antarctica has areas which are more similar to the Moon and Mars than anywhere else on Earth. The physical similarities, coupled with strong parallels between the general nature of crew activities, functional requirements and operational constraints on antarctic stations and prospective lunar/planetary bases offer instructive comparisons. These include operations and logistics; facility planning, design and construction; utility system selection and design; and development of automatic and telerobotic systems.

The Sasakawa International Center for Space Architecture (SICSA) is planning an international research and technology demonstration facility in Antarctica to take advantage of these many analog opportunities. The Antarctic Planetary Testbed (APT) program will provide a basis for new insights into planning for Moon and/or Mars missions.

Introduction

People and activities on small antarctic bases have many attributes in common with astronauts on space missions. Highly motivated crews cut off from outside resources and services perform roles of recognized importance under difficult, sometimes dangerous situations. Such circumstances are difficult or impossible to recreate in an artificial environment. By complimenting laboratory experimentation and simulations with testing in a harsh,

[1]Director, Sasakawa International Center for Space Architecture (SICSA), College of Architecture, University of Houston, Houston, TX 77204-4431.

[2]Research Scientist, Sasakawa International Center for Space Architecture (SICSA), College of Architecture, University of Houston, Houston, TX 77204-4431.

remote environment, a broader picture of operational and mechanical performance may be painted.

A previous paper describes the APT program and the extensive data which may be transferred from antarctic bases to space mission planning (Bell, 1988). The following paper concentrates on some of the operational and technological areas that will be investigated by APT.

Operations and Logistics

Antarctic exploration and science activities have important parallels to lunar/planetary missions. Both involve dedicated people who must conduct difficult tasks with limited means, under conditions that often pose hardships and personal risks. Objectives sometimes require traverses over rugged terrain to locations that are remote from the main base where accidents, equipment failures or unforeseen events (e.g. snow storms and large solar proton events) can put people and missions in peril. Stresses resulting from long periods of isolation, fatigue, and interpersonal conflicts in small groups can degrade performance and judgement at critical times. Emergency countermeasures, including backup systems, operation protocols, search and rescue procedures and medical interventions must be planned well in advance.

The harsh antarctic environment is unpredictable and unforgiving. While personnel receive extensive survival training and instruction in emergency procedures, life-threatening situations may not be avoidable. Totally fail-safe systems are not practical, particularly when equipment is subjected to severely low temperatures. High winds and blowing/drifting snow threaten structures, travel and obscure visibility, sometimes cutting off survey teams and making rescue attempts difficult or impossible. These conditions bring true realism to space analog lessons that enable a better understanding of requirements for risk assessment and management.

1. Antarctic/Space Field Operation Similarities

Many typical field missions in Antarctica are similar in nature and in requirements to field studies that are likely to occur on future lunar/planetary bases. Some require individuals to journey distances away from the close protection and support of a main station. Examples are:

- Extended traverses associated with geological survey and sample returns (e.g. meteorite/rock collection, core drilling and seismic recording).

- Exploration to seek and observe extant and extinct organisms and ecosystems (on Mars as well as various Antarctic regions).

- Expeditions to set up or repair equipment at unmanned weather/magnetosphere observation posts (Antarctica) and radiotelescope installations (Moon).

- Search and rescue missions associated with all of the above activities.

2. Logistic Support Requirements and Constraints

Delivery of people and materials to and from bases in Antarctica/space represents the most difficult and costly programmatic aspect of these initiatives. Ice conditions make the continent inaccessible by ship during several months of the year. Extremely low temperatures and major winter storms frequently interrupt air service. The U.S. South Pole Station for example, is cut off from the large McMurdo Station between March and October. Air strips and roadbeds are properly absent in more environmentally sensitive locations such as dry valley areas. Access there is limited to helicopter range distances from a main station during moderate weather conditions.

Rigorous logistic support constraints lend realism to antarctic space operations analogs. Such restrictions force careful planning of crew changes, material and weather-dependent construction schedules and emergency rescue strategies. They also raise the importance of measures to make efficient use of consumable materials. Transportation of large quantities of fuel required for heat and power is particularly costly. Applications of state of the art wind, solar, waste utilization and other technologies might reduce this burden. Such technologies may also reduce pollution associated with diesel generators commonly used now.

Facility Planning, Design and Construction

Facility planning priorities and considerations for polar and extraterrestrial settlements are similar in many aspects. Important types of influences common to both are transportation constraints, harsh environmental conditions, and critical safety requirements. The APT offers useful means to transfer lessons which will benefit polar and space programs.

Research and concepts that have come to light through the course of planning for *Space Station Freedom* and other space habitats can be applied to upgrade outdated and inadequate facility designs that characterize some existing polar stations. In doing so, these advanced concepts and general insights can be tested in Antarctica to provide a valuable experience database to guide planning for future lunar/planetary settlements. Listed below are a few considerations in planning for these facilities:

- It is likely that the first lunar shelters will be self-contained, preconstructed modules brought from Earth to minimize necessary construction/set-up time, labor and equipment. Modular techniques can demonstrate similar benefits in Antarctica.

- The maximum space station module size established by the Space Shuttle payload capacity is approximately 42 ft. long (12.8 m) and 14.5 ft. (4.4 m) in diameter. The maximum construction element size for transportation to Antarctica by U.S. Navy ships is 20 x 8 x 8 ft. (6.1 x 2.4 x 2.4 m); by Hercules or Starlifter aircraft is 8.33 x 6.66 x 7 ft. (2.5 x 2 x 2.1 m); and by Sikorsky S-64 Skycrane helicopter is 27 x 8 x 8 ft. (8.2 x 2.4 x 2.4 m). (SICSA has designed a system to create approximately space station module equivalent volumes from 4 ship/helicopter-sized containers.)

- In Antarctica, opportunities for construction in and with snow are limited because temperature sensitivity prohibits extensive use for sophisticated applications. Some habitats have been placed in tunnels for protection from harsh surface conditions. Snow loads on tunnel roofs, however, cause considerable structural deformations and maintenance problems. Placement of lunar modules in trenches, then covering them with soil (regolith) is often proposed as a radiation protection measure. Parallels between these Antarctic and lunar approaches pose similar construction considerations.

- Low temperatures in Antarctica and on the Moon require efficient, safe heating systems and good insulation. Facilities need to be well sealed for pressurization on the Moon, and from the ingress of high winds and blowing snow in Antarctica.

- Overheating due to near perfect insulation and no air infiltration represents a major problem at some antarctic stations (e.g., Scott Base). Evacuation of excess heat is a vital consideration for facilities in space.

- Fire is a great hazard and concern in Antarctica. Since humidity is very low, wood dries out and becomes very combustible. Accordingly, metal framing/construction is preferred. Stringent fire prevention methods include sprinklers and alarm systems. Sprinklers demand ample reserve water storage and reliable, self-contained pressure systems. Similar, but even more critical concerns apply to extraterrestrial bases where emergency evacuation and fire fighting resources (e.g., water) are much more restricted.

- Sewage and waste disposal is a critical environmental concern in Antarctica. Treatment plants with provisions for recycling materials and/or removing sludge are needed. Similarly, effective waste management/recycling methods are vital for safety and resource conservation in space.

- Reliable, efficient, non-polluting power systems are needed in Antarctica and in space. Power storage systems are necessary to supplement solar systems during long antarctic/lunar nights.

Work is in progress to plan and develop special APT habitats (Transhab Modules) which will provide living and research accommodations for crews. Individual units will be modest in size, but efficient in character, to facilitate air transportation to the selected site. Severe transportation constraints posed by launch vehicles and surface landing systems will similarly limit the unit size of future lunar/planetary habitats.

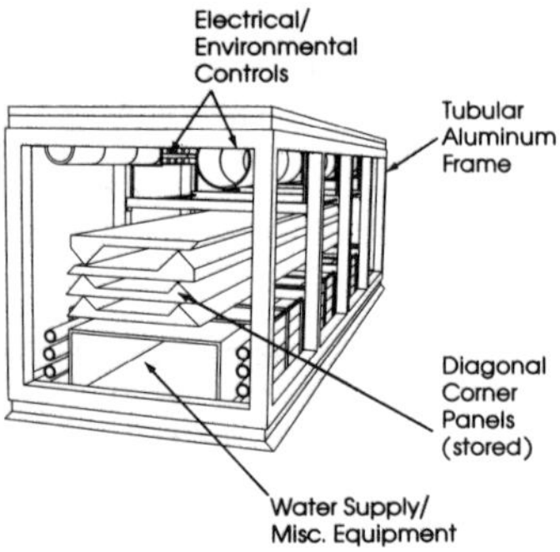

Figure 1. Module Core Package

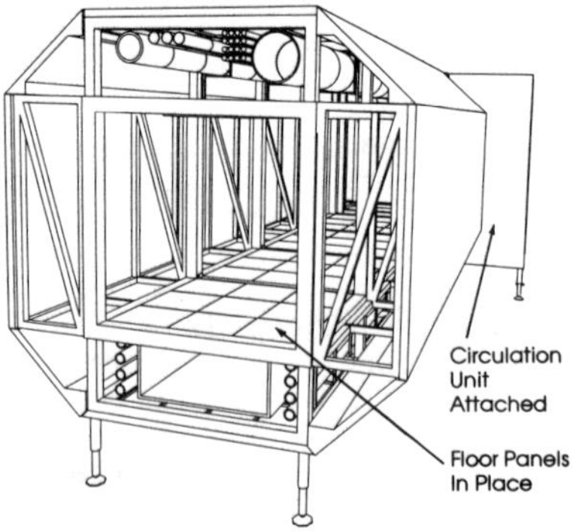

Figure 2. Module Deployed

APT crew sizes will be restricted to the smallest practical populations necessary to accomplish essential tasks. This is mandated to maximize space analog realism and to minimize adverse human impacts upon the site and any natural ecosystems that may be present. Specific numbers of people to be accommodated will be influenced by seasonal weather constraints and scientific priorities dictated by APT site conditions and applications.

The Transhab Modules are being designed to have interior volumes and furnishings which are very similar to those that will exist on *Space Station Freedom*. Basic living and research laboratory elements are modular racks for equipment, and functional units for crew occupancy. These elements can be added, subtracted and repositioned in response to changing needs and conditions over time. The space station-type racks will contain health maintenance equipment, food preparation systems, storage units, and laboratory experiment devices. The functional units will offer private enclosures for beds, showers, personal hygiene accommodations and toilets. Utility connections are being designed to facilitate easy access and removal for periodic equipment modifications and routine/emergency servicing.

SICSA is also investigating requirements and approaches using resin foam-rigidized structures to create large, insulated enclosures for space and terrestrial applications. Such construction offers an attractive alternative to move conventional rigid systems for specialized uses where simple, spacious volumes are needed. A key advantage is the ability to transport these structures in a compact, transport-efficient and easy to deploy form.

The APT program will provide diverse opportunities to test new materials, construction processes and assembly techniques under harsh environmental and restricted operational conditions. Advantages and limitations of alternative methods can be compared in terms of actual labor time and problems encountered, durability under temperature extremes, and other performance features. Construction of large towers, and antennas at the APT site can provide opportunities to demonstrate and evaluate advanced methods to create erectable and deployable frames. Representative space applications for such systems include major space station infrastructure keel and mast elements as well as expansive frameworks for proposed solar space satellites to beam power to Earth.

Utility Systems Selection and Design

As in space, high transportation costs, restrictions on cargo payload volumes and limitations on periods of site accessibility pose serious logistic constraints. These constraints impact the design of facilities and major equipment elements, resupply of consumables and crew rotation cycles. Stringent conservation measures are necessary to conserve fuel for power and heating (a dominant logistic burden) and to protect the environment.

1. Waste Management and Recovery

APT can serve as an operational testbed for new/improved waste management and recovery technologies which can be beneficial in space and remote terrestrial settings. Types of systems to recover water that are being given consideration include: vapor compression distillation,, thermoelectric integrated membrane evaporation, reverse osmosis and multifiltration.

Although current baseline equipment for the space station is limited to waste stabilization and does not include waste recycling systems, consideration is being given to waste processing in an augmented life support system to be added at a future time. Waste processing will be necessary for extended interplanetary missions.

All waste treatment techniques involve the oxidation of the organic material either by microbial action (as practiced on Earth in municipal water treatment facilities and waste dumps) or by high temperature (incineration and wet oxidation). Waste treatment systems for an augmented space station life support system fall into two categories, those based on application of "proven" technology and those based on new technology. Examples of the first category include dry incineration and wet oxidation, while supercritical water oxidation and waste electrolysis represent emerging technology options.

2. Power Systems

Important power system selection objectives are to minimize dependence upon imported fuel sources, environmental pollution and ease of maintenance. Systems with potential value for planetary applications will be tested on a continuing basis and used when feasible. A variety of technologies are being considered, including biomass systems to provide fuel from organic wastes; wind turbines, solar systems, and nuclear power systems for power generation; and advanced batteries and fuel cells for energy storage.

The APT program will apply advanced power technologies developed by many different nations when and where possible. NASA programs such as the Civilian Space Technology Initiative (CSTI) and the Pathfinder program, for example, are supporting aggressive efforts to develop advanced power systems for the future. Improved systems created through these and other enterprises may offer clean and efficient options to replace the fossil fuel-driven systems used throughout Antarctica today.

Automatic and Telerobotic Systems

Limitations on human operations imposed by small crews, premiums on labor time, harsh working conditions and difficulties in traveling to remote activity sites dictate an emphasis on automated and teleoperated systems for many activities in space. Similar conditions can apply to field research activities and the support of unmanned stations in Antarctica. Automatic/telerobotic requirements common both to space and Antarctica can include:

- Teleoperated/robotic construction and repair systems for habitats and ancillary structures.
- Automation of remote unmanned Earth and planetary science outposts.
- Automated/robotic sample detection, collection, and analysis systems for lunar/planetary and polar exploration research.

1. Construction and Repair Systems

Technologies currently exist to erect and repair "tinker toy" type space structures using teleoperated systems. Possible future methods might automatically fabricate frame structures in space using "beam builders" that apply manufacturing processes presently used in industry. Both of these approaches might be useful for constructing large towers and antenna systems on lunar/planetary bases as well as elements for space stations. Demonstrations in Antarctica can afford training opportunities for future space mission crew candidates under realistic operational and environmental conditions. Resulting structures can provide utilitarian benefits in Antarctica.

Fully automated manufacturing processes might eventually transfer existing terrestrial technology to the space environment. Proposed beam builders, for example, would transform metal strips contained on spools to triangular beam sections complete with struts and ties. While necessary equipment to accomplish this operation might be complex and bulky, the process could be valuable to create structures using extraterrestrial materials obtained from the Moon, Mars or asteroids.

2. Deployment and Operation of Unmanned Observatories

Unmanned observatories in Antarctica provide important atmospheric and astronomical data. Strong arguments favor a future unmanned lunar observatory on the far side of the Moon which would present similar conditions and requirements. Challenges common to the antarctic and lunar facilities include overcoming difficulties in deploying and servicing remote instruments, avoiding structural thermal strains, and avoid distortions to accomplish precise pointing control. Advanced automated and telerobotic devices demonstrated in Antarctica can address these needs.

Facilities presently in use in Antarctica have automatic low-power systems that provide for the acquisition, preprocessing and transmission of data from remote areas. Deployment and continuous operation of high-resolution arrays for such stations to monitor magnetic perturbations and meteorological conditions around the South Pole are an essential part of cusp and polar cap studies and measurements of the atmosphere boundary layer. Automatic stations also serve as floating buoys on the antarctic pack ice belt.

The U.S. Polar Research Board (PRB) recommends that new clusters of remote observatories with real-time data transmission to a central site and/or orbiting spacecraft are needed for atmospheric and upper atmospheric studies. Such accommodations, along with large very low frequency array (VLFA) radio transmitters and receivers also apply to a future lunar observatory.

Several new construction and automation technologies for a lunar observatory might be demonstrated in Antarctica. Examples are:

- Large, stable structures to support observatory components, including geometrically precise elements using advanced materials such as metal matrix composites with very low coefficients of thermal expansion.

- Steerable dish radio telescopes including means to move and align very large structures for high precision tracking (i.e., to milliarc-seconds) with controls to prevent/damp out structure interaction interferences.

- Technologies to automatically measure and maintain surface accuracies of millimeter and submillimeter radioastronomy parabolic dishes to within 5-10 microns under changing thermal conditions.

- Means to assemble and disassemble structures without loss of accuracy.
- Remote communications and control systems. (Observatories on the far side of the Moon and in Antarctica will need communication antennas for uplink and downlink to orbiting satellites.)

3. Collection and Analysis of Surface/Subsurface Samples

Manned and unmanned lunar/planetary missions will rely upon automated devices to seek, obtain and analyze surface and subsurface material samples. Some initiatives, and particularly those which precede human settlements, may utilize unmanned survey rovers to explore regions of interest in search of useful resources. These robotic prospectors might be pre-programmed to follow designated routes, using artificial intelligence (AI) to avoid obstacles and hazardous terrain conditions. Alternatively, they might also be guided by remote human observers on Earth or in space who view the surroundings via television monitors, issuing commands through relay satellites. Automated or teleoperated scoops and coring apparati incorporated into the prospectors would collect samples to be returned for study or to be analyzed by an automated onboard laboratory.

Key elements of these advanced systems can be developed for beneficial antarctic survey applications as well as lunar/planetary simulations in Antarctica. Important components can include lightweight, durable and efficient means to obtain deep core samples; compact, portable specimen analysis equipment; and effective, easy to operate teleoperated manipulators and remote monitoring/control devices.

Summary

Severe environmental conditions on the Moon and Mars will pose hardships for people and machines. Failure to validate procedures and systems under realistic conditions is likely to be costly in terms of human life and/or failed missions. The APT facility will afford a valuable environment for realistic simulations and assessments.

References

Bell, L., and G. Trotti. 1988. Earth-based analogs of lunar and planetary facilities. Paper presented at the Second Symposium of Lunar Bases and Space Activities of the 21st Century. Proceedings in press. Houston: Lunar and Planetary Institute.

A CENTER FOR EXTRATERRESTRIAL ENGINEERING AND CONSTRUCTION (CETEC)

Peter A. Hart[1]
Steven D. Howe[2]
Stewart W. Johnson[3]
Gerald G. Leigh[4]
Raymond S. Leonard[5]

ABSTRACT: The Center for Extraterrestrial Engineering and Construction (CETEC) has been envisioned as an organization, a center of knowledge and information, and a set of experimental test facilities. The primary purpose would be to deal with the technologies related to engineering and construction on extraterrestrial bodies. The organization responsible for CETEC is proposed to be a consortium of universities, government laboratories, government agencies, private industries, and small businesses. An international cooperative approach to the research facility is envisioned, which will benefit space engineering researchers from all over the world.

BACKGROUND: The CETEC concept was initiated through an awareness on the part of members from the founding organizations of the emerging need and tremendous opportunities associated with such a center (Figure 1. Artist Conception - CETEC Facility).

In the fall of 1988, the CETEC concept was presented to the Office of Exploration at NASA headquarters. Discussions following that presentation indicated a need to obtain a consensus from a larger body of knowledgeable persons regarding the need for such a center, who would be the user community, and what types of research, development, and testing would be accomplished in such facilities. It was decided that the best way to accumulate this information would be to hold a workshop dedicated to that purpose. The workshop was held in Albuquerque, New Mexico, in February 1989. The purpose was to define a Center for Extraterrestrial Engineering and Construction (CETEC). The concept presented for discussion was that CETEC would be a center of knowledge and information and would operate a group of test facilities where engineering, construction, mining, and materials processing problems that may be encountered on extraterrestrial bodies could be simulated realistically. A goal of the CETEC workshop was to organize and build upon the extensive body of knowledge that resides with the individuals invited to attend the workshop. The workshop was seen as a means through which a number of very knowledgeable people could be brought together to identify and discuss what problems would be encountered, what technologies would need to be developed, and what test facilities would be required to support such efforts. Many of the attendees have worked with space related activities for many years and are considered experts in their respective

[1]Peter A. Hart, NMERI, The University of New Mexico, Albuquerque, NM 87131

[2]Steven D. Howe, Los Alamos National Laboratory, P.O. Box 1663, Los Alamos, NM 87545

[3]Stewart W. Johnson, BDM International, Inc., 1801 Randolph Road, S.E., Albuquerque, NM 87106

[4]Gerald G. Leigh, NMERI, The University of New Mexico, Albuquerque, NM 87131

[5]Raymond S. Leonard, Ad Astra, Ltd., Rt. 1, Box 92 LL, Santa Fe, NM 87501

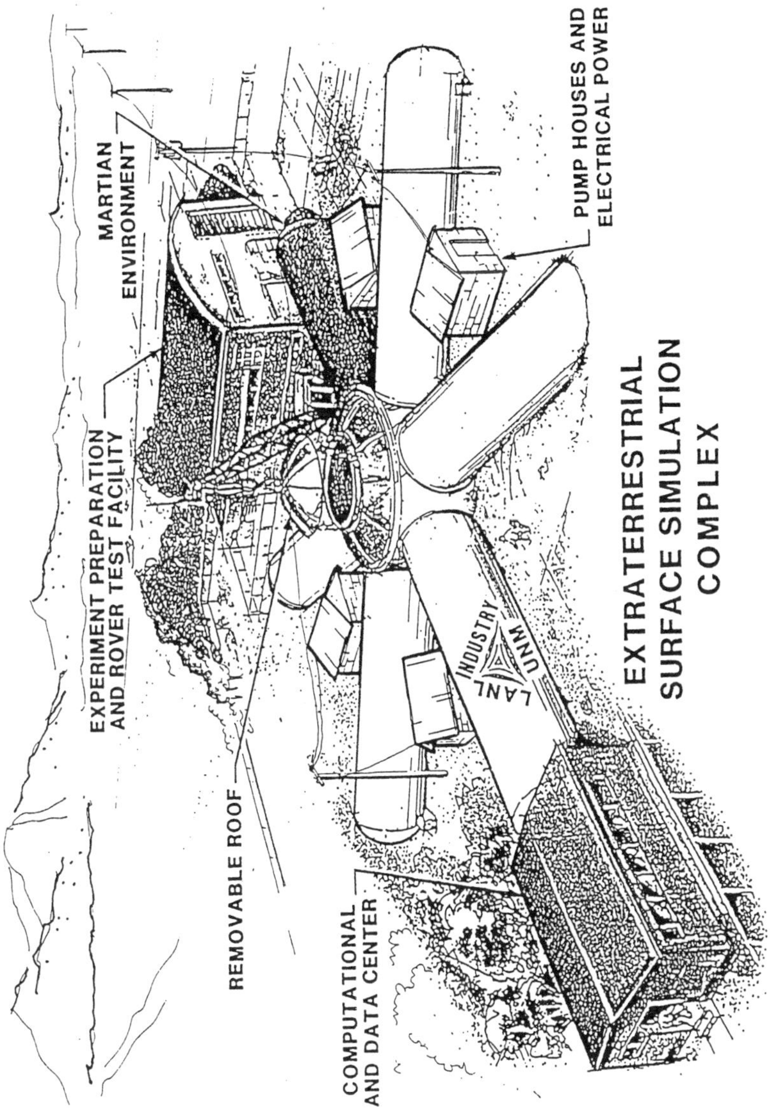

Figure 1. Artist Conception - CETEC Facility

fields. The collection, distillation, and collation of their thoughts on these topics constituted the main effort of the CETEC workshop, and are the key points to be presented in this paper. A more comprehensive report of the workshop and list of attendees are available from the authors of this paper.

INTRODUCTION: Why Does the US Need an Extraterrestrial Surface Simulator? In 1969, man first set foot on the surface of the Moon. Three and one-half years later the astronauts from the Apollo 17 flight departed the Moon and no human has returned since. The total accumulated time spent by humans on the surface was three weeks. For the 15 years following the Apollo program, there was little national interest in returning to the Moon. Instead, the nation concentrated on creating a space transportation system and designing a space station.

In the mid-1980s, this situation began to change. Discussions of future lunar outposts and Mars missions have increased throughout the international space community. A number of research and development projects strongly oriented toward a return to the Moon have been funded. Several well-attended conferences, such as the Lunar Base Symposia (1984, Washington; 1988, Houston), and Space 88 and Space 90 (Engineering, Construction, and Operations in Space--1988, 1990, Albuquerque) are among the signs of an awakening national interest in space exploration and in the use of materials from the Moon, Mars, and the satellites of Mars. Recent milestone documents, such as *The Report of the National Commission on Space* and *The Sally Ride Report*, and statements by presidents of the United States on national space policy emphasize the need for US leadership in the exploration and development of space. The Japanese have set forth their plans for a lunar base. In order to explore the Moon and to make a mission to Mars possible soon after the year 2000, it will be necessary to set goals for extraterrestrial development by 1992. Preparation for 1992 goal setting will require a center such as CETEC.

The enabling technologies required for existence on the Moon are being developed and the groundwork is being laid for a return to the Moon--and to stay. This movement of humankind toward a new frontier follows clear precedents such as the voyages to the Americas, the settling of the West, and the establishment of outposts in the Antarctic.

Advances in engineering technology and the application of science to societal needs are generally goal- or market-driven. Many nations have chosen space exploration as one of their national priorities for promoting technology development. The anticipated need for engineering, construction, and operations in space and for facilities such as lunar bases provides the focuses and goals for advancing US construction technology and capabilities. Past experiences with Apollo and Skylab and the present Space Transportation System (STS) have repeatedly demonstrated the need to test and validate equipment and processes in simulated environments.

NASA has for more than 25 years conducted "paper studies" on the creation of lunar and martian bases and colonies. In addition, a number of organizations and companies have conducted both funded and unfunded studies to develop concepts for processing extraterrestrial resources. Some of the very early studies defined the needs and requirements for the Apollo landings. These studies also led to experimentation and the development of actual flight-qualified hardware, such as the rover and a drill, for geophysical exploration. Since the Apollo landings, little experimental work in the areas of construction, mining, and processing in the environments of outer space has been pursued. CETEC will be instrumental in fostering such activities.

CETEC is envisioned to be both:

- A facility for conducting experiments needed to develop the technology pertinent to constructing facilities and operating equipment on other planetary surfaces.
- A clearinghouse of knowledge and expertise for conceiving, designing, and developing the systems needed to establish and maintain human presence on the planetary surfaces.

Why Build an Extraterrestrial Surface Simulator Now? Many years are required to proceed from the idea stage to proven concepts and working hardware. For example, in the case of the Hubble Space Telescope, this required more than 25 years.

Engineering tests are a vital part of the process of developing products that must work in new and ill-defined environments. These are the environments that the engineer and constructor will face when they must develop new and workable schemes for engineering and construction on the Moon and on planetary surfaces.

It appears that lunar surface construction will be a reality in the first decade of the twenty-first century, and perhaps as early as the late 1990s; therefore, it is essential that concepts be tested in the early 1990s. Concepts that permit testing to establish their validity must be selected. Early involvement of the test community will help focus planetary construction concepts in practical, cost-effective directions.

For What Will an Extraterrestrial Surface Simulator be Used? To design and build structures on extraterrestrial surfaces, we must understand not only the basic processes but the scaling laws involved in moving from the benchtop to an extraterrestrial construction site. An extraterrestrial surface environmental simulator in which pilot plants and equipment can be run is necessary in order to develop economically and safely the capability to live and work in space.

The simulator will also be used to provide benchmark tests for a wide variety of suggested extraterrestrial (ET) construction schemes. Candidate construction approaches on the Moon range from extensive use of native lunar materials to heavy reliance on imported materials and systems requiring assembly, checkout, start-up, and operation. Testing in a simulated ET environment is required in order to begin to sort out the critical factors associated with each proposed approach to extraterrestrial construction. These tests will allow engineering research to be focused with greater confidence and credibility in the most productive directions. The scientific and basic engineering research communities will proceed with a much greater level of confidence and productivity when they are able to test the theories and basic concepts supporting the design effort.

Candidate areas for research include:

- Deploying and erecting/inflating building materials and structural systems in ET environments
- Manipulating/processing aggregates on ET surfaces for construction and shielding
- Mining/boring/coring of ET mineral deposits and geological formations
- Maintenance and repair procedures for equipment working in ET environments
- Developing chemical process streams for extraction of key elements from soils and rock.

CHALLENGE: On January 5, 1988, then President Reagan approved a revised national space policy that set the direction of US efforts in space. The policy was the result of a five-month interagency review, which included analyses of previous presidential decisions, the National Commission on Space report, and the implications of the space shuttle and expendable launch vehicle accidents. The primary objective of this review was to consolidate and update presidential guidance of the US space program and to provide a broad policy framework to guide US space activities well into the future. The resulting presidential directive reaffirmed the national commitment to the exploration and use of space in support of our national well being.

GOALS AND COMMITMENT: On July 20, 1989, President George Bush called for a new bold step in America's space program. He said, "The time has come to look beyond brief encounters. We must commit ourselves anew to a sustained program of manned exploration of the solar system and yes, the permanent settlement of space." The president went on to say, "I'm proposing a long range continuing commitment. First, for the coming decade--for the 1990s, Space Station Freedom, our critical next step in all our future endeavors. And next, for the new century, back to the Moon. Back to the future, and this time to stay. And then a journey into tomorrow, a journey to another planet, a manned mission to Mars."

THE INFRASTRUCTURE TO IMPLEMENT POLICY: Two themes seem appropriate to any discussion of the infrastructure needed to implement national space policy--especially within the context of the CETEC goals. First, a human presence on a planetary surface requires technology similar to that found on Earth, but dissimilar to that needed in orbit. Just as three decades of technology base development have provided the basis for the US space station, so too must we begin now to establish the technology base and infrastructure to allow for the evolution of a similar basis for solar system exploration and colonization. Second, CETEC will provide a unique capability to nurture the technology integration, validation, and development required for larger-scale extraterrestrial surface operations in an environment similar to that to be encountered.

Specifically, CETEC will support the development of the needed infrastructure and take us a step closer to implementing national space policy by:

- Providing a center for developing and testing new technologies for expanding the human presence into the solar system. Through the unique capabilities of the CETEC complex, government, industry, and university users will be able to develop the techniques required to establish human habitation on other planetary bodies.
- Promoting the development of new technologies for scientific, technological, and economic benefits. These technologies could likely have terrestrial applications to robotics, mining, materials, and construction techniques.
- Expanding private sector participation by providing a testing center for industry to develop new technologies for future space exploration. CETEC would provide a facility where companies could learn to work in space.

We believe that a facility of this type, possessing the unique capabilities currently envisioned, is a necessity for successful planetary missions and can produce valuable benefits for its constituency and the nation.

CONCEPT OF OPERATION: As currently envisioned, CETEC will introduce experiments into the vacuum chambers via an airlock. Because of the extensive efforts that will be required to preprocess the lunar simulant in order to achieve a vacuum of 10^{-4} Torr or less, the tanks are not intended to be exposed to the atmosphere once they are prepared. Consequently, all experiments performed in the tanks, until they are acceptable for human habitation, must either be automated or teleoperated.

The following is a summary of the technical design requirements for CETEC:

- Lunar construction equipment design and development will require a large facility for simulation of the lunar environment, parametric research, and endurance testing of scaled models of the machines.
- Generic components, such as fasteners, bearings, seals, cables, sheaves, motors, etc., can be tested in vacuum chambers to ensure compatibility with lunar environments.
- Teleoperation remains a major issue that must be resolved and on-line when the first test article enters the program.

- A tank-rated robot will be needed to service the test articles and maintain the inside of the facility. It would be advantageous to design the robot much like the lunar version so that it might also yield relevant experience in remote maintenance and repair of the construction equipment and long-term experience with complex machines. A hybrid three-legged, wheeled robot with a robotic arm is suggested. A kit of spares and expendables would accompany each test article. This robot would be a candidate configuration of a lunar service robot for construction and other applications on the Moon.
- Ways of simulating reduced gravity for validating parameters such as traction and dust dispersion need to be refined, the development of which would occur at CETEC.

USES: CETEC can function effectively as a research, development, and test facility for teleoperated and automated robotic systems in a vacuum environment with fine-grained soils and thermal cycling. Such capability is important in lunar astronomical observatories as well as mining, construction, manufacturing, building systems, resource extraction and processing, and human factors.

Construction Technologies: One component of the human exploration of space could be a 16-meter optical telescope located on the Moon. This telescope would be used to search for planets around other suns and perhaps determine the probable atmospheric composition of any planets found. Techniques for construction of this telescope and of other observatory elements would occur at CETEC.

Technologies for deployment and construction should minimize the need for intervention by construction workers in space suits. Teleoperated cranes may serve as backup for automated off-loading of components from arriving payload packages. Methods will need to be developed to prevent the accumulation of fine-grained particulates from the lunar regolith on mating surfaces of contiguous elements. Deployment and erection sequences must be carefully preplanned so that components match up in spite of temperature variations from component to component over time. Confidence in deployment and erection technologies will be critical in determining the future success of the observatory and of other components of a lunar base.

Another challenge is that related to the design of radiators for equipment that are capable of rejecting heat during power generation. A CETEC simulator will allow component and subsystem testing of heat rejection capability. Different materials may be tested and approaches applicable to dusty environments may be validated.

A robust space exploration program on the Moon and eventually Mars will require a power rich environment if we are to realize the full benefit of our investments. Power rich implies the need for a dynamic power generation system which implies the need for large radiation systems. A lack of atmosphere on the Moon and the very thin Martian atmosphere imply that the heat rejection methods will be primarily radiative. Radiative heat transfer is very sensitive to surface absorptivity and emissivity, which are both dependent upon surface contamination. Consequently, developing and validating power systems for facilities and outposts on other planets will require testing the heat rejection subsystems. This type of testing will necessitate a fairly large area that can maintain extreme thermal differentials and simulate dust clouds.

SUMMARY: The CETEC concept has evolved and matured since its conception in mid-1988. The research complex, as currently envisioned, consists of a set of interlinked vacuum chambers which will be able to provide simulated lunar and martian environments with soil and rock simulants. Adjacent to the vacuum chambers will be a large, several acre plot of simulated lunar soils for testing construction equipment and concepts prior to vacuum testing. Also planned are a series of habitats for use in studying human factors, operating concepts, architectural layouts, and possibly closed environmental living system concepts developed elsewhere. The facility will be designed to facilitate the involvement of the public without impairing research activities or endangering visitors.

The evolving CETEC concept will truly represent a national asset for testing and validation of concepts and systems intended for service at a lunar outpost, a mission to Mars, and other extraterrestrial bodies.

REFERENCES

1. A Center for Extraterrestrial Engineering and Construction (CETEC): Findings and Recommendations of the CETEC workshop held in Albuquerque, New Mexico, February 27 - March 3, 1989, (128 pages, available from authors).

2. Johnson, Stewart W., and John P. Wetzel, editors. (1988). Engineering, Construction, and Operations in Space, ASCE, NY.

3. Johnson, Stewart W., and John P. Wetzel, editors. (1990). Engineering, Construction, and Operations in Space II, Volumes 1 and 2, ASCE, NY.

4. Mendell, Wendell, ed. (1985, 1990). Lunar Bases and Space Activities of the 21st Century, Vol. 1 and 2, LPI, Houston.

5. National Commission on Space (1986). Pioneering the Space Frontier, Bantam Books, NY.

6. Ride, Sally. (1987). Leadership and America's Future in Space, US Government Printing Office.

Subsection 4.2: LIFE SUPPORT SYSTEMS

BIOSPHERE II: CLOSED ECOLOGICAL SYSTEMS ENGINEERING

William F. Dempster[1]

Abstract

This paper provides a general description of the Biosphere II project — a 3.15 acre materially closed ecological system containing tropical rainforest, savannah, marsh, marine, desert, intensive agriculture and human habitat biomes — currently under construction and scheduled for completion in September 1990. It further describes the approach taken to solve particular problems that arise due to the fact that the system is completely materially closed and contains several complex ecosystems. Materials selection for avoidance of contamination and methods of controlling contamination are discussed. A method for pumping water which is benign to life forms in the water is described. Systems choices are strongly influenced by the criterion of no external resupply of equipment, spare parts, or consumables for very long durations. The Biosphere II project is aimed at creating regenerative ecological systems as life habitat prototypes for space locations, and to conduct ecological research applicable to the biosphere of the Earth.

General Description

Biosphere II is the first attempt to construct a large scale materially closed self-regenerative ecological system capable of indefinitely sustaining a human crew. The name Biosphere II is taken as a progression from the biosphere of planet Earth which, as the only observed biosphere in existence at present, is termed Biosphere I. All known life occurs within the biosphere of the Earth — a materially-closed, informationally and energetically open, stable, complex, evolving system of ecosystems.

---

[1]Director of Systems Engineering, Space Biospheres Ventures, Post Office Box 689, Oracle, Arizona 85623.

For both Biosphere I and II, material closure is true in the sense that no material transfers occur across the system boundary to a degree that has any practical effect on the operation of the system. Thus, although the planet Earth receives input of some meteoritic material and bombardment by the solar wind, the amounts are so incommensurably small that there is not a significant effect on the life sustaining ecological cycles. Similarly, for Biosphere II the containment will reduce external exchange beyond the point of significant influence on the operation of the system. The targeted external exchange of atmospheric gases is not to exceed 1% of the contained volume per year.

Informationally and energetically open simply states that there are no limitations on either information or energy crossing the boundary. Solar energy is the most obvious energy transmission. Biosphere II is fully glazed over some 16,000 square meters so that sunlight can provide the energy for plant growth. Further examples are electric power and thermal energy. Airtight electric feed-throughs provide for 2 Mw power transmission and closed loop hot and chilled water piping will carry heat in and out of Biosphere II. The energy source for electric power, hot and chilled water will be an energy center external to Biosphere II. Information transfers occur by computer data transfers, video, telephone, and radio transmissions.

Stability in ecosystems, although difficult to rigorously define, refers to similarity continuing over a time that is long compared to the cycle times of the ecosystem components. Most biomic regions of the Earth exhibit stability in this sense — in terms of types and densities of organisms, although numbers and proportions of component species may vary over time — and it is anticipated

Figure 1. Biosphere II Longitudinal Section showing wilderness biomes (from left to right): desert, marsh, marine, savannah and rainforest.

that the major biomic spaces of Biosphere II will retain their characteristics for decades.

Complexity is emphasized in Biosphere II by inclusion of seven biomes; five are representative of natural wilderness biomes and two are anthropogenic (of human origin). The five wilderness biomes — tropical rainforest, savannah, ocean, marsh and desert — represent major features beyond the usual approach to regenerative life support (Figure 1). There will be approximately 3,800 species of macroscopic plants and animals in Biosphere II at closure; some perennial plants or long-lived species, and some will be annuals or short-lived species. The presence of wilderness biomes adds a great deal of complexity and diversity to the entire system.

Species selection and configuration of the ecosystems — including 'habitats' or micro-ecosystems of varying environmental conditions within the major biomes — is being conducted by expert ecological consultants (see Acknowledgements, below). Documentation of species included in Biosphere II at closure is being recorded in a computer database developed by Space Biospheres Ventures for this purpose; this will serve as comparative data for ongoing evaluation and future research.

The initial period of closure is planned to be two years with a human crew of eight. The system is intended to be entirely recycling and self sustaining with regard to all materials requirements including food, water, and atmosphere. Nutrients for plant growth will be recycled from plant residues, animal, and human wastes through a composting system. There will be a selected stock of spare parts for the mechanical, electrical, and electronic systems to enable repairs while maintaining full closure. If no major problems surface during the initial two year occupancy, the crew can be exchanged through an airlock and operation continued indefinitely. The materials and design of Biosphere II have been oriented toward a 100 year operational lifetime.

The different biomes of Biosphere II occupy distinct areas (see Figures 2 and 3). Beginning at the north end of the wilderness area, the rainforest biome covers an area of 43.5 meters square. An artificial mountain is constructed under the pyramidal space frame and the area is planted with species representative of Amazonian rainforest. Extending southward from the rainforest, the area 84 meters long and 10 meters wide on the west side of the long wilderness section is planted with species from tropical savannah regions. A fresh water stream that originates in the rainforest flows southward through the savannah. The east side of the same long section is occupied by the 45 meter long x 20 meter wide ocean biome. The ocean comprises beach, lagoon, coral reef and deep ocean and is filled with a mixture of real sea water brought from the Pacific Ocean and artificial sea water prepared from salts plus fresh water for a total volume of about 4 million liters. The salts very closely approximate the salts naturally occurring in sea water. The deepest portion is 8 meters deep.

South of the ocean is the salt gradient marsh system. The savannah stream enters the southernmost and highest part of the marsh and flows through into the ocean. This freshwater flow maintains the gradient from fresh to salt water. Tidal exchange occurs between the ocean and marsh biomes by the action of a system that pumps water from the ocean to the marsh and then allows the risen marsh waters to flow back to the ocean by gravity.

The southern pyramidal space frame of the wilderness covers the desert area which is 37 meters square. It features species of tropical deserts such as are found in Baja California. A small salt lake occupies the center of the desert and will hold brackish water on a seasonal basis.

The wilderness areas described above comprise the entire east section of Biosphere II. It is connected by two passageways to the west section that

Intensive Agriculture Biome (foreground) Habitat (background) Rainforest Ocean

includes both the Intensive Agriculture Biome (IAB) and the Habitat. The IAB is 55 meters x 41 meters and provides the agricultural base for domestic animals and humans occupying Biosphere II. A wide variety of crops will be raised including wheat, sorghum, barley, rice, amaranth, malanga, yams, white and sweet potatoes, squashes, beans, broccoli, tropical fruits, strawberries, and numerous herbs and spices. Domestic animals include chickens, fish, pygmy goats and pygmy pigs. The chickens produce eggs as well as meat and the goats produce milk and meat.

The Habitat provides individual living space for each of the eight Biospherians (as the members of the human crew are called) plus kitchen facilities, maintenance shop, analytical laboratory, medical facility, recreation room, library

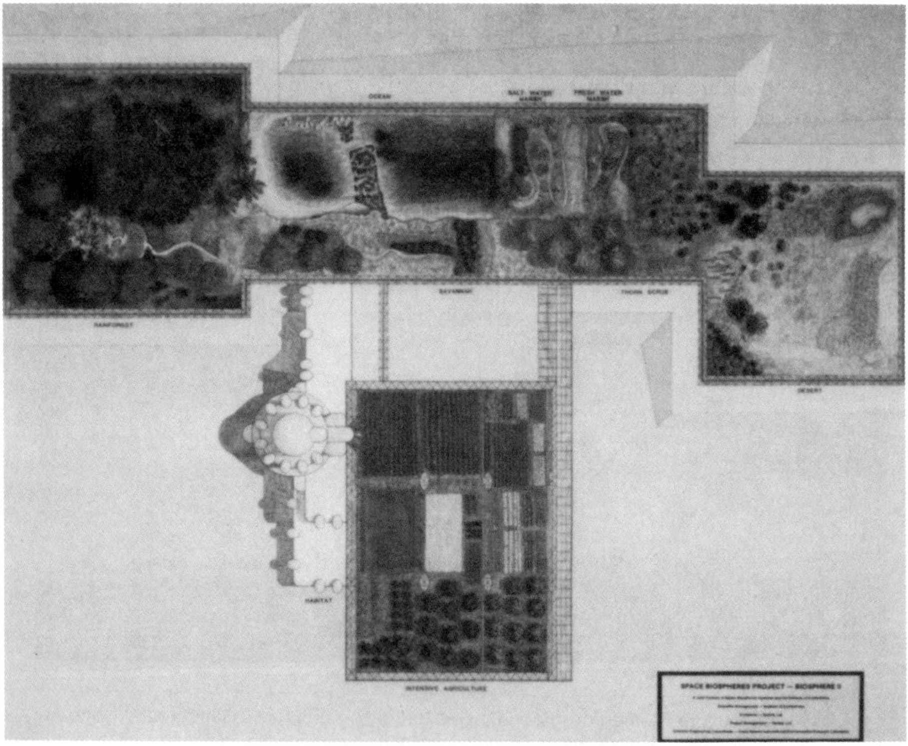

Figure 3. Biosphere II Floor Plan

and observation tower, and command room. The command room is the center of computer monitoring and control of Biosphere II. The Habitat stands against the full 55 meter north side of the IAB and has a curving front. The total footprint of the Habitat is 1077 square meters. It has two main floors plus two mezzanine levels and a library in the high central tower. The Habitat skin is predominantly steel paneling mounted within a space frame structure, with selected glazing as suitable to occupied buildings.

Outlying from the central structure described above are two variable volume chambers on the west and the south called 'lungs'. The purpose of the lungs is to provide pressure relief as the contained atmosphere expands and contracts with temperature inside Biosphere II. The combined variable volume of the two lungs is 43800 cubic meters. The lungs consist of a cylindrical steel tank 48 meters in diameter and 7.3 meters high. The top of the tank is sealed with a flexible membrane which also holds a central aluminum dish 28.5 meters in diameter. The weight of the dish imparts a slight pressure (about 1 mm Hg) to the air contained within. The result of this arrangement is that the internal air pressure of Biosphere II is greater than the ambient air pressure by about 1 mm Hg. Adjustment to internal temperature variation is passive and immediate. The connecting ducts between the lungs and the central part of Biosphere II are 2.4 meters in diameter. They are sized for a person to walk through rather than for airflow resistance which is negligible. Similarly, variation in the external barometric pressure results in lung volume change. If there is any leakage from the system, the internal overpressure will force it to be outward and preclude external contamination from entering. Also, leakage anywhere in the entire Biosphere II system will manifest as a slow descent of the membrane and dish as their weight forces the contained air out through the leak. The lungs therefore provide for system-wide verification and precise measurement of the degree of sealing achieved. The lungs are further protected by weathercover domes of space frame with painted steel paneling.

All of the spaces described above, with the exception of the external energy center, are one contiguous air volume. The lungs can be individually isolated by airtight doors for separate testing and in anticipation of membrane replacement after 30 to 50 years. In the construction sequencing, the IAB will be momentarily isolated together with one lung for airtight testing.

The Technosphere of Biosphere II

The principal purpose of the technical systems of Biosphere II is to support all of the life systems within the enclosure. The technical function is therefore subordinate to the life function and cannot override the life system requirements. There are several specific design areas where this consideration has made a substantial impact.

Materials

The materials of construction must be enduring and non-toxic. Heavy metals are of particular concern. Cadmium has specifically been excluded as a fastener coating. Copper piping has also been excluded.

It is important to realize that in a fully closed system long term accumulations can become critical after years or decades even if short term rates are within nominally accepted limits. A consequence of this consideration has been the exclusion of galvanized sheet metal for most of the air handling equipment and ductwork. In a very humid environment galvanized steel (i.e. a source of zinc) can be expected to become rusted steel in several years to a few decades. The zinc used in galvanizing also contains significant quantities of lead, cadmium and copper. The substitute used for galvanized steel in these applications has mainly been stainless steel.

Biosphere II is sealed on the lower side walls and floor slabs with a stainless steel liner of composition 20 Cr - 25 Ni - 6 Mo. Classed as a 'super stainless steel', this alloy is highly resistant to corrosion even in sea water service. There are, however, two separate problems which differ by orders of magnitude. The first and most obvious is corrosion which could threaten to penetrate the liner and break the airtight seal. The second is release of minute amounts of metals from the liner into Biosphere II which could accumulate and become toxic to living organisms. A testing program was conducted with samples of the liner material immersed in a recycling marine life system at the Smithsonian Institution and it was found that contamination by nickel, chromium, and molybdenum rose in the system at a relatively rapid rate. Although no corrosion that would threaten to penetrate the sample sheets was observable, there was a sufficient release of metal to threaten the marine life systems. In view of this data, an epoxy based coating is specified and applied on all liner surfaces that are immersed in water or overlain by soils.

Whereas metals have the potential to accumulate over long periods of time, organic contaminants are more susceptible to being broken down by a variety of processes. We have observed removal of airborne contaminants such as formaldehyde, carbon monoxide, methane, ethylene, solvent vapors, etc. by exposure to plants and soils. It appears probable that microbes metabolize many organic contaminants and reduce them to harmless compounds such as CO_2 and H_2O. This makes a strong case for inclusion of large amounts of microbial life forms in long duration space missions and in closed systems generally. Biosphere II incorporates some 15000 cubic meters of soil which will be fertile and rich in microbial life. A portion of the air circulation is forced through the IAB soil beds to increase the air-soil contact.

Sea Water Handling

The marine and marsh biomes contain water that is rich in plankton and microbial life forms. Conventional centrifugal pumps subject the pumped water to an instant of extreme pressure and acceleration which is lethal to these forms and an alternate method of moving the water was needed. The method adopted is to use vacuum to suck water to heights up to 2 meters and then release it to flow by gravity. In the ocean wave generator, the water is sucked up into a vertical chamber 17 meters wide by 0.5 meters deep by 2.4 meters high. Ambient air is then vented into the top of the chamber allowing the water to fall and surge out from the 17 meter wide mouth. The surge propagates across the ocean creating a wave that can be varied in height up to 0.7 meters and in frequencies up to once every 5 seconds. This wave action is a close approximation to natural waves which are crucial for the functioning of the living coral reef and other ocean biome life forms.

Tidal exchange between the marsh and ocean will be accomplished by pumping ocean water into the marsh allowing the marsh to rise 0.5 meters while the ocean is lowered about 0.25 meters. The timing will be synchronized with lunar cycles to simulate natural tidal frequencies. The pumps for creating the tides utilize the vacuum lift system. Within the marsh there are 6 subdivisions progressing from a fresh water pond (freshest) to a red mangrove area (saltiest). Each subdivision recirculates between 200 and 800 liters per minute by use of vacuum lift pumps.

Desalinator

The fresh water stream delivers an average of between 12 and 16 liters per minute to the ocean after passing through the marsh system. The Earth's oceans return analogously received fresh water by evaporation and rainfall, but in Biosphere II evaporation will be an order of magnitude too small to match the input rate of fresh water. This necessitates the use of artificial means to separate fresh water from the ocean and return the fresh water as a source for wilderness rainfall and irrigation. Processes available for this function are reverse osmosis, flash evaporation, vapor compression distillation, or just plain boiling. Reverse osmosis entails use of acids and bases for membrane cleaning and/or frequent membrane replacement and would therefore present a problem both for source and disposal of the chemicals and/or membranes. Vapor compression and boiling both will result in fouling of heat transfer surfaces by scale deposition of sea water salts which will in turn imply chemical cleaning or chemical methods of scale prevention. Only flash evaporation operated at low temperatures (40 C) and high vacuum (55 mm Hg absolute) offers a solution for long term operations without major chemical or replacement requirements. The heat requirement for operating the flash evaporation system will be taken from waste heat produced in the energy center that serves Biosphere II. The heat exchange surfaces of the flash evaporator will be of titanium which does not present a metals contamination problem.

Summary

Biosphere II is to be a completely material-closed self-regenerative life system capable of supporting a crew of eight humans for long durations. It includes areas representative of wilderness in addition to agricultural and spaces for human habitation. The exclusion of external material resupply has major impacts on the choice of materials of construction and on the types of mechanical systems internally employed.

Acknowledgements

An interdisciplinary team of scientific, management, architectural and engineering consultants are participating in the Biosphere II project, including Sarbid Ltd. (architecture, design management and coordination); the Institute of Ecotechnics (total systems ecological management); Carl Hodges, Director of the Environmental Research Laboratory of the University of Arizona (intensive agriculture biome design and some aspects of the scientific engineering), Dr. Ghillean Prance, Director of the Royal Botanic Gardens at Kew in association with the Institute of Economic Botany at the New York Botanical Garden (tropical rainforest biome design), Dr. Walter Adey, Director of the Marine Systems Laboratory of the Smithsonian Institution (marsh and marine biomes design), Dr. Peter Warshall of the University of Arizona Office of Arid Land Studies (tropical savannah biome design and vertebrate species selection), Dr. Tony Burgess of the University of Arizona Geochronology Laboratory (desert biome design), and Dr. Scott Miller of the Bishop Museum (entomological community design).

References

The references cited below provide further direct information on the Biosphere II project that are not fully elaborated within this paper.

Allen, J. and M. Nelson. *Space Biospheres,* Synergetic Press, Tucson, 1986. Second edition (revised) 1989.

Augustine, M. "Biosphere II: The Closed Ecology Project" in Burdett and Soffen (ed.) *The Human Quest in Space,* AAS 86-119, Vol. 65, Science and Technology Series, American Astronautical Society, pp. 243-254, 1987.

Dempster, W. "Biosphere II: Technical Overview of a Manned, Closed Ecological System", SAE Technical Paper Series no. 891599. 19th Intersociety Conference on Environmental Systems, 1989.

Dempster, W. "Biosphere II: Engineering of Manned Closed Ecological Systems", *Journal of Aerospace Engineering,* 1990, publication pending.

CONCEPTUAL DESIGN FOR A LUNAR-BASE CELSS

Steven H. Schwartzkopf[1] and Hatice S. Cullingford[2]

ABSTRACT

Future human exploration is key to the United States National Space Policy goal of maintaining a world leadership position in space. In the past, spacecraft life support systems have used open-loop technologies that were simple and sufficiently reliable to demonstrate the feasibility of spaceflight. A critical technology area needing development in support of both long duration missions and the establishment of lunar or planetary bases is regenerative life support. The information presented in this paper describes a conceptual design of a Lunar Base Controlled Ecological Life Support System (LCELSS) which supports a crew size ranging from 4 to 100. The system includes, or incorporates interfaces with, eight primary subsystems. An initial description of the Lunar-Base CELSS subsystems is provided within the framework of the conceptual design. The system design includes both plant (algae and higher plant) and animal species as potential food sources.

INTRODUCTION

We are reporting here on the initial results of an on-going study which will lead to a conceptual design for a Lunar-Base Controlled Ecological Life Support System (LCELSS). The conceptual design is intended to accommodate a crew size ranging from 4 to 100. The LCELSS study is driven by a recognition that bioregenerative life support technologies are likely to be needed for advanced missions. This need is based on mission requirements to: 1) provide a safe, reliable human life support system to accommodate long mission durations, 2) maximize the degree of self-sufficiency of the base, 3) minimize the economic costs associated with resupply and the accompanying complexity of logistics, and 4) maintain a familiar, Earth-like living environment to promote human productivity and psychological well-being.

STUDY PHILOSOPHY

Previous CELSS studies have generally tended to develop their designs around ecological trophic level diagrams, that is by establishing blocks for the crew, higher plants and/or algae, bacteria, and/or higher animals. In this study, the approach is to evaluate the system from a functional perspective to

---

[1] Lockheed Missiles & Space Co. Inc., Sunnyvale, CA 94088-3504

[2] NASA, Lyndon B. Johnson Space Center, Houston, TX 77058-3696

determine whether any new design drivers can be identified. Figure 1 summarizes the initial functional view of the system. In this diagram, catabolic activites are those reactions in which system waste materials are degraded into less organized raw materials, while the anabolic reactions are those which utilize the raw materials in the synthesis of life-sustaining compounds. Note that in this view, raw materials include not only inorganic elements but also simple organic compounds.

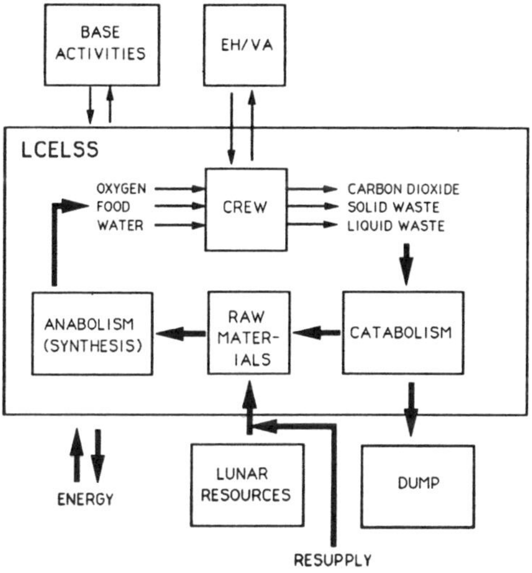

Figure 1. Functional block diagram of the LCELSS system and associated interfaces.

SYSTEM DEFINITION

Eight subsystem technology areas have been addressed in the LCELSS design effort. These subsystem areas include: 1) air revitalization, 2) water processing, 3) food production, 4) biomass production, 5) food processing, 6) waste processing, 7) in-situ resource utilization, and 8) extra habitat/vehicular activity. Of these eight areas, the first through the sixth are direct parts of the LCELSS design, while the last two are not directly involved in the design but rather are principally related through their interfaces to the LCELSS. Each of these subsystem technology areas, along with its associated bioregenerative technology candidates, is discussed below in more detail.

AIR REVITALIZATION

Air revitalization encompasses CO_2 removal, CO_2 reduction, O_2 production, and trace contaminant control. In physicochemical life support systems, each of these functions is normally handled by a separate machine. In a CELSS, these functions can be incorporated directly into the photosynthetic production of food.

Trace contaminant control can be handled either physicochemically or with bioregenerative technology. The study emphasis is on bioregenerative technologies, specifically bacterial reactors. Two different types of bacterial reactor have been proposed for trace contaminant control. The first uses soil beds, through which the atmosphere to be treated is pumped. Bacteria in these beds have been demonstrated to degrade organic compounds such as CO, H_2S, and SO_2 (1). Such beds could be established inside the LCELSS using lunar regolith seeded with bacterial innoculum brought from earth.

The second type of trace contaminant control utilizes a liquid bed reactor, in which the atmosphere is pumped through a shallow bacterial culture (2). The reactor requires a solid substrate onto which the bacterial community can adhere, and again could be made from lunar regolith. Advantages to both kinds of reactors include the capability to return to normal operation after poisoning and increased adaptability to new contaminant species.

WATER PROCESSING

Sources of water available for processing and recycling include humidity condensate, wash water, and urine. Water from fuel cells or from physicochemical CO_2 reduction can also be available (3). In past life support systems, physicochemical techniques have been used for water processing. Higher plants supply essentially pure water through the transpiration process. The transpired water must be condensed and collected with a condensing heat exchanger. However, condensate can become contaminated by trace amounts of materials from the heat exchanger, or possibly even by volatile organic compounds produced by the plants which condense along with the transpiration water (4). As a consequence, water processing with higher plants must incorporate a polishing step in which these trace contaminants are removed before the water is returned for use by the crew.

FOOD PRODUCTION

One of the key differences between physicochemical and bioregenerative systems involves the production of food. Although physicochemical food production has been accomplished, the materials provided by these technologies often present undesirable side effects such as nausea. Frequently, the raw materials required for physicochemical food production must be of very high purity, and would be difficult to obtain from life support system waste materials.

In a CELSS, food production can be based on terrestrial agricultural capabilities. There are several advantages to this approach, including crew familiarity with and acceptance of the foodstuffs produced, as well as the "bonus" of air processing (CO_2 removal and O_2 production) accomplished by higher plants as they grow (5,6). Current estimates of the area required to feed a person generally range from 20 to 30 square meters. The exact area depends on both the mix of crop species grown and on the productivity achieved by each. Hoff, et.al. (7) have proposed a mixture of 10 species which satisfy, although minimally, all human nutritional requirements. These species include soybean, peanut, wheat, rice, potato, carrot, chard, cabbage, lettuce and tomato, require about 24 m^2 of growing area to support one person.

The largest problem facing the use of higher plants in a CELSS is the need for light. If artificial lighting is used exclusively, analysis indicates that each square meter of growing area will require 0.5 to 1.0 kW of electrical energy to produce the minimum acceptable levels of photosynthetically active radiation (i.e., 300 to 600 $\mu mol/m^2/sec$) This study indicates that the electrical power required for illumination can be reduced by as much as 75% by using sunlight. This reduction would be achieved by using solar radiation during the lunar day and incorporating low-intensity artificialy lighting (approximately 10-15% of full Earth sunlight intensity) supplemented by higher atmospheric CO_2 concentrations during the lunar night.

Animals can provide additional food for the crew, and can feed on some of the waste plant materials. Thus, an additional protein source can be provided, while at the same time assisting in the processing of waste plant materials. Our current concept emphasizes the use of aquaculture, specifically Tilapia fish, as a means of producing animal protein for crew consumption.

BIOMASS PRODUCTION

Biomass production by higher plants can potentially contribute to a wide variety of life support needs. Normally inedible plant materials can be processed to produce human foodstuffs, for instance. One example of this application is alfalfa, which produces high protein vegetable material. Alfalfa protein content can reach 15-20% by dry mass, while wheat is usually 10-15% dry mass protein. In addition, alfalfa produces as much as 60% more biomass per unit area than does wheat (8). Thus, if alfalfa biomass was processed to remove the protein for human food, the generation of protein per unit of growing area could potentially be greater than twice that of wheat.

Other plant species can provide additional useful products. Jojoba could provide lubricating oil for LCELSS machinery. Guyaule might be used to provide rubber products for o-rings and/or seals. Tissues, wipes and writing paper could be produced from a variety of cellulose-containing plant materials. Finally, woody shrubs might be grown to provide raw materials for interior uses such as particle board partitions or shelves. Although there are safety issues involved in some such uses (e.g., flammability concerns), they could help produce a more psychologically acceptable environment for the crew.

FOOD PROCESSING

Food processing options may be grouped into two general categories, 1) processing of normally human-edible materials, and 2) conversion of normally inedibly materials into human-edible. Figure 2 summarizes our initial plan for laying out these processes. In this plan, edible materials may be eaten directly after washing, cooked for immediate consumption or for storage and later consumption, or processed to remove a specific component (either to enhance their digestibility or to obtain a component for specific uses). Normally inedible materials can be extracted or treated to produce compounds for human consumption (e.g., carbohydrate, protein, or oil), processed to produce feedstocks for animal consumption, or provided to the waste processing system if they had no food value.

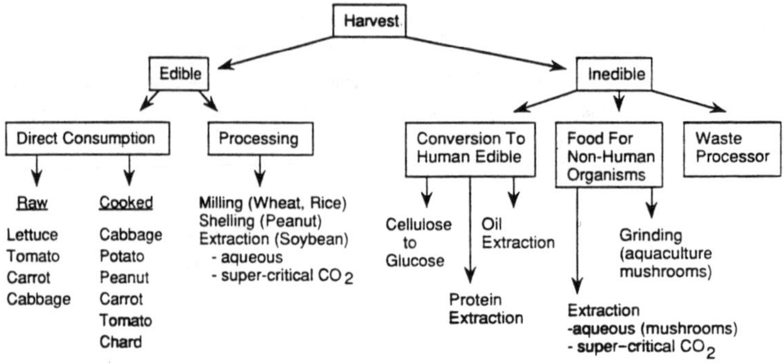

Figure 2. Food processing system flow chart.

WASTE PROCESSING

Waste processing must address both the biological waste materials produced by crew, animal and plant metabolic activities, and the "non-biological" waste materials derived from packaging materials, daily activities, etc. One definite advantage in the implementation of an LCELSS would be to ensure that expendable materials delivered to the lunar surface were as highly degradable as possible. In this fashion, for instance, wrapping or packaging materials could be fabricated from degradable plastics. After processing, such degradable materials would assist in filling the carbon, oxygen, and hydrogen reservoirs of the LCELSS.

Many previous CELSS concepts have incorporated high energy methods of waste degradation such as wet oxidation or super critical wet oxidation. These high energy methods convert all organic waste materials into inorganic salts, water and gases. In the process, all of the energy stored in the chemical bonds of the waste materials is lost. By using either bioregenerative technologies or appropriate physicochemical technologies (eg, low temperature wet oxidation or microwave processing) some of the chemical bond energy can be provided to the system by converting wastes into low complexity organic materials which can be used as feedstocks for bacteria, algae or higher plants.

IN-SITU RESOURCE UTILIZATION

One of the significant design advantages resulting from siting a base on the lunar surface involves the availability of in-situ resources. Table 1 summarizes the elemental composition determined for regolith samples obtained from the Apollo and Luna missions. Table 2 summarizes the elemental composition by percent of a corn plant (Zea mays), a human, carbohydrate (sucrose), fat and protein. From the plant composition values presented, Table 2 indicates that over 95% of plant biomass is composed of oxygen, carbon, hydrogen and nitrogen. Over 87% of human biomass is composed of these same four elements.

Element (%)	Mare High Ti A-11	Mare High Ti A-17	Mare Low Ti A-12	Mare Low Ti A-15	Mare Low Ti L-16	Basin Ejecta A-14	Basin Ejecta A-15	Basin Ejecta A-17
Al	7.29	5.80	7.25	5.46	8.21	9.21	9.28	10.90
Ca	8.66	7.59	7.54	6.96	8.63	7.71	6.27	9.19
Cr	0.21	0.31	0.24	0.36	0.20	0.15	0.19	0.18
Fe	12.20	13.60	12.00	15.30	12.70	10.30	9.00	6.68
K	0.12	0.06	0.22	0.08	0.08	0.46	0.14	0.13
Mg	4.93	5.80	5.98	6.81	5.30	5.71	6.28	6.21
Mn	0.16	0.19	0.17	0.19	0.16	0.11	0.12	0.08
Na	0.33	0.26	0.36	0.23	0.27	0.52	0.31	0.30
O	41.60	39.70	42.30	41.30	41.60	43.80	43.8	42.20
P	0.05	0.03	0.14	0.05	0.06	0.22	0.07	0.06
S	0.12	0.13	0.10	0.06	0.21	0.08	0.08	0.06
Si	19.80	18.60	21.60	21.5	20.50	22.40	21.70	21.00
Ti	4.60	5.65	1.84	2.11	2.11	1.02	0.79	0.97

Table 1. Elemental composition of Lunar regolith (After 9).

A comparative evaluation of these two tables shows that of the four biologically significant elements named above, only oxygen is available from regolith in significant amounts. Although trace amounts of carbon, hydrogen and nitrogen may be extracted from lunar regolith, initially it will probably be easiest to carry these elements from earth. As capabilities for resource extraction develop, there will be less need to rely on bringing even these major constituents from Earth.

EXTRA HABITAT/VEHICULAR ACTIVITY (EH/VA)

Although EH/VA technologies are not part of the LCELSS, it is necessary to include interfaces in the design to provide for regeneration of EH/VA materials by the LCELSS. As a consequence, it is also imperative that EH/VA technologies be designed to interface with CELSS technologies.

We have identified four types of EH/VA, including: 1) a self-contained suit and life support system, 2) a suit with an umbilical connection to a habitat or rover, 3) an open surface rover, and 4) a closed surface rover. Although it would be necessary to interface each of these with the LCELSS, bioregenerative life support technologies are particularly important in 2 through 4. In those options, an algal culture could be used to supply atmosphere recycling during lunar day. The use of an umbilical suit would make such an application especially practical. We have assumed that food, water, and waste processing would be accomplished with current suit or rover methods.

Element	Zea mays	Man	CHO(Sucrose)	Fat	Protein
O	44.43	14.62	51.42	11.33	24
C	43.57	55.99	42.10	76.54	52
H	6.24	7.46	6.48	12.13	7
N	1.46	9.33			16
Si	1.17	.005			
K	0.92	1.09			
Ca	0.23	4.67			
P	0.20	3.11			
Mg	0.18	0.16			
S	0.17	0.78			
Cl	0.14	0.47			
Al	0.11	-			
Fe	0.08	.012			
Mn	0.04	-			
Na	-	0.47			
Zn	-	0.01			
Rb	-	.005			

Table 2. Elemental composition of selected materials, by percent (After 10).

DATA BASE

The data base we have developed to support the LCELSS study includes data in five different areas. These include: 1) references used in the study, 2) technology summary and evaluation sheets, 3) lunar environment data, 4) human life support input/output data, and 5) definitions and weighting factors for the technology summary and evaluation sheets. The function of the data base is to serve as a centralized, computer-based handbook of the study data. The data base software provides capabilities for storing, retrieving (browsing or reference hard copy), and manipulating (search, sort, subset formats) data items. Table 3 lists human life support input/output data derived from the database.for crew sizes of 4 and 100 persons.

CONCEPTUAL DESIGN.

Figure 3 provides the block diagram for the LCELSS conceptual design. This design accommodates the mass in- and out-flows summarized in Table 3, and includes both animal and plant components as food sources for the crew. It also stresses the use of hardware modules to accomplish specific functions, thus decreasing development risk and increasing overall system reliability. For example, heat exchange, water recovery, and trace contaminant control will be implemented by the same hardware designs, regardless of whether they serve the crew, plant or animal compartments.

ITEM	DAILY MASS IN/OUT FLOWS FOR CREW SIZE (kg) CREW SIZE = 4	CREW SIZE = 100
INPUTS		
Oxygen	3.35	83.64
Food	2.47	61.82
Water in Food	2.00	50.00
Food Preparation Water	2.87	71.82
Drinking Water	7.44	185.91
Hand/Face Wash Water	7.27	181.82
Shower Water	14.55	363.64
Clothing	4.55	113.64
Clothes Wash Water	50.00	1250.00
TOTAL MASS	94.49	2362.27
OUTPUTS		
Carbon Dioxide	4.00	100.00
Respi./Persp. Water	7.31	182.73
Urine	6.02	150.45
Feces Water	0.36	9.09
Sweat Solids	0.07	1.82
Urine Solids	0.24	5.91
Feces Solids	0.13	3.18
Hygiene Water	21.82	545.45
Clothing	4.55	113.64
Clothes Wash Water	50.00	1250.00
TOTAL MASS	94.49	2362.27

Table 3. Daily life support system input/output data for a 4- and 100-person crews.

The design provides for separation and processing of plant and animal biomass as described earlier. All waste materials from the food processors are either used as foodstuffs for the animals or sent to the waste processing system for recovery of essential elements. The design also provides for isolation between both atmosphere and water of the crew, animal, and plant compartments. This isolation not only assists in preventing the possible spread of bacteria and/or volatile organic compounds between the compartments, but also makes possible independent control of atmospheric composition. This composition control capability is essential for managing the growth and productivity of higher plants, as they generally have different composition requirements from those of humans. The buffer reservoirs for atmospheric composition control (e.g., CO_2, O_2, and N_2) can be shared between the crew, plant,

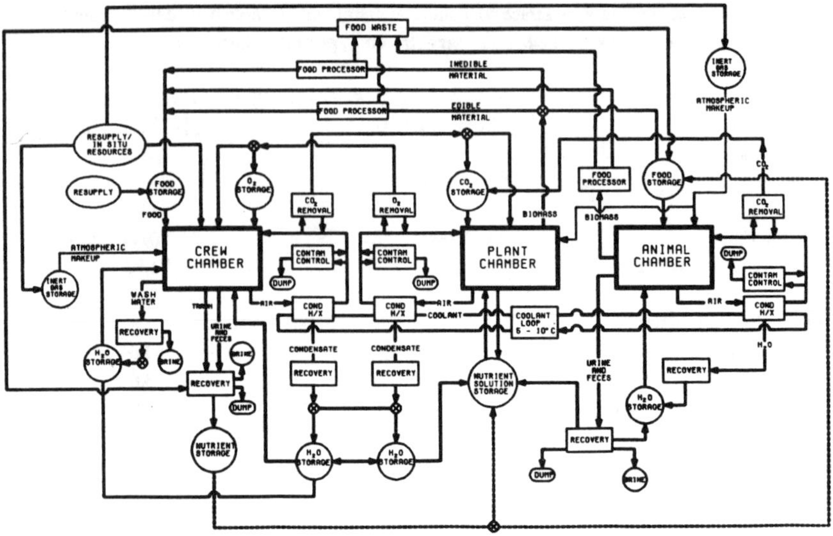

Figure 3. Block diagram of the LCELSS conceptual design.

and animal compartments. The water reservoir system for the animal compartment is isolated from that of the crew, however the crew water reservoir is replenished with plant transpiration water that has undergone a polishing step.

CONCLUSION

The focus of the LCELSS study is on developing a conceptual design for a safe, reliable life support system based on bioregenerative technologies. The initial conclusions derived from the study include: 1) the use of both plants and animals as food sources for the crew provides significant advantages in terms of constructing an adequate diet, maintaining a familiar living environment for the crew, and decreasing both engineering complexity and energy use in the higher plant growth unit, 2) by combining appropriate food processing technologies with the growth of higher plants not normally used for production of human food, productivity of essential nutritional compounds (e.g., protein) may be increased over that achievable with common food plants, 3) through the use of specific waste processing technologies, much of the chemical energy contained in waste materials can be recovered and used directly by the system, 4) although many of the elements abundant in Lunar regolith are useful to LCELSS, three of the elements most common to living biomass (N, H, & C) are only available in trace amounts and will have to be supplied from Earth in the initial stages of LCELSS development, 5) bioregenerative technology can be used for air revitalization in several EH/VA applications, and provides a simple interface to the overall LCELSS, 6) the conceptual design provides for isolation of the biotic components (crew, plants, animals) to minimize contamination problems and maximize the capability for independent control of the component environments, and 7) the conceptual design incorporates hardware modularity and enables sharing of some buffer reservoirs by the biotic components.

LITERATURE CITED

1. Bohn, Hinrich and R. Bohn. 1988. Soil Beds Weed Out Air Pollutants. Chem. Eng. April 25, 1988. p 73.

2. Binot, Roger A. and P.G. Paul. 1989. BAF-an Advanced Ecological Concept for Air Quality Control. Proc. 19th Intersociety Conference on Environmental Systems, July, 1989, San Diego, CA.

3. Cullingford, H.S., and M. Novara. 1988. Conceptual Design of a Piloted Mars Sprint Life Support System. Proc. 18th Intersociety Conference on Environmental Systems, July, 1988, San Francisco, CA.

4. Derendyayeva, T.A. 1977. Investigation of the Possibility of Using a Condensate of Transpiration Moisture of Ratata for Cultivating plants in Biological LIfe Support Systems. Joint Publ. Res. Ser. L/6859 Dept. Comm. Washington, D.C., Moscow Kosm. Biol. I Aviakosm. Med. 10(6). p 70.

5. Novara, M. and H.S. Cullingford. 1988. Bioisolation Analysis of Plants and Humans in a Piloted Mars Sprint. Proc. 18th Intersociety Conference on Environmental Systems, July, 1988, San Francisco, CA.

6. Volk. T. and H.S. Cullingford. 1988. Crop Growth and Associated Life Support for a Lunar Farm. Proc. Lunar Bases and Space Activities of the 21st Century, 1988.

7. Hoff, J.E., J.M. Howe, and C.A. Mitchell. 1982. Nutrititional and Cultural Aspects of Plant Species Selection for a Controlled Ecological Life Support System. NASA Contractor Report #166324, Ames Research Center, Moffett Field, CA.

8.Martin, J.H., W.H. Leonard, and D.L. Stamp. 1976. Principles of Field Crop Production. Macmillan Publiching Co., Inc., New York, N.Y.

9. Phinney, W.C., D. Criswell, E. Drexler, and J. Garmirian. 1977. Lunar Resources and their Utilization, in Space-Based Manufacturing from Non-terrestrial Materials. G. O'Neill, Ed.

10. Epstein, E. 1972. Mineral Nutrition of Plants: Principles and Perspectives. John Wiley and Sons, New York, N.Y.

Investigation of HVAC Systems
in
Low Gravity Application

David L. Sharp[1]

ABSTRACT

This paper has been developed to identify and summarize the principal areas of investigation necessary to design and install functional HVAC systems in low gravity applications. These investigations must consider design, construction and control details and restrictions for systems and components. Included in these investigations should be subjects such as the mechanics of prime movers in lunar gravity, temperature and pressure requirements, air distribution methods and physical size, transport velocity requirements, resources recovery interface with life support systems and both energy consumption and conservation. A suggestion of priorities for the investigations is presented. These priorities have been established to support development of basic design criteria and methods for functional HVAC systems. Further investigation of more detailed subjects can be performed at a later date.

INTRODUCTION

The existing literature on extraterrestrial, surface based facilities provides information on modular life support system, such as the Advanced Integrated Life Support System (AILSS) investigated at NASA Langley Research Center (Hall, 1969), and the Controlled Ecological Life Support Systems (CELSS) research conducted at NASA Ames Research Center (Majumdar, 1985). This research has concentrated mainly on military or exploratory missions ranging in size and duration from four men for 30 days to nine men for 500 days (Rockwell, 1969), under more or less primitive conditions. Much of the work has assumed weightless condition or artificially induced gravity, although work performed by Armstrong and King at the General Dynamics Convair Division (Armstrong et al, 1968) has outlined many of the areas of bioengineering life support system

---

[1] Senior Engineer, Advanced Technology Division, Fluor Daniel Inc., 3333 Michelson Drive, Irvine, CA 92730

design. The literature is, however, noticeably lacking in design information for practical air conditioning and distribution systems. There are few published examples detailing design or construction methods for HVAC systems in low gravity, space-restricted applications.

During the next few generations, the human race will move from a totally earthbound society, to a terrestrially supported orbiting space station, to permanently inhabited self-sufficient outposts on the moon, and perhaps farther. These outposts will require practical application of air conditioning and distribution techniques in combination with the environmental systems necessary to support life and productive activities in extremely hostile conditions. The basic principles of ventilation system design will need to incorporate the changes introduced into the mathematical model by the differences in the pull of gravity. The design of any such facilities support system requires determination of the mission requirements and consideration of reliability and maintenance. It has been repeatedly demonstrated that environmental conditions, both real and perceived, directly affect personnel ability to perform work or concentrate on a specific task (Rasmussen et al, 1973 and Johnson et al, 1977).

ENVIRONMENTAL DESIGNS

The practical aspects of controlling temperature, humidity, odors and other airborne contaminants at work stations, recreational areas, and living quarters need to be investigated. Methods used in ventilation system design for submarines, antarctic stations and other isolated and hostile environments must be studied to determine their applicability to a lunar colony. It is clear that military personnel, construction crews, and, to a large extent, scientific expeditions, may be expected to live and function for a limited time under conditions that most people would or could not tolerate. However, personnel arriving later cannot be expected to live and effectively produce under pioneering conditions, particularly if permanent habitation is the desired result.

Environmental requirements for the health and comfort of people are available from as convenient a source as the ASHRAE Fundamentals Handbook. However, determination should be made about the temperature control and ventilation requirements for special purpose areas such as exercise areas, large lounge/auditoriums or other common area, food preparation areas, and private quarters. These private quarters will be important to the physical and emotional well-being of the inhabitants. Large areas with their attendant large population concentration have quite different air conditioning requirements than work stations or private rooms. In large areas there is a need to move large volumes of moist air away from the people in the area. This moisture, produced by respiration or physical exertion, needs to be directed to the resource recovery plant so that the water can be removed, probably by condensation, and recycled. Temperature control in the large areas is not as critical as long as the volume of air remains adequate. Exercise rooms need to remove the heat of physical exertion by supplying air at cooler temperature than that required for large room ventilation. Office work stations may be controlled for a central location but the air must be distributed over relatively large areas, and should have sufficient volume for heat removal and odor control.

Private quarters may be the most difficult challenge to meet due to severe restrictions placed on the size of equipment. Each person's desire for a comfortable temperature should be addressed to the extent possible. A method for allowing this individualization of centralized systems should be investigated and incorporated into the design of the facility at the earliest time. Manufacturing and fabrication ventilation systems will need to be separated from the normally inhabited area systems to prevent cross-contamination and to reduce the overall size of equipment necessary to provide a shirt-sleeve working environment. This extra equipment will require additional building material and enclosed space, but the extremes at which these systems will operate suggest that high reliability and back-up capability will be needed. Having separate and independent systems for separate areas and functions is a method of achieving redundancy and spare capacity.

Any portion of the structure that is exposed to the lunar surface must be insulated to minimize the effects of the lunar temperature extremes on the heat transfer through the shell of the facility. Most of the structure is expected to be covered with regolith for radiation shielding (Richter et al, 1988). It must be determined if the same thickness necessary for shielding will provide thermal insulation so that the facility is relatively unaffected by changes in surface temperature. What type of special design features will be required to adjust to large swings in air conditioning needs if enough insulation can not be added? The final heat sink for unrecoverable heat must be carefully considered. This heat sink will be required to reverse its operation every two weeks in accordance with the moons day/night cycle, and will function under exposure to extreme conditions, ranging from 250 to -250° Fahrenheit.

Odor and fumes control will have a large impact in determining air distribution and volume. Methods for removing offensive odors produced by cooking or bodily function must be determined, as recent studies indicate that airflow of 15 ft^3/min for each person is necessary to control human odors. Industrial and/or toxic fumes created either by fabrication activities like welding fumes, or by processes such as chemical reduction of materials to obtain oxygen must be isolated and treated separately. Control of dangerous fumes is of vital importance, as contamination of the atmosphere in a closed environment can be disastrous. It is desirable to have people work without heavy suits, but the equipment necessary to remove fumes such as glove boxes and fume hoods have a large impact on ventilation system control, equipment space requirements and cost. Current filtration units used in atmospheric clean-up, such as charcoal adsorbers, are very large, high maintenance items seemingly unsuited for use in the lunar colony. A safe method for directing the flow of fumes, and their removal from the general environment must be determined. Even as simple and necessary an action as the operation of certain types of electric motors, which produce ozone, will have an affect on the breathable atmosphere.

Dust infiltration and contamination must also be minimized particularly in air locks and other high traffic areas. Moon dust is a fine, rough surfaced dust that stays suspended in the atmosphere for long periods of time. How much filtration will be required on return air systems to avoid machinery maintenance problems? The health effects of dust inhalation are available from medical research and published literature, but does it adequately address the effects of very dry moon dust and its environment? The behavior of moon dust in a moist atmosphere should be studied. Could the dust absorb enough moisture to have an affect on the habitability of the facility? How much dust can the inhabitants inhale before it becomes a nuisance or health risk? Particulate filtration units require equipment space and additional maintenance space.

PHYSICAL CONSIDERATIONS

The integration of physical delivery systems, ie., the supply and return ducts, into the facility design must be investigated. While it is possible to deliver large quantities of air through small diameter pipes, a large energy penalty is involved as well as unacceptable levels of noise or other vibration. Large horsepower motors are generally required for high pressure air distribution systems. These motors can consume considerable power, produce waste heat and take up large areas by themselves. Each area of the facility will have its own special requirements for distribution systems, just as each area will have its own needs for temperature control and ventilation rates. Studies should be conducted to determine the best compromise between comfort, space availability, energy conservation and reliability. Ideally, each such area would have a system designed specifically to meet its unique needs; however minimizing facility equipment space will be a major, if not controlling, factor.

Not only does the equipment need space for maintenance activities, critical components of any HVAC system need to be redundant, with several layers of independence and protection. Safety concerns with the accidental loss of pressure boundary must be included in the HVAC design. Equipment would need to be sized to provide excess capacity to overcome credible leaks and losses. Ventilation supply equipment must have at least two separate and independent connections to oxygen generation or supply systems to provide protection against common mode failure. Ducts must be constructed in such a manner and with appropriate materials to ensure that conditioned air will not leak into undesired locations and will not rupture if faced with excess pressure. The additional stress imposed by large, high pressure ventilation systems should be investigated in conjunction with structural design as the facility is expected to be pressurized to near earth normal (Armstrong et al, 1968) which presents some unique problems concerning differential thermal expansion and pressure induced stresses. In the event of catastrophic failure in an area, could sufficient capacity and rapid control response protect personnel from rapid depressurization; should the affected portion of the system be isolated, or should the entire system be shut down to limit atmospheric gas losses?

CALCULATIONS and THEORY

Many of the equations used in the basic design principles may need to be revised for use in low gravity design. The Bernoulli equation, basic to fluid flow analysis,

$$gz = V^2/2 + p/\rho$$

and the Grashof Number, used in convective heat transfer analysis,

$$gl^3\beta T/v^2$$

are examples of the wide range of potential impacts of changing the value of gravity. The calculated conditions will be different from the values used in previous design. For example, Coriolis forces on a space station that is spinning to induce gravity will have a significant effect on fluids when compared to the effects on the moon. Prime mover work requirements, therefore energy usage, may be reduced due to gravitational effects on duct friction losses, after all, the ever helpful velocity pressure equation

$$p_v = (V/4005)^2$$

is derived from

$$p_v = \rho V^2/2g_c C_p.$$

Simple, perhaps not very significant, things like the particulate entrainment velocity in the duct and the throw of air from a supply register will be affected, not just work and energy values. Low gravity will present as many different and challenging variations as did zero gravity, with a major difference being that the spin on the moon is already applied and is fixed.

Conclusion

Lastly, with the objective of a lunar colony to be self sufficient, a closed environmental system will most likely be used. Resource recovery methods and systems will probably be tested and used on the space station, but it is envisioned that the space station will still receive a significant fraction of its consumable from Earth. Recoverable materials will be present in the air returned through the ventilation system. Water present in the atmosphere from breathing, hygienic requirements, food preparation activities and industrial processes should be recovered. Heat recovery systems need to be studied. For example, one of the current technologies, the heat pipe, is affected by gravitational effects (Chapman, 1974). Carbon dioxide, as well as other airborne atmospheric contaminants, must be removed from the breathable atmosphere. While carbon dioxide is often considered to be a waste product, in a truly closed life support system there can be very little waste. It is supposed that space station technology for carbon dioxide scrubbing and oxygen recycling will be modified and used.

The items mentioned briefly in this article are just a few of the many areas of design and construction that will require re-examination when they are moved into low gravity applications. Many other related topics need to be investigated, such as infectious disease quarantine and radioactive materials handling. Because of the differences in gravity strength, rotation, type of environmental systems and mission objectives, these areas will require different solutions than those applied to the space station and will more than likely surprise the designers and builders in ways not even imaginable at this time. Some priorities must be set for establishing goals so that resources can be allotted and used efficiently. Ventilation and air conditioning techniques used in normal terrestrial facilities are well established and generally not given the degree of forethought that will be required for closed systems like the lunar facility. Early investigation should concentrate on those items which lead directly to a functional system. This should consist of at least the following steps.

1) Develop calculations necessary to support accurate equipment sizing and energy use estimates.
2) Incorporation of empirical data from space station and exploratory lunar missions as soon as possible.
3) Development of a computer model for the new facility design.
4) Test the model under actual conditions during the early stages of construction.

Other areas for investigation can then be opened as the basic requirements for life are met.

REFERENCES

Armstrong, R.C. and King, C.D., 1968, "Life Support Systems on the Moon", Technical Information Service, American Institute of Aeronautics and Astronautics, New York.

Beltz, W.A., (ed), Vol 9, No. 21, Oct 1989, "Occupational Safety and Health Reporter",

Chapman A. J., 1974, "Heat Transfer", Macmillan Publishing House, New York.

Hall, J.B. Jr, 1969, "Advanced Integrated Life Support Systems - A Status Report". National Aeronautics and Space Administration, Langley Research Center, Hampton, VA.

Johnson R.D. and Holbrow, C., (eds), 1977, "Space Settlements: A Design Study", NASA Scientific and Technical Information Office, Washington D. C.

Majumdar, Mousumi, 1985, "Controlled Ecological Life Support Systems (CELSS), NASA Scientific and Technical Information Center, Washington D. C.

North American Rockwell Corporation, 1969, "Review of the Extended Lunar Shelter Problem", Space Division, Downey.

Rasmussen, J.E. (ed.), 1973, "Man in Isolation and Confinement", Aldine Publishing Co., Chicago.

Richter, P.J., Anderson, T.L., Drake, R.M., and Kumar, R.P., 1988, "Strategy for Design and Construction of Lunar Facilities", ENGINEERING, CONSTRUCTION, AND OPERATIONS IN SPACE, S.W. Johnson and J.P. Wetzel (eds),.

COMPOSTING FOR LUNAR AGRICULTURE

Judith Fielder and Nickolaus Leggett[1]

Abstract

Alternative technologies for the microbially mediated biological oxidative transformation (composting) of human wastes and other organic wastes are presented. Batch mode and continuous process composters are proposed to convert waste materials into a form that is a useful input for lunar agriculture.

Each composting system provides environmental conditions necessary for composting. The waste materials are composted over a period of time and the resulting composted material is added to an engineered soil in plant growth chambers.

Composting technology used in conjunction with a regolith-derived soil for agriculture allows organic wastes to be effectively recovered for re-use in food production.

Introduction

Recycling organic waste materials for reuse is a serious consideration for a lunar base. Although there are several recycling technologies that can be utilized by a lunar base, most technologies require secondary processing facilities to get material back into the cycle in forms that can be utilized by the agricultural system. This is not the case with composting, the microbial biological oxidative transformation of organic waste materials. Composting systems convert waste organic materials into a form that is suitable and beneficial for use in a regolith-based agricultural system.

[1] Consultants, 1432 Northgate Square, #2, Reston, VA 22090.

This paper presents composting system design considerations that support a regolith-based system of lunar agriculture (Fielder and Leggett, 1988).

Composting Requirements

Certain environmental requirements must be met to ensure effective and optimum composting of organic waste materials in the systems we describe (Table 1).

Table 1. Composting System Requirements

Temperature control	50° - 60° C
Moisture control	$\geq$50%
Aeration environment	Variable-temperature dependent
Oxygen Requirement	5-15% v/v of available air
pH adjustment	5.5 - 8.0 pH
Carbon dioxide	<5% v/v of available air

An aerobic environment must be maintained with an adequate air flow to remove accumulated carbon-dioxide and control temperature. Considerably more air will be needed to control temperature than to provide an aerobic environment (Haug, 1986; Finstein and Miller, 1985). The temperature of the composting material must be maintained between 50 and 60° centigrade (Kuter et al. 1985; Sikora and Sowers, 1985; Verougstraete, et al., 1985). Moisture content of the composting material should be maintained as high as feasible above 50% without affecting the aerobic condition. The pH should be balanced if necessary at the onset of composting and appropriate microbial organisms introduced to initiate composting.

Composting Inputs

As the lunar base evolves from an initial, first generation lunar base to a second generation mature base, the input materials for composting will change. Initial lunar bases will most likely have rudimentary agricultural systems with quick turnaround, low carbon to nitrogen (C:N) ratio crop types and a low number of base personnel. The biomass to be composted will be predominantly human waste and fast composting crop wastes. As the lunar base evolves and lunar agriculture matures, however, there will be a diversification in the types of crops grown and a concurrent increase in the number of personnel. C:N ratios of plants grown will likely be higher and larger amounts of crop residue will be introduced into the composting system due to the differences in biomass production. There will also be greater amounts of human wastes. The wastes with a higher C:N ratio require a longer period of time for the composting process. C:N ratio limits the composting rate and cannot be averaged across plant materials. Some examples of representative C:N ratios are given in Table 2.

Table 2. Carbon:Nitrogen Ratios of Some Representative Plant Waste Materials

Material	C:N Ratio
Alfalfa	12:1
Oat straw	75:1
Corn stalks	80:1
Sawdust	300:1 - 400:1

For a variety of reasons, woody plant types are likely to be introduced into a mature lunar base environment. Composting time for this material increases significantly. The option of burning select types of high C:N ratio plant wastes may be considered since the ash can be introduced into the compost and thus recycled.

Composting Technologies

The following composting technologies are suitable for use in a lunar base environment singly or in combination. Materials used to construct the systems must be resistant to corrosion in the presence of high humidity and organic acids. It must also be resistant to abrasion by composting material. In addition, the materials used cannot have an inhibitory effect on the composting process nor may it contain elements toxic to plant growth or to humans. The composting process itself is not utilized to sterilize the medium of harmful microorganisms. There will be an initial sterilization stage for human wastes and select crop wastes. Preparation of organic wastes is a consideration. While initial low C:N ratio plant wastes may not need to be prepared for the composting process, this freedom rapidly disappears when composting plant wastes with higher C:N ratios. Waste material particle size has an appreciable effect on composting duration. Therefore, as soon as feasible a precomposting unit that will macerate, thoroughly mix sterilized human and crop wastes, and also adjust pH and introduce microbial agents should be installed. This could be centralized units which are connected to the following composting reactor systems with flexible tubing.

Composting Drum

Drum composters are proposed for small initial lunar bases. The drum composter is a cylinder that is mounted fixed in a horizontal position (Figure 1). The cylinder is designed to be rotated about its axis and is equipped with a door in its side.

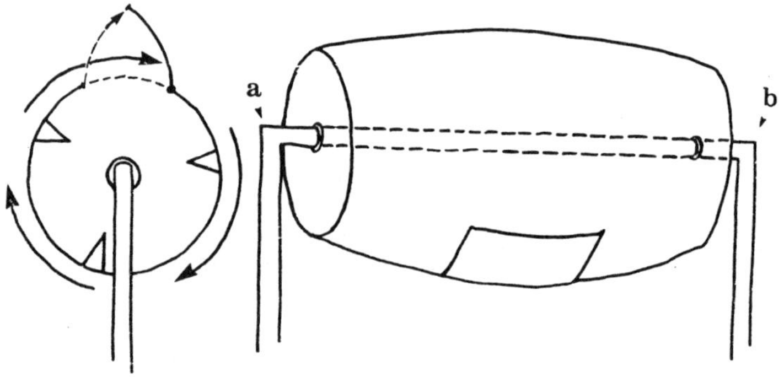

a. H_2O, O_2 Control; b. CO_2 Removal

Figure 1. Drum Composter

This composter is designed for batch-mode operation. The design is based on currently used drum composters (Kishimoto et al., 1987). At the beginning of the process, the cylinder is rotated so that the door on its circumference is facing upward. This door is opened and material to be composted is poured into the cylinder.

The door is closed and latched. The cylinder is slowly rotated during the 10-day to 2-week composting period or longer if necessary. The cylinder can be rotated continuously by a geared-down electric motor or it can be turned periodically by hand.

At the end of the composting process, the cylinder is rotated so that the door is facing downward. The door is unlatched and opened and the fully-composted material is poured into an output hopper for use.

This batch-mode composting technology uses a standard size drum one to two meters in diameter and several meters in length. Larger lunar base installations can be served by several of these standard-sized composters. Multiple composters offer redundancy as well as a more even composting process than that provided by a very large drum.

Archimedean Screw

The archimedean screw composter offers a continuous-action composting process that is well suited to a larger lunar base. The archimedean screw technology consists of a large screw mounted within a fixed cylinder (Figure 2). The cylinder itself is oriented in a horizontal or sloping position relative to the floor of the lunar base.

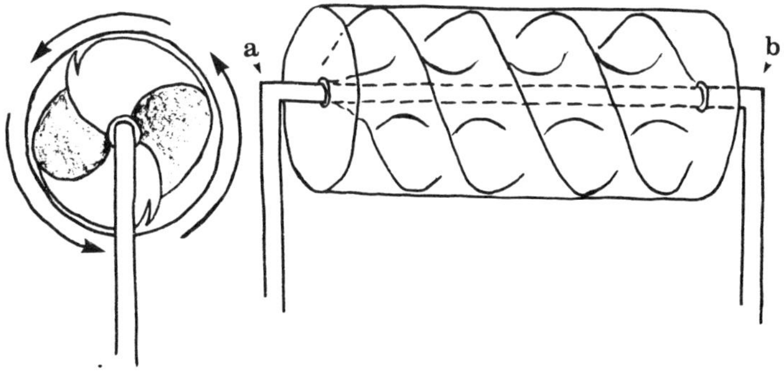

a. H_2O, O_2 Control; b. CO_2 Removal

Figure 2. Archimedean Screw Composter

Material to be composted is added at the input end of the composter. The slowly revolving screw moves the material down the length of the cylinder. The action of the screw stirs the composting material and makes sure that it is aerated by the air flowing through the cylinder. Successful action of this composter is dependent on the material volume being limited so that a percentage of free air space is available within the cylinder.

After a residence time of 10 days to 2 weeks, the compost arrives at the output end of the cylinder where it is collected for use.

In normal operation, the screw is rotated by an electric motor driving a gear system that delivers a very low rotational rate. In emergency situations, the composter's screw can be rotated periodically by hand.

Rotating Cylinder

The third composting technology is a hollow rotating cylinder with a helical dam installed within the cylinder (Figure 3). This cylinder is installed in a horizontal position relative to the floor of the lunar base. Material to be composted is input at one end of the cylinder. Fully-composted material is removed from the other end of the cylinder.

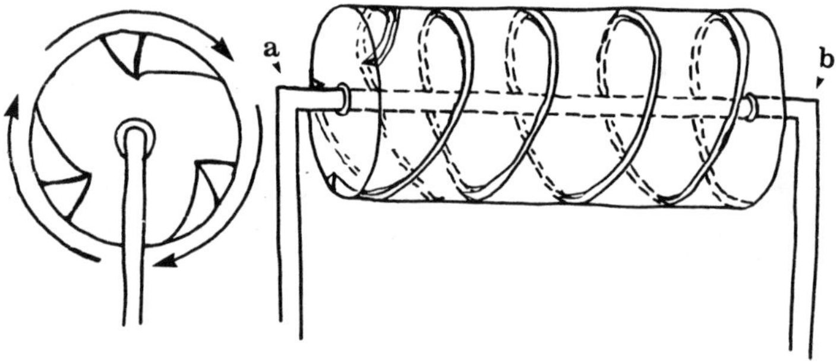

a. H_2O, O_2 Control; b. CO_2 Removal

Figure 3. Helical Dam Composter

The slow rotation of the cylinder moves the composting material down through the cylinder towards the output end. The throughput of material is controlled so there is ample empty volume within the cylinder. The dam carries the material along the axis of the cylinder. Material rotating upward within the cylinder tends to tumble downward insuring a continuous mixing of the composting material as it migrates towards the output end.

The rotation of the composter is established so that the composting material will take 10 days to 2 weeks to move from the input to the output. Residence time for the composting material can be controlled by controlling the rotation rate of the cylinder itself.

Mixing and Storage

Once the compost has been composted it can be mixed with lunar regolith by means of a small drum mixer or an equivalent system (it may be necessary to use a counter-rotating paddle system to optimize volume efficiency) and placed in the agricultural system's various plant beds and growth trays. If, however, there is a surplus of compost at the time of its completion of the composting cycle, the compost can be stored. Storage units should contain the means for keeping the compost aerated and moist.

Use of Composted Material

The composted material is added to lunar regolith material to form an engineered soil for the growth of plants. Organic matter in soil is one of the fundamental necessities for biological fertility (de Bertoldi, et al., 1985). It has been demonstrated that the addition of composted organic material will dramatically increase the fertility of relatively inert growing media such as pulverized fuel ash (Stentiford et al., 1985).

In addition, the composted material enhances the physical structure of the soil and provides passage for vital air flow to the plant roots. The composted material also stores water and nutrients for extended periods of time maintaining an optimum environment for plant growth.

Composting Variables

All of the composting technologies are equipped with air circulation systems to provide air flow through the compost. This air flow is used to control aeration; it is also used to control the temperature at which the composting material is maintained. Feed air to the compost systems should be oxygen rich. Exhaust air from the composting systems will be carbon dioxide rich and can be channeled to the plant growth areas (Golueke, 1986). This channeling system must be a controlled system capable of delivering the carbon dioxide rich exhaust air to only illuminated plant growth areas. This exhaust air may need to be filtered to remove objectionable odors.

It will be necessary to remove a substantial amount of waste heat generated by the composting process to keep the composting material within the optimum temperature range. To do this it may be necessary to supplement air cooling with some type of active refrigeration.

A continuous air flow through the composting system will tend to desiccate the composting material, active moisture replacement is designed into the units.

With judicious handling of aeration, moisture content, system spin rate, and particle size it is possible to control the rate of speed at which composting will occur. If necessary, the composting process can be slowed considerably. And while there are finite limits to the speed of completion due to C:N ratios, continuous composting should more than handle the needs of a lunar base.

Research Needs

Since composting is a microbially driven process research is required in the area of microbial life cycles, behavior, and efficiencies in the one-sixth gravity environment of a lunar base. In addition, the composting process generates large quantities of low-grade heat. Technology is needed to efficiently remove that heat from the composting area with a minimum energy expenditure.

Heat generated by composting is calculated roughly utilizing the bulk density and volatile solids contents of the waste material to be composted and the rate at which composting is proceeding (Finstein and Miller, 1985; Verougstraete, et al., 1985). Gaps exist in information for energy requirements involved in all phases of composting making reliable and accurate energy balances impossible to make (Golueke, 1986). Models of composting specifically tailored for lunar base waste flows and materials would be useful in predicting mass balances.

Conclusion

Composting provides a direct and practical means for recycling organic material in a lunar base that uses regolith-based agriculture. Several alternative composting technologies are available for this purpose.

REFERENCES

de Bertoldi, M. et al. (1985), "Technological Aspects of Composting Including Modeling and Microbiology," Composting of Agricultural and Other Wastes, Gasser, J.K.R., editor, Elsevier Applied Science Publishers, London, pp. 21-41.

Fielder, J., and Leggett, N., (1988), "Lunar Agricultural Requirements Definition," Engineering, Construction, and Operations in Space--Proceedings of Space 88, American Society of Civil Engineers, pp. 344-351.

Finstein, M.S. and Miller, F.C. (1985), "Principles of Composting Leading to Maximization of Decomposition Rate, Odor Control, and Cost Effectiveness," Composting of Agricultural and Other Waste, Gasser, J.K.R., editor, Elsevier Applied Science Publishers, London, pp. 13-26.

Golueke, C.G. (1986), "Compost Research Accomplishments and Needs," Biocycle, April 1986, pp. 40-43.

Haug, R.T. (1986), "Composting Process Design Criteria," Biocycle, October 1986, pp. 53-57.

Kishimoto, M. et al., (1987), "Simulation of an Aerobic Composting of Activated Sludge Using a Statistical Procedure," MIRCEN Journal of Applied Microbiology and Biotechnology, Vol 3(2), pp. 113-124.

Kuter, G.A. et al., (1985), "Effects of Aeration and Temperature on Composting of Municipal Sludge in a Full-Scale Vessel System," Journal of the Water Pollution Control Federation, Vol. 57(4), pp. 309-315.

Sikora, L.J., and Sowers, M.A., (1985), "Effect of Temperature Control on the Composting Process," Journal of Environmental Quality, Vol. 14(3), pp. 434-439.

Stentiford, E.I. et al. (1985), "Forced Aeration Co-Composting of Domestic Refuse and Sewage Sludge in Static Piles," Composting of Agricultural and Other Wastes, Gasser, J.K.R., editor, Elsevier Applied Science Publishers, London, pp. 42-55.

Verougstraete, A. et al. (1985), "Heat Recovery from Composting and Comparison with Energy from Anaerobic Digestion," Composting of Agricultural and Other Wastes, Gasser, J.K.R., editor, Elsevier Applied Science Publishers, London, pp. 135-146.

FLUID BEHAVIOR CONSIDERATIONS FOR WASTE MANGEMENT IN LOW-GRAVITY ENVIRONMENTS

Anthony M. Wachinski
Associate Professor of Civil Engineering
USAF Academy, Colorado

Kurt T. Preston
Ph.D. Candidate
Purdue University
West Lafayette, Indiana

ABSTRACT: Design of waste recycling systems for spacecraft requires a knowledge of fluid behavior in microgravity. As gravity is reduced, phenomena usually ignored in the One-G environment of earth can dominate physical or biological processes. This paper provides an explanation of Zero-G, microgravity terminology, and microgravity fluid behavior. Its purpose is to educate civil engineers and waste management professionals on Zero-G basics.

INTRODUCTION

In light of the recent 20th anniversary of the day American astronauts walked on the moon and President Bush's support of the U.S. space station, a manned lunar base, and piloted mission to Mars, consider the following: Not a drop of waste water has been treated by any spacecraft treatment/recycling system -- either by the Soviets or Americans. (Balebanov,1989)

Current plans for our Space Station are to recycle as much waste water as possible to reduce shuttle resupply requirements. A mission to Mars may require total recycle and regeneration of water, air, and food -- a planet-earth type ecosystem. However, fluid behavior in space differs markedly from that in a gravity-dominated environment. The purpose of this paper is to initiate the general environmental engineering sector on the challenge of waste water recycle in the microgravity or Zero-G environment.

DEFINITION OF ZERO-G

The term Zero-G is really a misnomer. It strictly means a total absence of gravity, i.e., the net contact force on the system under consideration is zero. This condition occurs in orbit when a spacecraft is affected only by gravity without any extraneous forces applied by thrust, lift, air drag, or ground contact. This condition rarely occurs while an object is in routine orbit. The spacecraft merely appears to be at Zero-G because it is following a curved free-fall path that never intersects the earth's surface or the denser levels of the atmosphere where increased air drag would slow it below orbital speed.

Let us now consider a collection of objects inside an orbiting spacecraft. The spacecraft and the objects inside remain in orbit because all are attracted toward the earth's center by its gravity. The attraction is uniformly distributed and influences every particle in the craft. To an astronaut inside the ship, objects seem to float freely as though gravity had been "turned-off" -- an illusion created because all of the objects, including the astronaut, are in free-fall together. The gravitational field acts uniformly on all objects in the orbiter and produces a uniform acceleration that causes relative movement between mass points.

In contrast, on earth, a body is supported by applying forces along its boundaries. The forces acting on the internal mass points of a body are the result of the gravitational and mechanical forces transmitted through the body from the supporting structure when on the earth's surface. In solids, the forces are transmitted through stresses; in liquids, forces are transmitted through hydrostatic pressure (Congelliere et al, 1963). On earth, we perceive the stresses and pressures that we interpret as "gravity" because of the reaction forces on the floor, chair, or any material object resisting earth's gravitational pull.

In orbit, a spacecraft may appear to be free of external force, but it is not. Even at orbital altitude there is a trace of atmosphere which applies some drag to a spacecraft's hull. As a spacecraft is slowed before reentry, objects floating inside appear to "fall" toward the "front" (the side toward the velocity vector), and objects, in contact with the bulkheads of the craft, experience tiny reaction forces that "feel" like gravity to the astronaut and the objects. It appears that a weak gravitational field is at work. Forces producing similar effects in space include thrusters firing, machinery or people moving in the spacecraft. There is even a "gravity-gradient force" because the earth's gravitational field is not perfectly uniform.

All of these effects contribute to a "microgravity" environment aboard orbiting spacecraft. Objects behave as they would in a very weak gravitational field, and space engineers must account for the effects. We assign the value of "One-G" to the gravitational acceleration we experience at the earth's surface; microgravity effects in space may vary from 10^{-6}(one millionth the surface value) due to trace atmospheric drag up to 10^{-2} g (one-hundredth g) due to thrusters firing. Although small, such acceleration can have important effects on fluid behavior in spacecraft.

BEHAVIOR OF FLUID IN MICROGRAVITY

Hundreds of articles have been written concerning fluid behavior in microgravity. The primary purpose of these is to answer questions concerning materials processing in microgravity. Little research has been conducted to specifically answer microgravity questions applied to waste treatment. Obviously, fluid behavior in microgravity is different than that in a gravity dominated environment. As gravity is reduced, phenomena usually ignored may dominate a physical process. Basic phenomena such as buoyancy, settling, convection, mixing and diffusion are considerably different in a microgravity environment. These and other differences in behavior may have significant influence in the choice of the waste treatment process chosen for a particular situation. Terminology is also different. Chemical engineers, NASA engineers, and microgravity researchers describe fluids using the concept of phases. A phase is a homogeneous substance viewed independently of shape and size. A single phase on earth might be raindrops or river water. Aboard the shuttle orbiter, Skylab, or the space station, single phase water is water with the characteristics of potable water without significant amounts of dissolved or suspended solids (Wachinski, 1987). Two-phase (or multiphase) water is water and a gas such as air, water and significant amounts of dissolved or suspended solids, or all three. Salt and a saturated brine is two-phase. Water at its triple point is multi-phase.

Current thinking of NASA engineers is that single phase water will pose no special design considerations in a microgravity environment (Brasseaux, 1988). Experience with potable systems (Sauer, 1970; Bustamante and Sauer, 1971) seem to support this. Multiphase systems, however, require special consideration (Brasseaux, 1988).

To understand these considerations, we must first review a few basics. In low gravity or microgravity, body forces approach zero. As a result, hydrostatic pressure approaches zero, buoyancy convection approaches zero, and sedimentation approaches zero (Bayuzick, 1987).

Hydrostatic Pressure

The earth's gravitational field does not cause bubbles to rise in a liquid, rather it is the result of a pressure field gradient within the liquid. This gradient is a result of the support given to the liquid to prevent its acceleration by the field. If the support is removed, i.e., it is dropped on earth or it is in a spacecraft experiencing microgravity, the hydrostatic pressure gradient within the fluid essentially disappears and, because the gas within the bubble is accelerated at the same rate as the liquid surrounding the bubble and the walls of the container, there will not be forces tending to create relative movement between the bubble and the liquid or the container (Congelliere et al, 1984). The pressure in an incompressible static fluid is given by

$$p = p_o + \rho g h$$

where h is the gravity direction; its magnitude is the distance from the free surface. ρ is the density of the fluid, g is the force of gravity, and p is the pressure. If g is zero, the hydrostatic component of the pressure is zero.

Buoyancy Forces

On earth, thermophysical processes are influenced by buoyant forces. In many physical processes, the magnitude of this buoyant force overwhelms other forces such as surface tension, capillarity and viscous shear. However, as the gravitational field weakens, these other forces begin to play complex and even competing roles (Labus, 1984).

In microgravity, buoyant forces approach zero. The buoyancy force, F, on a body of volume V in a liquid can be expressed as

$$\vec{F} = -\int \nabla p dv$$

where p is the pressure gradient. If no body forces exist, p is zero, and the buoyancy force is essentially zero. Therefore, in a low gravity environment, buoyancy convection is, for all practical purposes, absent. The absence of buoyancy convection means that all phenomena associated with density differences are also affected. One such phenomenon is the separation of liquid phases of different densities, liquid dispersions.

Dispersions

The stability of liquid dispersions in low gravity was studied by Lacy and Otto (1974). Three different dispersions of oil and water were prepared on Skylab 4 and compared with the same dispersions prepared on earth. The emulsions prepared in microgravity were stable for over ten hours, whereas the same fluids on earth separated completely in ten seconds, indicating a rate of coalescence in low gravity far lower than that on earth.

The reluctance of similar phases to separate is a serious design issue not only in physical processes but in biological issues as well. Separation of gases from liquid and separation of cell mass from liquid will necessitate suitable membrane transfer (never tried in microgravity) or centrifugal separation units (Peterson *et al.*, 1985).

Another phenomenon associated with waste management impacted by buoyancy convection is mass transfer. Whenever a gas and a liquid are in contact, mass transfer across the gas-liquid interface occurs as an equilibrium is established between the partial pressure of a species in the gas phase and the concentration of dissolved species in the liquid. The overall rate of gas transfer is a function of the coupled effects of molecular diffusion, surface phenomena and hydrodynamics.

On earth (One-G), the difference in phase densities (buoyancy) couples diffusional mass transfer with the hydrodynamics of buoyant motion. In microgravity, buoyant translation is essentially zero as is its effect on gas distribution in the liquid. Since density variations are also essentially eliminated, the free convection velocities which influence the overall mass transfer process are also lessened (DeWitt, *et al.*, 1988).

Bubble surface area and internal convection within the air bubbles control gas transfer to growing cells. Bubble size produced at the sparger orifice is controlled by a combination of gravitational forces and surface tension. In microgravity, bubble size will increase rapidly -- severely limiting the amount of oxygen transfer. With essentially no buoyant forces to enhance gas transfer by mixing, other ways of mixing require development (Peterson, *et al.*, 1985). The power requirements for mixing in microgravity have not been determined at this time.

Sedimentation/Flotation

In One-G, the settling of dilute suspensions of solids follows Stoke's Law:

$$V_x = \frac{g}{18\,\mu}\,(P_s - P)\,d^2$$

where V_x is settling velocity, d the diameter of solid particle (considered a sphere), Γ = mass density of the liquid, μ = dynamic viscosity, Γ_s is density of the particle and g = acceleration due to gravity. The elimination of gravity essentially obviates sedimentation as a mechanism for separation. Centrifuges can, of course, be used, but their use increases spacecraft vibration and related micro-G effects.

Surface Tension

Surface tension is the dominant force used to position liquids in microgravity. Figure 1 shows the microbehavior of water in a partially-filled container (Padday, 1986). Figure 2 shows the microgravity behavior of water and mercury in two-container configurations. Note the solid-liquid interface in Figure 2.

The behavior of liquids depend upon their boundary surfaces, i.e., where a liquid comes in contact with solids or gases. Liquids have cohesion and adhesion; both are forms of molecular attraction. Cohesion enables a liquid to resist tensile stress; adhesion enables it to adhere to another body. Cohesive forces pull the liquid molecules together, and adhesive forces pull the liquid molecules toward the solid. If the adhesive force is larger, small amounts of liquid will spread out smoothly on the surface (wetting). If the cohesive force is larger, the liquid will form little droplets on the solid surface. Capillarity is also due to both adhesion and cohesion.

SUMMARY

The characteristics of microgravity fluid behavior that impact waste management all stem from one fact, i.e., the absence of body forces. Absence of body forces translates to a lack of hydrostatic pressure and the absence of any buoyant force. Under these conditions, mass transport is controlled by diffusion. Sedimentation and flotation are eliminated, while surface tension and capillarity now play the major role in fluid behavior. The changes are significant for all fluid processes, but they are extremely critical in spacecraft environmental design.

Figure 1. Behavior of Water in a Partially-filled Container in (a) One-G, (b) Microgravity (Padday, 1986).

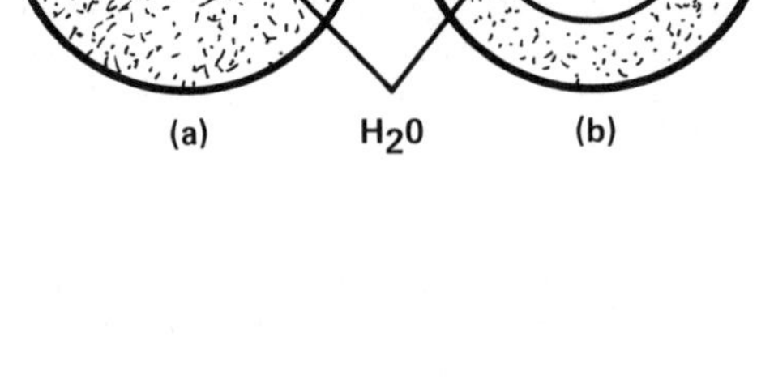

Figure 2. Positioning of Liquids in Microgravity (Murray *et al.*).

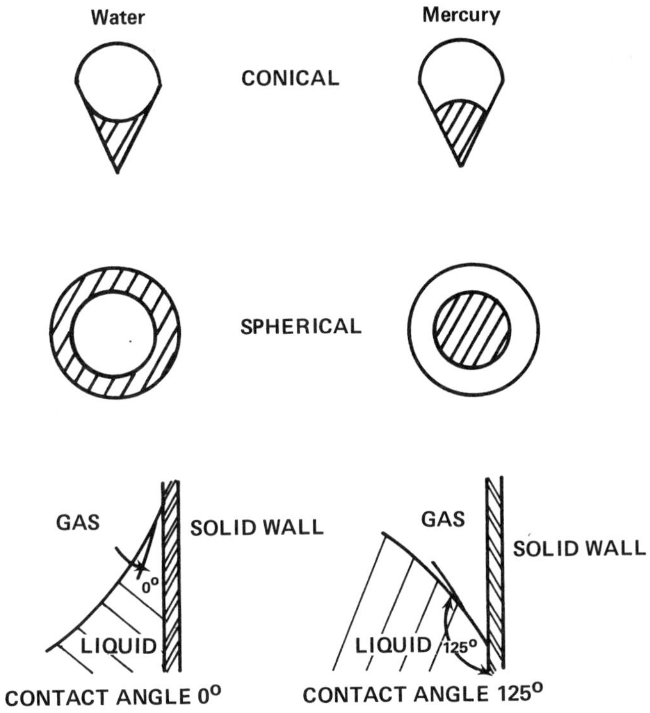

REFERENCES

1. Balebanov, Vyacheslav. (1989). ASCE Bi-lateral Technical Exchange with Academy of Sciences. USSR, Space Research Institute. Moscow. June 9, 1989.

2. Bayuzick, Robert J. (1987). "Space Processing of Metals and Alloys." Commercial Opportunities in Space, (Eds., F. Shahrokhi and K.E. Harwell), Progress in Astronautics and Aeronautics, AIAA, Washington, D.C., 137-146.

3. Berenson, P.J. (1963). "Fundamentals of Low Gravity Phenomena Relevant to Fluid System Design." Physical and Biological Phenomena in a Weightless State. (Eds., E.T. Benedikt and R.W. Halliburton).

4. Brasseaux, Hubert (1988). Telephone interview.

5. Congelliere, J.F., et al., "The Zero G Flow Loop: Steady-Flow, Zero-gravity Simulation for Investigation of Two-Phase Phenomena."

6. Congelliere, et al. (1963). "The Zero G Flow Loop: Steady-Flow, Zero-Gravity Simulation for Investigation of Two-Phase Phenomena." Physical and Biological Phenomena in a Weightless State. (Eds., E.T. Benedikt and R.W. Halliburton). Advances in the Astronautical Sciences, AAS, Los Angeles, California, 223-263.

7. DeWitt, Kenneth J., et al., (1988). "Mass Transport Phenomena Between Bubbles and Dissolved Gases in Liquids Under Reduced Gravity Conditions." 26th Aerospace Sciences Meeting, AIAA, Nevada.

8. Labus, Thomas L. (1984). "Fluid Dynamics in Microgravity." Proceedings of the Workshop on Microgravity Science and Applications. Pasadena, CA., December 3-4, 243-264.

9. Lacy, Lewis L. and Guenther, Otto H. (1974). "The Stability of Liquid Dispersions in Low Gravity." AIAA/AGU Conference on Scientific Experiments of SkyLab, AIAA Paper No. 74-1242, Huntsville, Alabama.

10. Lacy, Lewis L. and Guenther, Otto H. (1974). "Observations of the Liquid/Solid Interface in Low-Gravity Melting." AIAA Paper No. 74-1243, Huntsville, Alabama.

11. Morrison, Dennis R. (1977). "Behavior of Fluids in a Weightless Environment." Bioprocessing in Space Proceedings of the 1976 NASA Colloquim on Processing in Space, NASA, Houston, Texas.

12. Murray, Robert W., et al., "Engineering Aspects of Zero Gravity Personal Hygiene and Waste Management Systems." Source unknown.

13. Padday, J.F. (1986). "The Behavior and Management of Liquid Systems in Low Gravity." Proceedings of the 6th European Symposium on Material Sciences Microgravity Conditions, Bordeaux, France.

14. Peterson, et al. (1985). "Design Concepts for Bioreactors in Space." Ames Bioregenerative Life Support Systems Conference, NASA, Ames, N88-17179.

15. Sauer, R.L. and Bustamante, R.B. (1971). "Water Supply and Waste Management in Spacecraft -- Past, Present and Future." Twenty-sixth Purdue Industrial Waste Conference, Lafayette, Indiana.

16. Sauer, Richard L. (1970). "Spacecraft Metabolic Water Supply." Spacecraft Potable Water Symposium, ACS, Houston, Texas.

17. Sauer, R.L. and Westover, J.B. (1974). "The Potable Water System in Skylab." Intersociety Conference on Environmental Systems. ASME.

18. Wachinski, A.M. (1988). "Waste Management in the U.S. Space Program." Journal WPCF, Vol. 60, No. 10.

Subsection 4.3: LAUNCH FACILITIES AND SUPPORT

NATIONAL AERO-SPACE PLANE PIONEERING NEW FRONTIERS

Major Jess Sponable[1]
Mr Terry Kasten[2]

Abstract

Paper reviews the current status, technology and potential future space launch applications of the joint DOD/NASA National Aero-Space Plane (NASP) program. Key program goals include first flight of the X-30 experimental spacecraft in 1997 with a single-stage-to-orbit flight by 1999. A conceptual design of a first generation operational vehicle derived from the NASP technology is presented. Operating out of conventional runways with a substantial 20,000 lb payload to space station, future operational vehicles offer the hope of more responsive access to space at significantly lower costs.

Introduction

The National Commission on Space in its 1986 report identified three goals to help open the space frontier:

1. Advancing our understanding of our planet, our Solar System, and the Universe;

2. Exploring, prospecting, and settling the Solar System; and

3. Stimulating space enterprises for the direct benefit of the people on Earth.

The Commission further noted that accomplishing these goals economically depends on the **PREREQUISITE** implementation of two additional thrusts:

1. Advancing technology across a broad spectrum to assure timely availability of critical capabilities; and

2. Creating and operating systems and institutions to provide low-cost access to the space frontier.

---

[1]Chief, Space Applications Branch, AFSC/NAR, Wright-Patterson AFB OH 45433-6503
[2]Chief, Systems Study Engineer, AFSC/NAR, Wright-Patterson AFB OH 45433-6503

The latter two thrusts are what the National Aero-Space Plane (NASP) program is all about.

The NASP is pioneering both the technology base and launch systems required to support a strong U.S. space program. Our charter is to develop the technology required to field a fully reusable Single-Stage-To-Orbit (SSTO) manned craft called the X-30. Operated like an aircraft from conventional runways, the X-30 will use ramjets, scramjets and rockets to rapidly accelerate to orbital velocity. It will also be able to cruise like an aircraft from subsonic to hypersonic speeds.

The technology base under development within the NASP program can support a family of NASP Derived Vehicles (NDVs) with numerous mission applications. Although no programs to develop NDVs have been initiated, concepts for NDVs are being investigated. Possible NDV concepts range from a near term scaled up X-30 with a Medium Launch Vehicle class payload to a heavy lift launch capability in the out years. Conversely, scaling down the X-30 technology to an endo-atmospheric cruise configuration could support the development of aircraft flying in the Mach 5-8 range. Illustrated in Fig. 1, a first generation operational vehicle derived from NASP technology is expected to operate as much like an aircraft as possible.

Program Summary

To accommodate future NDV needs, the X-30 will demonstrate manned single-stage-to-orbit flight with a spin-off long range hypersonic cruise capability above Mach 5. Operated from conventional runways the X-30 will takeoff and land horizontally like an aircraft with a fully reusable long life structure. Designed for a 24 hr rapid turnaround time, the X-30 will require a minimum number of ground support equipment and personnel. Equally impressive, the X-30 will be able to ferry itself from an abort landing site back to its take-off point: Edwards AFB, California.

The NASP program is truly a national effort; the NASP Joint Program Office is composed of USAF, Navy, NASA, and DARPA personnel, supported by dozens of government organizations, laboratories, universities and contractors across the nation. The NASP Joint Program Office reports through the Air Force chain of command to a joint steering group that includes senior personnel from all of the participating organizations. The national scope of the NASP program is illustrated by the diversity of participants. From universities and research laboratories to a host of industrial concerns there are over five thousand people from throughout the United States working on the NASP program; all focused on achieving the NASP goals.

Figure 1 - Airplane-Like Operations/Supportability

The NASP team is focused on an ambitious schedule. Currently, we are over halfway through Phase 2 of the program. Phase 1 was the early conceptual studies undertaken by DARPA. Phase 2 began in early 1986 and brought with it significant funding and the stature of a major national program. Phase 3 will begin in 1993 and will initiate the construction of the X-30 flight vehicle. First flight of the X-30 is scheduled to occur in 1997 with the first orbital flight in 1999.

The NASP program has been an extremely competitive program from its inception. The original Phase 2 contracts let in early 1986 involved three engine companies: General Electric, Pratt & Whitney and Rocketdyne. The five airframe companies included Boeing, General Dynamics, Lockheed, McDonnell Douglas and Rockwell. An engine and airframe concept review in the Jul-Aug 87 timeframe marked a major downselect point that left two engine contractors (Pratt & Whitney and Rocketdyne) and three airframe contractors (General Dynamics, McDonnell Douglas and Rockwell). All of these companies are competing to build the X-30 aircraft during Phase 3. The acquisition strategy for Phase 3 is still being formulated. We are seeking a strategy that will assure a high quality technical product, capture the benefits of multiple contractor participation, and allow us to control the cost of the program.

The competitive nature of the NASP program is highlighted by the magnitude of contractor investment in the program. Attempting to win the Phase 3 contract to build the X-30, the contractors will invest approximately $700 million of their own money during Phase 2. The contractors recognize the value of NASP technology to the future livelihood of the aerospace industry. Recognizing the program's potential, NASA is also a strong contributor to the program. However, the DOD will fund the majority of the Phase 2 and 3 program because the X-30 will enable a family of exciting new flight vehicles.

A historical perspective will help illustrate why the DOD and NASA are willing to invest such major funds in the NASP program. The first U.S. jet flight of the XP-59 occurred in the early 1940s. This experimental vehicle enabled an entire family of jet aircraft that dramatically enhanced the capability of the United States to wage aerial warfare. Equally significant, it enabled new commercial transports that have brought the United States and the World much closer together.

Similarly, the X-30 will enable an equally capable family of operational vehicles. Ranging from SSTO vehicles no larger than conventional aircraft to extremely long range hypersonic cruise vehicles, the family of possible NASP Derived Vehicles (NDVs) portends an exciting aerospace future.

Beginning in 1986 the Joint Program Office has continued to take a close look at how the X-30 vehicle scales up to support large payloads to orbit. Seven different contractor studies using different designs and scaling relationships affirmed that vehicles supporting a shuttle class payload would have a take-off gross weight in the 600,000 to slightly over 1,000,000 lb range. These studies further affirmed that lower weights combined with aircraft-like operations could substantially reduce the operating and support costs of future space launch systems. Although very preliminary, operating cost estimates for a shuttle class NDV fall in the $1 to $9 million per flight range exclusive of any development or production costs. The Joint Program Office is putting increasing emphasis on cost estimating methodologies ; however, our initial conclusion is that an NDV can offer truly economical access to space.

Perhaps just as important as cost are the military needs (depicted in Table 1) an NDV might be able to satisfy. Benefits of military NDVs center around providing assured access to space, a responsive launch capability and flexible operations. An example of each is the aircraft level of reliability that must be built into an NDV (assured access), the rapid turn-around time that enables a surge flight capability (responsive access) and the all azimuth flight capability that enables an NDV to access all required orbits (flexibility).

Table 1

COMMERCIAL NEEDS SYNERGISTIC WITH MILITARY REQUIREMENTS

ASSURED ACCESS	RESPONSIVENESS	FLEXIBILITY
o SAFE ABORT	o RAPID TURN-AROUND	o ACCESS TO ALL ORBITS
o AIRCRAFT RELIABILITIES	o HORIZONTAL PROCESSING	o RENDEZVOUS & DOCK
o ALL WEATHER FLIGHT	o CONTAINERIZED PAYLOADS	o SELF FERRY
o STANDARD RUNWAY BASING		o ON-BOARD AUTONOMY

Technical Progress

The top technical challenges faced by the NASP program fall into three general categories: airframe, propulsion and vehicle integration. Construction of the airframe and engines is dependent on the development of advanced materials that combine the attributes of low weight, high strength and high temperature resistance. Long life cryogenic tankage and the development of actively cooled structural components round out the key technology needs that will enable construction of the X-30 airframe. From a propulsion perspective, integrating a ramjet and scramjet propulsion system into the vehicle while maintaining efficient scramjet performance at the highest Mach numbers is an exciting engineering challenge. Closely related is the development and validation of computational fluid dynamics (CFD) computer codes that are used to optimize vehicle design and performance. The final category of vehicle integration is a catch-all that primarily addresses the need to build the X-30 airframe around the engine performance. Essentially, the entire forebody of the X-30 acts as a compressor that slows and compresses air into the engine inlet. The rear of the aircraft is contoured like a rocket nozzle to efficiently expand the engine exhaust while maximizing thrust and propulsive efficiency. The last two technology challenges include development of a very sophisticated flight control system and the development of slush hydrogen with which to fuel the vehicle.

All of the critical technologies required to build the X-30 are being addressed in the Phase 2 ground test program. All three airframe contractors are fabricating and testing major airframe components in a large number of facilities nationwide. Over 20,000 hours of wind tunnel testing in 30+ facilities will be accomplished. Similarly, the propulsion contractors are developing extensive propulsion test capabilities that will culminate in the testing of near flight weight engines in two Mach 8, continuous flow wind tunnels modified for use by the NASP program.

Confidence in the X-30 designs has increased dramatically with the acquisition of ground test data. With the program's inception in 1986 there was much uncertainty with respect to several key technologies including the high speed scramjets and the advanced materials. Today, halfway through the Phase 2 program, our confidence in these key technologies is far greater than what it was in 1986. By the Phase 3 decision we will have ground test data in hand to demonstrate most of the technologies required to build the X-30. Two key exceptions where additional data is needed are scramjet operation at near orbital velocity and our ability to integrate all the technology pieces into a vehicle design with a reasonable weight. These two key areas must be demonstrated by the construction of the X-30.

As with any program reliant on extensive ground test data, all technical approaches may not succeed during development testing. To accommodate this component risk without impacting the overall program, the Joint Program Office is funding parallel development efforts in all the key risk areas. The best example of this is the development of advanced materials. A cooperative materials consortium, consisting of all five prime NASP contractors, is funded with $150 million to develop numerous material candidates. These candidates can be grouped in five primary classes: titanium aluminides, carbon-carbons, metal matrix composites, high creep strength materials and high conductivity materials; the latter two classes used for engine structure and cooling respectively. Only a small number of these materials need to be successfully developed to enable construction of an X-30 vehicle. Similarly, active cooling options under development include the use of hydrogen fuel, other fluids, heat pipes and transpiration cooling techniques. With respect to the scramjets, the test data we are getting looks quite impressive; however, if necessary we are evaluating turning the rockets on as low as Mach 8 -- the maximum velocity at which our wind tunnels operate. Alternatively, the rockets on the vehicle may be used to augment ramjet/scramjet thrust whenever necessary. Also considered is the use of liquid hydrogen instead of slush hydrogen on the vehicles. All these technology options grow the size of the vehicles; however, the resulting sizes are not unreasonably large.

The Next Generation

A preliminary concept for a first generation NDV might have a 20,000 lb payload to the space station orbit. Such a payload would lead to an NDV that could complement other expendable launch vehicles early in the next century. The vehicle that carries this payload may have a Take Off Gross Weight of approximately 400,000 lbs and an Operating Weight Empty (OWE) of approximately 90,000 lbs. A 24 hour turnaround time, relatively low maintenance requirements and a short mission duration combine to help insure a mission oriented, supportable transport craft design. In terms of physical dimensions, such a vehicle would be smaller than a DC-10 and fit within the confines of the space shuttle external tank as illustrated in Fig. 2.

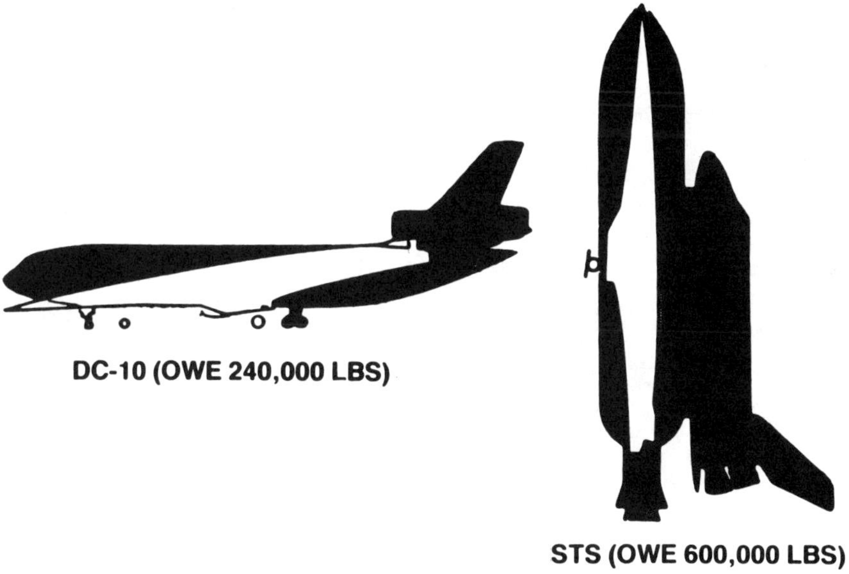

Figure 2 - NDV Size Comparison with DC-10 and Shuttle

The first generation NDV is anticipated to operate like a conventional aircraft from standard hangers using largely off the shelf ground support equipment. The two fueling carts shown in Fig. 3 are unique pieces of ground support equipment required to load hydrogen and oxygen into the vehicle. Access to the vehicle is through the wheel well with the cockpit located adjacent to the payload bay.

Figure 3 - NDV Operations Concept

Another important characteristic of the NDV concept is the containerized payload bay. The container weight is part of the NDV and includes standard interfaces to the vehicle. The payload developer can either utilize the standard interfaces available from the NDV through the container or install airborne support equipment within the container to accommodate unique payload needs. Standard payload modules could support missions such as on-orbit servicing, space station resupply, personnel transport and on-orbit remote manipulator arm operations.

The NDV concept complements heavy lift Expendable Launch Vehicles by supporting a large percentage of less massive space traffic. Equally significant, it would enhance the DOD capability for assured access to space by providing reliable access for smaller payloads, that can also be launched on expendable boosters in the event the NDV fleet were grounded. The payload bay mass and volume constraints still support 40 to 50% of DOD space traffic, nearly 90% of the near term SDI requirements (less of the long term SDI requirements) and around 90% of the civil and commercial launch requirements. These estimates do not account for possible new spacecraft designs optimized for NDV launch and on-orbit support. If this were to occur a larger percentage of payloads could be accommodated.

In terms of an annual flight rate, the NDV might fly 40 to 160 flights per year depending on the quantity of space traffic assumed. A key area of uncertainty in projecting future space traffic is the level of commercial launch activity. If the NDV lowers the cost of accessing space, we anticipate a surge in commercial launch traffic. Quantifying the magnitude of this increase requires an understanding of the demand elasticity of the commercial launch market. The increased demand may range from only slight increases in launch traffic to the creation and movement of entire industries into space.

A key question is when can the NDV be available to the government and to the commercial sector. Any schedule is dependent on the successful demonstration of the requisite technologies and the availability of adequate funding. The current X-30 schedule could support delivery of NDVs able to accommodate military space launch missions early in the next century. A second block-buy immediately thereafter could accommodate additional government requirements or conceivably, commercial demand for an NDV fleet modified for commercial operations.

The first generation NDV could support a high percentage of government civil and commercial space traffic. Military versions of the NDV will emphasize aircraft-like operations and relatively low operating costs. These attributes offer an exciting opportunity to accommodate commercial launch needs. However, the fundamental question of what commercial enterprise would buy the NDV, and how many, remains unanswered. We're not taking orders yet, but we are a national program with a national charter and the needs of the commercial launch industry will be considered as we formulate plans and schedules.

Concluding Remarks

It is worth reemphasizing that the NASP program is a national effort, involving thousands of US citizens. NASP is rapidly developing and maturing the key technologies required to build the X-30 and NDVs. We have already demonstrated significant technical progress and anticipate delivery of much more quantitative test data over the next several years. Although the conceptual NDV designs discussed in this paper are not part of an approved development program, the current NASP schedule could support delivery of an operational space launch capability early in the next century. The NASP program will enable hypersonic flight to space and leverage the American economy to take full advantage of commercial space opportunities in the twenty first century.

Commercial utilization of the resources and environment of the space frontier will begin a new chapter in the aviation saga. As with all past frontiers, opening space to mankind will come only at tremendous cost, both human and financial. The spirit of the men and women who exploit and settle the frontier will forge a new generation dedicated to mankind's neverending quest to fly: from the Greek flyer Icarus, to the Wright brothers at Kitty Hawk, and someday to the distant stars.

COMMERCIAL LAUNCH FACILITY ECONOMICS

by Mike Trial*, member ASCE

ABSTRACT:

Launch facility construction, maintenance and modification costs remain a largely invisible part of the cost of putting mass into orbit. The government can lease launch facilities to commercial users, but since these facilities typically were constructed in the sixties and for other missions, they often require considerable modification before they are useable for commercial launch services. Due to the criteria which drove their original design, many operational and maintenance inefficiencies remain. The cost of these inefficiencies will necessarily be passed along to launch service customers. By designing a commercial launch facility from scratch considerable savings can be realized.

EXISTING FACILITIES:

It is typically more expeditious for the government to remodel existing facilities rather than to build new ones if DOD is funding all or part of the facility. Remodelling can often be carried out as part of the vehicle acquisition program rather than as a new construction program. Often facility criteria are subject to major changes during the design process as vehicle configurations change. Initial operational capability dates are seldom slipped so facility design sometimes must incorporate compromises to prevent having to start the design process over again each time the vehicle changes.

Unfortunately, the parts of the launch facility at or below ground are the most expensive, and least changeable once design has begun. There are three

* Program Manager, US Army Corps of Engineers
20 Massachusetts Ave, Washington DC 20314

primary design elements in the below ground work: flame bucket and exhaust duct, launch deck and utilities. These are also among the first elements that must be constructed. So final facility configuration, final facility cost, and final facility flexibility (among vehicle types) must be decided at the outset. The cost for a launch facility for a heavy launch vehicle such as Titan IV can be upwards of $200 million and take in excess of 3 years, excluding installed equipment. If design and/or construction times are compressed to meet launch need dates, costs rise, and the risk of design errors or compromises increases. Although no government agency is studying constructing new medium launch facilities, costs to the government to construct a new facility for Delta or Atlas vehicles might be expected to be $150 million.

Designers must also decide whether vehicles will be integrated on the pads or integrated off-pad and transported to the pads for launch. Except for the Space Shuttle, American launchers are designed for integration on pad. There is considerable interest in designing future pads for off-pad integration, however, to increase the launch rates per pad. A new facility should ideally be able to accomodate either mode of operation.

GOALS FOR A NEW COMMERCIAL LAUNCH FACILITY:

A reasonable goal should be to construct a resilient,efficient, flexible launch facility for less than $50 million and in under three years time. The design should be maximized for operational efficiency, high launch rate, and low construction costs. The new facility should be flexible among medium launch vehicles and should provide fuel facilities and integration areas to service alternative small launch vehicles such as Orbital Science Corps' Pegasus or small truck-mounted launchers such as that proposed by Space Commerce Corp. A new launch facility should allow small and medium lift launch service providers to establish a presence in that market in the face of competition from both foreign launch services and leased government facilities. A secondary goal could be to allow for expansion into heavy lift vehicle services.

Advanced Launch System goals (Scott, 1989) mirror these goals: operational flexibility, responsiveness to payload manifest changes, reliability, resilience of the facility to launch catastrophes, and quick launch turnaround times.

Others (Logsdon, 1989) note that for American launch service providers to exploit what is expected to be a $1 billion market, costs must come down and readiness must improve. Additional launch facilities provide both and improve American ability to participate in international cooperative missions (Banks, 1989) by having better launch schedule flexibility.

THE MARKET IN THE NEAR TERM:

Launch facilities comprise a complementary good to the launch vehicles they service. To date the government has constructed all launch facilities in this country, making it difficult to evaluate supply and demand for launch facilities. It appears there is an emerging demand for more launch facilities, based on aggregate demand for kilograms to orbit, and on the desire for increased flexibility in launch vehicles, manifesting, and launch windows. Small, single payload vehicles launched from simple launch facilities can more easily meet flexibility and reliability demands than larger vehicles launched from government-constructed facilities.

Assessments of methods for reducing the cost of mass to orbit generally focus on larger launch vehicles (OTA July 1988, September 1988). Medium or small launch vehicles become more competitive when total costs are addressed. Costs such as delays to payloads are seldom addressed in total cost figures even though those costs can be hundreds of thousands of dollars daily. Flexibility in scheduling payloads can reduce delays. Smaller payloads or single payloads also reduce liability in case of launch failure, thus reducing insurance costs. Shorter on-pad timelines result in lower costs to the launch site operator and thus to the user. A launch facility should be able to accomodate modularity in vehicles to accomplish rapid changes in vehicle configuration to meet payload manifest changes. A high launch rate reduces facility fixed costs per kilogram to orbit.

Since the direct costs of earthwork and concrete for

the launch support structure (flame bucket and exhaust duct) can be expected to comprise almost 50% of the cost of facility construction (Sverdrup, 1988), careful attention to the design of these components is essential.

LAUNCH FACILITY CONSTRUCTION:

The launch support structure comprises the largest part of fixed facility construction costs. Two features of the structure are critical to both cost and flexibility among vehicles to be launched. These are the depth of the flame bucket and the size of the opening in the launch deck. A typical structure is shown in section view in Figure 1. If flame bucket depth can be held to no deeper than 20 meters and launch deck opening to no larger than 9 meters square, construction cost of the launch support structure could be held to $12 million. This compares very

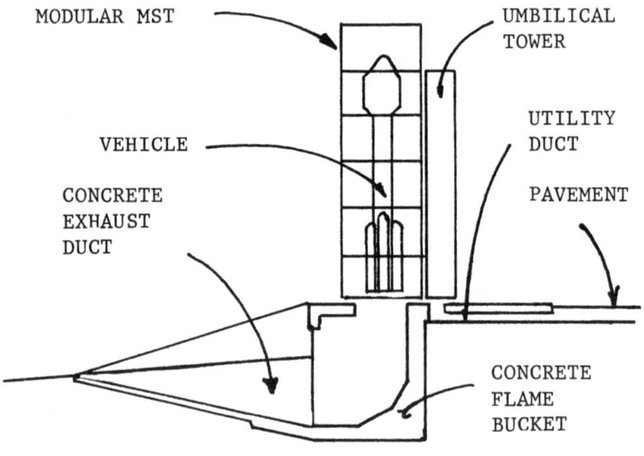

FIGURE 1: LAUNCH STRUCTURE SECTION

favorably with estimated costs of $32 million for a proposed government facility for Titan IV's, with a 30 meter deep flame bucket and a 12.3 meter square opening. This also illustrates the fact that costs for large vertical concrete structures such as flame buckets increase at a more than linear rate as depth is increased. This is due to the increasing complexity of formwork and scaffolding as wall height increases. In addition, the wall thickness must increase as height increases, resulting in more concrete and more rebar. A 20 meter depth appears adequate for proposed Delta and Atlas vehicles without sound suppression water.

The launch deck opening size is also a critical element in launch facility design and cost. Figure 2 shows a plan view of the launch deck with a 9 meter square opening. The exhaust opening should be sized to a minimum consistent with all proposed vehicles. This is difficult since strap-on boosters, requiring an increased opening size, are the preferred method of increasing vehicle lift capacity. Too large an opening is not optimal since exhaust overpressures are then able to affect the vehicle during the early seconds of ignition. The use of a removeable launch platform, configured to a specific vehicle and set into place by a crane prior to positioning of the MST offers one alternative. The opening can then be made large enough to accomodate all proposed vehicles.

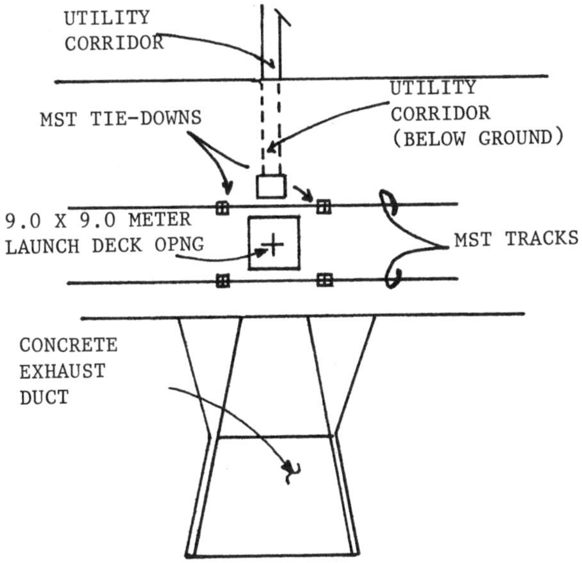

FIGURE 2: LAUNCH STRUCTURE PLAN VIEW

The second largest cost in launch facility construction is the mobile service tower (MST). Typical towers are self-propelled structures over 30 meters in height with an integral crane for vehicle integration. The tower provides access to the vehicle at all heights, the crane provides handling for the vehicle stacking operation, and the structure often provides payload and sometimes vehicle environmental control. Utilities to the vehicle are normally provided by the umbilical tower. Just prior to launch the mobile service tower is rolled back. Costs and fabrication times for large MSTs are considerable. A replacement MST for space launch complex 4 at Vandenberg approached $90 million and one year fabrication and erection time. The size and weight of the MST directly affects the track and tie-downs for the tower. If the MST can be reduced in weight there are direct savings in construction costs. Two further savings could be realized if the integral bridge crane in the MST could be eliminated and if the self-propulsion mechanisms could be eliminated.

Conventional MST's are typically designed around a single vehicle so flexibility is a problem. Multiple MST's for different vehicles is too capital intensive. A viable alternative for flexibility in MST's is to make them modular so that they can be configured for specific vehicles.

Figure 3 shows an alternative to the conventional MST: a modular MST, track-mounted, but consisting of modules constructed of specialized tube and coupler members (like that used in scaffolding) rather than the bolted structural steel currently used. The integral crane would be elminated. The same construction crane used to assemble the MST can serve as the vehicle assembly crane. Remote controls exist for modern 100 to 200 metric ton cranes allowing the operator to stand inside the MST, at the work point, during stacking. During launch the crane would be driven to a safe distance in the utility area and boomed down. The MST would be moved by pulling it

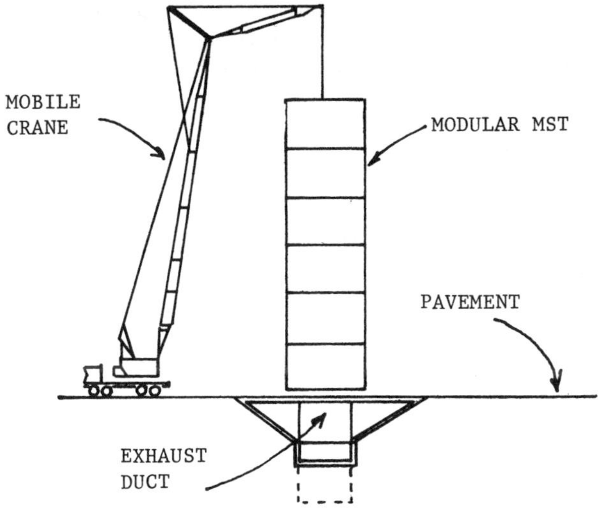

FIGURE 3: MST SIDE VIEW

with construction equipment such as a large front-end loader or dozer. Rolling gear on the bottom module of the MST and tie-downs to concrete anchors in the concrete would be much the same as conventional MST's but considerably lighter since the MST will be lighter.

The third major cost component of launch facilities is the utility work, most of which is below ground. Most existing facility designs were vehicle-specific and construction plans were of a conventional nature which emphasized lowest first cost rather than life-cycle costing. Utilities, including electrical power, water, sewage, propellants, oxidizers, inert gasses and inlet air routing is typcially done straight line from source to launch point, direct buried in the ground, with various lines running above and below each other. This makes changes and maintenance to the lines exceptionally difficult. If utilities are corridorized, in concrete ducts large enough to accomodate inspection perconnel and to accomodate line replacement, there is a significant increase in flexibility, maintainability, and in constructability. Figure 2 shows a concrete utility duct below the paved surface and an open concrete-lined, trench above ground. Utilities would reach the vehicle through the umbilical tower deck opening. If a future change called for conversion to transporter, this same opening could serve as the connection point to the transporter.

FUTURE EXPANSION:

As the commercial launch facility gains acceptance and as demand increases, a second launch support structure can be considered. Some components of the utility service and storage area could be common to both launchers. A degree of redundancy in support facilities should be maintained however, since one of the goals is to provide a highly reliable facility. Individual air conditioning plants, fuel handling, and power facilities provide significant backup value. Interchangeability among fans, pumps, motors, valves and the like can increase reliability since parts can be cannibalized in emergencies. Use of off-the-shelf componentry should be emphasized. Additional savings in spare parts stockage and in reduced employee

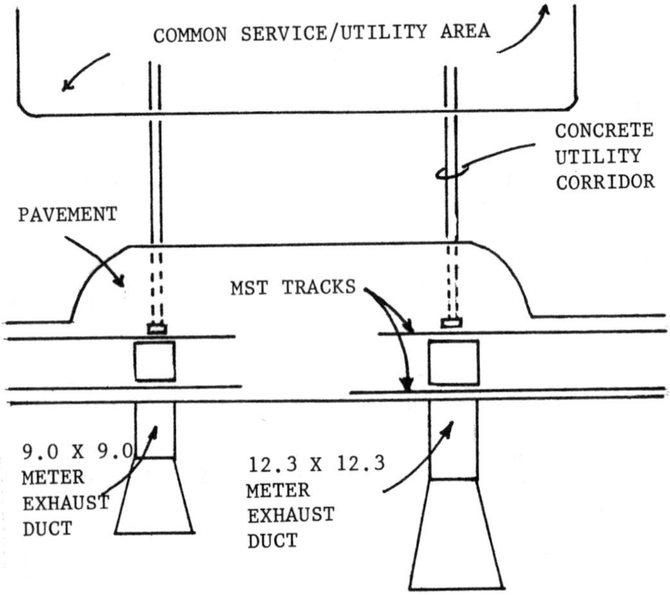

FIGURE 4: DUAL LAUNCHERS PLAN VIEW

training can also be realized by standardizing. Figure 4 shows a large launch support structure, suitable for Titan IV adjacent to a medium launch vehicle launch facility. Defense Explosives Safety Board quantity distances would require the facility work points be separated by approximately 500 m. however, based on current Titan IV configurations.

CONCLUSIONS:

Utilizing concepts that incoporate good construction practice and attention to the reliability, maintainability and operability goals of launch service operators, a commercial launch facility could be designed and constructed for considerably less that existing government facilities. A new launch facility would be convertible between mobile service tower and transporter modes of vehicle servicing. This commercially constructed facility could provide very frequent launches of small and medium launch vehicles with good maintainability and flexibility among vehicles.

APPENDIX

Banks, Peter M. and Sally K. Ride "Soviets in Space." Scientific American Number 2 Volume 260, February 1989: 32-40.

Logsdon, John M. and Ray A. Williamson, "U.S. Access to Space" Scientific American Number 3 Volume 260, March 1989: 34-40.

Office of Technology Assessment, Launch options for The Future: Buyers Guide OTA-ISC-383, July 1988: 8, 11.

Office of Technology Assessment, Reducing Launch Operations Costs OTA-TM-ISC -28, September 1988: 35.

Scott, William B., "ALS Cost, Efficiency to Depend Heavily on Process Improvements." Aviation Week and Space Technology, October 23, 1989: 41-43.

Sverdrup Corporation, Criteria Validation Study, Launch Complex Facilities, Space Launch Complex 7, Package II, Part 2. November 4, 1988.

Considerations for the Development of a Commercial Spaceport

Ralph W. Christie, Jr.[1] Member, ASCE

Abstract

Several states and countries are currently reviewing the feasibility of developing a commercial spaceport in their homeland. United Engineers & Constructors Inc. completed an extensive feasibility study in February 1989, which evaluated the development of a commercial spaceport in Florida. The author, who was the Project Manager for Spaceport Florida, provides an overview of the issues, methods and findings which are contained in the study. In addition, he presents ideas on what analyses and implementation steps should be considered when developing plans for a commercial spaceport.

Introduction

In 1984, President Reagan issued Executive Order 12465, entitled "Commercial Expendable Launch Vehicle (ELV) Activities", "...encouraging and facilitating commercial ELV activities by the United States private sector." Soon after, Congress enacted the Commercial Space Launch Act of 1984, signifying the government's endorsement of a U.S. commercial ELV industry. This legislation, coupled with the President's 1988 National Space Policy, has sparked considerable interest and activity from commercial launch vehicle providers; in addition, various states and countries have been evaluating how they can participate in the commercial launch market. Some of these governmental entities have investigated the possibility of developing a commercial spaceport - a site(s) which could provide commercial launchers a place to launch on a commercial basis.

Already established as a premier launch site, the State of Florida has undertaken studies to investigate exciting new possibilities in commercial space. As a portion of this continuing process, in July 1988, the Florida Department of Commerce awarded the "Space-

---

[1]Project Manager, United Engineers & Constructors Inc., 700 S. Ash, Denver, Colorado 80222

port Florida Feasibility Study" to United Engineers & Constructors Inc. (UE&C), Stearns-Roger Division, and its teammates.

UE&C and the study team executed from July 1988 to February 1989 a comprehensive assessment of the operational, economical and financial feasibility of a commercial spaceport located in the State of Florida. The project included assessment of complementary space-related opportunities and preparation of a blueprint Business Plan to take Spaceport Florida from a concept to commercial reality.

This paper gives an overview of the issues, methods and findings which are contained in the extensive three-part report. In addition, this paper presents ideas on what analysis and implementation steps should be considered when developing plans for a commercial spaceport.

Operational and Technical Considerations

Site Selection

The selection of a site for a commercial launch facility involves the consideration of many complex issues and criteria which are often interrelated. The screening process to select acceptable sites is a multidiscipline effort which considers such things as operational safety, space vehicle mission requirements, environmental risks, and infrastructure requirements. The final selection of the "chosen" site(s) from this candidate list, however, could also consider additional issues such as socioeconomic impacts, capital and operating costs, and financial feasibility.

The list of candidate sites in Florida was developed through a progressive screening process as shown in Figure 1. During the Florida study, this process was initiated on the county level with portions of the State being eliminated based on not satisfying certain criteria. As the number of acceptable sites became smaller, the focus shifted to evaluating specific areas in the county.

A modified version of the Kepner-Tregoe method was utilized in evaluating site alternatives. The criteria for the screening process was two types: (1) criteria which must be satisfied ("must" criteria) where each alternative site is evaluated on a "go/no

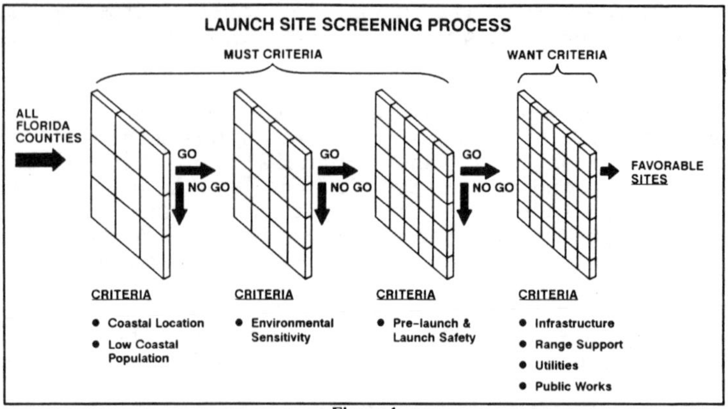

Figure 1

go" basis against these criteria, or (2) criteria which express relative desirability and which and are associated with relative advantages and disadvantages ("want" criteria). The "must" criteria includes items which relate primarily to issues of personal safety, property damage prevention, and protecting the environment. The "want" criteria, on the other hand, include items which concern issues regarding operational flexibility, availability of land, existence of necessary utilities and infrastructure, and the adequacy of community services.

The favorable launch sites that were identified in Florida were inactive pads 14, 15 and 16 at Cape Canaveral and an inactive suborbital site at Cape San Blas in the panhandle region of Florida.

Mission Considerations

When determining the suitability of a particular site for commercial launches, it is necessary to consider the different types of space missions (i.e. Low Earth Equatorial, Low Earth Polar, Geostationary, Suborbital, etc.) and their unique requirements and constraints. The various missions place special requirements on the location of the launch site, such as, flight azimuths, site latitude, vehicle velocity, and launch vehicle size.

Launch Vehicle Requirements

A number of expendable launch vehicles (ELVs) which have been launched on numerous military and civilian space missions are suitable for use in commercial space flights. These vehicles include ones that are a part of active space programs, e.g., the Scout, the Delta, the Atlas Centaur, and the Titan. In addition there are launch vehicles under development or proposed by Government agencies or private organizations. A challenge in planning a commercial spaceport is to design a launch facility which can service and launch as many different vehicles as the market dictates, in order to minimize the capital and operating costs required for launching. A particular exception could be a country/state who guarantees the launching of its preferred vehicles(s).

From study of the physical size and present/future pad requirements and operations of existing vehicles, it was concluded that Spaceport Florida should initially serve vehicles with suborbital capability through the range of those capable of delivering 7,000 pound payloads into Geosynchronous Transfer Orbit (GTO). In the planning of any potential commercial spaceport, one needs to identify the class of launch vehicles it will serve and define the launch facilities required. Based on review of the marketplace and the requirements of existing, as well as planned launch vehicles, the study team concluded that Spaceport Florida should initially focus on the following launch vehicles:

- ° Suborbital (sounding rockets)
- ° Small-class vehicles (up to 2,000 pounds to Low Earth Orbit (LEO)
- ° Medium-class vehicles (up to 7,000 pounds to GTO)

The study team suggested that Spaceport Florida should initially focus on servicing existing launch vehicles, considering that development of new vehicle can take several years.

One of the Spaceport's critical functions in competing for the commercial launch services market will be to provide cost-effective operations. Multi-vehicle launch capability has been proposed by United Engineers & Constructors to achieve this competitiveness. Each pad will be able to accommodate a variety of launch vehicles, reducing the overall capital and operating cost requirements from those of conventional single-vehicle pads. Adjustable platforms would be provided in the integration and assembly buildings, mobile service tower (MST), umbilical tower (UT), as appropriate, as well as provide adjustable clean room enclosures. This operational approach is a change from current U.S. practice, but many industry experts believe it is achievable.

Launch Pads

Small vehicle pads, which service suborbital and small LEO vehicles would normally require a minimum of construction. Core vehicle assembly and payload encapsulation could be performed off-site. Since these vehicles are normally propelled by solid rocket motors, liquid propellant storage and handling systems would not be required. Major support installations could include a launch control center, tracking, and telemetry capabilities. Sounding rockets (suborbital) could also be launched off of these small pads.

Medium-vehicle pads could service a variety of vehicles which deliver payloads to LEO and GTO orbits. Because most vehicles today in this class require integration on-pad (IOP) of boosters, core vehicle and payload, a mobile service tower (MST) with a payload clean room is required. The MST could contain movable and adjustable platforms with inserts to fit around specific vehicles, and will function as a gantry crane during assembly operations.

An umbilical tower would be required to provide fueling and electrical connections. Other major installations required are a launch control center, tracking and telemetry capabilities, and liquid propellant storage tanks.

The trend of future launch complexes is to provide an ITL (Integrate-Transfer-Launch) approach to operations where the vehicle and payload are integrated in facilities away from the launch pad, thereby greatly reducing the time required for a launch vehicle to be on the pad prior to launch.

Other new launch vehicle concepts, such as the air-launched Pegasus, could radically change the requirements for launch facilities.

Market Considerations

In order to create a commercially viable enterprise, it is necessary to have an understanding of the marketplace which it will serve. A commercial spaceport's market "products" are launch services, and the required infrastructure associated with space vehicle launches.

In 1988 most Western World market watchers forecasted 15 to 20 satellite launches per year on medium- to large-class vehicles through the mid-1990's. Also in 1988 Arianespace (the European Space Agency's commercial launch organization) predicted somewhat higher rates--20 to 28 launches per year.

The small orbital launch market (payloads weighing less than 2,000 pounds delivered to LEO) is not as well defined, but there appears to be a growing interest in this market. During the Spaceport Florida Study, estimates ranged from ten to 40 domestic launches per year through the mid-1990's and international sounding rocket launch requirements were projected to be between 70 and 100 per year during the same period.

During the planning of a commercial spaceport, an entity would need to evaluate their geographic location, funding capacity, infrastructure and land availability to determine what are the most likely target markets. An estimate of market share would also be required to determine the economic feasibility of the operation.

Organizational Structure and Strategy

There are several existing launch complexes in the world that provide launch facilities and services to commercial launchers: Kourou, French Guiana (operated by the European Space Agency), Cape Canaveral Air Force Station, Vandenberg Air Force Base, Xi Chang China, etc. The European Space Agency facility in Kourou currently captures about 50 percent of the commercial launch market and is intended to operate truly as a commercial spaceport. Cape Canaveral and Vandenberg launch services focus mainly on Department of Defense missions. Considering the many existing launch complexes, it is important in the planning and development of a commercial spaceport to provide launch services and operations that are commercially focused in order to be competitive.

A "Spaceport Authority" was recommended as the appropriate legal structure to provide leadership and coordination of Spaceport Florida's development. Such an Authority was established by general law in June 1989.

An appropriate role for the Spaceport Authority is to act as a facilitator and provider of the necessary site infrastructure to develop a spaceport. The Authority must take the lead in obtaining the necessary environmental permits, and assembling the land required to meet current and future market needs.

The Authority's management and operational approach could be very similar to an Airport Authority. For instance, at an airport, the Authority attracts airline companies to its facilities by providing the general services and infrastructure required--an airline company does not build runways and terminal buildings. Similarly, the Spaceport Authority would provide launch pads, and operations support buildings. (See Figure 2).

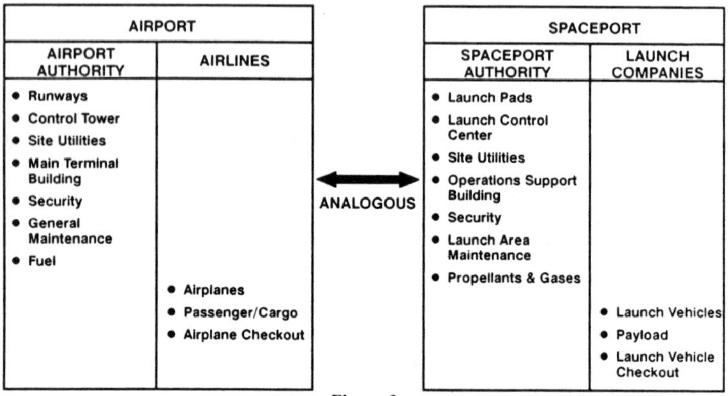

Figure 2

Although airline operations face different technical and managerial constraints than would a space launch operation, the analogy holds for many services and facilities. The Spaceport Authority must develop a commercial relationship with the launchers very much like an airport arrangement. In both cases, the private sector could possibly assist in building many of the required support facilities (for instance, hotels for airports, and payload processing facilities for a spaceport).

Potential Launch Services Revenues

Launch revenues for a commercial spaceport will be dependent on many factors, including but not limited to vehicle size, insurance requirements, the services and facilities required by launchers, and business conditions. Again, the analogy to airport operations holds true: airport landing and use fees vary with similar factors.

Launch rates will be influenced by the ability of a new spaceport to penetrate the commercial launch market and the potential for more launches from the reduction of launch cost per pound of payload or from the recognitions of other technology applications/ research in space.

The Spaceport Florida Study Team reviewed the launch cost that could be directly influenced by a commercial spaceport. From this analysis it was determined that approximately 15% to 35% (depending on vehicle size) of the launch vehicle charge (approximately $10 Million to about $100 Million, depending on the launch vehicle size and mission) can be affected by a spaceport. The costs that could be controlled by a spaceport are for launch operations support (i.e., range tracking, telemetry, documentation, propellant storage, launch pad usage fees, launch pad operation and maintenance, and possibly insurance coverage (third party and property damage).

Complementary Opportunities

The Spaceport Florida study team evaluated complementary space-related activities with the potential to provide economic development and other benefits to the state. Based on numerous discussions with industry professionals, the study team concluded that there are three specific areas of opportunity: space-related tourism, a Space Business Incubator, and fostering educational involvement in Spaceport Florida.

Tourism is Florida's largest industry, and space tourism could be a major stimulus and integral portion of the overall tourism infrastructure. Other government entities planning commercial spaceports could also benefit from space tourism, depending on their geographic location and accessibility.

Business incubators provide an opportunity for small, startup businesses to be nurtured into "going concerns". A Space Business Incubator, located near a commercial spaceport could provide specific support and encouragement for companies that might eventually become spaceport customers, suppliers or service providers.

The Governor's Commission on Space has indicated that Florida must strengthen its educational system to meet the current and future demands of the space industry. The entire Spaceport Florida scenario could stimulate the state's academic institutions at all levels. Spaceport activities can encourage participation in joint research and experimental programs, as well as development of specialized educational programs to meet industry's demand for qualified personnel. A Space Business Incubator can include companies which will rely on universities to carry out contract research and development. Even the space-related tourist attractions can emphasize education about space and its technological advancements.

Financial Feasibility and Benefits

The Spaceport Florida Feasibility Report addressed two major issues: financial projections and available financing methods. A main concern of any potential spaceport owner is whether the benefits of the spaceport will be greater than the costs. Figure 3 shows the potential funds inflows and outflows of Spaceport Florida. Without considering specific dollar amounts, two types of economic benefits are available to the stage. First, there is the profit potential (or net revenues) of the operations. Second, indirect benefits to the state include higher employment, an increased tax base, economic growth, etc.

The financial feasibility of a commercial launch center is extremely sensitive to the number of launches and the fees charged.

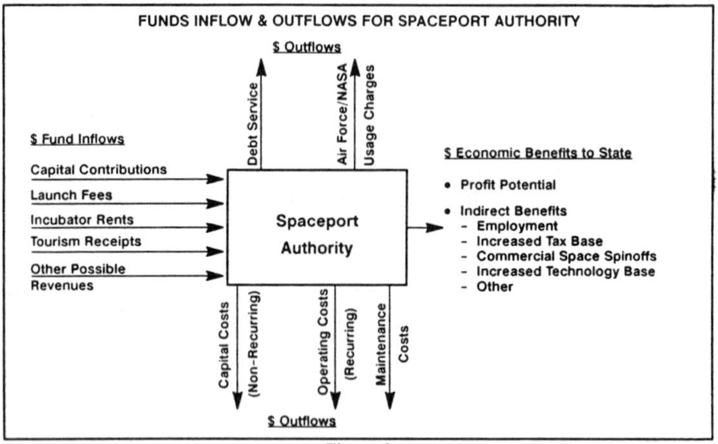

Figure 3

It was recommended that an off-site impact analysis be performed to identify all potential benefits to a state/country from commercial space-related industries.

The Spaceport Florida study team recommended four debt financing options as follows:

- Tax exempt general revenue bond, with credit support
- Tax exempt special facility revenue bond
- Taxable special facility revenue bond
- Corporate debt privatized facility

Summary

The development of a commercial spaceport requires the analysis of several important factors, such as operational/technical feasibility (including safety, environmental, infrastructure requirements, etc.), market potential, complementary opportunities and financial feasibility. The State of Florida has completed such an analysis and is preparing to build a small launch center at Cape Canaveral and Cape San Blas, Florida, as well as creating a space-business incubator and new space-related tourist facilities. Other states and countries, pursuing the development of commercial spaceports will require similar analysis; their follow-up steps will depend on their geographic location, faith in market penetration and their satisfaction of the financial benefits.

References

Spaceport Florida Feasibility Study (Parts I, II & III and the Executive Summary) completed by United Engineers & Constructors Inc. in February 1989.

PRELIMINARY ASSESSMENT OF THE POWER REQUIREMENTS OF A MANNED ROVER FOR MARS MISSIONS

MOHAMED S. EL-GENK and NICHOLAS J. MORLEY[1]

ROBERT CATALDO and HARVEY BLOOMFIELD[2]

Abstract

This paper presents the results of a preliminary study to determine the total mass and power requirements of a Manned Mars Rover. The primary power for the Rover is provided by a Nuclear Reactor Power System, while the Auxiliary and Emergency Power system is a hybrid of Solar Photovoltaics and Oxygen-Hydrogen Fuel Cells. Estimates of the Power requirements for the Nuclear Reactor Power System are determined as functions of the number of astronauts on board; the scenario of the emergency return trip in case of a complete malfunction of the reactor system; Specific Mass of the Power System; and the range and cruising speed of the Mars Rover vehicle. Results show that the cruising speed of the Rover and the soil traction factor strongly affect the traversing power requirement, and hence the mass of the nuclear power system. In order for the Nuclear power system mass not to exceed 50% of the total mass of the Rover vehicle, the cruising speed must be limited to 14.5 and 24 km/hr for a Power System Specific Mass of 150 kg/kWe and 50 Kg/kWe, respectively. The results of this study will help identify the primary power level for subsequent development of a point design and the integration of a nuclear power system with the Rover vehicle.

Introduction

Throughout the next three decades the focus of the U.S. space program will be directed towards manned outposts in space. As currently being defined by NASA's Office of Exploration (OEXP), the successful initiation of human exploration of the solar system will require system capabilities to evolve beyond their current status.

[1]Institute For Space Nuclear Power Studies, The University of New Mexico, Albuquerque, New Mexico, 87131

[2]NASA Lewis Research Center, Cleveland, OH, 44135

Innovative designs of both spacecraft and surface elements will be required along with thoughtful approaches to the problems of long-term space power [Andrews, 1989].

NASA's Office of Exploration is currently conducting studies that will yield a better understanding of the strategies necessary to conduct human exploration of the solar system [OEXP, 1989]. A permanent human presence on the moon or Mars is one of the ultimate goals. Precursor missions, both robotic and human expeditions, will determine optimal sites to establish such bases. Preferred sites would be abundant with resources, thus reducing the dependency on resupply of processing materials and life support supplies.

An evolutionary strategy is currently planned for manned missions to Mars, beginning with an initial exploration of Mars and building towards the establishment of an in situ propellant facility and ultimately a permanent manned base. Some of the mission design guidelines include an initial crew size of 4 growing to 8, reusability of selected vehicle elements, and aerobraking at both Mars and Earth. The program is to be carried out through a series of unmanned cargo flights and manned exploration missions. Expeditions to Mars, as well as permanent bases, are envisioned to require enhanced piloted vehicles to conduct science and exploration. Piloted Rovers, with extended range, have been identified as a viable means of achieving global access of Mars; a mission requirement for surface systems operations.

For these missions, a Rover vehicle with 30 kWe steady-state user net power is being considered for the exploration and sampling of the Martian surface. This power level is based on the projected electric power needs for operations by the Rover on the Martian surface and does not include traversing power. These operations include drilling, sampling and sample analysis, on board computer and instrumentation, vehicle thermal management, and the astronauts' life support system. The Rover may consist of a train configuration with up to four cars (see figure 1) and allows up to six astronauts [El-Genk and Morley, 1988].

In addition to the operation and the traversing power requirements, the Rover vehicle will be equipped with regenerative fuel cells to meet possible emergency power needs for utilities and life support for up to a full week. A total of three days worth of fuel cells is mounted on board the Control Vehicle, while four days worth of fuel cells is mounted on board the experimental unit car. During an emergency return, following an assumed complete malfunction of the main power supply, the

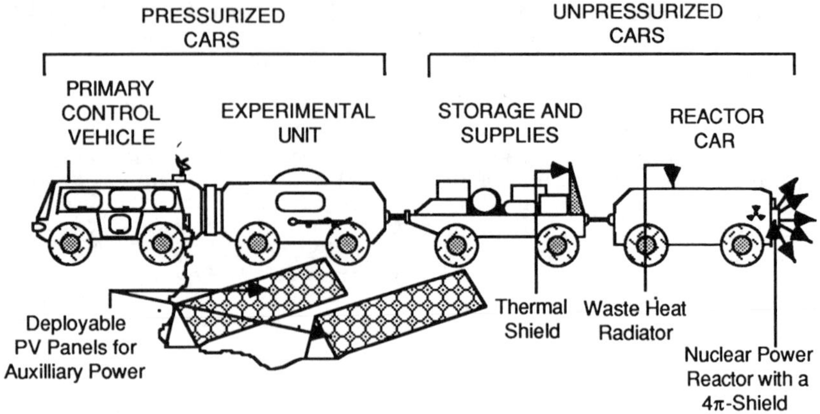

. FIGURE 1:LAYOUT OF A TRAIN TYPE MANNED MARS ROVER

Control Vehicle will be separated from the rest of the Rover. The fuel cells, together with deployable Photovoltaic (PV) arrays, will provide auxiliary power for both the life support and traversing. Both the primary power (including operation and traversing) and the emergency power requirements depend on the projected daily range and cruising speed in each case, as well as the number of astronauts on board. While the former depends on the total mass of the various cars of the Rover Vehicle [EEI, 1989], the latter depends on the mass of the Primary Control Vehicle, the Regenerative Fuel Cells and the PV panels (see figure 1).

Although there are many candidate power system options available for primary power (i.e. Solar Photovoltaic, Solar Dynamics, Radioisotope Thermoelectric Generators (RTG's), Dynamic Isotope Power Systems (DIPS)), each has inherent limitations which limits it's usefulness for a Manned Mars Rover.

The low solar insolation makes the use of a solar primary power system not feasible because of the large size, and hence, the mass of the solar panels. Such high mass will significantly impact the launch cost. In addition, the intensity of the dust storms on Mars will not only degrade the operation of the solar collectors, but also limit their useful lifetime [Kaplan,1988]. Also, an energy storage system (i.e., batteries or regenerative fuel cells) is required to meet the power requirements for life support, instrumentation, thermal management, and data analysis equipment at night.

RTG's are also excluded because of their high specific mass (~200 kg/kWe to 250 kg/kWe), the large inventory of Pu-238 required and the additional heat rejection requirements during the trip to Mars (10 to 16 months) [Bennett et al,1988].

Although they posses a much higher conversion efficiency and lower specific mass than RTG's, DIPS (specific mass of 80-100 kg/kWe) still suffer from many of the same limitations as RTG's. RTG's and DIPS are also costly due to the limited availability of Pu-238 for the General Purpose Heat Source for meeting the primary power needs. However, the use of DIPS may be considered for emergency power use.

For the projected power needs for the Manned Mars Rover mission (30 kWe plus), the low mass and high power density of nuclear reactor power systems make them an attractive option for meeting the Rover primary power requirements [Bloomfield et al., 1988]. Another advantage of a reactor power system is that it will not be operated until it is deployed on the Martian surface, thus eliminating any cooling requirements during transit to Mars. However, a reactor power system represents a technological challenge to the designer and the developer of the Mars Rover. Some of the critical technology issues include: (a) the reactor system configuration and integration with the Rover vehicle, and (b) the integrated shielding design and configuration to protect the astronauts from both reactor and natural radiations. The resolution of these (and other) issues is the subject of ongoing research at the University of New Mexico's Institute for Space Nuclear Power Studies in collaboration with NASA Lewis Research Center. This paper reports the result of the first phase of this research, which focused on identifying the primary power level for subsequent development of a Nuclear Reactor Power System, and its integration with a Manned Mars Rover vehicle [El-Genk and Morley, 1989].

Methodology and Model

In order to determine the Reactor power needed, it was first necessary to estimate the total mass of the Rover. This was completed in two successive steps. First, the configuration of the Rover was set, as in figure 1, and a component mass breakdown for each of the cars was made. This yielded masses of 5500 kg and 3127 kg for the Experimental Unit (EU), and the Supply Car (SC), respectively. For further details on the Rover vehicle design see El-Genk and Morley, 1989.

The mass of the Primary Control vehicle can not be explicitly determined because it depends in part on the number of astronauts and the mass of the Auxiliary power system, a hybrid of deployable PV panels and regenerative fuel cells. The mass of the Primary Control vehicle is then determined as a function of the scenario for the emergency return (cruising speed and day versus night travel). For the base case of 4 astronauts and 5km/hr, the mass of the Primary Control Vehicle (excluding that of the PV panels) was 5860 kg [El-Genk and Morley, 1989].

The second step involved determining the total Rover mass and power requirement. Because the mass of the Reactor Car (RC) depends on the total power required, which in turn depends on the total mass of the Rover and the specific mass of the reactor system, determining the power requirements is an iterative process.

As indicated earlier,the emergency return mission scenario calls for detaching the EU, SC, and RC from the Primary Control Vehicle and making the return trip at a reduced speed of 5 km/hr. The mass of the deployable PV arrays can be given as:

$$M_{pv} = \frac{n_a p_{1s}\left[t_c + \left(t_d - t_c\right)/\eta_{fc}\right] + \chi\left(8.0 \times 10^{-5}\right) M_{pcv} v\left(t_m/\eta_{fc}\right)}{\left[t_c/SM_{pv} - \chi\left(8.0 \times 10^{-5}\right) v\left(t_m/\eta_{fc}\right)\right]} \qquad (1)$$

In equation (1), t_c is the collection time , t_d is the average number of hours/day, t_m is the mobility time during both the day and night, n_a is the number of astronauts, v is the cruising speed, χ is the traction factor modifier used to compensate for the difference in gravitational acceleration and soil consistency from the moon to Mars, η_{fc} is the efficiency of the fuel cells, and SM_{pv} is the specific mass of the PV's. The total mass of the PCV then becomes the sum of the 5860 kg and the calculated mass for the PV's by equation (1). Figure 2 represents the emergency return mission scenario for the base case used in calculating the mass of the PV panels.

An analytical model was developed to assess the total mass and power requirements for the Mars Rover vehicle as functions of the cruising speed, number of astronauts,the Rover utility car mass, and the specific mass of the nuclear reactor power system. Unlike the other three cars of the Rover the mass of the Reactor Car (RC) cannot be determined by simply tabulating the components and summing their masses. This is because mobility power depends upon the total mass of the Rover (including the RC) and the mass of the reactor system is a function of

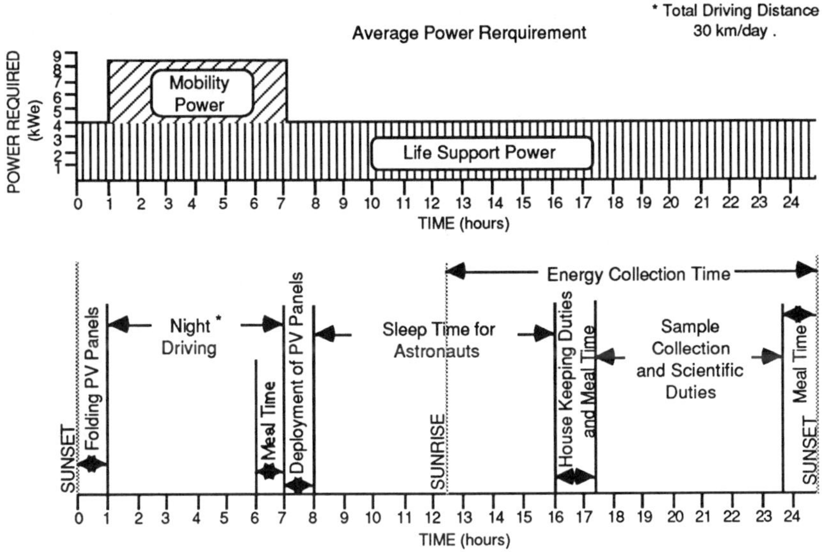

FIGURE 2:EMERGENCY RETURN MISSION SCENARIO AND POWER REQUIREMENT

the total power needed. An iterative approach, similar to that taken for the PV auxiliary system, is used to determine the Rover's total mass and Power requirement.

Although a good estimate of the power requirement is essential for reactor design, this is not the only quantity of interest to mission planners. The total mass of the vehicle is also important when considering launch and deployment costs. Therefore, the model is also set up to calculate the total mass of the Rover, including the Primary Power System (PPS). The power delivered by the PPS, P_t, is given as:

$$P_t = P_a + \chi(8.0\text{x}10^{-5})M_t \cdot v, \qquad (2)$$

where P_a is the net user power(30 kWe), M_t is the total mass of the Rover (kg) and V is the cruising speed (km/hr). While the mass of the reactor car, which is proportional to nuclear reactor system's power output, P_t, and specific mass, SM_r, can be given as:

$$M_{rc} = 1.1(SM_r \cdot P_t) \qquad (3)$$

The factor of 1.1 accounts for the mass of the frame and drive system of the car, which is taken as 10% of the total mass of the Reactor power system. Substituting the

the masses of the individual Rover cars into Equation (2), and rearranging the results yields the following expression for the total power requirement.

$$P_t = \frac{P_a + \chi\left(8.0 \times 10^{-5}\right)\left[M_{pcv} + M_{eu} + M_{sc}\right] V}{1 - \chi\left(8.8 \times 10^{-5}\right)\left(SM_r\right) V} \quad (4)$$

It can be seen from equation (4) that the total power requirement for the Manned Mars Rover depends on five primary variables: P_a, χ, SM_r, V, and the mass of the Utility Vehicles (M_{pcv}, M_{eu} and M_{sc}). The total mass of the Primary Control Vehicle is given by Equation(1). Equations (3) and (4) are solved iteratively to calculate the total mass and primary power requirements as functions of the mission five primary variables indicated earlier. The results of the parametric analysis including these variables are presented and discussed in the following sections.

Results and Discussion

A parametric analysis was conducted to determine the sensitivity of the total Rover mass and the primary power requirements to mission variables. In figure 3, the total reactor power requirements as a function of cruising speed, is shown for three different specific mass values (50, 100, and 150 kg/kWe). For all three curves, the minimum power is 30 kWe (user net power) at 0 km/hr. At low cruising speeds (< 10 km/hr) the reactor system's specific mass has virtually no effect on the primary power requirements. Once the speed begins to increase beyond 10 km/hr, the specific mass begins to show a significant effect on the total power. For example, at 30 km/hr the total power requirements double from 183 kWe at 100 kg/kWe to 378 kWe at 150 kg/kWe. For the base case of 100 kWe , the maximum cruising speed attainable for the Rover would be 23, 19, and 16.5 km/hr, for specific masses of 50, 100, and 150 kg/kWe, respectively.

As figure 4 shows, for 100 kWe of primary power, increasing the system's specific mass from 100 to 150 kg/kWe decreases the maximum cruising speed by 2.5 km/hr, but increases the Rover mass from 26 to 32 metric tonnes (~12% increase). This 6,000 kg increase in the Rover mass is significant, considering the launch costs to Mars. The results presented in figure 4 demonstrate the effect of the cruising speed on the mass of the reactor power system. As this figure indicates, for a specific mass of 150 kg/kWe, the mass of the power system will equal 50% of the total mass of the Rover vehicle at a cruising speed of about 14.5 km/hr. However, this mass becomes only 21.5%

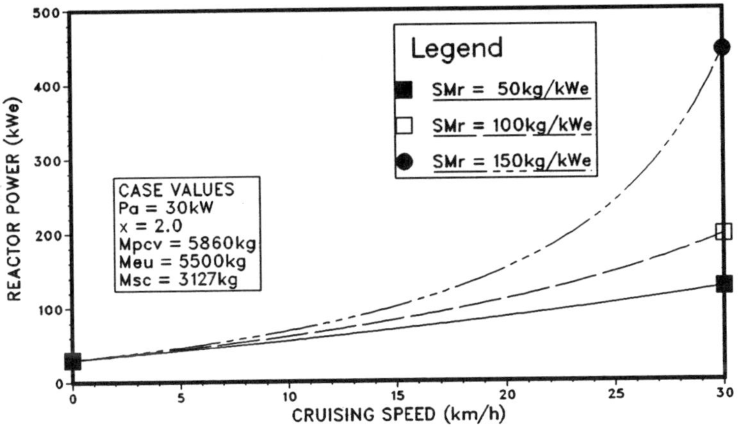

FIGURE 3: EFFECT OF POWER SYSTEM SPECIFIC MASS ON REACTOR POWER

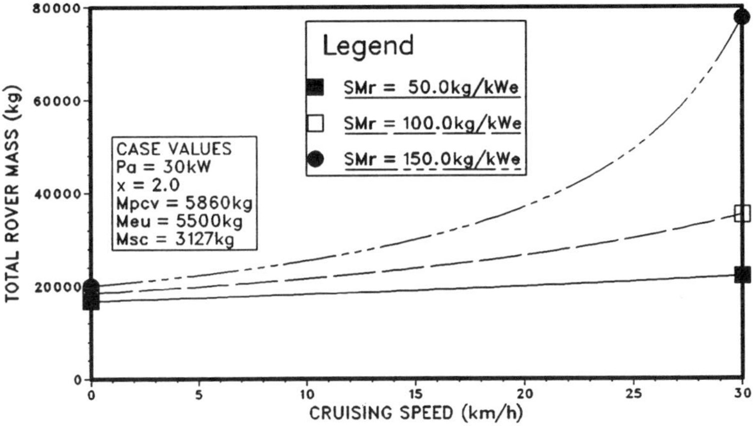

FIGURE 4: EFFECT OF POWER SYSTEM SPECIFIC MASS ON TOTAL ROVER MASS

of the total mass of the Rover when the system's specific mass decreases to 50kg/kWe. Also, at a cruising speed of 30 km/hr the mass of the nuclear power system equals approximately 81% and 33% of the total mass of the Rover vehicle for a system specific mass of 150 kg/kWe and 50 kg/kWe, respectively. The results delineated in figure 4 show the strong dependence of the total Rover mass on the cruising speed.

Figure 5 illustrates the effects of the soil traction parameter coefficient, χ, over a range of cruising speeds up to 30 km/hr. Since the gravitational

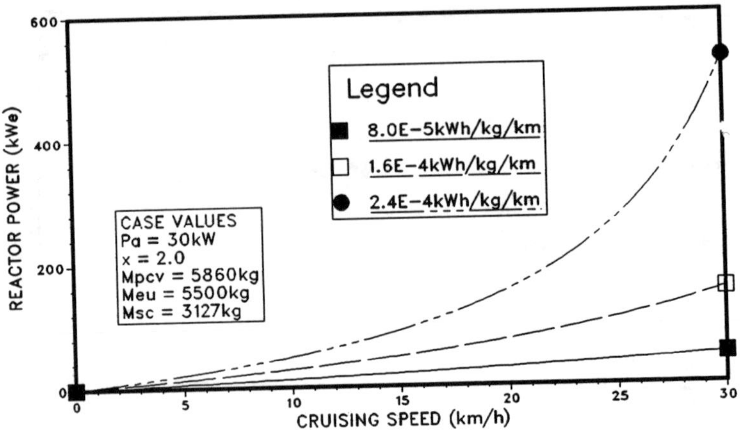

FIGURE 5: EFFECT OF SOIL TRACTION PARAMETER ON REACTOR POWER

acceleration on Mars is greater than that on the moon (soil traction parameter is based on data from Lunar Rover [EEI, 1988]) by ~2 times, the weight force will increase proportionally and the vehicle will exhibit a higher rolling resistance. However, the wheel design and soil consistency may vary significantly which in turn will affect the mobility power requirements.

The number of astronauts and the mass of the utility cars were also varied, but showed only moderate effects on the total Rover mass and the primary power requirements.

Summary and Conclusions

The results show that the mobility power requirement is a strong function of the cruising speed of the Rover vehicle. Therefore, the selection of a cruising speed for the Mars Manned Rover vehicle will be constrained by the mobility power requirement and the total mass of the nuclear power system. It has also been shown that reducing the specific mass of the power system through efficient integration of the nuclear reactor with the Rover Vehicle and minimizing the radiation shield mass would effectively reduce the total mass and the power requirements of the Rover Vehicle.

The results indicate that a 100 kWe primary power system will be sufficient to operate the Rover vehicle as long as the specific mass of the PPS can be held below 150 kg/kWe.

Acknowledgements

This research is performed under NASA Grant No. NAG 3-992 to the University of New Mexico's Institute for Space Nuclear Power Studies.

References

Andrews, D. G., "Near-Term Nuclear Space Missions", J. Space Power, vol. 8, No. 3, Proc. of IAF International Conference on Space Power, IAF-ICOSP 89, Carfax Publishing Company, Oxfordshire, United Kingdom, Cleveland,OH,5-7 June,1989.

Bennett, G. L., J. A. Janzen, J. J. Lombardo, and A. S.Mehner, "The Dynamic Isotope Power System Technology Program", Proc.3rd Intersociety Energy Conversion Engineering Conference,vol.3,pp. 131, Denver, Co, 1988.

Bloomfield, H. S. , et al., "Extended SP-100 Nuclear Reactor Power Systems Capability", Proc. 23rd Intersociety Energy Conversion Engineering Conference, vol. 3,pp. 671, Denver, Co, 1988.

Eagle Engineering, Inc., Lunar Surface Transportation System Conceptual Design - Lunar Base System Study Task 5.2, EEI Report 88-188, Houston , Texas, July, 1988.

El-Genk, M.S., and N.J. Morley, Conceptual Studies on the Integration of Nuclear Reactor Systems to Manned Mars Rover Missions, Report ISNPS-NAG 3-992/89, Institute for Space Nuclear Power Studies, University of New Mexico, Albuquerque, New Mexico, 1989.

Koplan, D.,Environment of Mars 1988, NASA Technical Memorandum 100470, U.S. Geological Survey, Flagstaff, Arizona,1988.

NASA (1988), OEXP Final Report, TM 4075, Office of Exploration, National Aeronautics and Space Administration, Washington, DC, 1988.

LUNAR BASE POWER REQUIREMENTS, OPTIONS & GROWTH

Robert D. Waldron[1]

ABSTRACT

A preliminary study of lunar base scenarios and associated power requirements has recently been completed. Minimum power levels of 50-100 kW are required for "science-only" activities, but power levels in excess of 1 megawatt are needed for propellant oxygen production from lunar materials at a useful rate. Early availability of lunar oxygen can favorably impact base transportation costs by reducing propellant cargo from earth to low earth orbit.

Main base and mobile power options for lunar surface operations include nuclear, solar (photovoltaic) -- with or without large scale storage, gravitational, and electrochemical systems. A parametric analysis for minimum import mass for energy intensive lunar industries is presented. For solar options, minimum energy storage requirements will require additional research on cyclic illumination effects on greenhouse operations.

Three energy storage methods requiring little or no earth import mass have been identified.

INTRODUCTION

Recent discussions (NCOS 1986, Ride Report, 1987) on long range goals for the US Space Program and the President's affirmation have included consideration of permanently occupied lunar bases for science, space resource utilization, and proving ground for long-duration space habitation. Detailed analysis of lunar surface activities in science, technology development, and applications projects have been reported (Battelle 1987) and facility import mass, volume, power, and human resource requirements have been estimated. Our studies have expanded upon this data base and developed a scenario for establishing and developing a multipurpose lunar base.

SCENARIO DEVELOPMENT

Our scenario was developed using 67 of 74 lunar related activities from the Battelle Study (4 with revisions) and 14 new activities plus 11 essential core facilities deemed mandatory for safety, life support and comfort of base inhabitants. These 92 facilities and activities would require delivered mass from Earth exceeding 500 tonnes (excluding return propellant and vehicles) in the absence of appreciable lunar surface parts production from lunar resources.

[1] Rockwell International, Downey, CA

The capacities of Earth-to-Low Earth Orbit (LEO) and LEO-to-lunar surface transportation systems will probably constrain base buildup to a period of at least 5 to 7 years. This makes it desirable to develop a rational process for prioritization of lunar activities to permit rank ordering of facility delivery to the moon.
We have first classified the 92 facilities and activities into 6 major categories, four of which are subdivided into 17 subgroups as shown in Table 1. Also listed are 8 ranked priority classes based on subjective benefit: cost ratios.

In the absence of either: 1) dramatic "quantum-leap" improvements beyond current chemical propellant methods or performance, or 2) lunar production of propellants, the logistics demands for the above lunar transport model would require delivery to LEO of earth supplied propellant with at least 4 or 5 times the net payload mass delivered to the moon.

The potentially enormous savings in transport costs which could accrue from lunar surface (or orbital) production of propellant oxygen (and optionally lunar or hybrid fuels) has prompted studies of potential production systems for lunar base propellant manufacture. Preliminary estimates of production facility and particularly power system masses needed to produce lunar oxygen at a meaningful level have been generally discouraging (Stump et al 1985, Stern 1988). However, the recent development of ultralight amorphous photovoltaic solar power converters (Hanak 1985, Landis 1989) has prompted a reexamination of the feasibility of production of oxygen early in lunar base deployment (Waldron 1988a).

ENERGY/POWER OPTIONS

Power needs for a lunar base can be divided into 5 categories:
1. Main base daytime power
2. Main base (minimum) nighttime power
3. Mobile power (mining, local & long distance surface transport)
4. Outpost power (unmanned or infrequently manned)
5. Short term (pulse) power (external propulsion devices)

Large scale power needed remote from the original main base as for specialty mining or processing, would be defined as a second main base.

The distinction between classes 1 & 2 is unimportant for nuclear power, but class 2 is important for analysis of solar primary power options since there exists some minimum level of essential (non-deferrable) power needed during the lunar night.

Figure 1 shows a classification of power and energy storage options available for lunar base operations. Analysis of conventional methods for high-level energy storage has shown that the mass of earth-imported equipment for electromechanical or electrochemical systems to provide 50 to 100 kW or more of mean power during an entire lunar night ($\sim$ 350 h) would be prohibitive. However, at least three energy storage methods dependent almost exclusively on locally derived

Table 1 LUNAR BASE ACTIVITIES & FACILITIES CLASSIFICATION

MAJOR CLASS		SUB GROUPS	NUMBER OF ACTIVITIES	PRIORITY ASSIGNMENT 1	2	3	4	5	6	7	8
(GOS) General and Observational Science	(PVE)	Space Plasma, Vacuum and Environmental Physics	3					3			
	(OPA)	Optical Physics & Astronomy	3								3
	(HEP)	High Energy Phys. & Astronomy	5							1	4
	(RFA)	RF and Misc. Astronomy	4							2	2
(BS) Biosciences	(MFR)	Medical, Fitness & Recreational	2	2							
	(BFP)	Biology & Food Production	2			1			1		
	(MWR)	Microbiology & Waste Management	4		3				1		
(M&LS) Materials and Lunar Science	(GRD)	Geophysics & Resource Devel.	7				3	3		1	
	(LFS)	Other Lunar Field Science	4				4				
	(SCP)	Surface Chemistry & Physics	5				1	4			
	(CMP)	Chem./Biochem. Matls. Prodn.	8		1				7		
(LT) Lunar Technology	(PPT)	Prime Power Technology	2	1?	1						
	(ESR)	Energy Storage & Retrieval	5		4					1	
	(BPT)	Beam Power Technology	2							1	1
	(TCT)	Thermal Control & Transport Tech.	6		5						1
	(CT)	Communications Technology	2							2	
	(M&A)	Manufacturing and Assembly Tech.	15		10				2		3
Miscellaneous		(Habitat Related)	2			2					
(PF) Prerequisite Facilities	(PF)	(Not generally Common)	11	11							

TRANSPORTATION PRIORITIES (LEO-MOON) FOR LUNAR BASE FACILITIES

1. Emplace essential core facilities (human safety, survival/rescue, base growth)
2. Lower transport unit cost or raise transport capacity(base growth rate) (Primarily: A. Propellant transport reduction, B. Payload mass replacement)
3. Improve or upgrade resident amenities
4. Establish or develop lunar-unique science (Primarily internal facilities)
5. Establish or develop lunar-unique science (Primarily external facilities)
6. Emplace moderate mass facilities for other science & tech. programs (Prim. internal)
7. Emplace moderate mass facilities for other science & tech. programs (Prim. external)
8. Develop selective priority rating system to schedule delivery of high import mass science or technology systems not included above

Table 2 MODE OPTIONS FOR ENERGY INTENSIVE LUNAR INDUSTRIES

OPTION	PLANT OPERATION FULL TIME	PLANT OPERATION DAYTIME ONLY	RATE	PRIME POWER OPER. FULL TIME	PRIME POWER OPER. DAYTIME ONLY	RATE	INPUT POWER RATE FOR ENERGY STORAGE
1	X		R_o	X		R_o	NONE
2	X		R_o		X	$R_o\ 1+\frac{1}{\eta}(\frac{2}{\gamma}-1)$	$\frac{R_o}{\eta}(\frac{2}{\gamma}-1)$
3		X	$\frac{2R_o}{\gamma}$		X	$\frac{2R_o}{\gamma}$	NONE

γ = Equivalent fraction of daylight period for power generation ($\gamma \sim .95$)
η = Round trip efficiency of energy storage method ($.2 < \eta < .9$)

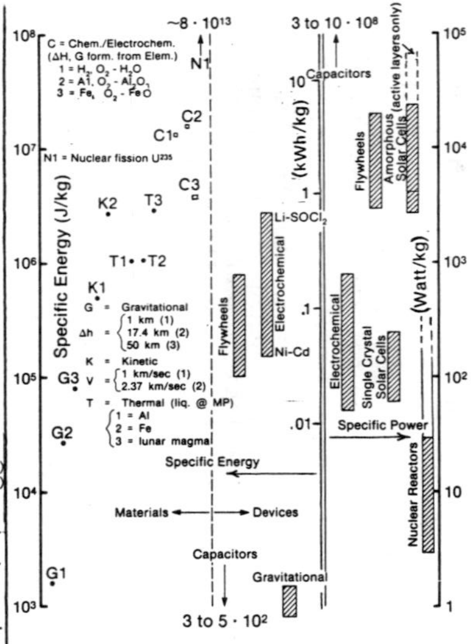

Figure 2 SPECIFIC ENERGY AND POWER

materials are available and could be used if required. These are shown with double borders in Figure 1.

Energy Storage Using Lunar Mass

In common with terrestrial practice, energy storage on the moon might employ gravitational, thermal, electrochemical or kinetic methods to permit nighttime power generation from solar radiation prime conversion. The specific energy capacity or J/kg for various storage methods are shown on the left section of Figure 2. Gravitational energy storage requires very large masses (of lunar soil) using the lunar analog of terrestrial pumped hydrostorage. Most of the materials handling equipment involves mature technologies and could probably be produced locally. (The same technology applied to down-hill mass transport only can be used to provide full-time (day/nite) power).

Thermal storage would use solar concentrators or solar electric systems to heat or melt lunar rock or soil. During the lunar night, the heat would be extracted to operate a heat engine subject to Carnot cycle limitations. Heat could be rejected to the night sky or to radiation cooled lunar soil (Waldron 1988a). The Thermal energy reservoir could be made entirely from lunar materials and probably a portion of the heat engine, generator and heat exchange equipment as well.

The preferred electrochemical energy storage system would use the elements of an iron-air (or iron-oxygen) battery (Demczyk, 1980) which may be described as a semi-fuel cell. To conserve earth import mass, it would be desirable to reprocess the discharged anodes & cell electrolytes to recover water and KOH during the lunar night to transfer to fully charged cells. The oxygen cathodes could also be made transferable. The electrochemical reactants, Fe & O_2, can be entirely derived from lunar sources.

The specific mass kg/kW based on mean nighttime power, $\bar{p}$, over t = 370 hr is given by:

$$SM = \frac{MASS(M)}{\bar{p}} = (1 + \alpha)\ SMB + K \cdot (ESM) \cdot t$$ Where SMB is the specific

mass of battery (projected as 5.5 kg/kW), ESM is electrochemical energy specific mass of the $Fe\text{-}O_2$ couple (about 1.3 Kg/kWh), α is the ratio of cell reprocessing to cell discharge time, and K is a factor $>$ 1 for energy dependent mass terms above the theoretical minimum. The second term is probably entirely derivable from lunar materials while up to 75% of the first term may be so derived.
As an example, if α = 1 and K = 1.25, SM=2(5.5) + 1.25(1.3) 370 = 11+601.3= 612.3 Kg/kw. However assuming 75% of the first term could be locally derived, the import specific mass, ISM, for 95,98 or 100% lunar derived second terms would be:

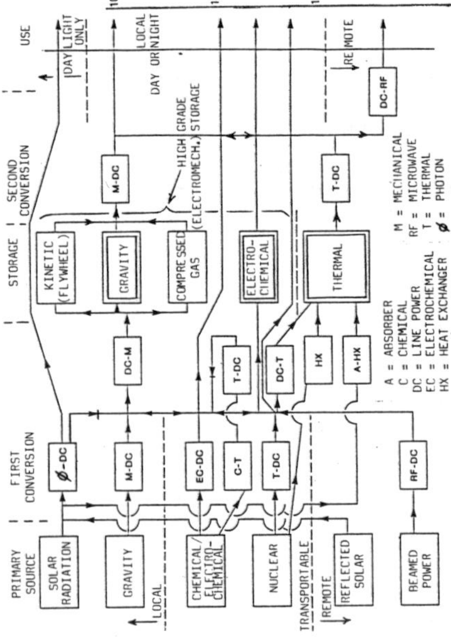

Figure 1 ENERGY/POWER OPTIONS FOR LUNAR BASES

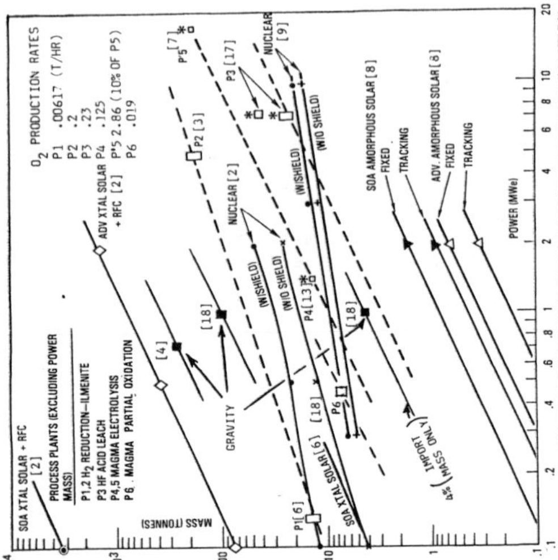

Figure 3 MASS & POWER REQUIREMENTS FOR LLOX PRODUCTION

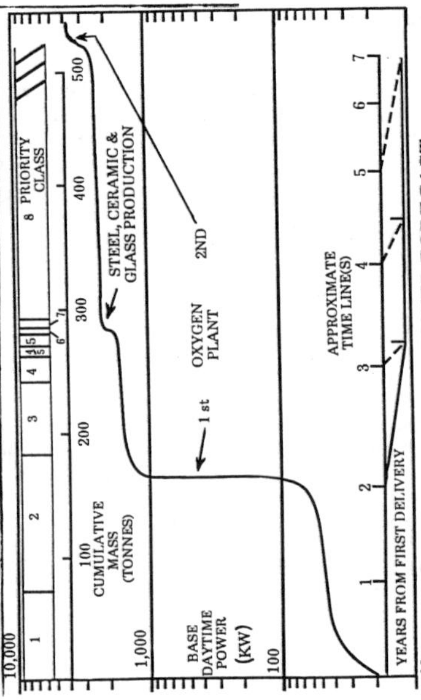

Fig. 4 LUNAR BASE POWER FORECAST

$$ISM = .25(11) + \begin{cases} .05 \\ .02 \\ 0 \end{cases} \cdot 601.3 = 2.75 + \begin{cases} 30 \\ 12 \\ 0 \end{cases} = \begin{cases} 32.75 \\ 14.75 \\ 2.75 \end{cases} \text{ kg/kw respectively.}$$

MAIN BASE POWER

Operating Mode Options

Most of the energy intensive operations on the lunar base will be associated with oxygen or other propellant production, materials refining, and hot forming operations such as metals casting, forging glass or ceramics production. Most of these operations are not manpower intensive and could be operated equally well either during lunar daytime only or during both lunar day and night periods.

Assuming that meaningful production levels on an annualized basis are sought, there are three options for production based on plant operation, prime power operation and optional energy storage usage which are summarized in Table 2.

Option 1 represents a reference or baseline system using either nuclear or gravitational power with reference rates R_o. Option 2 requires the same plant size as Option 1, but more than twice the collecting power and storage capacity for more than 15 days equivalent full time output. Option 3 requires a plant larger than the first two by a factor $2/\gamma$ and similarly a power plant capacity larger by the same factor.

The net import mass advantages for the respective options depend on the specific masses of photovoltaic versus nuclear or gravitational power systems (kg/kW) and for option 3 on the specific mass of the production plant(s).

The total import masses for the respective options are:

$$M(1) = \bar{p} \cdot (SMP + SMR)$$
$$M(2) = \bar{p} \cdot (SMP + [1 + (1/\eta)(2/\gamma - 1) \quad SMS)$$
$$M(3) = \bar{p} \cdot ((2/\gamma) SMP + (2/\gamma) SMS)$$

where $\bar{p}$ is Mean Operating Power Rate for 8,000 h/y operation,
SMP is Specific Mass of Production Plant (kg/kW),
SMR is Specific Mass of Reference Power Plant (kg/kW),
SMS is Specific Mass of Solar Power Plant (kg/kW) (including import mass for energy storage if required),
η is Round Trip efficiency for energy storage/recovery, and
γ is Equivalent fraction of daylight hours for full solar power output

The selection of preferred operating mode would be based on a combination of factors and not solely on minimum import mass.

However, we have attempted to estimate specific mass factors both from prior studies and our current evaluation to forecast possible trends.

Figure 3 shows State of the Art (SOA) actual values or current or advanced technology (ADV) estimates for nuclear, gravitational, or solar power and oxygen plant masses vs mean power level from which values of SMP, SMS, & SMR can be derived. One notes that there is a remarkably large spread in specific mass values for photovoltaic power systems between crystalline and amorphous converters. Note also the appreciable departure of mass linearity with power for nuclear systems.

For energy storage systems as required for option 2, it is clear that totally imported storage systems are much more massive than equivalent nuclear power systems and option 1 is preferable--however if energy storage can be accomplished primarily with lunar mass facilities, option 2 could be preferred if the import fractional mass for storage is less than a nuclear power system of equivalent power.

A comparison between options 1 & 3 shows the import mass advantage depends on relative specific masses for both power plants (reference vs solar) and production plant facilities. The latter depend on specific process selections.

From the previous mass expressions we may derive:

$$\Delta M(3\text{-}1) = M(3) - M(1) = R_o \left[(2/\gamma - 1)\ SMP + (2/\gamma)\ SMS - SMR \right] .$$

If the ratio $\frac{SMS}{SMR} = \zeta < 1$ and $\gamma \approx .95$, we obtain:

$$\Delta M(3\text{-}1) \quad R_o \left[1.1\ SMP + (2.1\zeta - 1)\ SMR \right]$$

. The import mass advantage would lie with option 3 if $SMP < \frac{1 - 2.1\zeta}{1.1} \cdot SMR$.

More refined estimates of specific masses, particularly for processing plants such as for propellant oxygen production and liquefaction must be made including detailed analyses of portions which could be fabricated from lunar materials before definitive assessments can be made for the respective options.

MINIMUM NIGHTTIME POWER REQUIREMENTS

The minimum nighttime power requirements are only of particular interest for solar powered bases without energy intensive lunar industries or where such activities are present and are operated in the option 3 mode previously described. For these cases the power requirements for essential (non-deferrable) activities can be estimated and will be dependent on a number of assumptions related to habitat structure, insulation, lighting, and heat rejection systems as well as life support systems and greenhouse management cycles.

For our analyses, the following assumptions were made:

1) Habitat will be thermally insulated with high efficiency due to 1 meter or more of lunar soil covering for radiation shielding.
2) Solar energy will be ducted in using advanced hollow light pipe technology for general interior and greenhouse lighting during lunar daylight. The artificial lighting for lunar night use can use the same light pipes to reduced air conditioning load due to lamp heating.
3) Radiant heat rejection for interior heat loads can be performed without heat pumping using geometrical shielding throughout the lunar cycle. (However, use of heat pumps could result in smaller & lighter radiators but higher power consumption).
4) Greenhouse power for illumination & transpiration can be reduced during the lunar night to cloudy or overcast earth equivalent levels (15-day maximum). This will lower mean photosynthetic rates and require a larger and heavier greenhouse.

Under the above assumptions, the life support power requirements should be less than for spacecraft or space station systems. We further estimate that greenhouse illumination & transpiration power can be lowered to 20% or less of that for area-intensive greenhouse systems using full artificial lighting. We estimate that for a base with full greenhouse facilities, the essential nighttime power can be held to 2.5 kW per inhabitant.

MOBILE POWER

Power for surface vehicles is needed for both short and long haul applications and for small to large (mining class) loads. Reliability factors become more important as distance from main base or service facilities increases.

Electrochemical systems appear to offer the most practical solutions to most vehicle power needs. The iron-oxygen semi-fuel cell or rechargeable battery can be constructed primarly from lunar materials and has a reasonably low specific mass. A slight variant of this system which provides fully storable electrochemical reactants is to use hydrogen peroxide to supply the oxygen using external catalytic dissociation to provide oxygen and water. The hydrogen peroxide would be remanufactured from water and oxygen extracted from $Fe(OH)_2$ formed on the anodes.

OUTPOST POWER

Reliable low level operations and standby power systems will be needed remote from the main base for emergency life support of long traverse vehicle sorties and transmission of scientific data such as for geophysical measurement probes. Nuclear power systems appear to be particularly suited to such applications.

PULSE POWER SYSTEMS

High power, short duration pulsed power systems are needed for several space transportation applications involving external propulsion.

Linear electromagnetic accelerators or mass drivers operating at acceleration levels of 1,000 G's or higher can propel payloads up to 1 tonne or more to orbital or escape speeds. The stored energy requirement is:

$E = mV^2/2\eta'$ where m = payload mass, V = terminal velocity and η' is conversion efficiency. The pulse duration is about 0.25 seconds @ 1,000G.

A second application involves using momentum transfer between ejected lunar soil and a descending vehicle to provide a low deceleration (lunar soft landing) maneuver designated as dustbraking. Peak power requirements are typically between 100 & 400 W/kg (vehicle gross weight) with total ejection times of 200 to 900 sec. depending on deceleration levels selected. Total ejection energy requirements are given by mean ejection power times ejection duration or approximately peak power times half ejection time (about 10^4 to 2.10^5 J/kg).

Both of the above applications require an energy storage system with high discharge rate such as a homopolar generator system. Duty cycles will generally be fairly low but interval times between successive operations will be limited by charging power capacity.

ALTERNATE (OFF SURFACE) POWER REQUIREMENTS FOR LUNAR MATERIALS PROCESSING

For export propellant needs and general earth orbital structural materials needs, return of lunar raw or refined materials to LEO is required. This can be accomplished more efficiently for payloads that can withstand high acceleration levels using external propulsion systems (mass drivers, suborbital ejectors, (Waldron 1988b) etc.). Availability of multimegawatt power systems in LEO or eccentric earth orbit combined with lunar materials processing facilities could then receive raw lunar soils and provide propellant, shielding, and refined metals and oxides for use in earth orbit.

Nuclear power systems appear to be the method of choice since solar systems are subject to exoatmospheric drag, eclipsing, excessive moment of inertia, and long term van-Allen radiant degradation. A possible alternative is to use microwave or laser beamed power from remote solar (or other source) converters.

ENERGY/POWER FORECAST

The rate of growth of lunar base power requirements will generally increase with the base build-up as constrained by transport schedules. However, the overall power requirements will be dominated by oxygen production once the propellant manufacturing system is in operation. We have attempted to forecast overall lunar base power growth based on an aggressive development scenario. This model was based on solar rather than nuclear power systems operated during lunar daylight only.

Figure 4 shows a projection of base power requirements growth, where oxygen production was initiated about 2 years after initial base occupancy and a second plant installed after about 5 years. The power levels were based on an initial capacity of 500 tonnes/y lunar liquid oxygen (LLOX) and a second duplicate plant operated during lunar daylight only. (Use of nuclear or gravitational power and operations for 8,000 h/y could either double annual capacity or alternatively permit power and plant hourly capacity to be halved.)

CONCLUSIONS

1. Power sources in the multimegawatt range are needed early in the lunar base deployment time frame for propellant production on a meaningful scale from lunar materials.
2. Nuclear power sources are convenient for lunar night time operations of energy intensive facilities but are neither mandatory, nor unequivocally the most cost effective option.
3. Large scale processing of lunar materials in earth orbit will be initiated following adoption of external propulsion systems for lunar mass return. Nuclear or possibly beamed power systems appear to be preferred sources.

REFERENCES

1) Battelle, Columbus (1987), "Advanced Space Transportation System Mission Analysis", NAS 9-17356
2) Bloomfield, H.S. (1988) "Nuclear Reactor Power Systems for Lunar & Planetary Bases", LBS-88-151, Second Symposium on Lunar Bases & Space Activities in the 21st Century, Houston
3) Davis, H.P. (1983) "Lunar Oxygen Impact Upon STS Effectiveness" Eagle Engineering, EEI83-63
4) Dalton, C. (1972) "Conceptual Design of a Lunar Colony, NASA CR-129164
5) Demczyk, B.G., Bryant, W. A. Liu, C.T., & Buzzelli, E.S. (1989) "Performance and Structural Characteristics of the Iron-Air Battery System" in Proceedings of the 15th Intersociety Energy Conversion Engineering Conf.
6) Eagle Engineering (1988) "Conceptual Design of a Lunar Oxygen Pilot Plant", EEI 88-182, NASA Contr. No. NAS 9 - 17878
7) General Dynamics, Convair Div. (1979) "Lunar Resources Utilization for Space Construction" Final Report, NASA Contr. No. NAS 9-15560
8) Hanak, J.J. (1985) "Ultralight Monolithic Photovoltaic Modules of Amorphous Silicon Alloys" in Proceedings of 18th Photovoltaic Specialists Conf., IEEE
9) Keaton, P. W. and Carlson, D. (1985) "Nuclear-Electric OTV" Memorandum to B. Roberts NASA JSC, quoted in Eagle Engrg. (1988)
10) Landis, G.A., Bailey, S.G., and Flood, D.J. (1989), "Adv. in Thin-Film Solar Cells for Lightweight Space Photovoltaic Power" NASA TM-102017.
11) National Commission on Space (1986), "Pioneering the Space Frontier", Bantam
12) Ride, S.K. (1987), "Leadership and America's Future in Space", NASA
13) Stern, M.O. (1988) "Advanced Propulsion for LEO-Moon Transport" Interim Progress Report, Calif. Space Institute, NASA Grant NAG 9-186
14) Stump, W.R. Christiansen, E. L., Babb, G., and Sullivan, D. (1985) "Analysis of Lunar Propellant Production, Rpt. No. 85-103B, Eagle Engineering
15) Waldron, R.D. (1988a) "Space Station Evolutionary Power Technology Study: Task II Report -- Lunar Mission Power Impacts" RI/RD 88-191 Rocketdyne Div./Rockwell International, NASA Contract No. NAS 3-24902
16) Waldron, R.D. (1988b) "Efficient Transportation Systems for Periodic Earth-Moon Rendezvous to Facilitate Repetitive Mass Transport" LBS-88-135, Second Symposium on Lunar Bases & Space Activities in the 21st Century, Houston
17) Waldron, R.D. (1988c) "Lunar Manufacturing: A Survey of Products and Processes" Acta Astronautica 17 (7) 691-708
18) Waldron, R. D. (1989) Unpublished study

POWER FOR MAGMA ELECTROLYSIS

Kenneth A. Owrey[1]

Abstract

This paper summarizes qualitative issues of electrical power distribution for near-term lunar facilities. The issues to be discussed were developed during a conceptual study of a Lunar Liquid Oxygen Plant (LLOX). Issues are discussed based on current terrestrial technology and construction techniques. The discussion identifies areas for further detailed study and evaluation.

Introduction

The Lunar Liquid Oxygen process studied was Ilmenite reduction by magma electrolysis (McCullough, 1990). The process studied proposes to use a low voltage high current dc power to produce molten magma from the lunar regolith. This yields oxygen and magma by-products at approximately 1300°C. The process cell and equipment would be housed inside a protective structure requiring power for lighting and ventilation. The process plant is assumed to be remote from the power source.

In performing the study to develop a Lunar Liquid Oxygen (LLOX) plant and supporting facilities, a conceptual design for the electrical power distribution system was required. Development of the conceptual electrical distribution design raised several issues which require further detailed evaluation. The issues raised involve the type of power, ac versus dc, distribution materials and installation techniques,

[1]Fluor Daniel Inc., 3333 Michelson Drive, Irvine, California 92730

electrical system availability and reliability, and system expansion capabilities for both loads and sources.

The conceptual distribution system for the LLOX process plant is shown in Fig. 1. The electrical criteria for this process require an estimated 2.5 megawatts of power at 6 volts dc, to supply the electrolysis process. This will yield an operating current of 400,000 amps.

Power sources were not included in the conceptual designs. However, photovoltaic supplies were assumed as the near-term power source, with eventual replacement by a nuclear source (Sovie, 1988). This assumption dictates a fourteen earth-day production based on the lunar day.

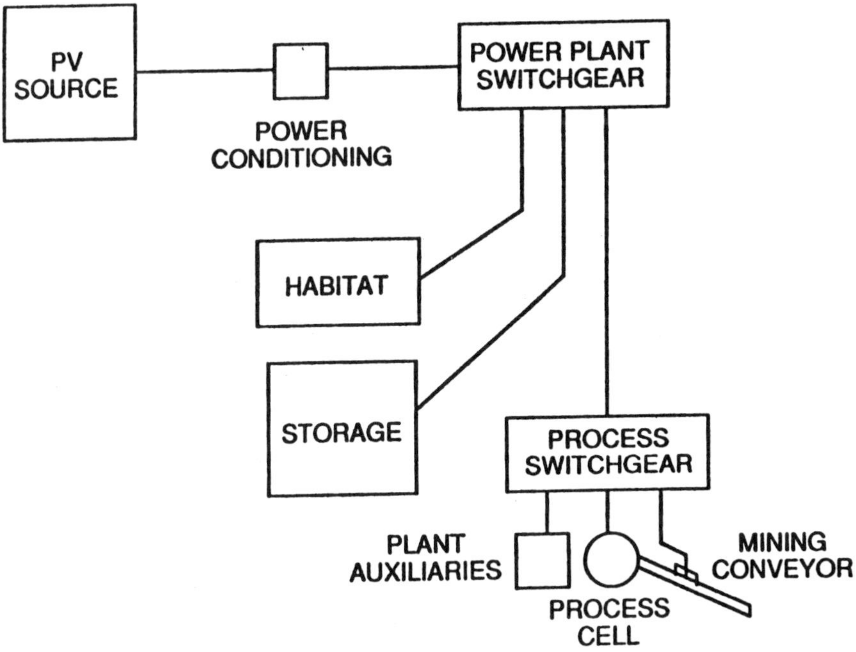

Figure 1: Conceptual Electrical Distribution

The elements of the lunar environment considered during the conceptual study which influence design of electrical distribution are as follows:

- Gravity: 1/6 of earth
- Atmospheric pressure: 0.0 kPA
- Regolith thickness: 3.0 meters
- Regolith characteristics: Electrically nonconductive, good thermal insulator.

Type of Power

Power can be distributed in either an ac or dc form at any selected voltage. Specific design criteria for the process and support facilities will influence the selection of type of power. As an example, rotating machinery characteristics may necessitate ac rather than dc motors. Heating or lighting loads will operate from either type of power. Lighting and motor loads may be more efficiently operated at voltages higher than the process 6 V dc. Variations in voltage levels and ac to dc power conversion will necessitate power conditioning equipment, including converters, inverters, transformers and associated circuit breakers and relays.

Dc distribution systems need to be evaluated for losses, conductor size and number of conductors required to carry the load. Low voltage dc will produce high current and losses requiring increased source capacity. High voltage dc will solve the high loss problem but will require complex conversion equipment, converters and inverters, at both the source and utilization points. This conversion equipment increases the mass to be transported to the lunar surface.

Ac distribution at low voltage has the same disadvantages as low voltage dc distribution. Higher ac voltages have the advantages of easy conversion between voltage levels and lower distribution losses. Ac power can be distributed as single phase or three phase configurations. Three phase power distribution will decrease conductor volume, and therefore mass; however, it will increase the number of conductors from two, for dc or single phase, to four for a three phase system with a neutral. Increased conductor quantity impacts installation complexity.

The difference in complexity and mass of transformer vs conversion equipment for high voltage ac vs dc requires further detailed analysis, including the subject of heat dissipation.

The selection of ac or dc power and the distribution voltage level is essential and influences the design of the entire distribution system both in the process plant and between the source and the plant. It affects size, quantity, and physical configuration of the conductors, and therefore the entire design.

Materials & Installation

The primary concern for all materials and equipment used in the electrical distribution system is the mass. Terrestrial conductors are predominately copper to provide high conductivity and minimize losses, it correspondingly has a high mass. The conventional alternative would be the use of aluminum which would reduce the mass, however, the volume of material to conduct the current for 2.5 mW will increase. Another consideration for ac systems would be to investigate high frequency applications combined with alternate conductor forms. This could reduce mass and volume and not sacrifice conductivity.

Another issue influencing material selection is the environmental conditions on the lunar surface. Temperatures ranging from -157°C to +121°C will create problems for exposed conductors involving expansion, contraction, and the resultant stresses. The magnitude of this problem will depend on the type of installation and the distance from the source to the load. The temperature range will affect the conductor support and terminations as well. Supports and terminations will require special design to compensate for their own and conductor stresses, and still permit simplified installation and testing. One other factor which will compound the high temperature condition on the conductors is conductor heating produced by the electrical losses. These losses are a function of the transmission voltage and type of power.

Some discussion of the LLOX plant conductors is necessary to indicate the problems to be confronted by the in-plant distribution system. As stated, the plant will operate at 6 V dc and 400,000 Amps. This magnitude of current will require parallel bus construction, again raising either mass or volume or both. This type of construction will necessitate the development of specialized connectors to simplify construction and testing, and to account for expansion and high stresses. Present technology and equipment do not meet these criteria.

Techniques for installation of the power distribution system outside the plant require evaluation of several variables to determine the optimum solution. Conventional techniques would involve abovegrade pole line distribution or an underground duct type network. Each of these has advantages and disadvantages.

The design of an abovegrade system, such as a pole line, requires accurate data on the terrain in order to calculate the span between poles and the tension on conductors. Data for sag and tension calculations using conventional conductor material is not currently available at the temperature extremes on the lunar surface. Additionally, abovegrade systems are subject to physical damage from machinery. This type of installation presents minimal hazards to personnel.

An underground direct burial technique provides excellent protection of personnel and protection of the conductors from physical damage. The disadvantage of this type of installation is the need for trenching equipment and subsequent location and repair of faulted conductors, if necessary. Another drawback is the poor heat dissipation provided by the lunar regolith. This will compound the conductor sizing problem.

Another approach feasible for lunar installation is to lay the conductors directly on the lunar surface. This is the simplest construction requiring no support hardware or trenching machinery. The heat dissipation is not of the magnitude of the abovegrade or underground installation. The major drawback is that the conductors are subject to physical damage and personnel may be exposed to high voltage and current.

Regardless of the type of installation determined best for the lunar surface, all types will require evaluation and development. Connections and terminations must be designed which will permit simplified installation and testing, and will withstand the voltage, current and physical stresses imposed. All materials used for conductors and support hardware must withstand thermal expansion and the stresses developed by high electrical current.

Availability/Reliability

A Lunar Liquid Oxygen plant and supporting facilities will require a highly reliable electrical system. To achieve the required availability the system must be designed to minimize physical damage to equipment and conductors, and to provide alternate circuits to back-up the system in the event of electrical faults. This applies to both the LLOX plant and the transmission system. The distribution network and circuit breakers, relays and sensing devices must be designed and constructed to sense faults and automatically transfer to alternate circuits to prevent equipment damage and provide adequate safety for personnel.

To provide uninterrupted production, the electrical system design will have to duplicate process redundancies, as shown in Fig. 2 and provide redundant circuits for all support facilities. Redundancy can be achieved by a variety of systems, ranging from one operating system with a 100% spare system, to multiple smaller units with a single small unit as spare. The specific scheme needs to be matched to the process requirements. Redundant circuits, however, will require increased number of conductors, hardware, and protective devices. This translates into increased mass to be transported to the lunar surface. Additionally, protective devices, circuit breakers and protective relaying, must be developed to interrupt currents of the magnitude, 400 KA, needed for the process. Terrestrial circuit breakers and switching networks for the voltage levels and currents indicated are complex, high mass mechanisms. These devices must withstand the electrical

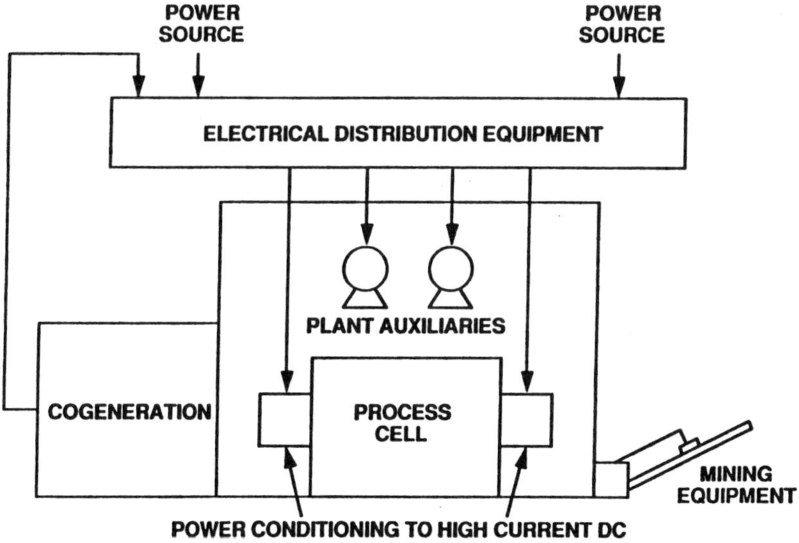

Figure 2: LLOX Plant Power Distribution

and magnetic forces generated by high current, while accounting for the lower gravitational force and poor mounting. Low mass switching mechanisms and protective devices with sufficient mountings must be developed to insure an electrical distribution system which meets all availability goals.

During the study, the heat from the magma by-products was recommended as a source of waste heat for cogeneration of electric power. Conceptual designs using closed systems with organic fluids were proposed. Feasible techniques of cogeneration, utilizing waste heat should be investigated and evaluated during further studies.

<u>System Expansion</u>

The LLOX plant study provided a design which establishes an initial production plant and minimum support facilities, as shown in Fig. 3. Increases are planned in stages adding plants, facilities and larger power supplies. The electrical system design needs the capacity and capability to supply the final, indicated in Fig. 3, as well as the initial installation. Two approaches to future expansion exist, a modular or parallel design, or initial full capacity.

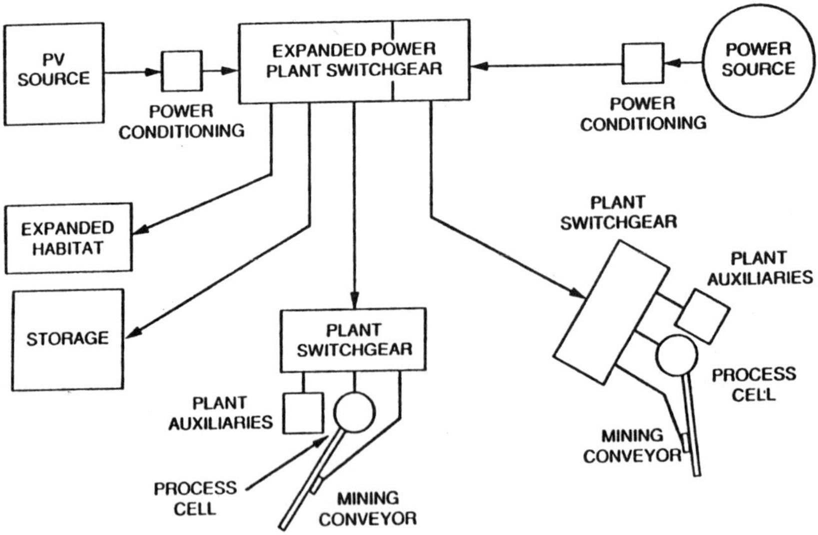

Figure 3: Expanded LLOX Distribution System

Initial full capacity requires designing the system only once. Disadvantages of this approach are a lack of flexibility, the total capacity is fixed by initial design, and a high initial mass and equipment expense relative to plant output.

The more desirable approach is to install modular components and parallel circuits to meet each phase of construction. This will yield a lower initial mass and equipment cost and a more flexible system once installed. The modular design applies to all components and materials including the source, the power conditioning equipment, distribution switchgear, transmission lines and the process plant distribution components and loads. Current technology however, does not produce equipment which will meet the LLOX electrical needs in the lunar environment and meet low mass requirements.

Summary

The LLOX plant conceptual study identified several areas of concern for the selected process. These areas of concern also apply to the distribution system between the source and the LLOX plant. The design of any lunar power distribution system will incur technological problems, and the issues are not limited to those presented. The magnitude of each depends on the process and the facility electrical criteria. As indicated in the discussion, the issues are interrelated and the solution to one may adversly impact another. For these reasons further quantitative analysis needs to be performed to determine optimum systems and materials.

Present terrestrial equipment needs evaluation relative to the lunar environment to determine development requirements. The design and development considerations identified apply to any lunar or planetary installation, not just the LLOX plant studied.

References

Sovie, Ronald J., "Power Systems for Production, Construction, Life Support and Operation in Space," Engineering, Construction, and Operations in Space, Edited by Stewart W. Johnson and John P. Wetzel, 1988.

McCullough, Edward and Mariz, Carl, "Lunar Oxygen Production via Magma Electrolysis," Space 90 Proceedings.

Design Considerations for a Ceramic Fabric Radiator

Keith A. Pauley[1], Brent J. Webb[2], and Andrew C. Klein, PhD[3]

Abstract

The design of an Advanced Ceramic Fabric (ACF) thermal management device for use in both interplanetary and near-earth space must consider several important aspects of the environment. First, the radiation field at various locations is dominated by a proton component which deposits its energy on the surface of the device. Second, various liner materials must be compatible with the working fluids selected for the system. Third, the fluid dynamics and heat transfer characteristics of this device should be adequately characterized. With the selection of materials that are compatible with the operating environment, the Bubble Membrane Radiator (BMR), which uses ACF materials, may be utilized for several future space missions.

Introduction

The successful operation of the spacecraft and other secondary equipment is related to the effectiveness and survivability of the thermal management system. One such advanced thermal management system, the Bubble Membrane Radiator (BMR), exhibits several key attributes such as effectiveness, light weight, and a small launch volume.

Figure 1 illustrates the design of the Bubble Membrane Radiator (BMR) as described by Webb (1988). This device rotates about its central axis to induce a small artificial gravity. The working fluid enters the radiator via feed lines incorporated in the main axis of rotation and is then sprayed from a central nozzle into the spherical cavity of the BMR. This vapor condenses on the inner surface and flows toward the equator of the BMR, where it is collected for return. While primarily designed to reject tens of kilowatts at operating temperatures from 70 - 200°C, the design may be scaled to fit a broad range of conditions by altering the working fluid and total surface area.

----

[1]Development Engineer, Pacific Northwest Laboratory, P.O. Box 999, Richland, WA 99352.

[2]Senior Research Engineer, Pacific Northwest Laboratory, P.O. Box 999, Richland, WA 99352.

[3]Assistant Professor, Dept. of Nuclear Engineering, Oregon State University, Corvallis, OR 97331.

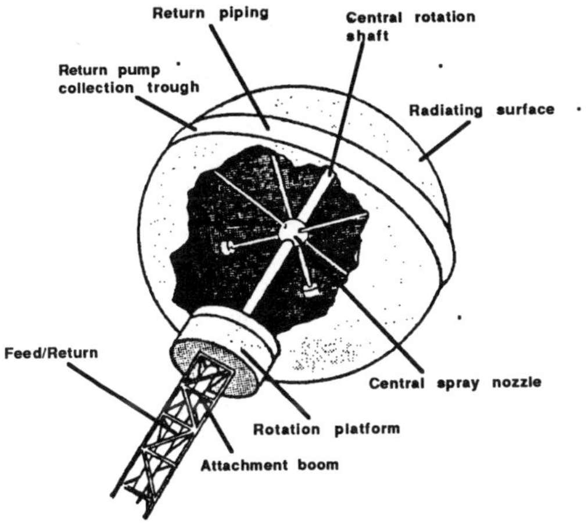

Figure 1. Bubble Membrane Radiator Design

The BMR design incorporates an innovative composite material using Advanced Ceramic Fabrics (ACF). An ACF is a textile made from continuous fibers of metallic oxides and other low-density materials. The BMR shall utilize a layered structure as shown in Figure 2. The outer ACF is the structural load-bearing component of the wall, while the thin metallic or plastic inner liner acts as the permeability and pressure boundary to contain the working fluids of interest.

Several critical design considerations have been assessed for the BMR, as presented in Table 1. This paper, however, will be limited to a review of space environment, materials compatability, and thermal hydraulic issues.

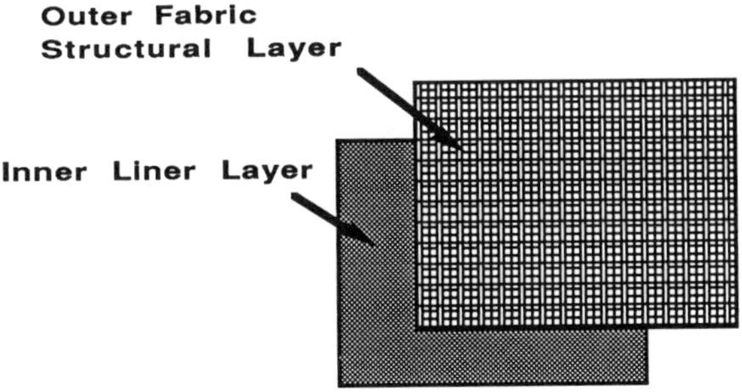

Figure 2. ACF Composite Structure

TABLE 1. Key Technology Issues

Issue	Status (current)	Comments
mass	demonstrated experimentally	high pressure tests prove that experimentally low mass design is achievable
performance	demonstrated experimentally	strength and emittance are acceptable
emittance	demonstrated experimentally	0.70 effective emittance are already acceptable
survivability	demonstrated	must be verified for various configurations
transitioning	demonstrated	molding design able to take pressure
bonding	demonstrated experimentally	requires verification in space environment
flexibility	demonstrated experimentally	requires verification in space environment
startup	not tested	only experiments can verify startup/deployment
space environment	testing underway	reviewed, requires proper selection of materials; experimentation at NASA Lewis and aboard shuttle
compatibility	testing underway	reviewed, requires proper selection of materials; experimentation at Oregon State University
thermal hydraulics	testing underway	condensation film model under development

Space Radiation Enviroment

The space radiation environment is commonly divided into interplanetary and low-Earth space. While these environments are substantially different in terms of the fluxes and energies of constituents, the radiation interaction modes with the BMR materials are not.

Interplanetary Space Environment

The radiation field in interplanetary space comes from two primary sources: solar and cosmic.

The solar wind is a fully ionized, electrically neutral plasma that travels along magnetic field lines from the inner solar corona. During quiet periods of solar activity, the composition of the solar wind is approximately 95% protons, 4.47% alpha particles, and 0.53% heavier charged particles. These radiations typically are generated at high flux levels and relatively low energies (Jursa 1985).

The second large source of radiation in interplanetary space is galactic particles. Galactic cosmic radiation is composed of high energy nuclei at low fluxes. The primary cosmic radiation observed in the vicinity of Earth is composed of approxiamtely 83% protons, 13% alphas, and 1% nuclie of atomic number $Z > 2$, and 3% electrons. These display a steady mean flux intensity from outside the solar system (Smith and West 1983). With large kinetic energies, these particles have the potential to induce significant, yet localized, damage in the materials (Jursa 1985).

Low-Earth Space Environment

Since 1958, the radiation field in the earth's magnetosphere called the "Van Allen Belt" has been researched and well documented. Within the Van Allen belt, two main regions exist. The inner and outer belts are composed primarily of low energy protons. The proton flux variation over a distance of approximately 7 Earth radii is shown in Figure 3.

The inner belt ranging from 1.2 to 2 earth radii is comprised principly of protons with about 0.1% alpha particles. The doses associated with these fluxes often exceed 10^7 rads/day at the outer surface of a spacecraft (Durcanin et al. 1987).

The outer belt lies at approximately 3 to 4 earth radii, the majority of the particles in this region are electrons. The flux and energy spectrum of the particles in this belt exhibit large variations associated with solar activity and geomagnetic storms (Condon and Odishaw 1967).

Atomic oxygen is the most significant ambient constituent of the low earth orbital environment. The flux of monatomic oxygen has been observed to rapidly degrade materials by oxidation. Space Shuttle experimentation has found that interactions between spacecraft surfaces and high-energy oxygen atoms can cause thermo-optical property variation and mass loss.

Space Radiation Interactions

Three principal interaction modes exist between the environmental radiation and the BMR: hydrogen deposition, materials charging, and radiation-induced sputtering. When an energetic charged particle is incident on a material, an electron is commonly stripped from the surrounding lattice resulting in two of these interactions. If protons are

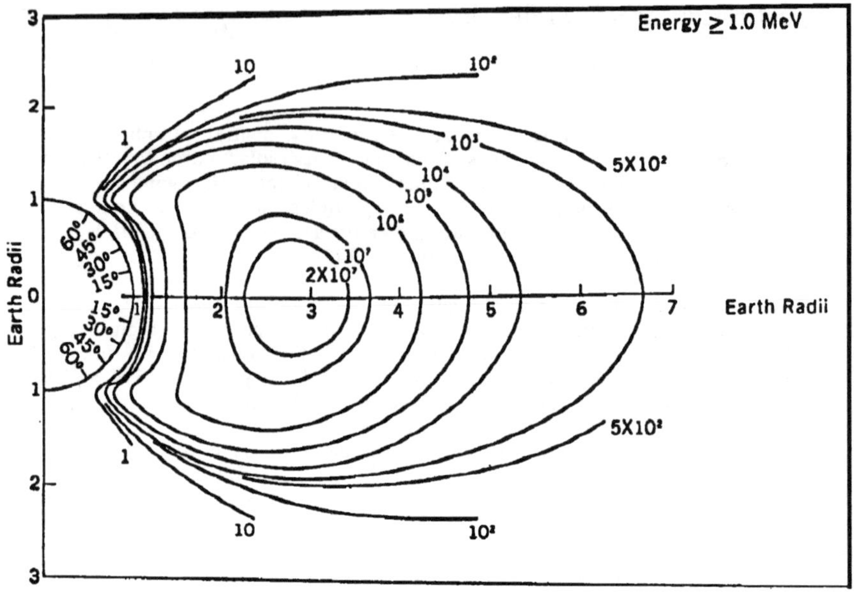

Figure 3. Constant Proton Intensity Flux Contours to 7 Earth Radii (Smith and West 1982)

involved, hydrogen is deposited below the surface of the material. The loss of charge in the vicinity of the deposition allows for a buildup of positive charge, therefore creating the possibility of arcing between surfaces of unequal charge distribution.

Sputtering is the erosion of atoms from the surface of a materials induced by a radiation field. This interaction mode is limited to high energy particles which strike surface atoms, thereby displacing them from their lattice sites. The energy of the incoming particle must exceed the binding energy of the lowest valence shell electron to produce this effect.

Though all of these effects are observed in the more severe low Earth orbit (LEO) applications, hydrogen deposition due to the incident proton field will dominate the deep space environmental consequences. In the LEO environment, material oxidation due to monatomic oxygen will be of principal concern. The overall damage sustained to any ACF component can be minimized in either of these regions, however, by the selection of materials. These materials must be able to exhaust the captured hydrogen and be resistant to oxidation in the presence of free oxygen.

Continuing research is being conducted on the behavior of ACF materials in the space radiation environment. Experiments at the NASA Lewis

Research Center (November 1989) and on board the space shuttle (August 1990) will further characterize the ACF materials in the monatomic oxygen environment.

Materials Compatability

A thorough consideration of the materials used in this design is also mandatory in order to achieve a high probability of successful mission completion. Three different types of materials are proposed to be used in the BMR as the working fluid, liner, and load-bearing shell. Materials considered are found in Table 2.

Due to the operating conditions, the working fluid must be compatible with the liner material. Of specific concern are issues associated with the degradation of the materials properties of the liner in certain working fluids which may cause inefficient operation or system failure. These issues include mass depreciation, surface roughing, chemical byproduct reactions, and transmissivity losses.

A review of materials compatability information yields several considerations in the selection of liner/working fluid combinations. First, most metals oxidize in the presence of oxygen-bearing liquids, causing corrosion of inner liner surfaces. Second, some organic fluids are solvents for several hydrocarbon-based liners. However, specific data on any of these combinations of materials is limited over the appropriate range of operating temperatures.

Further materials compatability research is now being conducted for PNL at Oregon State University. The primary test matrix exposes the aluminum, copper, and titanium samples to several proposed working fluids (water, methanol, acetone, and toluene) for various durations. Mass fluctuations are then observed as the indicator of compatability.

TABLE 2. BMR Materials

Working Fluid	Liner	ACF Shell
Water	Zirconium	Graphite
Ammonia	Niobium	SiC
Toluene	Titanium	SiO2
Freon-type organics	Mylar[(a)]	Nextel[(b)]
	Aluminum	Nicalon[(c)]

(a) Mylar is a registered trademark of E.I. du Pont de Nemours & Co. (Inc.), Wilmington, Delaware.
(b) Nextel is a registered trademark of the 3M Corporation, St. Paul, Minnesota.
(c) Nicalon is a registered trademark of Nippon Carbon, Japan.

Thermal Hydraulics

The fluid dynamics and heat transfer characteristics of the BMR are being studied both analytically and numerically. The problem which has been posed can be visualized in Figure 4, where R is the inner radius of the sphere, r is the distance from the center to the inner surface of the condensate layer, d is the condensate film thickness, w is the angular velocity of the system, t is the tangential coordinate direction, and a is the radial coordinate direction. While substantial work has been conducted on these thermal hydraulics models, results are yet forthcoming. Several effects have been considered in the study of the fluid dynamics of the BMR. The problem as posed requires the incorporation of: 1) a variable gravitational field, 2) variable centripetal forces, 3) variable wall surface temperatures, and 5) variable Coriolis forces. The film thickness, local heat transfer coefficient, the heat flux at the liquid-solid interface, and the laminar-turbulent transition can be predicted from the general solution of the Navier-Stokes equations for the boundary layer.

These parameters are found in the following way. The film thickness, temperature and velocity distributions are the direct results of the conservation equations. The heat flux at the wall is found by assuming pure conduction at the interface node.

$$q_w(\Theta,\phi) = -k_l \frac{dT}{dr} \Big| r = \text{wall} \qquad (1)$$

By examining the temperature distribution through the condensate film, the local heat transfer coefficient may show to be

$$h(\Theta,\phi) = \frac{q_w(\Theta,\phi)}{T_w(\Theta,\phi) - T_{in}(\Theta,\phi)} \qquad (2)$$

By integrating over Θ and ϕ, the total heat transfer coefficient may be found.

$$h_{total} = \int_o^{2\Pi} \int_o^{2\Pi} h(\Theta,\phi)\, d\Theta d\phi \qquad (3)$$

The heat transfer aspect of this problem also incorporates several interesting features. The heat transfer mechanisms are coupled as shown in Figure 5. These include: 1) condensation, 2) forced convection, 3) conduction through the liner, 4) conduction through the fabric, 5) terminal radiation from the liner surface, and 6) terminal radiation from the fabric surface. From this analysis, the temperature distribution, heat flux, and overall heat transfer coefficent may be assessed by employing the Fourier and Newton rate equations.

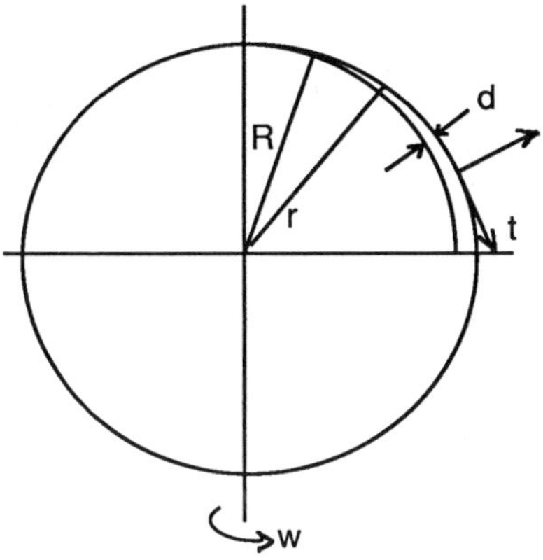

Figure 4. Spherical Film Condensation Model

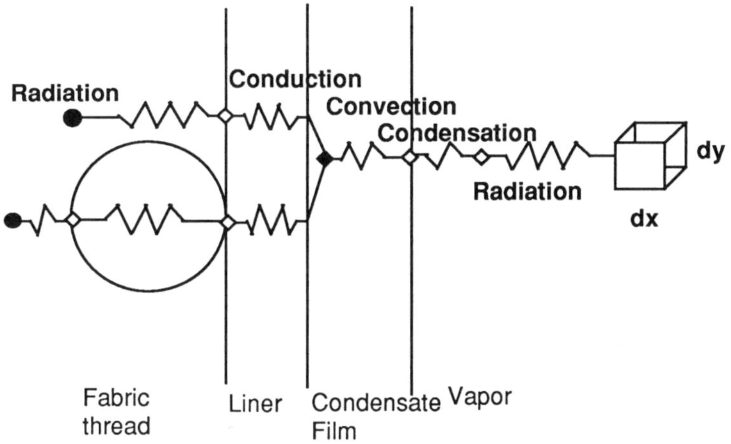

Figure 5. Heat Transfer Model

The numerical formulation will utilize a two-fluid model for the full coupled continuity, motion, and energy equations. The condensation rate will therefore be predicted by an analysis of the energy of the vapor as it approaches the inner film boundary. The analytical and numerical models will then be compared in order to evaluate their relative accuracy.

Conclusions

Design of the BMR concept is dependent on a number of parameters, including the radiation space environment, materials compatability, and thermal hydraulics. The use of an integrated design approach is required by the long mission lifetimes, up to 10 years, and high success probabilities which are dictated by the various missions applications.

The radiation field in interplanetary space is dominated by the proton component. The two primary sources of these particles are the solar wind and cosmic sources. The main interaction mechanism of the proton flux is the deposition of hydrogen in the lattice of the surface materials. While this has little effect on the mechanical properties of the ACF materials, the thermal and optical properties may vary significantly. The effects of this variation have yet to be determined.

The LEO environment contains a large monatomic oxygen component. This radiation field may significantly oxidize and corrode BMR components. While the oxygen-saturated ceramic materials are inert to this attack, further testing and verification is being conducted at the NASA Lewis Research Center and on the space shuttle through FY90.

Materials compatibility issues center around the use of oxygen-bearing fluids. The lack of specific and applicable degradation data on the materials and working fluids has necessitated a series of experimentation currently underway at Oregon State University to assess the spectrum of compatibility issues.

The fluid dynamics and heat transfer characteristics of the BMR are also being studied. The analytical and computer models which are being developed will predict the condensate film thickness, velocity and temperature distributions, heat flux at the liquid-solid interface, effective heat transfer coefficient, and the laminar-turbulent transition point. These will provide a well characterized set of thermal and flow parameters for the spherical BMR design.

References

Condon, E. U. and H. Odishaw, eds. 1967. Handbook of Physics. McGraw-Hill Book Company, New York, New York. pp. 9-317,318.

Durcanin, J. T., et al. 1987. "The Definition of the Low Earth Orbital Environment and its Effects on Thermal Control Material." Proceedings of AIAA 22nd Thermophysics Conf. AIAA-87-1599. American Institute of Aeronautics and Astronautics, New York. pp. 1-12.

Jursa, Adolph S., ed. 1985. Handbook of Geophysics and the Space Environment. Air Force Systems Command, Air Force Geophysics Laboratory, Cambridge, Massachusetts.

Smith, Robert E. and George S. West, compilers. 1983. Space and Planetary Environment Criteria for Use in Space Vehicle Development. NASA TM-82478, NASA Science and Technical Information Branch, Marshall Space Flight Center, Alabama.

Webb, B. J. 1988. Preliminary Design and Analysis of a Rotating Bubble Membrane Radiator for Space-Based Nuclear Power Systems. OSU-NE-8805, Oregon State University, Corvallis, Oregon.

NOMENCLATURE

Variables

k - material thermal conductivity
h - convective heat transfer coefficient
q - heat flux
T - temperature

Subscripts

l - liquid state
w - (surface of inner liner) at $r = R_0$
in - (vapor-liquid interface) at $r = R_1$

The Optimum Dimensions of Radiative Spines

A. Razani and H. Zohoor [1]
Mechanical Engineering Department
The University of New Mexico, Albuquerque, N.M. 87131

Abstract

Radiative pin fins (spines) are important in conjunction with heat rejection systems for space applications. In the design of radiators for space applications, spines possessing minimum mass are of great interest. In this study a conducting-radiating spine with an arbitrary profile is considered. The profile is assumed to be a function of the temperature in the spines by introducing a free parameter. With this choice of profile, the nonlinear heat transfer equation for the conducting-radiating spine can be integrated analytically. Using this method, a nonlinear integral equation for the temperature of the spine is obtained. This equation can be conveniently solved numerically to find the optimum dimensions of the spine.

Introduction

Fins can be used to increase the rate of heat transfer from a solid object. The finned surfaces are widely used in heat exchange devices. Most of these fins are purely convective where the rate of heat transfer from the surface to the surrounding fluid is linearly proportional to the temperature difference. This leads to a linear differential equation for which an analytical solution is available (Kern and Kraus, 1972; Schneider, 1955).

The optimization of convective fins has been considered by several authors (Jones and Smith, 1970; Cobble, 1971; Dhar and Arora, 1976). A general optimization problem of convective fins has been previously analyzed (Maday, 1975; Razelos, 1983). Recently, optimization of convective fins of constant

---

[1] On sabbatical leave from Sharif University of Technology, Tehran, Iran.

cross section has been considered by Sonn and Bar-Cohen (1981). The effect of curvature on optimum straight fins has been investigated as well (Chung and Kan, 1987). Optimization of radiative fins for space application has also been studied (Reynolds, 1963; Schnurr, et al., 1976). The optimization problem, then is to find the shape and dimensions of a fin which would minimize the volume or mass for a given amount of heat dissipation, or, alternatively, to maximize the heat dissipation for a given volume (or mass). One way to analyze the optimization problem is to select a suitable profile and then determine the dimensions of the fin to yield maximum heat dissipation for a given fin volume.

The radiating fins are important in conjunction with heat rejection systems for space applications. The radiating fins for space application are also of interest in that they permit operation of equipment at reasonable temperatures. Optimization of radiative fins is particularly important for space applications. The resulting differential equations for radiative fins are nonlinear; therefore, the analysis of these fins is more complex and in most cases analytical solutions are not possible. This is especially true when optimization of radiative fins is being considered. For fin optimization problems, it is natural to assume that the profile should be correlated with the temperature of the fin. Radiative and convective fins become less effective as their temperatures are reduced. Therefore, at the point in the fin where the temperature is low, the volume or mass is not effectively used. This demonstrates the fact that optimum profiles and temperatures should be correlated. In this work, we assume that the fin profile is related to the temperature of the fin by introducing a free parameter. Using the profile, the nonlinear differential equation can be integrated and the resulting equation can be solved numerically to find the optimum dimensions of the fin.

Theory

The steady state energy equation for a conducting-radiating spine can be written as

$$\frac{d}{dx}\left(kA\frac{dT}{dx}\right) - \sigma P(\epsilon T^4 - \alpha T_e^4) = 0, \qquad (1)$$

where σ is the Stefan-Boltzmann constant and the cross-sectional area A and the perimeter P are a function of x (the distance from the base of the spine). T_e is the effective temperature of radiating surfaces seen by the spine. In order to include the effect of radiation heat transfer to the spine from other surfaces, the usual assumption of $t_e = 0$ is not considered in this work. The emissivity, ϵ, and absorptivity, α, are not taken to be equal to include the effect of possible differences in the emitting and absorbing spectra. As pointed out before, based on physical considerations, the temperature and the perimeter for the optimum spine are correlated. The equation for the conducting-radiating spine can be integrated analytically when this correlation is considered by introducing a free unknown parameter, β. The dimensions of the spine will be optimized with respect to β as a design parameter. The thermal conductivity, k, can be any polynomial function of T.

Figure 1 shows the geometry and coordinate system for a radiating fin with an arbitrary profile. The origin of the coordinate system is taken at the center of the spine base. The distance from the spine axis to any point on its surface is denoted by Y.

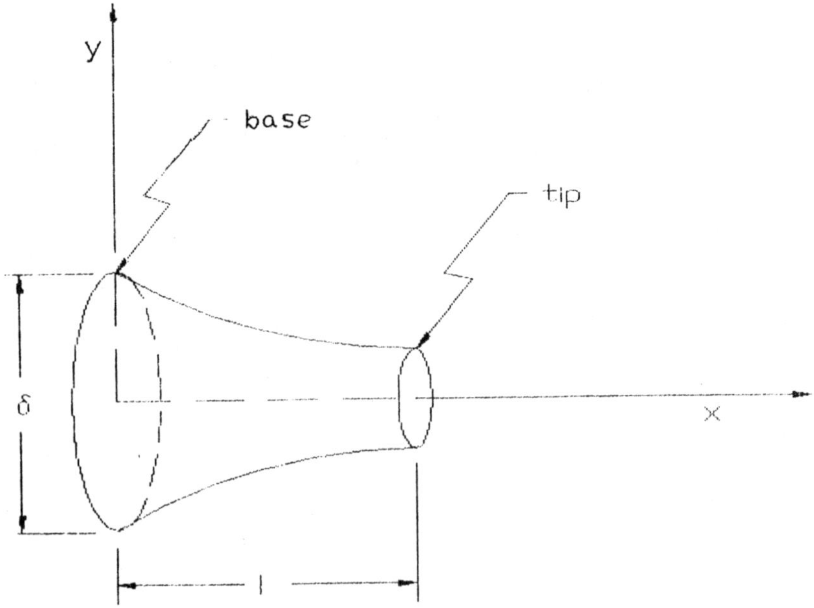

Fig. 1. The spine and its coordinate system

For the symmetrical spine, its volume and surface can be written respectively as

$$V = \int_0^1 \pi Y^2 dx, \qquad (2)$$

$$S = \int_0^1 2\pi Y dx, \qquad (3)$$

where l is the length of the spine.

In this work, we assume that the profile is correlated with the temperature and can be represented by

$$Y = \frac{\delta}{2}\left(\frac{T}{T_o}\right)^{\beta}, \qquad (4)$$

where δ is the spine base diameter and T_o is the base temperature.

The temperature is given in degrees Kelvin. The free parameter β controls the shape of the spine profile and consequently its volume, cross-sectional area, and perimeter. For example, we have a cylindrical spine with constant cross-section, where $\beta = 0$. With this choice of profile, equation (1) can be written as

$$\frac{d}{dx}\left[k\delta^2\left(\frac{T}{T_o}\right)^{\beta}\frac{dT}{dx}\right] = 2\sigma\delta\left(\frac{T}{T_o}\right)^{\beta}(\epsilon T^4 - \alpha T_e^4). \qquad (5)$$

This equation should be solved together with boundary conditions at the base and the tip of the spine. These boundary conditions are given by:

$$T = T_o \text{ at } x = 0 \text{ and}$$

$$\frac{dT}{dx} = 0 \text{ at } x = l, \qquad (6)$$

where an adiabatic boundary condition at the tip of the spine in used. If we multiply equation (5) by $(kA\frac{dT}{dx})$ and integrate the resulting equation, we get

$$\left(\frac{dT}{dx}\right)^2 = \frac{8\sigma}{k\delta}\left(\frac{\epsilon}{5+3\beta}\, T^5\theta^{3\beta} - \frac{\alpha T_e^4}{1+3\beta}\, T\theta^{3\beta}\right) + C \qquad (7)$$

where $\theta = T/T_o$ is a dimensionless temperature, and C is the constant of integration. This constant can be obtained from the adiabatic boundary condition given by equation (6).

If we denote the temperature of the tip of the spine by T_1, equation (7) can be written as

$$\left(\frac{dT}{dx}\right)^2 = \frac{8\sigma}{k\delta}\Big[\frac{\epsilon}{5+3\beta}\left(T^5\theta^{3\beta} - T_1{}^5\theta_1{}^{3\beta}\right) - \frac{\alpha T_e^4}{1+3\beta}\left(T\theta^{3\beta} - T_1\theta_1^{3\beta}\right)\Big] \tag{8}$$

where $\theta_1 = T_1/T_o$.

Equation (8) can be used to obtain the temperature distribution in the spine. Taking the square root of equation (8) and integrating the resulting first order differential equation, one obtains

$$\int_{T_o}^{T}\Big\{\frac{8\sigma}{k\delta}\Big[\frac{\epsilon}{5+3\beta}\left(T^5\theta^{3\beta} - T_1{}^5\theta_1{}^{3\beta}\right) - \frac{\alpha T_e^4}{1+3\beta}\left(T\theta^{3\beta} - T_1\theta_1^{3\beta}\right)\Big]\Big\}^{-1/2}dT = \int_0^x dx. \tag{9}$$

From equation (7), the rate of heat transfer from the spine, q_o, can be calculated.

$$q_o^2 = \frac{k\sigma\beta^2\delta^3}{2}\Big[\frac{\epsilon}{5+3\beta}\left(T_o^5 - T_1^5\theta_1^{3\beta}\right) - \frac{\alpha T_e^4}{1+3\beta}\left(T_o - T_1\theta_1^{3\beta}\right)\Big], \tag{10}$$

The required optimization problem is to relate the heat transfer rate from the base of the spine to its volume. From equations (2), (3), (4) and (10) one obtains

$$\frac{1}{V}\left(\frac{q_o^5}{k\sigma^4 T_o^{17}}\right)^{1/3} = (256\pi^2)^{1/3}\Big[\frac{\epsilon}{5+3\beta}(1 - \theta_1^5) - \frac{\alpha\theta_e^4}{1+3\beta}(1 -$$

$$\theta_1)]^{5/6}\{\int_1^{\theta_1}\theta^{4\beta}[\frac{\epsilon}{5+3\beta}(\theta^{5+3\beta} - \theta_1^{5+3\beta}) - \frac{\alpha\theta_e^4}{1+3\beta}(\theta^{1+3\beta} - \theta_1^{1+3\beta})]^{-1/2}d\theta\}^{-1}. \qquad (11)$$

where $\theta_e = T_e/T_o$.

In the present formulation, the important quantity is the temperature of the spine, T_1. Now, for a given volume we should find maximum q_o by varying the spine tip temperature T_1. A computer program has been written to find, T_1 for the optimum spine. It should be pointed out that for each β an optimum spine can be selected which corresponds to the particular profile. Therefore, β is a design parameter which can be changed to find a profile suitable for structure and design considerations. Once T_1 is obtained from equation (11), the optimum q_o can be calculated. Equation (10) can then be used to determine the optimum base diameter of the spine, δ. The temperature distribution in the optimum spine is found from equation (9). Using the temperature distribution of the optimum spine, its profile is determined from equation (4).

Results and Discussion

In order to show the effect of different optimum spine profiles represented by equation (4), the optimization problem of a radiating fin for different values of β is presented. In the radiating fin optimization problems for space applications usually a simplifying assumption of $t_e = 0$ is made (Wilkins, 1960). In this work, an effective temperature of the surrounding medium of $t_e = 300^o k$ is assumed. Any constant surrounding temperature may be considered in this model such that one can analyze the effect of radiating surfaces seen by the spine. For the examples given below, the following numerical values are used:

$V = 5x10^{-5}\ m^3$,

$\epsilon = \alpha = 1$,

$k = 100\ w/m^o k$.

Figure 2 shows the optimum dimensionless parameter

$N = \frac{1}{V}(\frac{q_o^5}{k\sigma^4 T_o^{17}})^{1/3}$ as a function of the profile parameter β for three different base temperatures of $T_o = 500^o k$, $T_o = 1000^o k$, and $T_o = 1500^o k$. From this figure, it can be seen that N reaches a maximum value of approximately $\beta = 8$. This figure is important for design analysis. Thus, one can find the best profile for a given volume V and base temperature T_o. It should be pointed out that the best value of β does not necessarily mean that the spine would be a structurally good design. However, the figure can be used to find the optimum spine that meets given design criteria.

Figure 3 shows the effect of β on the ratio of the optimum spine tip temperature to its base temperature. This figure indicates that the dimensionless spine tip temperature is an important parameter for optimization. This can also be seen from equation (11). It is interesting that this ratio shows little variation for values of base temperature $T_o > 1000^o k$.

The ratio of the optimum spine length to its diameter as a function of the profile parameter, β, is given in figure 4 for the three different spine base temperatures, T_o. It should be pointed out that for the large base temperatures, the optimum spine is very short. For example, when $\beta = 0$ (cylindrical spine with constant cross section) the length and diameter of the optimum spine are equal. In this case, the adiabatic boundary condition used in equation (4) is not accurate and the heat loss from the tip of the spine should be taken into consideration. However, this complicates the mathematics involved and the model is capable of taking a general boundary condition into account.

In all cases considered, the "length-of-arc" assumption is used. This is a very good assumption for a small value of β. For large values of β, the error introduced may be a few percent, but these cases are of little practical significance. For the model under consideration, the effect of the profile curvature should be approximated in such a way that the advantage of analytical integration would not be lost.

Conclusions

The effectiveness of a fin decreases as the temperature of the fin relative to the temperature of the environment is reduced. This is, naturally, an important consideration in the design of optimum fins. In this work, radiative spines are optimized using a temperature correlated profile. This is accomplished by introducing a free parameter so that different profiles may be analyzed for optimization. The relation of the temperature correlated profiles has an added advantage for the analysis of radiative fins. This is due to the fact that the heat transfer equation can be integrated analytically. As a result of the analytical integration a dimensionless parameter for the analysis of optimum radiative spines has been developed. This dimensionless number, $N = \frac{1}{V}\left(\frac{q_o^5}{k\sigma^4 T_o^{17}}\right)^{1/3}$ is important in the design analysis of optimum radiative spines. The dimensionless number N is obtained in terms of an integral equation with only one variable, the dimensionless tip temperature of the spine. Therefore, the tip temperature of the spine becomes the optimization variable. This does not mean that the tip temperature must be measured or must be given so that the radiative spines can be optimized. The tip temperature only serves as an optimization parameter for numerical calculation and it is obtained as a part of the solution.

References

Cobble, M. H., 1971, "Optimum Fin Shape," Journal of the Franklin Institute, Vol. 291, No. 4, pp. 630-639.

Chung, B. T. F. and Kan, R. W., 1987, "The Optimum Dimensions of Convective Straight Fins with Profile Curvature Effect," AIChE Heat Transfer Symposium Series, edited by R. W. Lychowski, pp. 77-83.

Dhar, L. P. and Arora, P. C., 1976, "Optimum Design of Finned Surfaces," Journal of the Franklin Institute, Vol. 301, no. 4, pp. 379-392.

Jones, D. C. and Smith, F. L., 1970 "Optimization Arrangement of Rectangular Fins of Horizontal Surfaces for Free-Convection Heat Transfer," ASME Journal of Heat Transfer, Vol. 92, pp. 6-10.

Kern, Q. D. and Kraus, D. A., 1972, "Extended Surface Heat Transfer," McGraw-Hill, New York.

Maday, C. J., 1975,"The Minimum Weight One-Dimensional Straight Fin," ASME Journal of Engineering for Industry, Vol. 97, pp. 1190-1193.

Schneider, P. J., 1955, "Conduction Heat Transfer," Addison-Wesley, Reading, Mass..

Razelos, P., 1970, "The Optimum Dimensions of Convective Pin Fin," ASME Journal of Heat Transfer, Vol. 105, pp. 411-413.

Reynolds, W.C., 1963, "A Design-Oriented Optimization of Simple Tapered Radiating Fins," Journal of Heat Transfer, pp. 193-202.

Schnurr, N.M., Shapiro, A.B., and Townsend, M.A., 1976, "Optimization of Radiating Fin Array with Respect to Weight," Journal of Heat Transfer, pp.643-648.

Sonn, A., and Bar-Cohen, A., 1981, "Optimum Cylindrical Pin Fin," Journal of Heat Transfer, Trans. ASME, vol. 103, pp. 814-815.

Wilkins, J. E., Jr. 1960, "Minimum-mass Thin Fins Which Transfer Heat Only by Radiation to Surroundings at Absolute Zero," Journal of Society Industrial Applied Mathematicas, Vol. 8, No. 4, pp. 630-639.

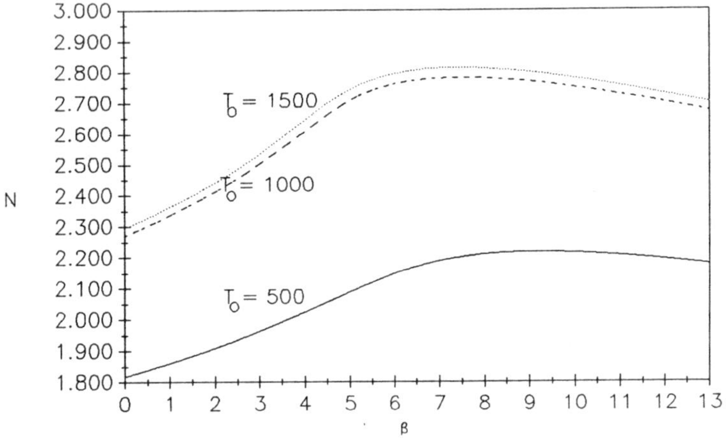

Fig. 2. The optimum dimensionless parameter, N as a function of the profile parameter, β

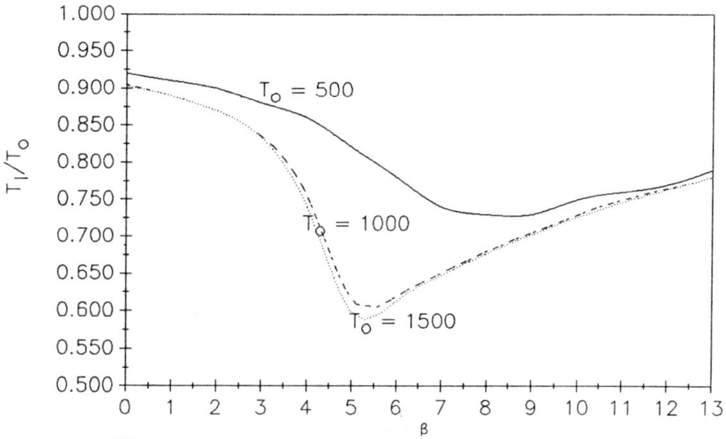

Fig. 3. The ratio of the optimum spine tip temperature to its base temperature as a function of the profile parameter, β

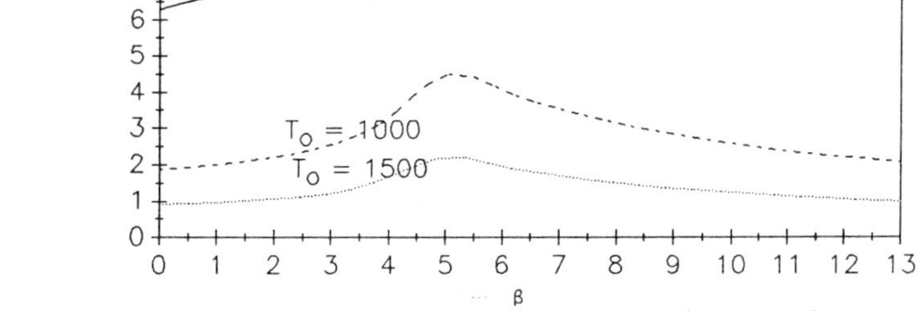

Fig. 4. The ratio of optimum spine length to its base diameter as a function of the profile parameter, β

Subsection 4.5: HUMAN FACTORS

Implications of the New Radiation Exposure Limits on Space Station Freedom Crews.

M. Stanford, McDonnell Douglas Space Systems Company - Space Station Division, Houston, TX 77058, and
D. S. Nachtwey, NASA - Johnson Space Center, Houston, TX 77058.

Abstract

Spaceflight unavoidably increases the exposure of astronauts to natural ionizing radiation. Any increase in radiation exposure increases the risk of contracting cancer or inducing genetic mutations. Since risk avoidance is equivalent to dose avoidance and since complete dose avoidance in space is not possible, levels of acceptable risk must be established. Research over the last decade indicates that the cancer risk per dose-equivalent has increased and that the relative carcinogenic effectiveness of certain types of space radiations may be much higher than previously thought. Therefore, new dose-equivalent limits have been recommended to NASA by the National Council on Radiation Protection and Measurements (NCRP).

For Space Station Freedom, the major radiation component which contributes to the total dose is trapped protons form the South Atlantic Anomaly (SAA). Based upon the new annual limits, an astronaut may not receive more than a depth-dose equivalent of 50 rem/year. Calculations indicate that a 180-day stay aboard Freedom could result in a worst-case depth-dose of roughly 30 rem. A 180-day mission in a nominally shielded spacecraft (1 g/cm^2) in a nominal, constant atmospheric density orbit with a varying altitude could result in a depth-dose equivalent of roughly 10 rem. This is twice the annual allowable dose-equivalent for terrestrial radiation workers. For the first time in the US Manned Space Program, crewmembers will be approaching a significant portion of their established exposure limits. Because of this, the measurement and management of radiation exposures and the assessment of the health risks involved will play a significant role in manned operations.

Space Station Freedom Radiation Environment at LEO

Low Earth Orbit (LEO) space flight involves exposure to a complex mixture of primary radiation types (Jursa, 1985). These include trapped protons (mostly from the South Atlantic Anomaly) and electrons, galactic cosmic radiation (not a significant contributor to overall dose at Space Station Freedom's orbital parameters), plus secondary radiation products produced by collisions of the primary radiation with Space Station Freedom (SSF), its furnishings and within the crew's bodies. This secondary radiation consists mainly of fragments of heavy ion projectiles and fragments of nuclei hit by energetic particles, including neutrons, electrons, spallation products, and gamma rays. Additionally, the radiation environment will vary over the length of the entire Solar Cycle.

Therefore, the radiation environment within Space Station Freedom (which will fly in a 28.5 deg. inclination between 150-250 nautical miles) will be constantly changing in intensity and composition and may vary considerably over the life of the Station. The radiation experienced at a given point within the Station is strongly dependent upon the external radiation impinging on the structures of the spacecraft (as described above) and upon the amount and composition of these materials (spacecraft walls, furnishings, stowage, electronics, etc.) in the radiation path. Structural materials (the final configuration of Freedom's hull construction has yet to be decided) both attenuate the incident radiation and serve as a source of complex secondary radiations. Because of these manifold interactions, the radiation within Freedom will be considerably more complex than the primary radiation incident upon it; and it is highly time-variable due to the changing orientation of the structural materials and the varying incident radiation field. This complex multicomponent radiation field presents a unique and different radiation measurement problem for any space radiation protection program.

Previous Radiation Exposures on Manned Spacecraft

Space Shuttle flights to date have involved relatively low career doses. Details of exposures encountered during STS missions can be seen in Figure 1 (Hardy, 1989). It is important to note the influence altitude plays in exposures received by STS crews. The lowest dose rate encountered in these missions occurred on flight STS-5, which flew at a low altitude and at 28.5 degree inclination. The enhanced effect of altitude can be seen from the data from flight STS-51J. The duration of the flight also plays a significant part. Skylab 4, a mission of 84-days at an average orbital altitude of 435-km and at 50 degree inclination, resulted in an average total bone marrow dose equivalent of about 77 mSv (7.7 rem) (Bailey 1977).

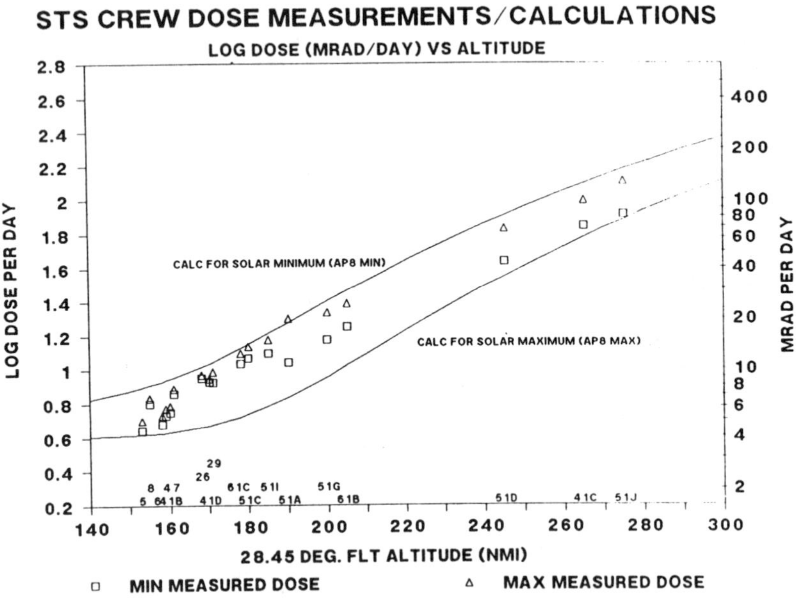

Figure 1. Theoretical and Measured Doses for Representative STS Flights. Note the Effect of Altitude on Dose.

Uncertainties in Current Knowledge Concerning Space Radiation and Its Effects

A major uncertainty facing those who assess the risks associated with exposures to space radiation involves the densely ionizing, high LET (Linear Energy Transfer) radiation (Fry, et al, 1985). This type of radiation is much more effective in inducing biological damage than low LET radiation (certain species of high LET radiation may be 20 to 50 times more damaging then low LET radiations). Since high LET radiations have greater effectiveness in inducing damage, the measured dose (in rad or gray) must be weighted by a quality factor (Q) to obtain the dose-equivalent (in rem or sievert) for risk assessments.

The challenge ahead for Space Station Freedom dosimetry is to provide an accurate determination of LET, particle charge, mass, energy and direction. The suite of instruments proposed for Freedom to accomplish these tasks include personal thermoluminescent dosimters (TLDs similar to those routinely used in Shuttle flights), a Charged Particle Directional Spectrometer (CPDS) and a Tissue Equivalent Proportional Counter (TEPC). Combining the data from this instrument package will greatly enhance our understanding of the secondary radiation produced inside Freedom and will enable the avoidance of overly conservative radiological health risk assessments. Additionally, the dose projection models now in use must employ decades-old, inaccurate magnetic field models and must be updated to account for current measurements made during Shuttle flights. As understanding of the dynamics of the geomagnetosphere increases, modeling the temporal, seasonal and solar cycle variations at LEO will enable a better assessment of the natural environment than the current statistical (static) models do. The proposed suite of radiation measuring instruments for Freedom will provide most of the needed data to upgrade these dose projection models.

The National Council on Radiation Protection and Measurement (NCRP) Recommendations

In 1970, the current radiation protection guidelines were adopted by the National Aeronautics and Space Administration (NASA). The career limit was set at 4.0 Sv (400 rem). Since that time, new risk assessments (Sinclair, 1983) have been obtained for radiation exposures as well as a better understanding of the radiation environment for different mission scenarios. Additionally, the nature of the astronaut corps has changed (women now are members) and many crewmembers will be exposed to repeat missions of different time lengths and orbital parameters. For these and other reasons, NASA requested the NCRP to re-examine NASA's radiation protection guidelines. This task was undertaken by NCRP Scientific Committee 75.

The new recommendations for career dose limits are illustrated in Figure 2. They are based on a 3% lifetime excess risk of cancer mortality and take into account the age at first exposure and the sex of the crewmember. The new career limits range form 1.0 Sv (100 rem) for a 24-year-old female up to 4.0 Sv (400 rem) for a 55-year-old male, compared with the previous single limit of 4.0 Sv (400 rem). The career limit for the lens of the eye has been reduced from 6.0 Sv (600 rem) to 4.0 Sv (400 rem), while the skin exposure level has been reduced from 12 Sv (1200 rem) to 6.0 Sv (600 rem).

IONIZING RADIATION EXPOSURE LIMITS[1]

Exposure interval	Depth (5 cm)	Eye (0.3 cm)	Skin (0.01 cm)
30 days	25 rem[2]	100 rem	150 rem
Annual	50	200	300
Career	100 to 400[3]	400	600

CAREER EXPOSURE BY AGE AND SEX

Sex	Age			
	25	35	45	55
Male	150	250	325	400
Female	100	175	250	300

FOOTNOTES:

[1]These dose-equivalent limits are being recommended to NASA by the National Council on Radiation Protection and Measurements (NCRP). Approval by the Administrator is pending.

[2]This table is expressed in conventional units due to common usage by the discipline. The SI unit is the sievert (Sv), which is equivalent to 100 rem.

[3]The career depth dose-equivalent limit is based upon a maximum three-percent lifetime excess risk of cancer mortality. The total dose-equivalent yielding this risk depends on age at start of exposure and sex. The career dose-equivalent limit is approximately equal to: 200 + 7.5 (age - 30) rem for males, up to 400 rem maximum, and 200 + 7.5 (age -38) rem for females, up to 400 rem maximum.

Figure 2. The New Recommendations for Career Dose Limits.

Astronaut Radiation Exposure Aboard Space Station Freedom

Figure 3 illustrates the worst case as well as the nominal worst case accumulated deep dose equivalent in rem for an astronaut who experiences repeated tours of duty aboard Freedom. Compare these rates of dose equivalent accumulation with terrestrial workers and one clearly sees the nature of the problem in assuring radiation health protection for astronaut crews who experience multiple tours of duty in the Station.

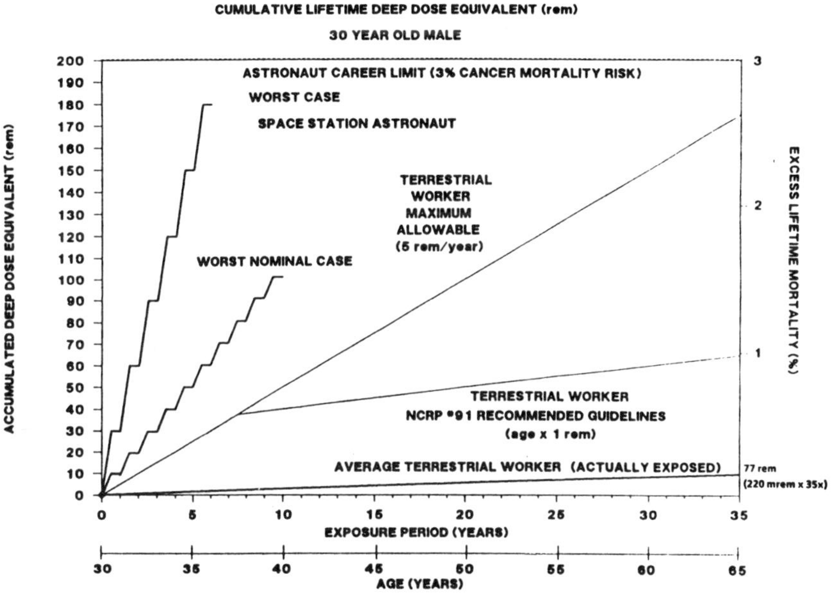

Figure 3. Theoretical Worst-Case and Nominal Worst-Case Accumulated Deep Dose Equivalent (rem) for an Astronaut Flying Repeated Tours Aboard Space Station Freedom.

To respond to this challenge a robust radiation health program must be implemented. The elements of such a robust program can be seen in Figure 4. It is a multifaceted program integrating the elements of active and passive dosimetry, modeling and simulation of the radiation environment and a comprehensive assessment of the health risks associated with exposure to the natural space radiation environment. In addition, mission

RADIOLOGICAL HEALTH PROGRAM

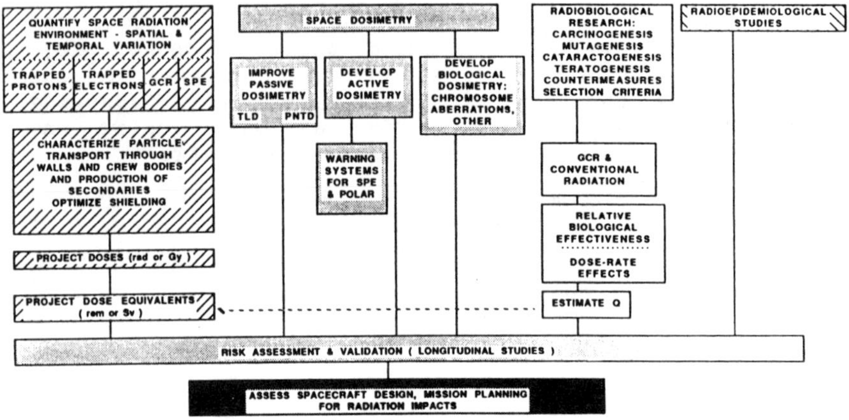

Figure 4. An Integrated Robust Radiation Health Program Porposed for the Space Station Era and Beyond.

planning, space craft design and operations must include analyses to assess radiation impacts on the overall utilization of Freedom. One promising method of reducing the radiation exposure of Freedom would be to fly the station in a constant drag orbit that takes full advantage of atmospheric removal of trapped particles (Pfitzer, 1989). Another method would be to fully characterize the radiation environment within Freedom to locate radiation "hot spots" and rotate the crew to minimize exposures in those areas. In all cases, the ALARA (As Low As Reasonably Achievable) principle will be employed to optimize as much as possible the many different strategies associated with management of radiation exposures over the careers of the astronauts on Freedom.

Conclusions

Radiological health concerns should not preclude the extended stay times envisioned for Space Station

Freedom. However, radiological health is a significant issue for all extended duration missions.

To ensure the health and safety of Freedom astronauts, accurate records of all radiation exposures (medical as well as space exposures) are being maintained.

A tumor registry for the entire astronaut population is being established and longitudinal studies are being conducted on this group.

A vigorous research program should be supported to assess the biological consequences of exposures to high LET radiations. Our present understanding of the biological effectiveness of high LET radiation is not adequate for accurate health risk assessments.

Constant improvements in radiation dosimetry, modeling and simulation used in predicting dose projection, and continued research into the biological effects of chronic low level exposure to the space natural environment will ensure the safety and well being of all future users of Space Station Freedom.

References

Bailey, J. V. Dosimetry During Space Missions. IEEE Trans. Nucl. Sci. NS-23 4:1379; 1977.

Fry, R. J. M.; Powers-Risius, P.; Alpen; Ainsworth, E. J. High-LET radiation carcinogenesis. Radiat. Res. 104:S188-S195; 1985.

Hardy, A. C. Office of Space Exploration, NASA Johnson space Center personal communication.

Jursa, A. S. (Ed.) Handbook of Geophysics and the Space Environment. Air Force Geophysics Laboratory, Air Force Systems Command; 1985.

Pfitzer, K. A. Radiation dose to man and hardware as a function of atmospheric density in the 28.5 deg. Space Station orbit. McDonnell Douglas doc. No. MDC-H5387; 1989.

Sinclair, W. K. Radiation safety standards, space hazards vs. terrestrial hazards. Adv. Space Res. 3:151-159; 1983.

A Prototype Model for Human/Automation Trade-offs

Richard Johnson[1]

Abstract

Future space construction missions will involve both human and machine constructors. Selection of the optimum constructor mix requires a model of constructor capabilities and requirements. The database for that model is developed via extrapolation from current literature. Optimization is done via minimization of total mission cost using a linear programming approach. This prototype is the first cut at producing a general tool for choosing a near-optimum constructor mix for any space construction mission. The linear programming optimization model illuminates several significant representational and data-gathering problems.

Introduction

Past space programs have not required the assembly or construction of vehicles in space or on other planetary bodies. Ambitious future space missions, such as those detailed in recent reports (Ride, 1987) (NCoS, 1987) and even Presidential speeches (Bush, 1989), will involve spacecraft, stations, and bases too large to launch in a single piece.

The creation and maintenance of such structures will involve on-orbit or on-surface construction and testing. A whole suite of constructors including humans, teleoperated robots, self-deploying structures, and fully autonomous robots might be available. Each type of constructor will have associated advantages and disadvantages depending upon the specific construction tasks. The primary question is exactly how much and what sorts of automation will best

[1] PhD Candidate - University of Colorado at Boulder, Center for Space Construction, Campus Box 429, Boulder, CO 80309

augment human capabilities. A prototype of the CSC constructor trade-off model may provide some initial answers, but most importantly it will allow us to test our modelling approaches.

General Construction Simulation

This constructor selection model is part of a larger effort in modelling the entire construction process. The overall Construction Simulation (SIMCON) model will include vehicle selection for construction material delivery and varied program construction sequences among other construction planning issues. Ideally, the prototype constructor selection model should be designed to interact smoothly with the rest of the modules in the SIMCON model. One side benefit of producing the constructor selection model is the insight it will provide on just what those module interfaces should be. In order to get some useful work done initially, however, the prototype will start by assuming input of construction location and provide outputs of near-optimum constructor selection and the attendant minimized constructor cost.

Constructor and Job Representation

The word constructor means one who, or that which constructs. Constructors include humans, humans augmented with tools, teleoperated machines, dull & smart robots, and even self-deploying structures. When constructors are used in a project, they can be referred to as a construction team.

Determining a near-optimum construction team for a given task or set of tasks can be viewed as a 4-step process (table 1). The constructor selection model is intended, when given information on tasks and constructor capabilities, to resolve the comparison and optimization step.

1)	Identification of task requirements
2)	Identification of constructor capabilities
3)	Match-up of requirements with candidate construction systems.
4)	Comparison of competing systems and selection of optimum [over single- or multi-mission program].
	(Helleckson, 1989)

Table 1 - Constructor Selection Process

Modelling Assumptions

As with all models, this one is somewhat abstract, making simplifying assumptions about constructors, tasks, and optimization criteria. Determination of requirements and capabilities can be handled at levels from primitives (move, grasp, etc.) to general tasks (assemble truss, dig trench, etc.). Given a single constructor and a simple construction task, the constructor's productivity can easily be represented in terms of the time required to perform the necessary sequences of Elementary Construction Functions (ECFs). A generalized set of ECFs, gleaned and generalized from many sources, includes Move, Rotate, Align, Grasp/Release, and Apply Force (Helleckson, 1989). Note that this sort of description requires detailed knowledge of the assembly choreography and constructor ECF capabilities. Determining such knowledge for all missions of interest would require a detailed design for each mission - a prohibitive level of effort. When more than one constructor is assigned to a task, the effort becomes even more severe.

A coarser view is obviously required to produce a workable general model. Representing the construction tasks on a higher level allows us to forego the detailed mission choreography and concentrate instead on cost based constructor selection. The higher level tasks (table 2) were chosen mainly because we have some reference data available on work rates at this level, such as that for pressurized module (can) attachment, for EVA astronauts using Space Transportation System (STS) 4 psi suits (NASA, 1989).

As mentioned earlier, the model uses overall mission cost as it's optimization variable. This ignores safety constraints, construction time, and many other important trades. Such a simplification does map well into the initial stages of mission design, where safety and other issues are not usually the primary focus. However, in future models such criteria will need to be included, perhaps in the form of an empirically determined trade function.

Unload		Transport	
	Truss Parts		Cans
	Can Support Parts		Truss Parts
	Solar Array Parts		Solar Array Parts
	Radiator Parts		Palettes
	Can		Can Support Parts
	Palettes		Shroud Parts
	Shroud Parts		
Build		Attach	
	Truss Bay		Radiator
	Can Support		Solar Arrays
	Solar Array		Palette
	Radiator		Shroud
	Shroud		Can
	Palettes		
	Make Electrical/Optical/Fluid Connections		

Table 2 - Reference Mission Construction Tasks

In addition, the prototype model is not all-inclusive in time. It is limited to a single near-term Low Earth Orbit (LEO) Design Reference Mission (DRM) - assembly of a satellite servicing facility (Sparn *et al*, 1989). The representative mission is composed of multiple phases with different constructor teams available for each.

Available Constructors

The constructors available for the mission (table 3) are limited to those currently in use, under development, or available via simple application of current technology. The most basic constructors are deployable structures. For example, self-deploying solar panels are commonly used on manned and unmanned satellites. The expected EVA astronauts will also be available. Finally, teleoperated machines such as the shuttle's Remote Manipulator System (RMS) and the planned Flight Telerobotic Servicer (FTS) will also be in the constructor pool.

Self Deploying Truss (SDT)
Self Deploying Solar Array (SDSA)
Self Deploying Radiator (SDR)
Extra-Vehicular Activity Astronaut using Space Transportation System suit (EVA/STS)
Extra-Vehicular Activity Astronaut using new Space Station Freedom hard suit (EVA/SSF)
Flight Telerobotic Servicer under STS control (FTS/STS)
Flight Telerobotic Servicer under ground control (FTS/GC)
Flight Telerobotic Servicer under SSF control (FTS/SSF)
Remote Manipulator System under STS control (RMS/STS)
Remote Manipulator System under SSF control (RMS/SSF)

Table 3 - The Constructors

Model Specifics

The prototype model uses a Linear Programming (LP) approach for optimization of constructor selection. The optimization will initially be performed solely on the basis of cost.

Objective Function

As with most LP models, this one has a single objective function and a number of constraint equations. It will initially be solved with the standard LP Simplex algorithm. The objective function is:

$$\text{Minimize } \mathbb{C} = \sum_{i=1}^{m} \sum_{j=1}^{n} X_{ij} C_i ,$$

Equation 1

where C is the total cost of the constructor team, X_{ij} the time that the ith constructor W_i is assigned to do task T_j, and C_i the cost per unit time of using the given i^{th} constructor.

Constraining Functions

There are two basic constraints on the selection of the independent variables.

The first is the requirement that all units of a particular task be completed (Total Work):

$$\sum_{i=1}^{m} X_{ij}P_{ij} >= \text{Total Work required to do task } T_j$$

Equation 2

Each task has its own Total Work equation where P_{ij} is the productivity of the constructor in task units per minute.

The second constraint equation ensures that that the constructors are not over-subscribed (Time Available):

$$\sum_{j=1}^{n} X_{ij} <= \text{Time Available for constructor } W_i$$

Equation 3

Each constructor has its own Time Available equation.

Additional constraints on constructor selection are not included in the model (due to lack of data and/or a method of representation). Such constraints involve constructor interference and environmental exposure limits, among others.

Database

The data for the matrices is kept in a HyperCard stack, which provides a quick prototyping environment. In addition, extracting the data into a more powerful database is easy. For a more complex model, the use of classes and inheritance in an object oriented database will be helpful, but for the limited set of constructors in the prototype it was unnecessary. A simple flat database, with one record for each constructor type, was adequate for the prototype.

Most of the test-case numbers used in the prototype model (tables 4 and 5) were educated guesses. Such estimations were adequate to test the model. Of course, hard data will be necessary as the modelling work proceeds - when reasonably accurate results are needed.

	Unload truss	Transport truss	Assemble truss	Unload cans	Transport cans	Attach cans
SDT	0	0	.2 bay/min (assumed)	0	0	0
EVA/ STS	.2 bay/min (derived NASA 1989)	.2 sect/min (derived NASA 1989)	.05 bay/min (NASA 1989)	.1 can/min (derived NASA 1989)	.1 can/min (derived NASA 1989)	.0167 can/min (NASA 1989)
FTS/ STS	.1 bay/min (assumed)	.1 sect/min (assumed)	.033 bay/min (assumed)	.067 can/min (assumed)	.067 can/min (assumed)	.0125 can/min (assumed)
RMS/ STS	.05 bay/min (assumed)	.1 sect/min (assumed)	.01 bay/min (assumed)	.067 can/min (assumed)	.1 can/min (assumed)	.01 can/min (assumed)

Table 4 - Productivities of Constructors on Tasks

Results & Conclusions

The full prototype model would not run under the limited linear programming analysis tools available at this stage of our modelling project. The initial trials of the model were thus restricted to a subset of the full DRM; *ie*, assembly of the 10-bay truss by the constructors shown in tables 4 and 5. Truss assembly has been the subject of two Shuttle flight demonstrations, EASE (Akin, 1986) and ACCESS (Levin, 1988), as well as ground-based studies (Heard *et al*, 1988). It thus provided the most sensible subset for investigation. As one might assume for such a simple mission without any larger mission context, a single constructor was selected for the job. An EVA astronaut would work 200 minutes at a "cost" of over $400,000 to complete the 10 bay truss.

Selection of the single cheapest constructor is inevitable in this case since the job didn't require any of the constructors to work beyond their time limits. Of course, with more accurate numbers for costs and productivities and additional tasks, other outcomes are possible. The prototype model thus produced untrustworthy constructor selection results - a good example of "garbage in, garbage out." However, the prototype did fulfill its main function; the model's problems provide direction and focus for further constructor selection modelling.

Problems and Model Extensions

The model has one fairly important conceptual problem. Launch costs and task set-up times are not represented well. All costs in the model are amortized and represented in dollars per minute of operation. For development and production costs, this reasonable, standard practice doesn't produce significant errors.

	Technology Develop. $	Production/ Training $	Launch $	Technology Useful Lifetime	Constructor Useful Lifetime	Mission Useful Lifetime
SDT	10 mill. $ (assumed)	800 k$ (assumed)	300 kg 1.32 M$ (assumed - to LEO)	1000 hrs (assumed)	1 hr (assumed)	1 hr ∞ hr for avail. time (assumed)
EVA/ STS	0 (assumed - already available)	700 k$ training (derived Havens 1986)	300 kg 1.32 M$ (assumed - to LEO)	1000 hrs (assumed)	36 hrs (assumed)	12 hrs (NASA 1989)
FTS/ STS	500 M$ (assumed)	1.4 M$ (assumed)	250 kg 1.1 M$ (assumed - to LEO)	10000 hrs (assumed)	100 hrs (assumed)	40 hrs (assumed)
RMS/ STS	0 (assumed - already available)	500 k$ training (assumed)	400 kg 1.76 M$ (assumed)	10000 hrs (assumed)	200 hrs (assumed)	40 hrs (assumed)

Table 5 - Costs of Constructors

For launch costs, however, amortization introduces noticeable perturbations in the final solution. If a given constructor is present at the assembly location, full launch costs must be paid no matter how many minutes it works. On the other hand, if it isn't present the launch costs are zero.

For missions combining constructors with launch costs differing by an order of magnitude, amortizing those costs can lead to a false-optimum solution. Launched constructors, especially those with high costs, will work minimal amounts if they're used at all. In many such cases, the result will be an "optimal" cost much less than the honest value (when full launch costs are taken into account). If an alternative work breakdown would produce a result between the false optimum and the honest costs, that solution is not presented.

A similar problem occurs on the task time side of things. Most constructors don't come "out of the box" ready to do their tasks. Significant amounts of time must be spent in setting up for the task, and in cleaning up afterwards. The Assembly EVA Task Menu (NASA, 1989) recommends adding 20% of the task time to account for such ancillary activities. With more detailed data on EVA and robotic operations, we may be able to produce a more accurate representation.

Our first priority will thus be to use more advanced modelling techniques (or a LP technique that has some limited exception handling) to accurately model the non-linear set-up costs and times.

Simultaneously, the problem of finding optimal solutions for these complex models will be investigated. Tabu search is currently being considered (Glover, 1989). Tabu search is a higher level heuristic, or meta-strategy, designed to find near-optimal solutions to optimization problems. The Tabu Search method has been used in a number of problem settings, and has been able to find solutions superior to the best results previously obtained by any method. Specifically, Tabu Search has been applied to complex scheduling problems where it has achieved results within 98-99% optimality running on a micro-computer.

Some subsequent modelling efforts will most likely use non-linear techniques. Handling constructor costs more accurately using step functions based on X_{ij} is one possibility. If a constructor is used, then the step functions causes the full launch, production and development costs to be added in. If it isn't used, then the step functions are zero. However, we're not yet aware of solution techniques that can handle such step functions.

In addition, we need more hard data on constructor capabilities and requirements. The majority of the numbers used in the prototype constructor selection model are extrapolations and educated guesses. Additional data is available in the literature and via contacts with NASA and contractor personnel working in the field.

Since the model represents each constructor as a separate entity it doesn't deal with coordinated constructors (of the same or different types) working together on a task. Data on the synergistic benefits and problems produced by such constructor mixes needs to be developed. Once such data is available, each likely mix can be entered as a new constructor in the database.

Finally, overall mission cost is not a sufficient optimization criteria for a general model. Choice of a near-optimum constructor team needs to be based on some sort of composite criteria, perhaps using an empirically derived optimization function.

When completed, the constructor selection model should provide an accurate and quick tool for mission and system designers. It will allow them to select and cost constructors for any given construction mission, singly and as a part of larger programs.

Acknowledgements

Dr. George Morgenthaler for advice and direction
Diane Stockton for exposure to new techniques.
Chuck Palmer for wild ideas that might actually work.

References

Akin, D.L. (1986), "EASE - Experimental Assembly of Structures in EVA - Overview of Selected Results", Space Construction, Proceedings of NASA Langley conference, August 6-7 1986, J.A. Hagaman, ed, NASA Conference Publication 2490, pp. 199-218.

Bush, G.H.W. (1989), "Remarks by the President at 20th Anniversary of Apollo Moon Landing", The White House Office of the Press Secretary, July 20, 1989, Washington, D.C.

Glover, F. (1989), Graduate School of Business and Administration, University of Colorado at Boulder, Personal communication.

Havens, K.A. (1986), "A Synopsis of the EVA Training Conducted on EASE/ACCESS for STS-61-B", Space Construction, Proceedings of NASA Langley conference, August 6-7 1986, J.A. Hagaman, ed, NASA Conference Publication 2490, pp. 153-182.

Heard, W.L., H.G. Bush, J.J. Watson, (1988) "Space Truss Construction Studies", Space '88: Engineering Construction and Operations in Space, S.W. Johnson, J.P. Wetzel, eds., ASCE, New York, pp 722-732.

Helleckson, B. (1989), "The Selection of Constructors for Space Construction", Proceedings of Space '90, ASCE, in press.

Levin, G.C., J.L. Ross, S.C. Spring (1988), "An Overview of the EASE/ACCESS Space Construction Demonstration", Space '88: Engineering Construction and Operations in Space, S.W. Johnson, J.P. Wetzel, eds., ASCE, New York, pp 712-721.

NASA (1989), "Assembly EVA Task Menu", unpublished NASA working document, Craig Fondling (703)438-5868, NASA JSC, June 5, 1989.

NCoS (1986), (National Commission on Space), Pioneering the Space Frontier, New York, Bantam Books.

Ride, S. (1987), Leadership and America's Future in Space, Report to the Administrator, NASA.

Sparn, T. *et al* (1989), "Design Reference Mission Initial Report", unpublished CSC white paper.

Managing Cultural Diversity in Orbital Habitats

James A. Wise[1], Kristiina McConville[2], and Nasr Al-Sahhaf[3]

Abstract

This paper presents the background, outline, and initial findings of an international project on "Managing Cultural Diversity in Orbital Habitats". The purpose of this project is to investigate significant habitability design and operations issues that arise with respect to multicultural crews working for long time durations in space facilities. These issues are being explored in the context of near Earth and lunar operations through expanded surveys and interviews with cosmonauts and astronauts who have had significant spaceflight experience.

Introduction

This paper is a first look at delineating potential problems of culturally diverse spacecrews. While this has not been a significant issue in the crew factors of spaceflight to date, we expect the situation to change dramatically over the next decade. At the time of this writing, both Soviet and U.S.space programs have orbited 19 foreign astro/cosmonauts who performed specialized research duties for their countries. The Soviets are currently training two Britons as that country's first spacefarers, and design for the *Freedom* space station continues as a collaboration of the U.S., Canada, Europe, and Japan.

For future missions, private associations like the Planetary Society have argued for a joint U.S-Soviet Mars Mission (Planetary, 1987) in their "Mars Declaration". This last summer, the International Space University met for its second year in Strasbourg, with future sessions planned for Toronto, Moscow, and Tokyo. The ISU annually enrolls over a hundred students from over 20 countries who are engaged in or planning space projects. As they return home to continue their work, the exploration of space will become a truly international endeavor.

Extended missions and staffed space bases will require international crews to live and work together under isolated, potentially hazardous and confined conditions for

----

1.Visiting Professor, Dept. of Industrial & Operations Engineering, University of Michigan, Ann Arbor, MI 48109.
2.College of Engineering, University of Toronto, Toronto, Canada.
3.King Abdulaziz City of Science and Technology, Riyadh, Saudi Arabia.

prolonged periods of time. Six months is the current planned crew rotation cycle in the LEO *Freedom* station, while Soviet cosmonauts have already orbited a year in *Mir.* A Mars journey is expected to take up to three years. The psychological and social data accumulated from laboratory studies, analog settings (submarines, oil platforms, Antarctic bases) and spaceflight experience show that such conditions of isolation and confinement in technology dominated environments frequently produce deteriorated crew relations as well as very significant psychological disturbances and losses in productive work (Connors, Harrison, and Akin, 1985). What would be the added effects of ethnically diverse crews?

These questions were seriously posed by a working group at NASA's Man-Systems Integration Standards Workshop at Johnson Space Center in January of 1988. The first and second authors of this paper subsequently undertook both a literature search and a cross-cultural survey of students at the first International Space University (ISU) in the summer of that year (Wise & McConville, 1989). This was followed by a second ISU survey (class of '89), and interviews with astronauts and cosmonauts attending the 5th Planetary Congress of the Association of Space Explorers in Riyadh during November of 1989, where 10 of the 19 foreign spacefarers were present. These surveys and interviews have been aimed at "scoping the field" of cultural diversity as it may pertain to space human factors, and producing some needed documentation for the considerable number of anecdotes concerning foreign crewmembers. This paper presents a summary of the results so far, organized in terms of the relevant habitability and operations issues.

Background

Ethnic differences have long been recognized in Human Factors, particularly as they pertain to anthropometrics of different populations. The recently completed NASA STD-3000 (1988) requires anthropometric accommodation for spacecrews from the 5th percentile Oriental female to the 95th percentile American male. In 1975, Chapanis edited an instructive volume illustrating how ethnic differences can affect other commonly accepted Human Factors results. Some of these occur on the level of 'control-display relationships', and involve behavioral stereotypies in the ways that members of different ethnic populations expect controls and their associated displays to respond. In the United States, for example, light switches turn up for 'on' and down for 'off', opposite to their action in the UK. Perhaps more problematic are sensors that may display the same results in either metric or English equivalents in different locations. This condition has already caused confusion in SpaceLab operations.

Wong and Lyman (1988) investigated cultural stereotypes that might affect the design of space station systems, and found that out of 24 tested control-display relationships, only one was responded to similarly and significantly by both Japanese and American subjects!

Such instrumental differences may also extend to the types of computer interfaces that control Expert Systems (Holland and Quinn, 1987). In the Artificial Intelligence field, one of the greatest hindrances is the elucidation of so-called "cultural knowledge" upon which presuppositions about the organization of the world is based. Such knowledge seems to define the structural characteristics of an Expert System's knowlege base, so that we should not expect astronauts from different cultures to access the system, seek information, and construe problems in the same ways. Yet the *Freedom* space station is planned to be critically dependent on different Expert Systems for its various maintenance and management tasks. What special characteristics of knowledge base construction and information search procedures must be encoded to ensure that these systems are equally useful to foreign as well as American astronauts?

Beyond the level of person-machine interaction, the aerospace research literature is almost nonexistent on cultural crew factors. In his pathfinding survey of 54 cosmo/astronauts on human communication in space, Kelly (1988) found only three prior articles that acknowledged the special language and communications problems of multicultural spacecrews.

Besides special interface and communications concerns within spacecrews and with Earth, there are many potential issues that are unique to multicultural crews. These divide themselves into two major categories of Habitability and Operations. Since none of these issues had been investigated before, the surveys and interviews of this project focused exclusively on habitability and operations questions above the person-machine interface level.

The ISU Surveys

The 1988 and '89 International Space University classes held at MIT and Strasbourg provided unprecedented opportunities to survey advanced students from 27 different countries who had spent nine weeks in a rigorous space studies program and intensive design exercises involving future space facilities. Many of these individuals were returning to roles in their countries' space agencies (or equivalents), and so were in both informed and privileged positions to comment on habitat design, operations and policy issues from their own cultural perspectives. Seventy percent of the surveys' respondents said they were either "committed" or "very interested" in going into space.

Both surveys were in the form of a questionnaire handed out to all students in the latter third of the summer session. Students completed the survey on their own time and returned them to an assigned mailbox before adjournment. From both '88 and '89 classes, there were 69 (35%) respondents(49 M and 20 F), of which 32 were North

American and 37 from other continents. Twenty-one different countries were represented in all.

For those questions that presented a rank order of options or ratings on a five point scale, the number of respondents to each option were multiplied by a factor (5-1, highest to lowest) indicating their ranked or rated level of preference. These products were then summed to create a rough overall desirability index (d.i.) for the option. Since there were about 70 respondents to each question, this means that a d.i. of approximately 350 is the highest score possible for an option, while a uniform distribution of rankings would generate a d.i.of approximately 210. If an option were always ranked last, it's d.i. would be aproximately 70. This index is only meant to be a shorthand indicant of an option's relative preference. No statistics were computed on d.i's.

Astronaut and Cosmonaut Interviews

All interviews were informally conducted by the first author at the Fifth Planetary Congress of the Association of Space Explorers, held in Riyadh, S.A., from Nov. 10-15, 1989. An interview schedule outlining the selected range of topics was first given to the respondent, who reviewed it prior to the interview. In deference to their wishes for privacy, none of the respondents will be uniquely identified as sources of specific information unless this was their explicit intent.

The remainder of this paper presents summary results of both ISU surveys and ASE interviews, organized by issue topics identified and abstracted from our study resources.

Food and Meal preparation.

Both U.S. and Soviet spaceprograms appear to deserve high marks for their attempts to satisfy crewmembers personal tastes in foods and to respond to culturally distinct food requirements. All foreign astronauts and cosmonauts interviewed indicated that they were able to take along selections of their own cultural cuisine on flights, except in one case which involved a visit to a Soviet crew going for an orbital endurance record, where there were concerns about upsetting their biological regimen. That flight was limited to new drinks only.

On U.S. flights there have been special food arrangements made for a foreign crewmember's religious proscriptions, and another's personal tastes that were well outside standard spacefare. These special meals appear to be met with good natured acceptance and benign curiousity by other crewmembers. Our overall impression at this point is that cultually distinct and even 'culturally themed' meals may be significant habitability enhancers on a long duration mission. There has already been success with such special meals on U.S. nuclear submarines.

The interviews echo the preferences of our survey respondents, where 40/69 or 58% indicated it was "important" or "very important" to have their own cultural cuisine available. The most preferred service style (d.i. = 232) was "cafeteria style with foods selected in portions from a serving station". The "NASA style of prepackaged foods in single serving portions" and "family style of foods served in one container and shared" were the least preferred options (d.i. = 173 and 171 respectively). Some survey respondents also strongly urged an occasional "space barbeque" while acknowledging its technical difficulties.

Alcohol and Drug Usage

Soviet and NASA policies agree in not allowing intoxicating beverages on spacecraft, but there have been exceptions. The U.S. has flown wine twice(to our knowledge) as part of religious Communion service. The much publicized wine that Patrick Baudry, the French 'spationaut', took along on STS Flight 51G in 1985 was not for his personal consumption. It was a vial of wine from the vineyards of a family friend, who received it unopened as a souvenir.

Informed sources relate that, infrequently, a very small bottle of cognac may find its way into resupplies for cosmonauts who are undergoing long duration stays in orbit. This is usually reserved for some special celebratory occasion, and is highly appreciated.

Our survey respondents (n=68) exhibited wide variability in their preferences, with the "U.K. Navy policy of limited alcohol rations/day" being a slightly desirable option (d.i.=217), and the least desirable (d.i.=117) being "alcohol only for religious purposes".

But the "no alcohol" policy option showed both the most 1st ranked preferences (n=27) and the most 5th ranked (least) preferences (n=25), showing the strong contrast of attitudes towards this policy. Where some alcohol usage was permissable, beer and wine were the preferred beverages (n=53 for each) with hard liquor additionally acceptable to 35 individuals. Only 3 of the 18 (17%) West Europeans responding to a proposed alcohol policy ranked the "no alcohol" policy first, whereas 10 of the 22 (45%) U.S. and 3 of the 9 (33%) Canadian respondents did so. Seven of 10 (70%) Middle East, Asiatic, South American, Indian and African respondents and all three East Europeans ranked "no alcohol" first.

The availability of "certain types of recreational drugs" was strongly rejected. Sixty-two of 69 (90%) respondents indicated a "no drugs allowed" policy, with some novel proposed sanctions for offenders!

Both interviews and survey responses show that there appears to be some more enlightened policy needed when it comes to alcoholic beverages in space. From included comments, the proposed policy of "limited rations/day" would have been much more popular if it had read "/week or month". While it was not unanimously expressed, several experienced spacefarers noted that an occasional glass of wine with meals, or a celebratory drink would be unlikely to hurt anything while being a certified morale enhancer.

From our perspective, it appears that the issue here is not so much having alcohol for its own sake, but rather having a socially sanctioned celebratory beverage that can be used to mark special times and occasions. Every culture has such drinks, and whether or not they contain alcohol, their value derives from their use in social ritual. If alcohol is to be banned or severely limited, some other types of social drinks must be substituted.

Personal Clothing and Interior Decor.

Present NASA and Soviet policies entail a uniform style of crew dress, although both permit flight crews to choose civilian styles of off-duty clothing. Each Particular NASA mission crew usually selects a style of shirt based on a common theme. While our survey respondents barely disapproved of the current policy of mission uniforms (d.i.=196), they also indicated that "personal design of uniform style" was near equally regarded (d.i.=204), and this had the greatest number of 1st rankings (n=19 of 62). Choice of clothing styles is clearly not a major issue with our survey respondents, nor was it ever indicated as an important personal decision in the interviews. But it may become a more important concern on long duration missions.

As for interior design and decor, our respondents (n=65) mildly felt that "decorations for shared spaces should be selected by a resident's committee" (d.i.=223) and that "individuals should be encouraged to add decor to shared spaces from their personal aesthetic views" (d.i.=213). Least popular was the view that "there should be a 'military style' approach" to decor (d.i.=98).

When asked to suggest interior design or decor options that crewmembers of their own cultural origin might prefer, respondents near unanimously asked for more natural and varied colors, plants, landscape pictures, and natural woods, regardless of their particular national origin. Human beings' love for nature and natural materials and forms, especially in high technology habitats, seems to transcend national boundaries. The Soviets extensive use of natural scenes and working gardens to maintain morale in their *Salyut* and *Mir* crews is already well documented, as was Skylab crews' appreciation of the pine-scented wipes they used to clean their spacesuit faceplates (See Wise and Rosenberg, 1988).

Sounds and Aromatics.

Both Soviet and U.S. spacecrews make considerable use of personally selected music tapes. This reflects the finding of our survey in that the most preferred background sounds were "personal selection of music in all areas" (d.i.=250) with "reproduced Earth sounds (wind, waves, etc.)" also scoring acceptably (d.i.=222). Least preferred were "sounds of life sustaining machinery (pumps, fans, etc.)" with a d.i. of 138.

The survey also asked about the acceptability of odors deliberately introduced into a lunar base. There was a high preference (d.i.=268) for "reproduced earth odors such as "candles, forest, farm, sea, rain", and to a lesser degree for cooking smells (d.i.=228). Only two individuals out of 69 rated reproduced Earth odors as unacceptable, and one of these reserved approval of "pine odor" specifically

These findings again mirror our interviews and published literature, where spacefarers of all countries appear to deeply miss the smells of their home planet.

Calendar and Religious Observances.

Different cultures and countries have different calendar events, sometimes reflecting contradictory or conflicting political and religious backgrounds. One wonders how an international crew might choose to observe the occurrences of Christmas, Rosh Hashanna, and Ramadan on the same mission, or perhaps May Day and the Fourth of July? From a set of eight policy options in our survey, the most preferred was a "common calendar representing a selection of events and record/timekeeping methods reflecting the different cultures" (d.i.=249). While "observance of one's own national and poliltical holidays" (d.i.=229) was slightly more preferable than "observance of political holidays from countries other than your own" (d.i.=199). Our interviewed spacefarers generally expressed an interest in observing a selected international calendar, particularly if it meant a few more holidays!

Although Sunday was considered the most common day set aside for religious observances (n=37 of 67), half as many (n=19) indicated no special day, and 4 showed greater than once per week. Twenty-eight asked for a separate area for religious activities, while another 5 simply indicated a quiet place or personal quarters. Twenty-seven felt no need for such a requirement. The majority felt that either once a week (n=21) or greater than once a week (n=19), there should be time scheduled for "reflection, meditation, or religious activity" in a space base. Approximately equal numbers (27 and 28) indicated that they did and did not utilize the services of a religious professional in their religious activities. Forty-six indicated they would be willing to share facilities and religious professionals with other denominational groups. The seven who showed reluctance in this regard were distributed across all cultural groups.

Meditation, prayer, and religious services of sorts have been practiced on both US and Soviet flights. On STS 51-G in 1985, Prince Sultan bin Salman Al-Saud became the first Muslim astronaut, maintaining his daily prayer schedule on the shuttle mid-deck, where fellow crewmembers "gave him whatever space was available for his needs".

Communications and Language.

The highest scores of the survey options occurred in this category, underscoring the importance of maintaining communications with friends and relatives on Earth.

They strongly felt (d.i.=320) that "all personal communication should be totally private and confidential", and that "personal communication should be voice and visual in real time" (d.i.=307). It was also desired that "the crew should be allowed to return home in emergencies" (d.i.=295), that "there should be scheduled visits home" (d.i.=279), and that "there should be personal packages from individuals on Earth" (d.i.=295). The policy option that "no personal information should be transmitted between Earth and the lunar base even in emergencies" was most unacceptable (d.i.=89). Restricting Earthside contact to "ground control" and "filtering" personal information was also quite unacceptable (d.i.=109).

Most (63 of 69) of the survey's respondents spoke two or more languages. But only 37 endorsed the option that there be "one chosen official language" for a base. If this was specified, it was unanimously English, or English and another tongue like French, German, or Russian.

The question of operationally approved languages is an important one, because it appears to be the root cause of a troubling incident that occurred on a Shuttle flight. While performing an on-board experiment with a foreign ground investigator, a foreign payload specialist found it necessary to communicate directly in the Earthside scientist's native language. Permission for this was denied, on the basis that the shuttle commander must know what is being communicated at all times. A ground translator was subsequently enlisted, although with much greater difficulty and time committment.

However, other foreign astronauts report that they have been able to engage in spontaneous, unrehearsed foreign language transmissions from the orbiter, when no one else aboard could have understood them. Foreign cosmonauts also communicate directly in their native tongue from Soviet spacecraft to ground stations, so this seems to be an acceptable practice. The incongruity of this one incident, then, is startling, and raises some question of whether the same proscription was enacted on other shuttle flights.

While recognizing the need for a single flight language, it seems that international crews will need more than one operational language when carrying on the diverse work within a space station or lunar base. Crewmembers must be able to communicate in their native tongues, both in personal and scientific transmissions, and a clear policy is needed to ensure this capability.

Real and Perceived Differences in Values and Attitudes.

Perhaps the most difficult cultural differences to characterize are those that deal with social values and attitudes. Yet it is just such potential conflicts of ethics and morals, preconceived cultural stereotypes and misattributions of meaning that are most deleterious to crew interaction in the long run.

Perceived differences do exist. In our surveys, 31 of the 37 non-North Americans reported they felt there were significant differences between persons of their culture and Americans in terms of nonsexual interpersonal touching, etiquette, speaking voice and distance, assertiveness or aggressiveness, and social and sexual values.

Despite perceived cultural differences, 48 of 61 respondents indicated that there was no cultural group from which they would not feel comfortable having a leader. The 13 who gave reservations here came from no particular cultural group(s), and mostly said their reluctance was based on perceived gender inequalities of certain cultures.

In this regard, we made a special effort in the ASE interviews to investigate a well known anecdotal story concerning the Saudi astronaut, Prince Sultan. Supposedly while on a KC-135 flight, he once declined to have a recording sensor placed on his body after it had been used on a woman crewmember. The supposed reason was attributed to his royal status. When one coauthor of this paper first heard this story, he felt it might rather involve a Moslem or Arabic restriction with regards to the genders. However, when inquiring about this alleged incident with Prince Sultan and other members of his shuttle flight, we discovered that *the incident had never occurred at all*, and that there was nothing in either his Islamic or cultural background that would ever require such a prohibition.

Like the widely misinterpreted and misreported "Frenchman's wine" of the same flight, we wonder whether this anecdote would have ever received the notoriety and prominence it had if the foreign astronaut had not been a member of a culture whose gender roles are widely misperceived and misunderstood in the Western world.

Herein lies the problem's crux in establishing international spacecrews: we all too easily think and act in terms of accepted cultural stereotypes based more on ignorance than understanding, and being technically well educated and capable in one's culture does not innoculate one against such cultural chauvinism.

Much work remains to be done in terms of training, crew selection procedures, and operations policies to ensure healthy, productive, and harmonious international spacecrews. We hope that the research reported here will assist this effort, and help develop space exploration as a cooperative endeavor of all nations.

REFERENCES

Chapanis, A., Ed. (1975). Ethnic Variables in Human Factors Engineering. Baltimore, Johns Hopkins University Press.

Connors, M.M., Harrison, A.A. & Akin, F.R. (1985) Living Aloft: Human Requirements for Extended Spaceflight. Wash.D.C.: NASA SP-483.

Holland, D. and Quinn, N., Eds. (1987) Cultural Models in Language and Thought. NY, Cambridge Univ.Press.

Kelly, A.D. (1988) Human Communication in Space. Unpub. Masters thesis, Stanford Univ., Palo Alto, CA.

National Aeronautics and Space Administration (1987) NASA-STD-3000 Man-System Integration Standards. Washington, D.C.: NASA.

The Planetary Society. (1987) The Mars Declaration. The Planetary Report, 7, (6).

Wise, James A. & Rosenberg, Erika. (1988) The Effects of Interior Treatments on Performance Stress in Three Types of Mental Tasks. CIFR Technical Report No: 002-02-1988.Grand Valley State University, Grand Rapids, MI. Submitted to Environment & Behavior.

Wong, C. & Lyman, J.(1988) American and Japanese Control Display Stereotypes: Possible Implications for Design of Space Station Systems. in Proc. Human Factors Society, 32nd Meeting, (pp.30-34) Santa Monica, CA: Human Factors Society.

Subsection 4.6: ARTIFICIAL GRAVITY

ARTIFICIAL GRAVITY RESEARCH FACILITY OPTIONS

Susan K. Rose, Timothy L. Stroup[1]
Lockheed Missiles & Space Co., Advanced Programs

ABSTRACT
On a long duration manned mission to Mars, the physiological changes caused by microgravity may be counteracted by artificial gravity. This paper evaluates several different classes of research options - a centrifuge, a free-flying animal facility, and a rotating manned spacecraft. Relative comparisons are made based on the initial constraints imposed on the facility, the operational restrictions for maintaining a healthy crew, based on research to date, and the science research requirements. The science requirements focus on the three primary physiological systems altered by microgravity - cardiovascular deconditioning, calcium loss, and muscle degradation. Significant design drivers, as well as high priority research areas and a recommended design approach are identified.

INTRODUCTION: GRAVITY
Humans are establishing a more permanent presence in space via the Soviet space station Mir and the developing Space Station Freedom. Advanced planning studies for lunar and Mars bases are underway. The scenarios of future manned planetary exploration involve extended duration exposure to microgravity during transit (up to two years for a Mars mission), extended duration habitation in low gravity environment (1/6 and 1/3 Earth's gravity) and exposure to high g-levels after the microgravity transit (up to 6 g's for an aerobrake return to earth).

Data from previous human space missions indicate that there are physiological changes that affect human performance and that could jeopardize mission success for a manned Mars mission. First, Soviet experiences of long duration stays in microgravity (one year or more) have shown that astronauts require weeks of physical rehabilitation once they return to earth's gravity environment (Atkov, 1989). Second, there may be physiological damage to the crew. Of the many known physiological changes induced by microgravity, three that are likely to have major impacts on a successful long duration manned Mars mission are cardiovascular deconditioning, calcium loss, and muscle degradation. Each of these could cause permanent physiological damage or impairment preventing the astronauts from performing their mission.

---

[1] O/53-51 B/580, LMSC, Sunnyvale, CA, 94088

There are many different options for minimizing the negative effects of microgravity on the human body: exercise, lower body pressure suit, drug therapy, and diet modification. Currently, exercise is most commonly used as a countermeasure. But, this is costly - approximately 4 crew-hours per day, 25% of the waking astronaut time. In addition, it has not been clearly shown that exercise eliminates the problems. It may, in fact, only slow the changes. Another option is to induce an artificial gravity analogous to that experienced on earth. Rotation induces centripetal acceleration - a radial acceleration within the rotating environment. This may load the physiological systems such that the degradation caused by microgravity does not occur. Unfortunately, there has only been minimal research of physiological systems in a rotating environment. There have been a few studies within ground-based centrifuges, most notably the large human centrifuge in Pensacola (Stone et al 1970) and rat research in a small centrifuge aboard the Kosmos 936 satellite. However, to understand the physiological impacts of an artificial gravity environment and then to be able to design a Mars vehicle, more research is required. This research must minimally address - is *there a configuration of radius, spin rate, and g-level in which humans could survive and perform their mission requirements on a manned mission to Mars*?

DESIGN PARAMETERS:SCOPE

For artificial gravity research there are three different classes of options based on the complexity of the facility: a centrifuge contained within an existing space station; a dedicated, free-flying, non-man-rated, rotating facility; and a dedicated, man-rated, rotating facility. Each option can be characterized and compared. based on initial restrictions, operational requirements, and science requirements. The initial restrictions imposed include low cost, short development time, and applied research directly applicable to a manned Mars mission. Maintenance of a healthy and productive crew is the only operational requirement considered. The science requirements consist of obtaining the data required to prepare for a manned Mars mission.

Cost and development time are difficult to quantify prior to a final design, but there are representative values. For the purposes of this paper, the mass to orbit (estimated ~$4400/kg to orbit), utilization of new technologies (examined qualitatively), and man-rated versus unmanned (the former raises the cost by approximately an order of magnitude and increases the development time of comparable vehicles by a factor of two (Stoewer 1989)) will be representative of the development cost. The moment of inertia and kinetic energy will be indicated as representative of the highest operational cost of such a facility. The time rate of change of kinetic energy is equal to the power required for each spin up and down of the facility.

DESIGN PARAMETERS: OPERATIONAL

Rotation induces centripetal acceleration, which is dependent only on radius(r) and spin rate(ω):

$$G \text{ (centripetal acceleration)} = \omega^2 r \text{ (Fig. 1)}.$$

This acceleration differs from the gravitational acceleration experienced here on earth, in that: 1)it can change in magnitude,

2)it will vary along the radius - and thus along the height of a specimen, 3)it will cause cross coupling effects with motion within the rotating environment: Coriolis force = 2ωxV (<=2ωV), where V is the local velocity within the rotating spacecraft.

There are few hard limits as to what is an acceptable artificial gravity environment. There exist many different optimal ranges for manned artificial gravity facilities (Table 1). Yet there is no concurrence as to the required radius, g-level, and spin rate. This is one of the primary reasons that artificial gravity research is required for the possible development of a rotating manned Mars vehicle.

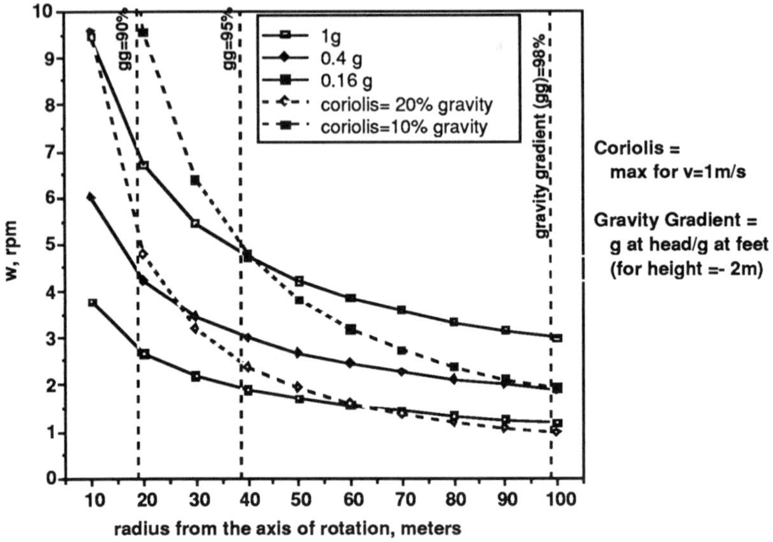

Fig.1 Artificial Gravity as a Function of Spin Rate and Radius w^2r

DESIGN PARAMETERS: BIOLOGICAL

On Earth, initial biological research in an emerging field is generally explored using animal models before examining humans. There is no reason to assume that artificial gravity research will progress any differently. Artificial gravity research on the three primary physiological systems affected by microgravity - cardiovascular deconditioning, muscle degradation, and calcium loss will study many of the same models that microgravity research uses. These animal models have been developed to varying degrees (Table 2). Other biological effects which may be of concern include immune responsiveness and hormonal changes. But because of limited data, it has not been determined if they are clinically significant (Steele 1989).

In microgravity cardiovascular research, rats have been used to study changes in heart biochemistry, changes in blood volume and sodium concentration, and the reduction of red blood cell mass (Morimoto et al 1987). Though it has not been unequivocally demonstrated that the rat is the proper model for cardiovascular artificial gravity research, there are data indicating that the rat's responses to microgravity are similar to those of humans. Blood volume researchers have also used dog, water immersion models to simulate microgravity (Miki et al 1987).

Microgravity exposure produces distinct, structural, metabolic and functional changes in the musculo-skeletal system. The Soviets found that a 7-day exposure was long enough to observe rat muscle atrophy (Gazenko et al 1987). Ground based microgravity research models using rats have included hindlimb suspension and whole body suspension (Steffen et al 1987). These have focused on muscle mass, protein levels, muscle unloading and nucleic acids. Calcium and bone loss studies have used rats, baboons and Rhesus monkeys to look at bone growth, formation, repair, metabolism and osteoporosis. Weightbearing bones have been compared with non-weightbearing bones in a microgravity environment (Vico et al 1987).

reference	spin rate (rpm)	g-level, g	radius, m	comments
Stone, 1970	**6**	0.6	16	"compromise between excessive radius and intolerable cross coupling" Pensacola Slow Rotation Room research
Thompson, 1965	**6 rpm**	0.58	14.4	"the lightest acceptable system for providing artificial gravity would be..." Slow Rotation Room research
Graybiel, 1960	**10 acceptable 6 recommended**			"... it is possible to adapt ... to velocities of 10rpm ... For levels below 10rpm, the problem becomes easier"
Galle, Yemel'yanov in Thompson,1989	**6.67**			" ... 6.67 rpm cannot be regarded as the maximum tolerable to man"
Loret, 1963	**4 max**	0.2-1g	54	max limit on gravity gradient=15%
NASA TM-4075, 1988	**6.5**	<=.64	(13.5)	
Wercinski,Searby, Tillman 1988	**0.4**	0-1.4	<=980	"A 5% tangential Coriolis force was calculated to determine the maximum radius and minimum spin rate"
Lemke, 1988	**1**	0-1.33	<=700	g gradient <0.01g/ft
I.S.U., 1989	**3**	<=1	100	
Schultz et al, 1987	**2**	1	223	

Table 1 Recommendations on Spin Rate, Radius, and G-level for Manned Artificial Gravity Research

From these experiments, some limitations on space based artificial gravity research techniques have been identified which impact study of the three physiological systems of concern. Most cardiovascular research measurements will need to be taken during or immediately after artificial gravity conditions. The samples cannot be brought back to Earth to be analyzed. For example, oxygen and carbon dioxide levels in blood degrade with in seconds after blood is withdrawn. Blood volume changes and shifts would occur within the time it takes to go from gravity to microgravity in the spin down of a rotating vehicle. In addition, because most of the instruments available to measure the cardiovascular changes are not currently automated, manual operation of the procedure is required. These limitations are not critical for most muscle degradation and calcium loss experiments. The muscle tissue or bone sample can be stabilized or fixed so that the pertinent measurements can be obtained in earth based laboratories.

Model	Type	Investigation	Ref.
CARDIOVASCULAR			
Rat	hind leg suspension	blood volume & sodium conc	Morimoto 1987
Dog	water immersion	blood volume	Miki 1986
Dog	water immersion	plasma volume increase	Miki 1987
Rat	suspension	red blood cell mass	Lange 1987
Rat	microgravity	cardiac deconditioning	Philpott 1985
BONE & CALCIUM			
Rat	microgravity	weight-bearing & non bones	Vico 1987
Rat	suspension	bone fracture repair	Sweeny 1985
Rat	pelvic suspension	bone formation	Turner 1985
Rat	suspension	bone metabolism	Halloran 1985
Rat	microgravity	incisor calcium levels	Rosenberg 1985
Rat	hindlimb unloading	bone formation	Bikle 1988
Baboon	in vivo	osteoporosis	Holt 1987
MUSCLE			
Rat	suspension	muscle mass, protein levels	Musacchia 1987
Rat	suspension	muscle atrophy	Tischer 1985
Rat	microgravity	morphological & biochemical muscle changes	Riley 1985
Rat	microgravity	muscle protein, RNA & DNA	Steffen 1985
Rat	body suspension	muscle unloading	Steffen 1987

Table 2 Animal models used in microgravity research

Other concerns that affect the design of an artificial gravity facility include the duration of each experiment, number of subjects, and gravity level. From a scientific perspective, the more subjects available for testing the reliable the data. Long duration experiments, from six months up to one year, are very important in understanding how artificial gravity affects the primary physiological systems. This length of time is needed to determine if there are any unexpected long term adverse effects of artificial gravity. The number of data points or samples taken, also needs to be statistically significant. This will enable scientists to obtain an understanding of physiological effects on the subjects. Also each experiment needs to be performed more than once to be sure accurate results are obtained.

DESIGN OPTIONS

There are three classes of design options for artificial gravity research based on the complexity of the facility - centrifuges, non-man-rated free-flying facilities, and manned free-flying facilities (Fig. 2).

system	radius,m	mass,kg	a=payload surface area, m^2 v=payload volume, m^3		approx. moments of inertia, kgm^2	conditions for 1 g*
Centrifuge						
COSMOS  (KOSMOS 936: Atkov)	0.5	-	~3	~0.3	-	w=42 cf=8.86
Space Station Centrifuge 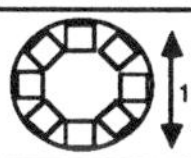 1.8 m (LMSC proposal, 1989)	~0.9	~600	~2.26 (2prh)	~0.7		w=31.5 gg=10.89 cf=6.6
Free Flying						
LifeSat spinning on axis 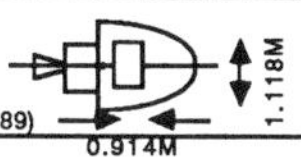 1.118M 0.914M (LMSC proposal, 1989)	0.45	1271 (@launch)	1.02	1.14	Ixx=Iyy=465.2 Izz=435.4	w=40.01 gg=17.56 cf=8.38
Space Station hab module rotating about axis 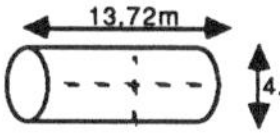 13.72m 4.57m	2.27 (transverse axis)	~22700	80.60	225.05	Ixx=Iyy=365,942 Izz=104,424.5	w=19.86 gg=4.33 cf=4.16
	6.96 (minor axis)					w=11.36 gg=1.42 cf=2.38
lifesat on variable length tether 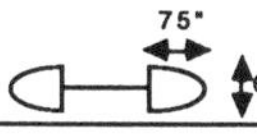 75" 61.8"	variable	2542	~2.04	~2.28	Ixx=Iyy=465.2 Izz=870.8	w>= 21.0 gg>4.84 cf=4.4
Manned Facility						
external tank rotating about axis 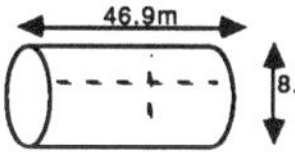 46.9m 8.4m	4.2 (transverse axis)	35425	506.38	2599.09	Ixx = Iyy = 41,819,040.5 Izz= 51,223,26	w=14.51 gg=2.31 cf=3.04
	23.45 (minor axis)		55.42 at any r	2599.09		
VGRF 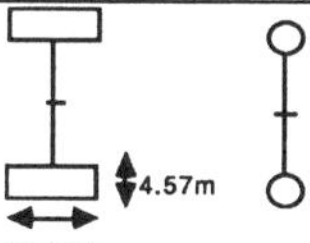 4.57m 13.72m I.S.U NEWTON facility, 1989	variable - max between 18 and 980m	~235,000		620.3	I_{xx} =2204x10^6 I_{yy} =58x10^6 I_{zz} =2209x10^6	w=0.95-6.97 cf=0.2-1.46

* FOR 1G:
w=spin rate(rpm); gg=gravity gradient(m/s^2/m); cf=coriolis force (m/s^2)-the maximum for v=1m/s

Figure 2 Design Option Classes for Artificial Gravity Options

To date there has been centrifuge research performed on earth, a centrifuge has flown aboard Kosmos 936, and a centrifuge is planned for the Space Station Freedom. The ground based research is limited as the radial acceleration is imposed over earth's gravity and as the radius is limited in total length by physical considerations. These are the reasons for moving into space based artificial gravity research. A space based centrifuge is considered the most simple of the three classes of facility options. It is contained within a larger host vehicle from which it can draw supplies - power and thermal control, greatly simplifying its own design. This also puts restrictions on the facility in dimension and duration of experimentation. Centrifuges are generally used for animal research and require a high spin rate for the nominal 1g environment. There do exist designs of small radius manned centrifuges, but current research indicates that dynamic loading is required to eliminate the negative effects of micro-gravity. This approach is thus not considered here as a viable design solution. The high spin rate centrifuges in turn may create excessive gravity gradient and Coriolis forces (within a 1.8 m centrifuge at 1g, a velocity of 1m/s induces an acceleration 6.6 newtons - perpendicular and 67% the magnitude of the radial acceleration). Long duration experiments and the removal of samples without changing the spin rate/ g-level are possible, but may be limited by physical constraints of the primary spacecraft.

A free-flying animal facility has the possibility for longer radii and thus is capable of providing a better environment with lower Coriolis Force and gravity gradient. In addition, provided autonomous life support can be supplied, long duration research is possible. Such a facility is a complete spacecraft and will require extensive development time. An unmanned facility would allow very limited animal research. Complex automated instrumentation would be required, as well as frequent crew visits to maintain equipment and remove animals to be studied after return to earth.

A large manned facility provides the means to study human subjects within a rotating environment. Such a facility is ideal for studying man for the transit to Mars. Such a facility is capable of optimizing all the operational and science restrictions - large radius, slow spin rate, continuous rotation, access to samples without changing the spin rate, human research. However, it is a very complex facility and may take ten times longer to develop (Stoewer 1989) and as much as one hundred times the cost of a similar unmanned facility.

CONCLUSIONS

If artificial gravity is to be considered a viable option for a manned mission to Mars, a research facility is required. Within current scenarios (NASA-TM-4075 1988), inconsistencies exist in the phasing of design and development of a man-rated, artificial gravity, Mars transit vehicle. A key problem is that the operational limits are not known. This is the primary objective of a rotating research facility - to determine a combination of spin rate, g-level, and radius in which humans could travel to Mars.

Certain general design drivers arise from the science requirements:

- All primary physiological subsystems can be examined using animal models
- A Mars mission will require long duration experimentation
- The musculo-skeletal and calcium subsystems can be examined after return to earth. The cardiovascular subsystem must be examined immediately.

Directed research addressing these questions would accommodate a large number of human and animal subjects in an artificial gravity environment for long durations. The optimal research tool would be a large, variable radius, variable gravity facility capable of housing both man and animals, with access while rotating. For three reasons the reality of such a facility is questionable. First, such a research facility is more complex than Space Station Freedom and thus requires more money and time to develop. Second, if such an effort were undertaken just after the operations of Freedom are started (1997), optimistically it would be 2008 before this facility was operational. Within any of the current schedules for a manned mission to Mars this is too late to begin design of the Mars vehicle. Third, the development of this vehicle would be in parallel with Freedom operations and (potentially) a lunar mission. For these reasons, smaller and less costly facility designs must also be examined as possible options to gain the knowledge to send humans to Mars.

Analogous to the manned facility, but smaller in scope is a free-flying animal facility. Such a facility has the potential of being an ideal research environment. However, by definition, there is no human research and the animal research that can be performed is greatly limited by the lack of access to the facility by researchers.

Centrifuge research has already begun aimed at addressing the primary concerns of the physiology involved in an unknown gravitational environment. A centrifuge is capable of performing all of the required research, except that involving humans, at lower cost and development time. A centrifuge with animal research can provide the necessary information to start defining the physiological boundaries imposed by artificial gravity and enough understanding to plan to send humans to Mars.

REFERENCES

Atkov, O., International Space University lecture Summer 1989.

Bikle, D. and B.P. Halloran (1988), "Effect of Skeletal Unloading on Bone Formation", 1987-1988 NASA Space/Gravitational Biology Accomplishments, T.W. Halstead Ed., NASA TM-4079, pg 101-103.

Gazenko, O.G., E.A. Ilyin, E.A. Savina, L.V. Serova, A.S. Kaplansky, I.A. Popova, V.S. Oganov, K.V. Smirnov, and I.V. Konstantinova (1987), "Study of the Initial Period of Adaptation to Microgravity in the Rat Experiment Onboard Kosmos-1667", The Physiologist 30(1): S-53 (Suppl.).

Graybiel, A., B. Clark, J. Zarriello (1960), "Observations on Human Subjects Living in a 'Slow Rotation Room' for Periods of Two Days". AMA Archives of Neurology, Vol. 3, p.55-73.

Halloran, B.P., D.D. Bikle, T.J. Wronski, R.K. Globus, M.J. Levens, and E. Morey-Holton (1985), "Effect of Simulated Weightlessness and Chronic 1,25-Dihydroxyvitamin D Administration on Bone Metabolism", The Physiologist 28(6): S-127 (Suppl.).

Holt, G.R., T.B. Aufdemorte, P.J. Sheridan, R.G. Triplett, B.F. Boyce, and G.R. Mundy (1987), "IN VIVO Biocompatible Implant Systems for Studies of Bone Pathophysiology under Earth Gravity and Microgravity Conditions", Three Decades of Life Science Research in Space, Abstracts, pg 88.

International Space University, Summer 1989 NEWTON design project.

Lange, R.D., J.B. Jones, and P.C. Johnson, Jr. (1987), "Comparative Aspects of Hemotological Responses in Animal and Human Models in Simulations of Weightlessness and Space Flight", The Physiologist 30(1): S-113 (Suppl.).

Lemke, L.G. (1988), "An Artificial Gravity Research Facility for Life Sciences", 18th Intersociety Conference on Environmental Systems, San Francisco, CA, July 11-13, 1988.

Lockheed Missiles & Space,Co. proposal for Space Station 1.8 m centrifuge, (1989).

Lockheed Missiles & Space,Co. LifeSat proposal, (1989).

Loret, B.J. (1963), "Optimization of Space Vehicle Design with Respect to Artificial Gravity". Aerospace Medicine, 34:430-441, .

Miki, K., G. Hajduczok, S.K. Hong and J.A. Krasney (1986), "Plasma Volume Changes During Head-out Water Immersion in Conscious Dogs", Am. J. Physiol. 251: R 582.

Miki, K. S.K. Hong, and J.A. Krasney (1987), "Mechanism of the Increase in Plasma Volume During Head-Out Water Immersion (WI) in Dogs", The Physiologist 30(1): S-100 (Suppl.).

Morimoto, T., E. Sugimoto, H. Nose, T. Okuno, T. Natsuyma, and M. Morita (1987), "Continuous Determination of Blood Volume and Blood Sodium Concentration on Conscious Rats: A Potential Tool For the Analysis of Water Balance During Weightlessness, The Physiologist 30(1): S-98 (Suppl.).

Musacchia, X.J., J.M. Steffen, R.D. Fell, and J. Dombrowski (1987), "Physiological Comparison of Rat Muscle in Body Suspension and Weightlessness", The Physiologist 30(1):S-102 (Suppl.).

NASA TM 4075 (1988), Exploration Studies Technical Report, Vol. II: Study Approach and Results, Office of Exploration.

Philpott, D.E., A. Fine, K. Kato, R. Egnor, L. Cheng, and M. Mednieks (1985), "Microgravity Changes in Heart Structure and Cyclic-AMP Metabolism", The Physiologist 28(6): S-209 (Suppl.).

Riley, D.A., S. Ellis, G.R. Slocum, T. Satyanarayana, J.L.W. Bain, and F.R. Sedlak (1985), " Morphological and Biochemical Changes in Soleus and Extensor Digitorum Longus Muscles of Rats Orbited in Spacelab-3", The Physiologist 28(6): S-207 (Suppl.).

Rosenberg, R.D., and D.J. Simmons (1985), "Electron Microprobe Analysis of Ca, S, Mg, and P Distribution in the Incisors of Spacelab-3 Rats.", The Physiologist 28(6): S-189 (Suppl.).

Schultz, D.N., C.C. Rupp, G.A. Hajos, and J.M. Butler Jr.(1987), "A Manned Mars Artificial Vehicle", Space Tethers for Science in the Space Station Era, pgs. 320-335.

Steele, K.J., (1989),"Zero Gravity and the Immune System: A Challenge to Man's Survival in Space", Analog Vol. 59 (4): 36.

Steffen, J.M. and X.J. Musacchia (1985),"Effect of Seven Days of Spaceflight on Hindlimb Muscle Protein, RNA and DNA in Adult Rats", The Physiologist 28(6): S-221 (Suppl.).

Steffen, J.M., R.D. Fell, and X.J. Musacchia (1987), "Physiological Responses During Whole Body Suspension of Adult Rats", The Physiologist 30(1): S-94 (Suppl.).

Stoewer, H.(1989), International Space University lecture, Summer 1989.

Stone, R.W., Jr., W.M. Piland, W. Letko (1970), "Certain Aspects of Onboard Centrifuges and Artificial Gravity". Fourth Symposium on the Role of the Vestibular Organs on Space Exploration. (Naval Aerospace Medical Institute, Pensacola, Florida). Office of Technology Utilization, NASA, p.331-345.

Sweeny, J.R., H.E. Gruber, M.E. Kirchen, and G.J. Marshall (1985), "Effects of Non Weight Bearing on Callus Formation", The Physiologist 28(6): S-63 (Suppl.).

Thompson, A.B. (1965), "Physiological Design Criteria for Artificial Gravity Environments in Manned Space Systems", presented at the Symposium on the Role of Vestibular Organs in the Exploration of Space, U.S. Naval School of Aviation Medicine, Pensacola, Fla, 20-22 Jan 1965.

Thompson,A.B. (1989), "Some Notes on Maximum Rotational Rate in Obtaining Centrifugal Artificial Gravity", unpublished Martin Marietta document.

Tischer, M.E., S.R. Jaspers, E.J. Henriksen and S. Jacob (1985), "Responses of Skeletal Muscle to Unloading - A Review", The Physiologist 28(6): S-13 (Suppl.).

Turner, R.T., G.K. Wakley, and B.W. Szukalski (1985), "Effects of Gravitational and Muscular Loading on Bone Formation in Growing Rats", The Physiologist 28(6): S-67 (Suppl.).

Vico, L., D. Chappard, A.V. Bakulin, V.E. Novikov, and C. Alexandre (1987), "Effects of 7-Day Space Flight on Weight-bearing and Non-Weight-bearing Bones In Rats (Kosmos 1667)", The Physiologist 30(1): S-45 (Suppl.).

Wercinski, P.F., N.D. Searby, B.W. Tillman (1988), "Space Artificial Gravity Facilities: An Approach to Their Construction", Engineering, Construction, and Operations in Space: Proceedings of Space 88, pgs. 734-749.

Variable Gravity Research Facility: A Concept

Paul F. Wercinski, Marcie A. Smith,
Robert G. Synnestvedt, and Robert G. Keller

Abstract

One of the most exciting missions NASA is currently studying is a human mission to Mars. Mission scenarios range from 20 day surface expeditions to permanent human bases on the martian surface. A mission to Mars, however, regardless of the length of stay on the surface, involves interplanetary trip times lasting over one year. Although humans have lived in spacecraft in Earth orbit for long durations, there are fundamental engineering and life sciences issues yet to be resolved in supporting an interplanetary manned mission. One such issue concerns artificial gravity: Is human exposure to artificial gravity necessary for the mission success, and if so, what is the optimum means of achieving artificial gravity? Answering these questions prior to the design of a Mars vehicle would require construction and operation of a dedicated spacecraft in low Earth orbit. This paper summarizes the study results of a conceptual design and operations scenario for such a spacecraft, called the Variable Gravity Research Facility (VGRF).

Introduction

Background

That gravity plays a role as a fundamental stimulus in the development of humans is evidenced by the physiological changes which develop in the absence of gravitational acceleration. Physiological systems which counteract gravity on Earth, such as muscles and bones, atrophy in the absence of gravitational acceleration, making normal performance difficult when humans return to Earth. Some deleterious effects resulting from long-term exposure to zero-gravity include: fluid loss, loss of muscle mass, cardiovascular deconditioning, and bone mineral loss.

Space Exploration Projects Office, NASA Ames Research Center,
Moffett Field, CA 94035

For over thirty years, scientists and planners have considered the development of a rotating manned habitat in space (Von Braun, 1952; Kramer et al, 1960; Loret,1963; NASA TR703,1971). One configuration for a manned centrifuge in orbit would create artificial gravity by rotating a habitat about a spin axis. The artificial gravity is a function of both the rotation rate and radius. The "floor" of this spacecraft is the habitat surface farthest from the spin axis on which the occupants in the habitat experience centripetal acceleration, or artificial gravity.

Artificial gravity does not entirely replicate true gravity and has unique accelerations which effect human performance in the artificial gravity environment. One effect is a centrifugal acceleration gradient which effects the human body fluid distribution and changes the weight of an object moved radially (Figure 1a). Another phenomenon associated with an artificial gravity environment is cross-coupled angular acceleration due to rotations, such as head motion about axes which are not parallel to the spin axis (Figure 1b). Motion of objects within a rotating environment can cause Coriolis accelerations, which are proportional to the velocity of the object and the vehicle rotation rate. Radial translations produce a Coriolis acceleration tangential to the spin axis. This effects walking, climbing "up" or "down", and the location where "dropped" objects fall (Fig. 1c). Translations tangential to the spin axis produce a Coriolis acceleration in the radial direction perpendicular to the spin axis. This force makes the weight of an object or a human lighter or heavier depending on the direction of movement relative to the direction of rotation (Fig. 1d).

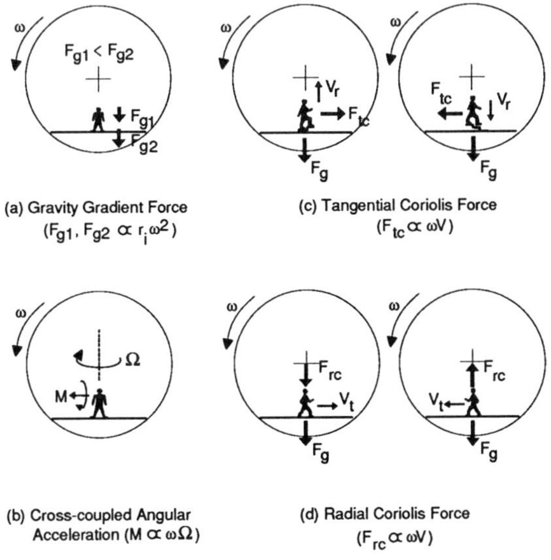

Figure 1. Artificial Gravity Related Forces and Accelerations

These variations in acceleration may adversely effect crew member motor performance and comfort. To understand these effects on the crew, artificial gravity should be induced over a range of radii and rotation rates. Therefore, the VGRF should be capable of providing a means to independently vary the radius and the rotation rate in order to allow investigation into the relationship between artificial gravity and human performance.

Study Initiative

The study summarized in this paper investigated the engineering and operational feasibility of an artificial gravity facility for humans in Earth orbit. The facility configuration under investigation is a manned habitat tethered to a counterweight. The facility is spun around the facility center of mass to create centrifugal acceleration which simulates artificial gravity at the habitat. The facility varies the spin rate by firing thrusters and varies the radius of rotation by changing the tether length between the habitat and counterweight.

The primary purpose of the facility is to study human performance, physiology, and comfort under various acceleration loads (g-levels) and rotation rates. The specific study objectives are to:

1. Develop requirements for a free-flying, tether connected, variable radius facility to perform artificial gravity research on humans.
2. Describe conceptual designs of such a facility, including the alternative of a passive counterweight.
3. Describe an operations plan for this facility.

Key configuration issues for the facility were specified as follows:

1. Nominal crew size is to be 3 humans.
2. Launch date is to be 1997, with a design lifetime of 3 years.
3. The facility will provide 1 g acceleration at the habitat at a rotation rate of 1 rpm, with operation possible at 2 g at the maximum radius.
4. Operational cycling will consist of 90 day missions, each mission consisting of up to five start/stop cycles.
5. The radius of rotation will be controllable during rotation.
6. Launch vehicle will be the National Space Transportation System (NSTS), or Shuttle.

Discussion

Many configurations for an artificial gravity spacecraft are possible. A "dumbbell configuration" with two end masses, a habitat and a counterweight, appears to be the most cost-effective and simplest design for a manned, Earth orbiting, variable gravity research facility and is discussed herein.

Habitat Module Options

For the VGRF, a standard Space Station Freedom (SSF) module can be outfitted to house a crew of three, however, there is no additional rack space for non-human research. It is estimated that 6 SSF double racks would be required for non-human gravitational biology research. Unless much of the equipment required for spacecraft operations is moved from inside the habitat module, volume would have to be added to the VGRF habitation module in order to accommodate both human and non-human research.

Different habitat module options were investigated in order to accommodate the need for additional pressurized volume including: 1) a single standard SSF habitat module, 2) two smaller habitat modules docked together, 3) a single, extended habitat module, 4) a standard habitat module with an attached node.

Selecting a standard SSF habitat module as the VGRF habitat reduces the development cost by using developed hardware which meets constraints imposed on the VGRF by the Shuttle launch vehicle. A detailed drawing of the habitat with the strongback truss structure, used to support externally mounted subsystems, is shown in Figure 2.

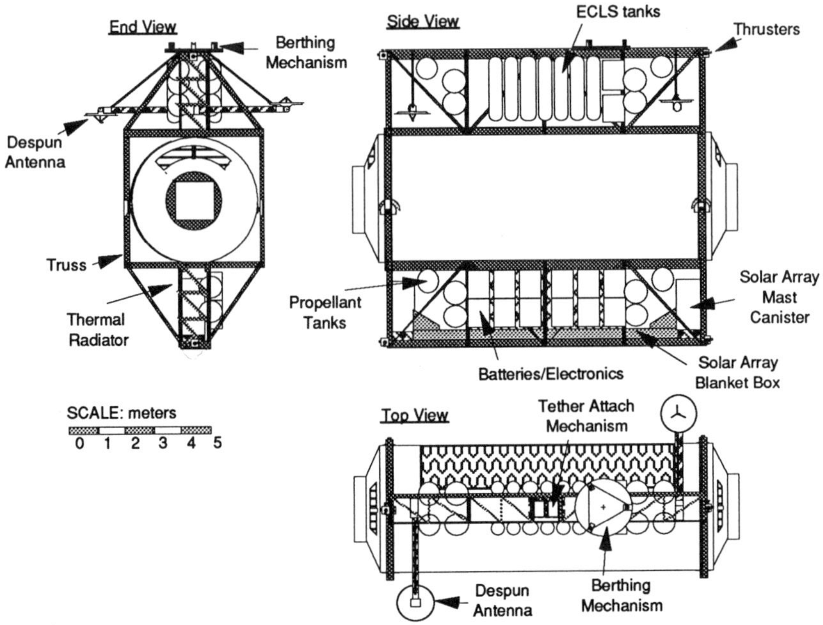

Figure 2. VGRF Habitat Module and Strongback Truss Structure

Counterweight Options

This study investigated three different types of counterweight options, shown in Figure 3, to determine which option best met design criteria for mass, operations, and mission suitability. The first option, an active module, performs all the propulsive maneuvers for spin up and spin down of the VGRF and controls tether deployment and retrieval. The second option is a "dead weight" launched into Earth orbit for another purpose, but capable, with modifications, to be used as a counterweight. The third option uses a second habitat module as a counterweight. In general, assembling the VGRF may require Extra-vehicular Activity (EVA) although the facility will be designed for assembly by the SSF or Shuttle Remote Manipulator System (RMS). The three counterweight options were driven by the assumption that everything the crew requires for power, life support, and communications is located at the habitat.

The active module option uses a derivative from an Orbital Maneuvering Vehicle (OMV) as the propulsion system for the counterweight for start/stop spin maneuvers. Propellant tanks are augmented with a higher capacity fuel capability and the OMV-derived module is connected to an Unpressurized Logistics Carrier (ULC) which houses the tether and tether reel device. The active module requires the longest tether of all the configurations, over 4000 m, when the counterweight is nearly empty of propellant. The disadvantages of the active counterweight are the vehicle complexity due to the necessity for complete subsystems of power, attitude control, communications, and propulsion system.

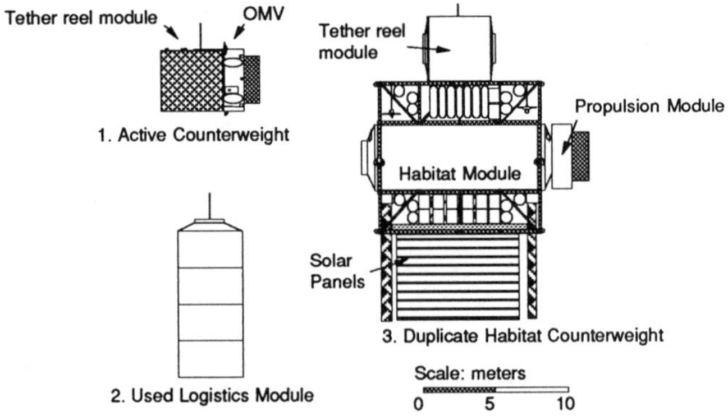

Figure 3. Counterweight Options

The second option, the "dead weight", has the advantage of using a "free" mass already in orbit to reduce VGRF launch mass. The ideal dead weight counterweight for the VGRF is a small massive object that can be easily attached to a tether and maneuvered around the Station and Shuttle. For the purpose of this study, a used Pressurized Logistics Carrier (PLC) from the SSF program was studied as a candidate dead weight.

The third option, a habitat counterweight, has the advantage of increasing the habitable volume of the VGRF and minimizing the tether length. However, disadvantages of the habitat counterweight are the added cost of hardware, the additional mass, and the operation complexity of resupplying two separated manned vehicles. To resupply each habitat, the Shuttle may have to dock separately with each module, increasing the complexity of the resupply procedure.

The active module was selected as the baseline counterweight for this study, due to the disadvantages of the other options. Using the habitat counterweight was rejected due to the cost of the extra mass to orbit and the PLC dead weight was not chosen because of the great cost in additional propellant required. Further study will be required to verify the operational capability of the OMV as a system in a hyper-gravity environment and the on-orbit lifetime of the OMV relevant to the VGRF mission.

VGRF Conceptual Design

An illustration of the VGRF in its fully outfitted, docked configuration is shown in Figure 4. The tradeoff studies associated with the subsystems used in the VGRF conceptual design included propulsion, tether, tether reel device, power, life support, communications, attitude control, and other secondary systems. A summary description of all the primary VGRF subsystems is given in Table 1. Trade studies for two of the key subsystems, propulsion and the tether, are described next.

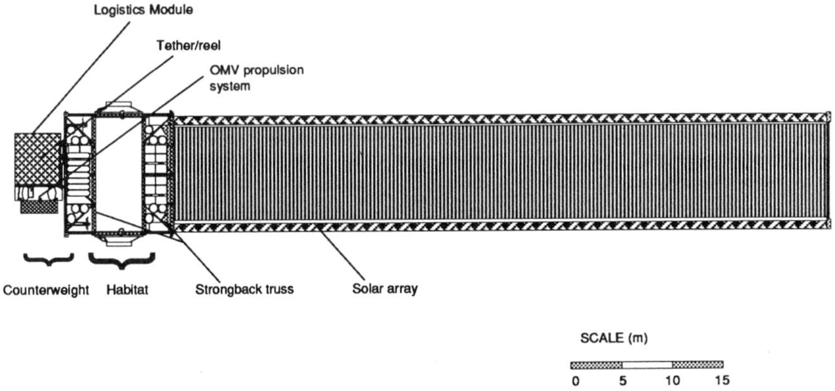

Figure 4. VGRF with habitat and counterweight docked

Table 1. VGRF Subsystem Description

COUNTERWEIGHT		HABITAT	
Mass (Dry)	9736 kg	Mass (Dry)	38938 kg
(Wet)	22000 kg	(Wet)	41100 kg
SYSTEMS		**SYSTEMS**	
Propulsion		Propulsion	
Bi-prop propellant	12250 kg	Bi-prop propellant	2150 kg
Thrusters (4)	580 N	Thrusters (16)	100 N
Power		Power	
Body mounted solar cells area	19.0 m^2	Deployed solar array	620 m^2
Power output	1.8 kW	Power output (peak)	30 kW
Ni-H2 batteries	45.6 A-hr	Power output (nominal)	21 kW
		Ni-H2 batteries	480 A-hr
Tether (Kevlar)		Attitude Control	
Cross-section	8.76 cm^2	Primary pitch control:	tether attachment point
Length	4100 m	Primary yaw control:	reaction wheels (4)
Mass	5166 kg	Secondary roll, pitch, yaw:	thrusters
Attitude Control		Life Support (90 day supply)	
Primary pitch control:	tether attachment point	Air	540 kg
Primary yaw control:	reaction wheels (2)	Water	2650 kg
Secondary roll, pitch, yaw:	thrusters	Food	490 kg
Communications		Communications	
Omni antenna (2)		Omni antenna (4)	
Parabolic (1.0 m), gimballed, despun antenna (2)		Parabolic (1.25 m), gimballed, despun antenna (2)	

When estimating the size and scope of the propulsion system, the start-stop cycle ΔV's, had the most significant impact in sizing the propulsion system engines and propellant tanks. A parametric study was performed to analyze the propellant requirements for a given set of VGRF mass properties over a range of g-levels and spin rates. At the beginning of a mission, the counterweight is full of propellant. As propellant is expended, the counterweight loses mass, and thus it must rotate at an increasing distance from the center of mass of the total VGRF system. As the counterweight becomes lighter and the tether longer, two opposing effects occur. First, the thrusting moment arm of the counterweight increases so the thrusters can impart a greater change in angular momentum of the system. Second, the greater radius increases the rotational moment of inertia of the system and works against the higher thruster capability. The VGRF will be capable of spinning up to 1 g/2 rpm five times, or to 1 g/1 rpm three times, or to 2 g/1.4 rpm only one time, in a given mission.

The tether design was another important subsystem in the VGRF design. The baseline design for the tether does not accommodate power transmission or data transfer between the habitat and counterweight modules. These features were not added to the tether because of the added weight and loss of flexibility to the tether. The cross-sectional area of the tether is determined by the ratio of maximum load to tensile strength. For the VGRF, the maximum load occurs when the habitat rotates at 2 g, providing a load of 806 kN. The area is increased by the safety factor, which for this study, was set at three. Kevlar was chosen as the tether material with a cross-sectional area of 8.76 cm^2 and density of 1.26 kg/m fulfilling the tether design requirements.

Launch/Assembly Operations

The VGRF will be launched in four launches on the Shuttle and assembled at SSF. For on orbit assembly at SSF, the first VGRF Shuttle docks to SSF and unloads the habitat module using the Shuttle RMS. Once the VGRF is docked with SSF, the astronauts can pressurize the facility and verify the subsystems. The second Shuttle launch delivers the truss structure and the remaining equipment to attach to the exterior of the habitat. The astronauts will assemble the truss structure and attach it to the habitat module using the SSF RMS or, if necessary, with EVA. The third Shuttle launch brings the counterweight module, to be docked with the habitat module. The fourth Shuttle launch carries a pressurized logistics module with the remaining equipment for the habitat module. Figure 5 shows the four Shuttle payload launch configurations.

Flight	Side View	End View	Main Elements	Mass (kg)
1	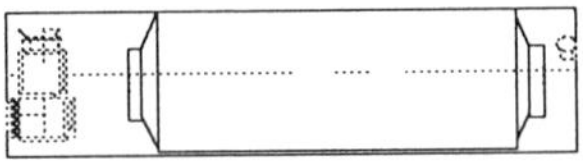		Hab module	17,600
2	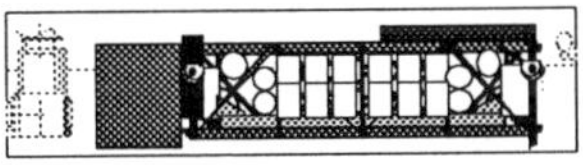		Strongback, Power system, Propellant	17,600
3	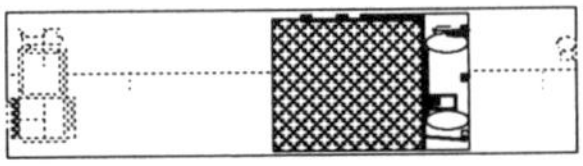		Counterweight	14,900
4	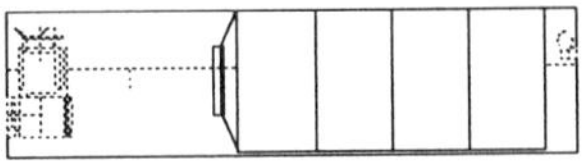		PLC, Outfitting, Crew	17,200

Figure 5. VGRF Launch Manifest

When the assembly and system verification are complete, the VGRF will separate from the SSF. Initiating its maiden check-out mission, the VGRF will deploy its solar panels and test the power, propulsion, and attitude control systems. The counterweight and habitat module will separate while the tether deploys to its longest length, stabilized in a gravity gradient orientation. The counterweight communications link will be verified and the attitude determination systems in the deployed tethered configurations will be tested. After initial system operations have been verified, they must be verified again while the facility is rotating. Initially, a very slow rotation rate and g-level will be used. The facility will then be run through a test profile of gravity levels and rotation rates to thoroughly test each subsystem in all configurations. At the end of the verification, the facility will be ready for the first crew of "subjects" for the artificial gravity experiments. The facility will despin and the tether will retract for docking with the Shuttle.

Nominal Operations

After being resupplied by the shuttle, the facility is ready to start a 90 day mission. After the shuttle departs, the facility is in the docked configuration, the habitat and counterweight are docked together, with a gravity gradient stabilized attitude corrected to allow power generation from the solar panels. The astronauts check the habitat module for loose articles and lock down all moving parts. Using thrusters, the facility is spun to a very low rotation rate about the desired spin axis orientation. The spin axis direction is chosen so that solar incidence on the solar panels is maximum in the middle of a given mission. Once the facility starts rotating, the tether is gradually deployed. As the tether deploys, the spin rate decreases and the propulsion module must apply thrust to maintain a minimum tension in the tether. The spin rate and radius are increased until the final radius/rotation rate combination is achieved.

Resupply and Crew Exchange

Prior to Shuttle resupply, the VGRF habitat and counterweight are docked and orient to the gravity gradient attitude. For docking with the Shuttle, however, the VGRF will need to reorient to allow the Shuttle to dock in a tail-down attitude (Figure 6). After docking, the crew transfers VGRF supplies from the logistics module in the Shuttle bay to the VGRF. Then, VGRF waste is loaded into the Shuttle logistics module. The RMS on the Shuttle is used to replace the VGRF propulsion module with a full tank. The crew is then exchanged, and the Shuttle can depart.

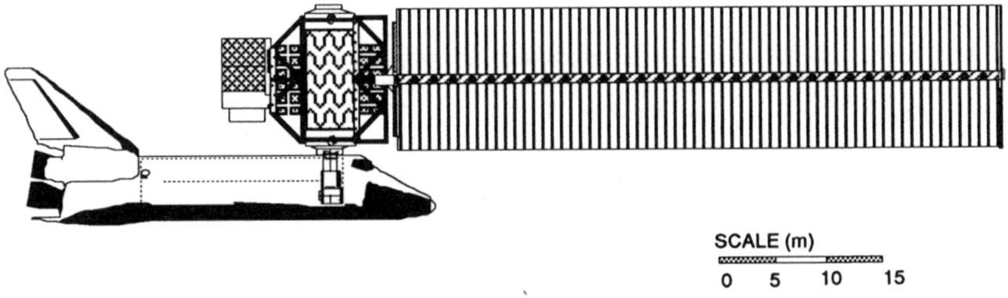

Figure 6. VGRF docked to Shuttle

Summary

This paper has summarized a spacecraft design and operations study for a Variable Gravity Research Facility. The VGRF is designed to accommodate a crew of 3 for 90 day missions with resupply to be conducted by the Shuttle. With a launch date in 1997 and a design lifetime of 3 years, the VGRF would provide the means to answer fundamental questions concerning the effect of artificial gravity on humans in space from a physiological and habitability perspective. With this knowledge, the possible implementation of artificial gravity can be assessed for long duration manned missions.

References

1. "Artificial Gravity Space Station Physiological Effects and Design Criteria", NASA Project Technical Report Task 703, 1971.

2. KRAMER, S. B., and BYERS, R. A.: A Modular Concept for a Multi-Manned Space Station. In "Proceedings of the Manned Space Stations symposium, Institute of the Aeronautical Sciences, New York, 1960.

3. LORET, B. J.: "Optimization of Space Vehicle Design with Respect to Artificial Gravity", Aerospace Medicine, Vol. 34, May 1963

4. VON BRAUN, W.,"Crossing the Last Frontier",Collier Magazine, March 22, 1952

ARTIFICIAL GRAVITY:

Human Factors Design Requirements

Stephen D. Capps*

Abstract

A major challenge of manned spaceflight is to assure the health, well-being and performance of the space inhabitants. As man ventures beyond Earth's orbit to settle the Moon, explore and settle Mars and its moons, and beyond, the issue of prolonged exposure to weightlessness on man becomes extremely important. This outward progression presents designers with the challenge of creating environments conducive to the average human as well as creating the means of human survival for extended spaceflight and settlement.

Introduction

If, from a human factors standpoint, we assume artificial gravity to be either necessary or highly desirable, many factors must be considered in designing a vehicle. The environment created by spinning a vehicle is very different than earth gravity, which suggests design limitations based on human tolerance and adaptation. From these design limitations, a basic set of human factors design requirements can be derived that will begin to drive vehicle configuration concepts and designs.

Environment

The *artificial gravity level* is the centripetal acceleration produced by rotation and is dependent on the angular velocity and the radius. The artificial gravity force can be described by the following equation.

$$Fg = \frac{Mw^2r}{9.81}$$

---

* Boeing Aerospace & Electronics, P.O. Box 240002, MS JX-23, Huntsville, AL 35824-6402.

Fg = Artificial gravity force, in "g's".
w = Angular velocity, in radians/second.
r = Radius, in meters.

This equation provides a fixed relationship between rotational speed and radius. If either the radius or the angular velocity is increased, the artificial gravity level will also increase. This translates into a trade-off between cost and complexity (which depend on the radius) versus physiological/psychological concerns (which depend on angular velocity) (Diamandis, 1987).

The *gravity gradient* is the change in gravity level due to the radius of the vehicle. The artificial gravity force always acts perpendicularly outward from the axis of rotation (Loret, 1963). There is a 100% gravity gradient between the center of rotation and the outer rim which causes objects, or people, on the outer rim to weigh more at the rim (feet) than they do towards the center (head) (Diamandis, 1987). Therefore, for a man standing in a rotating habitat, there exists a gravity gradient that can be calculated by the following equation (see figure 1).

$$G = \frac{\Delta r}{r}$$

G = Gravity gradient expressed in a percentage.
Δr = Change in radius in meters.
r = Radius to rim in meters.

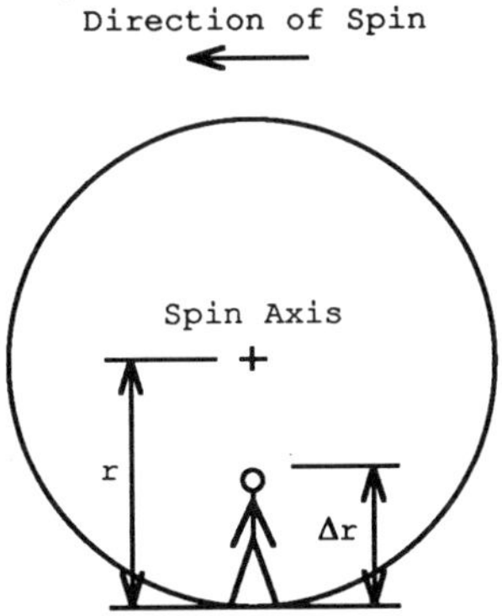

Figure 1: Head-to-toe Gravity Gradient (Loret, 1963)

For large values of r, the gravity gradient will be negligibly small; whereas, for small radii, the gradient will be relatively large. Depending on man's ability to tolerate this peculiarity, the gravity gradient becomes significant in design, particularly in small radius habitats (Loret, 1963).

A *Coriolis force* is an apparent force that is applied to an object moving linearly within a rotating system (Diamandis, 1987). The direction in which the Coriolis force acts is not constant as it is for the artificial gravity force, but varies both in magnitude and direction depending on the geometric relationship between the spin axis and the velocity vector of the crew member relative to the habitat, according to the following vector equation.

$$\overline{Fc} = -\frac{2M(\overline{w} \times \overline{v})}{9.81}$$

$\overline{Fc}$ = Coriolis force, in "g's".
$\overline{w}$ = Vector quantity, in radians/second.
$\overline{v}$ = Velocity of crew member, in meters/second.

For radial motion, a crew member ascending/descending a ladder, the Coriolis force acts to one side for motion toward the spin axis and to the other side for motion away from the spin axis (figure 2). For tangential motion, perpendicular to the spin axis, the Coriolis force acts along the same line as the gravity force, adding to it for motion in the direction of the spin and subtracting from it for motion against the direction of spin (figure 3). For axial motion, parallel to the spin axis, the Coriolis force is non-existent since the cross product of *w* and v is 0 (figure 4).

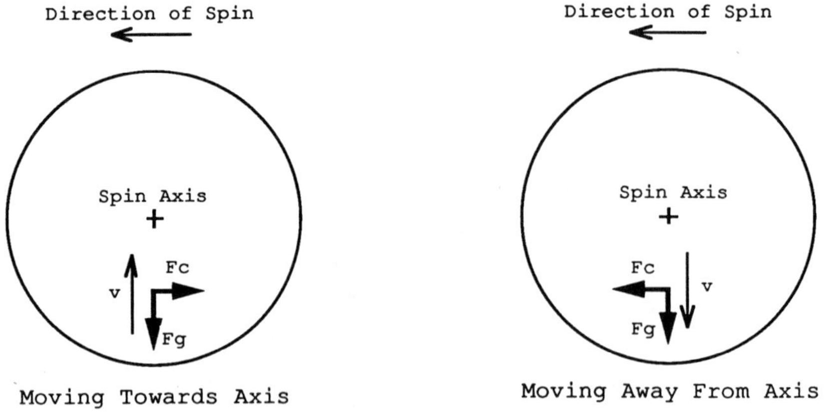

Figure 2: Radial Coriolis Forces (Loret, 1963)

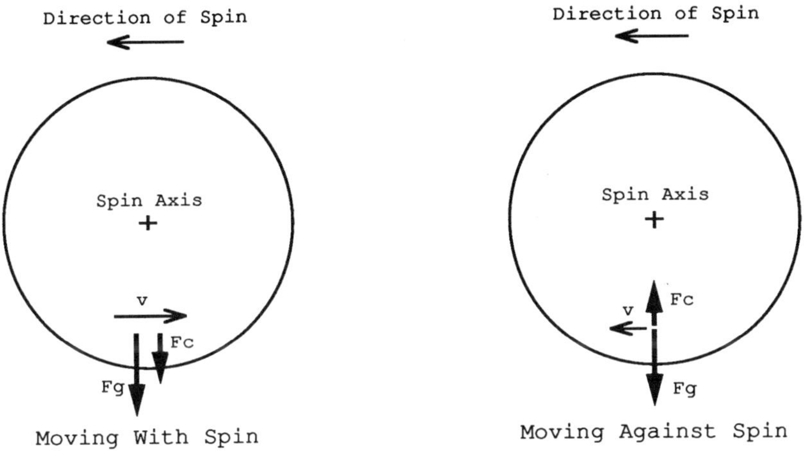

Figure 3: Tangential Coriolis Forces (Loret, 1963)

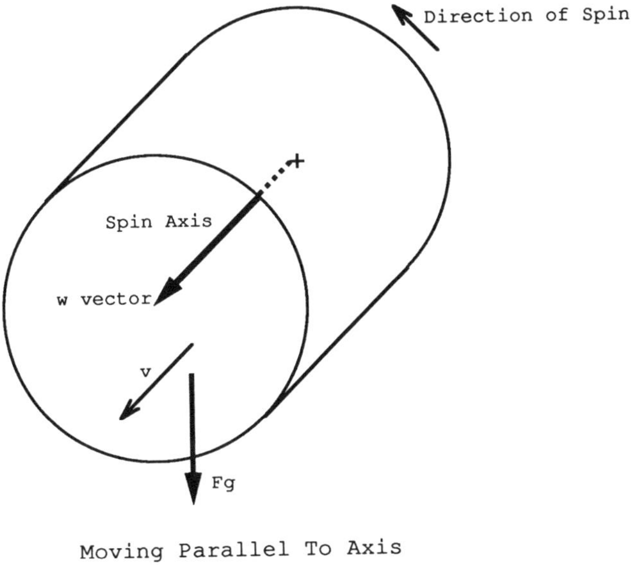

Figure 4: Axial Coriolis Forces (Loret, 1963)

Design Limitations

The *artificial gravity level* can change, due to Coriolis forces inherent in tangential locomotion, which makes it possible for a crew member to experience more than 1g if walking in the direction of spin. To avoid this occurrence, the "base" gravity level must be lower than 1g to allow a person walking at normal speeds to remain within the 1g limit.

Experiments have been conducted by Aerospace Medical Laboratory personnel at Wright-Patterson Air Force Base which indicate that a lower limit of 0.2g should be established (Loret, 1963). As in the upper limit, the lower limit is determined by the Coriolis force of tangential locomotion against the direction of spin, so crew members at no time go below the 0.2g limit during normal walking.

Gravity gradient design limitations have yet to be decided based on experimental data; however, speculations and assumptions have been made of 15% (Loret, 1963) and 20% (Diamandis, 1987). A 15% gravity gradient between head and feet suggests a minimum vehicle radius of 11.75m, while a 20% gravity gradient would require a minimum radius of 8.75m.

Coriolis forces due to radial motion in the vicinity of the axis of rotation cause a distortion of the environment due to change in resultant force both in magnitude and direction. This force is likely to cause visual illusions and mental confusion. Therefore, radial traffic should be kept to a minimum or even avoided, as well as the preclusion of movement across the spin axis.

Tangential motion, which causes a change in gravity level, poses a problem because there is no experimental evidence available to indicate the ability of man to discriminate between small gradations of gravity or on the maximum permissible deviation from local gravity levels which can be tolerated without adverse psychophysiological or locomotive effects. A maximum permissible limit of 50% (Dole, 1960) variation between tangential walking and stationary gravity levels has been estimated. However, for materials handling, 25% (Stone, 1970) and 20% (Thompson, 1989) has been recommended.

Because of the Coriolis forces inherent in moving tangentially in a rotating environment, the habitat should be designed to take full advantage of the situation. Since no Coriolis forces are felt when moving parallel to the vehicle spin axis, the long axes of the crew compartments should be placed parallel to the vehicle spin axis to avoid Coriolis forces along major circulation paths (Loret, 1963).

If head rotation takes place around an axis parallel to the spin axis, there is a minimum tendency for sickness to occur. Translated to design, the crew workstations should be oriented so that the head-to-toe axis is parallel to the spin axis. Therefore, display consoles should be oriented in an up-down position rather than a left-right position. In the same manner, crew bunks should be oriented axially with the long axis of the body parallel to the spin axis (Loret, 1963).

Deterioration in human performance and comfort can result due to angular accelerations (head motions); therefore, special attention must be given to vehicle design to prevent or minimize the possibility of sickness. Experiments indicate that an angular velocity of 0.4 (4 rpm) to 0.67 rad./sec. (6 rpm) is practical for an upper limit on w (Dole, 1960, Stone, 1970 and Thompson, 1989).

Locomotion experiments in the NASA Langley Rotating Space Station Simulator (a 12m dia. circular platform) were conducted to simulate human reaction to walking at varying levels of gravity. Rotation speeds varied from 3 to 10.5rpm to produce 0.05 to 0.75g. Results showed that walking at levels from 0.167 to 0.3g was very comfortable. From 0.3 to 0.5g, the subjects began to feel sensations of heaviness in the legs and body (Letko and Spady, 1970).

Adaptation

Adaptation schedule investigations in the Pensacola Slow Rotation Room (SRR) suggested that sickness could be minimized or avoided through periodic small velocity increases. Reason and Graybiel (1970) conducted an experiment designed to construct an adaptation schedule. Results suggested a 0.4 rad./sec. (4rpm) limit on angular velocity based on overall human tolerance.

In another study, Graybiel and Knepton (1972) analyzed characteristics of direction-specific adaptation effects acquired during an incremental adaptation schedule and revealed by symptoms of motion sickness during the execution of head movements after either a one-step return to zero velocity or a reversal of the direction of spin. Results of the study showed that both the short-term (direction-specific) and long-term (nondirection-specific) adaptation effects can be elicited during a standardized incremental adaptation schedule.

A *reach effectiveness* experiment, conducted by O'Laughlin, Brady and Newsom (1968), was designed to test reach movements requiring timely and precise contact with instrument control consoles that could be adversely affected by Coriolis forces. Under stress and exaggerated conditions, subjects were correct over 99% of the time. It was determined that Coriolis forces had no measurable effect on the precision and accuracy of reach activities in a rotating environment.

In 1969, the *Rockwell Rotational Test Facility* was developed by North American Rockwell which was capable of conducting artificial gravity tests at 3, 4 and 5rpm with a radius of 22 meters. A crew of 4 subjects were chosen to be rotated at 4rpm for seven days.

Locomotion results paralleled those of the NASA Langley Rotating Space Station Simulator. Conclusions of the test were that the crew could function proficiently without morale, cooperation or sickness problems.

The *Pensacola Slow Rotation Room* (SRR) was a nearly circular, windowless room with a diameter of 4.5 meters which contained laboratory and living facilities for 4 test subjects. Six separate runs were made to sample the range which might be used in generating an artificial gravity environment at the following velocities: 1.71, 2.22, 3.82, 5.44 and 10.00rpm (Clark and Graybiel, 1961).

Tests were conducted to determine muscular strength and coordination, eye-hand coordination and arithmetic computation. Results showed that the only deterioration was in the area of motivation.

Human Factors Design Requirements

The shaded area of figure 5 shows the human factors design envelope based on the design limitations inherent to an artificial gravity environment. This figure denotes all the previous design requirements to give a range of angular velocity and radius. The upper radius is shown at 250ft. (76.2m), but could be extended to any desired radius that fits the human factors zone.

Table 1 shows the result of this study compiled into a set of human factors design requirements. These are general requirements that could be expanded as mission objectives and requirements become more specific.

Conclusion

For future, long duration space missions, artificial gravity may prove essential for crew survival. In this case, a new set of design parameters are introduced that must be followed, from a human factors standpoint, in the design of rotating habitats. Further research is needed to determine the exact human factors envelope using the requirements of this paper as a departure point.

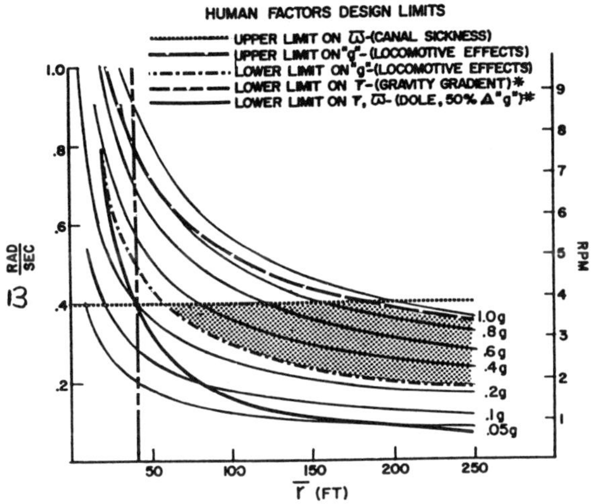

Figure 5: Human Factors Design Envelope (Loret, 1963)

Table 1: Human Factors Design Requirements

1. Upper gravity level should not exceed 1g.
2. Lower gravity level should not go under 0.2g (0.2 to 0.3g recommended).
3. Gravity gradient between head and feet should not exceed 15 to 20%.
4. Upper limit for angular velocity should be 0.67 rad./sec. (6 rpm).
5. Change in gravity level due to tangential walking should not exceed 20%.
6. Radial traffic should be kept to minimum.
7. Transport across spin axis and human activity at the hub should be prohibited unless hub is non-rotating.
8. Long axis of crew compartments should be placed parallel to spin axis.
9. Crew duty stations should be designed to minimize head movement out of the spin axis.
10. Sleeping bunks should be oriented with the long axis parallel to the spin axis.

Rotating habitats may be expensive to develop, but they may be the only comprehensive solution to the problem. Therefore, if humans are seriously considering extended spaceflight and settlement, it would be beneficial to investigate, in further detail, the effects of artificial gravity on humans and how a vehicle can be designed to satisfy human needs.

References

Clark, Brant and Graybiel, Ashton. February 1961. *Human performance during adaptation to stress in the Pensacola Slow Rotation Room*. Aerospace Medicine, Vol.32. pp.93-106.

Diamandis, Peter. 1987. *Reconsidering artificial gravity for twenty-first century space habitats*. M.I.T. Man-Vehicle Laboratory, Cambridge, Mass.

Dole, S.H. 1960. Design criteria for rotating space vehicles. The Rand Corp., Santa Monica, CA.

Graybiel, Ashton and Knepton, James. November 1972. *Direction-specific adaptation effects acquired in a slow rotation room*. Aerospace Medicine, Vol.43. pp.1179-1189.

Letko, W. and Spady, A.A. 1970. *Walking in simulated lunar gravity*. Fourth Symposium on the Role of the Vestibular Organs in Space Exploration, Naval Aerospace Medical Institute, Pensacola, Florida. Office of Technology Utilization, NASA. pp.347-351.

Loret, Benjamin J. May 1963. *Optimization of space vehicle design with respect to artificial gravity*. Aerospace Medicine, Vol.39. pp.430-441.

O'Laughlin, T.W.; Brady, J.F. and Newsom, B.D. May 1968. *Reach effectiveness in a rotating environment*. Aerospace Medicine, Vol.39. pp.505-508.

Reason, James T. and Graybiel, Ashton. January 1970. *Progressive adaptation to Coriolis accelerations associated with 1rpm increments in the velocity of the Slow Rotation Room*. Aerospace Medicine, Vol.41. pp.73-79.

Stone, Ralph W. Jr. August 19-21, 1970. *An overview of artificial gravity*. NASA/Langley Research Center, Hampton, Virginia. Presented at the 5th Symposium on the Role of the Vestibular Organs in Space Exploration, Pensacola, Florida, 1970.

Thompson, Allen B. 1989. *Some notes on maximum rotational rate in obtaining centrifugal artificial gravity*. Martin Marietta Corporation, Denver, Colorado.

PARTIAL GRAVITY:
Human Impacts on Facility Design

Stephen Capps* and Nathan Moore**

Abstract

Partial gravity affects the body differently than earth gravity and microgravity environments. The main difference from earth gravity is human locomotion; while the main difference from microgravity is the specific up-down orientation and reach envelopes which increase volume requirements. Much data is available on earth gravity and microgravity design; however, very little information is available on human reactions to reduced gravity levels in IVA situations (without pressure suits). Therefore, if humans commit to permanent lunar habitation, much research should be conducted in the area of partial gravity effects on habitat design.

Introduction

Proper design and layout of human habitation areas is imperative to the proper functioning and success of a lunar base. The reduced gravity level (0.16g) on the Moon will affect interior architecture by changing rack design as well as volumes due to human locomotion changes. Psychological issues will become very important, due to the extended mission lengths (6 months +), and the increased amount of isolation/confinement will be a problem because of the remoteness of the environment. These factors will affect the design and volumetric requirements needed for human habitation. The following study addresses lunar habitation in a Space Station *Freedom* size module with a crew of four to determine the proper architecture for a partial gravity environment.

---

* Boeing Aerospace & Electronics, P.O. Box 240002, MS JX-23, Huntsville, AL 35824-6402.

** NASA/Johnson Space Center, Man-Systems Division, MS SP33, Houston, TX 77058.

Gravity Level

Modular Rack - The Space Station *Freedom* interior architecture works well in a microgravity environment where walking surfaces are not required and large massive racks can easily be maneuvered. Crew facilities in *Freedom* are accommodated by a modular rack system where the crew can sleep in any orientation due to the lack of gravity. In partial gravity (0.16g), this system breaks down. The 680kg (earth weight) equipment racks on *Freedom* become less moveable on the Moon because they would weigh 110kg. Also, the volume contained by an empty *Freedom* rack is too small to accommodate a crew member on the Moon who must sleep horizontally as well as providing a stand-up volume for donning and doffing clothing. As a result, a new internal configuration and rack system must be proposed for lunar application to accommodate both laboratory and habitation uses.

A standard *Freedom* rack is 1.07m by 1m by 2.13m. A rack for the lunar surface can be reduced to a size of 0.75m by 1m by 2.13m, which would weigh approximately 75kg lunar weight when fully outfitted with scientific equipment. This will allow two astronauts to more easily maneuver the racks for installation and reconfiguration.

Walking and Running Gaits - Humans are designed to walk in a 1g environment and have adapted to a certain force and traction due to that gravity level. A change in the gravity changes the forces and traction acting against the human body and therefore, changes the gait. A comparison of earth gravity (1g) and lunar gravity (0.16g) walking and running gaits is shown as follows:

In *1 gravity walking*, the muscular energy is expended to lift the legs thus creating potential energy. This lifting of the legs offsets the center of gravity of the body in the forward direction, thus resulting in an acceleration in the forward direction (the transfer of potential energy into kinetic energy). In the case of walking, potential energy and kinetic are out of phase (Margaria and Cavagna, 1964). In other words, some of the kinetic energy is turned into potential energy when the body is lifted, thus causing forward motion.

In *partial gravity walking* (0.16g), less muscular energy is expended, thus making less potential energy available. Less acceleration in the forward direction makes walking velocities lower; therefore, the critical speed at which walking shifts to running becomes lower. In a lunar gravity situation, walking is impractical, slow and virtually impossible (Margaria and Cavagna, 1964).

Another aspect of walking in partial gravity is the fact that astronauts "bounce" higher because they are used to expending earth gravity forces to walk. Extended stay in 0.16g will probably result in a minimization of this "bouncing" due to muscle and bone atrophy and the astronauts adaptation to the 0.16g environment (Keller and Strauss, 1988). However, this "bouncing" will still be more than that on Earth, which in turn affects the interior architecture.

In *1 gravity running*, the shift from walking to running takes place at a speed of 8.5km/hr., at which point the potential energy, accumulated during the body lift, about equals the kinetic energy. Higher speeds require acceleration to be sustained by a direct muscular push, which increases both kinetic and potential energy. The transition from walking to running puts both potential and kinetic energy in phase (Margaria and Cavagna, 1964). In other words, the kinetic energy and potential energy become equal to sustain the forward speed changes in the step cycle.

In *partial gravity running*, as in walking, the maximum speed is lower because the low apparent weight of the astronaut reduces the vertical force component of traction producing movement (Margaria and Cavagna, 1964). Partial gravity running also has the same aspects regarding "bouncing" as walking, which results in greater "bouncing" heights due to the increased vertical force component.

Partial gravity jumping helps in traction by increasing the vertical force component of the maneuver. Lunar gravity locomotion has the advantage of having low energy cost of speed maintenance per distance traveled than that required in an earth gravity environment (Margaria and Cavagna, 1964). In a 0.16g simulation (Spady and Krasnow, 1966), test subjects were able to jump vertical heights six to seven times higher and perform standing broad jumps about two times farther than is possible in earth gravity.

Partial gravity loping is, according to 0.16g simulations, the most comfortable walking gait at about 3m/sec., which is much faster than the most comfortable walking gait on Earth of about 1.2m/sec. (Hewes, Spady and Harris, 1966). This is due to the reduced energy requirements needed to accelerate.

Posture - Human posture in lunar gravity also differs from posture in an earth gravity environment. In lunar gravity, as the speed is increased, the forward body inclination gets progressively larger (figures 1 and 2). For example, the inclination of a sprinting gait on Earth is 10° while the same gait on the Moon is 60° (Hewes, Spady and Harris, 1966).

Traction - The reduction of traction can make human balance and locomotion hazardous in any environment. With a reduction of gravity, a human experiences a reduction in the friction between himself/herself and the surface of the ground. This can become a serious hazard because astronauts cannot rapidly change positions to avoid moving or falling objects or gain a surer foothold or handhold while in a precarious position (Hewes and Spady, 1964). Another constraint affecting locomotion is that the inertial force required to start moving from a complete stop is the same as it is in 1g; therefore, the subject must overcome the same inertial force in lunar gravity as in earth gravity while utilizing less traction.

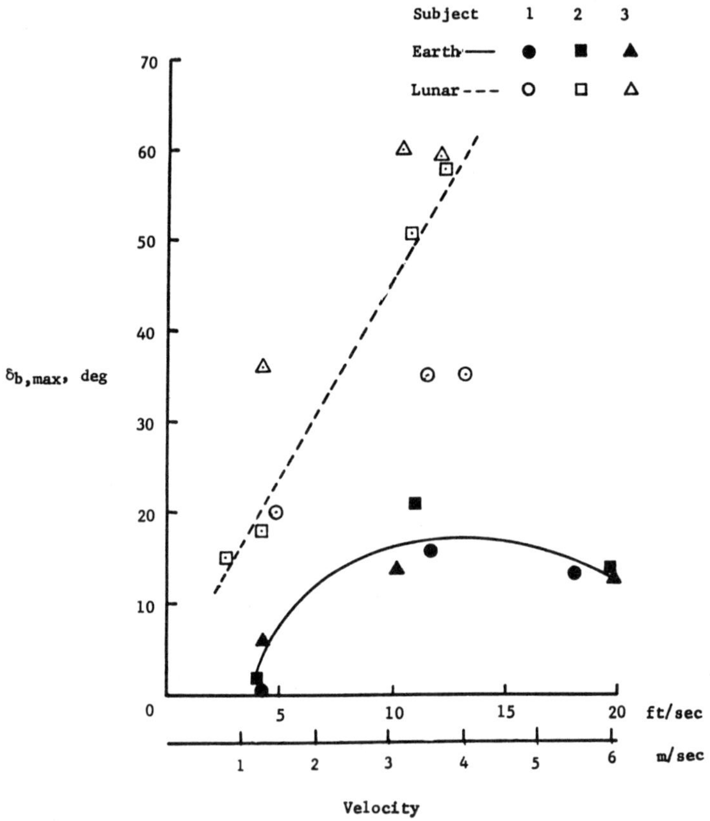

Figure 1: Body Inclination Against Velocity (Hewes et al., 1966)

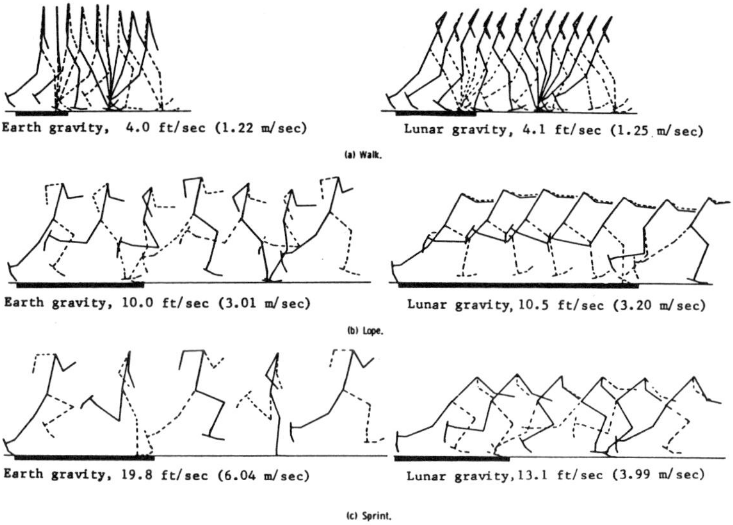

Figure 2: Body Inclinations(Hewes, Spady and Harris, 1966)

Psychology

Short Duration Mission - For mission durations of a few days to couple of weeks, crews can share personal quarters by rotating shifts, as is done when the Space Shuttle carries Spacelab. Crew members also do not need near as much volume for recreation, exercise, dining, etc. due to the time factor and the fact that crews can rotate shifts, which reduces redundancy of space.

Medium Duration Mission - For mission durations of less than six months, crews begin to require their own private personal quarters for sleeping as well as private recreation (reading and communication with relatives), and will require more volume for grooming and personal hygiene. Crews will also begin to work standard earth shifts which will result in more volume needed for dining, recreation, exercise and meeting areas.

Long Duration Mission - For mission durations of six months or more, crews begin to require all the necessary "comforts of home". Each crew member will need a private sleeping area with personal storage, a dressing area and a sitting area. Full recreational and exercise facilities will be required as well as a complete health maintenance facility.

Confinement/Isolation - A major aspect of human isolation for prolonged periods of time is the psychology of confinement. Breeze (1961) conducted a study to determine the minimum amount of space that humans need for survival for various lengths of isolation. He concluded that 1.42 m^3/person was adequate for one or two days of confinement, whereas 7.36m^3/person is needed for one to two months, and 17m^3/person is needed for periods exceeding two months. Fraser (1968) concluded that for two months confinement, a volume of 7.08 to 11.31m^3 was required to avoid psychological impairment of the crew. These volumes are absolute minimums and do not include divisions into functional units such as the work unit (operational tasks, facility management), public unit (dining, recreation, exercise), personal unit (sleeping, personal privacy and storage), and service unit (toilet, laundry, public storage) (Fraser, 1968).

Another factor which affects the volumes is the layout of the interior architecture. Volumes can be psychologically confining if they are narrow and equipment is difficult to access. Therefore, care should be taken in configuring volumes to ensure the proper functioning and psychological well-being of the crew.

Design Impacts

Corridors - Since starting/stopping distances increase due to reduced gravity, corridors should be clear of obstructions as well as employing hand mobility aids to control direction and speed. Therefore, optimum configuration would be an isolated corridor with no equipment or furnishings to interfere with circulation.

Floor Surfaces - The reduction of traction is a problem that can be overcome through careful design. Floor surfaces should be designed with a coarse surface that adheres to shoes with a traction sole. Figure 3 shows one possible concept; however, an "astroturf" type surface with rubber cleat soles may suffice. This will increase the traction to aid in forward locomotion.

Figure 3: Traction Surface for Reduced Gravity

Ceiling Heights - Humans bounce higher when walking and running in reduced gravity, thus requiring a sufficient ceiling height for head clearance. During normal walking in 0.16g (1.25m/sec.), humans will probably bounce no higher than 2.44m because of their body inclination of approximately 20°. At a sprinting speed in 0.16g (3.99m/sec.), the forward body inclination is approximately 60°. However, as seen from the Apollo program, loping at a speed of about 3m/sec. is the most comfortable speed which has a forward body inclination of about 45°. Figure 4 shows an analysis of both walking and loping gaits in 0.16g, which shows that a ceiling height of 2.44m will probably be sufficient. In small, enclosed spaces, such as the crew quarters and the personal hygiene facilities, a 2.13m ceiling height will probably by sufficient because of the minimum amount of locomotion within those spaces. One safety factor that can be employed is a padding system on the ceiling in case the astronauts do bounce too high initially, before adaptation occurs.

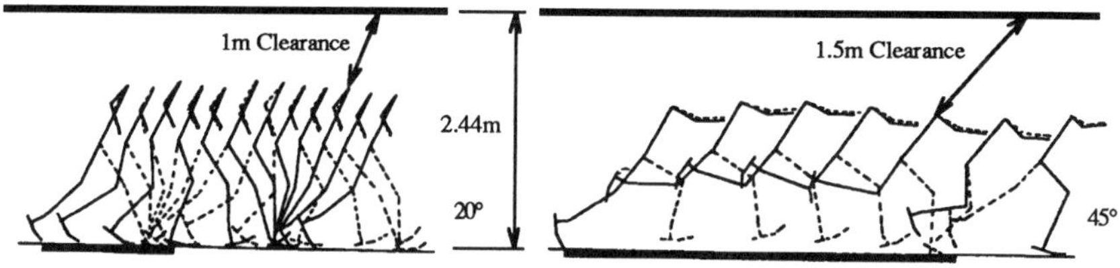

Figure 4: Ceiling Heights in Lunar Gravity

Staircases - Tests have shown that humans can jump 3.7 to 4.3m vertically when unencumbered by a spacesuit, while humans can only jump 51 to 56cm on Earth (Hewes and Spady, 1964 and Spady and Krasnow, 1966). Also, in the same series of tests, it was found that the easiest way to ascend/descend stairs with a landing height of 1.25m was to simply jump. These results are not surprising because a vertical distance of 1.25 m in 0.16g is equivalent to 20cm in earth gravity, or a standard stair riser height. If we assume a maximum strength deterioration of 67% (Keller and Strauss, 1988), then stair risers would be required at about 41cm spacings. Therefore, we can conclude that stairs should be designed with a stair riser spacing of somewhere between 41cm and 1.25m (figure 5), depending on the length of stay and deterioration, which is a subject for further experimentation.

Chairs - Due to reduced gravity, a normal and comfortable sitting position will probably be different than that of earth gravity. Earth gravity pulls the body to a near 90°

normal and comfortable sitting position (figure 6). In microgravity, the body assumes a neutral position (figure 7) that is very different from the position taken in earth gravity. Therefore, it can be assumed that the body will probably take a position somewhere in between the earth gravity and microgravity positions shown in figures 6 and 7. Figure 8 shows the design of a partial gravity chair which was designed in speculation of what the body might do in 0.16g. This is also a subject for further experimentation; however, it stands to reason that the body would take a position closer to the neutral body position which could be accommodated by a chair that simply supports the body at the knees and the buttocks/thigh region.

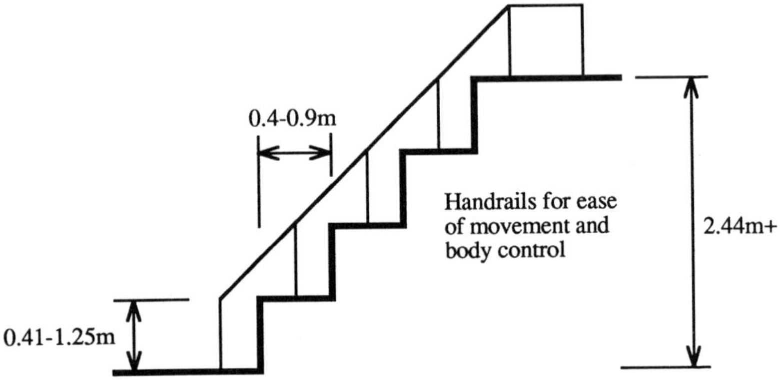

Figure 5: Stairs in Lunar Gravity

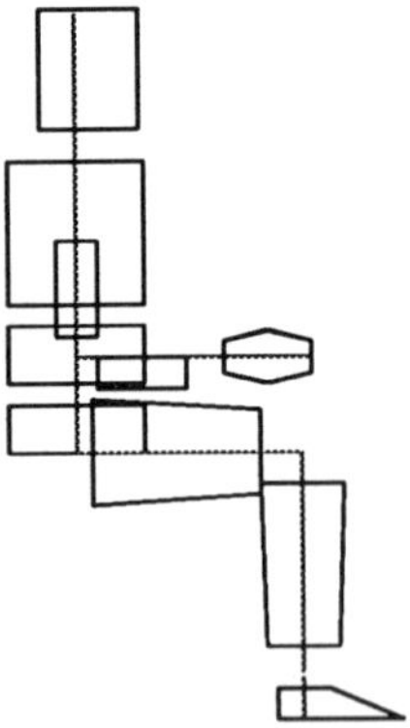

Figure 6: Earth Gravity Sitting

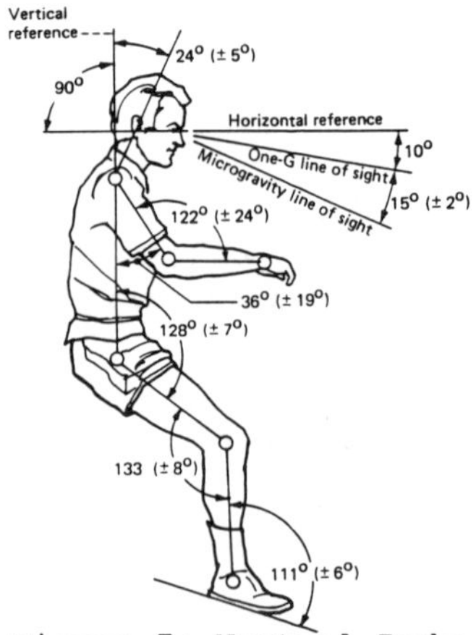

Figure 7: Neutral Body Posture (MSIS, 1987)

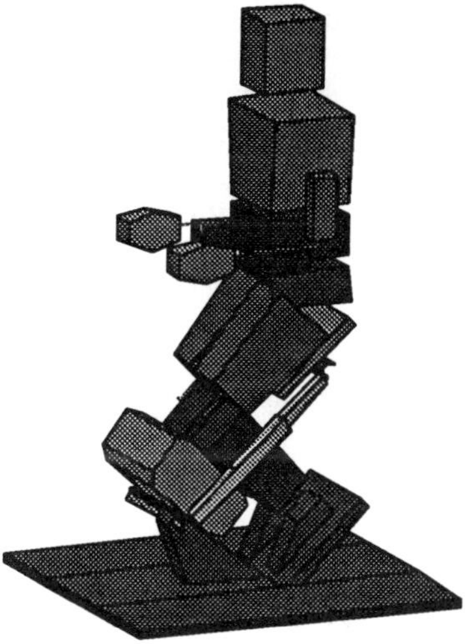

Figure 8: Lunar Gravity Sitting

Facility Design

Table 1 shows a summary of the volumes deemed necessary for a crew of 4 based on the previous requirements. The total volume of the module is approximately 200m^3, which is the volume for 8 people on Space Station *Freedom*.

Table 1: Lunar Base Habitation Volumes

Area	**Volume (m^3)**
Corridor	45.00
Crew Quarters	32.00
Health Maintenance Facility	24.00
Personal Hygiene/Waste Management	8.00
Galley/Wardroom/Experiments	48.00
Public Storage	7.00
ECLSS	3.31
Structure	--
Mechanical/Storage	32.69
Total	200.00

Conclusions

In conclusion, this study has addressed the various parameters of human habitation, including the change in gravity level and the psychological issues, and has derived design solutions that meet the criteria set forth.

Human habitation in the lunar environment poses many new challenges that have been addressed in this paper. However, many of the design parameters brought forth by this study warrant further investigation to test hypotheses. The design solutions in the paper do, however, provide a basis for speculation and testing of theories, which can be investigated further as mission goals and scenarios become more specific.

References

Breeze, R.K. November 1961. *Space vehicle environmental requirement based on equipment and psychological criteria.* ASD-TR-61-161, Pt.1. Wright-Patterson Air Force Base, Ohio.

Fraser, T.M. 1968. *Confinement and free volume requirements*. Space Life Sciences, Vol.1. pp.453-466.

Hewes, Donald E. and Spady, Amos A. March 1964. *Evaluation of a gravity-simulation technique for studies of man's self-locomotion in lunar environment*. NASA Technical Note D-2176. NASA/ Langley Research Center.

Hewes, Donald E.; Spady, Amos A. and Harris, Randall L. June 1966. *Comparative measurements of man's walking and running gaits in earth and simulated lunar gravity*. NASA Technical Note D-3363. NASA/Langley Research Center.

Keller, T.S. and Strauss, A.M. April 5, 1988. *Bone loss and human adaptation to lunar gravity*. Lunar Bases and Space Activities of the 21st Century, Westin Galleria Hotel, Houston, Texas. Paper No. LBS-88-092.

Margaria, R. and Cavagna, C.A. December 1964. *Human locomotion in subgravity*. Aerospace Medicine, Vol.35, No.12. pp.1140-1146.

Man-Systems Integration Standards. March 1987. NASA-STD-3000, Vol.1.

Spady, Amos A. and Krasnow, William D. July 1966. *Exploratory study of man's self-locomotion capabilities with a space suit in lunar gravity*. NASA Technical Note D-2641. NASA/Langley Research Center.

HEDRATECTURE AND ARTIFICIAL GRAVITY

Joseph J. Mangan, A.I.A.[1]

ABSTRACT

There is a need, according to the National Commission On Space, to develop three critical space facilities: A Space Base, A Variable-Gravity Research Facility, and a Lunar Surface Outpost. An attempt will be made here to make use of HEDRATECTURE, a sub-discipline of civil engineering, to develop special structural forms that could be of some value in approaching these goals. HEDRATECTURE is the art and science of building with structural frames that follow the form of the convex polyhedrons.

The Lunar Surface Outpost using tetrahedral frames as structural supports has been previously described in "The Expandable Platform As A Structure On the Moon" (Mangan, 1988). This paper will describe the addition of a ring construction near the Lunar Surface Outpost Base (the Expandable Platform). The rotating dwelling elements within this structure will provide artificial gravity along with a complete simulated Earth environment. The Space Base will also be developed as a Variable-Gravity Research Facility.

INTRODUCTION

It has been established that there are serious health problems to be overcome if people are to work in low gravity environments for even a limited duration. When one considers that along with psychological stress, humans can suffer from cardiovascular deconditioning, loss of muscle mass, demineralization of the bones, loss of red blood cells and impairment of the body's immune response, it would seem that consideration should be given to the construction of

---

1. Member of American Institute of Architects, P.O. Box 326, Ho-Ho-Kus, New Jersey, 07423.

special structures where these problems could be addressed.

For over 30 years there have been various concepts developed on the construction of rotating habitats that would produce an artificial gravity (Von Braun, 1952, Kramer et al, 1960, Loret, 1963). It will be shown here that the hedratectural forms of the tetrahedron and icosahedron, in a configuration to create an artificial gravity, provide simple redundant structures that give an effective response to the problem.

BACKGROUND

The need for an artificial environment that would be similar to that of Earth has been known for some time. There has been considerable attention given over the last thirty year to the effects of g-forces on the human body. This experimental work has been done mainly with rotating arms (set at variable speeds), with neutral buoyancy tanks (simulating low gravity), and with water braked rocket sleds.

We can produce forces in a zero-g environment that would have an effect on the human body that would seem very near what would be encountered in a true Earth gravity environment (9.81 m/sec^2).

We must find a way to provide a Variable - g Research Facility (VGRF) that could provide not only a laboratory in space, but a place where people could recover from the debilitating effect of living in a weightless condition for a long period of time. This facility could also provide a place in space or on the Moon where special surgery, requiring an Earth-like gravity, could be performed.

It is felt here that there could, eventually, be a need for two distinctly different types of VGRF facilities:

The first would be a rotating space base set at a position in LEO (low earth orbit), or in LLO (low lunar orbit), or possibly in one of the La Grange points.

The second would be a facility built on the Moon as a fixed circular ring constructed within the sloped side walls of a crater or on a selected minimally undulating regolith surface and equipped with one or more vehicular dwelling units. The vehicular dwelling units would attain an artificial gravity force by the velocity of their rotation.

HEDRATECTURE AND 3 GEOMETRIC FORMS

Since any construction in space will have to make use of structures that have a minimal number of redundant interfitting parts that can be robotically assembled, it is felt here that two of the five convex polyhedron forms used in Hedratecture can be used to advantage.

Hedratecture, which is a new sub-discipline of civil engineering, is the art and science of "building with hedrons". It makes use of the five convex polyhedrons as building forms. These five forms are the four sided tetrahedron, the six sided hexahedron, the eight sided octohedron, the twelve sided dodecahedron and the twenty sided icosahedron (Fig 1). They all are geometrically interrelated, and since building in space will eventually require that structures be made up of simple, redundant parts that are telerobotically assembled in space or on a planet, these forms used as structures seem to respond directly to that need.

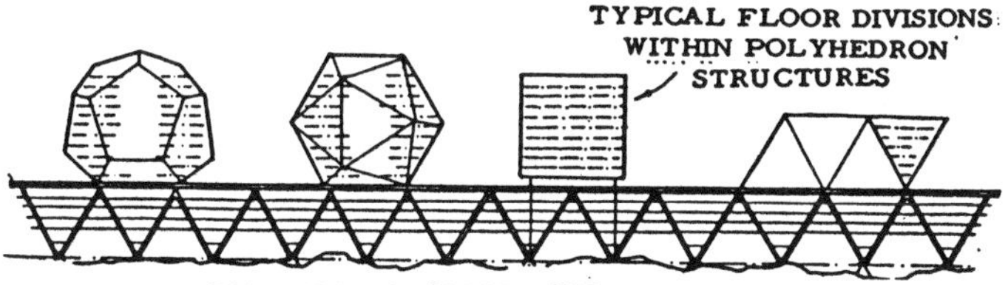

POLYHEDRONS PLACED UPON PLATFORM

FIG. 1

In the case of the rotating space base, the form of the icosahedron is used. For the fixed circular ring the tetrahedron is used.

THE ROTATING SPACE BASE

The rotating space base would be made up of a hub element, two twin structures constructed in the form of icosahedrons, and a tether with elevators connecting both icosahedral structures to the hub (Fig. 2). An advantage of the icosahedral form is that it is "a building" that can be highly pressurized and provide the very familiar interior ambiance of a building on earth. This type of structure would not only have an artificial gravity but it would also have a pressure and air composition very similar to that of earth.

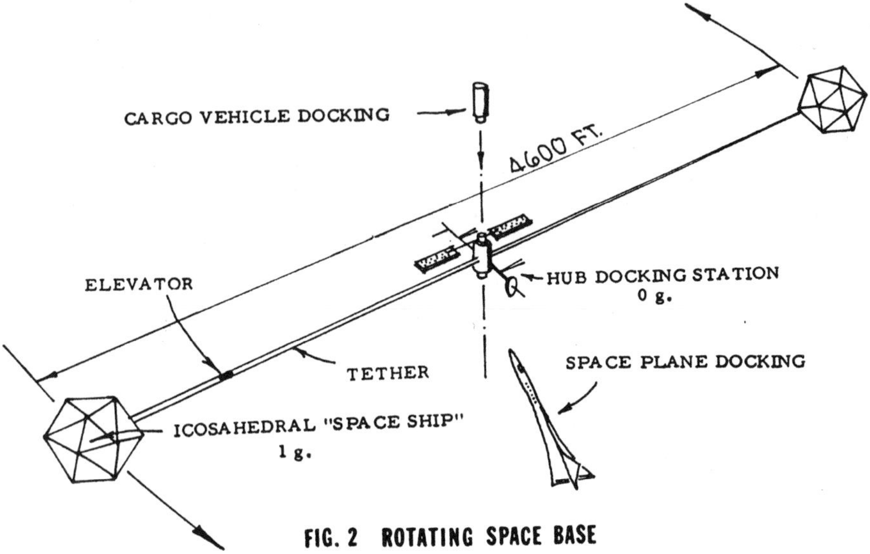

FIG. 2 ROTATING SPACE BASE

In this particular type of space base it would be possible to modulate the artificial gravity to 1/3 G or 1/6 G, if required, by modifying the radius or the rotating velocity.

The structural elements of the large icosahedral frame would, like all Hedratectural structures, be made up of a limited number of special epoxy coated columns, beams and connectors. These elements would be taken to LEO where they would be robotically assembled.

These twin icosahedral-framed structures would, in essence, be large spaceships set equidistant from each other. It would be necessary to have a tether connecting both through a hub. The tether would actually be made up of six cables made of low-weight polymeric material of high tensile strength. These cables would provide a shaftway for the elevators that would run from the hub to each of the icosahedral "space ships ".

The icosahedral-shaped structures at each end of the tether would be more like buildings than like space ships. They would have floors much like a conventional building and elevators connecting these floors. However they would have their own propulsion system to provide rotation and modulation of rotation (Fig 3).

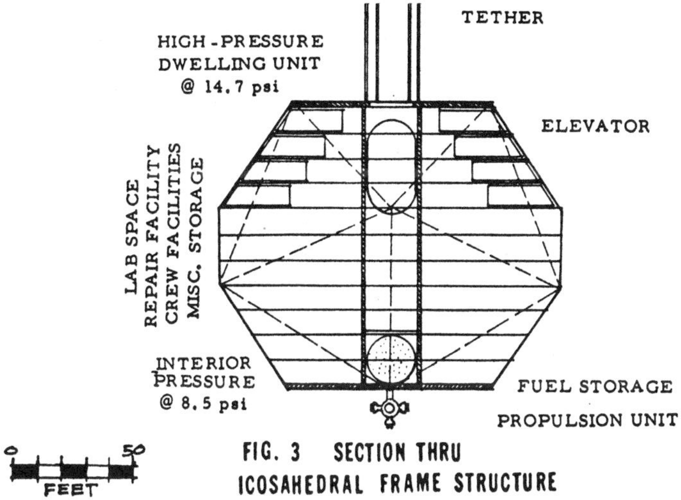

FIG. 3 SECTION THRU ICOSAHEDRAL FRAME STRUCTURE

One of the salient features of the icosahedral and dodecahedral configurations is that they lend themselves to pressurization. In space these structures would be pressurized in different parts of their interior with pressures ranging from 8.5 psi to 14.7 psi. These pressures can be extremely high- 14.7 psi is 2,116.8 lbs/sq. ft.- but the structural engineering can accommodate such pressures. In most cases the dwelling units (habitats) would be self contained and thereby not require the whole icosahedral frame to be highly pressurized. Also, the pressure in many areas of the frame would not need to be 14.7 lbs/sq. in. (1.04 kg/cm^2).

The hub is an important part of the Rotating Space Base. Though it would be constructed as a dodecahedron if a large volume were required it is shown in this paper as a simple epoxy coated container that would be taken directly from earth, pressurized and placed in orbit. It would provide the main docking ports for access to the station. It would contain a reel system for modulation and adjustment to the tether. It is the main elevator lobby for the base and it contains solar arrays and communication equipment (Fig 4). After the hub was in orbit the icosahedrons would be robotically constructed and the tether put in place. The propulsion system would then be activated to produce the required velocity.

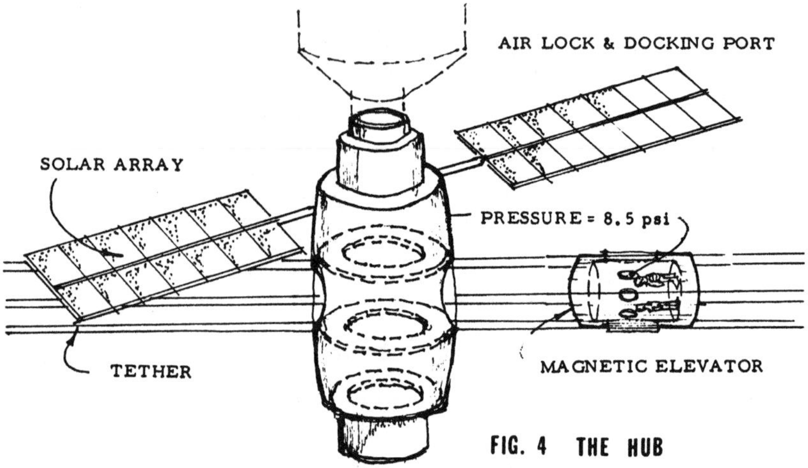

FIG. 4 THE HUB

It is not the purpose of this paper to discuss all the mechanical systems of this Rotating Space Base, such as the reel system, radiators, solar arrays, lighting, air conditioning, the propulsion system and air locks. It is intended to show that structures can be constructed out of simple convex polyhedral forms with a limited number of redundant parts providing all the amenities for a VGRF.

The special significance of the rotating space base with its hub and space ship-like buildings is that it could provide a host of necessary facilities other than the VGRF, e.g. repair shops, a hospital, gymnasium and possibly a swimming pool. Most of all, it would provide the special laboratories and astronomy observatories along with living quarters with an environment very similar to that of earth.

It is even possible that these icosadral-shaped structures could be detached and taken to Geostationary orbit and under their own propulsion system on to libration point (1). From there it could be possible to land and attach to a previously built Expandable Platform on the Moon.

If the Rotating Space Base were in low earth orbit (LEO), it would be possible for its hub to act as a dock for not only rocket powered cargo transport vehicles, but for an aerospace plane that could leave earth with horizontal take-off, discharge its passengers at the hub of the Rotating Space Base and return to Earth with a runway landing.

Though not fully described in this paper, it is not beyond present technology to accelerate the tethered rotating space ships and move the whole entity toward Mars. This would allow for the astronauts to keep in excellent physical condition on the long journey. The space ships could be so designed as to allow landing on Mars.

The large volume of the icosahedral frames could adequately carry all the required elements to construct a base along with a return vehicle and ample fuel for the two-way trip.

For the purpose of this presentation a dimension of 4,600 ft. (.7 km.) will be used for the diameter of revolution of the station, with the two icosahedrons intersecting the orbit at their centers. At that diameter, to produce the equivalent of terrestrial gravity, the revolutions per minute would be 1.13 rpm. The velocity is 272 ft/sec. or 185 mph. (299 km/hr).

The structural elements of the icosahedrons would be made of aluminum composite and be robotically assembled in space. Each icosahedron has a usable floor area of 150,000 sq.ft. (14,000 m^2), a weight of 670,000 lbs. (30,000 kg.), and a volume of 2,000,000 cu. ft. (58,960 m^3). Each has a weight of 2.86 lb/sq.ft. (14.44 kg/m^2) and a weight-to-volume ratio of .335 (.509 metric).

THE FIXED CRATER BASED RING

The fixed crater based ring would be an extension of the transportation trestle where structural tetrahedral frames are used to carry magnetically elevated vehicles along a given trajectory while providing a 3 m protective cover of regolith (See "Hedratecture and the Expandable Platform", Mangan, 1988).

Prior to construction, a selection of the site would be made using the most refined survey information. For the purpose of this paper a theoretical crater would be selected. The crater would be near the outpost and have a radius of approximately 2,300 ft (701 m).

There are innumerable craters on the moon of varied diameters and of varied composition. Fresh lunar craters up to 9.32 miles (15 km) in diameter have a consistent diameter/depth ratio of 5. Large craters are much shallower (Taylor, 86). Older craters, having been eroded by meteorite bombardment, have become wider and shallower, thereby increasing the diameter-to-depth ratio. Even the smoothest areas on the lunar surface are undulating plains, which makes the use of leapfrogging tetrahedral frames so viable.

In recent years it has become apparent that the bearing capacity of the regolith on the Moon is as high as some compacted Earth soils (Carrier, 87). This means that at a very short distance from the surface there would be sufficient bearing capacity to support special preformed footings (Mangan, 87). On the rim of the crater there would be areas that might require special pedestals in order to properly prepare the surface for a structure. The angle of repose for all small craters is close to 30° (Taylor 86).

Since there is such a multitude of small craters and since there is such consistency in their morphology it can be considered plausible that a hypotetical small crater with a radius of 2,300 ft. (.7 km) exists. This size crater will be used as the setting for the placement of an expandable platform. The angle of repose will be taken as 30° and the adjustable footings of the tetrahedral frames will allow the trestle to run around the rim of the crater in a perfect circle. The trestle will provide the attachments for magnetic track, the rotating vehicles and the protective cover (Fig. 5).

The objective is to provide a force of 1g at the perimeter of the ring. The speed of the rotating vehicles is the variable factor. The g-factor can vary with a change in the velocity of rotation.

In order to achieve 1g at the floor of the rotating habitats the basic formula for horizontal acceleration is used:. horiz. acc. = v^2 / r , solving for v (velocity in ft./sec). In order to derive the horizontal acceleration, a vector diagram is used.

This diagram also provides the angle of equilibrium for the rotating unit at 1g, at 9°-36'-36".

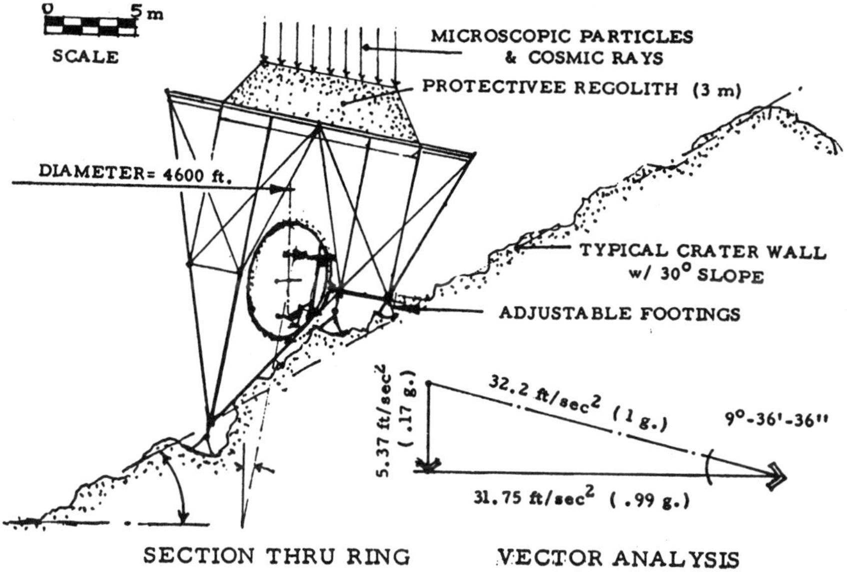

FIG. 5 ROTATING RING

If one sets the diameter of the crater at 4,600 ft. (1.4 km), or a radius at 2,300 feet, then the revolutions per minute will be 1.129. The speed at the perimeter of the crater would be 185.4 miles per hour (296.6 km/hr). At this speed the the floor of the habitat is almost perpendicular to the floor of the crater and one would have many of the sensations felt in terrestrial gravity.

It is true that artificial gravity could be produced by smoothing out the side of a given crater and building a roadway directly on the regolith. But it is felt here that the cost of cut and fill, including blasting and soil moving with heavy earth imported equipment, and application of the surfacing finishes would be many times more costly than the Hedratectural approach with simple redundant robotically-placed tetrahedral frames employing light equipment.

With the circular structure on the moon there would be a coriolis force, but it is felt that it would be negligible considering the relatively large diameter of the structure.

Unlike the Rotating Space Base, which moves in a O-gravity environment, the Fixed Crater Based Ring (the merry-go-round) moves over a track that must be designed to magnetically levitate the moving vehicles (Gillies, 86). The track must be kept in exact position, relatively free of lunar dust.

As with the rotating Space Base it is not the intent of this paper to present solutions for the many mechanical engineering problems that must be solved but only to present the considerable value that a Hedratectural approach provides as the simplest and most economical way to construct Variable Gravity Research Facilities in outer space and on the Moon.

REFERENCES

Carrier, W.D. lll: 1987, "On the Moon", Journal Florida Engineering Society.

Carrier, W.D. lll, J.K. Mitchell, A. Mahmood: 1973, "The Nature of Lunar Soil", Journal of Soil Mechanics and Foundations, Div. ASCE, Vol. 99, ASCE, New York, N.Y.

Gillies, G.T.: 1986, "A Review of a Precision Magnetic Suspension System", Symposium 86- Atlantic City, Lunar and Planetary Institute, Houston, TX., pp. S86 1a- S86 1g.

Johnson, S.W., R.S. Leonard: 1985, "Evolution of Concepts for Lunar Bases", Lunar Bases and Space Activities in the 21st Century, Lunar and Planetary Institute, Houston, TX., pp.47-45.

Mangan, J.J.: 1988, "The Expandable Platform as a Structure on the Moon", Engineering, Construction, and Operations in Space", ASCE, New York, N.Y., pp.375-387.

Mixon, R.W., W.W. Hankins lll, M.A. Wise: 1988, "Robotic Space Construction", Engineering, Construction, and Operations in Space, ASCE, New York, N.Y., pp.656-662.

Taylor, G.J.: 1986, "Geological Considerations for Lunar Telescopes", Lunar Bases and Space Activities in the 21st Century, Lunar and Planetary Institute, Houston, TX., pp. 189-194.

Wercinski, P.F., N.D. Searby, B.W. Tillman: 1988, "Space Artificial Gravity Facilities: An Approach to their Construction", Engineering, Construction, and Operations in Space, ASCE, New York, N.Y., pp. 734-749.

Subsection 4.7: EDUCATION

EDUCATION: THE FUTURE IS NOW

Stewart W. Johnson,[1] Member, ASCE
John P. Wetzel[2]
M. Anis Johnson[3]

ABSTRACT

The United States of America as a nation faces a major challenge in improving its entire educational system, especially education in mathematics, science, and engineering. Our future in this technologically competitive world rests on young people now in our elementary, middle, and high schools and in our universities. Members of technical and professional societies can relate to the general public in positive ways by working with the schools to enhance the education of our young people.

INTRODUCTION

The authors of this paper have had valuable opportunities to interact constructively with students and faculty at the elementary, middle, and high schools, and university levels. The Young Astronauts who enthusiastically met with Astronaut Dr. James van Hoften at the Space 88 conference in Albuquerque, NM in August 1988 were primarily elementary school students. They heard Astronaut van Hoften tell about the space shuttle mission on which he repaired the Solar Max satellite. At the middle school level our efforts are in a company join-a-school program with a middle school in Albuquerque. The join-a-school program pairs businesses and schools and facilitates partnerships that help both businesses and schools improve their programs.

---

[1]Principal Engineer, Advanced Basing Systems. BDM International, Inc., 1801 Randolph Road S.E., Albuquerque, NM 87106; and Adjunct Professor of Civil Engineering, University of New Mexico.

[2]Associate Staff Member, Space Systems. BDM International, Inc.

[3]Educational Consultant, Albuquerque, NM.

At the high school level we are developing a student paper session for the Space 90 conference, as we did for Space 88, as well as an extensive exhibit of student art related to space exploration. We have also participated in tutoring programs and in technical presentations for high school students. The Space 88 student program was recognized and praised by the Albuquerque Public Schools Board of Education and the Superintendent at a Board meeting.

At the university level we continue to work with faculty and graduate and undergraduate students in several departments at the University of New Mexico and New Mexico State University on research efforts relating to extraterrestrial facilities engineering. A special emphasis in this effort is on the construction in the 21st Century of an astronomical observatory on the lunar surface (Johnson et al., 1989; Burns et al., 1989). The focus of these efforts is productive working relationships among industry, faculty, and students in departments of engineering, geology, and astronomy. Continued support of education by all members of the engineering and scientific community is important to the essential technological advance of our country.

Importance of Technology

We must not forget that the real basis of world leadership is science and technology applied to medicine, energy, communications, transportation, agriculture, manufacturing, pharmaceuticals, space exploration, and defense (da Costa, 1989). Before Confucius, China was the world leader in science and technology. Chinese astronomers, world explorers, chemists, metal workers, and medical people were the leaders of their time. When followers of the humanitarian philosopher Confucius sought the ideal social system for China, they decided to expend energy on solving domestic problems and avoid wasting resources on exploration and science. The result was a decline of the power of the Chinese and eventual takeover by barbarian armies. This takeover occurred despite the fact that, early in the 15th Century, China led the world in gunpowder, printing, metallurgy, engineering, and medicine. China between 1405 and 1433 sent large fleets to open westward trade routes. An expedition of 62 ships reached the east coast of Africa by 1431. The Ming Emperors halted the voyages and, in an atmosphere of conservatism, Chinese science and technology faltered, trade initiatives ceased, and the Chinese became an exploited people (The Ming Dynasty, 1988).

We in the United States are at the crossroads. We are faced with a difficult challenge in maintaining in our citizenry the basic understanding of science needed to make informed decisions on technology and technology applications. This literacy in science is important so that the United States by default does not follow the path taken by the post-Confucius Chinese. A robust space science and technology program in the United States is critical to our future as a world leader.

Educational Testing Results

The Educational Testing Service (ETS), with funding from the U.S. Department of Education and the National Science Foundation, recently performed a study of 24,000 13-year-old students in the United States, the United Kingdom, Spain, Ireland, Korea, and four Canadian Provinces (Byrne, 1989). The tests given as a part of the study were translated from English into the native languages. Math achievement results were highest in Korea. The lowest scores in mathematics were by students from the United States and from French-speaking Quebec.

In science the highest achievements were by students from British Columbia and Korea. At the bottom in science were students from the United States, Ireland, and French-speaking populations of Ontario and New Brunswick. The American students were weakest in mathematics and science skill areas most often predictive of future careers in those fields. According to Archie Lapointe of ETS, the results are reasonable indicators of what 23-year-olds of 1999 will be able to accomplish in science and mathematics.

U.S. Educational System

The shortcomings of our United States educational system are widely recognized (Adams, 1989). Where these challenges are particularly significant for our future as a nation are in science and technology. In spite of efforts to help improve teaching and to upgrade courses, the National Science Foundation reports that "even our best mathematics students have the lowest achievement scores of any country for which comparable data are available." The National Assessment of Educational Progress (NAEP) has found that "only seven percent of U.S. 17-year-olds have the knowledge and skills to perform well in college level science courses." In mathematics, because of losses in the 1970's, recent gains in performance levels still leave us behind where we were twenty years ago. Blacks, Hispanics, and other minorities will constitute 35 percent of our population by 2020 (Tifft, 1989b). Blacks, Hispanics, and women continue to display lower performance levels in mathematics as they approach graduation from high schools.

The precollegiate preparation in our public schools has been described as dismal (Adams, 1989). That fact has apparently contributed to a degradation in the quality of curricula in our universities. In many fine liberal arts programs, the required science content has been reduced. Remaining science courses have often been reduced in rigor and mathematical content. The result of this watering down has led to a situation where maintaining strong graduate programs is heavily dependent on enrollments of students born in other nations. At present more than 50% of those earning doctorates in mathematics and engineering programs in the United States come from abroad. Only about one-half of these students will remain here after graduation.

Many students entering college have weak backgrounds in algebra, geometry, and trigonometry. Some calculus courses taught in high school give these students unwarranted confidence when they enroll in college calculus. The short supply of high school mathematics teachers exacerbates the problem. Computers and sophisticated hand-held calculators will force changes in the teaching of calculus (Cipra, 1988).

The teaching of physics has not been revolutionized in the way that the practice of physics has been revolutionized by the computer. Although research in physics is accustomed to rapid change, the rate of change in teaching is slow. Easier access to computers, better software, and faculty members prepared to use computers in teaching are needed. The content of physics courses and physics textbooks needs revision. Consideration is being given to application of artificial intelligence techniques and expert systems, use of simulations as lecture demonstrations, and computer data acquisition, programming, and modeling. These applications in physics lectures and laboratories could lead to substantial improvements in the teaching of physics in coming years (Wilson and Redish, 1989).

Public comprehension of national scientific and technological challenges is important for our future. We have strong research departments but making the best use of outputs of these departments is difficult. One reason for the difficulty is the pervasive illiteracy in science and technology in our society. Our civilization rests on scientific and technological underpinnings. Can we sustain our leadership as the numbers of our scientists and engineers fall and the literacy of the general population in science and engineering remains low? Richard C. Atkinson, President of the American Association for the Advancement of Science, believes that shortly after the year 2000 our demand for Ph.D.'s in the natural sciences and engineering in the academic, governmental, and industrial sectors will be less than 60% filled by our graduation rates (Holden, 1989).

Schooling in Other Nations

In the age of space, we live in a competitive world which relates to many nations and technical disciplines. Figure 1 demonstrates the relative importance of engineering education in the USSR and in Japan as compared to the U.S. (Quayle, 1988). The Japanese soil produces few things that would sustain a nation so the country chooses to cultivate schooling. Tomonao Hayoshi, a Japanese space scientist, notes that the long-term emphasis has been in the primary education system with an emphasis on technology studies (Triplett, 1989). The view is that technology studies are of obvious practical value and can be adopted and digested very easily. The Japanese are increasingly aware of educational systems in other nations (Tifft, 1989a).

COUNTRY	NUMBER	% OF TOTAL GRADUATES
USA	78,000	8%
USSR	335,000	39%
Japan	72,000	20%

Figure 1. 1985 Engineering Graduates.

Japan's National Space Development Agency (NSDA) has become a major partner in the international space station. Their Japan Experimental Module (JEM) will be manned by three Japanese astronauts who are already in training. Plans also include microgravity experiments, a free-flyer, an unmanned shuttle, and eventually a space plane. Their H-II rocket is to have a lift capability close to that of the Ariane 4 of the European Space Agency (ESA). Their planning seems to focus on the mid-1990's although the H-II is scheduled to be ready in 1992 when they could begin offering international launch services (Triplett, 1989). The Japanese continue to search for a program which has a consensus of popular support and will prove to be practical, profitable, and of benefit to everyone.

CHALLENGES TO IMPROVE EDUCATION

Education is a journey, not a destination. Education is lifelong. We need to motivate and guide potential engineering and science students from elementary school onward. There is a continuing need to attract more and better students to programs in science and engineering. There are many good programs that seek to attract young people toward the sciences and engineering. These include high school outreach and adopt-a-school or join-a-school programs.

In every major challenge the United States faces, the quality of education is important. International competitiveness requires the presence of a healthy and vital scientific and technical enterprise in the United States. The scientific and technical communities and the federal government can work together to improve the quality of education that is so vital to scientific and technical enterprise. Some bold initiatives are required (Massey, 1989):

1. Members of professional societies could ask their members to contribute several hours per week to enhance science and mathematics education in schools, museums, and other organizations.
2. Professional societies' members could work to encourage and establish programs for students to transition from precollege institutions to institutions where they earn degrees in science and engineering.
3. Writers of research proposals should identify funds for participation of undergraduates in research projects.
4. The general public needs to be educated on the nature of the problem.

The individual initiatives suggested above can be facilitated by federal government support. Improvement of all education, engineering and science education in particular, needs to be highlighted and raised to a level of national concern. The nation's greatest challenges - being competitive internationally, being literate in engineering and science, having a capable work force, and meeting the needs of the environment and the space programs - all require science, engineering, and education.

In the report "Educating Scientists and Engineers, Grade School to Grad School" published by the Office of Technology Assessment in 1988 and in recent reports from the National Science Foundation, the challenges we face as a nation are clear (Holden, 1989). Estimates by the National Science Foundation suggest that the dip in the population of 18- to 24-year-olds to 24 million in the mid-1990's from 30 million in 1980 will lead to a substantial shortage of scientists and engineers by the year 2006. The reasoning is that about 5% of the people age 22 will earn B.S. degrees in engineering and science and only 5% of the B.S. degree holders go on to earn the Ph.D. The numbers project a shortfall of 675,000 in the scientist/engineer population by 2006 and a shortfall in Ph.D.'s of 9000 by the year 2000. The Ph.D. shortfall is predicated on the retirement of about one-third of current university science and engineering faculty by the year 2000. Studies show that white males now constitute 47% of the total workforce but by the year 2010 they will make up less than one-third of the college-age population. Currently Blacks and Hispanics are 20% of the U.S. population, but they receive only 7% of U.S. B.S. degrees and only 4% of science and engineering doctorates. The percentage of Blacks and Hispanics earning doctorates in physical sciences, engineering, mathematics, and computer sciences is about 1%. Women receive only about 7% of the doctorates in engineering and 16% of the doctorates in the physical sciences.

Hilliard Williams, who is the research director at Monsanto Corporation, suggests that fears of a shortfall are overplayed and that market demand will work to draw needed people into the sciences and engineering (Holden, 1989). He believes that a strong effort is necessary to bring women and minorities into the programs where shortfalls exist. Large numbers of women and minorities will have to be attracted to science and engineering careers. The 8% of high school graduates who apparently have the grades and courses but do not go to college should be encouraged to go to college to help fill the gap. There are now large numbers of programs sponsored by government, universities, and industry which have the goal of raising participation of women and minorities in science. In Black and Hispanic college freshman, interest in engineering and physical science has increased over the past decade (Green, 1989). There remains the challenge of preparing and motivating greater numbers of these and other students in high school for college engineering and physical science programs (Tifft, 1989b).

From the viewpoint of an American high school student in a large Albuquerque high school in an affluent neighborhood, the following points were made (Haas, 1989):

> Sports in high school get the spotlight. At my high school, the football team hasn't had a winning season in years, yet the players are treated like stars or heroes. They are given pep rallies, special assemblies, focus on the morning announcements, etc. The winning entries at science fair who went on to state competition, one of whom went to international competition - were never given any public recognition at our school - only 7 of 1800 students even entered a project - that's less than 1% and if it reveals students' interest and motivation, we're all in trouble. Without recognition - the feeling of accomplishment, the feeling of belonging - it is unlikely that most teens will have any continuing interest in science and math.

George E. Brown, Jr. (1989), Member, U.S. House of Representatives, states that "scientific literacy is as vital as language, historical, or cultural literacy. Those who master science have the potential to wield great power over those who do not. A democratic society may flounder unless all citizens understand the spirit, character, and values of the science that empowers so much of society." Science and technology are the economic and cultural engines for much of the world. As we live and work in such a world, we need a high quality science education program so that we do not become a nation of science illiterates. Brown calls for encouraging women and minorities to enter technical careers both out of fairness and to augment a shrinking number of white males available for careers in science and engineering. The educational system must encourage students from kindergarten through high school to consider and prepare for such careers.

Ways must be found to encourage bright young people to follow careers in science and technology. The long and rigorous training required for careers in science and engineering and the greater economic incentives offered by other careers discourage young people from entering science and engineering. In spite of a steady demand for engineers, the overall interest of freshman in engineering careers has shown a surprising decline over the past six years (Green, 1989).

Science education must again be made a rewarding career. The professional value system must place proper emphasis on the task of imparting knowledge to the young. The educational process for scientists and engineers must be accorded a priority at least equal to the priority given to faculty research.

EDUCATION AND ITS FUTURE IN SPACE

On July 20, 1989 at the Air and Space Museum in Washington, D.C., George Bush, President of the United States, spoke on the 20th anniversary of the Apollo Moon Landing:

> ....Space is the inescapable challenge to all the advanced nations of the Earth. And there's little question that, in the 21st century, humans will again leave their home planet for voyages of discovery and exploration. What was once improbable is now inevitable. The time has come to look beyond brief encounters. We must commit ourselves anew to a sustained program of manned exploration of the solar system - and yes - the permanent settlement of space. We must commit ourselves to a future where Americans and citizens of all nations will live and work in space.

This vision of the future, which encompasses people of all nations living and working in space, must begin with our young people. People that are now thirteen years old will be twenty-four in the year 2000 when a return to the Moon is forecast. Previously we described the results of the tests of our 13-year-olds versus those of other nations. The results are not a basis for optimism that we as a nation can meet the scientific and engineering challenges of the next century because our young students scored near the bottom in skills related to future success in science and engineering.

In a September 1988 article in Defense Science, Vice President (then Senator) Dan Quayle noted that an ambitious space policy could be in jeopardy because of a lack of qualified people. The NASA/International Space Station Freedom, Mission to Planet Earth, an outpost and science station on the Moon, voyages to Mars, the National Aerospace Plane (NASP), and national security missions in space will require young people with expertise in many disciplines. Quayle noted that even if we have the financial backing to proceed with all of these programs, we lack the trained people to pursue energetically the required technological development. He called for a coordinated effort to provide the necessary resources and incentives for better education of more young people at all levels.

In August 1988 at the Space 88 Conference held in Albuquerque, New Mexico, students of Albuquerque Public High Schools participated in the conference by presenting papers and preparing manuscripts for publication in the proceedings. The student session in which their papers were presented was attended by conference delegates and thirty high school students from throughout New Mexico. Awards were presented to the students by Astronaut, Dr. Mary Cleave. Also, elementary school age Young Astronauts enthusiastically attended a special presentation by Dr. James van Hoften on his space shuttle mission to repair the Solar Max satellite.

We, the scientific and engineering community, must be willing to motivate and serve as mentors for our elementary, middle, high school, and university students on a larger scale. Students must be rewarded for their efforts and achievements in science and technical areas in order for us to promote their continued interest and education.

ACKNOWLEDGMENTS

The constructive comments from various people are appreciated in the preparation of this manuscript, especially reviews from Phil Richter and Ed O'Donnell.

APPENDIX I. REFERENCES

Adams, Robert McC. (1989). "Smithsonian Horizons," Smithsonian Magazine, Vol. 20, No. 4, page 8.

Brown, George E., Jr. (1989). "Project 2061: A Congressional View," Science, Vol. 245, No. 4916, page 340.

Burns, Jack O., Nebojsa Duric, G. Jeffrey Taylor, and Stewart W. Johnson (1989). "Astronomical Observatories on the Moon," Scientific American, Vol. 262, No. 3.

Byrne, Gregory (1989). "U.S. Students Flunk Math, Science," Science, Vol. 243, No. 4892, page 729.

Cipra, Barry A. (1988). "Calculus: Crisis Looms in Mathematics' Future," Science, Vol. 239, No. 4847, page 1491.

da Costa, Robert (1989). "The Currency of Leadership," Defense Science, Vol. 8, No. 4, May 1989, page 5.

Green, Kenneth C. (1989). "A Profile of Undergraduates in the Sciences," American Scientist, Vol. 77, No. 5, pages 475-480.

Haas, Jennifer (1989). "Concerns About the State of Mathematics and Science Education," Presentation to Education Outreach Committee, Albuquerque, NM, June 12, 1989.

Holden, Constance (1989). "Wanted: 675,000 Future Scientists and Engineers," Science, Vol. 244, No. 4912, pages 1536-1537.

Johnson, Stewart W., et al. (1989). "Lunar Astronomical Observatories: Design Studies at the University of New Mexico" in Engineering Excellence: People Make the Difference, Proceedings, 1989 Annual Conference in Lincoln, NE, American Society for Engineering Education, Washington, D.C., pages 12-14.

Massey, Walter E. (1989). "Science Education in the United States," Science, Vol. 245, No. 4921, pages 915-921.

Quayle, Dan (1988). "Can the U.S. Win the Other Space Race?: Space Science Education," Defense Science, Vol. 7, No. 9, pages 52-53.

Tifft, Susan (1989a). "Japan's Search for U.S. Colleges," Time, Vol. 134, No. 4, page 57.

Tifft, Susan (1989b). "The Search for Minorities," Time, Vol. 134, No. 8, pages 64-65.

"The Ming Dynasty," (1988). Time, Vol. 132, No. 16, pages 10-11.

Triplett, William (1989). "Japan is Knocking at Heaven's Door," Air and Space Smithsonian, Vol. 4, No. 3, pages 52-63.

Wilson, Jack M., and Edward F. Redish (1989). "Using Computers in Teaching Physics," Physics Today, Vol. 42, No. 1, pages 34-41.

APPENDIX II. ADDITIONAL SOURCES

"AAAS Seeks Members to Work with Teachers," The AAAS Observer, Supplement to Science, July 7, 1989, No. 6, page 15.

Allis, Sam (1989). "Lessons from on High: Project STAR," Time, Vol. 133, No. 2, page 65.

Ardis, Colby V. (1989). "Teaching Civil Engineering in Space," Civil Engineering, Vol. 59, No. 9, page 6.

Banks, Peter M. and Sally K. Ride (1989). "Soviets in Space," Scientific American, Vol. 260, No. 2, pages 32-40.

Boyle, Charles (1989). "We Can Help Fix Science Teaching By Fixing Tax Policy," The AAAS Observer, Supplement to Science, July 7, 1989, No. 6, page 4.

Combemale, Jean-Loup R. (1989). "Research in Technology and the Federal Government, Defense Science, Vol. 5, No. 8, pages 19-22.

Comber, Mary Jo (1989). "Program Making Teachers High-Tech Resource," Albuquerque Journal, July 16, page D-2.

DeJarnette, Fred R. (1988). "Establishing a University-Based Mars Mission Research Center," Engineering Education, Vol. 78, No. 9, pages 23-25.

Ellis, Richard A. (1989). "Engineering and Technology Degrees, 1988," Engineering Education, Vol. 79, No. 4, pages 511-521.

Hartmann, Melvin J., Frederick P. Povinelli, and Steven C. Hartman (1988). "NASA's University Space Engineering Research Centers," Engineering Education, Vol. 78, No. 9, pages 19-22.

Kluger, Jeffrey (1989). "2001 Habitat: Tokyo Takes to the Sea," Discover, Vol. 9, No. 11, page 84.

Toffler, Alvin (1970). Future Shock, Random House, New York, NY, page 485 (561 pages).

New Interdisciplinary Engineering Design Course in Planetary Materials and Resource Utilization

W.I. Rose[1], J.B. Paces[1], C.A. Chesner[1], B.J. Pletka[2], A. Hellawell[2], S.K. Kawatra[2] and J.E. Pilling[2]

Abstract

A new senior/graduate-level course addressing topics of planetary materials and resource utilization was given at Michigan Technological University in the spring of '89. The purpose of the course was to increase interest in the future of space colonization and related engineering, to galvanize a working group of faculty from diverse fields interested in collaborative research, and to initiate an interdisciplinary undergraduate/graduate program in space-related studies. Development of the course required the concerted, coordinated effort of a core group of faculty from several departments. Since no similar course could be found to use as a guide, we concentrated on addressing topics fundamental to permanent, self-sustained extraterrestrial bases, with particular emphasis on utilizing indigenous materials and resources, and on processing these materials under low-gravity and vacuum conditions. The success of the course was largely due to 13 visiting speakers from NASA, USGS, universities and private companies. They each gave one or two classroom lectures in their field of specialty, spent considerable time with students and faculty participating in the course, and gave an evening lecture which was part of a 10 week, publicized lecture series open to the whole university and community-at-large. Ten internal faculty from six different departments (Geological Engineering, Metallurgical Engineering, Mechanical Engineering, Biological Sciences, Mathematics, and Civil Engineering) also presented classroom lectures. Twenty two senior and graduate students from a variety of engineering programs were registered for the course. They were divided into six interdisciplinary groups whose mission was to work on design projects which culminated in papers and oral pre-

[1]Department of Geological Engineering, [2]Department of Metallurgical Engineering, Michigan Technological University, Houghton, MI 49931

sentations. Several of the student design projects have led to further investigations and off-campus presentations. The entire lecture sequence, about 50 hours in all, was videotaped. This resource has been edited and is available for distribution.

A university grant of $50,000 covered expenses: a post-doctoral fellow who acted as the course coordinator, travel expenses for 13 visitors, and video production costs. The course helped us develop an effective interdisciplinary working group, and establish the framework for future space-related research and study programs.

Introduction

The original motivation for the course was our perception that NASA's near-future space exploration goals would foster a substantial number of career opportunities for engineers with an appreciation for the unique problems associated with extra-terrestrial conditions. We also perceived an opportunity for possible funding of the needed innovative space-related research. As one of the countries largest engineering schools, we felt that these career and research opportunities should be emphasized. This course was therefore initiated as a means of introducing faculty, graduate students, undergraduates, and the community, to some of the engineering challenges of planetary exploration and space manufacturing. As an additional motivation, many of the faculty with space-related interests saw the benefit of stimulating interdisciplinary engineering efforts as a vital means of expanding specialized perspectives. After attending an April 1988 NASA Symposium on Lunar Bases and Space Activities in the 21st Century, we felt that solutions to many of the engineering problems facing lunar and space outposts demanded an interdisciplinary approach which had not to date been utilized in our engineering curricula.

Initial Efforts

Although we felt that a new course would be the best way to initially address the space-related engineering needs of the university, deciding on the form and content of such a first-time interdisciplinary course was difficult. While existing courses in several different departments each contained some relevant material, an appropriate combination of interdepartmental courses had never been taken by students due to the lack of any established degree program. Furthermore, no individual faculty member had the proper overview for establishing the framework for the interdisciplinary course. We also perceived that a new program should be introduced gradually and allowed to expand as we gained more experience. Therefore, we decided to combine efforts on a single initial course which could galvanize existing interests and stimulate students and faculty to investigate a new interdisciplinary curriculum.

We devised a course for engineering students which gave them enough background on extraterrestrial materials and conditions so that specific design problems could be addressed. Because the core group of faculty came from materials science, mineral processing and geological backgrounds, we started with a strong orientation toward planetary materials. We decided that the new course should be accessible to advanced undergraduates, who already have basic engineering and science training, and to graduate students at different levels who would provide more academic maturity to the class and its design groups. We also hoped that a broader group of faculty would contribute appropriate lectures from their own specialties.

We tried to find similar courses on other campuses to use as a guideline. Courses in lunar and planetary geology exist on many campuses, but we wanted a strong engineering design element rather than a purely scientific approach. We found specialized engineering courses in microgravity and materials processing, such as those given by Jean Koster at the University of Colorado. However, for our initial course, we were more interested in a broader scope which would involve a wider spectrum of participants. Although these other courses offered insights into organization of some of the important topics which we felt needed to be discussed, we were ultimately left without a clear structural model.

Our desire for an interdisciplinary format made it unclear in which department the course should reside. We were reluctant to assign the upper-division course to a single department because students from other departments might not elect it. University procedures for courses listed outside of an academic department are very cloudy and students typically feel ambivalent towards them. How can university faculty interested in interdisciplinary research topics encourage a broad spectrum of students to consider an upper division class outside of their normal curriculum? Ultimately, we opted to use "wild-card" course numbers designated for "Advanced Topics" in various departments. We also recognized the need for a strong publicity campaign, describing the course to undergraduate and graduate students.

Given the limited number of internal faculty with space-related experience, we realized that visiting speakers could play a crucial role, not only in the success of the course itself, but also by interacting with faculty and graduate students to stimulate further research projects. Although we had some ideas for appropriate external scientists and engineers, we had difficulty devising a comprehensive list. We were also concerned about the cost of having a number of external speakers. Furthermore, it was obvious that we needed an individual to organize the course, but it would be an overwhelming task for a faculty member with many ongoing responsibilities. Therefore, we went to the university in search of support.

We convinced university administrators to fund the new course on the grounds that it would help initiate a creative new engineering program for Michigan Tech which is not being duplicated at other universities. We also argued that this course would help develop a better interface between university faculty and various NASA programs. This argument was strengthened by NASA's initially positive review of a proposal from Michigan Tech for an Engineering Research Center. We also pointed out that the course would additionally benefit the university by being videotaped and distributed as both an up-linked, televideo course, and as a non-profit, educational package. The $50,000 funding was budgeted into three major categories: video-production costs, traveling expenses for 10-15 external speakers, and salary to hire a course coordinator. For our coordinator we chose Jim Paces, a new Ph.D. graduate from our Geology program who had been a NASA Graduate Student Trainee at Johnson Space Center, had many contacts, and a broad knowledge of planetary science.

Organizational Details Prior to the Class

The course coordinator worked steadily for five months prior to the beginning of the course in March 1989. His first task was organizing a course outline (Table 1) and a list of visiting speakers (Table 2). This fluid activity involved many meetings, phone calls and letters. At first, the course syllabus changed constantly. Initially, we were interested in a very broad focus, but this narrowed with time as the nucleus of interest came largely from geology and materials science/engineering. The course title evolved to "Planetary Materials and Resource Utilization: Design and Applications" to reflect the need for self-sufficiency in future space operations. Topics covered during the course reflected the interests and specialties of faculty members participating in it. There were three broad categories: non-terrestrial geological resources (lunar mineralogy, petrology, surface conditions, remote sensing); exploitation of non-terrestrial resources (geotechnical and mining, lunar power systems, lunar materials processing systems); and exploitation of non-terrestrial conditions (low gravity and high vacuum materials processing in space). Clearly, all of the topics, which were discussed for anywhere from one to maybe a few hours, could have been expanded greatly. We hope that the inspiration of the present course will stimulate more-specialized, technically-rigorous courses in the future.

Crucial help in developing the external speaker list came from Dr. Wendell Mendell of NASA-JSC. Given an early copy of the course outline, he provided a list of potential speakers qualified to discuss each of the topics we had included. The involvement of Dr. Mendell and other NASA personnel greatly contributed to our success with outside speakers. Most accepted our invitation, were well-prepared, and gave stimulating, professional lectures (Table 2). We usually asked for two different talks: a more-technically oriented lecture to be delivered to the class, and a more-generalized presentation open to the public.

The coordinator got both faculty and students interested in the new course. Internal faculty were enlisted to summarize fundamentals, to introduce the concepts discussed further by external speakers, and to bridge gaps not covered by the visiting speakers. Initially many faculty and students were skeptical about the new course. They were unfamiliar with NASA's near-future plans and confused by the country's indecisive stance on the societal importance of space issues. In spite of this skepticism, the new course attracted a number of students because of the interesting and exciting aspects of the subject matter. To publicize the course, we used articles in campus and local papers, bulletins, and flyers. These advertisements explicitly encouraged student participation from the widest possible spectrum of engineering and scientific backgrounds, with the only requirement being upper level standing. Perhaps the most effective publicity was through word-of-mouth. The efforts of the course coordinator were essential here in broadening participation and giving us a critical mass.

The Class Itself

Since the entire course was to be videotaped, special consideration had to be given to classroom logistics. We chose to hold class lectures in a videoclassroom, a small (30 seat capacity) room containing video-production equipment. The video-format compromised viewing conditions for the live audience, particularly in viewing slides and overhead projections which were transferred directly to TV monitors rather than being projected onto the wall. In addition the small room forced any overflow audience to watch the lecture in an adjoining room with a live video-feed. However, we felt that these compromises were necessary to obtain high quality videotape of the visiting lectures for use as a resource in future classes. We used a large lecture hall for the evening presentations to allow for a larger community audience. Overall, we felt that we were able to generate good participation. In addition to the 22 students enrolled for credit in the class, we had extra, non-enrolled faculty and grad students at class sessions, and often more than 100 attended evening lectures.

Half of the 22 enrolled students were undergraduates and half were graduates. Five different departments were represented. Although the course had no textbook which was closely tied to material presented in class, lecturers supplied us with pertinent readings and/or references. The symposium volume "Lunar Bases in the 21st Century" (LPI, 1984) was used for background reading: many of the included papers served as models for the design projects (Table 3) which formed an essential part of the class requirements. We divided the students into five groups of four or five each, taking care to maintain a diverse set of disciplines. An advanced graduate student (five doctoral candidates were enrolled) led each design group, and was assisted by one or more of the internal faculty. At the end of the quarter, each group turned in a written paper, gave a polished oral presentation and fielded questions from classmates and faculty. The presentations and papers formed the basis of assigned grades. The quality was impressive, and two of these design projects are being presented at Space 90.

Students were asked to evaluate the class. They generally reacted positively to the visiting lectures, the subject matter and the design emphasis. They were critical of the uneven coverage, gaps, and overlaps that resulted from the multi-lecturer format. The interdisciplinary nature of the participants also meant that, at almost every lecture, some students (and faculty) were not prepared with sufficient background material, while others listened to material with which they were already quite familiar. Lecture styles and content were, not surprisingly, uneven. Internal lecturers were more reluctant than the visitors to step outside their own safe disciplinary niches. The absence of a unifying "script" led to obscure and frustrating transitions between speakers. However the design group experiences were very positive, and overshadowed all of the other minor problems.

Summary

We feel that the objectives we had set forth for this course were met effectively. It did require a tremendous effort on the part of five faculty and the coordinator. We plan to offer the course again this year, building on our past experience. Eventually we expect that new, more specialized courses will be added to the curriculum and that a degree program in space-related engineering will be established. The video of 25 hours of visitors' lectures is being made available to other users. Individual lectures can be obtained and there are also "short course" groupings of lectures on topics such as: "Geology and Mineralogy of the Moon" and " Scientific Use of Microgravity". We also plan to edit a condensed 10 hour short course version of our experience which will span the entire breadth of the class. Several months of video editing should produce a high quality product with good continuity.

Recommendations for Others

1) Build a nucleus of flexible faculty who enjoy spending some time thinking about things outside of their secure specialty.

2) Try to span departments. This made the class richer, and students and faculty learned much about other disciplines and activities on their own campus.

3) Outside speakers were very willing to interrupt busy schedules to participate in such an effort, and made effective contributions.

4) Keep the momentum going. We are very worried that now, unless we keep up the effort, the benefits we got from the course will quickly evaporate.

5) Video classroom designs are critical. Try to select a room the right size with good viewing conditions for the graphics.

6) Community involvement was very worthwhile. The evening lectures attracted other faculty, high school teachers and students, families, even religious zealots and radicals, who all made the experience better and more interesting.

TABLE 1:
PLANETARY MATERIALS AND RESOURCE UTILIZATION, Spring 1989
Syllabus *

SPEAKER	TOPIC
WEEK 1	
W.I.Rose, GE	Introduction and Purpose
****Paul Spudis**	Geological History of the Moon
Paul Spudis	Exploring the Moon in the 21st Century
****Paul Spudis**	Lunar Base in Mare Smythii
WEEK 2	
****John Alred**	Engineering a Lunar Outpost
John Alred	NASA Office of Exploration: Future
J.B. Paces, GE	Lunar "Ore" deposits: Mineralogy
S.T. Bagley, BIO	Biological Extraction Processes
WEEK 3	
W.I. Rose, GE	Magmatic Processes and Rock Types
W.W. Predebon, ME	Surface Processes: Impacting
K. Kawatra, MY	Surface Processes: Radiation
****Larry Taylor**	Lunar Mineralogy and Petrology
Larry Taylor	Origin of the Lunar Soil
Larry Taylor	Thin Sections of Lunar Samples:
WEEK 4	
****J. Zimbelman**	Mars as a Planet
J. Zimbelman	Exploration of Mars: Past, Present and Future
W.I. Rose, GE	Remote Sensing - Polar Orbiter
Lab Session	Lunar thin section/remote sensing
WEEK 5	
B.D. Alkire, CE	Geotechnical Engineering

****David Carrier**	The Strength of Lunar Soil
David Carrier	Mining on the Moon
****Ron Sovie**	Power Systems Options for Lunar/ Space Base Applications
Ron Sovie	Power Systems in Space
WEEK 6	
****Dave McKay**	Processing Raw Materials
Dave McKay	Emerging Technology for Utilization of Lunar Resources
K. Kawatra, MY	Concentration/Separation
B.J. Pletka, MY	Ceramics Engineering
WEEK 7	
****James Blacic**	Properties, Production and Application of Lunar Glasses
James Blasic	Melt Tunneling: Unique Engineering
J.E. Pilling, MY	Cellular Structures
WEEK 8	
R.W. Kolkka, MA	G Jitter
A. Agrawal, ME	Two Phase Fluid Dynamics
R. Boudreault	Material Science and Manufacturing in Microgravity
R. Boudreault	Use Space to Help Diabetics?
WEEK 9	
****Jean Koster**	Low G Science: Challenge in Fluid Mechanics and Transport Physics
Jean Koster	Behavior of Fluids in Space
A. Hellawell, MY	Aspects of Materials Processing
****John Perepezko**	Drop Tubes & Low-G Simulation
John Perepezko	Containerless Processing in Space
WEEK 10	
****Martin Glicksman**	Solidification Experiments in Low-G Environments
Martin Glicksman	University Involvement in Space Programs
****Wendell Mendell**	Human Exploration and the Need for Space Resources
Wendell Mendell	Justifications, Policies, and Key Technological Problems for 21st Century Space Colonization

Class Project Presentations

* Names in bold letters=invited external speakers
Table 2 lists addresses and affiliations
For internal speakers, department key: GE=Geological Engineering, BIO= Biological Sciences, ME=Mechanical Engineering, MY=Metallurgical Engineering, CE=Civil and Environmental Engineering, MA=Mathematics.
** Video of lecture available from Michigan Tech, Dept of Education and Public Services (906) 487-2585

TABLE 2
PLANETARY MATERIALS AND RESOURCE UTILIZATION, 1989
Invited External Speakers

John W. Alred. Exploration Studies Office, Code 1Z2, Vangard Building, NASA Johnson Space Center, Houston, TX 77058 / Lunar Base Study System; Bootstrapping; General space systems engineering

Richard Boudreault. Canadian Astronautics Limited, 1050 Morrison Drive, Ottawa, Ontario K2H 8K7 / Microgravity processing

W. David Carrier. Bromwell and Carrier, Inc., P.O. Box 5467, Lakeland, Florida 33807 / Geotechnical; Mining; Civil Engineering

James Blacic. Geophysics Group, Division of Earth and Space Sciences, MS C335, Los Alamos National Laboratory, Los Alamos, New Mexico 87545 / Lunar Glasses

Martin E. Glicksman. Materials Engineering Department, Rensselaer Polytechnic Institute, Troy, New York 12181 / Solidification and microstructures; University involvement in space programs

Jean Koster. Associate Director, Center for Low Gravity Fluid Mechanics and Transport Phenomena, Dept. Aerospace Engineering Sciences, Engineering Center, Campus Box 429, University of Colorado at Boulder, Boulder, CO 80309-0429 / Microgravity fluid dynamics and materials science

David McKay. Planetary Science Branch, Mail Code SN2, NASA Johnson Space Center, Houston, TX 77058 / Space resources utilization; Lunar oxygen; Lunar materials processing

Wendell W. Mendell. Mail Code SN21, NASA Johnson Space Center, Houston, TX 77058 / Space resources utilizations; Lunar base concepts

John Perepezko. Metallurgical and Mineral Engineering, University of Wisconsin, Madison, WI 53706 / Containerless processing; Simulation of low gravity conditions; Drop tube experiments

Ronald Sovie. NASA-Lewis Research Center, Mail Code 301-5, 21000 Brookpark Road, Cleveland, OH 44135 / Lunar power sources

Paul Spudis. Branch of Astrogeology, U.S. Geological Survey, 2255 N. Gemini Drive, Flagstaff, AZ 86001: Planetary Geology and Petrology / Lunar origin theories

Lawrence A. Taylor. Dept. of Geological Sciences, Univ. of Tennessee, Knoxville, TN 37996-1410 / Mineralogy, geochemistry and petrology of lunar materials

James R. Zimbelman. Center for Earth and Planetary Studies, National Air and Space Museum, Smithsonian Institute, Washington, DC 20560 / Planetary Science; Viking results; Mars geology

TABLE 3

PLANETARY MATERIALS AND RESOURCE UTILIZATION, 1989
Class Design Projects

Purpose: Form small working groups in order to investigate a given design problem related to lunar base development. Use an integrated, multidisciplinary approach taking advantage of the individual backgrounds of each group member.

1) Define specific aspects of the problem
2) Determine appropriate requirements
3) Research background and principles
4) Propose solution

Goal: Prepare a concise paper modeled after those presented in Lunar Bases and Space Activities in the 21st Century, 1984, Lunar & Planetary Institute

Topics:

*I. REGOLITH BRICK PRODUCTION
Most appropriate raw materials; Brick design/geometry; Production techniques; Energy requirements

II. SOLAR POWER SYSTEMS FOR MATERIALS PROCESSING
Direct (thermal) vs. Passive (Photovoltaic); Furnace design; Transport and assembly on lunar surface; Site/geometry considerations

III. LUNAR 02: COMMINUTION
Bedrock (mining, fracturing, blasing) vs. Regolith (particle size distribution, screening); Maximum particle size requirements; Dry milling and handling techniques

*IV. LUNAR 02: SEPARATION AND CONCENTRATION OF ILMENITE
Distribution and form of oxide particles; Dry handling; Purity requirements

V. LUNAR 02: EXTRACTION OF 02
Mineral chemistry; Reduction reactions; Hydrogen source; by-products; cryostorage

VI. DEALING WITH DUST
Particle distribution; Cohesion; Filtration; Abrasion

*Papers based on Topics I. and IV. are being presented at Space 90--see Strenski et. al. and DeLa'O et. al.

Working on the Moon: The Apollo Experience

Eric M. Jones[1]

Abstract

Work experience gained during the Apollo missions suggests that lunar base productivity can be enhanced through careful attention to suit design, repair capabilities, equipment design and testing, and the dust environment.

Introduction

The successful completion of any scientific or engineering project on the Moon will depend, in part, on human abilities to do useful work under lunar conditions. In making informed decisions about such things as the use of humans rather than robots for specific tasks, the scheduling of valuable human time, and the design and selection of equipment and tools, good use can be made of the existing experience base.

During the six completed landing missions, Apollo lunar surface crews conducted 160 astronaut-hours of extra-vehicular activities (EVAs) and also spent a similar sum of waking hours working in the cramped confines of the Lunar Module (LM). The first three missions were primarily proof-tests of flight hardware and procedures. The ability to land equipment and consumables was very modest but, despite stay times of no more than 32 hours, the crews of Apollos 11, 12, and 14 were able to test their mobility and their capability of doing useful work outside the spacecraft. For the last three missions, thanks to LM modifications which enabled landings with significant amounts of cargo, stay times more than doubled to three days. The crews

[1] Earth and Environmental Sciences, Los Alamos National Laboratory, Los Alamos, NM 87545

were able to use Lunar Rovers to conduct extensive local exploration and to travel up to 8 kilometers away from their immediate landing sites. During these final missions, the astronauts spent enough time doing work of sufficient complexity that their experience should be particularly useful in the formulation early-stage lunar base operating plans.

Apollo Constraints

Time constraints were the dominant fact of life in Apollo, and were ultimately the result of political rather than of technical factors. Of particular importance were decisions that, (1), the Apollo hardware would be designed to land crews; and, (2), there would be only a finite number of flight opportunities.

As defined in 1961, the basic Apollo mission was to land astronauts on the Moon and return them safely to Earth before the end of the decade. Consequently, NASA had to design the sparest possible set of flight hardware capable of satisfying the Kennedy criteria (e.g. Cortright 1975). Even with the LM upgrades, other than the mass of the descent stage engines and structures, NASA could only land about 6.0 metric tons on the lunar surface. Because the vehicle had to carry a crew, little of that tonnage could be spared for stocks of consumables and non-flight related equipment.

Had there been an opportunity to build an unmanned, cargo version of the LM descent stage and to use it for repeated landings at a single site, NASA could have accumulated assets to support very long stays and, eventually, a lunar base. However, because NASA's first priority was a manned landing and because the cost of the core program was high enough to strain political support, development of a cargo lander could not have been contemplated until design and testing of the piloted vehicle was virtually complete. At mid-decade, as spending on the development of the Saturn V, the LM, and the Command and Service Module began to decline, there was some hope that funds could then be allocated to a lunar base program; but political support never emerged and, in 1970, NASA was even forced to cancel three flights for which hardware had already been built (e.g. McDougall 1985)

During Apollo, NASA was never in a position to contemplate an accumulation of lunar assets and was only able to conduct a broad survey of lunar geologic history by sending minimally-equipped crews to a half dozen widely-scattered sites. Because stay times were short, almost every available minute was carefully allocated. There was some flexibility in the EVA time-lines and some allowance was made for unexpected difficulties and/or discoveries; but the slack never amounted to more than a few minutes per EVA. When serious difficulties were experienced, completion of a task came only at the expense of dropping something else. For example, on Apollo 15 flaws in equipment design and procedures forced the crew to spend an extra hour completing drilling tasks, an hour which then had to be deducted from their final geology traverse. Indeed, it was a credit to the planners and the crews that, because of careful use of simulation and training and careful assessment of the work loads, few serious problems were encountered.

In a lunar base era, operational flexibility will be greatly enhanced because of stays of several months duration and the availability of hundreds of tons of equipment, supplies, and structures. Nonetheless, although time pressures will be less important at a lunar base than during Apollo, the number and types of tasks which can be undertaken and the rate with which they can be completed will depend on many of the same basic factors - mobility, communications, team work, vulnerability to dust abrasion, and the like - which were important during Apollo.

Astronaut Mobility and Metabolic Rates

An astronaut's speed, confidence, and efficiency while moving will determine such things as the optimum mix of walking and riding time during traverses and will set limits on excursions ranges in the interest of a safe return to base following a vehicle breakdown. With each successive Apollo mission, crews increased their range of operations, accumulated data on mobility and metabolic rates, and gained the confidence and efficiency that comes with experience.

Learning how to walk and run effectively required adaptation not only to lunar gravity but also to the stiffness of the inflated suit, the bulk of the boots, the dramatic upward shift in center-of-mass caused by the backpacks, and, to the astronauts' inability to see their feet and such hazards as rocks, soft crater rims, and foot-grabbing cables. Because of training exercises which allowed partial simulation of the effects of 1/6-g, basic walking skills could be acquired in a matter of a few minutes. On Apollo 11, Armstrong and Aldrin quickly confirmed that they could walk short distances, carry and deploy equipment, and gather samples. On the longer missions, crews had time enough that gaits and hazard avoidance techniques (such as backing up as seldom as possible) became semi-automatic. On Apollo 15, for example, even though Scott and Irwin never made a traverse of more than 200 meters, during EVA-1 they increased their average walking speed from about 1 km/hr to about 1.5 km/hr, and then to 2.0 km/hr during EVA 2.

The first information relevant to the conduct of long traverses and emergency returns came on Apollo 12. During their second EVA, Conrad and Bean covered one 300-meter stretch up a shallow, three-percent grade at a speed of about 4 km/hr but then had to rest for a few minutes to get their heart rates and breathing under control. On Apollo 14, Shepard and Mitchell made the most difficult of the long Apollo traverses, climbing a ten-degree slope during the second half of their 1.4 km walk to the rim of Cone Crater. Although they tried to maintain a deliberate pace, they found they could only walk for a few minutes at a time before their breathing became labored and they had to stop to rest. Not including science stops, their average outbound speed was only about 1.7 km/hr. Not surprisingly, on the trip back to the LM, the walking got to be fun again and, going downhill, Shepard and Mitchell doubled their average speed.

Using this experience, NASA estimated that, with some care taken in avoiding long uphill slopes, crews making an emergency return to the LM could average 3.6 km/hr for periods of up to an hour, and 2.5 km/hr for longer periods. A return from a distance of 10 kilometers could, then, be completed in about 4 hours, provided only that the astronauts had sufficient oxygen and cooling water available. Consumption rates were related to equivalent metabolic rates. Oxygen use was about 0.16 lb/hr per 1000 BTU/hr; and, in addition to 0.3 lbs/hr of water used for equipment cooling, water use was about 1.2 lbs/hr per 1000 BTU/hr. Typical metabolic rates were 1000 BTU/hr while working around the LM or Rover, 550 BTU/hr while riding, 1290 BTU/hr while walking at 2.5 km/hr, and 1560 BTU/hr at 3.6 km/hr.

Working in Stiff Suits

Other than time constraints, getting work done during the Apollo EVAs was controlled primarily by the stiffness of the inflated suits. The suits would bend reasonably well at the knees and waist but, because of both the stiffness and the high center-of-mass, getting low was best accomplished if the astronaut had some means of support. Neither Armstrong nor Aldrin tried to bend over far enough to pick objects with their hands nor tried to kneel. As planned, they used tongs and scoops to grab samples and dropped equipment. Conrad and Bean discovered such tricks as using their tools and their tool rack as supports and using straps attached to the backpacks as a kind of toddler's harness. By the last few missions, astronauts were occasionally seen kneeling next to equipment or reaching down to pick up rocks. If there was no support close at hand, getting back up was then most easily accomplished by doing a strenuous push up followed by a frantic running-in-place as the astronaut tried to get his feet under his high center of mass. The astronauts had a number of long handled tools; but there were always cases when the right tool wasn't close at hand, an interesting sample was too big to grab with a tool, or a bolt was in an awkward position. At such times, the significant effort necessary to get within arms-length of the ground sometimes seemed worthwhile.

The stiffness of the suit arms was particularly irksome when an astronaut tried to work alone. For example, during solo attempts to sample soil, the astronaut would have to hold the end of the loaded scoop in one hand, an open sample bag in the other, and then, with both arms extended, try to pour the soil into the bag. For two people, soil sampling was easier: one person manipulated the scoop, the other the bag. Indeed many tasks could be done most efficiently if the astronauts worked as a team, each with a practised set of actions to perform.

Suit stiffness also produced severe forearm fatigue. Any movement or positioning of a leg, arm, hand, or finger away from the "rest" configuration of the pressurized suit required constant muscle tension. Because of nearly constant use, the forearm muscles suffered most. As Jack Schmitt has described the problem, "it was like squeezing a tennis ball repetitively. Within a half hour or so, the forearm muscles were sufficiently fatigued to ache and you reached a much lower level of productivity using your hands than when you started. Eventually, by pacing yourself, you reached a constant level of forearm pain such that you could tolerate it and still do the job and not drop things and still apply sufficient grip to work; and that then went on for the rest of the EVA. Fortunately, after a night's rest and sleep, that soreness went away. There was no residual soreness; and I think that's related to the lower gravity environment. On the Moon, your heart and cardiovascular system are so much more efficient in removing the products of metabolism such as lactic acid that you never damage any muscle fiber. You fatigue it, but you don't damage it. And the important thing was that, with rest, the soreness went away. It came back when you went out again and started work, but it wasn't compounded."

And finally, there was physical trauma that resulted from repeated reaching within the gloves. According to Schmitt "as you reached in the suit and just got a little bit of scraping from the rubber bladder, it grabbed at your fingernail and, eventually, lifted the nail right off the quick. It was a problem we knew about before the mission, because others had experienced it. Knowing that, I wore some nylon liners. I still had the problem, but not as rapidly as Gene Cernan, who didn't wear any liners. Ultimately, all my nails were lifted off the quick. And that was just continuous, traumatic soreness which faded into the background and you didn't worry about it. I don't recall having rough or damaged fingertips, but I think Gene and a lot of the other guys did."

Clearly, on missions of just a few days' duration, the consequences of stiff suits were tolerable. But, for longer stays, attention needs to be paid both to suit flexibility and to reduction of internal scraping and pulling. Indeed, careful attention to equipment design and testing will be critical to long-term efficiency and productivity.

Inadequate Equipment Testing/Faulty Designs

Under any novel circumstances or conditions of first use, some equipment is not going to work as expected. Indeed, it would have been surprising had there been no equipment problems during the missions. Because time was at such a premium, failures could cause serious losses of productivity and there were at least a few cases when a bit more attention to the design of particular pieces of lunar surface equipment could have produced a substantial return.

Understandably, NASA devoted most of its Apollo effort to the design and testing of equipment essential to the achievement of the landings and the safe return of crews. Flight hardware came first; suits and life support systems came second; and then everything else. Even for such prominent items as the Apollo Lunar Surface Experiment Package (ALSEP) and the Lunar Rover, development and production contracts were not let until 1966 and 1969 respectively; and, although representatives of the astronaut corps were sometimes involved in design activities right from the beginning (as with the Rover), some equipment became available for testing and familiarization only late in a mission training cycle. Crews had to be insistent to keep unfamiliar equipment off a flight manifest. As Pete Conrad implied during his mission, it was pleasure to open the LM equipment bay and find only "things I've seen before." Familiarity reduced the chance for surprises and the waste of extremely valuable time.

Some equipment problems were due to fundamental design flaws. On Apollo 11, a short core tube which Aldrin was supposed to push and then hammer into the ground, had been mistakenly designed with an internal bevel which compacted soil entering the tube. If the soil was fairly loose, the compaction wasn't severe enough to clog the core tube; but, on the Moon, natural soil compaction tends to increase dramatically with depth and, after pushing the tube in about 3 to 5 inches, Aldrin was able to hammer it only about 2 inches deeper. On Apollo 12, one of the experiment modules, its weight much lower on the Moon than on Earth, wouldn't sit upright but, rather, was flipped on its side by residual spring in the cables connecting it with the ALSEP central station. On an Apollo 15 heat flow experiment, drill stems which had been designed primarily with thermal characteristics in mind would not penetrate the compact soils encountered at depth. And, finally, on Apollo 17, a gravity-wave-detector which Schmitt has described as "the only Nobel-class experiment yet taken to the Moon" - failed because of the manufacture's insistence that one-sixth-g, pre-launch tests would reveal proprietary information and, therefore, should not be performed. As it turned out, such tests would have revealed an error in the design of a balance beam at the heart of the device, an error which prevented the beam from uncaging under lunar gravity conditions. These problems and others caused a considerable loss of data, both directly and through the collective, unplanned consumption of well over an hour of EVA time.

Other problems can be better ascribed to insufficient testing and training, most prominently in the case of the Apollo 15 deep core hole. Here, design of the drill stem was excellent but Scott and others had been working as late as the week before launch to devise procedures so that he could manually extract the core from compact soils. The effort was unsuccessful and Scott and Irwin extracted the core - sometimes described as as one of the most important samples yet returned from the Moon - only at a cost of about 15 extra minutes of EVA time and a severe (but unremarked) shoulder sprain suffered by Scott. (Later crews had a treadle and jack which made the job a great deal easier.) And then, to add insult to the literal injury, Scott also discovered that the vise mounted on the back of the Rover would not grip the core sections when he tried to separate them for transport back to Earth. As it turned out, engineering drawings used for Rover assembly showed the vise mounted backwards from its proper orientation. During assembly of the training unit used on Earth, the vise had been installed backwards from the faulty drawing and, therefore, worked properly. But the flight unit was assembled in accordance with the drawings and, therefore, cost Scott and Irwin a good deal of time and aggravation.

Repairs

In a lunar base era, an accumulation of experience with lunar conditions will aid in equipment design and testing and, with proper attention, should reduce the relative frequency with which inappropriately designed and/or inadequately tested equipment is delivered. However, because local fabrication capabilities will emerge only slowly, there will be a on-going need to modify and repair gear brought to the Moon.

Even during Apollo, astronauts were able to make some emergency repairs. In particular, on Apollo 17, Cernan and Schmitt, using a design devised and tested overnight in Houston, were able to make a replacement Rover fender using spare maps, clamps, and grey duct tape. The original fender had been damaged when Cernan inadvertently walked too close to it with a hammer dangling out of a suit pocket. The hammer caught under the fender and, because Cernan could not feel what was happening through the thick suit, the fender was torn off. At that point, Cernan attempted to re-attach the fender with tape and, not surprisingly, found that the dust - which tended to cover everything - prevented a good bond. The temporary repair lasted for about an hour's worth of driving; but a great deal of the ultimate success of the mission can be ascribed to the simple fact that Cernan and Schmitt had the opportunity to go back inside the LM, get out of the suits and gloves, and, with help from Houston, take time to fabricate a replacement fender in a relatively dust-free environment.

At a lunar base, the ability to make repairs and modifications in a clean, shirt sleeve environment will be essential to productivity. The repair shop will need an adequate supply of tools - particularly a lathe - and a good supply of spare parts, including such things as wire made from lunar metals. The feasibility of repair and modification will be greatly aided if equipment designs are made as simple as possible and with repair and modification in mind.

Dust In Everything

The problem of lunar dust is pervasive. The surface layers contain a high proportion of very fine particles which, because of their inherent dryness, tend to maintain a high electrostatic charge derived from their interaction with the solar wind. Almost any disturbance of the soil will loft a small cloud of particles which will rise much higher on the airless Moon than they would on Earth and which will tend to cling to anything they encounter. A walking astronaut lofts a particle cloud with every step and, because the particles fall so slowly under lunar gravity, the astronaut accumulates dust on his legs as he walks through the cloud. Before long, he will look like someone who has just crawled out of a coal bin. Similarly, even with four good fenders, a Rover will quickly acquire at least a fine coat of dust.

The dust creates at least two potential hazards to lunar operations. First, dust-coated surfaces tend to absorb more sunlight, increasing thermal stress on equipment. With regard to the Rover, the severity of the problem was fully appreciated only as a result of Apollo 15. On the last two missions, each of the commanders spent a few minutes at the beginning of each of the geology stops brushing dust off of thermally sensitive equipment. The second dust hazard is abrasion and mechanical clogging; and it was only on the Rover missions, particularly Apollo 17, that equipment was exposed long enough for problems to emerge. Toward the end of the third Apollo 17 EVA, the replacement fender began to fail because of more than twenty kilometers of steady sand-blasting; and a latch on the back gate was so hopelessly dust-clogged that it stopped working. The astronauts' visors had become severely scratched; frame count indicators on the cameras were unreadable because of the dust coating; and, in general, Schmitt thought that they were, at last, looking like a real geology expedition "all covered with dust."

At one point - and with little exaggeration - Cernan said that his dust-brush was his "most important tool;" and there is a general sense created by the collective Apollo experience that dust may well be the most serious impediment to productivity. In an effort to maximize the useful life of equipment intended for use outdoors, careful design attention should be paid to reducing vulnerability to abrasion, clogging, and thermal loading. Equipment should be designed for easy cleaning and, of course, roads and other prepared surface will be vital. Because lunar dust clings so easily to almost anything, careful attention will also have to be paid to keeping dust out of work and living areas.

The Fun of It All

In some ways, it is a shame that most people who watched or heard any part of the Apollo missions experienced only Apollo 11. As probably befit such an historic occasion, Armstrong and Aldrin did not crack a single joke for public consumption; nor did they sing or hum while they worked. However, working on the Moon can be an enormous amount of fun. On the later missions, with the spotlights largely turned off, people like Pete Conrad,

Al Bean, Dave Scott, Charlie Duke, Gene Cernan, and Jack Schmitt could relax a little bit and share their exhilaration.

With a little practice moonwalking was fun, almost effortless. The scenery was striking and, particularly at the mountainous sites, was even spectacular. At Hadley Rille and, at Taurus-Littrow, crews did their work against the backdrop of mountains which rise abruptly from the valley floors to summits of up to four kilometers. Because of the soft contours of the dust-covered lunar mountains, skiing terms came easily to the crews of Apollo 15 and 17. At one point late in his mission, Jack Schmitt imagined himself skiing off of the Sculptured Hills, and made "whoosh"-ing sounds as he hopped through the powder. And Gene Cernan looked out over the valley and the surrounding mountains and wished there were some air so that he could come roaring through in an airplane.

It was fun. It was spectacular. And it was little wonder that Pete Conrad was delighted to find that, even at 3.7 psi, he could whistle while he worked.

Acknowledgements and Sources

The material presented in this paper has been extracted almost entirely from the unpublished Lunar Surface Procedures volumes, Mission Reports, and Air-to-Ground Transcripts and audio tapes prepared by NASA at the time of the missions. The remaining material has been derived from interviews with Harrison H. Schmitt.

The assistance of Jack Schmitt and, as well, Janet Kovacevich, Joey Pellarin, and Dianna Ormsbee, all of the Johnson Spaceflight Center, is gratefully acknowledged.

This work has been supported by the Los Alamos National Laboratory, which is operated for the U.S. Department of Energy by the University of California.

References

Cortright, E. M. (ed) 1975, *Apollo Expeditions to the Moon*, NASA Publication SP-350.

McDougall, W. A. 1985, ... *the Heavens and the Earth,* (Basic: New York)

DESIGNED EXPERIMENTS: AN INTERDISCIPLINARY APPROACH TO SPACE EDUCATION

David O. Swint,* S. E. Jones,** Stephen R. Schmidt,†
and Stanley P. Rader‡

The challenge of space is forcing engineering faculties to re-examine their roles in preparing college graduates to contribute to the space exploration and colonization effort. One of the greatest obstacles to conquering space is lack of funding. Poor designs, cost overruns, rework and faulty products can effectively derail our space program. We need an increased educational emphasis on developing technical and managerial skills designed to improve the quality, reliability, maintainability and cost-effectiveness of our space-related engineering systems. This emphasis must be interdisciplinary in nature in order to have the impact required to substantially influence the way the space industry does business. This paper addresses a powerful interdisciplinary method of blending engineering theory with statistics in order to improve the cost-effectiveness of space-related experimentation and design of engineering systems.

INTRODUCTION AND BACKGROUND

This paper evolved from the USAF Academy's poster presentation entry in the Space 88 Conference on lunar based construction. The senior-level, semester-long core engineering design class project (titled Concrete: A Lunar Construction Material) which produced this poster presentation was highly successful (3rd place award at Space 88). However, it was also a very frustrating design experience for the students because of the large number of experiments required and uncertainty that the results obtained were indeed optimized. This prompted a repeat of this semester-long class project with

*Professor and Head, Department of Civil Engineering, United States Air Force Academy, CO 80840-5701

**Professor of Engineering Mechanics and Materials Science, The University of Alabama, and Visiting Professor of Engineering Mechanics, United States Air Force Academy

†Associate Professor of Mathematical Sciences, United States Air Force Academy

‡Associate Professor of Civil Engineering, United States Air Force Academy

different students . . . this time meshing with a senior-level applied statistics class. A senior-level engineering systems design class and a senior-level statistics class formed an alliance to develop a mathematical prediction model (predicting optimal engineering properties) for use by the students in the engineering design class. This orderly approach to experimental design reduced the students' frustration with the "hit-or-miss" strategy while reducing the 80 possible experiments to 18 experiments. These 18 experiments yielded a prediction equation for the remaining experimental conditions. A third semester offering of this class project produced confirmation tests for the optimum properties which supported the premise that the mathematical prediction model was a viable alternative to costly, time consuming tests of all possible combinations of factors (full-factorial). This experience convinced the students that the use of statistically designed experiments will maximize the amount of information while minimizing the number of experimental trials. An expansion of this experiment is discussed in the section titled Optimizing Lunar Concrete.

The success of these classes prompted the senior Ph.D.'s in the USAF Academy's four civil engineering tracks (environmental engineering, structural engineering, construction management, and geotechnical engineering) to audit the senior-level design of experiments class. The class emphasizes statistical modeling of products and processes not easily solved with traditional engineering techniques. The use of an in-class experiment using a catapult demonstrated the power of blending engineering theory with statistics. The engineering faculty found this approach to be easy to master. In fact, their group won the catapult modeling competition. This project is discussed in greater detail in the section titled Catapult Modeling. The designed experiments approach used in both the lunar concrete and catapult projects is presented in detail in the recent text by Schmidt and Launsby (1989).

OPTIMIZING LUNAR CONCRETE

Problem Statement

The Spring 1989 Engineering 410 class at the U.S. Air Force Academy studied lunar concrete. Their task was to design, fabricate, test and analyze concrete mixtures in order to maximize engineering properties. The design and testing of the specimens were performed in accordance with American Society for Testing Materials (ASTM) standards.

Assuming the feasibility of lunar concrete as a construction material, the selection of the optimum concrete mix was addressed. To accomplish this task, the Engineering 410 group had to determine the best of 80 possible combinations. The group used a D-optimal design to test only 18 different mix combinations. Variables included type of cement, environment, plasticizer, wetting agent, and reinforcement. From the results of the testing, a prediction equation was formulated, from which the characteristics of the remaining 62 combinations were estimated. The generation of these 18 test runs is discussed below.

Test Set-up and Operation

Cement – Two types of cement were used for the experiment, Calcium Aluminate and Portland Type III.

Environment – The two mixing environments used during this experiment were ambient and semi-evacuated (mixed at 200 Torr). This was done to determine the effect of atmospheric pressure on the porosity of the concrete.

Plasticizer/Wetting Agent – Concrete specimens with and without each of these additives were tested.

Reinforcement – Steel wire, aluminum wire, glass fiber, simulated lunar soil, and no additive. Though aluminum is known to react with the calcium oxide present in the cements, the group was asked to investigate its feasibility as a reinforcement. If a feasible reinforcement, using aluminum already in space will eliminate the cost of shipping it to the moon.

With the five factors and their possible variations mentioned above, the group had 80 possible combinations of ingredients to test. However, due to time and monetary constraints, it was not feasible to test all eight combinations. It was desired to find a subset of these 80 combinations, specifically 18 test runs, that would provide information on the overall effect of each of the 5 variables plus 3 two-way interactions between environment, plasticizer, and wetting agent. To satisfy this requirement a computer generated D-optimal design was constructed which would also provide the necessary information to construct a mathematical Taylor Series model which could be used to predict the response for the remaining 62 untested combinations.

For each of the 18 combinations discussed above, one beam and three column specimens were produced. For every concrete batch made, the following were held constant:

Sand – Ottawa sand was used in the concrete mixtures because its particle size distribution is similar to that of a lunar soil sample (Lin, 1987).

Water – Distilled water was used.

Mix – A sand-cement-water ratio of 1.75:1:0.435 was used for all specimens (Swint, 1989).

Reinforcement Additives – The reinforcement was added by volume after removing 2% of the sand from the mixture. This was done because of the different densities of the various reinforcement additives.

Mixing Time – Each sample was mixed for 3 minutes before being placed in the molds. According to the Fall 1988 class, a 3-minute mix resulted in the strongest concrete and removed a high percentage of voids from the specimens.

Curing Time – A 3 to 4-day cure of the Calcium Aluminate cement resulted in optimal strength during Dr. Lin's tests (Lin, Love and Stark, 1987). For this reason, the Engineering 410 class allowed the Calcium Aluminate cement mixes to cure for 4 days. The Portland cement cured for 7 days in accordance with standard testing procedures. The first day of curing took place in the molds at 100% humidity. The molds were removed after 24 hours and the remaining three days of curing took place at 100% humidity also.

Specimen dimensions – matched those used in Dr. Lin's tests:

Cylindrical Column: Diameter = 1″, Length = 2″
Square Cross Section Beam: Side dimension = 1.57″
Length = 6.69″

From each test run, the specimens were tested for compressive and flexural strength. A three-point load was applied to the beams (according to ASTM C 78-64). A load rate of 0.0025 in/min was used, consistent with previous semester tests, and the beams were loaded to failure. The columns were capped with sulfur (for uniform load application), and strain gauges were mounted. The load rate was set at 0.0012 in/min. The lower load rate (as compared with the beam loading) was done to facilitate data collection. As with the beam, the columns were loaded to failure.

From these tests it was possible to determine maximum flexural strength (modulus of rupture), maximum compressive strength, modulus of elasticity and Poisson's ratio for each mix. The compressive strength only is discussed below.

Analysis and Confirmation

Two sets of specimens were prepared for each of the 18 experimental conditions shown in Appendix A. Each specimen was tested for compressive strength resulting in two response values (A and B) for each row of the design matrix. The average compressive strength per experimental condition is displayed in the Appendix A column labeled "Compressive Actual".

Once all the specimens were tested, the data was used to determine the performance of the specimens in terms of compressive strength. Using the data from the 18 test combinations, the prediction equation below was used to estimate the value of the compressive strength for the other 62 untested combinations.

$$\begin{aligned}\widehat{Y} = \overline{T} + \; & (\overline{C}_i - \overline{T}) + (\overline{E}_i - \overline{T}) + (\overline{P}_i - \overline{T}) + (\overline{A}_i - \overline{T}) \\ & +[(\overline{P_i W_i} - \overline{T}) - (\overline{P}_i - \overline{T}) - (\overline{W}_i - \overline{T})] \\ & +[(\overline{W_i E_i} - \overline{T}) - (\overline{W}_i - \overline{T}) - (\overline{E}_i - \overline{T})] \\ & +[(\overline{E_i P_i} - \overline{T}) - (\overline{E}_i - \overline{T}) - (\overline{P}_i - \overline{T})] \\ & \text{(Schmidt, 1989)} \qquad (1)\end{aligned}$$

In this equation, "$\widehat{Y}$" represents the predicted compressive strength for any concrete sample with a given combination of ingredients. "$\overline{C}_i$" represents the average response when C is at level i. The "i" subscript allows for variation of the type of cement used. For example, $\overline{C}_1$ represents the average response for specimens made with Portland Type III cement, while $\overline{C}_2$ represents the average response for samples made with Calcium Aluminate cement. The other factors are interpreted in a similar fashion. The marginal averages can be located in Appendix A.

Just as the single factors provide marginal averages (average response), the double factor combinations provide information on interactions. An interaction is very similar to a marginal average except that instead of averaging the value of the property for experimental specimens containing one specific ingredient, one must average the value of the property for the experimental specimens containing the two ingredients in the interaction. For example, for "$\hat{Y}$" representing compressive strength, $\overline{E}_1\overline{P}_1$ is calculated by finding the average of the compressive strength for all of the 18 combinations containing both E_1 and P_1 (see Appendix A). "$\overline{T}$" is the average value of compressive strength for all 18 of the experimental runs. Applying the prediction equation to any of the 80 possible combinations of ingredients, one would be able to determine the estimated compressive strength for a concrete specimen containing the given combination. For instance, if one wishes to determine the compressive strength for the specimen with the ingredients Calcium Aluminate (C_2), wetting agent (W_1), plasticizer (P_1) and no additives (A_1) mixed under ambient conditions (E_1), the equation would be as follows.

$$\begin{aligned}\hat{Y} = \overline{T} + \; & (\overline{C}_2 - \overline{T}) + (\overline{E}_1 - \overline{T}) + (\overline{W}_1 - \overline{T}) + (\overline{P}_1 - \overline{T}) + (\overline{A}_1 - \overline{T}) \\ & + [(\overline{P}_1\overline{W}_1 - \overline{T}) - (\overline{P}_1 - \overline{T}) - (\overline{W}_1 - \overline{T})] \\ & + [(\overline{W}_1\overline{E}_1 - \overline{T}) - (\overline{W}_1 - \overline{T}) - (\overline{E}_1 - \overline{T})] \\ & + [(\overline{E}_1\overline{P}_1 - \overline{T}) - (\overline{E}_1 - \overline{T}) - (\overline{P}_1 - \overline{T})] \qquad (2)\end{aligned}$$

Using the mathematical model, the group was able to predict the compressive strength for all of the 80 combinations. In determining the optimal combinations, the graphical results of the prediction algorithm were analyzed, revealing the following three possible combinations to maximize compressive strength:

1. $C_2\ E_2\ W_1\ P_1\ A_1$
2. $C_2\ E_2\ W_1\ P_2\ A_1$
3. $C_2\ E_2\ W_2\ P_1\ A_1$

Once the optimal combinations were chosen, they were remixed and tested in order to validate the prediction algorithm. One mix, consisting of three cylinders and a beam, was done for each of the combinations of interest. The mixing and testing procedures were carried out exactly as they had been for the original 18 specimens. The combinations of interest predicted and actual results are shown below:

1. $C_2\ E_2\ W_1\ P_1\ A_1$, predicted average 5862.0 psi, actual average 6694.6 psi
2. $C_2\ E_2\ W_1\ P_2\ A_1$, predicted average 6119.0 psi, actual average 6941.6 psi

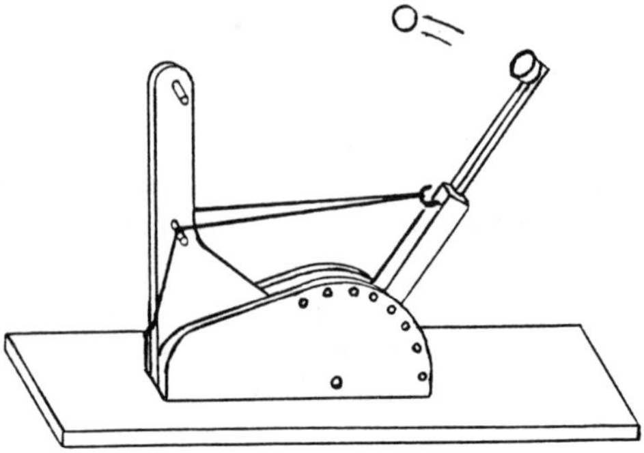

Figure 1: The Catapult

3. $C_2\ E_2\ W_2\ P_1\ A_1$, predicted average 6163.7 psi, actual average 6811.0 psi

The confirmation results exceeded predictions (from 647 psi to 833 psi). Although the difference in predicted and confirmation results is in a direction advantageous to the objective, the group was still concerned as to what caused this difference. One strong possibility is that an important interaction was not included in this design. To verify this thesis it is recommended that a follow on experiment be conducted allowing for all two-way interactions to be evaluated.

Summary

The purpose of this experiment was to find the optimum concrete mixture for lunar construction. It used the prediction equation as a statistical tool in this process. The confirmation data support the algorithm's predictions and thus show that the design of experiments approach can dramatically reduce the experimenter's work. For actual lunar construction, many more inputs must be analyzed to obtain the optimal combination. This increase in the number of input variables demands the use of the prediction equation in order to decrease time and money spent while still obtaining the optimum concrete mixture.

CATAPULT MODELING

The exploration and colonization of space will require development of products and processes that are not easily modelled using engineering knowledge alone. Other tools, such as statistics, can be used to supplement the engineering design process and thereby speed up product and process development times. The use of a mechanical device such as the catapult (see Figure 1) discussed below can be a very effective educational tool for demonstrating the power of blending engineering knowledge with statistically designed experiments. This rather simple example can provide motivation for applying the designed experiments approach to challenging space-related engineering problems.

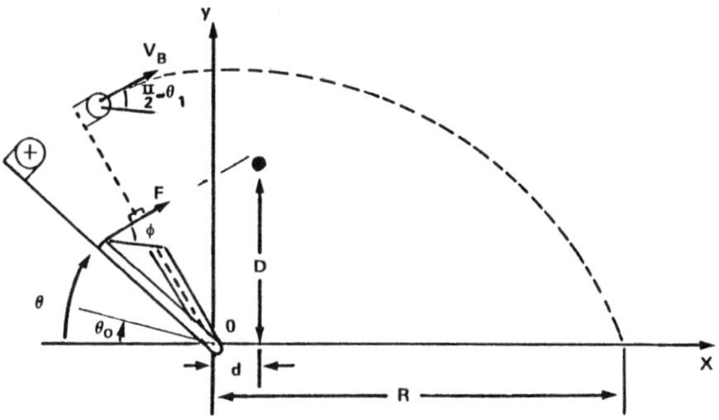

Figure 2: Catapult Parameters

Problem Statement

The Catapult is a device for launching a small projectile (ball) with initial speed v_B toward a predetermined, downrange impact site R (see Figure 2). The force F, which rotates the extensible moment arm of the catapult, is provided by a stretched rubber band. The rubber band may be attached to several points on the superstructure of the device and may assume several stretch lengths for a given configuration. The force exerted by the rubber band on the moment arm of the Catapult can be a complicated function of the stretch length. However, we assume that the deformed length is a function of the current position of the moment arm, $F = F(\theta)$.

Problem Solving Strategy: Traditional Engineering Approach

To strike the target at R, the following relationship between the variables in the problem must be satisfied (see Schmidt and Launsby (1989) for the specific details of the derivation).

$$\frac{gI_0}{4r_B} \frac{(R + r_B\cos\theta_1)^2}{\cos^2\left[\frac{\pi}{2} - \theta_1\right]\left[r_B\sin\theta_1 + (R + r_B\cos\theta_1)\tan\left[\frac{\pi}{2} - \theta_1\right]\right]}$$

$$= r_F\int_{\theta_0}^{\theta_1} F(\theta)\sin\theta\cos\theta d\theta - (Mgr_G + mgr_B)(\sin\theta_1 - \sin\theta_0) \qquad (3)$$

In this equation I_0 is the mass moments of inertia of the compound moment arm and ball assembly, r_F is the distance from 0 to the rubber band connection on the compound moment arm, r_G is the distance from 0 to the mass center of moment arm, r_B is the distance from 0 to the center of mass of the ball. M is the mass of the moment arm, m is the mass of the ball, and g is the local gravitational constant. θ_0 and θ_1 are the initial and launching angles for the catapult. The launch angle θ_1 is provided by a stop in the mechanism. The angle ϕ is given by

$$\tan\phi = \frac{D - r_F \sin\theta}{d + r_F \cos\theta} \tag{4}$$

The integral on the right hand side of equation (3) cannot be evaluated in terms of elementary functions. However, it can be numerically integrated routinely.

When equation (3) can be satisfied, the conditions for impacting the downrange site at R are met. However, before this can be accomplished, a force law for the rubber band $F(\theta)$, the moment of inertia I_0, and the distance to the mass center of the extensible compound moment arm r_G must be specified. The easiest way to accomplish this is through a series of simple experiments. Notice that a force law must be determined for each rubber band that is used with the device. For large extensions, these force laws will likely be nonlinear.

Extreme care must be taken to eliminate any sources of friction. The effects of friction have not been included in the derivation of equation (3) and the predictive capability of the equation is dependent on the degree to which friction can be minimized. The pin at 0 and the pegs over which the rubber band many pass may need to be lubricated.

Even when care is exercised and the unknown quantities are specified, there may be no solution to equation (3) for a given distance R. In this case, the force in the rubber band must be increased. On the other hand, there may be multiple solutions for a given distance R. The force $F(\theta)$ can be such that numerous combinations of initial and stop angles, θ_0 and θ_1, will satisfy equation (3).

Alternate Problem Solving Strategy: Blending Engineering and Statistics

It is going to be difficult at best to model the catapult using only the engineering approach just presented. Thus, it is advisable to seek an alternative approximate technique to speed up the modeling process. Assume that we know the true functional relationship of distance with each of the variables contained in the catapult. This function could be approximated with a Taylor Series model through the use of calculus. For the catapult as well as today's complex products and processes, the problem is that this functional relationship is typically unknown or difficult to determine. A Taylor Series model, however, can still be constructed using empirical data. For the catapult problem, a two level experimental design matrix can be used to generate the appropriate empirical data required to build a first order Taylor Series approximation which can include desired 2-way linear interactions. Less than 4 hours of instruction are normally required to explain a simple approach to accomplish this type of modeling. Air Force Academy Faculty and cadets attending an experimental design class recently used this approach for catapult modeling. They used engineering knowledge to brainstorm the four most likely variables that affect distance. These four variables were then included in an eight shot design matrix of highs and lows for each variable. The distance for these eight shots and the design matrix highs and lows were used to construct a first-order Taylor Series model. Confirmation tests of the model indicated that

± 2 inch accuracy can be achieved over the operational range of the device. After initial training, the entire experimental approach was accomplished in two hours.

Summary

The obvious conclusions are: 1) The Taylor Series modeling approach is fast and accurate; 2) in today's tight budget environment, engineers can increase their ability to reduce space-related product and process development times through the use of statistically designed experiments blended with existing engineering knowledge; and 3) the use of training devices such as the Catapult not only provide hands-on training but also make the learning process enjoyable.

CONCLUSIONS

New products such as lunar concrete and complex processes not unlike the catapult typify today's space-related engineering challenges. To meet these challenges in a resource-constrained environment, engineering graduates need to be exposed to the most broad-based and powerful problem solving methods available. The blending of engineering theory with statistics featured in the designed experiments approach is such a method. Its application to space-related engineering problems and processes may prove to be a key factor in making significant space exploration and colonization a reality.

REFERENCES

ENGR 410 class, Fall 1988. Final Report.

ENGR 410 class, Spring 1988. Display.

Flexural Strength of Concrete (Using Simple Beam with Third-Point Loading), ASTM C 78-64, pp. 40-42.

Lin, T. D. "Concrete for Lunar Base Construction." The Portland Cement Association, 1987.

Lin, T. D., H. Love and D. Stark. "Physical Properties of Concrete Made with Apollo 16 Lunar Soil Sample." Construction Technology Laboratories, 1987.

Schmidt, Stephen R., Lt. Col. (USAF) and Robert G. Launsby. Understanding Industrial Design Experiments, 1989.

Swint, David O., Col. (USAF). "Optimizing Lunar Concrete," Report to American Concrete Institute Committee 125, Lunar Concrete. 1989.

APPENDIX A

MARGINAL AVERAGES

Variable	Compressive Strength (psi)
C_1	3364.46
C_2	4464.06
E_1	3482.94
E_2	4369.28
W_1	4080.38
W_2	3870.31
P_1	3861.52
P_2	4089.17
A_1	5104.74
A_2	3777.05
A_3	2556.51
A_4	4232.34
A_5	4368.70

INTERACTIONS

Variable	Compressive Strength (psi)
W_1, E_1	3751.76
W_1, E_2	4343.28
W_2, E_1	3214.11
W_2, E_2	4395.28
E_1, P_1	3197.74
E_1, P_2	3768.13
E_2, P_1	4392.55
E_2, P_2	4346.01
W_1, P_1	3843.91
W_1, P_2	4375.97
W_2, P_1	3883.54
W_2, P_2	3859.73

COMPRESSIVE STRENGTH (psi)

	Design Matrix					Response		Compressive Strength (psi)		Deviation
Run	C	E	W	P	A	A	B	Actual	Predicted	(%)
1	1	1	2	2	1	4132.94	3428.83	3780.89	3893.40	- 2.98
2	2	1	1	1	1	5067.50	6042.79	5555.15	4962.02	+ 10.68
3	1	2	2	1	1	5207.55	4004.34	4605.95	5064.10	- 9.95
4	2	2	1	2	1	6944.25	6009.69	6476.97	6119.87	+ 5.51
5	1	2	1	1	3	3101.61	2792.21	2946.91	2214.17	+ 24.86
6	1	1	1	2	3	2272.73	2639.43	2456.08	2188.99	+ 10.87
7	1	1	2	1	4	2317.30	4278.08	3297.69	2702.07	+ 18.06
8	1	2	2	2	5	3997.97	4507.27	4252.62	4030.06	+ 5.23
9	2	1	2	2	2	4010.00	3670.00	3840.00	3665.31	+ 4.55
10	2	2	1	1	4	4260.26	5458.38	4859.32	4989.60	- 2.68
11	2	1	2	1	3	1914.95	1960.79	1937.87	2125.84	- 9.70
12	2	1	1	2	5	4318.83	5672.28	4995.56	4993.30	- 2.11
13	2	2	2	2	4	4870.00	4210.00	4540.00	4993.30	- 9.98
14	2	2	2	1	2	5692.65	5692.65	5692.65	4836.01	+ 15.05
15	1	1	1	1	2	1706.14	2294.38	2000.26	2534.73	- 26.72
16	2	2	1	1	5	3496.32	4219.52	3857.92	5125.98	- 24.74
17	1	2	1	2	2	3808.30	3342.25	3575.28	3692.58	- 3.28
18	2	2	2	2	3	2591.04	3179.28	2885.161	3317.47	- 14.98

T = 3975.35 psi **Actual - Predicted**

MIXING ASSIGNMENTS

C_1 - Portland Type III cement
C_2 - Calcium Aluminate cement

E_1 - Ambient mixing
E_2 - Semi-Evacuated

W_1 - Wetting Agent - 1 drop
W_2 - No Wetting Agent

P_1 - Plasticizer - 1ml
P_2 - No Plasticizer

A_1 - No Reinforcement
A_2 - Steel
A_3 - Aluminum
A_4 - Glass Fiber
A_5 - Simulated Lunar Soil

PROBABLE IMPACTS OF SPACE OPERATIONS ON AIR FORCE CIVIL ENGINEERING

John W. Mogge, Jr., Lt Colonel, USAF[1]

Abstract

Consider the bridge from today's space programs and systems to tomorrow's space reality. Current space policy is revolutionary in nature. Selected space related programs such as transportation systems, space control, space weapons and power have potential to effect change. Revolutionary characteristics are present in many of these programs and activities. Support concepts, specifically space logistics, engineering and space facilities, relate to these space programs and have evolutionary and revolutionary potential. Future considerations include an approach for civil engineers to develop their space support missions. The approach has four basic tenets: visualization of space as the basis of global power, focusing attention on the linkages between space policy and technology, acceptance of a bias for action in space, and the normalization of space functions in civil engineering mission areas.

Vision of Tomorrow

Today's vision can be tomorrow's reality. The bridge that takes us from visionary pursuits to realistic capabilities can be identified if one analytically studies the components of one's vision and systematically reassembles them in such a way that their influencing and controlling factors are brought to light and better understood.

One area for understanding how that bridge connects our vision of space today with the reality of space tomorrow is in the broad area of Space Logistics and its related component, civil engineering. This study has been undertaken to provide a separate analysis and assessment of the impact of space on future civil engineering missions. With a path leading to tomorrow, the impact of space on Air Force Civil Engineering can be estimated.

For this study, evolutionary change is defined as change occurring from the momentum of current trends or through change in awareness or knowledge. Revolutionary change is defined as change occurring as a result of abrupt significant events.

Policy Overview

In the area of policy formulation, space is quickly becoming a critical frontier in terms of national security. More important is the growing realization that space relates directly to nation-state security. Space then becomes an undeniable element

---

[1]Air Force Engineering and Services Center, Tyndall AFB, FL, May 1989.

of national security and has the potential to assume a critical role as routine access to space becomes more available.

Throughout the 1980s, expenditures for space programs by the Departments of Defense, Transportation, and Commerce, as well as many government agencies, have grown remarkably. In fact, the DOD currently accounts for over 70 percent of federal expenditures in space.

In brief, the US National, Department of Defense, and US Air Force policies emphasize the importance of space and recognize three broad objectives: assuring mission capability, exploring the potential of military-man-in-space and developing new technology to increase performance, and developing a general space control capability. The US Air Force Engineering and Services Space Support Policy can be summarized as continuing to support Air Force Operations in the traditional support roles, both from the earth and in space.

Enabling Technologies

NASA recognizes five areas of space technology which are enabling technologies for many programs with a wide variety of applications. These five areas of space technology are also recognized as key areas by the DOD and include: propulsion, entry technologies, precision control of large flexible structures, power, and space operations. In many research and development circles these are known as technology pulls; NASA refers to them as driver missions.

In the area of transportation, driver mission technologies are expected to make possible space-based orbital transfer vehicles (OTV), an advanced earth to orbit vehicle, translunar OTVs and, perhaps most importantly, a new generation of heavy lift vehicles. In the area of spacecraft the earth observation systems culminating in continuous geosynchronous coverage are the key programs. Perhaps most closely connected to civil engineering are the space station, large deployable reflectors, geosynchronous platforms and eventually the lunar base that the large space system technologies will make possible.

By 2010 the capability to routinely assemble large complex active structures will exist. The evolution of space-based high capacity utility power systems can be projected in decades by orders of magnitude in actual capability. Current SDI research may accelerate these time lines and bring nuclear generation capabilities into the realm of possibility much earlier than the NASA projections. The last mission driver technology area, space operations, has implications for routine manned and adaptive unmanned operations. As with the complex structures and power technologies, space operations have important long-term implications for civil engineering.

Space Station

NASA expects to be operating routinely in the space station by 1998-99. The configuration of the space station has been established to allow growth elements and unmanned platforms. It is conceivable that the station, or a duplicate of it, be configured to perform some military functions, such as exploring the potential of military-man-in-space. The implications of such capabilities, when tied to firm military requirements, presents enormous challenges to the DOD, especially when one considers that the US is less than 8 years away from permanent space station operations. However, as with the enabling technologies, the requirements for military capabilities on the space station are not firmly established in the DOD.

Functionally, NASA considers the station a multipurpose facility. The immediate and near-term uses include a national laboratory in space, a permanent observatory, a servicing facility, a transportation node, an assembly facility, a manufacturing facility, a storage depot and a staging base. Conceptually, NASA envisions the station to be an evolutionary development process beyond its initial operating capability. As more and more of the enabling technologies, developmental engineering and program definitions are mastered, the station program will become the baseline for virtually every aspect of space operations. From this standpoint the program has enormous implications for defense space systems and deserves very close attention.

Technology and Systems

This section looks at the three technological areas which have the greatest potential to shape the future for DOD. The three technology areas are: Transportation, Control and Weaponry, and Power.

Transportation, as a pacing technology driver, is shaped largely by the National Aerospace Plane (NASP) program for revolutionary impact and the Advanced Launch System (ALS) for evolutionary impact. In a sense, the evolution of Expendable Launch Vehicles (ELV) and the space shuttle can almost be assumed due to national recognition of space access imperatives and by current technological capabilities. Therefore, the NASP program becomes the focus since it offers the only real opportunity to make access to space far less expensive and immediate, thus revolutionizing space access.

Space control and its vital role in national security will demand technically advanced solutions. The programs that are underway, such as "Brilliant Pebbles," will be refined as their utility and urgency are debated by defense officials. Revolutionary technologies exist and with successful implementation, they will help shape force structure decisions and military activities well through the year 2010.

The third technological area is as vital to the Air Force's future in space as the first two areas; however, successful application of power as a space technology is necessarily dependent on the first two. Power, while not a fundamental technological area of space research, is at the very minimum a necessary co-requisite technology. In regards to probable impacts of space on Air Force civil engineering, it is in fact a vital area to review, since power generation or the provision of power for Air Force missions is a traditional mission support role for civil engineering.

Space stations, manned and unmanned platforms, directed energy weapons, satellites and space-based information systems are just a few space activities that will require power. Turn of the century power requirements for the space station are projected to be in excess of 200 kW. Future systems will be either nuclear or solar and are evolutionary with respect to earth technologies. There are solar arrays which deliver power at relatively low voltages and with unacceptable fluctuations. Research is under way to develop power processing systems which can help by synthesizing alternating current from direct currents supplied by solar arrays. Power storage systems are also being developed along with power controls to provide greater and more stable capacities. In the area of nuclear generation the issues concern efficiency in converting thermal power to electrical power, radiation shielding of manned facilities, and possible contamination due to inadvertent early re-entry of space based plants. None of these issues are beyond the reach of technological solution. Therefore it is possible to characterize this area of technology as clearly evolutionary.

Support Concepts

The notions of platforms, stations, and facilities in space suggest that space support engineering is quickly transforming itself to a more direct operational role. Unlike airbases, the stations and platforms are not currently envisioned as simply launch platforms for weapons, but observation stations, logistics nodes, and technical research laboratories. Also, peculiar to space is the notion that unlike terrestrial facilities the life support systems required in space are in and of themselves complex, technically demanding, and vital for manned presence. All of these aspects of facilities in space necessitate that the space support engineering communities prepare themselves for key roles in helping the US realize its full potential in space.

For the purposes of this study, space facilities are defined as everything in space that is not a satellite or an element of transportation. In current technical language, this would exclude the Space Transportation System (STS), NASP Derived Vehicles (NDV), Orbital Maneuvering Vehicles (OMV), Orbital Transfer Vehicles (OTV), and most satellite systems. Some elements of satellites, such as solar arrays, antennas, structural components, and housekeeping Orbital Service Units (OSU) of the future can be considered as exceptions to this definition. In addition to space facilities, or space based facilities, there is a set of space related facilities based on the earth. These are most often simply described as ground-based facilities. Regardless of the location of the facilities, all are the object of three distinct operations which military facility engineers perform.

First is the provision of the facility governed by planning, design, construction, or fabrication and acceptance functions. Second is the operation of such facilities, which generally entails the provision of utilities and actions which activate, engage, disengage, regulate or otherwise control systems and subsystems of the facility. Third is the maintenance of facilities which include a range of functions from simple part repair or replacement to depot level major system and subsystem exchange, repair or remanufacture.

Currently, there are four basic categories of space assets: vehicles, satellites, weaponry, and platforms. Traditional thought indicates that the latter category, that of platforms, will eventually become the province of facility engineers especially if they function as habitats, working space or laboratories for humans and in the case of unmanned platforms are clearly not vehicles, satellites or weapons. The space logisticians and engineers have a very significant role to play.

There have been technological advances which should not be overlooked. In a NASA report prepared by a contractor regarding human performance issues and space facilities, a number of important concerns were identified. Some of them are the lack of multipurpose tools, the omission of simple approaches such as double sided sticky tape, dissimilarities in training due to gravitational force (1G versus zero-G), blackbox level maintenance, complexity of maintenance instructions, routine maintenance by humans versus redundant and automated systems, and numerous others. Other issues even more universal include the need for privacy time by crews, the aesthetic needs of man in constant interface with machines, crew members and fitness in zero-G environments, safety hazards and design of space suits, and the full utilization of a crew member's intellect versus task rigidity and lock step methodology. These concerns and issues indicate that the needed attention to space facilities and engineering in support operations is in its infancy. In another technical report by the Air Force Rocket Propulsion Laboratory on large space system design, it was suggested that space facility engineers would need basic requisite skills in advanced computer-aided design and theory of structures in microgravity.

Some important implications can be drawn from the available studies:

First, the dynamic nature of space structures involves complex engineering calculations very different from the structural design of facilities on earth.

Second, the nature of zero gravity structural analysis requires construction or assembly methods dramatically different from these on earth.

Third, space facilities design must start from a point in space about which total dynamic flexibility exists versus fixed-land assets.

Fourth, vibration analysis and torque control have radically different theories in relation to space.

These are but a few examples of how different space structures and facilities are from a design engineer's perspective. What this may, in fact, suggest is that facility engineers in space will have to equip themselves with a highly specialized set of additional skills to those needed for terrestrial based requirements.

Space logistics as a support concept has in the past year come to grips with many of the challenges inhibiting evolutionary growth parallel with operational space programs. The imperatives of space support logistics have been recognized, and the continuing technology transfer of support architecture from NASA will lead to a normalized role in the Air Force logistics communities. The military-man-in-space issue may finally find a source of resolution through the demonstrated need of man-in-the-loop space logistics and space construction. Finally, in the area of space facilities, enormous potential for change may revolutionize facility engineering by creating a requirement for a special cadre of space facility engineers with specialized skills and education unique to the planning, design, construction, assembly, fabrication, operation, maintenance, and sustenance of space facilities.

Assessment

Three fundamental activities that comprise the facility engineer's mission are to provide, operate, and sustain facilities for the Air Force. Simply stated, facility engineers of the future will act on earth and in space.

Space policy recognizes the national security imperatives of space control as a fundamental element of future global power. Space control encompasses space defensive and offensive systems. Space defensive systems may include assets deployed for periods up to 20 years. Long deployment periods of space assets necessitates orbital maintenance. Orbital maintenance may include requirements for manned and unmanned platforms or facilities in space. Therefore, the relationship of space control to the provision of space facilities is revolutionary since presently, there are no such facilities.

Another relationship is that of transportation and providing ground based facilities. The National Aerospace Plane (NASP) offers revolutionary potential for assuring access to space in accordance with National Space Policy. Mach 12 and above propulsion is possible, but may not be in place by 1995. Once in place, NASP Derived Vehicles (NDV) will be built. The operational concept of NDVs requires specialized support facilities for cargo containerization, fuel, maintenance, etc. It follows that a successful NASP program will require large complex facilities at more than one operating location and, therefore, offers revolutionary impact potential in terms of the provisioning of ground-based facilities and systems.

A third relationship concerns DOD force enhancement activities in space. DOD space policy recognizes the services' dependencies on space in the broad area of enhancing or multiplying the effectiveness of existing force structure with the aid of space assets. Satellites represent common force enhancement assets. Projections exist that state the number of satellites used by the DOD will triple by the year

2010. Therefore, a revolutionary impact of providing such high-technology facilities throughout the DOD can be anticipated. Taking this assessment one step farther, the ability to operate, maintain or sustain the projected satellite constellations may in fact, require facilities in space much like the maintenance facilities outlined for on-orbit maintenance of a space defense system. Taken as a whole the scope and timing of just one aspect of force enhancement suggests more than revolutionary impact, and can be said to characterize the relationship of force enhancement and virtually all of the facility engineer's fundamental activities as highly revolutionary.

Future Considerations

Space, as the final frontier, is here. How Air Force civil engineering chooses to invest in space can be proactive or reactive. Air Force civil engineers should adopt a set of approaches to space that form the basis of policy and long-term involvement designed to insure ability to continue to execute missions and guarantee long term health. One such set of proactive approaches follows.

The first tenet of the set is to visualize the trend of the United States and the Soviet Union to turn toward space as their basis of global power. In doing this, civil engineers should see the space control mission clearly and start developing their roles in support of that mission. Civil engineers should understand that the distinctions between war and peace are different in space and consequently focus on their warrior roles early. Civil engineers should see space from a warrior's perspective and develop ways to exploit its efficiencies in combat. Civil engineers should establish their linkage to the weapon systems of space early as they have with war planes today.

The second tenet of the set is to focus attention on the linkages between space policy and space technology. As indicated the relationships of these areas have the potential to impact in dramatic ways. In doing this, civil engineers should continually stay abreast of developments in technology. This approach should be a broad-based effort with linkage to every space related research and development activity. This should necessarily lead towards moving space out of its very narrow channel in civil engineering and help develop a better understanding of the unique space force structure/facility relationship. Civil engineers should establish liaison contacts with NASA activities involved in space station research, space flight and science and technology research involving any type of future space facility as a means to generate early information about mission support requirements. Civil engineers should develop specialized focuses on space power, logistics, facilities and military men in space missions. Last, civil engineers should learn from the late start of space logisticians and develop ways to utilize space logistics technologies and coopt such technologies for their missions.

The third tenet is to accept a bias for action for space as our frontier. One such action is to recognize the unique role of the communicators in space. Air Force civil engineers should develop new understandings of those roles and respond to the need to develop specialty areas skilled in current and future operations involving knowledge systems, high-tech supportability, power and connectivity. Civil engineers should accept the premise that while space is not an exclusive club, the Air Force has been the service leader in space for over 30 years and will continue to be so. As such, civil engineers should take a leadership role in the design and construction of all future military space facilities. Air Force civil engineers should fashion a space policy that establishes this space construction agent role firmly in the DOD. In addition to this military construction agent role, Air Force civil engineers should seek to establish a space corps with a specialized mission in space parallel to that of the Corps of Engineers civil works districts on earth.

The fourth, and last, tenet is to normalize and specialize space where appropriate in civil engineering. This could start by continually redefining space in terms of the traditional. Parallel with the logistics community, the Air Force civil engineers should attempt to normalize space by reducing it to routine functions. This might be the development of dual hatted astronauts who perform space facility engineering missions once in space. Air Force civil engineers should build a cadre of space facility engineers specially trained in the provision, operation and sustainment of space-based facilities. These engineers should begin developing operational bridges to space parallel with space requirements and technology. This can be done by strengthening all engineers knowledge of space in entry, middle and senior continuing professional education courses at the Air Force Institute of Technology (AFIT). Last, and in keeping with the first and third tenets, civil engineers should act in visionary ways. As space support concepts and capabilities are developed, organizations should be changed to fully exploit those capabilities. If specialty capabilities are needed, dedicated organizations should be developed to support such requirements. In all, corporate commitment, broad involvement and vision should be the by-words of this tenet.

How civil engineering is impacted by space can be answered to some extent by outlining avenues of proaction for civil engineering in space. It is time for civil engineers to wholeheartedly embrace space and find ways in which they can contribute to the development of the final frontier in support of national security. "To explore space is a noble goal, but to preserve the national security, to protect U. S. assets, to ensure the right of free passage and to safeguard future explorers is a solemn obligation. It is an unavoidable prerequisite to pursuing more exciting goals" (Gen Piotrowski).

REFERENCES

1. Air War College, Air University, Maxwell AFB, Alabama. USAF Blue Ribbon Panel on Space, Final Report, November 1988.
2. Secretary of the Air Force, Chief of Staff USAF. Letter to all major commands and separate operating agencies. Subject: Air Force Space Policy, 2 Dec. 1988.

3. Ellis, George, Major General, "Briefing to the Space 88 Conference, Engineering and Services Liaison Group," 31 Aug 88. (In this briefing Gen Ellis emphasized the importance of the three traditional functions for civil engineers: to provide, to operate, and to sustain (maintain) facilities for the USAF.

4. Director of Engineering and Services HQ USAF. Letter to all Major Commands DE. Subject: Engineering and Services Space Support Policy, 2 Dec. 1987.

5. Knutson, Duane E., New Mexico Engineering Research Institute (NMERI), The University of New Mexico, Editor.

6. "Large Space System Design," Technical Report USAF Rocket Propulsion Laboratory, Edwards AFB, CA, Oct 1986, pp. 1-5.

7. NASA, "Human Performance Issues Arising From Manned space Station Missions," George C. Marshal Flight Center Contract NAS 835770, Oct 1986, pp. 1-58.

8. Piotrowski, John L., General, USAF, "Space Leadership: Vision or Vanity?" Signal, May 1988, p. 27.

9. Scott, James K., Colonel, HQ USAF/LEEX. Telephone interview, 23 Aug 1988.

10. Secretary of the Air Force, Chief of Staff USAF. Letter to all Major Commands and separate operating agencies. Subject: air Force Space Policy, 2 Dec 1988.

11. US Congress. House Committee on Science and Technology, NASA's Long Range Plans. Hearings before the subcommittee on Space Science and Applications. 99th Congress, 1st Session, September 17-19, 1985; Washington D. C., US Government Printing Office, 1986, p. 54.

12. Woodcock, Gordon R., Space Stations and Platforms. Foreword by Edward G. Gibson (Malabar, FL. Orbit Book Co., 1986), Pg VII: "Potential Department of Defense Use of Permanently Manned Space Station," p. 266.

SYNTHESIS OF A SPACE INFRASTRUCTURE

Ronald M. Jones [1]

Abstract

The Rockwell International "Integrated Space Plan" consolidates a broad range of near-term and long-range objectives for the development of space. Beginning with the Space Shuttle and the establishment of Space Station *Freedom*, it logically extrapolates a course: back to the Moon, on to Mars, laying the ground work for the emergence of space commerce and in-space industrial activity, exploitation of solar resources, development of human communities in space, and eventually planetary terraforming and human interstellar migration. The hundreds of elements necessary to support this very-long-range vision are chronologically presented in a 120-plus year projection which identifies supporting, evolutionary, and synergistic technology and program interrelationships. At it's heart is a long-term growth strategy that emphasizes resiliency and robustness in any future space transportation system architecture and a philosophical approach to infrastructure development which would assure that the elements we emplace to satisfy the recently defined lunar-Mars exploration initiative, establish a foundation that does not preclude a longer range strategic objective of tapping into the vast material and energy wealth of the inner solar system.

Introduction

On the 20th anniversary of the first lunar landing, President George Bush proclaimed a new American space agenda. Looking forward to the 21st century, the President asserted

---

[1] Advanced Programs, Rockwell International, Space Transportation System Division, 12214 Lakewood Blvd. D/385-210, MC AA-96, Downey, CA
Member AIAA

"...the time has come to look beyond brief encounters. We must commit ourselves anew to a sustained program of manned exploration of the solar system and... the permanent settlement of space." He continued, "I'm proposing a long-range continuing commitment. First, for the coming decade, ...Space Station *Freedom*, our next critical step in all our space endeavors. And next, for the new century, back to the Moon... and this time back to stay." "...and then, a journey into tomorrow, a journey to another planet, a manned mission to Mars."

This announcement followed four other important recent political developments relating to the future of the civilian space program. First, the establishment of a comprehensive national space policy that would for the first time "..expand human presence and activity beyond earth orbit into the solar system." Second, the National Aeronautics and Space Act of 1958 was amended, legitimizing the settlement of space as a long-range national goal. Third, the NASA Office of Exploration (OEXP) was created to provide recommendations and alternatives to the President for a focused program of human exploration of the solar system, principally directed at the Moon and Mars. And fourth, the President reestablished the National Space Council whose primary function will be to advise and assist the President on national space policy matters including long-range goals and space development strategy for national space activities.

This is a critically important period for the U.S. space program. Decisions made in the next 12 months in Washington could shape a future of unparalleled human opportunity and prosperity. Or, we could just as easily repeat the mistakes of the past and begin a major new initiative without a long-range strategy and unwittingly assure its eventual demise. Today we can't afford the possibility of another Apollo abandonment. Now is the time to develop a clear, bold, long-range strategy for space development to preclude this from happening.

In response to the rapidly changing civil space environment, an investigation was initiated at Rockwell to assess studies to date relating to long-range space objectives and their impact on likely avenues of human expansion into the solar system. Over two dozen post-Apollo era studies were assessed for future objectives, required capabilities, technologies, elements to attain those objectives, and credible mission scenarios to focus further study. An array of goals for both our civilian manned and planetary space programs were identified. Out of this assessment, a single very long-range blueprint for future space activities was synthesized along with a strategy for space development.

The Integrated Space Plan

The Rockwell "Integrated Space Plan" (ISP), is a comprehensive synthesis of capabilities, technologies, program elements and mission scenarios (figure 1). All elements are segregated by discipline and chronologically arranged within numerous evolutionary control-sequences that synergistically support a centerline critical path of space objectives. The critical path is sequenced to support an evolutionary program of human expansion into the solar system, progressively extending to greater orbital distances from low Earth orbit (LEO). This evolutionary growth sequence also becomes the foundation for the ISP's step-by-step "building block" approach to infrastructure development.

The ISP consolidates the genius and vision of many of the world's finest minds into a single comprehensive plan.

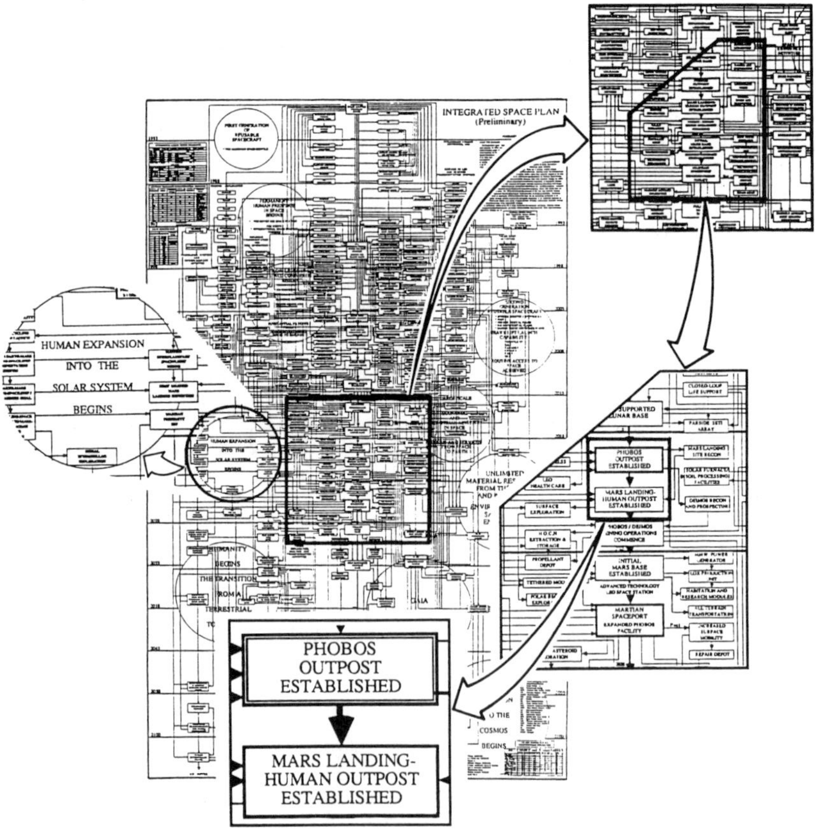

Figure 1. The Integrated Space Plan is a Comprehensive Synthesis of Hundreds of Space Program Elements Synergistically Arranged to Support Long-Range Space Program Development

It graphically depicts how humanity, through our space program, could grow beyond this planet, explore, exploit and settle the solar system and eventually migrate to the stars. The ISP identifies and organizes the fundamental steps to reach these goals, sets the steps to a rational timetable, and graphically portrays the systematic interrelationship of all of the elements involved. In so doing, a sobering overview of the size and scope of a strategically conceived space program emerges along with a meaningful benign long-range focus for our national technological prowess.

Using the Integrated Space Plan

The Integrated Space Plan is a tool. Its purpose is to assist in the process of synthesizing a space infrastructure that will satisfy space policy objectives. It has been developed for space planners, space enthusiasts, policy makers, and anyone else who wants to know more about the space program and the interrelationship of its elements.

The ISP is not a definitive plan. It is a "living" document. It is a compilation assembled from many sources. It is still being modified in an ongoing effort to compile the gamut of foreseeable elements necessary to expand the boundaries of human existence into solar space. Like a street map, it contains many paths to any destination.

The ISP infrastructure can be used as a framework to develop perspective on any space program element. It's "big picture" scope allows the significance of any program to be gauged within a longer range strategic scheme. An element can be assessed by it's ability to fill an important need. Does it occupy a necessary "niche" in the interrelated infrastructure? Will its existence enable, enhance, or compliment the functional utility of other programs? From this perspective, the evolutionary integrity of our "next steps" (i.e. Space Station *Freedom*, Shuttle-C, Personnel Launch System, and many others) can be assessed from the broader perspective of a comprehensive infrastructure. Because of its identification of long-range objectives, the ISP can also be useful for implying preferred or favorable avenues for systematic growth. Minimally, the ISP can be used as a running "score card" to track our progress, program by program, as we press forward into space.

Scope

The Integrated Space Plan is international in scope. It is a long-range extrapolation of the World's space programmatic opportunities. Key infrastructure-developing elements from the United States, Germany, France, Italy, Britain, Japan, and Canada have been incorporated into the plan. We have attempted to integrate important global milestones with currently envisioned NASA programs that support the U.S. space policy objective of expanding human presence into the solar system. We assume that the lunar base, like the Space Station *Freedom*, will be an international effort. We also assume that most of the commercial activity depicted is international. However, version 1.1 of the ISP does not include Soviet or Chinese launch vehicles or programs from rapidly developing newcomers to space like Israel and India. It is our hope that a more comprehensive global strategic space plan will eventually identify key niches for all participants in the international space community.

The Critical Path

The ISP's centerline is the critical path of core program objectives. Its thrust is human expansion into space. It progresses along a stepping-stone sequence of activity, logically extrapolating a course from low-Earth-orbit (LEO), to geosynchronous orbit, to the Moon, Mars and beyond. It evolved from a synthesis of numerous studies including assorted works from the popular literature. The ISP critical path is remarkably similar to the OEXP Pathfinder Initiative Case Study 4 -- Lunar Outpost To Early Mars Evolution. This case study has recently been split into two successive scenarios, the Lunar Evolution Case Study and the Mars Evolution Case Study. These case studies are significant because they use in-situ resources to leverage infrastructure development and compliment the ISP's strategic development strategy.

Structure

The ISP is laid out as a matrix, in columnar fashion, with time running from top to bottom. Simultaneously operating programs run from left to right. Columns on each side of the centerline are time phased to support centerline critical path milestones. Element initial operational capabilities across a range of programs and disciplines are coordinated to support a long-term thrust in a synergistic (cost effective) manner.

Columns are arranged into specific disciplinary areas including: propulsion systems and interplanetary expansion activities, precursor robotic exploration, high-Earth-orbit and cislunar transportation and operations, Earth-to-orbit and in-space transportation systems, significant in-space research and development programs, polar orbit activities, space commerce activities, space settlement establishment activities, and very long-range terraforming activities on Venus and Mars.

All column elements are sequenced to support each others evolutionary development as well as the centerline evolutionary development. Reading across the ISP from left to right yields the total scope of required activity within a given period of time to support the total "integrated" plan.

Elements

Boxes on the ISP are primarily programmatic (hardware) elements, but they may also be capabilities or achievable milestones that support the overall integrated infrastructure. They can be existing programs, future programs, existing or future capabilities, programmatic goals, technical milestones or human milestones.

Key "enabling programs" are identified by boxes with bold outlines. These programs are important because their development affects the development of many other programs. As "corner stone" elements, they are key to the infrastructure-building process.

In-space transportation nodes are identified by boxes with double lines. Nodes are important because they represent critical "stepping stone" milestones in the human expansion process.

Program Chronology

The ISP elements are laid out chronologically across five-year intervals up to the mid-21st century. Although the dates depicted are best estimates and have an uncertainty that increases with time, element chronologies are intended to be specific. Five-year increments were selected because that is a practical period of time for long-range planning and overall program depiction.

Program & Element Relationships

A number of important program relationships are depicted on the ISP. The physical connection of elements as with the Shuttle and its Remote Manipulator System or the Space Station and its supporting habitation or logistics modules, are depicted by one-way arrowed lines.

Direct supporting relationships as in the Space Station's servicing and tending capability for a harbored Space Transfer Vehicle (STV), or exploitation of the External Tank's utility as a pressure vessel to support the development of the lunar base, are depicted by arrowed lines.

Evolutionary relationships are also depicted with arrowed lines as in the Space Shuttle's evolution into an Extended Duration Orbiter (EDO), or the lunar outpost's evolution into the lunar base.

Synergistic relationships as in the mating of SpaceHab's unique capabilities with the STV to produce a manned STV, or mating the Flight Telerobotic Servicer to the Orbital Maneuvering Vehicle (or STV) to produce a maneuverable telerobotic servicing capability, or an External Tank mated to a Commercially Developed Space Facility (CDSF) to produce a volume intensive space station capability, are also depicted by arrowed lines.

Communications and data links between orbiting or coorbiting elements as in geostationary communications platforms and the elements they serve, or the Space Station and its coorbiting platforms, are depicted by two-way arrows.

Epoch Bubbles

The ISP's 120-plus year range is highlighted by a series of evolutionary bubbles or "eras of future history". Their purpose is to articulate the synergistic summation of the individual elements of a given time span. They represent the highest level groups of goals; plateaus of technological and human achievement attainable only through a comprehensive execution of the depicted ISP elements. Some attempt to articulate underlying philosophical motivations that direct the pragmatic aspects of humanity's expansion into space. Motivations like long term human survival, proliferation of the species, and human prosperity are all aspects of our future in space. These long-term cultural drivers, in reach within the span of a human lifetime through an effective channeling of existing knowledge and resources, should become topics for discussion and consideration as we grapple for purpose and meaning in our near-term space activities. They suggest an underlying strategy for space development that must eventually be addressed by the space community at large and understood by the tax paying public.

The Integrated Space Plan Strategy

Emergence of Purpose

Collectively, the bubbles on the ISP suggest a more significant purpose for our space program than exploring either Mars or the Moon. They suggest that the infrastructure we eventually emplace to extend human presence beyond Earth, be robust enough to exploit the solar system's abundant wealth to assure that we (humanity) sustain long-term evolutionary growth. This "terrestrially independent" focus, (as opposed to a "Mars" focus), would end our reliance on Earth's limited resource reservoir by finally breaking us out of our closed terrestrial system. This focus is the emphasis of the ISP's long-range growth strategy. If coupled with an evolutionary lunar-Mars expansion scenario, this stratagem promises a boundless frontier of unlimited wealth for a protracted national investment in space, not simply the planting of another flag on a distant world.

If the U.S. is serious about becoming a spacefaring nation, then it must develop a politically acceptable strategy to justify the expense of it's space program. This is the challenge before the National Space Council. The strategy they choose must maximize the return on our national investment. An important point that must be stressed is that the long-term returns from President Bush's Mars initiative could be far more significant than the goal itself. If properly conceived and implemented, the lunar-Mars infrastructure can become the mechanism for tapping into limitless material and energy resources that could fuel the fires of economic growth and human expansion in the 21st century.

Terrestrial Independence Strategy

This is the thrust of the ISP. In the spirit of our new space policy, the ISP uses Space Station *Freedom* and a Moon-to-Mars stepping stone evolution to guide infrastructure growth. The ISP strategy is to maximize the use of existing assets, common elements, and built-in modularity for extended utility and growth of capability. Infrastructure is built-up over a 30 year period. Evolvable elements are utilized as part of a "building block" architecture in a strategically conceived infrastructure that satisfies national objectives (i.e sending humans to the the Moon and Mars), but is specifically tailored for a longer range evolutionary purpose. That purpose we have termed "terrestrial independence" (TI). TI will end the era of human dependence on Earth's resources for sustenance and open the inner solar system, with it's unlimited space, energy, and material wealth, to human expansion, exploitation, and proliferation.

This philosophical approach to infrastructure development, coupled with current national space policy to facilitate international involvement in space, offers the opportunity to focus a significant amount of the World's resources toward achieving a near-term goal of "evolutionary" human significance. Biplanetary civilization, a major step towards TI, could be less than two decades away.

The Role of Space Commerce

The strategy of the ISP is to seize the advantage of our new national mandate and tap the vigor of the commercial market place to secure the long term economic development of space. An important aspect of the ISP as an "integrated" plan is that it combines commercial projects with NASA's lunar/Mars exploration program. This must be done because in the long run, space commerce will be the catalyst that enables very large scale projects like space settlements or solar power satellites. This will take time, however. The bottom line is, even though some commercial ventures will see profit during the early phases of orbital infrastructure development, it will take 25 to 30 years before we realize any significant payback from our space commitment. But if we approach the Mars initiative wisely, that payback will dwarf the front-end cost of infrastructure establishment once we open the inner solar system and assume command of its energy and material wealth. The long term interests of our profit-oriented system suggest that the government will focus on the pot of gold at the end of the rainbow. This is the logic incorporated in the ISP and is why the role of space commerce is so important in the long term development of space.

To initiate this process, ways must be identified to maximize free market forces and create near-term opportunities for commerce in space as mandated by the new Space Policy Directive. Integral to space policy and ISP evolution is the use of existing resources and capabilities to stimulate commercial activity. Included in our arsenal of assets and resources, the Space Shuttle and it's evolutionary elements should play a pivotal role in the development of on-orbit commercial activity. No element however, offers the range of commercial opportunity as the currently expended Space Shuttle External Tank (ET).

The Role of an Orbital Resource Base

Integral to the ISP growth strategy is the evolution of commercial activities leading eventually to large scale industrial capability in space. To begin this process, the near-term deployment of Space Shuttle External Tanks is employed. Creation of a raw material resource base and momentum reservoir would maximize future commercial opportunities and increase infrastructure growth options. Each ET in orbit would not only make available an abundant quantity of raw material (68,000 lbs. of mostly high purity aluminum) for eventual commercial exploitation, but it's "strongback" configuration can be utilized as either a platform (over 2/3 of an acre of potentially usable interior and exterior surface area) or pressure vessel of opportunity. Additionally, each ET contains over 75,000 cubic feet of potentially inhabitable volume.

A National Strategy for Space Development

To maximize the effectiveness of limited resources and make the most of the new lunar/Mars national mandate, the United States needs a national strategy to develop the space frontier. It is the intent of the Integrated Space Plan to suggest such a strategy. The ISP would focus on the long-range objective of terrestrial independence. Doing so would satisfy our national need to lead in space and reestablish our global preeminence in the development of advanced technology. It would create a framework for international cooperative efforts to peacefully develop our solar neighborhood. It would build an optimistic open-ended future for humanity by creating the opportunity for prosperous growth and proliferation in an environment unconstrained by energy, space, or material bounds.

What is now needed is community discussion of strategic options for space development and a consensus that TI, or a similar strategy, is the wisest approach to long term infrastructure development if we are to become a spacefaring nation and eventually, a solar civilization.

Summary

Twenty years after Apollo, the American space program has once again been been given a focus. The president has dictated a far-reaching plan to expand human presence beyond Earth orbit to surface of the planet Mars. We now have the opportunity to design and construct a comprehensive space infrastructure. If we use Space Station *Freedom*, the lunar base, and a Mars base as stepping stones, we could open our solar system, with it's limitless material and energy wealth, to human expansion, exploitation, and settlement. What we need is a consensus that this is the proper approach to space development and a strategy to achieve this objective.

A strawman space infrastructure has been developed called the Integrated Space Plan. The ISP graphically depicts an American consensus on the direction and growth of future space activities. It delineates, step-by-step, the critical programs that must be undertaken to construct a resilient and robust space infrastructure and offers a specific growth strategy, called Terrestrial Independence, for the development of space.

Acknowledgements

I would like to thank R. Fred Pfost who had faith in my perception of "the vision" and insisted I record it on paper. Fred was the catalyst that made the ISP happen.

I would like to thank Rockwell International management for their support of this project. Their vision is helping to propel our industry into the 21st century.

I also thank Anita Gale, Paul Turner, H. Robert Wasiuta, Matt Marshall, Scott Pace and Jonathan Post for their editorial assistance during the preparation of this paper.

And finally, I must thank the hundreds of scientists, engineers, and visionaries whose previous works contributed to the development of the Integrated Space Plan. I must single out Dr. Gerard O'Neill, Dr. Thomas Paine and the members of the National Commission on Space and Dr. Sally Ride and her colleagues, whose work collectively became the heart of the ISP. Their vision is the vision of the Integrated Space Plan. Their work has allowed a new generation to aspire to a greater purpose. As the individual who simply put their pieces together, I can honestly say that "I stand on the shoulders of giants."

APPLICATION OF STATISTICAL DISTRIBUTION THEORY TO LAUNCH-ON-TIME FOR SPACE CONSTRUCTION LOGISTIC SUPPORT

George W. Morgenthaler[1,2]

Abstract

The ability to launch-on-time and to send payloads into space has progressed dramatically since the days of the earliest missile and space programs. Causes for delay during launch, i.e., unplanned "holds", are attributable to several sources: weather, range activities, vehicle failures, human error, etc. Recent developments in the Space Program have placed increased emphasis on understanding and mastering every aspect of launch vehicle operations. There is a particular need for highly reliable logistic support of space construction and the subsequent planned operation of space stations, large unmanned space structures, Lunar and Mars bases, and providing "guaranteed" commercial launches.

The Center for Space Construction (CSC) has acquired historical launch vehicle data for analyzing space launch vehicle logistic support of space construction. This analysis will include development of a better understanding of launch-on-time capability and simulation of required support systems for vehicle assembly and launch which are necessary to support national Space Program construction schedules.

In this paper, we present actual unscheduled "hold" data of various launch vehicles. The data have been supplied by Industrial Associates of CSC. We seek to determine suitable probability models which can describe historical data and can be used for: a) inputs to broader simulations of launch vehicle logistic space construction support processes, and b) to determine which launch operations cause the majority of the unscheduled "holds", and can be changed to improve launch-on-time. In particular, the paper investigates the ability of a compound distribution probability model to fit the data. This model is manipulated to show how launch system changes can improve launch-on-time statistics.

Introduction

The major launch vehicle families of today have their roots in the sounding rocket and tactical and ballistic missiles programs of earlier decades. An ever-improving record of launch reliabilities has been established, and improvement of design and operational practices to increase the probability of successful launch, P, still continues (Ref. 1, Leonard, et al., 1989). As payloads become larger and more expensive and as the number of astronauts involved in a space project increases, further improvement of P is mandatory.

Efforts to improve launch-on-time capability as well as P have also continued. It is more critical now than ever before to focus on launch-on-time for several reasons:

---

[1] Professor, Aerospace Engineering Sciences Dept. and Center for Space Construction, and Associate Dean, College of Engineering and Applied Sciences, University of Colorado at Boulder, 80309-0429.

[2] The author gratefully acknowledges the assistance of the following Center for Space Construction undergraduate students: Catherine Gonet, Ecole Polytechnique Feminine, Paris, France; and Kadett Chan and Kendall Nii, Department of Aerospace Engineering Sciences, University of Colorado.

The Interruptability Problems of Space Construction

The successful assembly of a large space structure such as Space Station Freedom (SSF) depends upon launching 20 or more Shuttle payloads of construction material and astronauts on a planned schedule. If interruption of the delivery schedule occurs, the partially built SSF could be jeopardized because it might not be structurally, attitudinally, thermally, or power-wise stable in its partially constructed state. Other construction interruptions, such as unexpected astronaut illness and space debris impact of SSF, might occur. All "holds" are costly. Collectively, these problems are known as the Interruptability Problem of space construction. To reduce the consequences, cargo deliveries must be on schedule and emergency rescue launches be completely controllable.

Commercial Competition

France, the USSR, Japan, China, and the United States, are now competing vigorously for commercial space launch business. The market discriminators will not only be payload size, launch reliability, ride quality, and cost, but also the ability to launch-on-time. Most U.S. launch vehicles are based upon 20-30 year old technology and will be hard-pressed to meet guaranteed "launch-on-time" offers by competitors.

Meeting Space Launch Windows

The recent Soviet and European inspections of Halley's Comet and the ultra-precision planetary tour of Voyager II remind us of the criticality of meeting launch "windows." In 1975, when two Viking landers had to be launched within 30 days of each other for Mars surface exploration, there was difficulty in achieving this feat. In fact, to achieve the first launch, the first payload had to be switched to the second launch vehicle. In the future, the stakes will be very high and launch vehicle and launch operations technology must be improved still further to assure that "windows" are met.

Space Rescue, Police Action, and Military Mission Success

As space activity becomes more widespread, the necessity for police action, for military space forces, and for precision rescue missions will increase. In all of these operations, launch-on-time will be a necessary capability of the systems involved.

Systems Analyses of Launch-On-Time

The launch-on-time problem in the Space Construction era has at least three foci:

(1) Statistical Analysis of Operations

Analyze the statistical nature of launch operations, specifically, unscheduled "holds." How do we minimize hold time once a hold has occured? Maybe we need: a) more self diagnostic Artificial Intelligence (AI) equipment for rapidly locating the problem causing the hold, or, b) modular vehicle designs for more rapid replacement of defective parts on the launch pad.

(2) Improved Design of the Vehicle

U.S. designs of the past corrected failures when they occurred, but did not apply the new Total Quality Management (TQM) program of the U.S. Air Force, i.e., constantly improving the quality of all aspects of the system: hardware, software, and operational procedures. Analyze past subsystems causing "holds" and improve the engineering of those vehicle subsystems. Increase the test spectrum to assure performance. To the extent weather and range are causes of unscheduled "holds", more accurate weather models and improved range equipment must be developed. Devotion to a TQM goal will produce results.

(3) Modelling and Simulation of Space Program Logistics Supply as a Total Delivery System

Figure 1 shows the current U.S. Launch Vehicle Fleet and their delivery capabilities.

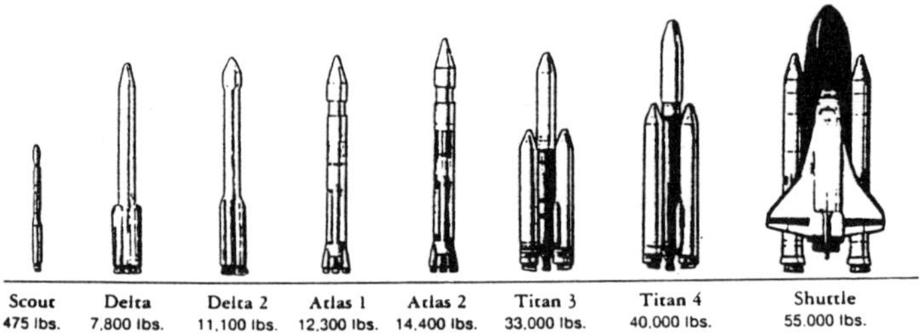

Fig. 1 Current U.S. Launch Vehicle Fleet Capabilities

Figure 2 depicts two potential new NASA cargo launch vehicles, Shuttle C and Shuttle Z, and their payloads, and also the U.S. Air Force Advanced Launch System (ALS), which may boost 150K to 280K lbs. to Low Earth Orbit (LEO).

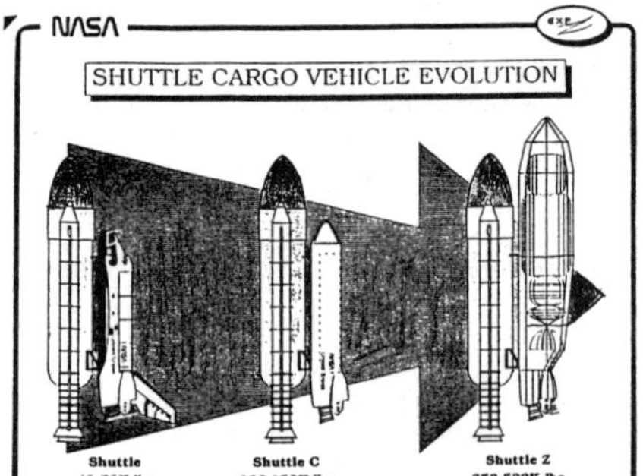

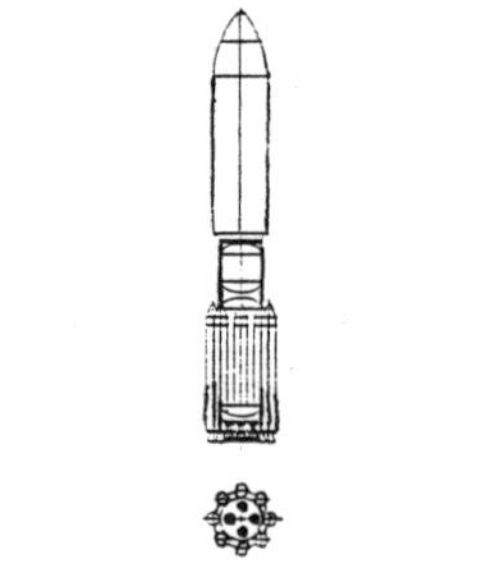

(a) Shuttle Family **(b) ALS 280,000 lbs. to LEO**

Fig. 2 Potential New Cargo Launch Vehicles

Figure 3 depicts a future orbital logistics support delivery system. If the ambitious Space Programs described in the National Space Commission Report (Ref. 2, 1986) and the report of Dr. Sally Ride (Ref. 3, 1987), or the Lunar and Mars missions described in the recent Office of Exploration Tech. Memo 4750 (Ref. 4, 1988) are to be achieved, a steady, predictable, and controlled sequence of launches using several launch vehicle families will be necessary.

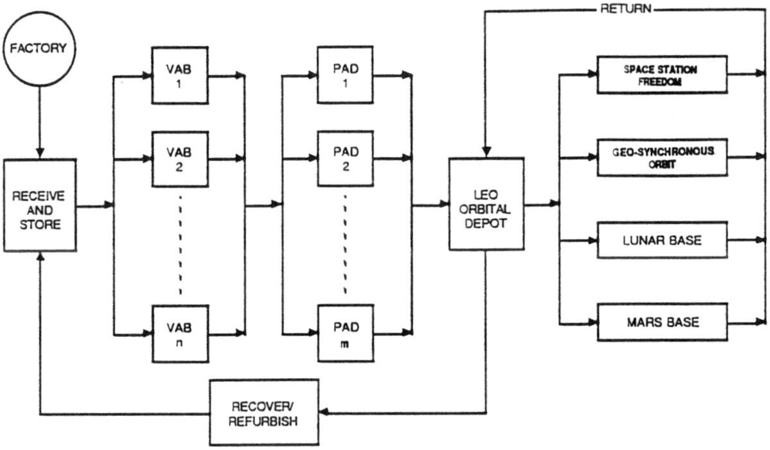

Fig. 3 Orbital Logistic Support Delivery Systems

If launch operating characteristics of the above vehicles are statistically characterized and embodied in a computer simulation model, it will be possible to define launch operating rules and set design specifications and select range equipment for the new cargo vehicles which will reduce Space Program logistic supply cost and improve launch-on-time probabilities. It is the limited objective of this paper to develop a better understanding of the causes and statistical behavior of "hold" times and to make recommendations for improvements to assure launch-on-time.

Compound Distribution Model of Launch Delay Time

Model for Calculating the Probability of "Making the Launch Window"

A launch vehicle is being "counted down" for launch. An unscheduled "hold" occurs and a repair time, t_1 is required. Another "hold" occurs and a repair time, t_2, is required, etc. The total unscheduled hold time is

$$S_N = t_1 + t_2 + \ldots + t_N,$$

where N is a random integer valued variable and each t_k has a service time distribution independent of N and of each other. The quantity S_N is a random variable which is the sum of a random number of random variables. We seek the density function h(t) of S_N, a so called "compound distribution", which can be used to calculate the probability of making the launch time, t :

(1) $H(t) = P[S_N \leq t] = \int_0^t h(u)\,du.$

Here H(t) is the cumulative distribution function (CDF) for S_N. We shall study data from four launch vehicle systems and find a functional form for h(t) and a formula for calculating H(t). We shall first find the distributions of N and of t.

N has a Geometric Distribution

The geometric density, g(N), can quite accurately represent the distribution of N, the number of unscheduled holds per launch. Let p be the probability that no unscheduled hold occurs, i.e., the probability that the vehicle launches on time as planned. Then, assuming independence from trial to trial and with q = (1 - p), the probability of a hold is:

(2) g(N) = P[N unscheduled holds, followed by launch] = $q^N p$, N = 0,1,2,3 ..., p > o.

Since $\sum_{N=0}^{\infty} q^N p = 1$, g(N) is a probability density function. Also, it can be readily verified that the mean of N, E[N], is $\left(\frac{q}{p}\right)$ and the variance, V[N], is $\frac{q}{p^2}$.

Through the cooperation of several Industrial Associates of the Center for Space Construction, the author has received the following historical launch data: the first 30 Shuttle (STS) launches from NASA, JSC; the most recent 89 Delta launches from McDonnell-Douglas; 38 Atlas-Centaur launches from General Dynamics; and the 12 Titan-Gemini launches from Martin Marietta. Figure 4 shows the fit of the geometric density, g(N), to these data from Delta, Shuttle, Atlas-Centaur, and the overall combined data, including the 12 Titan-Gemini launches.
The value of p in each case is calculated from $E[N] = \frac{q}{p} = \frac{1-p}{p}$ = sample mean of the data.

Two other distributions were also tried:

(3) the Poisson density, $f_1(N) = \frac{e^{-\lambda}\lambda^N}{N!}$, N=0,1,2 . . ., with $E(N) = \lambda$ and $Var[N] = \lambda$; and

(4) the Binomial density with m=6, the largest number of observed holds per launch

$$f_2(N) = \binom{6}{N} p^N q^{6-N}, \quad N=1,2,3,4,5,6$$

For $f_2(N)$, $E[N] = 6p$ and $Var[N] = 6p(1-p)$.

The geometric density function g(N) gave the best fit overall. (See Figure 4.)

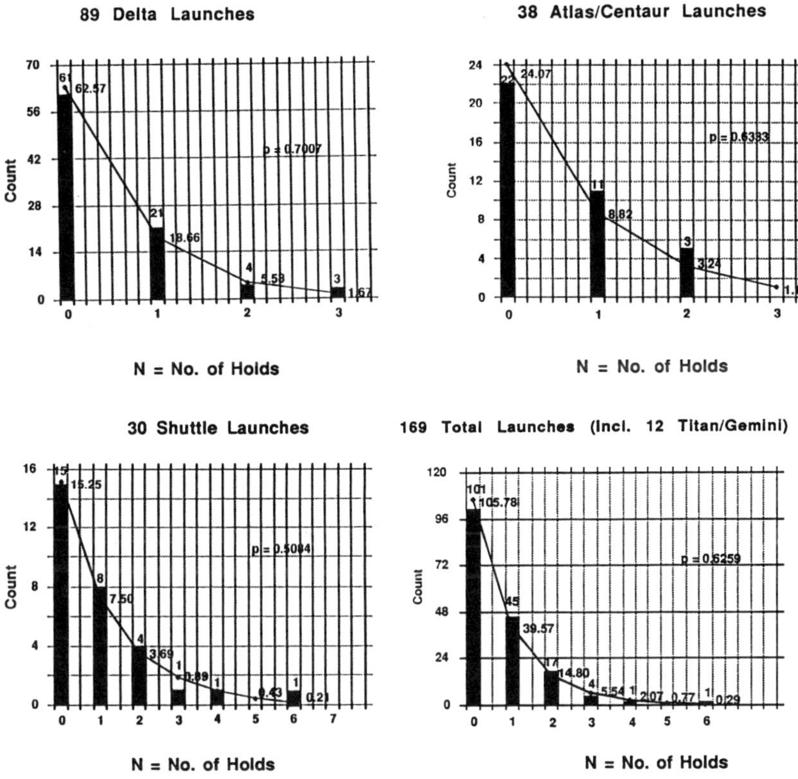

Fig. 4 Comparison of Geometric Density Function $g(N)=q^N p$, N=0,1,2... With Histograms of Number of Unscheduled Holds from Delta, Shuttle, Atlas-Centaur, and Total Combined Data, including 12 Titan-Gemini Launches.

The Distribution of Hold Times Per Hold is Exponential

We assume that the number of hold-causing problems that can be repaired in time Δt is proportional to Δt. This causes a reduction in the density of holds in effect at time, $y(t)$, by Δy. To first order, we assume that the percent reduction in y is thus also proportional to Δt, i.e.,

(5) $\frac{\Delta y}{y} = -\lambda\ \Delta t$, or $\frac{dy}{y} = -\lambda\ dt$. Then

(6) $y = Ae^{-\lambda t}$, and, since $\int_0^{\infty} Ae^{-\lambda t}\,dt = 1$, $A = \lambda$.

The density of hold times then is

(7) $f(t) = \lambda e^{-\lambda t}$, $t>0$, with $E[t] = \frac{1}{\lambda}$ and $Var[t] = \frac{1}{\lambda^2}$, and

(8) $F(t) = P[\text{hold-time} < t] = \int_0^t he^{-\lambda u}\,du = 1 - e^{-\lambda t}$.

Figure 5 shows that the Delta and Shuttle (STS) data are well-fitted by the exponential Cumulative Distribution Function (CDF):

$$F(t) = 1 - e^{-\lambda t},\ t > 0,\ \text{with } \lambda \text{ estimated from } \frac{1}{\lambda} = \frac{\sum t_i}{(\text{No. of holds})}.$$

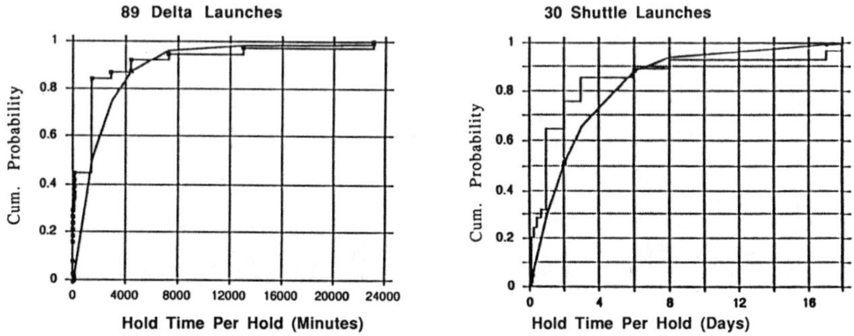

Fig. 5 Fitting of Exponential Distribution (8) to Cumulative Distribution Data of Hold-Time Per Hold for: a) 89 Delta Launches and b) 30 Shuttle (STS) Launches.

Distribution of $S_N=\Sigma t_i$. Total Hold Time Per Launch

Given the density for N and the density for the $\{t_i\}$, the distribution of

$S_N = \sum_{i=1}^{N} t_i$ is found by the theory of compound probability distributions.

The following theorem is proved in Feller (Ref. 5, 1966, pp 53,54):

Theorem. If N has the geometric density (1), $g(N)=q^N p$, N=0,1,2,3..., and if each $\{t_i\}$ has the exponential density (7), with parameter λ, then, if N and the $\{t_i\}$ are independent random variables, the compound random variable

$$S_N = \sum_{i=1}^{N} t_i$$

has the exponential distribution with constant λp, i.e., the density and CDF are:

$$\text{(9)} \qquad h(t) = \lambda p e^{-\lambda p t},\ t>0, \text{ where } E[S_N] = \frac{1}{\lambda p}, \text{ and } Var[S_N] = \frac{1}{\lambda^2 p^2}. \text{ Also,}$$

$$\text{(10)} \qquad H(t) = P[S_N \le t] = \int_0^t \lambda p e^{-\lambda p u}\, du = 1 - e^{-\lambda p t}.$$

In Fig. 6 we compare the Cumulative Distribution Function H(t) in (10) with the Delta and Shuttle (STS) total hold time data per launch, and observe that we have quite good fits.

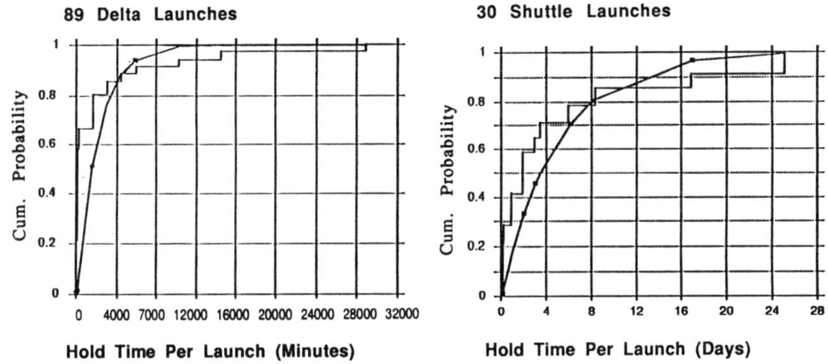

Fig. 6 Fitting of Exponential Cumulative Distribution (10) to Cumulative Distribution Data of Total Hold Time Per Launch for 89 Delta Launches and 30 Shuttle (STS) Launches.

Improving Launch-On-Time Probability

From the theorem,

$$P[S_N \le t] = 1 - e^{\lambda p t},\ t > 0, \text{ and } E[S_N] = \frac{1}{\lambda p} = \text{average unscheduled "hold" time.}$$

Here p is the probability of no holds per launch and $1/\lambda$ is the average hold-time per hold so that λ may be interpreted as the number of hold problems "fixed" per unit time. Increasing λ thereby reduces hold time.

The average hold time per launch, $E[S_N]$, can be used to measure the improvement in launch-on-time. We can decrease $E[S_N]$ by increasing p, increasing λ, or both.

<u>Reduction in Number of Holds, N (increase in p):</u>

Figure 7 shows the distribution of causes for unscheduled "holds" between range holds, launch vehicle holds, weather holds, payload holds, etc. Reduction in the average number of holds can be made by: improving the range equipment so that range holds disappear; improving weather prediction (by the use of parallel computing techniques and improving the accuracy of weather prediction models); improving the design of those elements of launch vehicles which have historically caused launch vehicle holds; or improving the design of such similar elements on payloads.

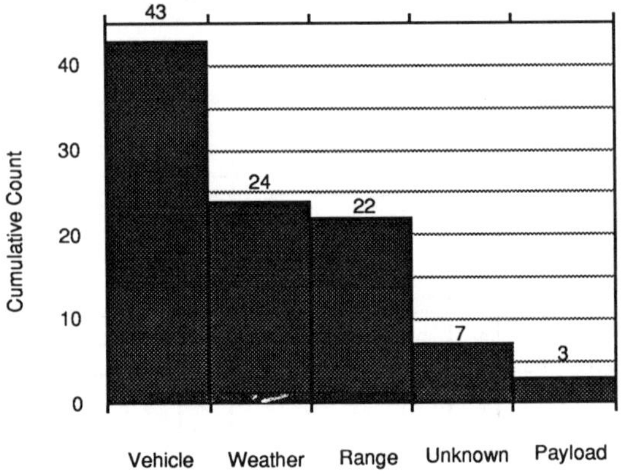

Fig. 7 Frequency of Unscheduled Holds By Cause for Total Data on 169 Launches. (Delta (89), Atlas/Centaur (38), Titan/Gemini (12), Shuttle (30))

<u>Reduction in Hold-Time per Hold (increase in λ):</u>

Hold times, given a hold, can be reduced by: a) using Artificial Intelligence diagnostic fault location methods to quickly find faults; b) keeping sufficient spares on hand to immediately fix launch vehicle problems; c) using modular designs so that failed items can be quickly replaced; and d) establishing better training programs for launch facility technicians.

What is the Relative Effect of increasing p or increasing λ as regards $E[S_N]$?

Let $f = E[S_N] = \frac{1}{p\lambda}$. Then we have:

$$df = d\{E[S_N]\} = \frac{\partial f}{\partial p} dp + \frac{\partial f}{\partial \lambda} d\lambda = -\frac{dp}{p^2\lambda} - \frac{d\lambda}{p\lambda^2}$$

$$= -\frac{1}{p\lambda}\left[\frac{dp}{p} + \frac{d\lambda}{\lambda}\right] = -E[S_N]\left[\frac{dp}{p} + \frac{d\lambda}{\lambda}\right].$$

Hence,

(11) $$\frac{dE[S_N]}{E[S_N]} = -\left[\frac{dp}{p} + \frac{d\lambda}{\lambda}\right].$$

Thus, a 5% increase in p alone, or a 5 % increase in λ alone will each give a 5% decrease in $E[S_N]$. Of course, the technical difficulty and cost of increasing p by 5% vs. increasing λ by 5% may be quite different. The value of formula (11) when combined with the data in Figure 7 is that an organized, predictable launch-on-time improvement campaign can be undertaken in a cost-efficient manner.

Conclusion

The historic approaches to launch-on-time must be improved. We must develop assured launch-on-time capability. This means understanding launch statistics better and designing and testing for decreased probability, $q = (1-p)$, of experiencing an unscheduled hold. We must also use Artificial Intelligence fault diagnostic techniques, design the launch system in a modular manner, and have spares on hand so that hold-time, when it occurs, is minimized. The author is currently expanding formula (11) and obtaining causal fault data to include a variety of specific sources of improvement in p and λ.

References

1. Leonard, Peter B., and Kisko, William A., "Predicting Launch Vehicle Failure", Aerospace America, September, 1989, pp. 36-38, and 46.

2. National Commission on Space Report, Pioneering the Space Frontier, Bantam Books, New York, N.Y., May, 1986.

3. Ride, Sally., Leadership and America's Future in Space, U.S. Government Printing Office, A Report to the NASA Administrator, August 1987.

4. NASA Office of Exploration, Exploration Studies Technical Report, FY 1988 Status (Tech. Memo 4750), October, 1988.

5. William Feller, An Introduction to Probability Theory and its Applications, Vol. II, John Wiley and Sons, Inc., New York, N.Y., 1966.

COLUMBUS GENERIC ELEMENT MANAGEMENT CONCEPT

John Sved(1)

ABSTRACT

The concept for largely autonomous onboard management of the ESA Columbus project elements relies on the automation of operational procedures. Reconfigurable items are commanded in sequence during execution of Automated Procedures that ensure failure management during indeterminate system states. At other times a new form of generalized decision tables will deal with incomplete failure symptoms. Unacceptable operations are avoided by the management of execution of a planned timeline of predefined Actions. The concept supports both unmanned and manned systems.

INTRODUCTION

The operation of systems, facilities and bases in a cis-lunar industrial infrastructure will be characterized by much greater autonomy. The recurring cost of the Earth based Mission Control Center will be reduced through automation of sustaining flight operations. The major routine tasks such as mission planning, the inevitable replanning when activities do not go exactly as planned and the diagnosis of anomaly symptoms that lead to deviations will be performed by onboard mission manager software applications.

This paper describes a concept developed in Europe for enhanced autonomous onboard management of space systems. Under the auspices of the ESA Columbus project, the operations and software community has developed a concept for a so-called Generic Element Manager (GEM). Representatives of the Columbus Operations Team come from the ESA Space Technology Center (ESTEC), European Space

---

(1) Member American Institute of Aeronautics and Astronautics, Flight Operations Engineer, MBB/ERNO Raumfahrttechnik GmbH, Bremen, FRG.

Operations Center (ESOC) and the Columbus industrial consortia led by prime contractor MBB/ERNO, Bremen.

The concept has its genesis in the evolving spacecraft avionics field where computer power in distributed microprocessor architectures will enable more elaborate software to be carried. ESA has funded a Standard Generic Approach to Spacecraft Autonomy and Automation (SGACAA) study (ESTEC 1987) that considered ways of implementing more onboard autonomy. A hierarchical architecture was identified where a Mission Manager computer delegates supervisory tasks to subordinate subsystem and payload managers as indicated in Figure 1. More recently, practical implementations that go into these manager functions have emerged for operational evaluation.

AUTONOMY CONCEPT

A spacecraft system is a complex electro-mechanical assembly controlled via a data distribution network that supports monitor and control software. The system has many potential configurations and corresponding states. It may be regarded as being in a stable state if there is no transition from one state description vector to another. Oscillating processes that are nominal may be described as stable modes of operation.

Classic unmanned spacecraft are reconfigured from one operational mode to another by sequences of primary commands uplinked for immediate or perhaps time tagged execution. Manned space missions have had the advantage of crew who were proficient in the execution of many tasks and who could retrieve documentation to guide them exactly through checklists and reconfiguration switch sequences. In the event of anomalies the crews' expertise has often been integral in Failure Detection Isolation and Recovery (FDIR) processes. Unmanned spacecraft have typically carried a limited FDIR facility that could autonomously switch to redundant equipment or a minimal activity safe mode to wait for ground intervention; with loss of payload productivity.

More onboard memory permits the procedural knowledge to be stored in databases for retrieval when required. Ground support focuses on the maintenance of the onboard database so that it reflects the system configuration and operational characteristics. Mission planning also originates from the Mission Control Center (MCC) which ensures that productivity remains as high as possible.

CONFIGURATION END ITEMS

From the operations viewpoint, the configuration of a space system may be described down to the level of

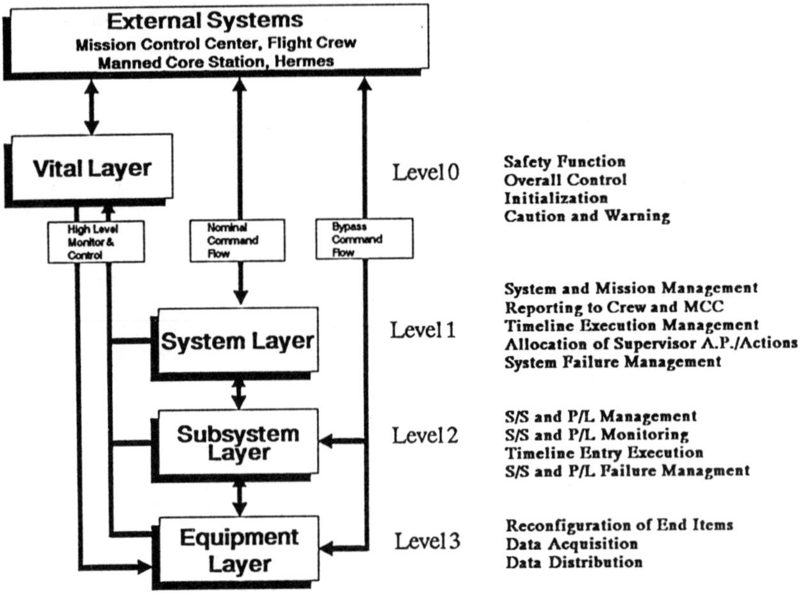

Figure 1. Hierarchical distributed management concept: Subordinate managers execute Automated Procedures allocated by the system manager.

individual re-configurable END ITEMS such as switches, valves, sensors, mechanical actuators, software modules, software control parameters, data tables to support software functions and similar entities.

The End Item naming convention will be a tree branch path system which allows for an understandable database structure, from the user viewpoint. It will be closely related to the engineering assembly break-down of the system.

Each discrete End Item command has a unique configuration database entry name such as CAL/ECLSS/Avionics_Loop/Fan1/OFF. The configuration database contains the Data Management System compatible discrete End Item Command. It may be a unique bit pattern containing the logical address of a DMS functional node, the specific I/O port and related parameters for signal conditioning. If the End Item is "smart" the signal conditioning is replaced by the bit pattern packet recognized by the End Item.

The intent of an End Item Command is to change the status or configuration of just one uniquely identified End Item. The general procedure for performing such a reconfiguration includes the steps shown in Figure 2. Such single objective procedures are potentially highly re-usable when included in various sequences of small procedures that make up a monolithic block activity for scheduling purposes.

AUTOMATED PROCEDURES

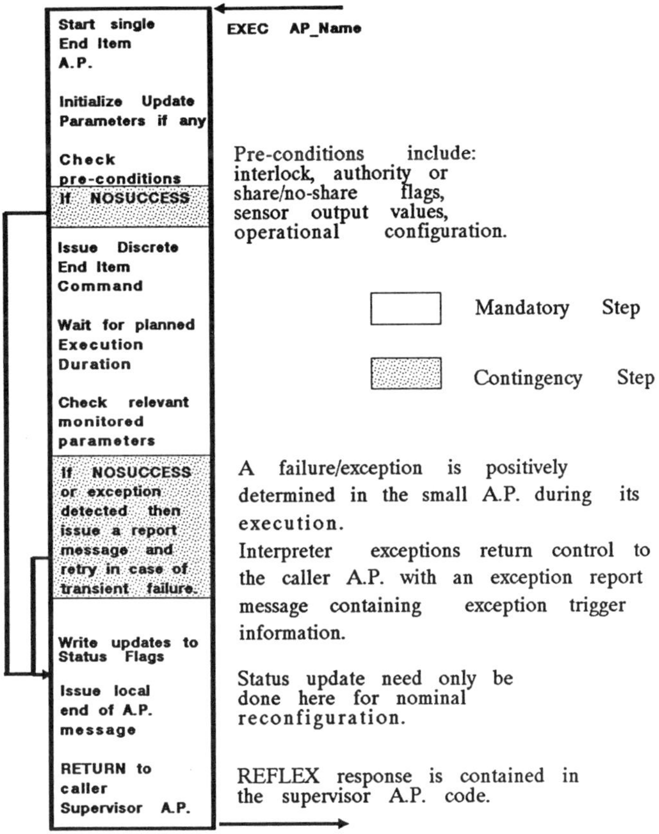

Figure 2. General End Item Reconfiguration Automated Procedure Structure.

A completely unmanned space system will rely on machine executable procedures while a system that is manned would also benefit from automation of the routine tasks ranging from complex reconfiguration sequences to gathering and presentation of information at a crew work station for interactive control tasks. A User Control Language (UCL) syntax is being defined in the Columbus project which can be used to encode the Automated Procedures (A.P.) and be executed either as interpreted source code or more efficient object code. The UCL source code has the appearance of a modern programming language

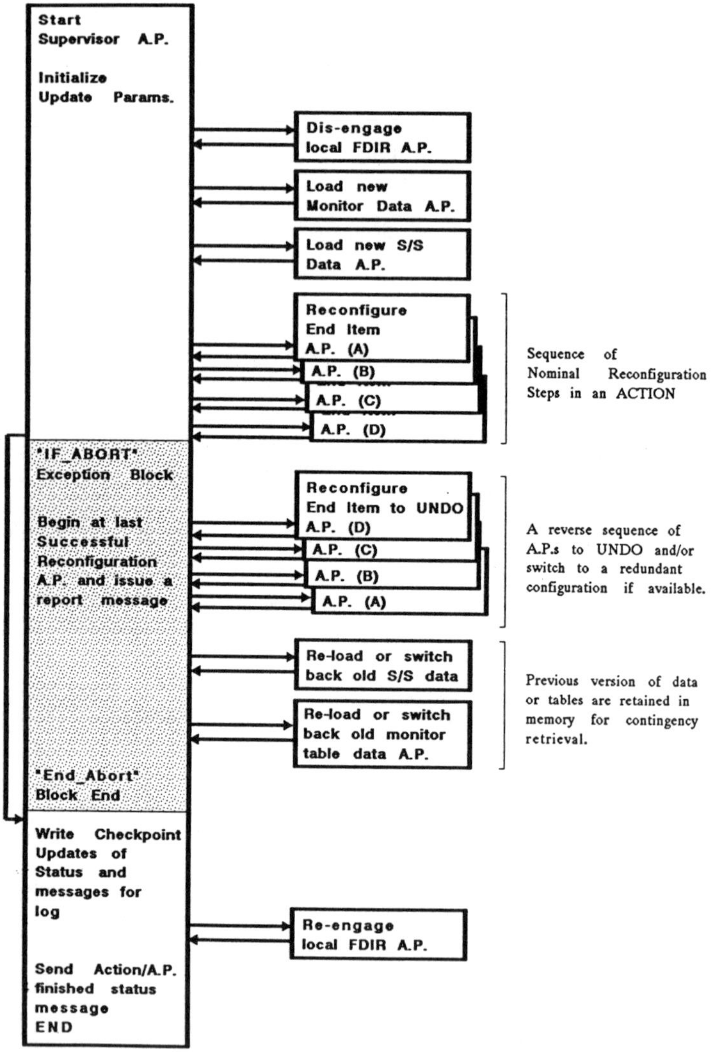

Figure 3. General Supervisor Automated Procedure Structure.

such as Modula-2 but with specific enhancements for space operations.

A procedural block of small A.P.s can be encoded as one supervisor A.P.. The supervisor A.P. essentially describes the overall procedure in terms of the subordinate A.P.s.. This is a hierarchical partition for ease of representation and timeline generation. Typically a supervisor A.P. may perform the generic procedures shown in Figure 3.

Run time exception handling is explicitly written into A.P. constructs with the operational objective of restoring the local configuration to a known safe or at least stable state on which further FDIR processes can operate.

The above procedure execution anomaly management function is one aspect of FDIR where a fast "REFLEX" response is necessary to stabilize the state of the system. A fast response time is required by operational criteria applied to critical functions that may have a minimal outage tolerance. This is commonly seen in "hard wired" designs.

STEADY STATE FAILURES

During steady state operations, where no reconfiguration sequences are being executed, the monitoring function can detect exceptions by table comparison cyclic checks. The set of exceptions need to be matched to predetermined exception patterns that define the symptoms of known potential failures, and hence, point to appropriate recovery procedures. Simple exact pattern matching mechanisms have proved inadequate in dealing with partial sets of exceptions. Sensors may be defective, the symptoms may be weakly manifested or dispersed in time. Simultaneous but largely independent failures may manifest conflicting symptoms that cancel an exception or lead to uncertainty with the failure isolation.

The MBB/ERNO developed "CONNEX" decision engine represents generalized decision tables in "Connection or Implication Matrices" in the sense that entries are "Implication Strengths". These range between zero and unity representing the degree of evidence a symptom bears to an anomaly.

Another important aspect is the question of "knowledge management", i.e. the capability to support decentralized generation of fault management databases, easy inspection and validation, easy updating and compact internal representation as well as high processing performance. Moreover, a strict separation of the fault management databases and the processing kernel has been achieved.

No numerical values of "Implication Strengths" need to be predetermined for each failure symptom. The relative significance of one exception and another is computed by CONNEX based on the instances in the matrix. The determination of symptoms to list on one axis of the matrix forces a thorough examination of instrumentation provisions (Kueke, 1989).

FAILURE MANAGEMENT HIERARCHY

During Steady State operations when there are no reconfigurations underway, the following processes operate.

- Exception messages are processed by CONNEX first at sub-system level with the relevant local knowledge domain of exceptions.
- The output recommended recovery Action, or options with computed "Proximity Ratios", are advised to the higher level (System Mission Manager).
- The SMM system level CONNEX processes the coincidental recovery recommendations from all subordinate managers. Low level sensor exception messages are not considered at system level but the supporting details are needed for explanation presentation.
- The SMM level Fault Manager determines the most "certain", or prudent, recovery response. It may be one of the lower level manager recommendations or it may be another contingency Action in the SMM CONNEX matrix/database.
- Explanations are provided to the crew who will select the recovery Action from the recommended options presented by the Fault Manager display/interface application at the Module Control Workstation.

  The Fault Management application will display explanations of why a particular recovery Action is recommended. Backtracking techniques will be displayed in a comprehensible way to the crew. The same information is passed to the MCC so that the ground based equivalent tools, plus other expert systems, can also analyze the anomaly.
- Unilateral subsystem/payload manager fault management action is possible if the hierarchical chain is broken but the response may preserve less mission productivity.

PRO-ACTIVE CHECKS

In some cases the failure may be avoided when a reconfiguration is constrained by a prerequisite system state. Checks can be performed to ensure that a reconfiguration procedure sequence, implemented in a supervisor A.P., is compatible with the current state.

The information about supervisor A.P. prerequisites is not contained within the executable code of the A.P. but in a representation of the supervisor A.P. and its subordinates. This is the so-called Action predicate which contains information such as:

- An informative name for the procedure block's operational goal (in path name syntax).
- The duration of the procedure block.
- The scheduled start time for the procedure block.
- The permissible slack time or time window.
- Resource usage characteristics as a simple step change that is associated with each reconfiguration.
- Authorization tokens required.
- Special parameter values other than default values used by A.P.'s.

ACTIONS

An Action is a rather abstract entity used to represent the most detailed layer of a schedule or timeline. A typical break down of concurrent Sub-Tasks into Actions and the corresponding executable supervisor A.P.s is shown in Figure 4.

The terms Tasks and Sub-Tasks for high level planning objectives and the term Action for the entities that must be arranged in a timeline is recommended. An "Action Database" exists within the Columbus Mission Preparation System that also uses a tree-branch-path name system. With a mission relevant copy of the Action Database onboard, only the Action Name, its planned time constraints and any changes to default parameter values, for A.P.s, need be uplinked for commanding. The onboard database entry is retrieved to assemble a complete Action predicate for checking planned resource usage against current resource availability before an automated System Mission Manager application invokes the execution of a supervisor Automated procedure. Temporal incompatibilities can be resubmitted until the prerequisite state is established by other reconfiguration sequences or the closure of the planned start time window.

TIMELINE GENERATION

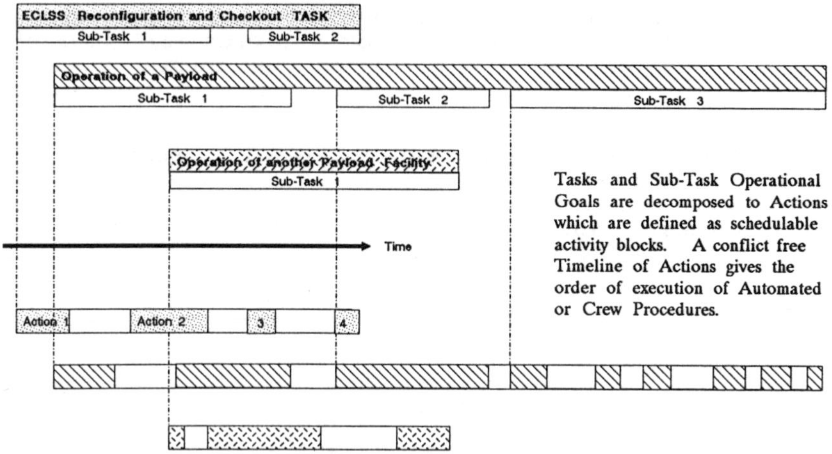

Figure 4. Timeline Decomposition.

The MBB/ERNO planning method and software tool is considered to be an improvement over the Spacelab implementation.

The planning process is based on a tool called Mission Activities and Resources Scheduler (MARS) which maintains a database of high level goals or Tasks which can be divided into sequential Sub-Tasks which consist of sequential Activities to be scheduled. Task, Sub-Task and Action/Activity antecedent and successor relationships plus their timing characteristics and scheduling constraints are efficiently built into a feasible schedule of multiplexed Actions. Interactive planning with the planning engineer is usual in order to select an activity when more than one could be next scheduled (Ohlendorf, 1989). Fully automated onboard planning or replanning by MARS for missions with poor contact opportunities with their MCC and a MARS based New Expert Planning Tool for User Networking Environment (NEPTUNE) for Space Station multi-user payload operations planning are under development.

MASTER TIMELINE OVERVIEW

The planning process output is a time ordered set of Actions. The tree path Action naming convention facilitates filter processes that can sort the Actions into their Sub-Task groups and thereby indicate the scheduled start and end of a Sub-Task; and similarly the time extent of Tasks.

Sorting processes can also reveal Crew Activity Plans for each crew member. Actions can be planned that reconfigure the Crew Workstation Displays in preparation for a new Sub-Task.

For the automated Mission Manager function the Action name convention also facilities contingency changes to the timeline as part of failure management operations. A Sub-Task that is aborted after partial execution of its Action sequence will need to be edited out of the timeline after reconfiguration Actions have established a stable state. Such a contingency Action may contain the Timeline edit A.P.s for the Mission manager to execute.

A timeline edit may be required in response to a reduction of available or allocated resources such as power or communications data rates and data storage capacity. Replanning could be performed by the MCC or in-situ crew or by fully automated tools.

GEM ANALYSIS

The determination of detailed functional or operational requirements is progressing by construction of dynamic functional simulation models. This is viewed as the safest way of ensuring that the total end-to-end process of System Mission Manager and the Generic Sub-system Manager functions are completely defined. The simulation tool selected is based on Petri Net theory which is well suited to modelling concurrent processes that can be described in terms of a predicate transition network.

The GEM concept is applicable to the entire range of foreseeable space missions. Working group level technical meetings are underway between ESA and NASA Operations teams concerned with the interfaces to the Space Station Freedom Onboard Management Application.

REFERENCES

ESTEC (1987) Standard Generic Approach to Spacecraft Autonomy and Automation (Phase 3); ESTEC Contract N6358/85/NL/PP(SC).

R. Kueke, A. Kellner and J. Sved (1989) Columbus Onboard FDIR Concept, Second European In-Orbit Operations Technology Symposium, Toulouse, Sept. 1989.

G. Ohlendorf, N. Schielow, A. Kellner (1989) Distributed Mission Planning Technology for Columbus, Second European In-Orbit Operations Technology Symposium, Toulouse, Sept 1989.

Normal Accidents and Logistics in Space Operations

Alexander K Bocast[1]

Abstract

Unpredictable systemic accidents, normal accidents, will occur in highly-complex large-scale space systems. To mitigate the catastrophic consequences of normal accidents, containment engineering will need to rely upon adequate logistics infrastructures provided by transform logistics channels to minimize consequential risk to target systems. The resource requirements of such an *in situ* logistics infrastructure to support the primary logistics channels of target systems may be orders of magnitude larger than current models estimate.

Introduction

The nature of space operations requires a careful re-evaluation of traditional concepts of logistics systems planning. In the light of normal accidents (Perrow, 1984), much more stringent demands will be placed upon logistics systems for permanent space operations than would be needed to support analogous terrestrial operations. Normal accidents are unpredictable, most often caused by complex interactions among otherwise independent failures, and ill-understood as they occur. Normal accidents will occur but as they are singular events, they fall outside our normal modes of analysis.

Space systems are singularly vulnerable to normal accidents. Existing in a hostile environment with long supply channels, there are few of the ameliorating and cushioning conditions found on Earth. Normal accidents

[1]Special Assistant for Technology, Office of the Assistant Secretary of Defense for Production and Logistics [OASD(P&L)S/AS&T], Rm. 1B466, The Pentagon, Washington D.C. 20301-8000.

in space systems will cause profoundly catastrophic results, unless containment engineering is used to reduce consequential risk. Containment engineering, in turn, will rely upon adequate logistics infrastructures.

Normal Accidents and Containment Engineering

Building upon Perrow's notion of normal accidents, systemic risk refers to the probability of catastrophic system failure precipitated by normal accidents. Consequential risk refers to the risk of catastrophic consequences stemming from normal accidents. By definition, systemic risk can not be reduced by engineering solutions; however, consequential risk can be reduced by engineering practice.

Containment engineering is engineering to contain the consequences of normal accidents, that is, to minimize consequential risk and to maximize system survivability. This is in contrast to conventional engineering that designs to prevent empirically known causes of failure. Containment engineering must assume that the source(s) of systemic failure will be unpredictable and probably unrecognizable when failure does occurs. Containment engineering can not minimize systemic risk but rather minimizes the risk of catastrophic consequences of normal accidents once they occur.

Engineering for catastrophic systemic failures is not well understood. Safety and human factors engineering speak to elements of containment, as do standard engineering principles for failsafe designs, e.g. redundant functionality. To date, these approaches have tended to focus on the component level rather than the system level or to concentrate on the protection of human lives rather than whole system sustainability.

Alternative Paths

The difference between predictive engineering and containment engineering can be illustrated by considering the difference between predictive engineering's use of physical redundancy and containment engineering's use of logical redundancy or alternative paths. Conventional fail-safe design often uses redundant elements as back-ups, hot spares, or even cooperating, voting partners. When one component fails, another identical component immediately replaces the failed component, thus preserving system functions. The critical assumption made is that the failure of one unit is not correlated with possible failure of redundant units.

Containment engineering suggests that systemic failure arises from the interconnection of complex components failing. The probability of failure of the individual components in a redundant set can not be seen as independent. In situations where unforeseen failure cascades from one place to another in a system, the original failure in the first component will be identically replicated in the redundant components.

Containment engineering's notion of alternative paths suggests that functional equivalence rather than duplicate physical implementation may be more reliable. At least one airplane program, the Aerobus, has taken the route of alternative paths; rather than provide redundant computers, different computers running different operating systems with software prepared by different contractors using different languages and different software architectures are used. Rather than three identical implementations, this program uses alternative paths which, while functionally identical, are as unlike as possible in their physical realization.

However, because normal accidents are both inevitable and unpredictable, containment strategies require that a wide spectrum of resources be distributed through a system so that these resources -- personnel and information as well as materiel -- may be available at the time(s) and place(s) of systemic failure (see Schmitt, 1988).

Logistics

From a design perspective, it is thus critical to understand the properties of the logistics channels that provide this spectrum of resources needed by containment engineering. A simple characterization of logistics channels relevant to the design of long-term space operations is suggested here. Relevant channel attributes will be developed to define the critical parameter of a space systems logistics channel, its latency.

Logistics is the process that maintains the operations of a target system. The target system is an assembly of resources -- machines, personnel, procedures -- that does something that it has been designed to do. The logistics that support the target system ensure that these resources are at the right place at the right time, in the proper amounts, and functioning properly. Thus logistics, as used here, broadly covers everything from training to resupply to maintenance and repair.

Looked at another way, logistics is the

transformation of a generalized resource into a target system specific resource. A logistics channel carries the resource from its generalized source to its specialized application.

These transforms, from generalized to specific, have a certain dimensional quality. A channel may act in any one or in any combination of these ways: across space (3-dimensional transform), across time, and/or across form. For example, a refrigerator transforms across time, maintaining the form and integrity of items it contains. A transportation system transforms across spatial coordinates -- with a necessary temporal component. Other transforms are physical transforms. While a storage or maintenance channel succeeds as long as no physical transform (a null transform) occurs over time, a manufacturing channel explicitly alters the physical characteristics of materiel that flows through it.

Characterization of Logistics Channels

A natural logistics channel is one that is provided by the natural terrestrial environment. These are commonly seen as life support functions when the target system is outside the biosphere of the Earth. The designer of a commercial product distribution system seldom worries about providing atmosphere, food, water, gravity, pressure, temperature, humidity, air flow, and other naturally occurring resources/processes. No specific action need be taken to ensure that these resources are available as required by the target system; they are provided by the natural environment.

An autonomous logistics channel is a channel that, while an artificial construct created by human activity, can be assumed to operate by the terrestrial system designer. They are autonomous in the sense that the target system can assume their presence and operation. These channels provide generalized resources that can be adapted to any number of specific systems.

Autonomous logistics channels are generally the result of commercial activities. For example, autonomous logistics channels provide gasoline, water and waste services, napkins, and paper. Generally, what is available for purchase through routine commercial activity falls into this category. A target system may simply assume that the products and services of autonomous logistics channels are readily available, on demand.

A primary logistics channel directly supports a

target system. This is the logistics channel normally considered by logisticians. A primary logistics channel transforms a resource from a generalized (natural or autonomous) resource to a resource specifically destined for and identified with a given target system.

The primary channels for space systems must provide the support generally assumed by the designers of terrestrial systems to be provided by natural and autonomous systems. For example, the logistics planning documents for the Space Station Freedom Program are quite careful and exhaustive in enumerating such requirements.

A <u>transform</u> channel is a channel that supports the support function provided by primary logistics channels. These are the channels that link autonomous and natural channels to the primary channels of a target system. They can be seen together as the <u>infrastructure</u> that supports primary channels. For terrestrial systems, there may be no distinction between natural or autonomous channels and transform channels. However, as we will see, the assurance of normal accidents will require explicitly designed and constructed transform channels within space environments.

<u>Logistics Channel Parametrics</u>

We also require terms for some logistics channel parameters. <u>Throughput</u> refers to how much of some resource flows through a channel during some unit time. Complementing this, <u>frequency</u> refers to how often channel output arrives at some destination. <u>Payload</u> refers to the amount delivered at a given frequency. <u>Drawdown</u> refers to how rapidly a given delivered stock is consumed.

At any given frequency, a payload may be calculated to provide a throughput sufficient to meet the demands of drawdown. The larger the payload, given a constant drawdown, the lower the required frequency to maintain a given throughput. Business logistics has looked extensively at the optimization of frequency and payload to obtain desired throughputs.

<u>Logistics Channel Latency</u>

In standard commercial and military practice, the logistics "ilities" -- reliability, maintainability, availability, etc. -- can be developed to provide a statistical model to drive such an optimization. Here we are primarily concerned with containment of consequential risk; normal accidents are the exceptional case rather than the normal or statistically expected case. For this we need to develop the additional concept of channel latency.

Latency refers to the time required for a channel to respond to and satisfy an ad hoc requirement. A low latency channel responds quickly to an ad hoc demand; a high latency channel responds slowly to such a demand. Under normal terrestrial conditions, natural channels have a latency that approaches zero: they provide their support continuously. Commercial logistics simply does not worry about the supply of gravity.

Again, the frequency of autonomous channels approaches continuous and, in the general case, their latency approaches zero. Requirements are satisfied upon demand. If I need milk, I obtain it from the nearest convenience store at any time. If I need medical assistance, an ambulance is only a few minutes away.

As a generalization, natural and autonomous channels are characterized by low latency; the resources they provide are constantly in those channels. Primary logistics channels for space systems, however, tend to high latency. The turnaround time from generation of an ad hoc requirement to its satisfaction -- at the place in space where the requirement was generated -- tends to be long, simply due to queue time, flight preparation, and transit times.

Channel latency links logistics system design to containment engineering. In particular, hostile space environments lack the natural and autonomous channels that may be assumed by terrestrial engineering. Containment engineering must compensate for these inherently low-latency logistics channels by providing low-latency primary channels. This can only be done by relying on an intermediate infrastructure of transform channels to absorb high-latency delays.

Recursion and Logistics Systems Design

Low-latency logistics channels linking a target system's primary channels to natural and autonomous channels through a constrained supply pipeline (e.g., Earth-to-Orbit vehicles) turn out to be recursively related: support systems are required to support the support systems. The form of this recursive relation is innocently linear but the consequents are exponential.

When primary channel latency exceeds the time constants of the target system, this recursion effect emerges. At this point, the primary logistics channels can no longer adequately supply the target system by providing a direct link to terrestrial natural and autonomous logistics channels. The primary channels can no longer cope with peaks in time-critical demand.

An infrastructure must be established as a buffer between terrestrial low-latency channels and high-latency primary channels supporting the target system. This infrastructure's primary purpose is to alter the latency of the primary channels, from high latency to low latency.

Consider one aspect of the logistics scenario given by Vajk, Engel, and Shettler (1979). In their logistics plan, they identify a fleet of orbital transfer vehicles (OTVs) sized to the transportation needs of the workforce for a space manufacturing facility. Certain elements of an OTV transportation network are not included in the Vajk scenario. These elements include pilots and crews, maintenance personnel, maintenance facilities, and service, maintenance, and repair time.

Given that an OTV fleet requires people and facilities to operate, these resources need to be moved to the sites where they will be needed throughout such a distributed transportation system. The OTV vehicles themselves are the most likely way to transport these resources.

Thus the OTV fleet (or flights) must be increased to handle the additional personnel required to operate the OTV fleet itself. These additional OTV's themselves require personnel to operate and maintain, for which additional OTV's will be required. These additional OTV's themselves require personnel to operate and maintain, for which additional OTV's will be required. We continue to repeat ourselves, in diminishing amounts, until the incremental recursive requirement for additional OTV's is terminated by existing slack in unit capacity, i.e., the last increment of personnel fills the payload capacity left over when the previous increment was satisfied.

The recursively-growing requirements for any given single system element, considered in isolation, normally rise quickly in an asymptotic approach to a steady-state value. For example, if a target system requires **A** units of a resource but the act of providing **A** itself requires an additional fraction **Q** of **A**, such recursion terminates after 4 iterations where **A**=100 and **Q**=0.1:

	Requirement Increments	Generated Total
r0	100	100
r1	100+10	110
r2	100+10+1	111
r3	100+10+1+.1	112
r4	100+10+1+.1+.01	112

Unfortunately, such a recursive requirement does not normally increase along only one factor. Assume for a moment that these are 112 OTV crews, up from 100. These additional personnel add 12% to the life-support requirements for OTV personnel. Somewhere along the chain of life-support, additional personnel will be required to provide and maintain the physical structures and processes to support these additional OTV crew requirements. These support personnel, in turn, will require more OTV support, which requires more life support, which requires more OTV support...

This indirect recursion is not analytically tractable. Unlike the simple recursion given above, a neat, clean analytic formulation of such indirect recursive interactions across a distributed system is not available. Numerical simulation using indirectly recursive algorithms gives surprising results: the total lift or payload requirements grow explosively and system deployment time lengthens considerably[2]. Convergence is not found in the neighborhood of **A+(Q*A)**, as might be expected.

This indirect recursion, this tracing through the web of interrelationships of resources in a space system, is analytically different in a fundamental way from the first-order logistics models that are customarily used (see for example Lewis, 1988). This change in mode of analysis from linear to recursive is somewhat analogous to moving from Newtonian to Relativistic analysis, not from wrong to right but from simple to more comprehensive.

Computationally-linear logistics models have sufficed for the small, simple systems that we have operated so far only because the effects of indirect recursion have been so small that they have not generated additional requirements beyond the unit capacities of existing logistics channels. The slack in unit channels

---

[2]This work is described in my Soon-To-Be-Completed doctoral dissertation on logistics requirements for space settlements.

has been sufficient, so far, to manage recursively generated requirements.

Even with simple systems, Lewis suggests that "the mathematics implies that missions will always experience shortages." Lewis argues that logistics will dominate certain aspects of space installation design and that excess capacity (a containment engineering strategy) will become an important design guideline. He concludes that "Long term space installations appear feasible [only] if...on demand or very frequent resupply is possible."

Lewis' conclusions are directly relevant to containment engineering and normal accidents because he shows that routine space operations can only be sustained in the presence of low-latency logistics channels. If low-latency primary channels are seen to be necessary using first-order analyses to achieve routine operating conditions in simple systems, how critical must they then be for responding to systemic failure, to the containment of the consequences of normal accidents, where instantaneous response must be demanded in a fog of confusion and a hostile environment?

Life-Cycle Resource Requirements

A containment engineering approach appears likely to increase the total estimate of life-cycle resource requirements for any given system. Containment engineering incorporates the hidden costs of systemic failure through the explicit costs of containing the consequences of those failures. Due to the recursive interactions among supporting transform channels, the infrastructure of logistics channels dictated by containment engineering may drive up life-cycle resource requirements. For permanent space operations -- complex technological systems existing within hostile environments and an inherently constrained supply pipeline to terrestrial sources -- requirements may rise by orders of magnitude over what has been expected using conventional estimation models.

This suggests that the logistics requirements for space operations are <u>not</u> likely to be dominated by the primary logistics channels identified by classical first-order systems engineering, upon which the design of space operations scenarios are typically based (for example, see Vajk, Engel, and Shettler, 1979). Rather, logistics requirements will be driven by transform channels that, taken recursively together, construct an infrastructure for a target space system.

Conclusions

The tendency of space planners is to accentuate the positive. They prefer to plan for the steady state, to assume that our systems will operate as we intend them to operate, within a systemically benign environment to support them. We are loathe to envision the inherent potentials for their destruction and the reality of normal accidents.

We need to face this unpleasant truth: if we sell space programs on the basis of the cost of the steady state, we are hiding the true costs of our systems. If we ignore the explicit costs for prevention and containment of catastrophe, normal accidents will still erupt, in spite of us, to reveal the inevitable hidden costs, in lives and resources, of uncontained system destruction and systemic reconstruction.

It is better policy to begin with sound advice about the relation between systemic risk, consequential risk, and the resources that are committed to space programs. If we fail to begin honestly, the political and policy backlash to the costs of recovery from catastrophe may be overwhelming.

APPENDIX: References

Lewis, William C (1988). "Logistics Support of Space Facilities," Engineering, Construction, and Operations in Space. ASCE, New York.

Perrow, Charles (1984). Normal Accidents: Living With High-Risk Technologies. Basic Books, New York.

Schmitt, Harrison H. (1988). "Constraints on Lunar Base Construction," Engineering, Construction, and Operations in Space. ASCE, New York.

Vajk, J. Peter, Joseph H. Engel, and John A. Shettler (1979). "Habitat and Logistics Support Requirements for the Initiation of a Space Manufacturing Enterprise," Space Resources and Space Settlements. NASA SP-428, NASA, Washington DC.

DYCAM
A Computer Aided Engineering Tool for Space Construction Planning

Ugo Racheli, Ph.D., P.E.[1]

Abstract

Planning and management of construction activities in space have not received sufficient attention up to the present as compared to research in propulsion, guidance and automation, and traditional aerospace disciplines. Space Station Freedom, a number of large platforms and extraterrestrial bases currently on the drawing board[3] require that a more focused research effort be directed at this novel type of space activity. A computing environment is needed which can provide assurance that space vehicles, platforms and extraterrestrial bases are, at all times during construction, viable systems with regard to structural and environmental stability, including life support in the case of manned operations. The Dynamic Construction and Assembly Module (DYCAM) described herein is intended to relate the configuration database of a specific mission to the mission's construction and operation plans in order to determine the stability of the system at a fine-grained set of node points, and to suggest corrective measures to ensure the survivability of the system if failures cause it to deviate from acceptable levels of safety and/or operability.

Introduction

On the twentieth anniversary of the first Moonwalk, celebrations have been held at several NASA Centers to commemorate the launch and successful completion of the Apollo 11 mission, by far the most spectacular and captivating venture (in or out of space) ever witnessed by the human race. But, as part of those celebrations, President Bush has outlined a much more ambitious program for NASA as we move into the age of space utilization:

---

[1]Research Associate, University of Colorado, Center for Space Construction, CB 429, Boulder, Colorado 80309

Space Station Freedom (SSF); manned Lunar bases; Mission to Planet Earth; exploratory manned missions to Mars.

Contrast the Apollo 11 mission with one of the Mars manned mission scenarios[2]. In lieu of launching a completely preassembled and tested spacecraft, the assembly of many new platforms and vehicles will be accomplished on orbit in evolutionary transportation nodes (perhaps on or near SSF), to a large extent by robotic workers and with a space crane; huge aerobrakes built up of thousands of structural elements and panels; subsystems assembled in Low Earth Orbit (LEO) after being delivered there by a variety of vehicles; the interfaces between subassemblies mated and tested in space with limited personnel and computer support.

Future missions, such as the manned exploration of Mars, will require a totally new view of the relationship of the remote mission components with Mission Control on Earth (or on a space station in LEO). This view centers on two novel concepts: distributed supervisory control and artificial intelligence to effectively manage data and provide "expert system" support to the explorers.

A direct comparison between the two missions outlined above dramatizes the challenge offered by space construction. Many future missions will require the assembly in a hostile environment of structures, space flight systems and life support systems that differ from earlier spacecraft both quantitatively and qualitatively in significant ways[6,8,9].

Constructability and survivability are the key factors at all stages of each mission from its conceptual design to space operation. Constructability analysis must provide feedback to the design effort by taking into account the very special conditions that will prevail once on orbit and on an extraterrestrial surface. At the same time, constructability analysis can profoundly influence the planning and execution of construction activities and operations in space by simulating such activities and performing in-depth failure modes analyses; this is what has been called the "feedforward" function.

We can now look at the space environment where future mission will be assembled and will perform and characterize its most salient features.

Novelty

Control-Structure Interaction (CSI) - Structures will require controls for stability that are not limited to attitude control, but will contribute to the integrity and viability of the spacecraft during construction and operation.

Lack of ground pretesting - The performance of critical components will not always be testable on the ground.

Longer operating life expectancy - The life-cycle planned for future missions is generally far longer than that of spacecraft currently in operation.

Complexity

Of the end articles - While the Apollo program dwarfs all other space missions flown to date, some future missions promise to be far more ambitious, complex, and demanding along practically all dimensions, calling for decentralized supervision and artificial intelligence support in the decision making process during construction and operation.

Of the construction environment - If, for instance, we look at any Shuttle mission dedicated to the construction of Space Station Freedom, we see an extremely complex environment[7]; it comprises a Shuttle coupled to the structure left on-orbit after the previous mission, astronauts engaged in EVA and provided with articulated positioning footholds, the Shuttle's RMS arm with an FTS unit acting as an end effector, and a mobile servicing platform with some combination of articulated arms.

Of supply logistics - Looking at future missions, the supply/logistics problem becomes ever more complex[4] due to uncertainties about the configuration of the to-be-constructed structure as well as the state-of-the-art and/or availability of the delivery systems.

Criticality

Failure modes analysis - When failure occurs in space it can propagate swiftly through a system with very little time to react and implement corrective measures.

One common criticism of the current space program is that while a very large effort is made to assure adherence to specifications[5], the analysis of failure modes and of the criticality of specific items to mission reliability is often performed after the design is complete and only because "it is called for in the contract."

Analyzing catastrophic failures - The survivability of a complex system in space, particularly one that includes humans, depends on a thin web of interactions that includes orbital dynamics, structural dynamics, power distribution, thermal balance, and life support. The hostility of the environment can be defined in terms of the isolation of the system from the reservoir of resources needed to maintain or repair its operational integrity.

Reaction times - Typically, reaction times in the event of failure on-orbit are very short. A pressurized compartment pierced by a micrometeorite is an example of an abnormality that can rapidly destroy the operability of a system.

Cost and political realities

At a time when the world faces unprecedented crises related to environmental pollution, overpopulation and other far reaching problems, a large percentage of mankind both in the U.S. and abroad expresses opinions that the price tag of planned future space missions is unacceptably high.

Notwithstanding the above, the supporters of our space program have succeeded in maintaining a reasonably strong level of funding. Bot there are signs that such support is being eroded by lack of confidence in the reliability of cost estimates for very large space projects—and construction contributes substantially to both the cost and the uncertainty of such projects.

Existing planning, construction, and operations management systems

On most construction projects there is traditionally a dichotomy between the technical information, which describes the configuration of the system to be built, and the scheduling information which describes how, when, and where it is to be built. When serious problems develop during the mission, they must be resolved in conference between the technical people and the schedulers. This scenario has been extended to the space program where Mission Control acts as the scheduler while the technical people are in the background, to be called into the picture wherever serious problems develop after countdown begins and throughout the mission.

Typically, bar charts and PERT/CPM[2] based systems are used for construction planning and management[1]. They address the basic problem of optimally allocating resources and minimizing time and cost functions in the acquisition phase of projects. There are, however serious problems in the use of standard PERT/CPM techniques for space construction planning.

Inaccessibility of historical data - All the systems we have reviewed start with a Work Breakdown Structure (WBS) prepared by some knowledgeable person (or team). By "knowledgeable" we mean privy to the detailed design of the physical objects to be built—a knowledge that is then filed

[2]PERT/CPM is the joint application of Program Evaluation Review Technique (PERT) and Critical Path Method (CPM).

away and not made part of the control system's database. Thereafter, the WBS generally remains static: the only real variables in the program are the level-of-effort durations of planned activities and their translation along the time axis when slack is available.

Difficulty and cost of upkeep - One common problem with scheduling programs, particularly those based on PERT techniques, is their very high upkeep cost[10].

Little replanning capability - For most programs, replanning, especially where it involves a change in the physical configuration of the system, involves a multi-step process where each database (i.e., configuration and WBS) must be revised in turn.

Risk assessment - PERT networks are based on variables (time or cost) which can assume values that depend on some analytic or statistical distribution. This level of risk assessment only allows the determination of probabilistic time frames for the overall performance of the task. In an analysis of failure modes, however, while probabilistic models can and are commonly developed for the occurrence of failures, the most important aspect is the propagation of the failure and the impact of such degradation on the operability and survivability of the total system.

Real-time operations management support - Most existing operation management program are not cognitively simple to use to assist in the real-time performance of the planned activities.

No clear pathway to configuration - While working on conventional construction tasks, the configuration can be kept in the background; on-orbit operations, on the other hand, will depend critically on a knowledge of the configuration (which includes the workforce and the temporary equipment as part of the construction environment).

In fact, each activity (or group of activities) causes a transformation of the configuration of some object (component, subsystem, or the entire system) which is generally not defined in the scheduling process. This can have disastrous consequences in the hostile space environment where mishaps during on-orbit construction can become catastrophic in short order.

Proposed DYCAM System

A section of the construction task for a typical space platform is shown in Figure 1. In this particular case it is assumed that the Shuttle will be used as the primary supply and construction base; the section of the total task shown in Figure 1 diagrams the conditions to be encountered between successive Shuttle missions.

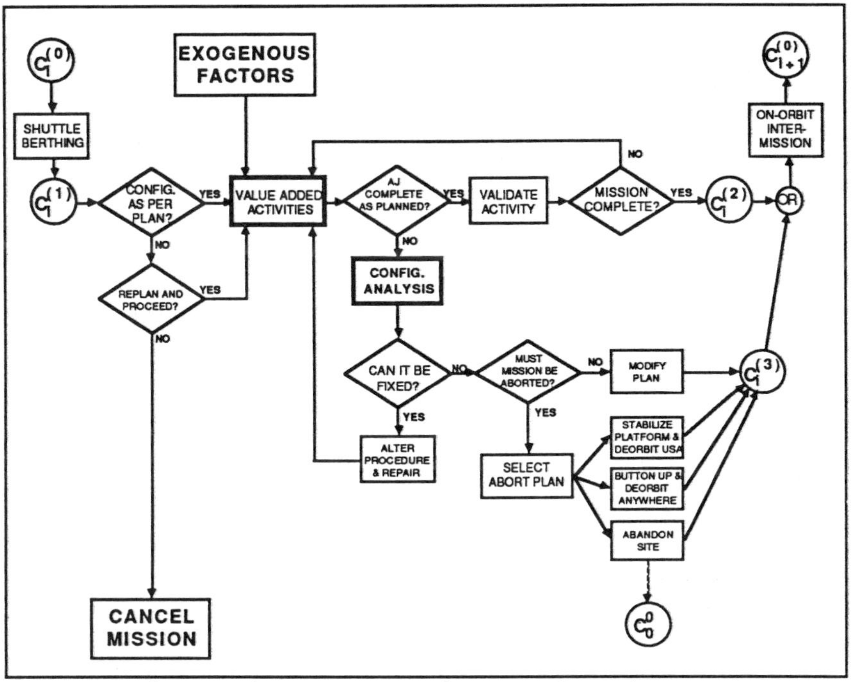

Figure 1. Construction Cycle Diagram

Prior to the berthing of the i-th Shuttle, the platform will consist of various subsystems arranged in some presumably known configuration $C_i(0)$. After berthing and inspection of the worksite, the total configuration must be revised to a new $C_i(1)$ level which reflects the configuration of the platform (revised to account for unforseen changes which might have occurred since the deorbiting of the prior Shuttle) plus the added Shuttle, its cargo and the workforce. Obviously, if $C_i(1)$ is different from the "calculated" configuration, $C_i(0)$, used during ground planning, some decision must be made on the proper course of action.

Each activity planned for the mission will be performed until it is either completed or interrupted; in the first case, an Activity Junction (AJ) is achieved and if it completes the mission objectives as planned, a new configuration $C_i(2)$ is achieved. If, on the other hand, an acceptable AJ cannot be reached, a subordinate configuration analysis must be conducted to determine the effect of the anomaly detected during the construction subtask. In some cases the anomaly may be corrected by ad-hoc repair procedures that reestablish the integrity of the platform so that the construction process can be resumed, albeit with some deviation from the planned activities schedule.

If the anomaly is sufficiently severe, a decision must be made whether to abort the mission. If so, three options are available depending on the severity of the disruption:

a. The most desirable aborting process includes the stabilization of the platform and a deorbiting plan that assures a Shuttle landing at one of the U.S. landing sites;
b. It may be necessary to deorbit as soon as possible and land the Shuttle at any suitable emergency site after some "buttoning up" of the platform to assure its survival until the next mission;
c. In extreme cases, the Shuttle will retrieve EVA personnel and deorbit as rapidly as possible. This may give rise to a $C_0(0)$ configuration, i.e., the total loss of the platform.

In all abnormal cases except $C_0(0)$, the platform will survive in some configuration, $C_i(3)$, which is different from the planned $C_i(2)$ configuration.

Finally, we must consider that the configuration remaining in orbit at the completion of the i-th mission—either $C_i(2)$ or $C_i(3)$—will undergo further changes due to a combination of environmental effects, autonomous controls, and ground commands. The (i+1)st mission will therefore encounter a new configuration, $C_{i+1}(0)$, and the process will be resumed for the (i+1)st cycle.

It must be noted that although the above scenario has parallels in all construction tasks on Earth as well as in space, the isolation of the construction site makes the knowledge of the configuration at each point of the process vital to the ability to cope with those unplanned deviations that, according to Murphy's law, are unavoidable. An operations management system for construction and assembly in space must incorporate a database which contains knowledge of the configuration at each point of the process. In addition, it must be able to access a database derived from the WBS which describes the activities to be performed in a conventional way (i.e. by references to resources, schedules and constraints) and another database which reflects a failure modes analysis. Finally it must be suitable for distributed supervision and control processes in order to assure the safe and effective performance of the construction task.

One goal of the current research program at the Center for Space Construction, is to develop a computer aided planning tool (DYCAM) capable of supporting construction and long-range operation of complex space missions. The overall task is divided into three phases: (1) the conceptual development of DYCAM, (2) the identification of a computer environment which contains a database management system of sufficient scope and with hooks for specialized analysis/design software packages, and (3) the development and programming of DYCAM in such a way that it interacts with the configuration database for a particular spacecraft and provides an interactive environment to relate planned and unplanned activities to the evolution of the spacecraft's configuration. In other words, DYCAM must display, analyze, modify and control the configuration of an evolving space structure which deviates from the original base-line configuration in response to random interruptions and problem resolution.

DYCAM implementation

Computing Environment - IDEAS**2 is a computing environment developed by the Structural Dynamics Research Corporation (SDRC) under the guidance of the NASA Langley Research Center (LaRC). It is essentially composed of a commercial Computer Aided Engineering (CAE) package (I-DEAS by SDRC) plus a number of enhancements introduced by LaRC and others. It is currently used to manage the configuration of the Space Station Freedom (SSF) at the Space Station Freedom Program Office in Reston, Virginia, and to assist in integrated planning and simulations of SSF construction missions at the Johnson Space Center. In addition, IDEAS**2 is used at LaRC to assist in preliminary engineering activities for a number of evolutionary space stations, orbital platforms, and planetary missions.

Hypertext architecture - A crucial aspect of DYCAM will be its interface with the planners/controllers of construction activities. Most of the programs running in the IDEAS**2 environment are intended to be used by technical people well versed in the use of the analytical tools; a sophisticated syntax is therefore not objectionable as a user interface. In the planning and execution of a complex program of construction activities, however, the user cannot be expected to be skilled in the process of populating databases and running engineering analyses in a CAE environment. It is believed that the most effective interactive environment can be provided by a hypertext architecture with liberal use of graphic navigational displays.

Interface - There are several key problems that need to be handled by the DYCAM interface:

a. There are, in effect, three databases that control the construction and operation of a complex system: (1) the configuration database, (2) the activities database, and (3) the failure modes analysis database. A key requirement of the DYCAM module is that it must provide bridges between these databases. The ability to display graphically a navigation map of each mission and to allow the expansion of each configuration or activity program is one of the principal features of an hypertext system. In addition, hypertext facilitates the interactive setting of parameters and the calling of programs to do analyses, modelling, simulations and the implementation of corrective actions. The above can be implemented by means of graphic displays as shown in Figure 2 in conjunction with a variety of input devices (e.g., mouse, light pen, touch screen).

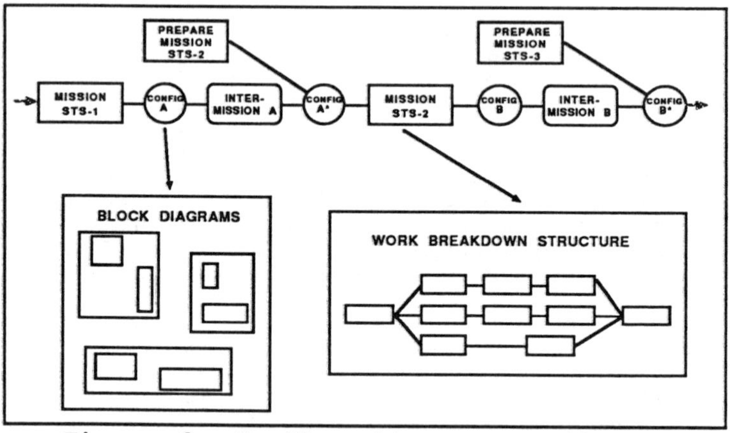

Figure 2. Configurations vs. Milestones

b. The interface must be sufficiently "smart" to be able to buffer data generated by sensors and by the software, chunk them into meaningful packets, interpret trends in their variation over time, determine their relative value versus established limits, and finally display them in a format that achieves the desired results (i.e., information, warning, alert, override) within the cognitive limitations of the user.

c. By applying statistically and heuristically generated risk factors, failure modes branching can be reasonably bounded. Using Hypertext techniques, the operator/supervisor can interactively control the depth of an analysis or the direction of a search without being confined to a sequential preprogrammed diagnostic routine. This is extremely important in real-time operations when confronted with critical conditions under extremely tight response time constraints.

d. When failure modes and the attendant configurations have been analyzed, checklists can be interactively generated to assist in planning (or managing) activities.

Conclusion

According to current plans, we see in the future large spacecraft, platforms and bases being constructed in extraterrestrial environments. The space science community, however, still thinks largely in terms of one-shot missions; but continued space exploration and utilization must include humans in greater numbers as active participants in space and will consequently be dominated by problems of life-support, communication, and logistics. We must now start to build the infrastructures that will make such an effort productive, hence economically and politically justifiable. Infrastructures are not self-terminating missions: they have to be built, assembled in-situ, operated, and maintained. They have to be reliable and evolutionary in support of long-term objectives, some of which will not be completely defined when the infrastructures are deployed.

Now, the implementation of infrastructures is the stuff that Engineering is made of. Bold thinkers and forward-looking scientists will always be needed to push the state-of-the-art to new achievements, but it will be up to the engineering profession to turn ideas into working infrastructures. To date, the engineering/academic community has only taken tentative steps in the direction of developing educational curricula and methodological tools for this enterprise. We need to extrapolate from the experience gained on Earth (particularly in the construction of large projects and infrastructures designed to operate in hostile environments) to meet the novel challenges in the space environment.

DYCAM is being proposed as a constructability analysis tool best suited for system engineering work. Its main purposes are: (1) to expose engineering students to the requirements for integrated system analysis by validation of spacecraft (and extraterrestrial base) design in terms of constructability, and (2) to research and apply computer aided tools in the planning and control of the construction phase of complex space systems. Its goals are to accomplish those tasks by relating in a logical and effective manner through a cognitively productive interface three distinct knowledge structures: (1) the physical configuration of a space system, (2) a Work Breakdown

Structure that represents the planned sequence of construction activities and (3) a structured analysis of failure modes.

Appendix A

References

[1] Blanchard, B. S. *System Engineering and Analysis*. Prentice-Hall, Englewood Cliffs, N.J., 1981

[2] Collins, M. *Mission to Mars*. National Geographic, Vol. 174, No. 5, November 1988, pp 733-764.

[3] Morgenthaler G. W. *Space Exploration and Utilization: The Next Frontier*. University of Colorado, CSC Report #1, November 1988.

[4] Morgenthaler G. W. *Optimal Selection of Space Transportation Fleet to Meet Multi-Mission Space Program Needs.* University of Colorado, CSC Systems Cluster Tech Note #3, March1989.

[5] NASA Publication, *Instructions for the Preparation of Failure Modes and Effects Analysis (FMEA) and Critical Items List*, NSTS 22206, August 1988

[6] National Commission on Space Report. *Pioneering the Space Frontier.* Bantam Books, New York, May 1986.

[7] Price, C. R. *Space Teleoperations, Now and Future*. IN-SPACE Technology Experiments Workshop, December 1988.

[8] *Project Pathfinder*, "Technology Benefits Assessment". NASA Office of Aeronautics and Space Technology Report D-5031, November 1987.

[9] Ride, S. *Leadership and America's Future in Space*. U.S. Government Printing Office, A Report to the NASA Administrator, August 1987.

[10] Silverman, M. *Project Management for the 90's.* Teleconference Manual, August 1989.

The Selection of Constructors for Space Construction

Brent Helleckson [1]
Elaine Hansen [1]
John Blanco [1]
Chris Echohawk [1]

Abstract

The process of space construction is defined by the construction goal and by the construction system selected to accomplish that goal. The goal provides the tasks that must be executed, while the selected system provides the capabilities to execute them. This paper outlines a high-level model of space construction and illustrates the relationship between constructor (i.e. "worker") and task. The construction process model is then utilized to investigate the process of constructor selection.

OVERVIEW

Space Construction Requirements

Historically, the ability to assemble or construct vehicles and structures in space or on an extraterrestrial surface has not been a major requirement of the U.S. space program. However, recent studies and commission reports such as Exploration Studies Technical Report (NASA, 1988a), Leadership and America's Future in Space (Ride, 1987), and Pioneering the Space Frontier (National Commission on Space, 1986) advocate an entirely new class of missions requiring permanently occupied Lunar bases, Martian exploration posts, large and complex space stations, and possibly Lunar and Martian settlement. Missions of this magnitude will require extensive on-orbit and on-surface assembly, construction, manufacture, and test. Programs such as the Pathfinder Technologies Program (NASA 1986), and the requirements of

[1] University of Colorado at Boulder, Center for Space Construction, LASP Engineering, Campus Box 392, Boulder, CO 80309

Space Station Freedom (SSF) attest to the growing importance of on-orbit and extraterrestrial surface assembly and construction as an enabling technology for the continued exploration and development of space.

Human/Robotic Dichotomy

To date, much of the literature regarding space construction has focused upon either human Extra-Vehicular Activity (EVA) capabilities or the application of industrial robotic technology to space. Excellent work has been done by NASA and various research institutions on the former (Hagaman, 1987; Fundling, 1989; Levin, 1988; Heard, 1988), and by NASA and the National Institute of Standards and Testing (NIST), among others, on the latter (Albus, 1987). While the dichotomy between humans and machines is a useful method by which to approach the problem of space construction, the optimal solution (as on earth) will likely result from a blend of both humans and machines in a mixed construction ensemble.

Research Efforts

One important aspect of research is the understanding of the appropriate mix of humans, teleoperated robots, supervised robots, and autonomous robots within a construction system. It is expected that this understanding will yield a method by which functional capabilities can be defined, thus allowing the comparison of multiple construction systems and selection of an appropriate mix of constructors.

GENERALIZED CONSTRUCTION PROCESS MODEL

Overview

In order to address the problem of an appropriate human/machine mix, it is necessary to understand the construction process. **Figure 1** illustrates a four-level model of the process of space construction. The outer three levels are wholly task-dependent processes, while the fourth level (i.e. the inner ring) is influenced by both task and construction system.

Level I: Construction Goal

The *Construction Goal* is the highest level of abstraction. Depending upon the level of construction detail at which the model is being applied, a construction goal could range from "assemble starboard solar dynamic system" to "install orbital replacement unit (ORU)." The construction goal is the expected result of the construction activity.

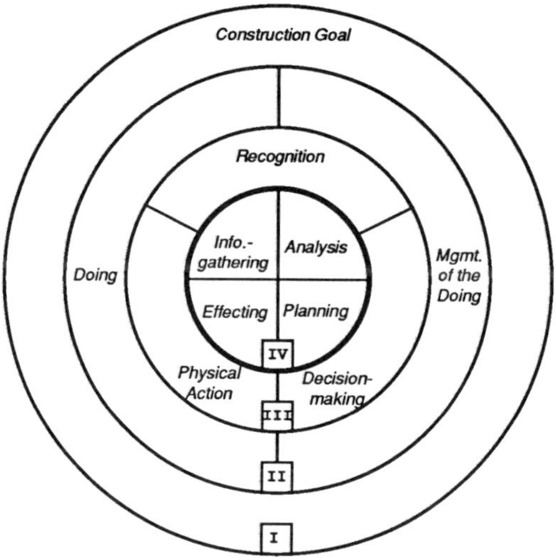

Figure 1: Generalized Construction Process Model

Level II: Doing/Management of the Doing

The construction goal can be viewed, at Level II, as a combination of the actual *Doing* of physical actions, and the *Management of the Doing* of those actions. Thus "install ORU" can be divided into those physical tasks required to *Do* the installation (e.g. attach ORU to truss structure, attach fluid lines, visually inspect fluid lines, etc.) and the *Management* of those tasks (e.g. monitor and apportion power usage, coordinate delivery of the ORU to the worksite with the onset of construction activity, retry upon failure, etc.).

Level III: Physical Action/Recognition/ Decision-making

Doing may require both *Physical Action* (e.g. grasp fluid connection) and *Recognition* (e.g. identify cross-hatch pattern indicating confirmed fluid connection). Management may require both *Recognition* (e.g. recognize power consumption levels) and *Decision-making* (e.g. choose between alternatives,). Therefore, Level III can be viewed as the combination of the *Physical Action,*

Recognition, and *Decision-making* requirements driven by the construction goal. Note that Recognition consists mainly of pattern recognition, be it spatial, temporal, situational, chromatic, logical, etc. Levels I through III can be examined without reference to the type of construction system used. The characteristics of each is dependent only on the task itself.

Level IV: Information-gathering/Analysis/Planning/ Effecting

Level III can be described in terms of another set of Level IV functions. This is the level at which the required functions become both task and constructor dependent. Level IV is essentially an iterative process consisting of *Information-gathering*, *Analysis*, *Planning*, and *Effecting* (i.e. the execution of elementary construction functions). As **Figure 1** indicates, *Recognition* comprises *Information-gathering* and *Analysis*, *Decision-making* comprises *Analysis* and *Planning*, and *Physical Actions* include *Effecting* and those actions required to gather information. It is worth noting again that at this level, the type and sequence of operations is highly dependent upon the constructor as well as the task.

Information gathering: *Information-gathering* can be accomplished in one of three ways; sensing, database query, or simulation. A relatively strong inverse relationship exists between the quality and completeness of the information in the database and the amount of sensing and/or simulation required. In the hypothetical case where the database contains perfect, complete representations of the world, constructor, and task, no simulation or sensing would be required. As the database becomes less than complete, less certain, or less accurate, the requirement for sensing and simulation increases. Note that the type and frequency of *Information-gathering* is highly dependent upon the constructor chosen as well as the task at hand.

Analysis: Within the context of space construction, *Analysis* is essentially the comparison of actual to the predicted results, and/or the synthesis of separate pieces of information to generate new knowledge. Analysis is also used to develop alternative responses.

Planning: *Planning*, in this context, is defined as a multi-step process consisting of the identification of sub-tasks needed to accomplish the construction goal, the sequencing of those tasks, the scheduling of those tasks into a time-line, and if necessary, replanning. Re-planning is distinguished from planning by the source of the construction goal. Planning implies a construction goal derived externally to the assembly system. Re-planning is done in response to a local situation, often to adapt to a deviation from expected conditions. Depending upon the severity of the situation, re-planning can involve re-scheduling, re-sequencing or even re-identification of sub-tasks. If Effecting, Information-gathering, and Analysis indicate that the construction goal has not been met, a re-definition of the construction goal or a re-selection of construction system may be forced.

Effecting: Elementary Construction Functions, and their arguments represent a language of primitives by which the physical act of construction can be described. While the language itself is neither task nor constructor dependent, the type of function chosen, the order of execution, and the arguments assigned to them are highly task and constructor dependent. With reference to the NASREM Telerobotic Control System Architecture (Albus, 1987), this language provides the interface between task-specific and constructor specific decompositions. Examples of collections of these construction primitives are available for many applications and/or specific robots (Mitzan, 1988; Price, 1988; Computer Technology Associates,1988; NASA 1988b).

Situational concerns

A physical description of the elementary functions required to accomplish a construction goal is not sufficient to allow an overall construction system (i.e. group of constructors) to be selected. The context or situation in which these functions are to be executed must also be considered as part of the task description. For example, a given constructor may well be able to fulfill each individual requirement, but may not be capable of executing them in the sequence required to achieve the construction goal. Other situational requirements include safety, reliability, consequences of failure, etc. The definition and categorization of situational influences are not yet well understood and will be addressed in future research.

Consolidation

The Generalized Construction Process Model, as presented, affords a framework within which the relationships between construction goal definition, task decomposition, and constructor selection can be examined. The model itself is in a developmental stage, yet will be utilized as is to examine the process of constructor selection.

PRELIMINARY INVESTIGATION INTO THE CONSTRUCTOR SELECTION PROCESS

Overview

The goal of the selection process is to create a construction system (i.e. an ensemble of constructors and capabilities) appropriate for the construction task(s) at hand. Assuming Space Station Freedom (SSF) represents the state-of-the-art in space construction, the task of theoretically and empirically defining the variables and criteria utilized for the match-up between on-orbit constructors and tasks has only recently begun in earnest (NASA 1988c). Prior to National Space Transportation System (NSTS) operations, the available methods of assembly were essentially self-deployment or space-suited-human manipulation. These may be viewed as the bounds of a continuum of space construction methods (see **Figure 2**). By way of example, the currently proposed Flight Telerobotic Servicer(NASA, 1988b) represents a local telerobotic method, while the recently launched Magellan probe illustrates a self-deploying method.

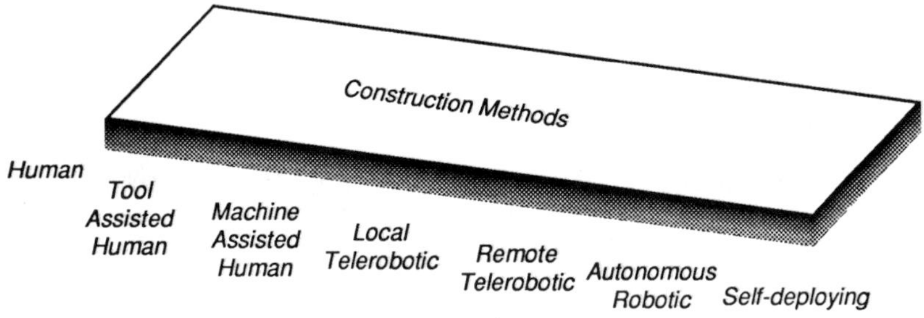

Figure 2: Continuum of Human and Robotic Construction Methods

The selection of an appropriate constructor is a four-step process, as summarized below and illustrated in **Figure 3**.

1. Identification of task requirements, in terms of Physical Actions, Recognition, and Decision-making requirements

2. Identification of constructor capabilities in these areas

3. Match-up of requirements with candidate constructor capabilities

4. Comparison of competing candidates and selection of the optimum system

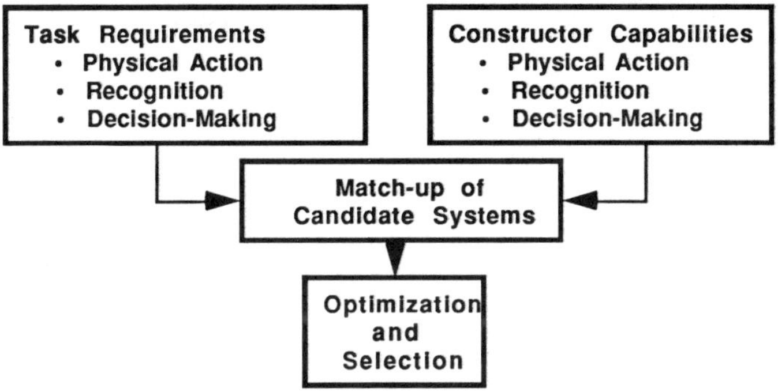

Figure 3: Constructor Selection Process

Task Requirements

As noted in the construction model, a construction goal can be described solely in task-dependent terms to the level of defining the requirements for Physical Action, *Recognition* and *Decision-making* (i.e. Level III). In general, *Physical Action* and *Recognition* requirements are driven by the physical requirements of the task and its environment. Requirements for *Decision-making* are driven by the situational context in which the task is to be performed and by the quality and completeness of the information available.

Constructor Capabilities

The identification of constructor capabilities serves two important functions. It allows the matching of tasks with those constructors capable of accomplishing them, and allows constructors to be compared and ranked on the basis of several variables.

Match-up of Candidate Systems

Without regard to optimization, a constructor is considered to be a candidate for selection if:

1. It is capable of executing the Level IV functions needed to accomplish the task as it is defined in Level III.

2. It is physically capable of executing the Level IV functions in the sequence dictated by the

task and environment (e.g. 15.5 ft-lbs torque at position (25, 104, -16) without supplemental lighting or active stabilization, and torquing must occur as pin B is rotated 60 degrees clockwise).

At this point in the process, no attempt is made to qualify selection based upon cost, time, safety, etc. These are optimization variables as opposed to "filtering" variables and are examined in the optimization step.

Implicit in this analysis is the assumption that constructors can be evaluated in ways that provide information useful for a matching process. For simple, well-defined tasks and individual constructors, such information is obtainable. For complex, ill-defined tasks with ensembles of multiple individual constructors, such information may not be obtainable.

Optimization

Once a group of candidate constructors has been identified, selection of the "optimal" choice remains. **Figure 4** is an idealized optimization scenario where information relative to each variable is available and the relationship between variables is known.

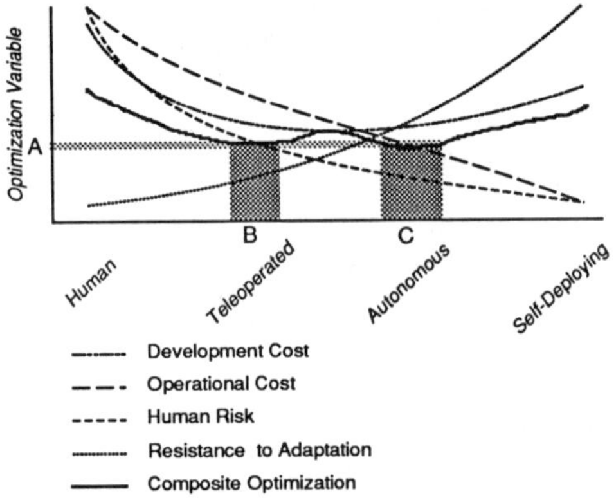

Range A = an acceptable range of optimization.
Ranges B and C = configurations capable of completing the construction goal

Figure 4: Idealized Optimization Scenario

Optimization on variables such as cost, time, safety, reliability, and consequences of failure, may well lead to a different constructor selection for each variable. Absolute mathematical optimization may be impossible because of the difficulty in obtaining information and in defining "optimal." Although absolute optimization may not be feasible, it is quite probable that a sub-optimization, resulting in a range of acceptable performance, can be identified.

Conclusion

From the preceding work, three conclusions can be drawn:

1. Task decomposition cannot be adequately accomplished using solely task-dependent methods. Constructor capabilities and methods of execution play an important role in determining how the task is decomposed.

2. Logical, defensible selection between construction methods is possible. Heuristical, "seat of the pants", and individually-biased selection methods can thus be avoided.

3. The following general models are required as a basis for constructor selection:

   A **Construction Process Model**, such as the four-level model presented here, to provide an understanding of what must be done for a given construction task.

   A **Selection Process Model**, such as that illustrated in **Figure 3**, which separates evaluation, selection and optimization into distinct steps.

   A **Constructor Management Model** to examine the interaction of the members of the construction system with each other, with the construction task, and with the work environment. This model will be developed in associated studies.

References

Albus, J.S., McCain, H.G., Lumia, R. (1987), "NASA/NBS Standard Reference Model for Telerobot Control System Architecture," NBS(NIST) Technical Note 1235.

Computer Technology Associates (1988), "FTS Detailed Servicing Scenario for an ORU Exchange," NASA Contract NAS5-30195.

Fundling, C. (1989), "Assembly EVA Task Menu," NASA-Headquarters working document.

Hagaman, J.A.,ed.,(1987), Space Construction, NASA Conference Publication 2490.

Heard, W.L., Bush, H.G., Watson, J.J. (1988), "Space Truss Construction Studies," Engineering, Construction, and Operations in Space, S.W. Johnson, J.P. Wetzel, eds., 722-733.

Levin. G.C., Ross, J.L, Spring, S.C. (1988), "An Overview of the EASE/ACCESS Space Construction Demonstration," Engineering, Construction, and Operations in Space, S.W. Johnson, J.P. Wetzel, eds., 712-721.

Mitzan, D.(1988), "Three-Dimensional Vision Structure for Robot Applications," IEEE Transactions on Pattern Analysis and Machine Intelligence, May 1988, Vol. 10, No. 3, 292.

National Commission on Space (1986), Pioneering the Space Frontier, New York, Bantam Books.

Ride, S. K. (1987), Leadership and America's Future in Space, Report to the Administrator, NASA.

NASA (1986), Project Pathfinder Research and Technology to Enable Future Space Missions, Office of Aeronautics and Space Technology.

NASA (1988a), Exploration Studies Technical Report, Vols. 1-4, Office of Exploration, FY 1988 Status, NASA Technical Memorandum 4075.

NASA (1988b), "Flight Telerobotic Servicer (FTS)-Phase C/D," Request for Proposal Solicitation Number RFP5-61629/302, NASA-Goddard, 11-15, 36-56.

NASA (1988c), "Space Station Assembly and Maintenance Implementation Definition Document (AMIDD)," NASA-JSC, Space Station Program Office.

Price, C.R. (1988), "Space Teleoperation, Now and Future," In-Space Technology Experiments Workshop presentation.

A Language Model for Space Construction Control

Brent Helleckson [1]
John Blanco [1]

Abstract

The optimal construction system for many space construction projects is a mixed construction ensemble. Such an ensemble consists of humans, robots with varying degrees of teleoperation and autonomy, and self-deploying structures. The methods of realtime management of such an ensemble have yet to be developed and proven. Musical theory is presented as a paradigm to facilitate the study of such a control methodology. A limited application of music theory and notation to telerobotic control is presented.

OVERVIEW

Space Construction Requirements

Throughout the Mercury, Gemini, Apollo, Skylab, and early Space Shuttle programs, launch vehicle capability and mission requirements allowed vehicles and structures to be assembled on earth and launched in an essentially fully-functional (albeit stowed) condition. Recent studies and commission reports such as Exploration Studies Technical Report (NASA, 1988), Leadership and America's Future in Space (Ride, 1987), and Pioneering the Space Frontier (National Commission on Space,1986) advocate an entirely new class of missions. This new class includes permanently occupied Lunar bases, Martian exploration posts, large and complex space stations, and Lunar and Martian settlements. Missions of this magnitude will require extensive on-orbit and on-surface assembly, construction, manufacture, and test. Recognizing this, NASA has instituted the Pathfinder

---

[1] University of Colorado at Boulder, Center for Space Construction, LASP Engineering, Campus Box 392, Boulder, CO 80309

Technologies Program (NASA 1986), one sub-program of which is tasked with the development of the techniques and equipment to undertake large-scale, in-space construction.

The Construction Ensemble

While it may be possible to accomplish the required construction and assembly tasks using only one type of constructor (e.g. only humans or only automated robots), cost, time, safety, and reliability considerations will dictate that the optimal construction system comprise a mixed ensemble of constructors working in concert. Each constructor in the continuum shown in **Figure 1** offers skills and advantages not available in the others. Also, the synergy of multiple constructor operations must not be overlooked.

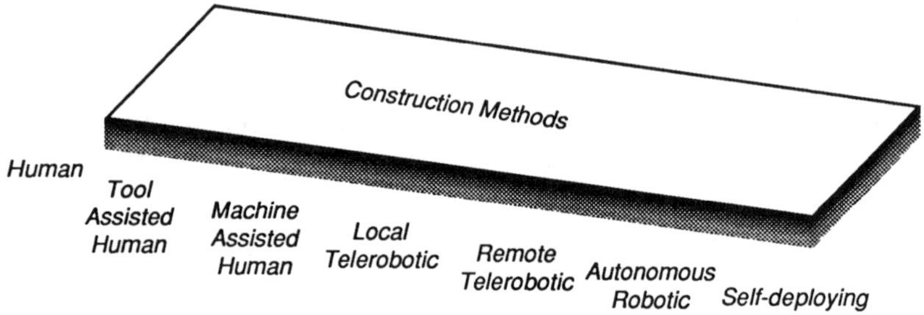

Figure 1: A Continuum of Space Constructors

ENSEMBLE MANAGEMENT

Paradigm

A study of the management and coordination of a distributed ensemble of constructors, each with differing skills, requirements, intelligence, consequences of failure, etc., can be approached through a model or paradigm. Such a model must be capable of describing lower level functions such as joint positions and task requirements, as well as higher level functions such as constructor coordination, task coordination, and resource utilization. Historically, civil engineering has utilized the Gantt, CPM, and Flow-Diagram methods of plan representation (Barrie, 1984). These methods are useful for "off-line" planning but have limitations when planning and execution occur simultaneously or in quick succession. It is also difficult to make sufficiently large amounts of detailed information available to the supervisor.

Several other methods of representing time and event-ordered sequences exist. Many have been undergoing refinement and testing for several centuries. The humanities offer language, choreography, and musical theory, among others, as methods of organizing, representing, and displaying events and relationships. NASA often refers to the sequencing of Extra Vehicular Activity (EVA) as "choreography" (NASA 1989). Similarly the NASA/NBS Standard Reference Model for Telerobot Control System Architecture (NASREM) devotes considerable attention to the problem of planning and the representation of tasks and constructor activities (Albus, 1987). In this application the Gantt system assumes "...the form of a musical score. This form suggests a possible notation for programming multiple cooperative tasks." Musical notation and theory offers a useful language for the display and organization of information relating to realtime space construction activities.

Ensemble Description

Consider an ensemble consisting of 15 separate two-armed constructors and a supervisor. Each constructor has between 5 and 20 controllable joints on each arm. The constructors are highly trained, and a high degree of intelligence is distributed throughout the hierarchy. Each constructor is capable of individual or closely coordinated activity.

The supervisor relies on line-of-sight optical communications as well as visual and aural feedback. A highly structured, yet highly flexible master sequence (i.e. plan), is available to the supervisor. The appropriate portions of the plan are available to the appropriate constructors. The master sequence consists of a series of ordered, coordinated graphical representations of construction activities. Despite the structured framework, the method of accomplishment is open to a great deal of individual interpretation. The preceding system description applies to a complex, mixed ensemble of constructors, as well as a 15-member orchestra and it's conductor. In the latter case, the conductor's score utilizes musical theory and musical constructs to fulfill the role of the master sequence. The score is a representation of the overall performance plan, and the assignments of the instruments with respect to time and to one another.

Musical Theory and Notation

Musical performance is a collaboration between performer, conductor, and composer. Each has exclusive, as well as overlapping functions to perform. The composer is responsible for the derivation of the end goal, the form and structure of the path, and the rough boundaries of interpretation. The conductor is responsible for a thorough knowledge of the composer's work, a thorough knowledge of the capabilities of the performers, and for the coordination of execution, interpretation, and rehearsal. The performer must execute the individual events while maintaining an awareness of the overall goal and the other performers. The link connecting this hierarchy is the musical score and its supporting theory. Musical constructs allow communication between the composer, conductor and performer. **Table 1** illustrates a partial listing of musical constructs organized according to responsibility of the musical collaborators (Ottman, 1970, Ottman, 1972).

Performer	Conductor	Composer
Pitch	Tempo	Cadence
Timbre	Phrase	Mode
Volume	Rhythm	Tonality
Duration	Interpretation	Harmony
Chord	Execution	Instrumentation

Table 1: Musical Constructs

These three collaborators and their responsibilities can be roughly mapped to a three-level view of a space construction system consisting of collaboration between a system manager, a local ("on-site") manager, and functional constructors. The music theory and notation system is highly complex and therefore offers a large number of ways in which to model construction operations. Tonality, for example, may be used to indicate safe modes of operation (e.g. consonance = safety, dissonance = danger). Key signature may be used to indicate the type of constructor. Phrasing and tempo may be used to represent the coordination of multiple constructors on a task.

By way of an example at the performer level, consider a teleoperated robot tasked with the removal of a connector. The mapping from musical notation to robot specific control may be done as in **Figure 2**.

Pitch	→ Single Joint Motion → Direction
Volume	→ Speed
Duration	→ Duration
Chord	→ Multiple Joint Motion
Formatta	→ Visual Alignment and Inspection

Figure 2: Music Theory to Telerobotic Control Mapping Example

Specifically, assume a five Degree Of Freedom (DOF) telerobot with closeable gripper under the following mapping:

DOF	Direction	Pitch
Base	Right	D (2)
	Left	D*b*
Shoulder	Up	E
	Down	E*b*
Elbow	Up	G
	Down	G*b*
Wrist	Up	A
	Down	A*b*
Wrist Roll	Up	B
	Down	B*b*
Gripper	Open	D (3)
	Close	D*b*

Note: D(2) is the second occurrence of D above concert A (440 Hz), D(3) an octave above D(2), etc.

Table 2: Specific Example of Telerobot Mapping

Under the initial geometry illustrated in **Figure 3,** the task of removing an RS-232 connector can be represented in the sequence in **Figure 4**.

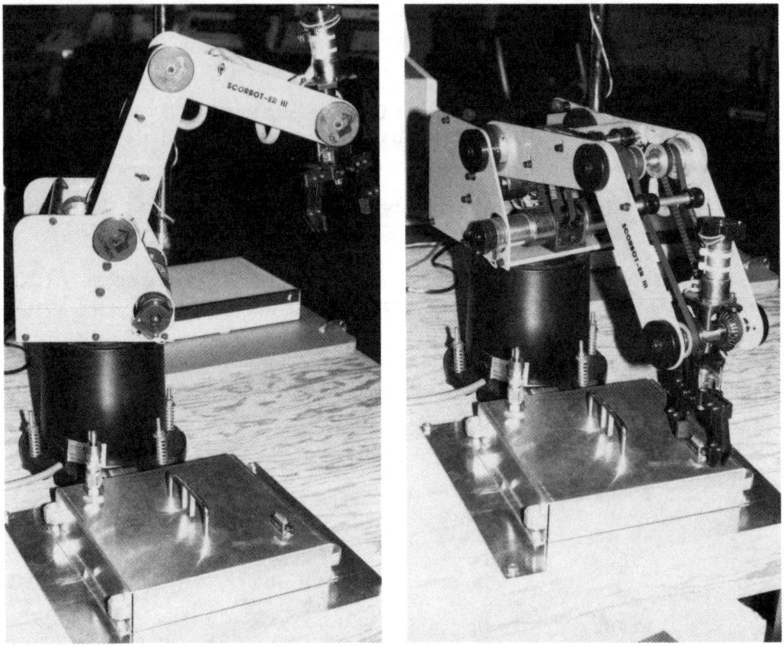

Figure 3: Telerobot/Task Geometry

Figure 4: Master Schedule Example

The system of musical notation and theory readily accommodates the description of the task in terms that are easily understandable by a trained operator.

A grouping of several of these sequences generates a master sequence (i.e. score) is illustrated in **Figure 5.**

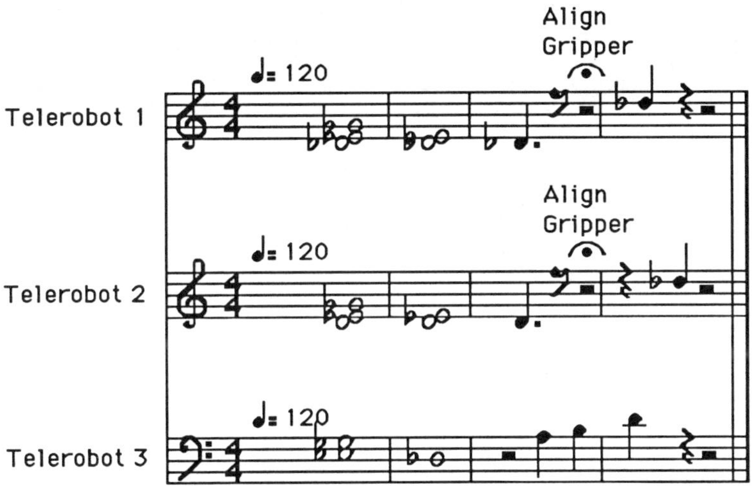

Figure 5: Typical Master Sequence or Score Used by the Local Manager

Application of Musical Theory

The preliminary application of musical notation and theory to the control of a simple telerobot has been accomplished at the University of Colorado Center for Space Construction (CSC). A musical keyboard equipped with standard Musical Instrument Digital Interface (MIDI) output is interfaced to a robot arm (see **Figure 6**).

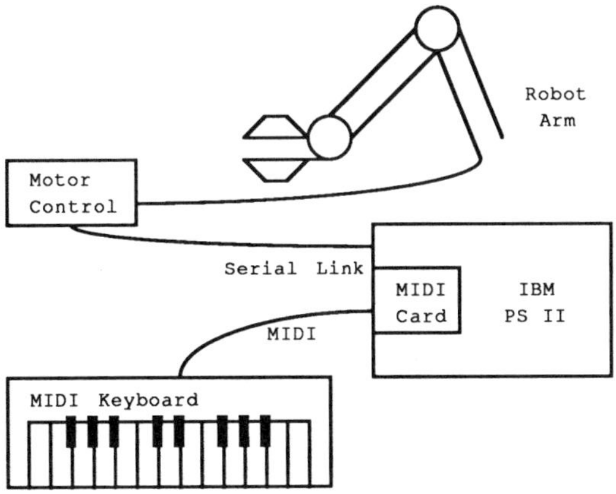

Figure 6: Connection of MIDI keyboard to robot arm.

The robot system used is a five DOF Eshed SCORBOT-ER III Educational Robot equipped with a gripper. The entire system consists of the mechanical robot arm, a controller box to activate motors in the robot arm, and an IBM PS II computer, The PS II supports the SCORBOT utilities software allowing control of the robot, and a MIDI coprocessor card, built by Music Quest Inc. MIDI protocol is a standardized method for connecting electronic musical instruments to one another, and to computer equipment (Yamaha, undated a; Yamaha undated b)). It is generally used to create modular synthesizer groups, to simplify the programming of synthesizers, and to write and record music.

MIDI uses a bi-directional serial interface that operates at 31.25 kbits/sec. Generally, there are two main types of messages that are sent and received over the MIDI interface. Channel messages are encoded with channel numbers meant for a specific destination. They generally comprise control information including messages that indicate whether a key on the keyboard has been pressed or released. The velocity at which the action has taken place is also encoded and transmitted. Other controller function information, such as pitch or modulation can also be transmitted as a channel message. System messages are not encoded with channel numbers and are intended for all units in a system. An exclusive message is a type of system message (i.e. not encoded with a channel number) that allows for moving large blocks of data across the interface.

The MIDI system shown in **Figure 6** maps two keys on the keyboard to each degree of freedom in the robot as indicated in **Table 2**. Software, written on the PS II in C and assembly language, receives messages from the MIDI card and maps the information to commands controlling the robot arm. This software makes use of a set of functions, provided with SCORBOT utilities, to control various aspects of the robot's behavior. Incoming MIDI messages are decoded and used to update a state variable containing fields representing each of the robot's axes or degrees of freedom. Each axis has a field for the direction and speed of motion. This state variable is continuously examined and messages are sent to the SCORBOT motor controller indicating which axis or axes are to move. Velocity information from the MIDI key-press messages are used to determine the speed at which the axes should move.

The score shown previously in **Figure 4** is an excerpt of the actual score used during the experiment. The first three bars, executed at 120 quarter notes per minute, result in the change in position from the left picture to the right. The formatta (𝄐) allows fine positioning, and the high D♭ closes the gripper. The remainder of the piece returns the SCORBOT to the starting position.

Summary

Musical notation and theory offers an existing, well-developed language for the study of coordinated, multiple constructor control. The simple application presented here represents a small sample of the ways in which musical theory and notation can be applied to the study of space construction operations. This application indicates that the keyboard and the musical score are useful methods of telerobotic task specification and control. Testing of productivity vs. mapping function, the role of tonality in boundary definition, and the required alterations in representation, among others, offer interesting areas for continued research.

References

Albus, J.S., McCain, H.G., Lumia, R. (1987), "NASA/NBS Standard Reference Model for Telerobot Control System Architecture," NBS(NIST) Technical Note 1235.

Barrie, D.S., and Poulson, B.C.,(1984) Professional Construction Management 2nd ed., New York, McGraw-Hill, 218-251, 488-513.

Fundling, C. (1989), "Assembly EVA Task Menu," NASA-Headquarters working document.

National Commission on Space (1986), Pioneering the Space Frontier, New York, Bantam Books.

Ride, S. K. (1987), Leadership and America's Future in Space, Report to the Administrator, NASA.

NASA (1986?), Project Pathfinder Research and Technology to Enable Future Space Missions, Office of Aeronautics and Space Technology.

NASA (1988), Exploration Studies Technical Report, Vols. 1-4, Office of Exploration, FY 1988 Status, NASA Technical Memorandum 4075.

Ottman, R.W. (1970), Elementary Harmony: Theory and Practice 2nd ed., Englewood Cliffs, New Jersey, Prentice-Hall.

Ottman, R.W. (1972), Advanced Harmony 2nd ed., Englewood Cliffs, New Jersey, Prentice-Hall.

Yamaha, (undated a), "Digital Programmable Algorithm Synthesizer DX7 MIDI Data Format," Hamamatsu, Japan, Nippon Gakki Co., Ltd.

Yamaha, (undated b), "What's Digital, What's MIDI," Hamamatsu, Japan, Nippon Gakki Co., Ltd.

A COST STUDY OF CONCRETE PRODUCTION ON THE MOON

Hiroshi Kanamori[1], Shinji Matsumoto[2]
Haruo Fujishiro[1], Kou Yatsuyanagi[1]

Abstract

In proposing and evaluating a construction system for a lunar base, the question of cost would be the most important one. Although it is not possible to conduct a detailed cost study at the present time, it is necessary to consider macroscopically the types of building materials from a cost point of view.

This paper presents results of cost study on concrete to be produced on the moon. A concrete production system was quantitatively analyzed and a cost model was formulated in this paper.

1. Introduction

Since the construction cost of a lunar base increases with the increased dependence on terrestrial materials, cost reduction can be achieved by producing some of the materials from existing lunar resources. Concrete, steel, and aluminum have been considered as the potential candidate materials. For the proposed cost model, we assumed an optimal production process for each material for calculation, and then compared the results.

2. Cost Model for Lunar Construction Materials

2.1 Assumptions

1: Researcher, Institute of Technology, Shimizu Corp.
3-4-17 Etchujima, Koto-ku, Tokyo 135 Japan
2: General Manager, Construction Engineering Dept.
Institute of Technology, Shimizu Corp.

(1) Cement and aggregate, basic components of concrete, can be produced on the moon.

(2) Water can be produced by hydrogen reduction of ilmenite. Hydrogen is the substance required to be transported from the earth.

(3) Iron can be extracted from the soil and rock of the moon.

(4) Carbon, which improves the mechanical properties of steel, can be transported from the earth.

(5) The size and weight of concrete plants are assumed to be similar to those on the earth.

(6) Energy required for production of lunar construction materials was calculated in reference with that of similar operations on the earth.

2.2 Formulated Cost Model

In order to calculate production costs of construction materials on the moon, the following four items are the primary variables:

(1) Cost for transporting from the earth to the moon the component materials of the plants for producing the construction materials.

(2) Cost for the construction of the plants on the moon. (proportional to the weight of the plants)

(3) Cost for transporting elements such as hydrogen and carbon from the earth to the moon.

(4) Cost of the energy to produce the construction materials on the moon. (including cost for running the plants)

Taking these conditions into consideration, the cost of the construction materials on the moon can be expressed by the following formula:

$$Cc = \frac{Wf \cdot Ct + K \cdot Wf}{Mp \cdot Pa} + Wt \cdot Ct + Cp \qquad (1)$$

Where
Cc: unit cost of the construction materials to be produced at lunar plants ($/ton)
Ct: cost for the transportation of one ton of materials

from the earth to the moon ($/ton)

Cp: cost of the energy to produce one ton of the materials at the lunar plants ($/ton)

Wt: weight of elements which must be transported from the earth to the moon in order to produce one ton of construction materials (tons/ton)

Wf: weight of the plants (tons)

Mp: annual production of the materials (tons/year)

Pa: plant life (years)

K : cost of constructing the lunar plants ($/ton)

3. Cost of Concrete Produced on The Moon

3.1 Production Process of Concrete

The process for producing concrete on the moon is shown in Figure 1.

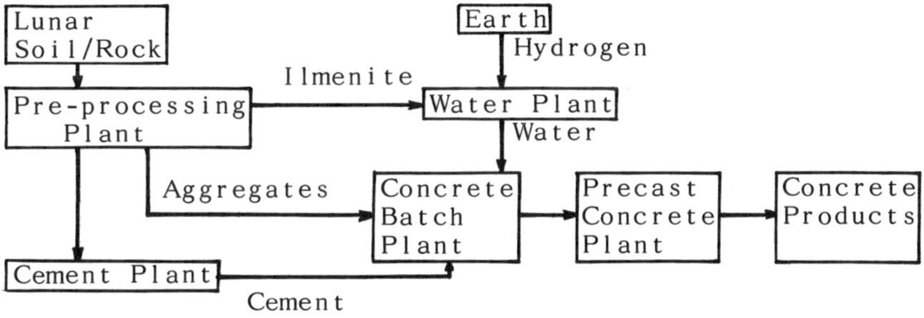

Figure 1 Concrete Production Process on the Moon

Pre-processing plant constitutes a plant for storing and processing lunar resources. Lunar soils and rocks are beneficiated and the ilmenite is separated using a magnetic separator in this plant [McKay et al. 1979].

Cement plant constitutes a plant for extracting cementitious elements from materials carried from the pre-processing plant. Cement with a composition close to that of high-alumina cement is produced in this plant [Lin 1987, etc.].

Water plant constitutes a plant for producing water by hydrogen reduction of ilmenite [Gibson et al. 1988].

At the Concrete batch plant, cement, water and aggregates are suitably blended and mixed to meet the individual quality requirements [Ishikawa et al. 1988].

Precast concrete plant constitutes a plant for making concrete products such as concrete panels and frames, and for curing to attain the prescribed qualities.

3.2 The Amount of Hydrogen and Ilmenite Necessary for The Concrete Production

Because water is a very precious material on the moon, as small a quantity of the water as possible should be contained in the concrete. To achieve this, water content is set at 150kg/m^3 in this paper. This means that the amount of water required to produce one ton of concrete products on the moon is approximately 64kg. The amount of hydrogen to be transported from the earth and the amount of ilmenite, in order to produce one ton of concrete, should be approximately 8kg and 600kg respectively, assuming the reduction efficiency of water to be 90%. Therefore, the Wt is set at 0.008 tons/ton.

3.3 Scale of Concrete Production Plants

The hexagonal concrete module we have proposed has approximately 40m^3 of living space. The weight of the concrete used is approximately 240 ton [Namba et al. 1988]. If a 100,000m^3 inner capacity concrete living facility is to be constructed within ten years, the annual production of concrete, Mp, is approximately 70,000 tons/year, including the concrete loss. The total weight of the plants, Wt, for this size of concrete production is shown in Table 1.

Table 1 Calculated Weight of The Concrete Plant

Plant	Weight (tons)	Major Facilities
Pre-processing Plant	10	Stock and transportation of raw materials
Cement Plant	200	Kiln, clinker crusher etc.
Water Plant	200	Reduction reactor, Ilmenite refining facilities etc.
Concrete Batch Plant	20	Material weighing system, Concrete mixer etc.
Precast Concrete Plant	100	Forms, Curing facilities etc.
Total Weight (Wt)	530	

By assuming the weight of the facilities, which are not influenced by the annual production of concrete, to be 20% of the total weight in Table 1 (when Mp = 0, Wt = 106, and when Mp = 70,000, then Wt = 530), weight of the plants, Wt, is expressed in the following formula:

$$Wt = 106 + 0.006057Mp \quad (2)$$

3.4 Other Factors for The Cost Calculation

Because the transportation cost is expected to decrease with the development of space transportation systems, it is set at Ct = 3.3×10^6 \$/ton and 6.7×10^7 \$/ton.

The plant life is set at Pa = 10 years and 30 years.

The construction cost of the lunar plant is considered to be nearly the same as that of space stations and therefore it is set at K = 6.7×10^6 \$/ton.

Table 2 shows the energy and production cost necessary for producing one ton of concrete products.

Table 2 Energy and Production Cost Necessary for Producing One Ton of Concrete Products

Production Process	Production (kg)	Energy Consumption (GJ)	Production Cost ($)	Remarks
Aggregate and Cement Materials	1167	0.105	23	80% of production efficiency on earth
Cement Production	170	1.162	29	Raw materials heated to 1800°C
Water Production	64	3.541	88	Raw materials heated to 1000°C
Concrete Mixing	1000	0.002	28	50% of mixing efficiency on earth
Concrete Products	1000	0.158	142	Control curing temperatures
Total	---	4.968	310	

From the above factors, the production cost is set at Cp = 310 \$/ton.

3.5 Concrete Production Cost

The result of the concrete cost on the moon as a function of the annual production is shown in Figure 2. Curves (1) through (4) in Figure 2 represent the values calculated in the four cases shown in Table 3.

Table 3 Cases for The Cost Calculation

	(1)	(2)	(3)	(4)
Transportation Cost (\$/ton)	3.3 × 10^6	3.3 × 10^6	6.7 × 10^7	6.7 × 10^7
Plant Life (years)	30	10	30	10

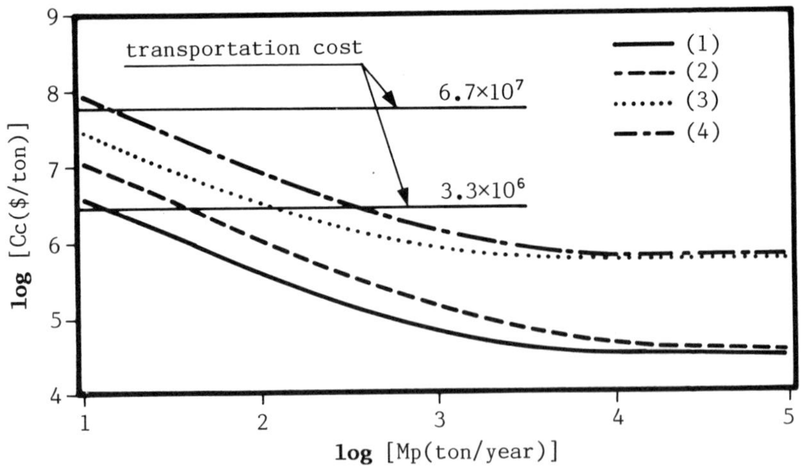

Fig. 2 Relationship Between Concrete Cost and Annual Production

From Figure 2, under condition (1), the production cost of concrete on the moon, when the annual production is more than 11 tons, is lower than that of transporting concrete products from the earth. In the same manner, under conditions (2), (3) and (4), the production cost is lower when the annual production is more than 32 tons, 4 tons and 12 tons respectively.

With the increase of the annual production and plant life, the fraction of plant cost included in the concrete cost decreases, and therefore, the cost of transporting hydrogen from the earth becomes prominent. For example, when Mp = 70,000 tons/year, Pa = 30 years, and Ct = 6.7 $\times 10^7$ \$/ton, concrete cost is 554,910 \$/ton, and about 97% of this is encompassed by the cost of transporting hydrogen.

4. Costs of Other Construction Materials on the Moon

4.1 Steel

4.1.1 Steel Production Process

Because a relatively large amount of iron is contained in lunar resources, several methods for extracting iron from lunar resources have been proposed [Haskin 1985, etc.]. Figure 3 shows one of the steel production methods on the moon.

Iron-rich materials separated in the water plant are regulated in their composition at the composition regulating plant, so that the mechanical properties of iron are improved. Several iron extracting methods are presently being considered for application at this plant.

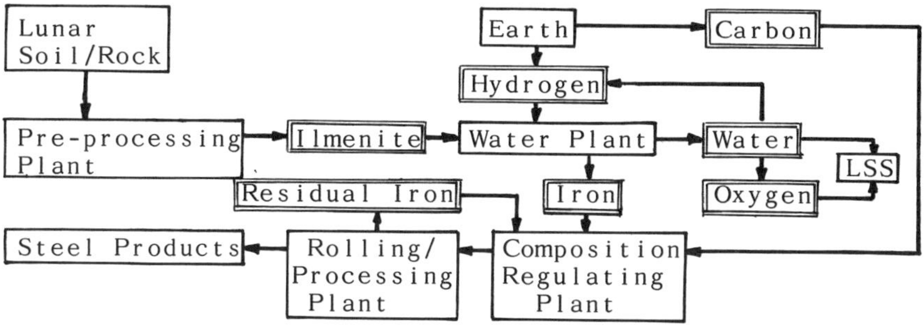

Figure 3 Steel Production Process

4.1.2 Factors for The Calculation of Steel Cost

Factors for the calculation of steel cost are set in a similar manner to those for concrete. The weight of the materials to be transported from the earth (Wt) for producing one ton of steel products is 0.017 tons. Of this weight, 0.007 tons is the weight of carbon. The cost of energy consumed in the production (Cp) is 770 \$/ton. The total weight of the plants is set at 1,000 tons when the annual production (Mp) is 25,000 tons/year, and 20% of this weight is assumed to be independent of the annual production. The values of K, Ct and Pa are set to the same values as those set for the production of concrete (Conditions (1) to (4)).

4.2 Aluminum

Since a lot of aluminum is contained in lunar resources, it is possible to produce aluminum products. However, since refining of aluminum using existing techniques uses a lot of water and electricity, aluminum products are expected to be transported from the earth. Therefore, the cost of aluminum products will be proportional to the transportation cost (Ct).

4.3 Cost Calculation for Steel Products

From the results of the cost calculation for steel products, it is found that the steel cost on the moon is lower than that of steel products transported from the earth when the annual production of steel products on the moon is at least 63 tons/year.

The relationship between annual production or plant life and steel cost is similar to that for concrete. For example, when Mp = 25,000 tons/year, Pa = 30 years, and Ct = 6.7×10^7 \$/ton, the steel cost is 1,238,036 \$/ton and about 92% of this cost is encompassed by the cost for transporting hydrogen and carbon.

5. Cost Comparison of The Materials

Costs of these materials should be studied based on the specific strength of each material. Table 4 shows the specific strength of each material. From this table, the weight of steel and aluminum necessary for

Table 4 Specific Strength of Each Material

Material	Strength (MPa)	Density (tons/m^3)	Specific Strength
Concrete	30	2.35	12.77
Steel	300	7.86	38.17
Aluminum	100	2.69	37.17

obtaining a strength corresponding to that of one ton of concrete is 0.335 tons and 0.344 tons, respectively.

If the annual production of concrete products is 70,000 tons, the annual production of steel products and aluminum products is 23,450 tons and 24,080 tons, respectively. The cost of each material at each condition is shown in Table 5.

Table 5 Material Cost Based on Annual Production Obtained from Specific Strength

	Material Cost ($/ton) by Condition (1)	(2)	(3)	(4)
Concrete	2.9×10^4	3.4×10^4	5.5×10^5	5.9×10^5
Steel	7.0×10^4	9.7×10^4	1.2×10^6	1.4×10^6
Aluminum	3.3×10^6	3.3×10^6	6.7×10^7	6.7×10^7

From Table 5, the cost of concrete products is 35% to 45% of that of steel products and 0.8% to 1% of that of aluminum products.

Table 6 shows the cost ratios of each material. The amount of each material in the table is equivalent to the amount which is calculated from the specific strength of each material. In each condition, the cost ratio of the concrete is estimated as 1.

As shown in Table 6, the cost of steel to be required to resist the load which concrete can carry is slightly lower than that of concrete itself. Because each material would be applied in distinctly different functional ways depending on its own characteristics, serviceability of the materials cannot be evaluated from this result alone. However, from an economical standpoint, the construction of steel reinforced concrete structures on the moon seems to be a realistic program.

Table 6 Material Cost Ratio

	Material Cost Ratio by Condition (1)	(2)	(3)	(4)
Concrete	1	1	1	1
Steel	0.81	0.96	0.73	0.79
Aluminum	39.1	33.4	41.9	39.1

6. Conclusion

We assumed a possible production process on the moon in order to obtain the cost of construction materials. The energy cost necessary for producing construction materials, the transportation cost from the earth to the moon, and the plant cost for producing construction materials were all set.

When the annual production is small, the depreciation cost of the production facilities exerts a great influence on the cost of construction materials on the moon. With the increase in the annual production, the transportation cost exerts a large influence.

Concrete, steel, and aluminum have been analyzed as possible lunar construction materials and the cost for each was calculated. As a result, when the annual production obtained from the specific strength is used as the basis, the cost of concrete products is approximately 35% to 45% of that of steel products, and approximately 0.8% to 1% of that of aluminum products. Because the amount of steel required to resist the load which concrete can carry is approximately 33.5% of that required for concrete, the actual cost of concrete and steel on the moon is considered to be almost the same. This means that the construction of steel reinforced concrete structures would be a realistic program.

The concept of the cost model proposed here can be applied to the cost evaluation for other future lunar construction materials as well.

REFERENCES

Gibson,M.A. et al.," LUNAR OXYGEN PRODUCTION FROM ILMENITE", Proceedings of Lunar Bases and Space Activities in the 21st Century, No.LBS-88-200, 1988.

Haskin,L.A.," TOWARD A SPARTAN SCENARIO FOR USE OF LUNAR MATERIALS", Lunar Bases and Space Activities of the 21st Century, p435, 1985.

Ishikawa,N. et al.," MAKING CONCRETE FOR LUNAR STRUCTURES", Abstract from the Symposium on Lunar Bases and Space Activities of the 21st Century, HOUSTON, 1988.

Lin,T.D., " CONCRETE FOR LUNAR BASE CONSTRUCTION", Concrete International, July, 1987, etc..

McKay,D.S. et al., " A GEOLOGIC ASSESSMENT OF POTENTIAL LUNAR ORES", Space Resources and Space Settlements, NASA SP-428, p243, 1979, etc..

Namba,H. et al.," CONCRETE HABITABLE STRUCTURE ON THE MOON", Engineering, Construction and Operations in Space, ASCE, Aug., 1988.

Level Of Recycling Effectiveness

Dick B. Parker[1]

Abstract

The Level Of Recycling Effectiveness (LORE) scale is presented. It is a conceptual framework for evaluating recycling and other alternative strategies of meeting needs in space activities. The LORE scale is developed, defined, and examples of its use are given. The LORE scale should help reduce planning time by serving as a framework within which initial thinking can take place when selecting between alternatives.

Introduction

A conceptual framework is proposed here for evaluating recycling and other alternative strategies of meeting needs in staffed space activities where resupply is required, or on the Moon where local materials may be used. The Level Of Recycling Effectiveness (LORE) scale provides a framework within which initial thinking can take place when selecting between strategies. The LORE scale can thus reduce planning time. It will tend to identify, at an early stage, items which might otherwise not be identified until substantially later, after considerable funds have been expended upon less optimal alternatives. Though most directly applicable to closed environment systems it can be applied to the entire range of space activities.

Background

All of us are interested in carrying out various functions to meet specific needs. We are not directly interested in raw materials such as iron ore or wheat fields. Our interest in these raw materials derives from

----

[1]Associate Professor, Department of Space Studies, University of North Dakota, Grand Forks, ND 58201.

their ability to be converted into functional products to meet our needs. Normally this conversion is at least a two step process. We do not eat wheat, we eat bread and pasta. Iron ore is converted to iron and steel before being made into cars. These useful items are converted to wastes rapidly or slowly: bread to exhaled carbon dioxide, water, and feces, and cars to junked cars plus some rust lost along the side of the road. This is typically a one way path.

As shown in Table 1, on earth products typically follow a one way path starting out as a RESOURCE which is extracted and converted to a RAW MATERIAL then to a FUNCTIONAL PRODUCT which is eventually converted to WASTE OR LOST INTO THE ENVIRONMENT.

TABLE 1

TYPICAL ONE WAY MATERIAL FLOW

ON EARTH	IN SPACE
RESOURCE (usually mineral) iron ore trees sunlight	RESOURCE (when available) aluminum in regolith oxygen in regolith sunlight
↓	↓
RAW MATERIAL steel logs electricity	STORES OR PRODUCTS FROM RECYCLING EQUIPMENT metallic aluminum liquid oxygen electricity
↓	↓
FUNCTIONAL PRODUCT cars paper fan operation	FUNCTIONAL PRODUCT habitats breathable air fan operation
↓	↓
LOST INTO ENVIRONMENT OR UNPROCESSABLE WASTE OR WASTE (RECYCLABLE) rust / junked cars lint / waste paper heat	LOST INTO SPACE OR UNPROCESSABLE WASTE OR WASTE (RECYCLABLE) eroded aluminum leakage / CO_2 heat

In space few resources are available so the raw material equivalent is often "STORES" which may be derived from earth or from recycling equipment. These become FUNCTIONAL PRODUCTS and are eventually converted to WASTE OR LOST INTO SPACE perhaps via leakage. The item of interest is still the FUNCTIONAL PRODUCT. We wish to achieve this FUNCTIONAL PRODUCT with minimal cost and because of high launch costs this often means minimal launch mass.

Recycling is a way to increase the amount of FUNCTIONAL PRODUCT for a given launch mass. On Earth there are other ways of viewing our interest in conservation of resources: as a process of slowing the movement of material from the RESOURCE state to the DISPERSED state; or as a process for increasing the length of time that material is in the FUNCTIONAL PRODUCT state and delaying its entry into the UNPROCESSABLE WASTE state. We also have some interest in seeing that resources are still available for our children.

Recycling and Alternatives

The normal recycling process tries to trap material in the waste stream before it becomes LOST INTO THE ENVIRONMENT and feed it back in place of a RESOURCE into a RAW MATERIAL. This is important and can be effective.

The most effective solution is, however, not normal recycling. For space activity the most effective (least mass) solution is to launch nothing to carry out that activity. Carried to the extreme this means we do nothing but at lesser levels many things can be left undone. Because launch costs are so expensive this solution is often used in space activities. In Skylab astronauts did not shower as often as on Earth thus saving water launch mass. Every item which is proposed for space flight should be examined with the question "Could we get along without it?"

There are often ways of getting the same FUNCTION out of a smaller amount of equipment / mass. There are other cases where we could use less of an item, or a smaller one of the same thing. This is perhaps a matter of looking at each piece and asking "Can we get along without this piece?" For items which wear out we can make them last longer. We can replace some items with different types of items of lesser total weight and accomplish the same functions. We can use an item which has served its function for an entirely different function, particularly if we design it to be converted.

Let us not forget that any change has ramifications on other aspects of a project. Where the function serves crew members then social and psychological impacts must also be considered. The trade offs are not best made in isolation from a larger context.

The least mass solution is DO NOT USE THE PRODUCT. Where we have people we must have air, but do we need nitrogen? After the Apollo 1 accident it was concluded that we do. Do those reasons still hold? Probably they do, but can we USE LESS OF THE PRODUCT? Do we need 780 mb of nitrogen or could we do with 400, or 100, or 600 mb? For that matter do we need 210 mb of oxygen? People in Denver don't seem to think so; they get along with 180 mb. Do we even need that? How about 150 mb of oxygen or the equivalent of 2,500 meters altitude? Any pressure reduction reduces the leakage rate and thus saves on the amount of gas needed to replace that irretrievably LOST INTO SPACE. Pressure reduction also increases the difficulty of cooling electronic equipment.

Another way of reducing the leakage rate is by improving the seals so that at a given pressure leakage is reduced. This in effect REPLACES air WITH A LONG LIVED sealant PRODUCT. That is we are substituting sealant and welds for air. Space station Freedom specifications are calling for lower leakage rates than have previously been tolerated.

Another means of using less of a product is to INCREASE PRODUCT LIFE. Clean clothing is perhaps an example of this. Clothes washing facilities are planned for Freedom. We can use less water for washing clothes if we wear the clothes longer between washings. This requires modifying our definition of clean clothing, and perhaps minor attitudinal changes.

A fourth means of using less of a product is to use an item for more than one purpose. An example of this might be the use of the External Tank (ET) as a strongback or as a pressurizable volume in LEO (Gimarc, 1985). This eliminates the need for a dedicated system to be launched. This type of double usage is simplified if the concept is in place during the design stage but may be possible with small modifications or even without modification for existing systems.

Proposed LORE Scale

Table 2 presents the Level of Recycling Effectiveness (LORE) scale. This is a hierarchical arrangement of the effectiveness of various systems. The high LORE numbers represent more effective reduction of material needs than the low numbers. The four means discussed above reduce the amount of material required to carry out activities. These methods are not as effective in eliminating mass launched as not doing the activity at all. Therefore, they are considered to be a lower level of effectiveness and do not reduce launch mass as much. Not using an item has a LORE value of 5 while the four means of reducing requirements have a LORE value of 4. Note that 100% efficient recycling is well down the effectiveness scale with a LORE value of 3.

TABLE 2

LORE SCALE

```
DO NOT USE PRODUCT
.
.
.    USE LESS OF PRODUCT
.    .
.    .
.    INCREASE PRODUCT LIFE
.    .
.    .
.    REPLACE WITH A LONG LIVED PRODUCT
.    .
.    .
.    USE SAME PRODUCT IN DIFFERENT MANNER
.    .
.    .
.    .    RECYCLE INTO THE SAME PRODUCT
.    .    .
.    .    .
.    .    REPLACE WITH MORE-AVAILABLE-RESOURCE PRODUCT
.    .    .
.    .    .
.    .    .    RECYCLE INTO A LOWER LEVEL PRODUCT
.    .    .    .
.    .    .    .
.    .    .    .    RECYCLE INTO AN END PRODUCT
.    .    .    .    .
.    .    .    .    .
.    .    .    .    .    LOST INTO SPACE OR
.    .    .    .    .    .  UNPROCESSABLE WASTE
.    .    .    .    .    .
5    4    3    2    1    0
```

LORE Scale

Level Of Recycling Effectiveness

The LORE scale values are based upon the following reasoning. DO NOT USE PRODUCT saves the entire required material and receives the highest LORE value of 5. Several approaches which use less material than normal save some of the material in a single functional cycle and carry a LORE value of 4. Standard recycling or substitution is rated at 3 because it does not reduce the amount of material required to carry out a function. A LORE value of 2 is given to a recycling activity which does not permit reprocessed material to go back into the same use from which the waste came. This is typically due to contamination causing the inability to maintain raw material quality. A recycling activity functioning in a manner from which material can not again be recycled receives a LORE value of 1. The LORE value of 0 is given to one way flow of material with no reuse of the material.

No recycling system is 100% efficient. There is always some loss in the system. Theoretically, oxygen recovery from exhaled carbon dioxide and water on space station Freedom can not be 100% efficient because of the inevitable leakage of gasses out of the habitation modules. On Earth converting junked cars back into metal to make new cars can not achieve 100% efficiency since it does not include the iron which is lost as rust along the highway.

Material substitution with "free" materials is ranked at the same level of effectiveness, LORE value of 3, as 100% efficient recycling. Almost free material is almost as good. Lunar oxygen on the moon is almost free at a lunar oxygen factory so burning imported hydrogen to produce water is a good idea.

The LORE scale is not capable of always selecting the optimal solution to a problem. For example recycling material back into the same product at 70% recovery may not be as desirable as recycling it into another product at 90% material recovery. Note, however, that 70% recycling material into the same product may be the more desirable of the two strategies as it can be repeatedly recycled recovering 49% of the original material the second time it is recycled (for a recovery rate of 119% in two cycles) with additional recovery the third time etc., while the 90% to a different product may not recycle very well again. What is ultimately of interest is the overall cost reduction which in space activities is often represented by total launch mass reduction needing to be launched. The LORE scale does provide an initial evaluation of relative ranking between alternatives.

Applying the LORE Scale

On Earth we do not usually recycle material back into the same product. Automobile bodies when melted down are contaminated with copper from the electrical wiring so can not be made into steel for automobile manufacturing. The steel is used in less demanding products. Newspaper typically is recycled into grey cardboard not more newsprint.

Some recycled material goes into uses from which it is recognized that it can not be recycled, that is into END PRODUCTS. A case of this nature in space might be using algae to produce oxygen from carbon dioxide and water. The algae does this in the process of using some of the material to make additional algae and we end up with algae biomass as an END PRODUCT as part of the process.

The least effective process from the point of view of recycling is to treat things as a one way flow and end up with wastes which are unprocessable. This may be the most cost effective way of operating. On Earth one way flow often is the most cost effective method, at least in the short run, but is sometimes cheap only because of how costs are defined. In space activity, as well, NASA has often been forced to make decisions on current costs rather than long-term costs.

With almost any item which one examines one can think of various ways to handle it at many of the levels of effectiveness. These methods are not necessarily cost effective. One must keep in mind that all design choices have consequences on other aspects of the system, including, of course, those choices made without considering their consequences.

The items the author is most familiar with are life support functions and since they are directly related to people, our social and psychological reactions must also be taken into consideration in making design choices.

As an example let us look at clothing.

DO NOT USE PRODUCT - LORE = 5: Not using clothing is not psychologically acceptable to most of our society so is not an option for NASA programs. In addition it might be a health problem.

USE LESS OF PRODUCT - LORE = 4: I personally hope that we can continue to avoid using neckties in space habitats. Perhaps we can reduce the average quantity of clothing to typical warm weather wear. Note that this choice is cultural and is not necessarily agreed to by individuals with other social standards. When we permit administrators to travel in space neckties may be reintroduced.

INCREASE PRODUCT LIFE - LORE = 4: Wearing clothes longer, thus getting them dirtier, might reduce the quantity required if more than two sets are needed between washings. On very long missions one might accept a longer wear definition of "worn out." Again this has social ramifications.

REPLACE WITH A LONG LIVED PRODUCT - LORE = 4: By increasing the temperature of the living space the quantity of clothing worn might be reduced to that typical of warmer climates thus substituting hardware for clothing. In some cases permitting higher temperatures might reduce the amount of equipment required for cooling as well as reducing the amount of clothing required. It will probably increase the size of dehumidifiers and increase design problems for electronic equipment cooling.

USE SAME PRODUCT IN DIFFERENT MANNER - LORE = 4: Can we sleep in our clothing and dispense with sheets?

RECYCLE INTO THE SAME PRODUCT - LORE = 3: Can clothing be designed so that worn out clothing can be made into clothing? If agriculture is available for food production then, perhaps it would be simpler to use cotton clothing, burning worn out clothing to produce carbon dioxide for growing cotton, than to establish a plastic fiber industry for synthetic clothing.

REPLACE WITH MORE-AVAILABLE-RESOURCE PRODUCT - LORE = 3: At a Lunar base can we make clothing from glass or some other fibers made on the Moon?

RECYCLE INTO A LOWER LEVEL PRODUCT - LORE = 2: Clothing is often recycled into rags and is likely to be used this way on permanently staffed facilities.

RECYCLE INTO AN END PRODUCT - LORE = 1: The author can not think of a reasonable end product.

UNPROCESSABLE WASTE - LORE = 0: Finally we could just throw clothing away and order some more from Earth. Slavin, et.al. (1985) suggested that it takes 27.5 pounds of water to wash 2.5 pounds of cloths. If water recycling is only 90% efficient then it is better to bring in 2.5 pounds of clothing than 2.75 pounds (10%) of replacement water.

For items which affect people living in space we must also consider the psychological and social consequences of doing things differently. However, there are likely to be areas where crews could easily adapt to providing a given function in a different manner than that to which they are accustomed.

In other cases the psychological issues dictate familiar approaches. It is worth noting that with internationally mixed crews we are already assuming that not everyone on a mission will be operating in their native social environment. Perhaps in some cases it would save launch mass to adopt another country's means of providing a function than the typical U.S. means.

An example is toilet paper usage. In parts of Europe during World War II toilet paper was not available so people wiped with what they called "diapers" and washed them. In some parts of the world people prefer to wash and dry their bottoms. Neither of these alternatives involve a single use material like toilet paper.

As our activity gets more diverse we will have more opportunities to use products and items as raw materials and in different ways.

Because of the high costs of launching mass the issue of reducing launched mass, including by recycling, has received much interest in space-related activity (Briggs, Sacco, 1985). The variety of suggestions for using external tanks, such as for habitats (LORE = 3) or for aluminum for aluminum-oxygen engines (LORE = 5), is one area (Gimarc, 1985). These ideas have been developed with no theoretical framework within which to evaluate them. The LORE scale provides a preliminary framework within which this thinking can more easily take place.

Summary

The hierarchical Level Of Recycling Effectiveness (LORE) scale provides a five level framework within which initial thinking can take place when selecting between alternative strategies for meeting requirements for space missions. The LORE scale can thus reduce planning time by assisting in identifying, at an early stage, the most effective (often least mass) solution to meeting a given function, particularly in respect to supply requirements for staffed missions.

On earth products typically follow a one way path starting as a resource, converted to a raw material, to a functional product, and eventually to a waste. There are often more effective ways, than 100% efficient recycling of material, to reduce waste generation. However, where the function serves crew members then social and psychological impacts must also be considered. All trade offs have ramifications on other aspects of projects so are not best made in isolation. Some examples are presented to illustrate the application of the LORE scale.

References

Briggs, R. and A. Sacco Jr. 1985. Environmental considerations and waste planning on the lunar surface. pp.423-430. In: Mendell,W. Ed. Lunar bases and space activities of the 21st century. Lunar and Planetary Institute, Houston, TX.

Gimarc, J. A. 1985. Report on space shuttle external tank applications. Space Studies Institute, Princeton, N.J.

Slavin, T., F. Liening, M. Oleson, R. Olson. 1986. Controlled ecological life support system (CELSS) physiochemical waste management system evaluation. NASA contractor report 177422. NASA J.S.C., Houston, TX.

Subsection 4.9: STUDENT PAPERS

Black Holes

Thomas C. Cannon[1]
Anthony W. Cannon[1]

Abstract

For many years, scientists have searched the skies for black holes; stars even more massive than white dwarves. As of yet, there have been no 100% sure sightings. Since they do not emit light, we cannot detect black holes by ocular means; however, efforts have been made to detect them in certain binary systems where they sometimes give off x-rays and by other methods. Although, presently, there are no practical uses for a black hole, speculation has been made as to possible applications which may be possible many years in the future.

Introduction

In the paper, *Philosophical Transactions of the Royal Society of London*, John Michell suggested that if a star was sufficiently massive and compact then it would have such a strong gravitational field that neither matter nor radiation could escape it: therefore, any light emitted from the surface would be dragged back toward the star. Michell suggested that although we would not be able to see these stars, later to be called black holes, their gravitational fields would still be felt.

Origin of a Black hole

To understand how a black hole is formed, it is necessary to first have an understanding of the life of a star. A star forms when a large amount of hydrogen gas collapses in on itself due to gravitational attraction. As the gas cloud contracts, the atoms which make up the cloud collide with increased frequency at greater speeds, causing the gas to heat up. Eventually, the gas becomes so hot that the collision of two hydrogen atoms causes them to chemically combine to form helium. The heat given off by this reaction is similar to that of a controlled hydrogen bomb explosion. This is what makes the star shine. This additional heat also increases the pressure of the gas, counteracting the

[1]Students, La Cueva High School, 7801 Wilshire, NE, Albuquerque, NM 87122.

gravitational attraction of the atoms within the star. Stars remain in this stable condition for a long time, however, sooner or later the star will run out of hydrogen fuel. When this occurs, the star will cool off and contract. What happened after this was not fully understood until the 1920's when Subrahmanyan Chandrasekhar, an Indian graduate student, worked out how large a star must be in order for it to be able to support itself after its fuel had been used.

Chandrasekhar reasoned that, according to the Pauli exclusion principle, when matter particles of a collapsing star move closer together, they must have different velocities. Therefore, the particles must move away from each other, causing the star to expand. A star can then reach a stable radius with this repulsion of particles offsetting the gravity of the star. However, Chandrasekhar calculated that a cold star more than one and a half times the mass of the sun would not be able to support itself against its own gravity. This mass is now known as the Chandrasekhar limit.

If a star is less than the Chandrasekhar limit, it will eventually stop contracting and settle into a stable state, perhaps as a white dwarf, a star with a radius of a few thousand miles. If, however, a star's mass is greater than the Chandrasekhar limit, it has one last chance for survival. If the star can manage to blow off enough mass to put it under the limit, then it would continue to collapse and eventually stabilize. However, it is doubtful that this always happens with a star of any size. If a star cannot halt the pull of gravity dragging it into itself, it will become a black hole.

Characteristics of a Black Hole

The gravitational force on the surface of stars alters the path of light, bending nearby light rays toward them. As a star contracts, the gravitational field at its surface grows in strength, and light is bent further in toward the star. Eventually, the pull of gravity at a star's surface becomes so strong that it is impossible for light to escape its pull. According to general relativity, nothing can travel faster than light, therefore, nothing can escape the star's pull, creating an area of space around the star from which it is impossible for light to reach a distant observer. This region is the black hole, whose boundary, called the event horizon, acts as a one-way membrane from which neither matter nor radiation can escape.

While the event horizon can be considered to be the surface of a black hole, another outer surface called the stationary limit, which is a result of the tremendous spin of the black hole, will also form. A neutron star must spin over one hundred times a minute to conserve its angular momentum. The rotation of a black hole must be many times greater. This rotational force, when combined with the gravitational forces of the black hole, causes any matter or radiation inside the static limit to be dragged in the direction of the black hole's spin. The area between the event horizon and the static limit is called the ergosphere and, because the force is directed circumferentially around the hole as opposed to directly toward it, both matter and radiation can eventually spiral out and escape. Also because of the incredible rotation of the hole, it is entirely possible for the matter or radiation to emerge with more energy than when it entered.

Searching for a black hole

The assumption has been made that black holes do indeed exist. However, to this date no solid confirmation or sighting of a black hole has been made. Because a blackhole emits no light, locating one by ocular means is next to impossible; however, the extreme gravitational forces of a black hole give it certain characteristics which may be used to detect them. Astronomers have observed systems in which two stars orbit each other, but they have also observed stars which orbit some unseen object. Although this does not necessitate the presence of a black hole, some of these systems, such as Cygnus X-1, are also strong sources of X-rays. Scientists have concluded that in this system, the gravity of the unseen object causes the matter blown off the visible star to spiral in toward it, gaining heat as it does so. This spiral of gases, called an accretion disk, emits a special form of detectable x-rays. However, this system will only work if the object which the visible star orbits around is very small; a white dwarf, a neutron star, or a black hole. By examining the orbit of the star, the lowest possible mass of the unseen object can be determined. In the case of Cygnus X-1, this is about six solar masses, which is too great for the object to be a white dwarf or a neutron star. Therefore, it seems that the object must be a black hole. There are other explanations for systems such as Cygnus X-1 which do not include a black hole, but these do not seem to be as probable as the one mentioned above.

The extreme gravitational forces produced by a black hole may also in some way be detected. According to general relativity, the movement of massive objects produces gravtational waves which travel from the distubance at the speed of light. The exploding, vibrating, or collapsing of large amounts of matter will produce such waves. As the waves move away from their source, they create a disturbance in the fabric of space time, causing a distortion in the shape of that particular region of space. If the distortion of an isolated mass can be measured, then it may be possible to detect these waves. The gravitational waves produced by different phenomena have different characteristics. The waves produced by a supernova, for example, would be different from the waves of a binary stellar system. This difference in waves gives each possible source a signature, or fingerprint. This would enable scientists to detect a forming blackhole or the collision of two blackholes by their own individual signals.

New technology in the field of astronomy could possibly lead to the confirmation of the existence of black holes. For example, the Lunar Optical-Ultraviolet-Infrared Synthesis Array[1] , or LOUISA, which could be built on the dark side of the moon could give a clearer picture of the accretion disk near the Cygnus X-1 system.

Possible Application of the Ergosphere

In the distant future, the ergosphere of a rotating black hole may make it possible for man to harness the energy of a black hole. One such plan, proposed by physicists Charles W. Misner, Kip S. Thorne, and John A. Wheeler, consists of a rigid spherical framework placed at a safe distance around a black hole, upon which cities are built. During the day, trucks move through the cities, collecting refuse and transporting it to special depots, where it is loaded on to special spacecraft which carry it down through the stationary limit Once the ship has entered the ergosphere, its cargo is dumped into the black hole, and the ship leaves the ergosphere with more energy than it started with. This excess energy is collected with special turbines on the surface of the superstructure and is converted into electricity.

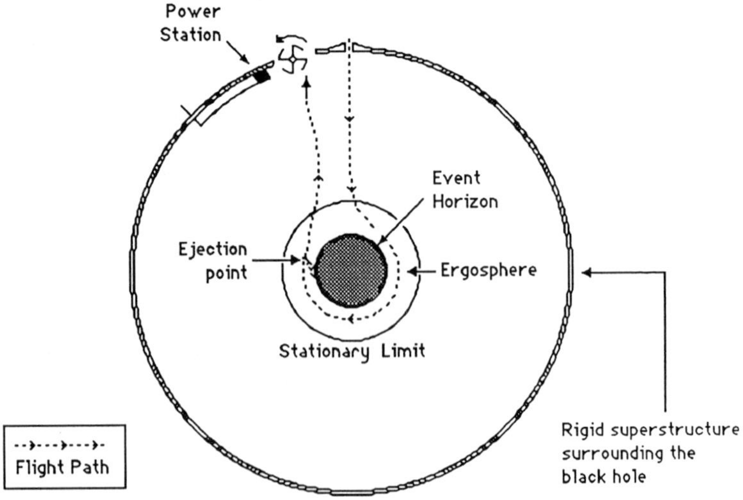

[1]proposed by Jack O. Burns, Nebojsa Duric, G. Jeffrey Taylor, and Stewart W. Johnson

If each depot on the surface converts one million tons of refuse a day, the amount of energy that five such depots could convert in one day would be equal to the amount of energy the earth receives from the sun in sixty-seven years.

Appendix I: Bibliography

Asimov, Isaac, *The Universe* (Walker and Company: New York, New York, 1980.)

Bartusiak, Marcia, *Thursday's Universe* (Times Books: New York, New York, 1986.)

Berry, Adrain, *The Iron Sun* (E.P. Dutton: New York, New York, 1977.)

Burns, Jack O., Nebojsa Duric, G. Jeffrey Taylor, and Stewart W. Johnson (1989). "Astronomical Observatories on the Moon," *Scientific American*, Vol. 262, No. 3.

Hawkings, Stephen W., *A Brief History of Time* (Bantam Books: New York, New York, 1988.)

Jeffries, Andrew d. and others, "Gravitational Wave Observatories" in *Scientific American*, (June 1987: vol 256. number 6) pp. 50-58

Motz, Lloyd, *The Universe: Its Beginning and End* (Charles Scribner's Sons: New York, New York, 1975.)

Sullivan, Walter, *Black Holes: The Edge of Space, The End of Time* (Anchor Press/DoubleDay: Garden City, New York, 1979.)

SPACE POWER SYSTEMS

Radiation: Energy Conversion

Bobby D. Jones[1]

ABSTRACT

Pair production is a way to change a photon, a gamma particle, into matter or energy. It is found that if a photon of energy greater than 1.02 MeV strikes a foil containing a large number of protons, high Z, the photon disappears and in its place an electron-positron pair is formed. A positron has the same mass as an electron, but the opposite charge, e^+.

It may be possible to utilize the photons of radiation and use the process of pair production to create useful energy. This conversion of radiation to energy would be inexpensive considering that radiation is readily available from a variety of sources.

INTRODUCTION TO ELECTRON AND POSITRON PAIRS

The concept of changing radiation into useful energy utilizes the process of changing a photon's energy into matter. This process is called **Pair Production.**

Photons: Mass, Energy, and Momentum

The photon is a relativistic particle - it travels at the speed of light, 3×10^8 meters/second in a vacuum. Thus, when dealing with its mass, energy, and momentum, relativistic formulas are used.

---

[1] Sophomore, West Mesa high School in Albuquerque, Albuquerque, NM

The mass, m, of any particle is shown by

$$m=m_0\Big/\sqrt{1-\frac{v^2}{c^2}} \qquad \text{Equation 1.}$$

where m_0 is the rest mass, v is the particle's velocity, and c is the speed of light. Since v=c for a photon, the denominator is zero. A photon is never at rest, so the rest mass, m_0, of a photon must also be zero, or its mass and energy would be infinite. The momentum of a photon, p, using the equation $E^2=p^2c^2+m^2_0c^4$* with $m_0=0$, is

$$p=\frac{E}{c} \qquad \text{Equation 2.}$$

The energy of a photon is equal to the Planck's constant, h, multiplied by the frenquency, f, is

$$E=hf \qquad \text{Equation 3.}$$

The momentum of a photon, p, is related to its wavelength by

$$p=\frac{hf}{c}=\frac{h}{\lambda} \qquad \text{Equation 4,}$$

where λ is the wavelength.

ELECTRON AND POSITRON PAIRS

Pair Production

Pair production is the creation of an electron-positron pair, matter, from a photon that strikes a foil containing a large number of protons, high Z, such as lead foil, Figure 1. A positron has the same mass as an electron, but the opposite charge, e^+. The positron is considered as antimatter. The photon disappears in the process of pair production.

Pair production is an example of rest mass being created from pure energy, and it occurs in accord with Einstein's equation, $E=mc^2$. A photon cannot create a single charge since electric charge would not then be conserved.

---

* The previous equation is used to show a useful relation between the total energy, E, of a particle and its momentum, p.

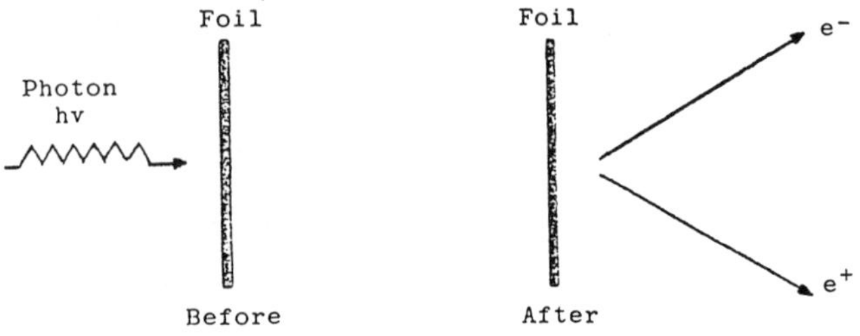

FIGURE 1 Electron-positron pair formation.

To produce an electron-positron pair, the photon must have energy equal to the electron-positron pair's mass multiplied by the speed of light squared.

$$E=mc^2=2(9.1\times10^{-31}kg)(3.0\times10^{8}m/s)^2= 1.64\times10^{-13}J, \text{ or } 1.02 \text{ MeV} \qquad \text{Equation 5.}$$

A photon with less energy cannot undergo pair production. If the energy of the photon is greater than 1.02 MeV, the excess energy goes into kinetic energy of the electron and positron.

Since $E=hf=hc/\lambda$, from Equation 4, the wavelength of a 1.02 MeV photon is

$$\lambda=\frac{hf}{E}=\frac{(6.6\times10^{-34}J\ x5)(3.0\times10^{8}m/s)}{(1.64\times10^{-13}J)}=1.2\times10^{-12}m \qquad \text{Equation 6,}$$

which is 0.0012 nm. Thus, the wavelength must be very short. Photons with such short wavelengths are in the gamma ray (or very short x-ray) region of the electromagnetic spectrum.

Pair production cannot occur in empty space for energy and momentum would not simultaneously be conserved. It can be shown that at any energy, an additional massive object, such as an atomic nucleus, must take part in the interaction to carry off some of the momentum. In Equation 2, for instance, energy is conserved, but the electron-positron pair has no momentum to carry away the initial momentum of the photon. If a pair is produced in a cloud chamber and a magnetic field is applied, the electron and the positron are deflected in opposite directions with equal curvature, Figure 2.

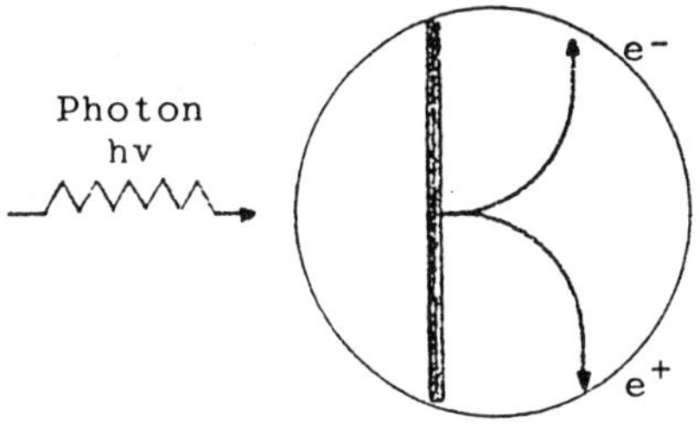

FIGURE 2 Electron-positron pair formation in a cloud chamber with a magnetic field.

Actually the pair formation takes place in the field of the nucleus and the conservation of energy gives

$$hf=2m_0c^2+E_++E_-+E_{nuc} \qquad \text{Equation 7,}$$

where hf is the energy of the incident photon, $2m_0c^2$ is the energy equivalent to the rest mass of the electron and the positron, and E_+, E_-, and E_{nuc} are the kinetic energies of the electron, positron, and the recoiling nucleus respectively. The mass of the recoiling nucleus is very large, therefore, it takes away a very small amount of kinetic energy and so E_{nuc} may be neglected. Thus, the equation would read

$$hf=2m_0c^2+E_++E_- \qquad \text{Equation 8,}$$

which clearly shows that the threshold for pair formation is $2m_0c^2$, or 1.02 MeV. Anything over 1.02 MeV, will go into the kinetic energy of the electron-positron pair.

Pair formation is also possible in fields of low Z, but the threshold in such cases are much higher. It can be shown that the pair formation in the field of an electron has a threshold of $4m_0c^2$ i.e., it will take place only if the photons have energies greater than 2.04 MeV. In this case, E_{nuc} would be greater.

Positron

The positron was discovered by Carl Anderson in 1932, in the cloud chamber photographs of cosmic radiation produced electron-positron pairs that were detected in the cloud chamber by the application of a magnetic field. The positron had already been predicted theoretically by Paul Dirac, who used a complicated set of quantum-mechanical equations for the electron. Positrons are not found in free space as are electrons. Positrons do not last long in nature because when they collide with an electron, both the electron and the positron produce two, or more photons, Figure 3. This pair annihilation is the process by which positrons are removed from circulation.

When a positron is created, it has some kinetic energy that it may lose by collisions with the atoms in its surroundings. After it has slowed down considerably, it may form a kind of atom with one of the electrons in the medium.

APPLICATION OF PAIR PRODUCTION

Pair production is a proven process. Harnessing the process to provide useful energy with minimal waste would give a leading edge to providing power for space systems.

The proposed system may utilize cosmic radiation as an emitter, or a commercial radiation source, (nutural radiation) such as cobolt 60. The system may then have collector plates, such as copper, which are filled with electrons. The electrons and positrons formed may strike these plates provided they have enough kinetic energy. The electron may strike one plate adding to the total number of electrons in that plate. The positron may then strike the other plate annihilating one of its electrons. Thus, one electron would be taken away fron the total number of electrons in that plate. If both plates contained the same number of electrons when the process began, a deficiency of two electrons may have been created between them. Current is caused by electrons flowing from the negative plate to the positive plate.

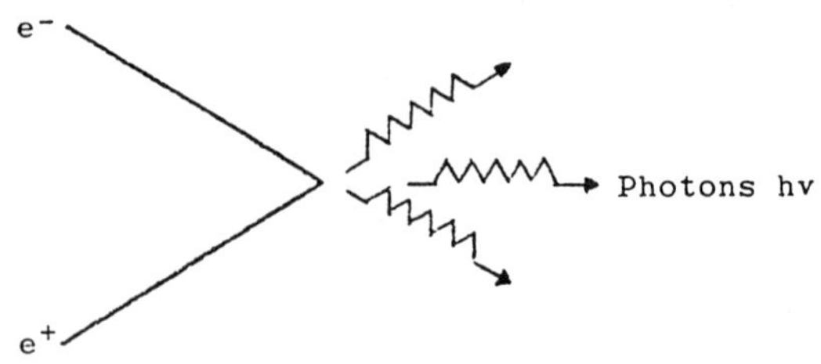

FIGURE 3 The collision of a positron and an electron result in gamma ray production.

REFERENCES

Arya, Atam P. Fundamentals of Nuclear Physics. Allyn and Bacon, Inc., Boston, Mass. 1968

Arya, Atam P. Fundamentals of Atomic Physics. Allyn and Bacon, Inc., Boston, Mass. 1971

Beiser, Arthur. Concepts of Modern Physics. McGraw-Hill Book, Boston, Mass. 1973

Close, Frank, Marten, Michael, and Sutton, Christine. The Particle Explosion. Oxford university Press, New York, New York. 1987

Giancoli, Douglas C. General Physics. Prentice-Hall, Inc., Englewood Cliffs, New Jersey. 1984

Hollon, James Max and Murphy, James T. and Zitzeeitz, Paul W. Physics Principles and Problems. Charles E. Merrill, Columbus, Ohio. 1986

Kaplan, Irving. Nuclear Physics. Addison-Wesley Publishing
Company, Inc., Reading, Mass. 1964

Mission to Planet Earth: The Quest to Save the Environment

Jody Jordan[1]

Abstract

The Mission to Planet Earth is a program designed to look at the earth as an integrated system. It is an internationally supported project. It was conceived because the current methods in use are inadequate and insufficient.

Background

Mission to Planet Earth is a project designed to provide detailed data and understanding of the climate changes relative to the global enviroment. The idea was first presented by a study group led by Sally Ride in 1987. Current methods for predicting climatic changes on a global scale are very limited, and the data being used is inadequate. The best models used today in the United States are the Global Circulation Models, which have limited usefulness because of insufficent information about the earth's enviroment.

There are a number of models that analyze the earth's enviromental problems independently, but none look at the earth as an integral system. The interaction on a global scale is far too complex to create a model given our limited understanding of the mechanics at work. It is hoped that a great deal can be learned by monitoring the earth as a whole from the unique perspective from space.

NASA has collected, over the years, a set of long-term data that has been used to create models that have proven consistent in their predictions. First, NASA currently has under construction three satellite projects that will gather data on global climate change. Secondly, NASA has a strong Earth Sciences program in place to support the Mission to Planet Earth Program.

The whole idea behind Mission to Planet Earth is to be able to foresee ecological disasters by understanding the earth and how it functions. With this information,

---

1Jordan, Student, La Cueva High School, 148 Whitetail NE, Albuquerque, NM 87122.

tragedies could be avoided. The information that would come out of this project would effect both national and international decisions about the enviroment.

The Reason for the Mission to Planet Earth Program

As a result of human activity since the beginning of the industrial age, the enviroment is in danger on a global scale. Some of the factors that are causing change are pollution, energy consumption, and overpopulation. The resulting problems are: the build up of greenhouse gases, the increase of methane in the atmosphere, the ozone layer depletion, the occasional warm ocean currents (El Nino), change in global weather patterns, and acid rain.[1]

The environment needs to be analyzed as an integrated system. This unique perspective can be obtained only from space, which is the purpose of Mission to Planet Earth. The Mission to Planet Earth Program will look at "worldwide enviromental threats and complex interactions between the atmosphere, oceans, and land,"[2] and will analyze the results of the interaction between human activities and the environment. The Mission to Planet Earth Project must be able to use the short term results to accurately predict the earth's enviromental future over a longer span of time.

The Mission will concentrate initially in two areas: the greenhouse effect and deforestation. There has already been evidence that the earth's climate is changing as the result of the greenhouse effect. The greenhouse effect is the trapping of solar energy and surface radiation. 70% of the incoming solar energy is absorbed by the earth's surface which is then reemitted at infrared wavelengths by the atmosphere and by the surface. This surface radiation is trapped by clouds and greenhouse gases and returned to the earth.[3] Some scientists think that the drought and the record high temperatures in the United States during the 1980's were a result of the developing greenhouse effect. The deforestation of Africa's and South America's chief rainforests caused 20% of the carbon dioxide build up in the atmosphere. Biologically, these rainforests also have many rare plant forms that are being destroyed daily. Economically, after destroying the trees, the ground has been found to be very barren, and it does not support crops well.

The Plans for the Planet to Earth Program

Mission to Planet Earth is an international project. If accomplished as currently conceived, it will be the largest cooperative space effort in history. The countries that will be involved are the United States, the Soviet Union, Japan, China, Australia, Brazil, Argentina, Pakistan, and eleven menbers of the European Space Agency (ESA). It also will be the largest civilian unmanned space effort ever attempted by the United States. The total program costs will be $15-30 billion dollars. The United States program will take 25 years to complete, with five years of preflight preparation and five years postflight of data analysis.

The components of the project are polar orbiting instrument platforms, sensors attached to the space station and small satellites called "earth probes."[4] Under the least expensive plan at $15 billion, 4-6 unmanned space platforms and a series of earth probes will be lauched. Under the expensive plan of $30 billion, five additional geosynchronous satellites will be launched.

If the program is funded, the polar orbiting instrument platforms would be the first to be launched. The first 15 ton, 40x15 foot spacecraft is scheduled to be launched in 1996, with launches continuing until 2011. The United States will provide two of the platforms, Japan will provide one to two, and the European Space Agency will provide one. The ESA and Japan are already in the "advanced definition phase."[5] The instruments on NASA's first Earth Observing System (EOS) will consist of 20 earth viewing sensors and 6-7 monitoring instruments that will be provided by the National Oceanic and Atmospheric Administration. NASA will conduct approximately 20 research projects using the data collected from the EOS sensors. These projects will give direction to the rest of the mission.

The earth probes are small satellites that will be launched every few years in equatorial orbits. They will collect data on rainfall, ozone depletion, magnetic fields, gravitaional fields, and radar altimeter data on the earth's topography. This data cannot be obtained from the polar orbiting platforms. The development of the earth probes has already begun. However, it has been determined that they would need replacing every five years, therefore tripling their cost.

The space station is an important part of Mission to Planet Earth. When it is launched in the late 1990's, it will carry a variety of instruments which will measure tropical rainfall, tropical winds, and solar radiation at the equator. It will also carry equipment for extensive photography of the earth.

The Geosynchronous-orbit platforms will only be used if NASA is given adequate funding. Two of these will be provided by the United States and two to three will be provided by Japan and Europe. These platforms would provide continuous environmental data on a whole hemisphere.

Conclusion

The purpose of Mission to Planet Earth is to observe the earth as an integrated system, from the unique perspective of space. The program already has the international support and the planning necessary to begin. The only area not confirmed is the budget.

This is a program that is vital to the survival of the earth. It is a project that will affect people internationally. People will be able to relate to this program because it will effect them personally, unlike a lunar base or a mission to Mars. The initial data received would enable scientists to begin to correct the enviromental problems that could seriously jeopardize life on earth. The present systems are fragmented and woefully inadequate for the magnitude of this problem.

REFERENCES

1. Ms. July/August, 1987. "Interview with Sally K. Ride."

2. Aviation Week and Space Technology, August 22, 1988, "International Space Effort to Assess Earth's Enviroment" Craig Covault

3. Scientific American, September 1989, "The Changing Climate" Stephen H. Schneider

4. United States Senate Report, March 8, 1989, Committee on Commerece, Science, and Transportation, Dr. Lennard A. Fisk.

5. Aviation Week and Space Technology, March 13, 1989, "Major Space Effort Mobilized To Blunt Enviromental Threat", Craig Covault

ADDITIONAL REFERENCES

Covault, Craig. "International Space Effort To Assess Earth's Enviroment." Aviation Week and Space Technology. August 22, 1988.

Covault, Craig. "Major Space Effort Mobilized To Blunt Enviromental Threat." Aviation Week and Space Technolgy. March 13, 1989.

Covault, Craig. "Ride Panal Calls for Aggressive Action To Assert U.S. Leadership in Space." Aviation Week And Space Technology. August 24, 1987.

Covault, Craig. "Ride Panal Will Urge Lunar Base, Earth Science as New Space Goals." Aviation Week and Space Technology. July 13, 1987.

Fisk, Lennard A. "The Space Debate, Mission to Planet Earth." Popular Science. July 1989.

Fisk, Lennard A. United States Senate Report to the Committee on Commerce, Science and Transportation. March 8, 1989.

"Interveiw with Sally K. Ride." Ms. July/August, 1987.

Schneider, Stephen H. "The Changing Climate." Scientific American. September 1989.

"Taking the Earth's Vital Signs." Time. June 5, 1989.

Van Allen, James A. "The Space Debate, Mission to Planet Earth." Popular Science. July 1989.

New Technologies for Unmanned Space Exploration

Chiem Whua Ma[1]

Abstract

There could be no better time than now for the development and implementation of new technologies such as V.L.S.I. and optoeraseable storage. Creative inventions from new mathematical concepts to the CRAY-XMP GaAs (sub-100-picosecond) processor are continuing to revolutionize our perception of the universe. New technology could be applied to the space exploration movement in ways only dreamed about previously. It can boost advancements in the next generation of unmanned space probes as well as encourage growth of the space exploration program.

Introduction

The idea of unmanned space exploration is not a new concept. Man has long dreamed of exploring the planets of the solar system and beyond. In the recent past, we have seen pictures of the outer planets from Voyager and soil analysis of Venus from the Soviet's Venera 13 and 14. As technology continues to advance, the possibilities for future unmanned space exploration will likely eclipse anything accomplished to date.

Improvements in current technology are beginning to make their ways into the space program. The rapidly changing method in which chips are manufactured has advanced both CMOS (Complementary Metal Oxide Silicone) and V.L.S.I. (Very Large Scale Integration) technology. Speed and density have gone up while cost and radiation sensitivity have fallen.

Newer technologies such as GaAs (Gallium Arsenide) are making it possible to do complex computations faster than ever before. Storage space has increased with the erasable compact disk. Higher resolution color C.C.D. (Charged Coupled Device) image sensors are available, and A.S.C.I. (Application Specific Integrated Circuit)/Standard Cell technology is making it possible to place full applications on single chips.

Possible future technological advances include transmission of power via microwave beams and advancements in artificial intelligence which could

[1]Student, Sandia High School and Career Enrichment Center; Internship Student (through C.E.C.) at Honeywell Defense Avionics Systems Division, Software Engineering - Dept. 1561. 7400 McNerney N.E.; Albuqueruqe, NM 87110

lead to such applications as automated building of bases in space or on other planets.

Technologies

CMOS, once a technology used mainly for its low power requirements, is now rivaling TTL (Transistor Transistor Logic) in speed and density. New V.L.S.I. CMOS chips flaunt gate capacities of up to 200,000 with extremely low power requirements. In addition, the problem of radiation sensitivity with CMOS is being overcome. High density CMOS has the tendency to become less sensitive to radiation as the transistor density increases. Special processing techniques are constantly being developed to increase total radiation dosage levels. At one time, large sums of money were paid for chips capable of withstanding 10,000 total rads[2]. Today, some technology with no special processing is capable of withstanding up to 100,000 rads. With special processing, that number can climb as high as 1,000,000 to 10,000,000 rads.

V.L.S.I. CMOS is ideal for many space projects in which the probe would be subjected to constant low radiation levels. Their low power consumption means longer processing times and shorter charge times. In addition to these attributes, the high density allows for very complex designs in small areas. V.L.S.I. chips, such as D.S.P.'s (Digital Signal Processor), allow real or near-real time signal processing, possible before only on mainframe computers.

Many algorithms which rely heavily upon the Fourier Transform[3] can now be performed using parallel processors in real time with data rates of 7.5 kilohertz[4]. (For comparison, the average I.B.M. 80386 with a math coprocessor at 24 megahertz can barely handle data rates approaching 900 hertz.)

In the search for new technology, GaAs has opened up a whole new concept in high speed computing. From GaAs comes ultra-high speed processors and laser diodes. GaAs processors are capable of operating in the gigahertz range, and laser diodes the size of a pin head can produce moderately powered laser beams.

---

[2]unit for measuring absorbed radiation doses, equivalent to 100 ergs of energy/gram. One erg is equivalent to the energy expenditure of one dyne acting through a distance of one centimeter.

[3]A mathematical tool used in frequency analysis, often used for applications such as voice and pattern recognition. Is also useful in predicting and analyzing many natural phenomena such as tides, waves, and chaos.

[4]A project the author is currently working on is a board capable of performing a 1024 point Continuous Fourier Transform at a rate of 40 kilohertz with 16 bit accuracy, yielding both high precision magnitude and phase information at any point within a signal.

GaAs processors often have a cycle rate[5] of anywhere from 2 nanoseconds to under 50 picoseconds (Bridge to V.L.S.I., 1990), setting them two orders of magnitudes higher on the speed scale than any other technology. Unfortunately GaAs processors, being relatively new, are currently very power hungry. GaAs processors are also not very dense. But to make up for the low density and high power consumption, the chips are insensitive to radiation, making them ideal for some special applications such as short bursts of intense computational power near areas of extremely high radiation. There is currently a large effort to improve power requirements and density for future generations of GaAs processors.

Laser diodes have already made their appearance publicly in the compact disk drive. Having the appearance of a rectangular L.E.D. (Light Emitting Diode), these small lasers can generate up to several watts of light-energy. Laser diode technology has made the optical erasable compact laser drive, a device capable of both writing and erasing compact disks, a reality. For years, laser disks have been used for high capacity[6] data storage. In a quality 120 megabyte magnetic media hard drive, approximately 1 megabytes can be stored per square inch. In optical storage, that number jumps to about 50 megabytes per square inch. Unfortunately, until only recently, compact disks could only be written to once, and then could not be erased. The new ability to erase them will probably lead to the eventual discontinuation of magnetic media when optical drive prices fall.

There are several advantages to using optical storage in space. The high capacity of optical storage means the availability of the megabytes of storage required for effective pipelining of data. Because erasable disks are only affected by magnetism for the short time that the surface plastic is in a molten state, these disks may be used near large sources of magnetic energy with low probability of data loss. Their small size and low mass make them ideal for space travel.

High resolution[7] color C.C.D. (Charged Coupled Device) image sensors are replacing high cost bulky imaging tubes and older lower resolution C.C.D. arrays. The size of a miniature match box, these small devices

---

[5]There are two measurements of chips speed relevant to determining overall processing throughput. The first is clock cycle rate which indicates how many times per second the clock line is cycled from logic "0" to logic "1". The second is cycles per instruction and is both processor and instruction dependant. It indicates how many cycles of the clock are required to execute one instruction. A 10 megahertz processor which can multiply in one cycle has a slightly faster computational throughput than a 100 megahertz processor which requires eleven clock cycles per multiply.

[6]An optical-erasable disc the size of a standard long play record approaches a storage space of five gigabytes per side.

[7]Lower resolution C.C.D. imagers are already in use in such projects as the Space Hubble Telescope, however their low resolution and single color specification dictate that three large arrays of image sensors be set up to obtain a reasonable color picture.

contain a bed of photosensors. For each pixel, there are three photodetectors, one each for red, green, and blue. C.C.D. sensors can contain as many as 640 by 350 pixels per unit, and do not suffer the problem of burning as encountered when using imaging tubes.

There are several proposed technologies for the more distant future. Microwave power transmission could help eliminate the need for a constantly running power generator in high power consumption applications. The possibility of advanced artificial intelligence could mean automating several dangerous tasks.

The proposed microwave power transmission device would consist of a sending dish in which high levels of energy would be converted to a narrow microwave beam. This microwave beam would then be pointed towards a distant probe or planet, the destination for the energy. The recipient of the beam would then reconvert the microwave energy back to electricity for use or storage.

New advances in the field of Artificial Intelligence could mean advanced automation. With the coming of faster real time digital signal capability, this field is quickly advancing. Pattern recognition is no longer a subject for science fiction; many algorithms already exist. They just run too slow or have yet to be implemented in some useful application.

Applications

There are many obvious advantages to unmanned space exploration. Many environments are too hostile for human occupation. Too many planets exist for all to be explored by humans. Many tasks are repetitive and/or prone to error. Humans are susceptible to the void of space. Human occupation requires the unnecessary expenditure of valuable payload space. Basically, machines are cheap, fast, light, and disposable. Humans are not.

In general, there are two types of exploratory probes. Those which conduct remote sensing, and those which land on a planet and do research on the planet surface. Remote sensing probes send back important information such as geography, average temperature, magnetic fields, and gaseous content. Probes which land upon the planet surface can perform more refined scientific study: soil sampling, seismological studies, and atmospheric sampling, for example.

Remote sensing probes are just beginning to scratch the possibilities of their usefulness. In the future, probes could be placed in to space filled with light-weight V.L.S.I. CMOS electronic scientific equipment, more than ten times what could previously be placed on board. As the probes passed the planets, they would continuously take C.C.D. pictures of each planet, storing the frames on laser disk for later retrieval. While in orbit, other equipment would constantly transmit data about the planet's atmosphere, temperature, and seismic activity. If more intense research was needed, a small sub-probe could be launched from the main probe to the planet surface. Useful only for a short period of time, this sub-probe would carry a low-power high-speed transmitter and contain GaAs technology. The high power and high computation rate of the GaAs technology is the most efficient use of the power and time available. While the mother probe would be in orbit around the planet, the sub-probe would quickly collect and analyze data, transmitting it back to the mother probe. The mother probe would then record the information on laser disk, and prepare it for long range transmission. After passing the planet, low level artificial intelligence could be used to decide which pictures or pieces of information would be most important. Using this information, the data could be sent back in descending order of importance, thus avoiding a flood of worthless, repetitious data. The sent data would dwindle in importance, but would be continuous, filling the time gap between two planets with information. Because of the low power CMOS technology, the probe would have a relatively large life expectancy.

Planetary probes pose the greatest potential for improvement by the new technologies. The possibility of automated research is exciting. In the past, almost all operations were carried out remotely from Earth; what operations were done by the equipment on board were handled by slow, dedicated[8] machines. The processing power of a mainframe computer allows probes to make real time decisions about their environment.

One application of new technologies might begin with the launching of a probe. Using onboard computers, the probe would guide itself to the destination point. Upon arrival, pictures would be taken of the planet surface. Using artificial intelligence, a suitable landing location would be selected. For reference, a small monitor probe might be released in to a simple decaying orbit to monitor the progress of the landing, sending important location information to the probe. The probe would then land itself using the onboard computer.

Upon landing, the computer would load in new instructions, changing the program from one of navigation to that of research. Then, using advanced artificial intelligence, the computer could use a robot to start assembling a small power generator. For power, until the generator was completed, microwave beams could be used to pipe power from another location. Upon finishing the power generator, the computer would assemble a communications system and a laboratory. When these two facilities had been completed, research could begin using robots to collect data. If facilities for human occupation would be needed, additional supply probes could be sent along with instruction encoded on an optical disk, and the robots could assemble a base suitable for human occupation. As with the remote sensing probes, data would be sent back according to its relative importance.

---

[8]Dedicated refers to machines which are built specifically for one and only one task. An I.B.M. and a V.C.R. are both computers, however a V.C.R. is a dedicated machine.

Conclusion

While the age of speed advancement started nearly ten years ago, we are only just beginning to see the impact on the space program. Since the invention of the integrated circuit with RTL (Resistor Transistor Logic) technology for guided missiles and satellites in the 1960s, improvements upon the basic idea have led to speeds of blinding proportion. Today, it is simply expected that current technology will be outdated within the month. With such a bright outlook upon the world of future technology, we are nearly assured that the growth will be applied to help satisfy our curiosity about space and beyond.

APPENDIX I: RESOURCES

Higgins, Richard J. (1990). "The Bridge to V.L.S.I.," in Digital Signal Processing in V.L.S.I., Prentice Hall, Englewood Cliffs, NJ

HITACHI (1988). "Operation Principles and Characteristics," in Optoelectronic Devices Data Book, Hitachi, Tokyo

Symposia on Space and Nuclear Power Studies, "A Mission to Mars", Albuquerque Convention Center, January 1990

Rigdon, Brian, discussion on opto-erasable drives, Electrical Engineer, Honeywell Defense Avionics Systems Division

Encouraging Students to Enter Math, Science, and Space-related Fields

Rhonda Morgan[1]

Introduction

When President Kennedy issued the goal to fly to the moon and back by the end of the decade, he stated: "This generation does not intend to flounder in the backwash of the coming age of space. We mean to be a part of it. We mean to lead it."

Surely this upcoming generation feels the same, but sadly America is losing its preeminence in space and in technological markets. Most people agree that education is the key to technological superiority, that our current state of education is prohibiting the United States from being as competitive as it once was, and that action must be taken soon to improve our schools to protect the future of our country economically and technologically.

Executing the following ideas is not going to convince every student to study and seek jobs in scientific fields. However, the ensuing programs may be able to sway the people on the edge, and further encourage and nourish those who already have an interest. Schools and industry can take action that will motivate and interest students in science, math, and space-related fields. This paper explores areas of improvement and possible means of motivation towards science, math, and space at the kindergarten, elementary, junior high, and high school levels.

Kindergarten

In the latest trend to improve education, some school systems across the country have structured kindergarten classes. Students copy numbers and letters from dittos and teachers' models, and use computers to learn reading basics. Jack Uphoff, professor of education at Wright State University, stated that "pushing academics at an early age increases chances of later flunking a grade, needing special tutoring, and developing learning disabilities." Consequently, an open ended approach should be taken in kindergarten education (Brophy, 1989). Some nonscience-oriented educators and administrators feel that science is a rigid discipline and must be taught with a very structured approach. However, science can be taught in an open atmosphere. The approach should be discussion oriented, the students should be encouraged to ask questions, and a meaningful response should be provided for each question. Principles taught should relate to what the students experience everyday. This in-class searching arouses their curiosity and causes them to

[1]Student, La Cueva High School, 7801 Wilshire NE, Albuquerque, New Mexico 87122

question how things work, to think about an answer, and to seek the correct answer, which is the basis of scientific discovery and space exploration.

Implementation of such a program could be relatively inexpensive and easy. A specialist can be hired for each school or several schools. Or each teacher could have a relatively small amount of training, be provided with a lesson book and still perform sufficiently.

Elementary Schools

Currently, students receive their real introduction to science at the elementary school level. A fun, comprehensive approach could excite and motivate many students. Sadly, many elementary school teachers do not have the training necessary to teach science or are afraid of science (Losing the Future, 1988). Jim Cook of the National Science Foundation commented, "Naturally [teachers] don't do a good job across the country of teaching science. And they turn these kids off." One possible solution is to increase the science course requirements for obtaining teaching certificates. Another answer could be workshops which teach basic scientific principles and how to teach these principles. Workshops should be filled with ideas, labs, ways to coordinate various media, contagious enthusiasm, and encouragement. Simple labs performed with inexpensive equipment can teach simple scientific principles. Computer simulations and different media, aside from being enjoyable to the students, could ease pressure on teachers not comfortable with science. In some areas of the country, more formal education may be necessary. In such cases, school districts and community businesses could defray course costs. Other incentives could be established and tailored within individual systems to encourage elementary school teachers to take science courses.

In classes, the approach should be hands-on. People learn by doing, and experiencing science is the best way to spark an interest in it. Activities can include labs, field trips, and demonstrations. Demonstrations are practical for experiments requiring specialized or costly equipment and/or materials. Demonstrations are also easier to manage in the elementary classroom environment and with the elementary age group. Herein lies potential for outside assistance. As part of their science courses, high school students could present demonstrations to various classes at different elementary schools. High Schoolers would provide a positive role model: older kids who are interested in and show enthusiasm for science. The demonstrations would benefit the teachers and elementary school students, and the high schoolers because of the research, planning, cooperation, and the reward of the children's reactions, thus filling two needs in the system simultaneously.

The goal is to produce a class of students entering junior high school that question the workings of the universe and enjoy problem solving. In essence, the effort is to raise the level of demand for science courses, science activities, and scientific answers.

Junior High School

The most frequently voiced complaint among junior high students concerning science courses is boring classes. Consequently, the students lose interest. This is no surprise considering the current state of the national junior high science class room. One third of all junior high school science teachers are misassigned, teaching subjects with which they are not very familiar (National Science Teachers Association, 1989). Eighty percent of all seventh grade students never take field trips and fifty percent of junior high students do not have access to lab equipment (Losing the Future, 1988). Obviously, the extent of the aforementioned problems vary regionally and between school systems. More undeviating is the general problem of boredom. Therefore, the next few paragraphs focus on enlivening junior high science classes in ways applicable to a wide range of systems. Many suggestions for science classes are also applicable to mathematics classes.

Teachers, with or without a specialty in science, need support to de-bore their classes effectively. The first logical step of support is a district science coordinator (for those districts without a coordinator). Innovative workshops, as previously mentioned, directed toward the junior high school teacher and curriculum could be helpful. Membership in organizations, such as the National Science Teachers' Association, informs teachers of opportunities for teachers and students, and offers many innovative ideas. Schools can encourage membership through endorsement or monetary assistance with membership fees. An inexpensive means of teacher support is to increase awareness of existing resources. Each of the NASA centers have teacher resource centers. Businesses within some districts also have special programs or are willing to provide speakers. In supporting teachers, the resources, or service of locating resources, should be provided. Then the teachers should be encouraged and motivated to share those resources with their students. Because students learn by example, teachers can be effective role models: the students may adopt the teacher's enthusiasm and concern for the subject.

Demonstrations are an excellent way to brighten the science class. They break regular routine and usually create excitement. Demonstrations by high school students are doubly beneficial, as previously discussed, but the level of sophistication and equipment usage may require professionals. This is a fine area for industry and businesses to assist in education. Other class activities could be field trips or guest speakers. Scientists, engineers, even astronauts, explaining the ways that they apply science in their work everyday could make science more real to the students and inspire them to further study science.

Quality, nationally-organized extracurricular activities exist in science, math, and space that could enthuse young people. With a science fair project, the student applies the scientific method and experiences science at work in an environment other than the classroom. MathCounts rewards kids for achievement in math through competition. Future Problem Solving competition teaches problem solving techniques. In the Young Astronauts, students learn about the space program and explore facets of space flight. To be effective, these programs need an enthusiastic and willing sponsor.

Alleviating boring classes requires enthusiasm, encouragement, curriculum diversity, and teacher support.

High School

High school is the last chance for the public school to motivate students to enter science, math, and space-related fields. It is also an important time to develop and nourish those who have become interested in science.

Numerous possibilities exist to assist teachers. For a school system without a communication network, such a network among schools concerning resources and ideas can prove invaluable. Workshops can also be useful to high school teachers. Besides presenting ideas in the teacher's specialty, workshops should stress the significance of briefly crossing into other areas of science. As a short case study: a solar eclipse is not directly related to chemistry, but a chemistry teacher not permitting class time to view the eclipse, by request of the students, only quenches curiosity. Teachers can have more in-depth labs utilizing high-tech equipment belonging to industry but not in current usage by the company. Teachers could also be motivated to encourage their students by the publicizing of teacher awards won through student's work.

Encouragement should not be reduced for high school students. However, students in regular, mandatory science classes regard courses with lack of interest or distaste. Industry demonstrations in areas rarely taught in regular classes, such as glass-blowing or plastics, could spark interest. Spontaneity and quick deviation from the curriculum, by the teacher, could help. In a personal experience, the chemistry class clowns chanted "thermite!" until the teacher performed a chemical reaction that yielded enough heat to melt iron. The class clowns may not remember the previous lab, but they still discuss the principles and powers behind the thermite reaction.

Extra-curricular activities offer a wide range of opportunities for the academist or the average student. Science fair can be a challenging learning experience which stresses careful use of the scientific method and experimentation, and involves awards. Grade incentives could encourage many students to participate. Numerous other contests exist at the high school level that call on diversified skills. Students should be encouraged to participate by teachers who are willing to help or find the necessary assistance. Another opportunity for school and industry cooperation is mentorships and internships. The students, by working in fields they are interested in entering, narrow their study and career choices, are given extra motivation to study science and math, are more aware of the requirements of potential fields, and receive "hands-on" experience not duplicatable in the classroom. Many opportunities exist for high school students to develop their interests and abilities. They merely need to be encouraged to take part.

Conclusion

Steps can and should be taken now to improve science education and to motivate students at all levels in public education to enter science, math, and space-related fields. The effectiveness of these steps relies on joint efforts by school administration, teachers, industry, and on quick but thorough instigation. Public education exists for the students, and any plan or program should be designed and executed by administrators, teachers, and businesses with thought for benefiting and furthering the students. The public schools need to produce students with scientific aptitude so that the students, as adults, will be able to adequately, or, hopefully, exceedingly fulfill their entrustment of the future. Improvement will be costly, requiring sacrifice of time and money, but neglect is more costly. As John F. Kennedy said, "We chose to go to the moon and do these other things not because they are easy, but because they are hard."

Appendix I: Resources

American Society for the Advancement of Science, "Science for All Americans", American Society for the Advancement of Science Inc., 1989 pp. 3-15

NBC News, Losing the Future, Dec, 1988

Nova "Can the Next President Win the Space Race?" Oct, 1989

Time "Man in Space", 1989
U.S. News & World Report

Brophy, Beth, "The Kindergarten Wars" April 10, 1989

Buckley, Jerry, "The Hard Lessons of School Reform, The Rochester Experiment." June 26, 1989 pp.58-60

Glastris, Paul, "Of More Than Parochial Interest" May 22, 1989 pp.62-62

Rachlin, Jill, "The Label That Sticks" July 3, 1989 pp.51-52

Rachlin, Jill, "Opening the School House Doors" May 8, 1989 p.62

Shapiro, Joseph P. "No Pass, No Drive." June 5, 1989 pp.49-51
"Opening a New Book on School Reform" Sept. 18, 1989 p.131

BENEFITS OF RADIOISOTOPE THERMOELECTRIC GENERATORS

Andrea L. Roth[1]

ABSTRACT

Since the 1960's the United States has had the capability to reach beyond the planet Earth with the aid of satellites and interplanetary space probes. As probes such as the Pioneer and Voyager series are venturing farther away from the sun, beyond the planet Mars, solar energy is no longer a feasible power source. The U.S. space program now utilizes nuclear power to fuel these missions. The nuclear power source is a radioisotope thermoelectric generator (RTG). A controversy has arisen concerning the environmental safety of the RTGs. This paper discusses the advantages of using nuclear power to fuel space vehicles and the possible environmental hazards involved.

INTRODUCTION

After the space shuttle Challenger explosion in January of 1986, the American space program was brought to a temporary halt. NASA programs were reevaluated. Safety testing and analysis were extended. "Foolproof" systems were modified to be even more secure. One of these modified systems was the nuclear power source used on many NASA missions: the radioisotope thermoelectric generator (RTG).

An RTG converts heat produced by the decay of Plutonium (Pu) - 238 into usable electric power. An RTG has two major parts: a heat source to produce thermal energy from the decay of Pu and a thermoelectric converter to convert this thermal energy into electricity. The current heat source used with the RTGs is the General-Purpose Heat Source (GPHS) (Cull, 1989).

---

[1]Student, Eldorado High School, 11300 Montgomery Boulevard, Albuquerque, New Mexico 87111.

HISTORY

Until the Pioneer spacecraft series, solar panels had been used as the primary space power system. However, solar energy is not effective as a power source beyond the orbit of Mars because it is too far away from the sun. Energy received from the sun is determined by the Inverse Square Law; that is, if the distance from the sun is doubled, the power will be reduced to 25% its original value (Smith, 1988). If the distance is tripled, the power is one-ninth the initial value, and so forth. Hence, the Pioneers became the first spacecraft to be powered solely by nuclear energy.

Just like the Pioneer project, the only feasible way to generate enough electricity for the Voyager probes was to use RTGs. The first of the Voyager space probes was launched on August 20, 1977. Voyager has traveled thousands of millions of km away from the sun (Smith, 1988), so solar panels were obviously not a feasible power source.

Voyager was equipped with three RTGs. They were located on a short boom, or extension, opposite the span platform. The RTGs were kept as far away as possible from the delicate instruments of the spacecraft as an additional safety precaution.

RTGs have also been used for the Apollo Lunar Surface experiments and the Viking Mars Lander. They have powered over 23 missions for NASA, the U.S. Navy, and the Air Force (Cull, 1989). They were most recently used to power the Galileo space probe, launched in October of 1989. The Galileo space probe's mission includes traveling past Jupiter and its moons (Myrabo, 1985). The upcoming Ulysses probe will also be fueled by RTGs. The RTGs have become an integral part of the American space program.

ADVANTAGES OF RTGs

The RTGs are strong, yet lightweight, and compact. They also contain no moving parts. These are many reasons why NASA chose to use RTGs. The previously mentioned heat source for the RTGs, the GPHS, was developed for use on the Galileo probe.

There are 18 separate GPHS modules in each GPHS-RTG. There is enough plutonium in each module to produce 250 watts of thermal energy. After this thermal energy is converted, each GPHS-RTG has a total of 285 watts of electrical energy. This is the amount of energy that the craft has at the beginning of the mission. The Galileo probe needs two RTGs to carry out its mission to Jupiter (Cull, 1989).

SAFETY OF THE RTGs

There are 144 plutonium pellets in each RTG. Each pellet is lined in an iridium shell, then two graphite shells, and finally an aluminum cover (Toufexis, 1989). This type of extensive protection is unique to the General Purpose Heat Source. The heat source used for the RTGs must be able to withstand the worst possible "accident environments", or disaster scenarios, before being utilized on a mission. There are two main types of "accident environments" that the GPHS-RTGs must survive. These are: 1) accidents occurring before launch, such as a fire or launch pad explosion, and 2) those accidents caused by failure of the actual spacecraft itself, such as during reentry or impact on Earth (Cull, 1989).

Because of the Challenger explosion, the safety testing for the GPHS-RTGs has been intensified. Before any RTG-powered craft can be launched, General Electric Company, the contractor for the RTGs, is required to give a safety report on the RTGs to the Department of Energy (DOE). Before flight, the reports must go from the contractor to the DOE, NASA, and/or the Department of Defense. Then, it goes to the Space Council, and finally the President (Cull, 1989).

From 1961 to 1977, at least 23 RTG-powered spacecraft have been launched. Of these 23 missions, three were aborted (Cull, 1989). The most widely known and criticized of these accidents occurred in 1964, when an Air Force satellite burned up during reentry into the Pacific Ocean. As a result, a harmful amount of radiation was released into the region. In all accidents to date which have resulted in fuel being released, the new GPHS-RTGs were not used. Only two accidents have occurred with the GPHS-RTGs, and neither resulted in the release of fuel into the environment. According to space officials, the probability of plutonium being released into the atmosphere in an aborted mission is no more than 1 in 1,428 (Toufexis, 1989).

TESTING OF THE RTGs

The RTGs underwent a variety of tests before being able to be used in the Galileo probe. They were shown to withstand the pressure from an explosion. In several tests conducted at Sandia Laboratories in Albuquerque, N.M., the RTGs themselves survived that pressure but the graphite covers were destroyed (Cull, 1989). Tests were also conducted to determine whether the capsules could withstand being hit by projectiles, such as fragments from an explosion. The fuel released when projectiles were fired at over 700 m/s at a capsule was minimal (Toufexis, 1989). Also, a 10 minute exposure to a solid propellant fire caused no fuel release in the capsule (Cull, 1989).

CONCLUSION

Overall, the GPHS-RTGs themselves have undergone almost ten years of testing (Cull, 1989). This length of testing is unique in the nuclear industry. Still, a spokesman from General Electric Co. said they receive many calls of concern from the public about the use of nuclear power. In October, a group of demonstrators gathered around the space shuttle shortly before its launch to express their concerns about the safety of the nuclear power source.

I believe nuclear power is the most efficient energy source for the future. I also feel there is a definite need for nuclear power in space travel. The human race must continue to reach into and explore the mysterious depths of outer space. RTGs allow space vehicles to travel beyond Mars and make new discoveries about the universe. I feel the benefits of using the RTGs as a power source for spacecraft outweigh the possible safety risks involved.

APPENDIX I. REFERENCES

Cull, Theresa A. (1989) "Safety Testing Program for the Galileo Mission Radioisotope Thermoelectric Generator." Proceedings of the Ninth International System Safety Conference.

Myrabo, Leik, et.al. (1985) The Future of Flight. Simon and Schuster, New York, pp. 63, 198.

Smith, Arthur. (1988) Planetary Exploration. Richard Clay Limited, London, pp. 116, 126, 151-153.

Toufexis, Anastasia. (1989) "Nuclear Fears About Galileo." Time, October 16, p. 76.

Alternatives for the U.S. Manned Space Program

Anna Maria Salazar*

Introduction

There are a number of directions the U.S. space program could take. Should a space station be built? Should a lunar outpost be constructed? Should the U.S. plan a manned trip to Mars? Each alternative requires a great deal of money, and to fund all three would be prohibitively expensive. The advantages of each alternative will be discussed in this paper to clearly define choices that the U.S. faces with the manned space program.

Space Station

A variety of uses have been planned for the space station. A space station will be used not only by NASA, but also by the commercial, defense, and scientific communities. The defense program requires security, however, which might limit private sector work. There is also a chance that the space station will not fulfill all of its goals because of the limit the U.S. has set on funds and time. The space station was estimated to cost approximately $8 billion in 1984, and will likely cost a great deal more as the work proceeds from early conceptual studies to design and construction (Simpson, 1985).

The primary advantage of a space station is, in comparison with the space shuttle, its ability to hold a larger crew for a longer time: 6 to 8 people in orbit for months instead of days (Simpson, 1985). Research and technological development from space can be greatly increased with manned laboratories. Scientists will have more time for experimentation, and will be able to conduct many more experiments in space without having to rely on the space shuttle. A space station also will provide permanent observatories for Earth and celestial observations. One of the most important functions of a

***Student, West Mesa High School, 6701 Fortuna NW, Albuq., New Mexico, 87121**

space station is to provide a transportation node for the delivery of payloads into other orbits, geosynchronous orbits, or interplanetary trajectories. Because the shuttle is limited to 600 miles altitude, it cannot provide this capability. Using a space station eliminates the element of time from the equation of selecting earth bound launch windows. Payloads are also able to be refueled and stored until an appropriate launch window opens. Another function of the space station is the continued maintenance and repair of orbiting satellites and space systems. Finally, the space station can be used to explore manufacturing of certain pharmaceutical and other materials in an environment where a near-perfect vacuum, a large range of temperatures, and microgravity exist to help materials be made with greater precision and less defects than when manufactured on Earth (Simpson, 1985).

NASA plans to build the space station in pieces on Earth, then launch them separately using the space shuttle and assemble the components in orbit. The structure of the space station will continuously change because of new advancements in technology, but certain elements are planned. The main sections of the space station are laboratories, observatories, transportation nodes, manufacturing facilities, and a storage depot. Different plans for the architecture of the space station are being considered, but most or all of the aforementioned facilities will be present.

The United States and the Soviet Union are both considering building permanently manned space stations. These space stations will be exploring an area new to humans. If the Soviet Union and the United States take their disagreements into space, where will it stop? If the United States and the Soviet Union want to start a new age, they should start it together with one space station that could serve the U.S. and Soviet scientific exploration. Not only will one space station serve as a testing ground for cooperative programs, but also will save time and money.

Lunar Outpost

One of the primary reasons for constructing a lunar outpost is for scientific use. Scientists can explore the geology, astronomy, and many other aspects of the moon for knowledge itself. Minerals or other natural resources can be mined from the moon, creating a new income. The final reason for building a lunar base is to, perhaps, start a colony on the moon.

The far side of the moon provides a much better location for observing the universe than the Earth. One reason for this is because the Earth has too much interference. Clouds, smog, and the reflection of lights combine to create an atmosphere that lowers the accuracy and clarity of telescopes. Many signals, such as microwave and satellite, hamper the reception of radio telescopes. The other reason that the moon is valuable to astronomers is that the Earth shines 60 times brighter than the full moon and, in doing so, washes out the stars in nighttime for optical astronomers (Alfred, Bufkin, Kennedy, Petro, Roberts, Stecklein, and Sturm, 1989).

Whether it is to study the universe, mine minerals, or start a colony, a lunar outpost could be very important. The economic aspects of the moon involve extracting natural resources and starting a colony. One resource, liquid oxygen fuel, may be very profitable. Lunar rocks have been found which contain oxygen, and the mining of these rocks can help to lessen the cost of colonization (Alfred, Bufkin, Kennedy, Petro, Roberts, Stecklein, and Sturm, 1989) Colonizing the moon could be profitable as natural resources are extracted.

Mission to Mars

A manned trip to Mars would advance the U.S. another major step in the exploration of the universe. The question is: Can one single nation afford the price of putting man on Mars? A manned flight to Mars is estimated to cost between five and twenty-five billion dollars (Oberg, 1982). As a comparison, the Apollo mission, which cost twenty-five to thirty-five billion dollars, was considerably more expensive than the mission to Mars is projected to be. It took much more basic research to get to the moon than it will to go to Mars. A great deal of the necessary technology for manned missions to Mars is already available because of previous space missions. Research money is needed only in isolated areas instead of the development of the new technologies that were needed for the moon mission.

There are a number of reasons for launching a manned mission to Mars. Scientists are still interested in the possibility of life on Mars. Evidence of previous life might be found, helping scientists to understand what happened. Another reason to send men and women to Mars is purely for exploration itself. What is Mars, or for that matter, any other planet like? Earth can be better understood if more knowledge about Mars ia accumulated, and humans may be able to discover what the probes missed.

A Mars mission involving a lunar base or a space station has been considered. In either one a great deal of time and money will be spent. The first step in both options is to build a structure, a moon base or a space station, which will be the focus of the mission. Then, using the moon base or space station as a liftoff point, the spaceship will journey to Mars. It is possible that a spaceship cannot reach Mars without using one of these options because of the difficulty in launching from Earth compared to launching from a lunar base or space station, so a space station or lunar base may have to be built first.

Other ideas are for a Soviet-United States joint mission

to Mars. The Mars mission is a very popular option because it is more of a risk and more exciting than the other projects. There may be some knowledge that can be discovered on Mars and elsewhere, and the sooner the United States and Soviet Union discover it, the better it will be for future citizens. If the Soviet Union and the United States work together, besides the aspects of peace, the United States would further its Mars project, making the manned trip occur sooner than expected.

Conclusion

The United States has three major options concerning the future direction of its space program: build a space station, develop a lunar base, or go to Mars. The nation does not have the resources to do all three in the near term, but they are all good ideas. The space station can further the scientific study of the universe and the Earth, and advance technology through basic research which can only be accomplished in space. A mission to Mars would mark the beginning of interplanetary exploration and possibly create a new commercial enterprise. A lunar base can help scientists to get a different view of the universe and develop a new income. Finally, working with the Soviet Union will promote peaceful relations and help the United States acheive more in space than it might otherwise accomplish on its own. All three options are important, but the United States must decide which is most cost effective, and what can best serve the interests of the U.S. space program.

References

Alfred, John; Bufkin, Ann; Petro, Andrew; Roberts, Michael; Stecklein, Jonette; Sturm, James; and Kennedy Kriss J. Lunar Outpost. 1989.

Oberg, James E. Mission To Mars: Plans and Concepts for the First Manned Landing. harrisburg, Pennsylvania: Stackpole Books, 1982.

Simpson, Theodore R. The Space Station: An Idea Whose Time Has Come. New York, New York: The Institute of Electrical and Electronics Engineers, Inc., 1985

SUBJECT INDEX

Page number refers to first page of paper.

AUTHOR INDEX

Page number refers to first page of paper.